T0334531

The Human Body

The Human Body
Linking Structure and Function

Bruce M. Carlson, MD, PhD
Department of Cell and Developmental Biology,
University of Michigan Medical School, Ann Arbor, MI, United States

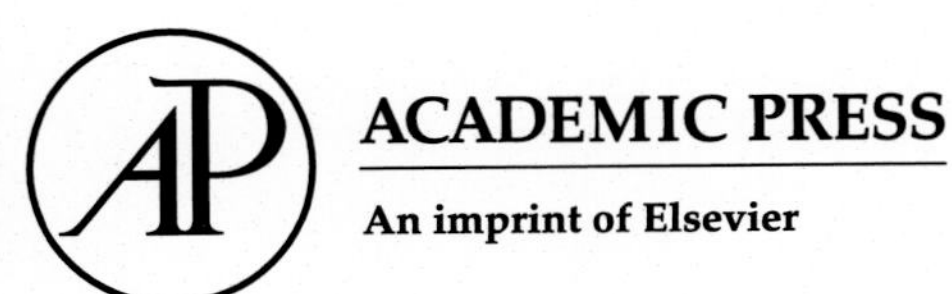

Academic Press is an imprint of Elsevier
125 London Wall, London EC2Y 5AS, United Kingdom
50 Hampshire Street, 5th Floor, Cambridge, MA 02139, United States
The Boulevard, Langford Lane, Kidlington, Oxford OX5 1GB, United Kingdom

British Library Cataloguing-in-Publication Data
A catalogue record for this book is available from the British Library

Library of Congress Cataloging-in-Publication Data
A catalog record for this book is available from the Library of Congress

ISBN: 978-0-12-804254-0

For Information on all Academic Press publications
visit our website at https://www.elsevier.com/books-and-journals

Publisher: Andre Gerhard Wolff
Acquisition Editor: Mary Preap
Editorial Project Manager: Carlos Rodriguez
Production Project Manager: Sreejith Viswanathan
Cover Designer: Christian Bilbow

Typeset by MPS Limited, Chennai, India

Dedication

To my grandson Isaac

Contents

Preface

The human body is complex, with lots of parts, both moving and nonmoving. Because of its complexity, the study of the human body has traditionally been broken down into a variety of individual disciplines, which occupy a good part of the first 2 years of a standard medical school curriculum. By the end of that time, a typical medical student has been confronted with a massive amount of detail, often presented within the context of the specific disciplines. What is difficult for the student is putting it all together to get an understanding of how all of the components of the human body work together in an integrated fashion.

For those who are not medical students, but who want to learn about how the body functions, the overwhelming amount of detail can be daunting to all but the most intrepid. A truly integrated coverage of what is known would require a multivolume treatise of many thousands of pages. One approach to this is the electronic Reference Module to the Biomedical Sciences published by Elsevier. There presently exist few resources that provide access to such information in less than encyclopedic fashion.

This book is focused on how structure and function work together. It places emphasis on bodily functions and how the structure of the body has evolved and is presently adapted to serve those functions. Over many decades of teaching at both the medical school and undergraduate levels, it has been my belief that by understanding how an organ system functions and by relating anatomical structures to these functions, one will develop and retain a much more integrative understanding of how the body is structured and why the various organs and tissues are organized the way they are. This is the way the body works anyway. There are few structures that do not have an obvious function, and functions cannot take place without the appropriate structures to support them.

The bulk of this book is organized according to functional systems. The principal focus is on the cellular and tissue level, but one cannot do justice to most functions without bringing some biochemistry or molecular biology into the discussion. At the other end of the spectrum, many functions are reflected in the configuration of gross anatomical structures. When appropriate, excursions into that domain will also occur in the text. The first two chapters (on cells and tissues) are designed specifically for those who have not had a background in cell biology or histology. The information presented in these chapters forms a basic vocabulary and conceptual base that will be important in understanding details of the later chapters.

It is my intent that individual chapters can be read through as a story, rather than in small chunks, as would be done with a typical textbook. I have included only the level of detail needed for an integrated understanding of the bodily functions covered. Therefore I do not want to burden the reader with lists of anatomical structures that are not specifically connected with functions. Similarly, I do not include details of physiological or biochemical processes that are not directly connected with the structures being covered. Many of the functions discussed often fall between the cracks of standard anatomy and physiology texts and are not discussed in either. Some references to pathological processes are made if they can help the reader to understand normal functions that are clarified by a description of what can go wrong if that function is disrupted. I also make brief excursions into evolutionary and developmental biology if these can clarify how important adult structures are organized. The text is heavily illustrated, often with figures borrowed from other fine texts, because figures are essential in obtaining a real sense for the topic being discussed.

My hope is that, upon reading through this book, you will emerge with a much better understanding and appreciation of our bodies—how they are put together and how they function. I always welcome feedback on my books—especially when the inevitable mistakes turn up. Feel free to contact me at brcarl@umich.edu.

Finally, I would like to thank those who have been instrumental in helping me to translate an idea into a book. First of all at Elsevier, to Janice Audet, who was willing to take a chance on a new idea for a textbook; to her successor, Sara Tenney for shepherding it through the publishing process; to Carlos Rodriguez who has spent many hours converting manuscript pages and raw illustrations into a real book. Finally, to my artist/illustrator Alex Baker who, as she has done with several of my other books, has converted rough ideas into wonderful illustrations. In addition, I thank both family members and friends, who read through early prototype chapters and encouraged me to pursue this project. Their support and encouragement have meant a lot to me.

Bruce M. Carlson

Sources of Borrowed Illustrations

Abbas AK, Lichtman AH: 2009 Basic Immunology 3/e, Saunders Elsevier, Philadelphia.
Abbas AK, Lichtman AH, Pillai S: 2016 Basic Immunology 5/e, Elsevier, St. Louis.
Adkison LR: 2012 Elsevier's Integrated Review Genetics 2/e, Elsevier Saunders, Philadelphia.
Bear MF, Connors BW, Paradiso MA: 2007 Neuroscience –Exploring the Brain 3/e, Lippincott Williams & Wilkins, Baltimore.
Berkovitz BKB, Holland GR, Moxham BJ: 2009 Oral Anatomy, Histology and Embryology 4/e, Mosby Elsevier, Edinburgh.
Bloom W, Fawcett DW: 1975 A Textbook of Histology 10/e, Saunders, Philadelphia.
Boron WF, Boulpaep EL: 2012 Medical Physiology 2/e, Elsevier Saunders, Philadelphia.
Bourne GH, ed.: 1956 The Biochemistry and Physiology of Bone, Academic Press, New York.
Browner et al., eds.: 2003 Skeletal Trauma 3/e, Saunders, Philadelphia.
Carlson BM: 2007 Principles of Regenerative Biology, Academic Press, San Diego.
Carlson BM: 2014 Human Embryology and Developmental Biology 5/e, Elsevier Saunders, Philadelphia.
Carroll RG: 2007 Elsevier's Integrated Physiology, Mosby Elsevier, Philadelphia.
Chew SL, Leslie D: 2006 Clinical Endocrinology and Diabetes, Churchill Livingstone Elsevier, Edinburgh.
Diaz MB, Herzig S, Vegiopoulos A: 2014 Metabolism, 63:1238–1249.
Drake RL, Vogl W, Mitchell AWM: 2005 Gray's Anatomy for Students, Elsevier Churchill Livingstone, Philadelphia.
England ME, Wakely J: 2006 Color Atlas of the Brain and Spinal Cord 2/e, Mosby Elsevier, Philadelphia.
Erlandsen SL, Magney JE: 1992 Color Atlas of Histology, Mosby-Year Book, St. Louis.
Fitzgerald MJT, Gruener G, Mtui E: 2012 Clinical Neuroanatomy and Neuroscience 6/e, Elsevier Saunders, Philadelphia.
Frost HM: 1973 Orthopaedic Biomechanics, Charles C. Thomas, Springfield, IL.
Gartner LP, Hiatt JL: 2007 Color Textbook of Histology 3/e, Saunders Elsevier, Philadelphia.
Gartner LP, Hiatt JL: 2011 Concise Histology, Saunders Elsevier, Philadelphia.
Guyton AC, Hall JE: 2006 Medical Physiology 11/e, Elsevier Saunders, Philadelphia.
Guyton AC, Hall JE: 2016 Textbook of Medical Physiology 13/e, Elsevier Saunders, Philadelphia.
Hall BK: 2005 Bones and Cartilage, Elsevier Academic Press, Amsterdam.
Hinz B, et al.: 2007 Am. J. Pathol., 170:1807–1816.
Johnson LR: 2014 Gastrointestinal Physiology 8/e, Elsevier Mosby, Philadelphia.
Jorde LB, Carey JC, Bamshad MJ: 2010 Medical Genetics 4/e, Mosby Elsevier, Philadelphia.
Kessel RG, Kardon RH: 1979 Tissues and Organs: A text atlas of scanning electron microscopy, W. H. Freeman, San Francisco.
Kierszenbaum AL, Tres LL: 2012 Histology and Cell Biology 3/e, Elsevier Saunders, Philadelphia.
Levy MN, Pappano AJ: 2007 Cardiovascular Physiology 9/e, Mosby Elsevier, Philadelphia.
Male D, Brostoff J, Roth DB, Roitt IM: 2013 Immunology 8/e, Elsevier Saunders, Philadelphia.
McBean D, van Wijck F: 2013 Applied Neurosciences for the Allied Health Professions, Churchill Livingstone Elsevier, Edinburgh.
Moore KL, Persaud TVN, Torchia MG: 2016 Before We Are Born, Elsevier, Philadelphia.
Myers TW: 2014 Anatomy Trains 3/e, Churchill Livingstone Elsevier, Edinburgh.
Nolte J: 2009 The Human Brain 6/e, Mosby Elsevier, Philadelphia.
Nussbaum RL, McInnes RR, Willard HF: 2016 Thompson & Thompson Genetics in Medicine 8/e, Elsevier, Philadelphia.
O'Neill RO, Murphy R: 2015 Endocrinology 4/e, Mosby Elsevier, Edinburgh.
Parham P: 2009 The Immune System 3/e, Garland Science, London.

Pollard TD, Earnshaw WC: 2004 Cell Biology, Saunders, Philadelphia.
Purves D, et al., eds.: 2008 Neuroscience 4/e, Sinauer Associates, Sunderland, MA.
Saladin KS: 2007 Anatomy & Physiology 4/e, McGraw-Hill, New York.
Siegel A, Sapru HN: 2011 Essential Neuroscience 2/e, Wolters Kluwer/Williams & Wilkins, Philadelphia.
Stevens A, Lowe J: 2005 Human Histology 3/e, Elsevier Mosby, Philadelphia.
Thibodeau GA, Patton KT: 2007 Anatomy & Physiology 6/e, Mosby Elsevier, St. Louis.
Turnpenny P, Ellard S: 2012 Emery's Elements of Medical Genetics 14/e, Elsevier Churchill Livingstone, Philadelphia.
Underwood E: 2015 Science 347:1186–1187.
Waugh A, Grant A: 2014 Ross and Wilson Anatomy & Physiology 12/e, Churchill Livingstone Elsevier, Edinburgh.
White BA, Porterfield SP: 2013 Endocrine and Reproductive Physiology 4/e, Elsevier Mosby, Philadelphia.
Young B, Loew JS, Stevens A, Heath JW: 2006 Wheater's Functional Histology 5/e, Churchill Livingstone Elsevier, Edinburgh.

Introduction

This book is about the human body, but I want to approach the subject in a special manner. The body of a human or any other animal is very complex. It contains a myriad of structures, each of which functions at many levels and in many ways. Large treatises have been written about almost every structure or function of the body, but they often treat the topic in isolation. These books may offer great detail about the structure or function in question, but what is often missing is a sense of how that organ relates to the rest of the body or how its structure and function are related.

From its length alone, it is obvious that the present book presents nowhere near the available information about the structure or function of any organ in the body, but what it does try to do is to view these organs as members of a highly integrated system. In the living body, integration occurs in two principal directions—vertical and horizontal.

Vertical integration involves dimensionality. For this book, the dimensions of vertical integration range from the ionic or molecular at the lowest level to that of the entire body at the upper level. This range of integration of both structure and function is well exemplified in the section on skeletal muscle, where physical movement of two bundles of protein chains past one another causes shortening of muscle cells, which then causes the shortening of a muscle, which then can cause a limb to move.

Horizontal integration can occur at almost any vertical level from the molecular to the gross. To belabor the muscle example once again, effective contraction at the molecular level depends upon a highly precise organization of the contractile and supporting proteins with respect to one another. This high level of molecular integration is reflected in the beautiful electron micrographs that highlight the repeating arrays of contractile proteins that make up the muscle cell (see Fig. 5.2). At the cellular level, the complex internal architecture of the liver (see Fig. 12.30) or spleen involves a variety of different cell types that are precisely organized to allow these organs to fulfill their numerous simultaneous cellular and biochemical functions. Even at the gross anatomical level, coordinated interconnections between brain, bones, muscles, tendons, and ligaments are required for normal posture or movements.

The more we learn, the more we realize that almost every structure has one or many functions and that at all levels these structures are remarkably well adapted to carry out their functions. A common axiom is that form follows function. This is accurate up to a point, but equally valid is the proposition is that function often takes advantage of existing form. An excellent example from the realm of comparative anatomy and evolutionary biology is our middle ear bones—the malleus, incus, and stapes. These auditory elements began their evolutionary journey as small bones supporting the jaw apparatus. Only later in evolution were they co-opted into a completely new function. Another example at the molecular level is conversion over evolutionary time of molecules that originally functioned as enzymes into crystallin proteins that lend transparency to the lens of the eye. Similarly, feathers on reptiles are thought to have first served for thermoregulation and only secondarily to have assisted in flight. Hairs may have first formed between scales for tactile purposes and were later co-opted for thermoregulation in mammals.

Most functions of the body begin with molecular events, in which a cell is exposed to a molecule or ion in its immediate environment. Through some structural specialization, that molecule or ion makes its way into the cell or triggers a molecular event within the cell. That event can cause a chain reaction that enters the nucleus of the cell and stimulates or represses the activity of a specific gene, which in turn leads to the production of a new type of protein. Or, the molecular event could stimulate the biochemical machinery within the cytoplasm of that cell to begin a cycle of biochemical synthesis or breakdown. Alternatively, very similar cellular responses can, in some cases, be stimulated by minute mechanical forces acting on the cell membrane or by the concentration of oxygen or the pH of the immediate environment of that cell.

Individual cellular responses can be confined largely to a single organ. At the other extreme, a response from one type of cell can influence far distant cells within the body. This is particularly true in the case of hormones or some antibodies, which are released from a local group of cells and are then carried by the blood to all parts of the body. In the case of nerves, a thought generated within the cerebral cortex can be quickly translated into movement of a distant toe through a specialized system of communication that utilizes networks of long nerve fibers much like telephone lines or electrical wiring.

Presenting multilevel structural and functional integration at an elementary level is a daunting task because acquiring a reasonable understanding of the structures or processes can involve some quite difficult material. One approach taken in this book is to encapsulate some of the more difficult content in boxes. The main text will provide a general outline of some of these complex processes, but the interested reader will have access to sufficient detail in the boxes to allow a fuller understanding of them.

The first two chapters introduce cells and tissues. Both of them are designed to provide the reader with basic concepts and vocabulary that are essential to understanding the remaining chapters. Because of this intent, much of the potential integrative treatment of these topics is left to the later chapters, where the structures and functions of individual cells and tissues are placed in a more functional context.

At an even more basic level, the introduction is an appropriate place to introduce structural terminology that is routinely used in anatomical descriptions. For animals except humans, the three cardinal axes are illustrated in Fig. I.1. The leading (front) end is called **anterior**, and the hind end is called **posterior**. Top—the back—is **dorsal**, and bottom—the belly—is **ventral**. The other axis is right–left. Because humans are bipedal, rather than quadrupedal (four-footed), human anatomists have declared that the top of the head is **superior** and the feet are **inferior** (Fig. I.2). Since we usually walk forward, our face or belly is considered anterior and our back, posterior. Right and left have not changed. Unfortunately, the terminology applied to human embryos is the same as that used for quadrupeds, so anterior in a human embryo means the top of the head or something closer to the head than a reference point. The back, then, is considered dorsal, even though posnatally the back is considered posterior. Confusing? Yes!

Another important set of reference terms is **proximal** and **distal**. Proximal means closer to a specific anatomical reference point, and distal means farther away. Therefore, the hand is distal to the elbow, and the shoulder is proximal to the elbow. By convention, in the digestive system, the intestines are considered distal to the stomach.

A last set of terminology refers to planes of section of the body. A **cross (transverse) section** refers to a cut through the body or a limb like a saw cutting through a tree trunk or a branch. A **sagittal** section (Fig. I.2) is one that splits the body or part of the body (e.g., the brain, see Fig. 6.11) into equal right and left halves, starting from the top of the head and down through the trunk. A **parasagittal** section is in the same plane, but not in the midline between left and right. Finally, a **frontal** or **coronal section** is one that is perpendicular to a sagittal section (see Fig. 11.4 for an example of a frontal section through the face).

Because this is not a strictly health-related book, I have placed the structure and workings of the human body in a broader biological context than is usual for a book of this kind. Succeeding chapters will make frequent reference to the evolutionary or embryological origins of important structures in the body. Although this is not a book on evolutionary biology or an embryological textbook, nevertheless understanding the **ontogenetic** (embryological) or **phylogenetic** (evolutionary) origins of structures or organs can allow a better understanding of why they are constructed the way they are. For some topics, e.g., vision or the bones of the middle ear, one can gain insight into their present structure and

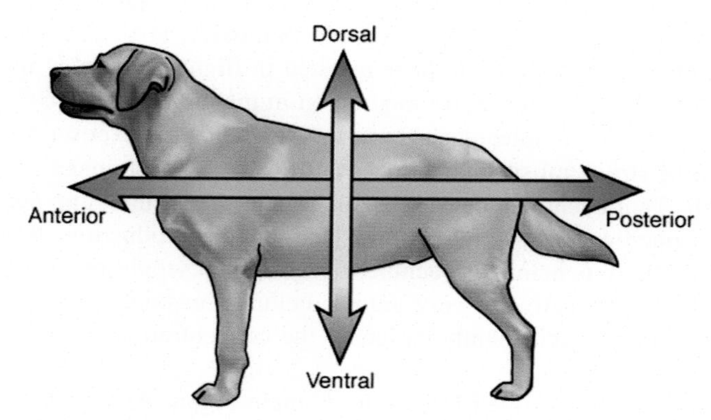

FIGURE I.1 Directions of the animal body.

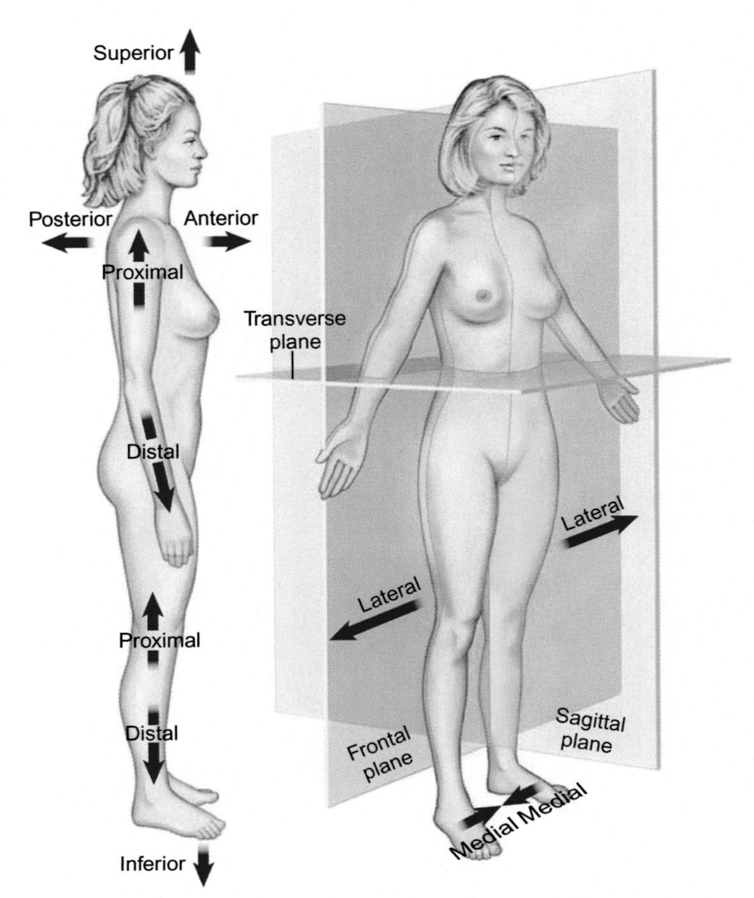

FIGURE I.2 Directions and planes of the human body. *From Thibodeau and Patton (2007), with permission.*

function by knowing their evolutionary history; for others, such as the skeleton, understanding their developmental history provides greater clarity.

The best way to read this book is to go through the chapters first as though you are reading a tour guide, in order to get the big picture. One can always fill in the details on a second reading.

Chapter 1

Cells

The cell is a fundamental unit of life. Many organisms consist of only a single cell, which is adequate for all physiological functions, including reproduction. Most plants and animals, however, are made up of enormous numbers of cells, which work in community. Not all cells are alike. The mature human body contains an estimated 200+ types of cells, each of which is specialized to fulfill a different function. Aggregates of similar cells form tissues and organs that seamlessly accommodate our functional needs. The number of cells that collectively constitute the human body is staggering. A recent careful estimate places that number at ~37.2 trillion.

Even more staggering is the number of co-inhabitants in our bodies—the bacteria and other microorganisms that constitute our **microbiome**. The bacteria, which may be 10 times as numerous as all the other cells of the body combined, share our body space in a mutually beneficial (**symbiotic**) relationship. All body surfaces that directly connect with the outside, for example, skin, digestive system, vagina, are home to significant populations of bacteria. Until very recently it has been assumed that our internal tissues are sterile, but bacteria are now being found in many tissues that are carefully analyzed.

In its most elemental representation, a **cell** consists of a nucleus surrounded by cytoplasm and is bounded on the outside by a thin plasma membrane. Each of these three is highly complex and contains many components. The **nucleus** is a repository for the genetic information that guides much of development and function. The **plasma membrane** serves as a gatekeeper that controls what comes into and what leaves the cell. Filling in the space between the nucleus and the plasma membrane, the **cytoplasm** contains a large number of internal structures (**organelles**), each of which serves functions as diverse as breaking down molecules and providing energy.

Despite their great variety, most cells are composed of a common set of organelles, but as cells specialize during development, certain sets of organelles that contribute to their particular function assume prominence. For example, muscle cells assemble arrays of contractile proteins that allow them to shorten upon demand, whereas islet cells of the pancreas develop the cellular infrastructure needed to synthesize and secrete insulin.

The aim of this chapter is to provide a general introduction to cellular structure and function. In order to provide a basic vocabulary, it outlines the most important structures found within cells and their roles in generic cellular functions (Fig. 1.1). The biology of many specific cell types will be covered later in the text, in the context of discussions of the organ systems in which they reside.

THE PLASMA MEMBRANE

The plasma membrane is not only a container for the contents, but it determines what passes in and out of the cell. Adjacent cells are joined to one another through specific cell adhesion molecules located on the surfaces of the plasma membranes. The plasma membrane also receives signals from its immediate environment that are transmitted to the interior of the cell, which then responds to the signals.

In essence, the plasma membrane consists of a **lipid bilayer** studded with a large variety of proteins, which play a variety of roles in the life of a cell (Fig. 1.2). About 50% of the plasma membrane is lipid, and the other 50% is protein. Some of the proteins are attached to either surface of the plasma membrane; others pass completely through it. Still others form channels of various sorts that allow the passage of water and a variety of **solutes** (ions and small molecules) through the membrane in either direction.

The Human Body. DOI: https://doi.org/10.1016/B978-0-12-804254-0.00001-6

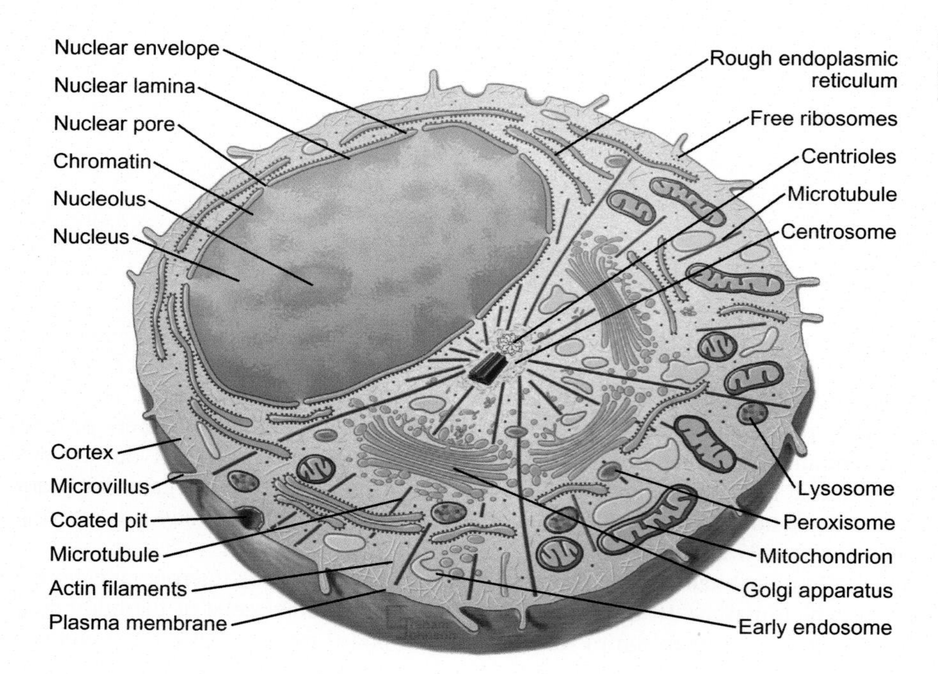

FIGURE 1.1 The major constituents of a generic mammalian cell. *From Pollard and Earnshaw (2004), with permission.*

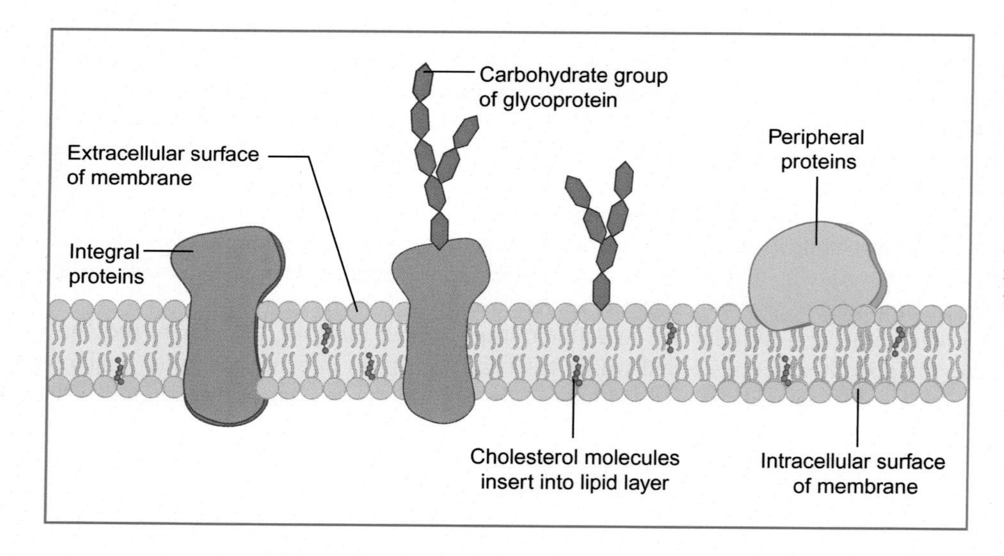

FIGURE 1.2 Types of molecules embedded in cell membranes. Cholesterol molecules are inserted within the lipid bilayer. Various types of proteins may extend entirely through the cell membrane or be partially inserted in the outer or inner face. Complex carbohydrates are often attached to membrane proteins or to the membrane itself. *From Carroll (2007), with permission.*

Lipid Bilayer

The lipid bilayer, which forms the framework of the plasma membrane, consists of two layers of phospholipid molecules. Each phospholipid molecule is shaped like a clothespin, with two fatty acids (the lipid) constituting the prongs and a combination of glycerol and phosphoric acid making up the head (see Fig. 1.2). Imagine two rows of these molecular clothespins, in which each layer lie side by side with the heads facing outward and the prongs facing inward. The fatty acid prongs are **hydrophobic** (do not mix with water), whereas the heads are **hydrophilic** and can interact with the watery fluids on either side of the membrane.

The plasma membrane contains a number of other forms of lipids, but in lesser amounts. Some of these molecules are connected with carbohydrate chains that protrude from the surface of the cell. The various carbohydrate chains on the outer surface form a fuzzy coat, often known as the **glycocalyx**. The glycocalyx of specific cell types, species and even individuals is unique, and it is a major contributor to the immunological identity of the cell and the individual (see

TABLE 1.1 Functions of Plasma Membrane Proteins

Type of Protein	Functions
Receptors	Receive chemical messages from nearby cells or other parts of the body. Trigger signal transduction response whereby the message influences gene expression within the nucleus.
Enzymes	Many membrane proteins possess enzymatic activity that cleaves chemical bonds in other molecules.
Channel proteins	Transmembrane proteins that form constantly open or gated channels that allow molecules to pass through the plasma membrane.
Carrier proteins	Proteins that bind to small molecules or ions and pass them through the plasma membrane.
Membrane pumps	Proteins that use energy from ATP to transport solutes across a membrane, even up a concentration gradient.
Cell adhesion molecules	Several groups of molecules that protrude from the cell surface and bind to similar molecules of neighboring cells.
Cell identity proteins	Glycoproteins whose characteristics differ from those of other cells and give the cells or the individual its immunological identity

Chapter 8). Human blood types (A, B, AB and O) are a reflection of biochemical differences in the glycocalyces of blood cells from different individuals.

An important internal component of the plasma membrane (~20% of plasma membrane lipids) is **cholesterol**, which stiffens the membranes by serving as anchors for the other molecules, but in higher concentrations actually contributes to the fluidity of the molecules within the membrane. Membrane fluidity is important in healing microwounds in the plasma membrane. At a practical level, this property allows one to puncture the plasma membrane to remove the nucleus from an egg and to introduce a new one into it, as is done in cloning experiments. In the field of reproductive medicine, the introduction of a sperm directly into an egg is one method for achieving in vitro fertilization. Even in normal fertilization, the plasma membrane of the sperm fuses with that of the egg as the sperm enters.

Plasma Membrane Proteins

Embedded within the plasma membrane are several categories of proteins, which serve many cellular functions (Table 1.1). Although the protein molecules themselves are stable, the membrane as a whole is a very dynamic structure, with the configuration of the lipid and protein molecules constantly changing. An early representation of membrane physiology was called the **fluid mosaic model**, in which the lipid component of the membrane was viewed as the surface of a soup with embedded protein molecules floating within it like dumplings. We now know that membrane physiology is much more complex than that and that strong connections exist between membrane proteins and some of the molecules located both inside and outside the cell. Nevertheless, membrane fluidity is real, and both protein and phospholipid molecules can often move freely within the plasma membrane.

Of the several categories of membrane proteins (see Fig. 1.2), one variety, **integral proteins**, consists of thousands of different types of proteins that are attached to the inner or outer surfaces of the plasma membrane or that pass completely through the membrane. These proteins have numerous functions. Some, called **receptors**, protrude from the outer surface and bind to signal molecules coming from other cells. After the binding event, the configuration of the receptor molecule typically changes. This change is then communicated to a series of proteins on the inside of the membrane, which are part of a second messenger (**signal transduction**) system that passes the signal through the cytoplasm to the nucleus, where it stimulates the expression of a gene within the DNA. Other integral proteins act principally as enzymes that break down molecules at the cell surface. For example, in the small intestine, they carry out the last stages of breakdown of carbohydrate and protein digestion. On muscle fibers, the membrane-bound enzyme **cholinesterase** breaks down the neurotransmitter acetylcholine (see p. 117) so that it does not continue to stimulate contraction of the muscle fiber long after it receives a signal from the nerve to contract.

Many intrinsic proteins contain carbohydrate chains that contribute to the immunological identity of the cell. Still others promote intercellular adhesion as they form stable bonds with other molecules of the same type on neighboring cells. One class of **cell adhesion molecules** (the Ig-CAMs) consists of members of the immunoglobulin family (see p. 215). Such adhesion molecules bind directly to other molecules of the same type on the next cell. Another large class of adhesion molecules, the **cadherins**, utilizes Ca^{++} as a binding agent between the two cadherins on opposite cell membranes.

Channels, Pumps, and Carriers

Most of the remaining membrane proteins are involved in the transport of materials across biological membranes (Fig. 1.3). There are three main categories of such proteins. The first category of membrane transport proteins is **membrane channels**, which allow the free transport of ions or small molecules along concentration gradients leading toward either the inside or the outside of the cell. These do not require great amounts of energy for their normal function. Membrane channels exist in a large variety of sizes and shapes, but their most important characteristic is that they allow the passage of both water and solutes through the cell membrane much faster than any other means. This is important in organs, such as the kidney, where large amounts of both water and ions must be continually processed in short amounts of time. Some channels are gated, meaning that part of the protein acts as a plug that opens or closes the channel depending upon the configuration of the protein. Gated channels in a nerve fiber allow the passage of Na^+ through the plasma membrane in a very short period of time (measured in milliseconds). Such rapid transport of Na^+ is required to generate an action potential, which serves as the basis of a nerve impulse (see p. 59). Channels specific for the passage of water are lined by **aquaporin** proteins. Although water molecules can diffuse through the phospholipid component of the plasma membrane, such passage is slow. Aquaporins allow very rapid transit of H_2O through the membrane.

Another variety of membrane protein forms membrane pumps (see Fig. 1.3). These pumps must utilize energy from one of several sources in order to function. The most common source of energy is the hydrolysis of **ATP** (adenosine triphosphate—the main source of internal energy in biological systems) into **ADP** (adenosine diphosphate) + phosphate, with the concomitant release of energy from the broken phosphate bond. In some systems, such as the eye, even light can serve as the source of energy. All cells have thousands of Na^+–K^+ ATPase pumps in their plasma membranes. Close to half of the ATP-derived energy used by a cell is used to power membrane pumps, which work almost continuously. The most important feature of membrane pumps is that they can transport materials up a concentration gradient—which is why so much energy is required. This advantage counters their intrinsic disadvantage, in that the overall flow of transport via pumps is considerably slower than that seen in channels. Nevertheless, things can be accomplished by means of pumps that cannot be done through channels. A good example is the inside of the stomach, where certain cells secrete hydrochloric acid in such concentrations that the pH can be as low as 1.0. A neutral pH is 7.0, and each pH point is a degree of magnitude greater

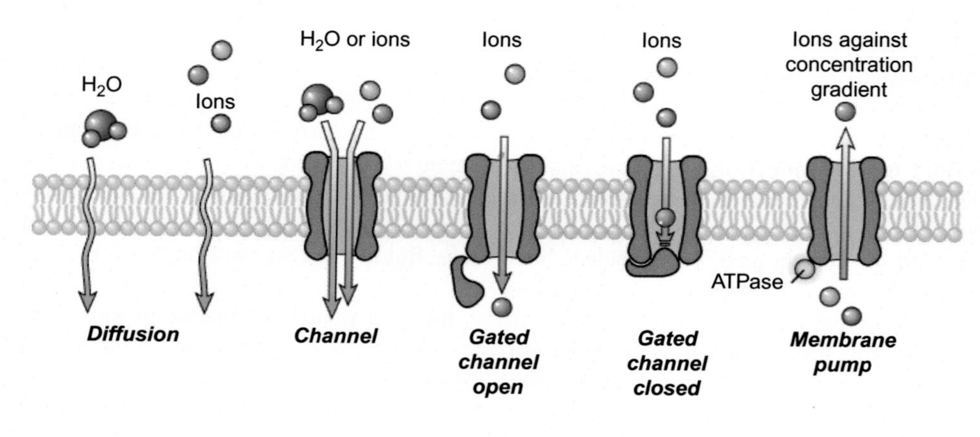

FIGURE 1.3 Upper row—Various means by which water and/or ions can be transported across a cell membrane. Lower row—Different means of ion or small molecule transport via membrane carriers across a cell membrane through channels with different properties.

than the previous one. Thus a pH of 1.0 is six degrees of magnitude greater than neutral (or 1,000,000 times as concentrated). Such a degree of concentration represents the upper limit of a membrane pump, but that alone is impressive.

The third category of transport proteins is **membrane carriers** (see Fig. 1.3). Although seemingly similar to channels and pumps, membrane carriers have a wide repertoire of functions. Some, called **uniporters**, provide low-resistance pathways for the transportation of simple sugars and amino acids into cells. Others, called **symporters**, are able to transport two or more kinds of simple molecules across a membrane at the same time. Still others, called **antiporters**, can drive an ion in one direction and another ion or molecule in the opposite direction. Some carrier proteins function by binding to a solute on one side of a membrane and then changing their configuration. As the configuration of the carrier protein is changing, the bound solute is brought to the other side of the membrane and then released. Insulin stimulates the insertion into the plasma membrane of many carrier proteins that bring glucose into cells.

Specialized portions of plasma membranes are used to hold cells together and to inhibit the passage of fluid and solutes between cells. Another type of membrane specialization, the **gap junction**, facilitates rapid communication between two cells. These specializations are discussed on p. 30 in the context of epithelial functions because they are most prominently represented in epithelia.

THE NUCLEUS

If the plasma membrane represents the gatekeeper of a cell, the nucleus (Fig. 1.4) represents the information library and server, because almost all of the genetic information of the individual is embedded in the DNA of the chromosomes. Upon appropriate processing of signals both intrinsic and extrinsic to the cell, the nucleus then unlocks specific bits of genetic information and produces molecular messages that are later acted on by other components of the cell.

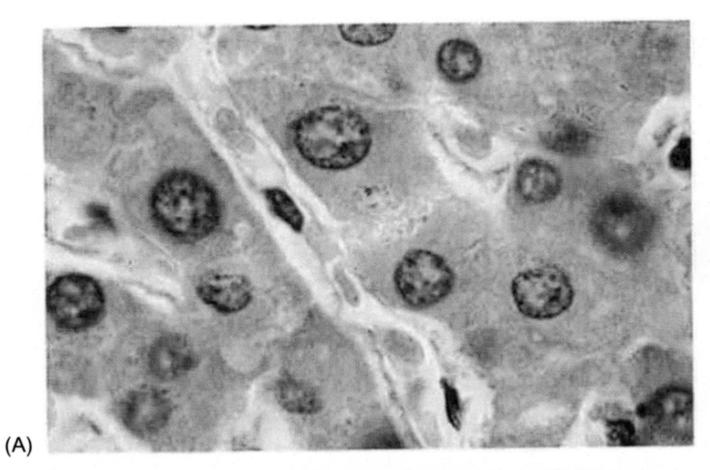

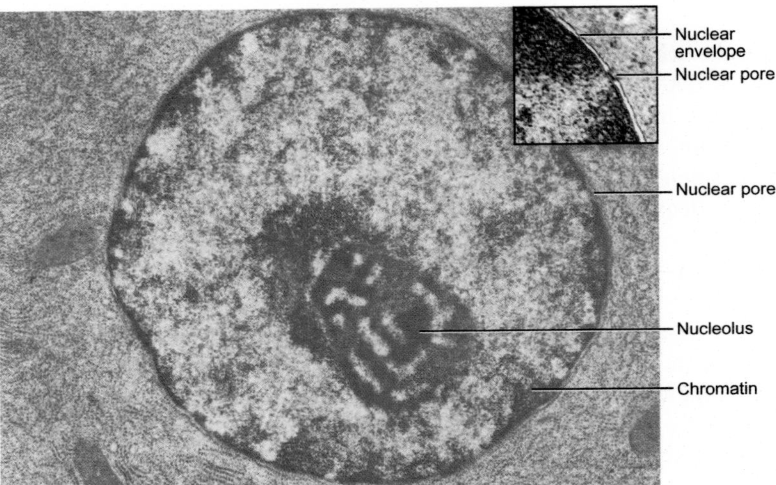

FIGURE 1.4 (A). Light micrograph of liver cells (hepatocytes), illustrating the appearance of the nucleus (purple) and cytoplasm (pink). (B). Electron micrograph of the nucleus of a cell, showing a large nucleolus and nuclear chromatin (condensed chromosomal material). The inset shows a nuclear pore, through which molecules enter and leave the nucleus. *(A) From Erlandsen and Magney (1992), with permission. (B) From Pollard and Earnshaw (2004), with permission.*

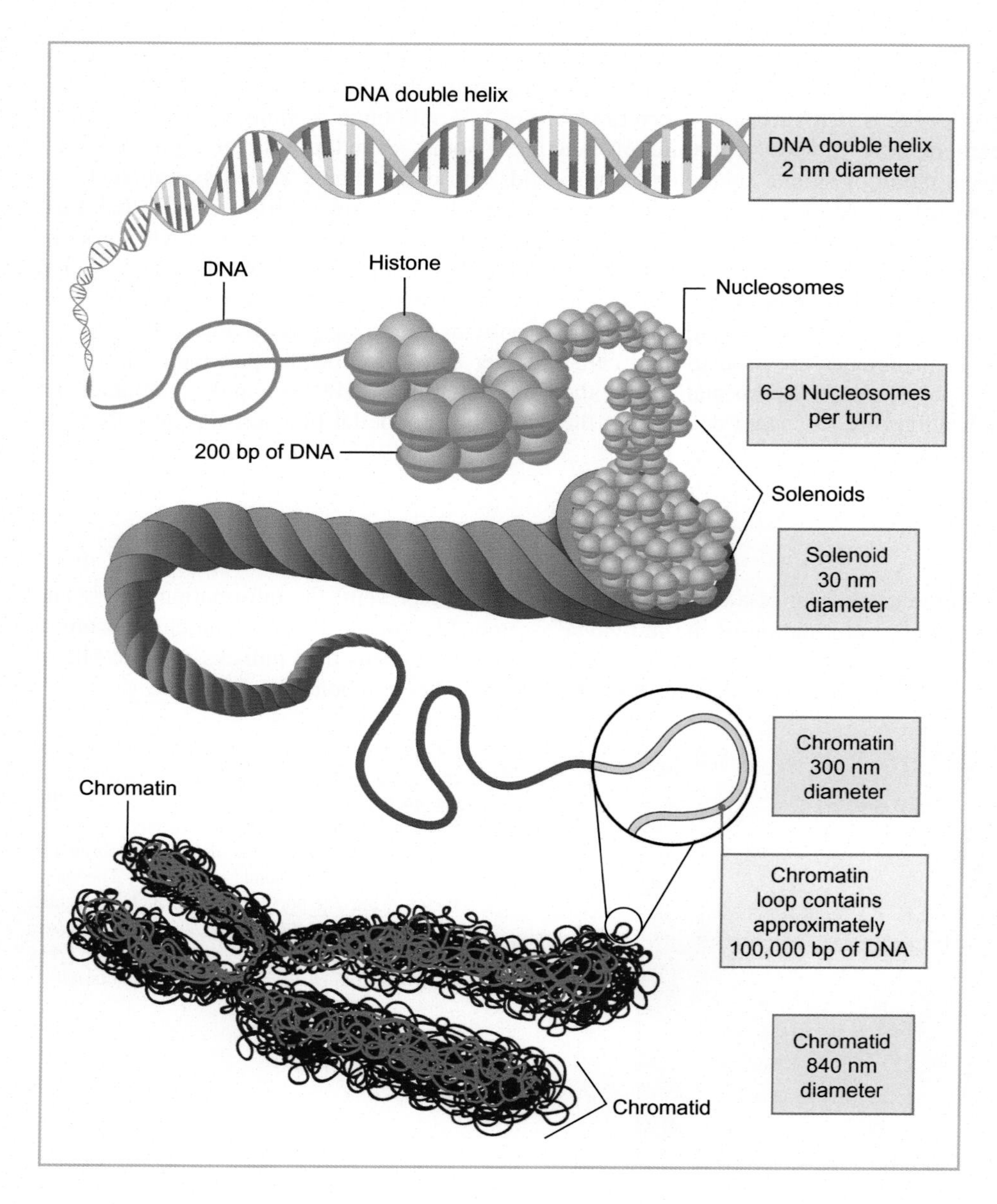

FIGURE 1.5 A chromosome, showing different levels of organization from a condensed chromosome containing two joined chromatids (bottom), as seen at the time of a mitotic division, down to the double helix of a DNA molecule. *bp*—base pair; *nm*—nanometer. *From Adkison (2012), with permission.*

The heart of the nucleus is the set of 23 pairs of **chromosomes**, which are the repository of the genetic information. The genetic information in the roughly 20,000–25,000 genes in the human genome is contained in the **DNA (deoxyribonucleic acid)** strands that form the core of the chromosome (Fig. 1.5). In any chromosome, the DNA exists as a coiled double helical structure composed of repeating units of four nucleotide bases adenine (A), cytosine (C), guanine (G), and thymine (T) in two strands of DNA. A single **nucleotide** consists of three components—a base, a sugar, and a phosphate group. For each nucleotide, the sugar (**deoxyribose**) and the phosphate group are the same. Like rungs of a ladder, the bases of the two DNA strands connect through chemical bonds in a very specific manner, so that only A and T or C and G can be bound together (see Fig. 1.8). A–Ts or C–Gs are called **base pairs**, and a human nucleus contains about 3×10^9 base pairs.

The DNA strands are very long and if laid end-to-end, the total amount of DNA in the nucleus would exceed 6 m in length. Packaging the DNA so that it will both fit into the nucleus and remain functional is complex. The first step is wrapping the DNA around several million tiny aggregates of eight small proteins called **histones** to form structures called **nucleosomes**. The DNA wraps itself roughly twice around each nucleosome, with some unwrapped (naked) DNA remaining between the nucleosomes, much like the string and beads of a necklace (see Fig. 1.5). This arrangement reduces the overall length of the DNA by roughly a factor of seven. Further packaging consists of folding of the

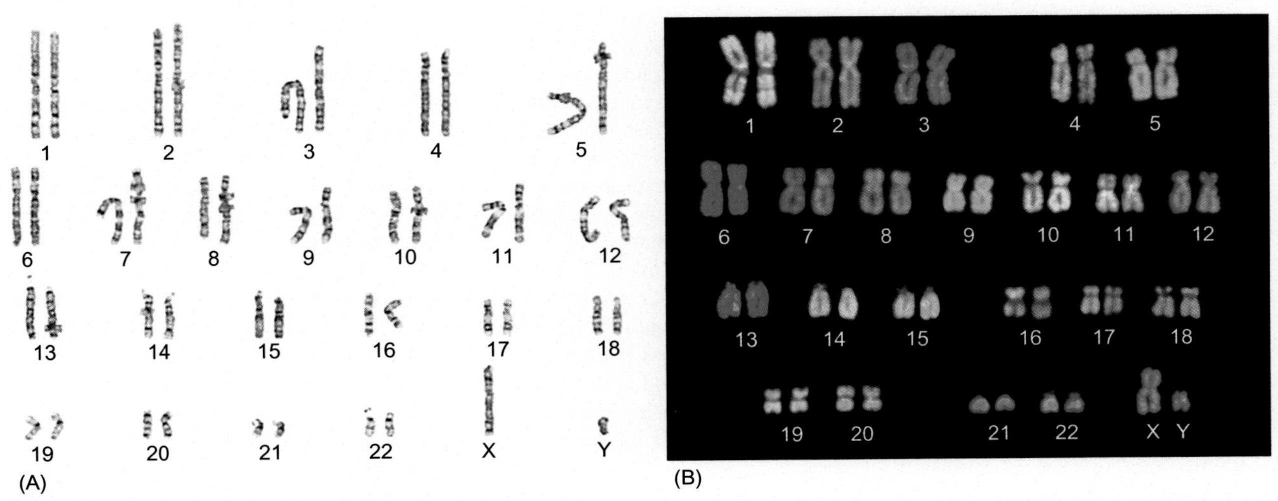

FIGURE 1.6 Condensed human chromosomes at the time of a mitotic division. (A) Individual chromosomes are matched with their homologous pair. They are arranged from the largest to the smallest and are assigned numbers. This chromosomal spread is from a male, and the X and Y chromosomes are quite different. (B) Spectral karyotyping and fluorescence *in situ* hybridization of human chromosomes. These techniques allow color-coding of individual chromosomes. *(A) From Jorde et al. (2010), with permission. (B) From Waugh and Grant (2014), with permission.*

necklace-like DNA strands, which reduces the overall length of the DNA by 40-fold. On top of that, additional folding and then a great deal of looping compresses the DNA to the point where it will fit into the nucleus. Despite all the compaction, the chromosomes within a resting nucleus are not identifiable as individual structures at the light microscopic level, although their presence is indicated by strands and clumps of stained material called **chromatin**. The fine strands of DNA, called **euchromatin**, are the most active, whereas the clumps (**heterochromatin**) represent inactive forms of DNA (see Fig. 1.4). Roughly 10% of the nuclear DNA is represented as euchromatin, and 90% is in the heterochromatin form. Only when a cell is preparing to divide, do the chromosomes further condense and become readily recognizable as individual entities (Fig. 1.6).

The chromosomes are contained within the **nuclear envelope**, which is a double membrane that is penetrated by many **nuclear pores** (see Fig. 1.4B) large enough to permit the passage of proteins and RNAs through the nuclear envelope. Nuclear pores are about 10 times the diameter of membrane channels, but protein meshworks across the pores control to some extent what passes through them. Within the nucleus, in addition to the chromosomes, is a prominent structure called the **nucleolus** (see Fig. 1.4). The nucleolus, which looks like a dark spot at the light microscopic level, is the site where **ribosomal RNA (rRNA)** molecules are synthesized and then assembled into ribosomes (see p. 11). Cells with prominent nucleoli are active in ribosome production, whereas in cells not actively producing RNAs or proteins the nucleolus is much less conspicuous. Because of their considerable activity in protein production, liver and developing muscle cells have prominent nucleoli.

THE CYTOPLASM

Organelles

If the plasma membrane represents the gatekeeper of a cell and the nucleus the information repository, then the cytoplasm would be analogous to the workshop, where things are made and broken down. The cytoplasm contains many organelles (see Fig. 1.1). Most cells contain almost all of the types of organelles, but their amounts and distribution vary considerably and reflect the locations and functions of the cells.

Production of materials within the cytoplasm begins with protein synthesis, and protein synthesis starts with molecules of **messenger RNA** (mRNA) in association with **ribosomes** (Box 1.1). The synthesis of proteins on an mRNA template is called **translation**. Although the molecular mechanics are essentially the same, protein synthesis follows two cellular pathways, depending upon whether the newly synthesized proteins will remain within the cell or if they are destined for export. Intracellular proteins are made on **polysomes—aggregates of free ribosomes in conjunction** with mRNAs. Proteins destined for export, on the other hand, are synthesized in the **rough endoplasmic reticulum (RER)**, which is a component of the endoplasmic reticulum.

BOX 1.1 Protein Synthesis

Synthesis of a protein begins in the nucleus, with a gene located along one of the DNA strands (Fig. 1.7). A **gene** is represented by a long series of **nucleotides** (Fig. 1.8), much like cars of a train, which carry specific information in what is called the **genetic code**. The information within the gene is transcribed to mRNA, which serves as the template for the synthesis of a protein. The sequence of amino acids within the protein precisely reflects the information carried in sequence of nucleotides within the gene.

Proteins consist of long strings of ~20 different **amino acids**. Each amino acid is represented by a series of three nucleotides (triplets) in both the DNA and RNA called **codons**. Through the codons, there is direct continuity between the sequence of nucleotides from DNA through mRNA to the amino acid sequence of a protein. The codons form the basis for the genetic code, which in total represents the 64 possible combinations of triplets of the four nucleotides (Fig. 1.9). For example, the DNA code for the amino acid methionine is TAC (the RNA code is AUG, see below). Other amino acids may be encoded by up to six different codons (triplets), but no codon encodes more than one amino acid. To complicate things further, RNA is organized in a slightly different manner from DNA. Whereas in DNA the bases are A, T, C, G, the T (thymidine) in DNA is represented in RNA by the nucleotide **uridine** (U) which, like T in DNA, can bind to the base A. Thus instead of the DNA code TAC, the RNA code for methionine is AUG. A string of nucleotide triplets, each representing a different amino acid, is the information base for forming a protein. (Another difference between DNA and RNA is the sugar component, which is **deoxyribose** in DNA and **ribose** in RNA.)

At any given moment, almost all genes in a cell are inactive (turned off). Genes are inactivated by a variety of complex mechanisms, but commonly by molecules that block access to the gene. The first step in protein synthesis is to turn on the gene in question through one of a variety of molecular mechanisms. This process often begins when a protein, called a **transcription factor**, binds to a segment of uncoiled DNA adjacent to the gene in question and effectively turns the gene on. The stage is now set for **transcription** of the gene, which involves the formation of a strand of RNA (in this example, **mRNA**) from the DNA template that constitutes the gene.

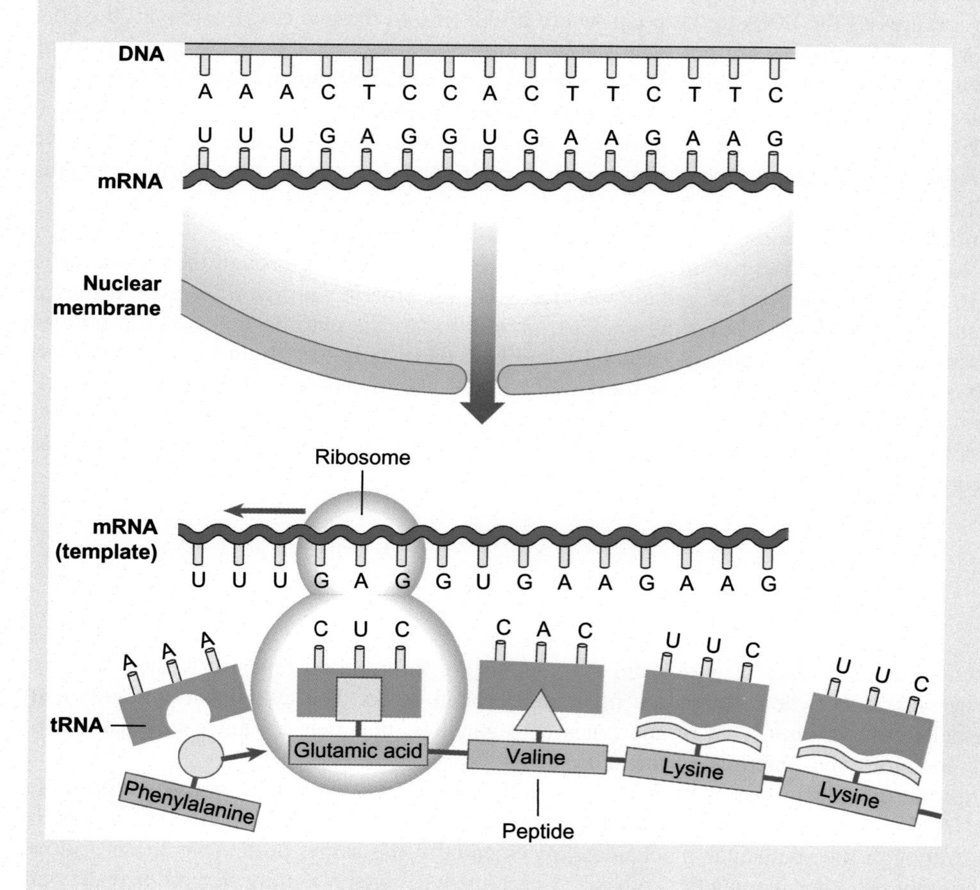

FIGURE 1.7 General scheme of the steps involved in protein synthesis, starting with the DNA molecule in the nucleus and ending with the assembly of a polypeptide chain within the cytoplasm. *From Turnpenny and Ellard (2012), with permission.*

(*Continued*)

BOX 1.1 (Continued)

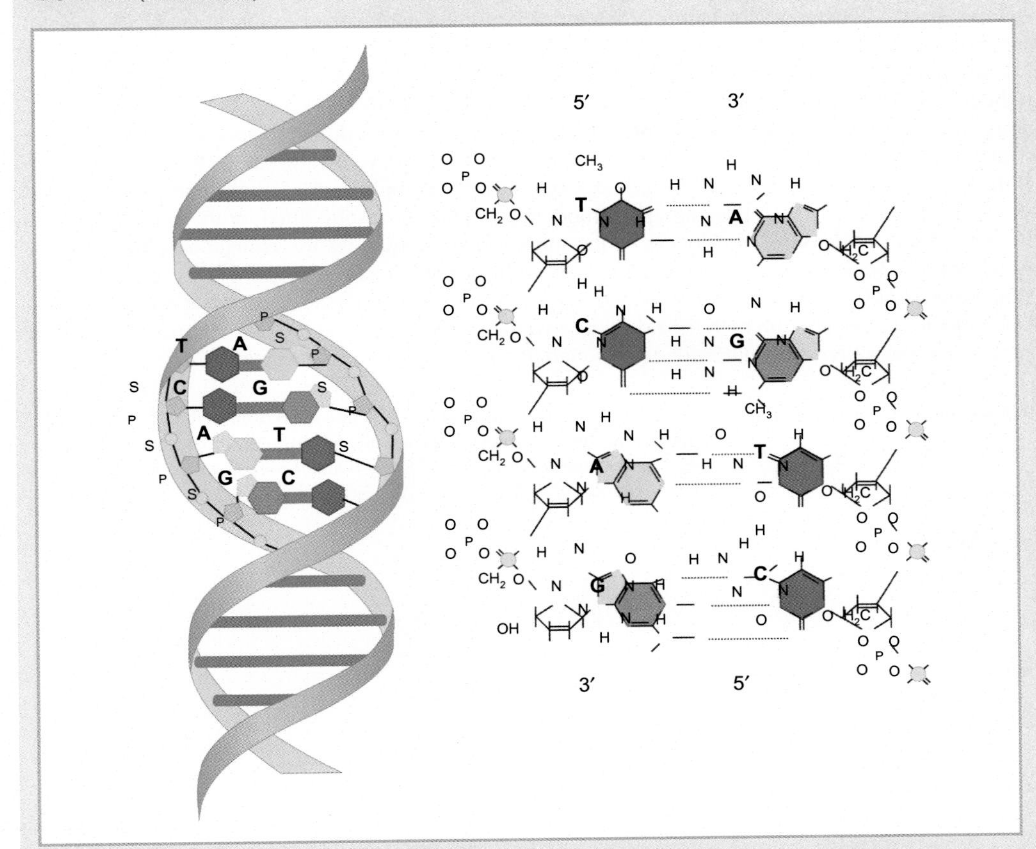

FIGURE 1.8 Structure of the DNA molecule, showing the organization of the nucleotides into individual chains of DNA, which are wrapped around each other in a double helix. A—adenosine, C—cytosine, G—guanosine, T—thymidine. Individual bases are connected to sugars (S), and the sugars are interconnected by phosphate (P) bonds. Overall, the DNA strands are organized in an antiparallel fashion. *From Adkison (2012), with permission.*

First position (5′ end)	Second position: U	Second position: C	Second position: A	Second position: G	Third position (3′ end)
U	UUU, UUC } Phe UUA, UUG } Leu	UCU, UCC, UCA, UCG } Ser	UAU, UAC } Tyr ● UAA, ● UAG } Stop	UGU, UGC } Cys ● UGA Stop UGG Trp	U C A G
C	CUU, CUC, CUA, CUG } Leu	CCU, CCC, CCA, CCG } Pro	CAU, CAC } His CAA, CAG } Gln	CGU, CGC, CGA, CGG } Arg	U C A G
A	AUU, AUC, AUA } Ile ● AUG Met	ACU, ACC, ACA, ACG } Thr	AAU, AAC } Asn AAA, AAG } Lys	AGU, AGC } Ser AGA, AGG } Arg	U C A G
G	GUU, GUC, GUA, GUG } Val	GCU, GCC, GCA, GCG } Ala	GAU, GAC } Asp GAA, GAG } Glu	GGU, GGC, GGA, GGG } Gly	U C A G

● = Chain-terminating codon
● = Initiation codon

FIGURE 1.9 The genetic code. The capital letters (A, C, G, U) represent the nucleotides in an mRNA molecule Three nucleotides represent a codon. The three-letter sequences are abbreviations for names of amino acids, for example, Phe—phenylalanine; Leu—leucine. The codons represented by green and red dots are special ones that initiate and terminate polypeptide chain formation. *From Pollard and Earnshaw (2004), with permission.*

(Continued)

BOX 1.1 (Continued)

RNA synthesis involves three stages—initiation, elongation, and termination. Initiation involves exposing the DNA segment constituting the gene so that a molecule of mRNA can be formed on the nucleotide template of the gene. In order for transcription to occur, not only must any protein molecules blocking access to the DNA be removed, but also the double helix of DNA itself must open up so that the mRNA can form along the split rungs of one strand of the DNA helix.

The key to transcription is an enzyme called **RNA polymerase**, which opens up the DNA double helix about 17 nucleotide pairs at a time and then facilitates the binding of complementary nucleotides to the open half-rungs of the exposed DNA (Fig. 1.10). For example, if the nucleotides CATG of the DNA were exposed, then the complementary nucleotides GUAC would be bound to the DNA in the form of a nascent RNA molecule. (Remember that in RNA, U is substituted for the T in DNA.) Thus the U of RNA binds to the A of DNA.) Addition of nucleotides to the mRNA chain constitutes the **elongation phase**. Like so many biochemical processes, RNA formation occurs at lightning speed, with the RNA chain elongating at a rate of 20–30 nucleotides per second. When the mRNA molecule has fully formed, the RNA polymerase reaches a **termination sequence** of DNA that causes the RNA polymerase to detach from both the newly formed mRNA molecule and the DNA strand. After release from the DNA and some processing, the mRNA is then transported through the nuclear pores into the cytoplasm.

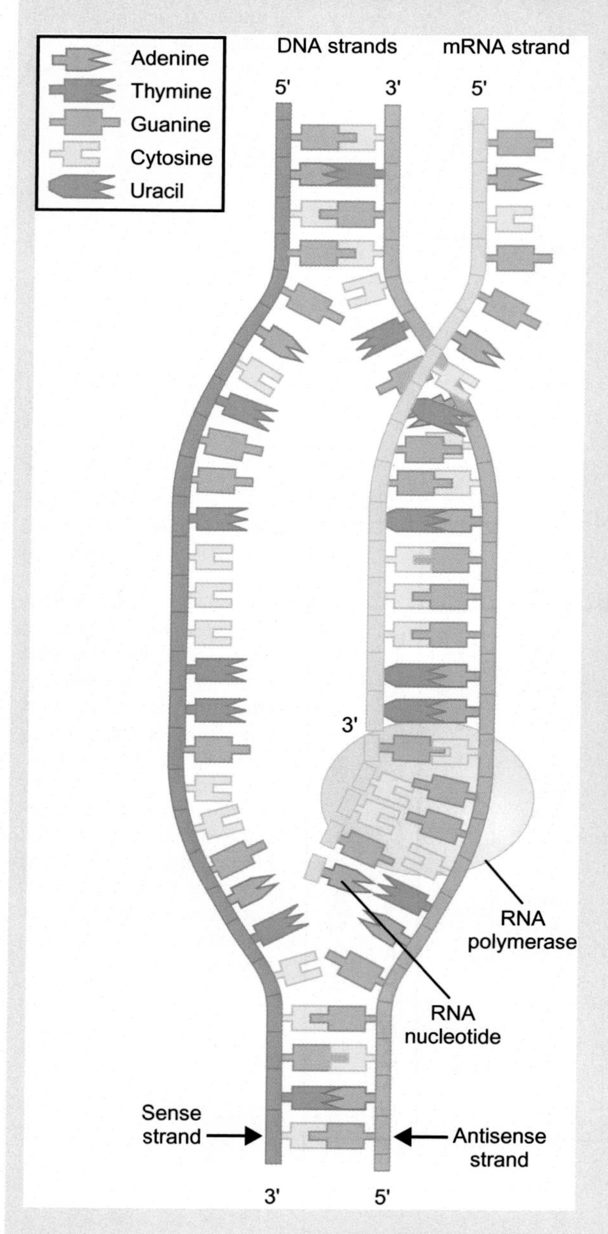

FIGURE 1.10 Transcription of DNA into mRNA through the action of RNA polymerase. mRNA nucleotides are assembled in an order complementary to that of the nucleotides in the DNA strand. *From Jorde et al. (2010), with permission.*

(*Continued*)

BOX 1.1 (Continued)

Despite the fact that there are only 20,000–25,000 genes in the human genome, the human body contains hundreds of thousands of different proteins. Such a disparity is possible because of **alternative splicing** of mRNA. The basis for alternative splicing lies in the organization of a gene (Fig. 1.11). The length of a gene in a DNA strand consists of coding regions, called **exons**, and noncoding regions, called **introns**. In order to make a functional mRNA molecule, the introns must be enzymatically removed from the newly formed mRNA and the exons joined together. In cases where there are multiple introns, the remaining exons, carrying the information for the construction of specific protein molecules, can be spliced together in various ways. Thus if a given gene has 5 or 6 exons, there are many ways in which they can be recombined to code for a protein.

Once it has left the nucleus and has entered the cytoplasm, the mRNA molecule then becomes associated with **ribosomes** in preparation for the next phase of protein synthesis—**translation**. Here the genetic information, now encoded in the mRNA molecule, is translated, much like a language, from a series of nucleotides to a series of connected amino acids, which constitute the backbone of the protein molecule. As explained more fully elsewhere in the text, translation can occur freely in the cytoplasm or can be connected with the RER.

Translation involves three essential elements—mRNA, ribosomes and **transfer RNAs** (**tRNA**) (see Fig. 1.7). In essence, a ribosome binds to the mRNA molecule. Amino acids are brought to that site by tRNAs, and the formation of a **polypeptide chain** (the backbone of a protein) is initiated. As amino acids are added to the chain, the mRNA passes through the ribosome in a manner analogous to a tape being pulled out from a tape measure.

Even at the electron microscopic level, ribosomes look like small dots, but in reality, they are very complex structures. A single ribosome is composed of two subunits (large and small) containing several types of rRNA molecules and over 80 different proteins. Initially, the end of an mRNA molecule, called the **leader sequence**, attaches to a special binding site of a small subunit of the ribosome. Then a special initiator tRNA attaches to the start codon on the mRNA, which is always AUG, the codon for the amino acid methionine. Finally, the large ribosomal subunit attaches to the small one to form a functional ribosomal–mRNA complex.

A key to this process is the structure and function of the tRNA molecule. To give a sense of proportionality, a tRNA molecule contains 70–90 nucleotide bases, an mRNA molecule up to 10,000 bases, and DNA up to 100 million base pairs. A tRNA molecule is shaped something like a clover leaf (Fig. 1.12). At the tip of the leaf is a series of three nucleotides that is called an **anticodon**. These three nucleotides are complementary to one codon of the mRNA. For example, an anticodon CCA of tRNA would bind specifically to the GGU codon of an mRNA. On the "stem" end of the tRNA molecule is a binding site for a specific amino acid corresponding to the codon of the mRNA and the anticodon of the tRNA. Thus for the tRNA with the anticodon CCA, the amino acid glycine, whose codon is GGU, would be attached. There are 61 possible different tRNAs, covering each of the possible amino acids. For many amino acids there exist up to six possible tRNAs because of redundancy in the code.

In the formation of a protein, a first step is binding of the anticodon end of the appropriate tRNA molecule to the corresponding codon of the mRNA (see Fig. 1.7). (As seen above, the formation of any protein begins with the codon AUG for the amino acid methionine.) Once formation of the polypeptide chain has begun, the amino acid at the stem end of the tRNA molecule becomes connected to the last amino acid formed in the growing polypeptide chain through enzymatic activity in the large subunit of the ribosome. During the process of polypeptide formation, the ribosome moves along the length of the mRNA molecule until it reaches the end. While this ribosome is en route, another ribosome attaches to the beginning of the mRNA and starts to produce another polypeptide. At any given time, several ribosomes may be attached to the same strand of mRNA. The complex of one mRNA molecule and several attached ribosomes is called a **polyribosome**. The process of making a polypeptide chain is repeated until the entire message in the mRNA has been read, ending with a specific **termination codon**. At that point, the completed polypeptide chain is released from the ribosome, and the ribosome breaks up into its subunits.

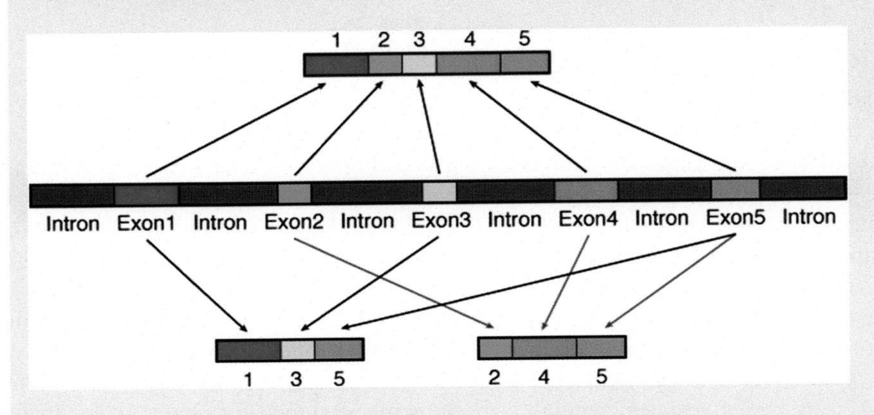

FIGURE 1.11 Alternative splicing, showing how different combinations of exons can be joined together to form the basis for the creation of different peptides.

(Continued)

BOX 1.1 (Continued)

A newly minted polypeptide chain is by no means complete. Even while the polypeptide chain is being synthesized, its free end begins to fold into what will become its final three-dimensional structure.[1] In almost all cases, small pieces are snipped off from the polypeptide chain at a later time in order to activate the protein. These pieces are called **signal peptides**, and before they are snipped off, they help direct proteins designed for export to the appropriate export pathways in the cell. Cytosolic proteins—those designed to remain within the cell—lack signal peptides. Often carbohydrate chains are added before a protein molecule is fully complete.

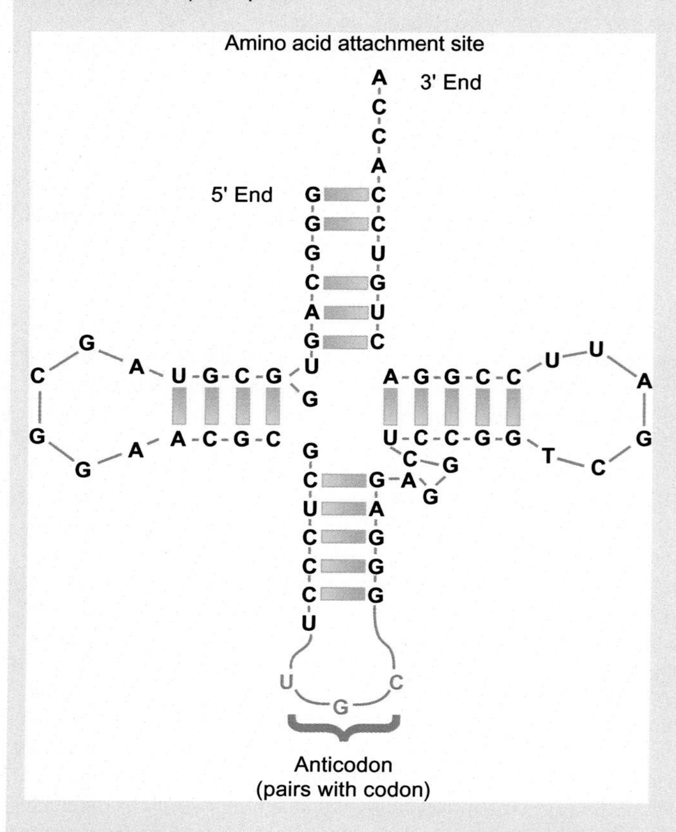

FIGURE 1.12 Structure of a tRNA molecule. Three nucleotides at the anticodon end of the molecule attach to three nucleotides of an mRNA codon (not pictured). The other (stem) end of the molecule is the amino acid attachment site, and the amino acid attached to the tRNA is added to the end of a forming polypeptide chain. *From Jorde et al. (2010), with permission.*

1. The three-dimensional structure of a protein is a function of its amino acid sequence. Among the amino acids, biophysical interactions determine a protein's final three-dimensional shape, and its shape is a major determinant of the functionality of any given protein.

Rough Endoplasmic Reticulum (RER)

The endoplasmic reticulum, which constitutes about 50% of the membranous surface within the cytoplasm, consists of three main regions that are involved in both synthetic and transport processes. The most prominent is the RER, which looks like a stack of flattened sacs studded with ribosomes on their outer surface (Fig. 1.13). The RER is continuous with the nuclear envelope. In fact, some cell biologists feel that the nuclear envelope arose evolutionarily as a modification of the RER. **Polypeptides** are synthesized on the outer surface of the RER and are then transferred into the interior of the sacs where they undergo **posttranslational modifications**. Folding of the polypeptide chains is probably a universal phenomenon within the sacs (cisternae) of the RER. In addition, it is common for small segments of newly formed polypeptide chains to be removed as they gradually take the form of the definitive proteins. Most proteins designed for export or inclusion into membranes add carbohydrate groups to the polypeptide backbone. For some of these proteins, the initial stages of this process occur in the RER. The modified proteins then move into a transitional region of the endoplasmic reticulum, where **transfer vesicles** containing the proteins, designed for export or for inclusion in membranes, bud off and are transferred to the Golgi apparatus (see Fig. 1.15).

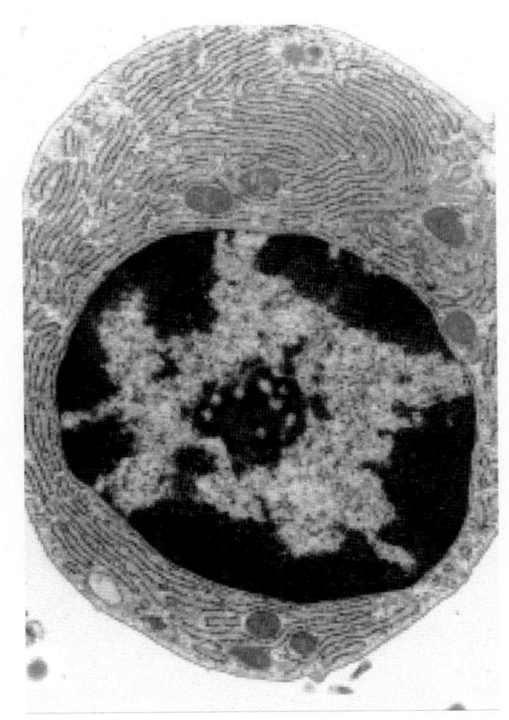

FIGURE 1.13 Electron micrograph of a plasma cell, showing a massive RER (rough endoplasmic reticulum) in the cytoplasm. In plasma cells, antibodies are synthesized in the RER. *From Pollard and Earnshaw (2004), with permission.*

Smooth Endoplasmic Reticulum

The next region of endoplasmic reticulum is an aggregation of tubules, often branched, called the **smooth endoplasmic reticulum** (Fig. 1.14). This area is heavily involved in the synthesis of fatty acids and steroids, certain aspects of carbohydrate metabolism, and detoxification reactions of lipid soluble drugs and alcohol. In skeletal muscle, the smooth endoplasmic reticulum is a reservoir of Ca^{++}, the release of which stimulates contraction of a muscle fiber. When chronically exposed to alcohol or barbiturates, the amount of smooth endoplasmic reticulum in a cell increases substantially. Expansion of the smooth endoplasmic reticulum is a major contributor to the tolerance of high concentrations of these substances seen in chronic over-users.

Golgi Apparatus

Once in the transitional region of the endoplasmic reticulum, the new proteins, enclosed in membranous vesicles, become transported to the **Golgi apparatus**, which in many cells consists of a stack of six flattened vesicles, called cisternae, located between the RER and the cell surface (Fig. 1.15; see Fig. 1.14). Each **cisterna** within the Golgi stack contains its own unique set of enzymes. Further posttranslational processing of proteins occurs as the proteins are passed from one Golgi cisterna to another until at the apical end (closest to the cell surface) of the Golgi apparatus, the finished proteins enter the trans-Golgi network, a series of interconnected tubules and vesicles that distribute the proteins to various cytoplasmic locations or to the exterior of the cell. The journey of a protein from RER to secretory vesicles takes about 45 minutes. One distribution pathway leads to the lysosomes (see below). Another directs finished proteins to the plasma membrane, where they are inserted. A third pathway carries proteins, enclosed in membrane-coated vesicles, from the Golgi apparatus to the surface of the cell. Individual vesicles fuse with the plasma membrane, open up and expel their contents into the extracellular space.

Lysosomes

Closely related to the Golgi apparatus are **lysosomes**. These membrane-bound organelles are the sites where most intracellular digestion occurs. They arise from and closely interact with the Golgi apparatus, and there is a brisk traffic of enzyme and membrane components between the two. Lysosomes contain $\sim$50 kinds of hydrolytic enzymes, which are maintained in an inactive form until they are needed. One of their main functions is to break down large molecules for

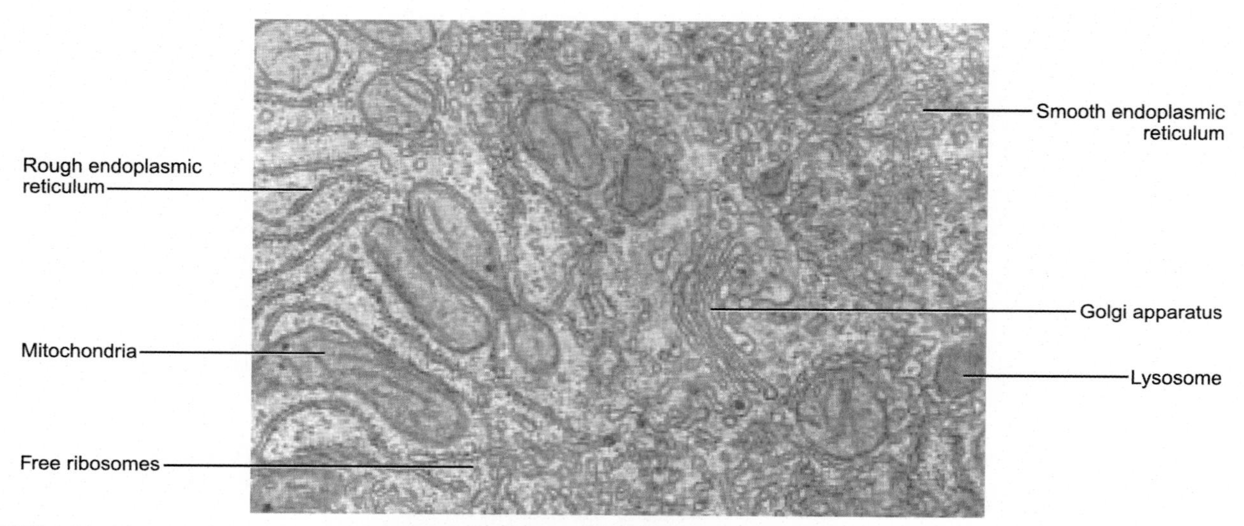

FIGURE 1.14 Electron micrograph of a thin section through a liver cell, showing a variety of cytoplasmic structures involved in synthetic and degradative activities. *From Pollard and Earnshaw (2004), with permission.*

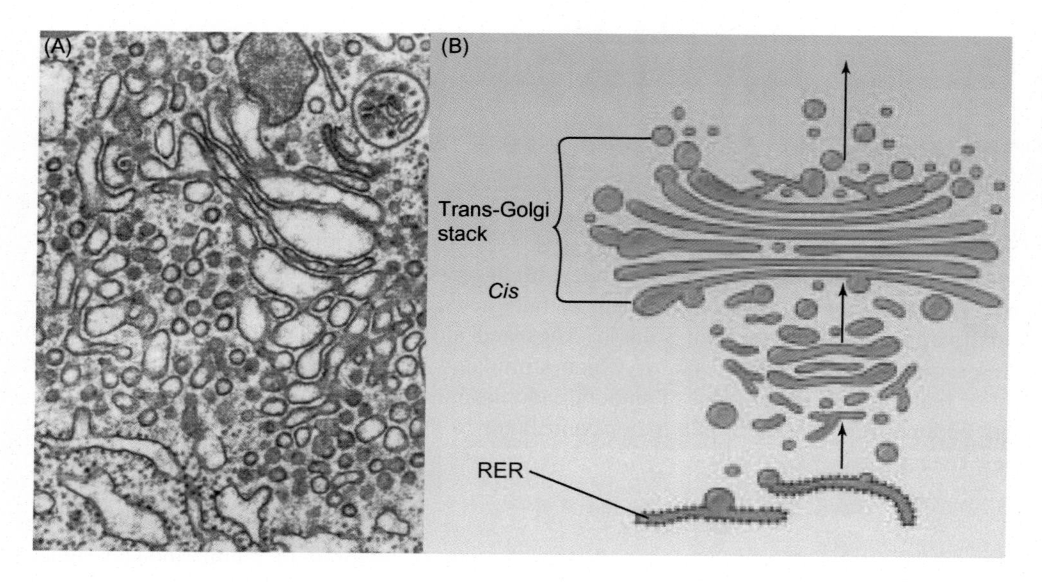

FIGURE 1.15 Electron micrograph (A) and drawing (B) of the Golgi apparatus. The arrows in B show the pathway of a protein, starting with its synthesis in the RER. *From Pollard and Earnshaw (2004), with permission.*

reutilization of their component parts in new synthetic activity by the cell. Because many of the lysosomal enzymes work best in acid conditions, the interior pH of a lysosome is 5.0. In order for this to occur without destroying the entire cell, the membrane surrounding a lysosome must be able to resist the enzymatic activity. In some types of cells, for example, white blood cells, lysosomes are actively involved in the phagocytosis and destruction of bacteria and extrinsic molecules or debris. At certain periods of life, cells die. Sometimes their death is programmed, a process called **apoptosis**, and lysosomes act as suicide bags that dissolve the cell when they release their enzymatic contents into the cytoplasm.

Peroxisomes

A related organelle is the **peroxisome**. Like lysosomes, they are membrane-bound and contain degradative enzymes. Unlike lysosomes, peroxisomal enzymes oxidize large molecules with the help of O_2. Drugs, free radicals, long-chain fatty acids, and alcohol are among the substances degraded within peroxisomes. During the process of degradation, **hydrogen peroxide** (H_2O_2) is produced, hence the name peroxisome. Further degradation by the enzyme **catalase** reduces the molecules broken down in peroxisomes to water and oxygen.

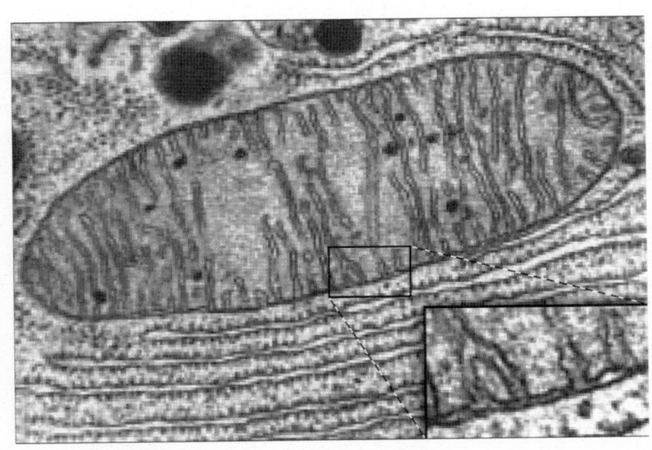

FIGURE 1.16 Electron micrograph of a mitochondrion, showing cristae protruding from the inner membrane (inset). The mitochondrion is surrounded by rough endoplasmic reticulum (RER). *From Pollard and Earnshaw (2004), with permission.*

Proteasomes

Proteasomes, barrel-shaped structures consisting of four rings of stacked proteins without a covering membrane, are important cytoplasmic players in the removal of damaged or misfolded proteins. Such proteins are recognized by other proteins, often called **chaperone proteins**, which bind to the defective proteins and lead them to the proteasomes, where they are degraded as they are fed through the central core of the barrel. In some important diseases, for example, Parkinsonism and Alzheimer's disease, abnormal proteins accumulate in the cytoplasm. How they escape degradation by the proteasomes remains a mystery.

Mitochondria

Often referred to as the powerhouses of the cell, **mitochondria** are small cytoplasmic organelles surrounded by a double membrane (Fig. 1.16). They are the only organelles other than the nucleus that contain their own DNA. In fact, through analysis of **mitochondrial DNA** (miDNA), researchers have concluded that mitochondria arose over a billion years ago when certain bacteria entered cells and took up residence in them in a symbiotic relationship that probably benefitted both the cells and the bacteria. Over the years, these bacterial invaders ultimately became incorporated into the cellular machinery to such an extent that they are now fully integrated into normal cellular structure and are critical for the functions of a cell. Nevertheless, as a residual reminder of their independent origin, mitochondria divide and fuse independently of other cell processes. In addition, their DNA, which encodes 37 genes, is circular, as is that of bacteria.

The most important function of a mitochondrion is to produce **ATP**, an energy-rich molecule that drives many biochemical reactions within a cell. The biochemical pathways leading to ATP formation are complex. One important aspect of mitochondrial function is that ATP is formed with minimal production of heat. Only in brown fat (see p. 45), through the action of a specific protein, is the metabolic pathway altered in favor of heat instead of energy production. Mitochondria also play a role in programmed cell death and in the production of steroids.

The outer and inner mitochondrial membranes have quite different properties. The outer membrane contains numerous pores that allow the ready passage of water and small molecules in and out from the space between the outer and inner membranes. The inner mitochondrial membrane is poorly permeable. It is packed with the highly ordered molecules involved in ATP synthesis. Extending inward from the inner mitochondrial membrane are numerous shelf- or tube-like projections (**cristae**) that increase the surface area available for synthesis and metabolic activity.

The numbers and ultrastructure of mitochondria reflect their activity. Highly metabolically active cells possess numerous mitochondria, which are large and contain many long cristae. For example a typical liver cell contains more than 1500 mitochondria. In metabolically inactive cells, not only are numbers and size reduced, but the cristae become much less prominent. An excellent example can be seen in kidney cells of frogs. In the spring and summer, when the frogs are busy reproducing and chasing food, the mitochondria have a very active appearance. But in the winter, when they are hibernating under the mud, the mitochondria become very reduced in number and size, and the cristae almost disappear. Mitochondria are also highly responsive to cellular activity. When a person trains for a running program, mitochondria accumulate in the muscle fibers of the legs.

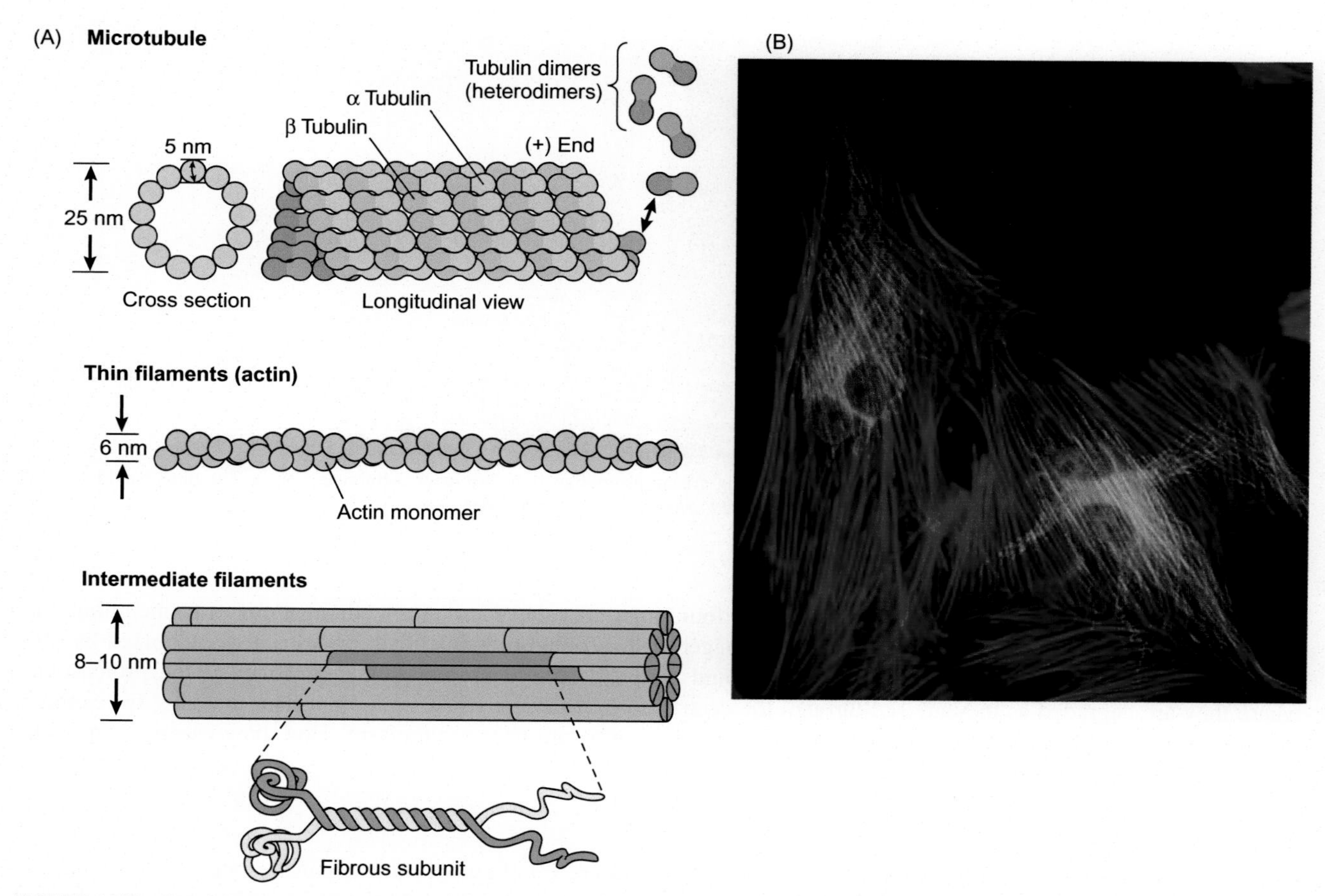

FIGURE 1.17 Cytoskeletal elements. (A) Drawings of organization of microtubules, thin filaments, and intermediate filaments. (B) An astrocyte (a glial cell) in the brain stained with fluorescent dyes. Nucleus—blue; actin filaments—red; microtubules—green. *(A) From Gartner and Hiatt (2007), with permission. (B) From Nolte (2009), with permission.*

The inner mitochondrial membrane surrounds the matrix, which contains the miDNA, as well as RNA, including ribosomes. These represent the machinery necessary for making mitochondrial proteins derived from the miDNA. In both sexes, all of the miDNA is maternally derived, because at fertilization, mitochondria of the sperm do not contribute to fertilized egg. Over hundreds of millions of years, the vast majority of the original miDNA gradually became transferred to the nucleus. Today miDNA encodes only about one percent of the proteins used by the mitochondria. The remainder of the proteins are synthesized in the cytoplasm from nuclear-derived mRNAs and make their way into the mitochondria. Both internally and externally derived proteins are arrayed in an orderly fashion within the inner membrane in order to take part in the pathways leading to ATP synthesis.

Cytoskeleton

Running through the cytoplasm are arrays of highly elongated proteins, collectively named the **cytoskeleton**. The cytoskeleton consists of three components—microtubules, microfilaments, and intermediate filaments—each with its own unique organization and functions (Fig. 1.17). Most of these functions involve the maintenance of or change in cell shape. Another important function is moving molecules or cells themselves from one place to another. An increasingly recognized function of the cytoskeleton is the transmission of extracellular mechanical forces to the nucleus, where deformation can influence the kinds and amounts of certain types of gene expression.

Microtubules

Microtubules constitute the most robust of the members of the cytoskeleton family. They are 25 nm in diameter and are constructed like a mini drinking straw, with a hollow center surrounded by a wall consisting of 13 filamentous units.

These units themselves are made up of end-to-end **tubulin** dimers (consisting of two molecules—α and β tubulin proteins) that form each of the 13 filamentous units (see Fig. 1.17). Microtubules are polarized, meaning that at one end tubulin dimers are added and at the other, dimers are removed. Like most intracellular processes, adding onto a molecule requires energy. In the case of microtubules, the energy is supplied by the breakdown of **GTP** (guanosine triphosphate) rather than by the more common ATP. A GTP molecule is attached to each tubulin protein.

Microtubules (see Fig. 1.17) play a wide variety of roles within cells, and because of their polarized structure they are involved in a variety of directional movements. For example, in nerve processes (axons and dendrites, see p. 51) they act almost as railroad tracks. Two carrier molecules, **kinesin** and cytoplasmic **dynein**, attach with one end to a molecule that needs transporting, and with the other, they ratchet themselves down the microtubules within the axon along with the molecule that they are transporting. During cell division—either mitosis or meiosis—microtubules are involved in the directional movements of entire chromosomes. Microtubules are also an important component of cilia or flagella (see p. 18), which either move materials over the surfaces of cells or move the cells themselves.

Microfilaments

Microfilaments are thin (7 nm) molecules composed principally of **actin** protein subunits, which polymerize to form elongated **actin filaments** (F-actin). Individual actin molecules, called G-actin, carry ATP to provide energy for the polymerization process. As G-actin molecules are added to one end of the F-actin chain, the ATP is reduced to ADP while it provides the energy for adding G-actin to the filament. Thus the G-actin units of the F-actin filament are bound to the lower energy ADP. Like microtubules, F-actin is a polarized molecule with a growing end, where G-actin units become attached and a retreating end, where they are removed. The process of adding units at one end and removing them from the other end is called **treadmilling**.

An important function of microfilaments is providing mechanical stability to cells. For example, in the microvilli (see p. 29) on the surface of intestinal epithelial cells or epithelial cells of the kidney tubules, microfilaments wind around one another to form bundles of filaments, thereby stabilizing the microvilli. In red blood cells, microfilaments interact with another protein, spectrin, to form a meshwork just beneath the plasma membrane that acts to maintain and preserve the overall shape of the cell even after it has been deformed by passing through a capillary (see p. 274).

Microfilaments play important roles in changing the shape of cells. When a cell divides, a ring of actin, working in concert with myosin molecules, constricts the dividing cell so that ultimately it forms a narrow waist, which ultimately completely breaks the connections between the two daughter cells. In embryonic development, one way an epithelium can change its shape from a flat plane to a tube is for a ring of actin filaments surrounding the apical ends of the cells to constrict. This changes the profile of the cell from a rectangle to something approaching a wedge.

A major function of actins is to partner with myosin proteins in causing entire cells to contract. Their role in muscle contraction is discussed in Chapter 5. Actins are similarly involved in less dramatic, but equally important cellular movements, such as the ameboid movements seen in some protozoa or the creeping of vertebrate white blood cells. Virtually any cell that moves under its own power uses actin filaments at some point. Branching actin filaments accumulate at the leading edge of a moving cell and become degraded at the trailing edge.

Intermediate Filaments

The third major category of cytoskeletal structures is the **intermediate filaments**, whose ~ 10 nm diameter lies between that of microfilaments and microtubules (see Fig. 1.17). In contrast to the other two cytoskeletal elements, intermediate filaments are not polar structures. They are put together through side by side polymerization between elongated individual monomer proteins.

Intermediate filaments are tough and resist substantial stretching. Their main function is mechanical support in a variety of cellular domains. One example is a class of proteins called **nuclear lamins**. Lamins form an orthogonal network just beneath the inner nuclear membrane and give it stability. In epidermal cells, bundles of keratins provide the strength for the epidermal cells to resist the mechanical trauma to which they are often subjected. Another important function for intermediate filaments is their attachment to hemidesmosomes, structures that are important in firmly connecting epithelial cells to their underlying basal laminae (see p. 28).

Centrosome

Among the cytoskeletal elements, microtubules exhibit the most highly ordered pathway of development. It all begins in the **centrosome**, a region near the center of the cell that serves as a microtubule organizing center. The most prominent feature of the centrosome is two small barrel-shaped **centrioles** (see Fig. 1.1), the walls of which are composed of nine triplets of short microtubules arrayed in a somewhat helical fashion. In a resting cell, the centrioles are oriented at

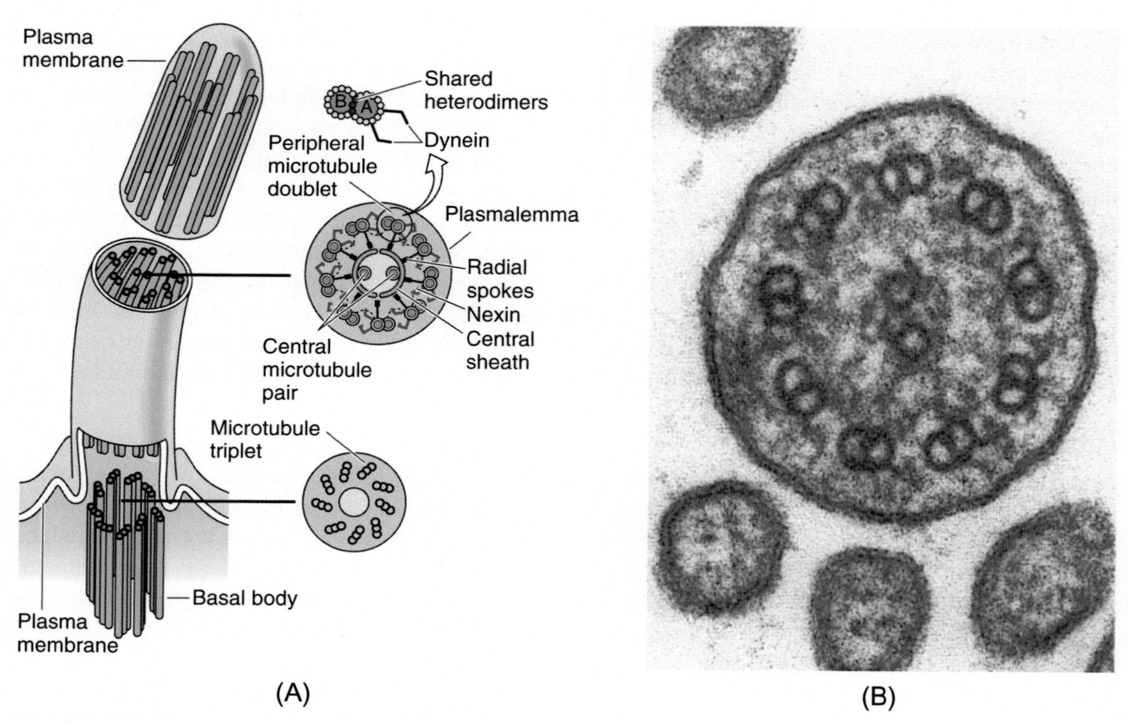

FIGURE 1.18 (A). Structure of a cilium. (B) Electron micrograph of a cross-section through a cilium, showing the nine pairs of microtubules surrounding a single central pair. *(A) From Gartner and Hiatt (2007), with permission; (B) From Stevens and Lowe (2005), with permission.*

right angles to each other. The major ongoing function of centrioles is to serve as the nucleus for the formation of microtubules, but in preparation for mitosis, the centrioles replicate. The resulting two pairs of centrioles form the basis for the two centrosomes that serve as the apices for each end of the mitotic spindle (see Fig. 1.22). In both dividing and nondividing cells, microtubules grow out from the centrioles.

Microtubules form the core (**axoneme**) of individual cilia and flagella, the hair-like structures that project from the surfaces of many epithelial cells and beat synchronously to move particles or sheets of mucus across the surface of the cells. The ciliary microtubules, however, are the products of a **basal body** which, as the name implies, lies at the base of each cilium or flagellum (see Fig. 1.18). Basal bodies are derived from centrioles, and a basal body then acts as an organizing center for the production of the microtubules found in the cilia. Through their microtubules, the cilia are firmly anchored to their respective basal bodies.

Cilia and Flagella

Cilia and **flagella** are phylogenetically ancient cellular organelles. Possibly as long as two billion years ago early protozoa developed cilia or flagella and used them to propel themselves through the water. The basic underlying structure of a cilium has not changed appreciably throughout this long evolutionary period. Cilia, which cover the free surface of some cells almost like blades of grass in a lawn, beat in unison at a rate as much as 100 beats per second. The surface of an epithelial cell contains from 50 to 200 cilia. Much like rowing a boat, ciliary beat has a power stroke that propels either the cell through a liquid or mucus over a cell that is fixed in place. In the latter case, however, a watery layer underlies the layer of mucus, and only the tips of the cilia contact the mucus itself. Without the watery layer, the cilia would become bogged down in the highly viscous mucus. The power stroke is followed by a slower recovery stroke. Flagella, such as the tail of a spermatozoon—the only human cell to contain a flagellum, are much longer than cilia, but their internal construction is very similar. The character of their beat differs from that of a cilium and resembles closely the way a snake swims in water.

The microtubular arrays of both cilia and flagella grow out of **basal bodies**, which are located in the cytoplasm just at the base of the cilia (Fig. 1.18). Through the action of transport proteins, tubulin subunits are carried along existing microtubules to their tips and then fuse with the growing tips of the microtubules. Through this mechanism, a cilium can grow to its full length in less than an hour. Like centrioles, basal bodies consist of nine helically arranged triplets of microtubules. Unlike either centrioles or basal bodies, microtubules within the shaft of a cilium consist of a cylinder of nine doublets of microtubules with an additional pair of microtubules in the center.

Over 200 accessory proteins are also associated with the microtubules. One of these proteins, **dynein**, is a major player in creating ciliary movement. Working with energy provided by the breakdown of ATP, dynein, in concert with other proteins, causes microtubule segments to slide past one another longitudinally, resulting in the bending of the cilium. The exact mechanism of ciliary motion is still the subject of considerable research. Exposure to cigarette smoke not only slows down the rate of ciliary beat in the respiratory system, but reduces the total length of individual cilia by about 10%. The shortened length further reduces the overall effectiveness of ciliary motion.

Cytoplasmic Inclusions

A final category of cytoplasmic structures consists of **inclusions**. These are not true organelles, but are aggregations of materials within the cytoplasm. A good example is the storage of neutral fat. In a typical fat cell, lipid accumulates in a large droplet that is not surrounded by a membrane. In common **white fat**, the droplet is so large that it occupies the vast majority of the volume within the plasma membrane so that when viewed through a microscope a fat cell looks like a signet ring, with a very thin rim of cytoplasm and a bulge caused by the nucleus. (In a normal microscopic preparation the fat dissolves out of the cell.) Brown fat cells (see p. 45) contain multiple small droplets of lipid rather than one large globule. Another common inclusion, especially in muscle fibers, liver, and brain cells, is **glycogen**, which is an efficient intracellular storage form for carbohydrates. Glycogen is also stored in the epithelial cells lining the vagina. When extruded from the cells, the glycogen is acted upon by resident vaginal bacteria and metabolized into lactic acid, which lowers the vaginal pH to a level that discourages the entry of most pathogenic bacteria. In most cells, the presence of inclusions, including fat, is an indication of a pathological process.

TRANSPORT ACROSS THE CELL MEMBRANE

One of the most prominent features of a living cell is transport of materials across membranes. We have already discussed transport of materials by diffusion or through various channels, transporters and pumps (see p. 4, Table 1.1). Another important means of transport is through membrane-coated vesicles. Such transport can be into a cell (**endocytosis**), out from a cell (**exocytosis**), or through a cell (**transcytosis**).

Endocytosis

Endocytosis comes in several varieties (Fig. 1.19). One of the simplest—**pinocytosis**—consists of the infolding of the plasma membrane to form a vesicle containing extracellular fluid with its dissolved solutes. The vesicle pinches off from the plasma membrane and migrates into the cytoplasm, where it fuses with a lysosome. The lysosomal enzymes digest any large molecules and release the digestion products into the cytoplasm for use in cellular metabolism.

Phagocytosis consists of a cell's engulfing a bacterium or some other large particle by folds of cytoplasm. The particle with its surrounding membrane becomes pinched off into the interior of the cell as a vesicle called a **phagosome**. The phagosome fuses with a lysosome to form a **phagolysosome**, which then digests the particle into its component parts. These are released into the cytoplasm for re-use, and the residue remains within the membrane-covered vesicle as a residual body. Phagocytosis is a common way by which protozoan amoebae take material into the cell, but it is also used by white blood cells to ingest bacteria and debris at sites of tissue damage or inflammation. In fact, macrophages (white blood cells residing in tissues) can ingest up to 25% of their mass in an hour.

A more specific process is **receptor-mediated endocytosis** (see Fig. 1.19). Here, ligand molecules, for example, cholesterol, bind to specific cell surface receptor proteins, which are located in **clathrin-coated pits**. Coated pits occupy 2%–5% of the surface of the plasma membrane. **Clathrin** is a protein located on the cytoplasmic side of the pit. After the cholesterol molecules are bound to the receptors, the pits invaginate to form spherical vesicles, which are coated on the outside by a meshwork of clathrin molecules. The budding of a single vesicle takes about 1 minute. The structure of clathrin molecules individually and communally promotes the formation of the spherical shape of the vesicles. Once the vesicle has formed and has pinched off into the cytoplasm, the clathrin molecules uncouple from the vesicle and are recycled back to the inner surface of the cell. Several vesicles then fuse into a larger structure called an **endosome**, within which the cholesterol molecules become released from their receptors. The receptors are also then recycled back to the plasma membrane, and smaller transport vesicles, which have budded off the endosome, fuse with lysosomes, where further degradation of the cholesterol molecules occurs.

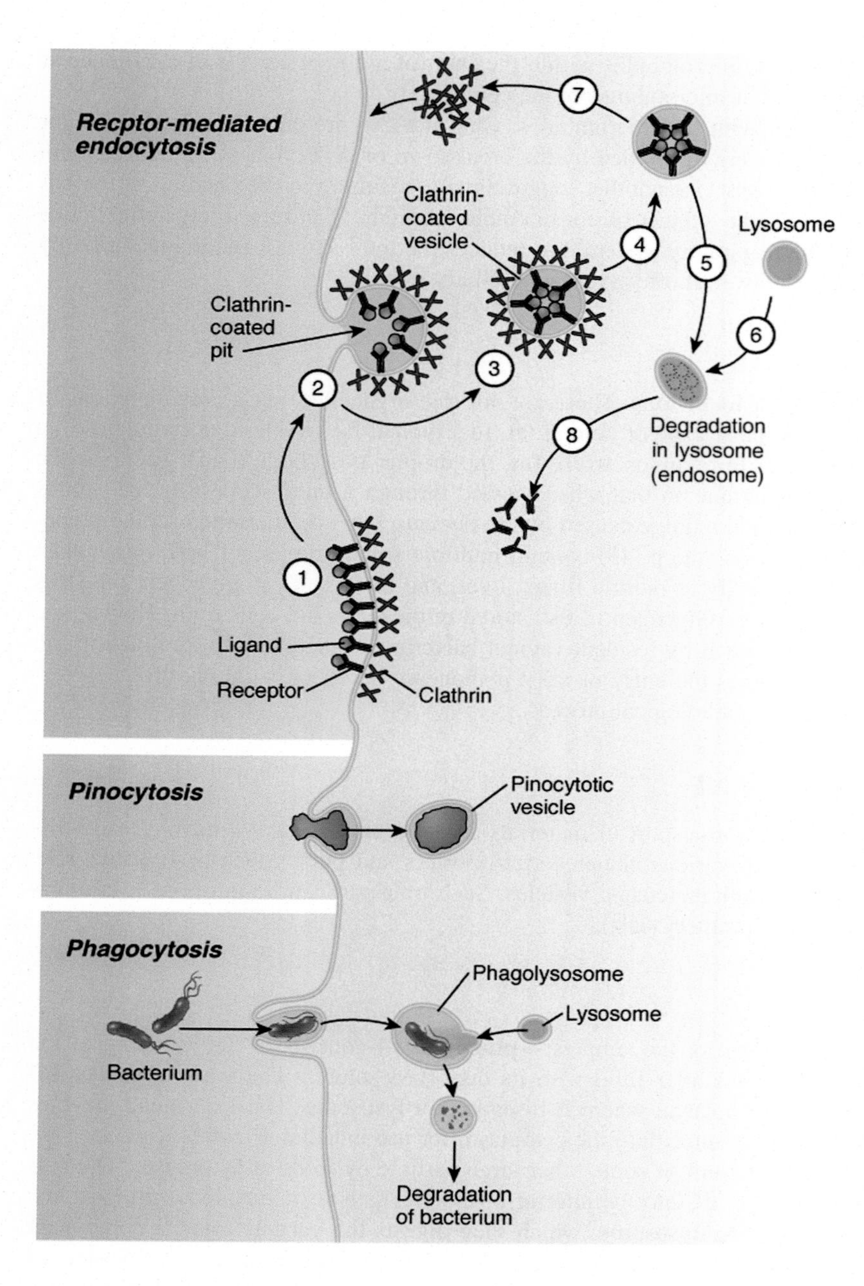

FIGURE 1.19 Three means by which material can be taken into a cell. Top—receptor-mediated endocytosis; middle—pinocytosis; bottom—phagocytosis.

Exocytosis

Exocytosis principally involves the budding off of membrane-coated vesicles from the terminal cisterna of the Golgi apparatus and their passage toward either the apical or lateral surface of a cell. These vesicles, which contain either proteins or carbohydrates designed for export, are not clathrin-coated. Once the vesicle contacts the plasma membrane, it fuses with it and opens to the outside, where the contents of the vesicle are deposited. Both exocytosis and the positioning of the Golgi apparatus occur on the side of the cell where the cellular product is most likely to be deposited. In most cells, this occurs between the nucleus and the apical surface, but in some cases, especially epithelial cells secreting a basal lamina, the Golgi apparatus is located in and exocytosis occurs from the basal side of the nucleus.

Transcytosis

Transcytosis is a process that takes material from one side of a cell and transports it in the form of a membrane-coated vesicle through the cell for its release on the other side of the cell. This mechanism is particularly prominent in the endothelial cells that line capillaries. It also occurs in the placenta, where there is an active exchange between the fetal and maternal sides of the placental barrier.

THE CELL CYCLE AND CELL DIVISION

According to recent calculations, the adult human body contains close to 40 trillion cells. Yet, the human body starts out as a single-cell fertilized egg. The number of cell divisions required to reach this number is almost unfathomable, but even more remarkable is the degree of cell production and turnover in the adult human body. With relatively few exceptions, such as nerve cells (neurons) and heart muscle cells, most cells in the body do not remain intact indefinitely. Many die at regular intervals and are replaced by newly formed cells, which arise by the division of precursor cells. Good examples are the epidermis of the skin and red blood cells. The outer surface of the epidermis is continuously shed, often as the result of friction, and must be replaced. Red blood cells, which in humans do not contain a nucleus, live between three and four months before they are removed from the circulation and replaced by new cells. In the adult human, ~2.4 million new red blood cells are produced each second. Thus the body is in a constant state of cellular flux. Central to all of this activity is cell division (**mitosis**), which is one component of the life cycle of a cell.

Cell Cycle

Until its last division, after which a cell dies, all cells go through a cycle of major events that terminate with the cell's division into two **daughter cells** (Fig. 1.20). The **cell cycle** begins with a long period, called **G_1 (gap-1)** during which the cell is growing in mass and is carrying out its normal functions. At some point, when the mass of the cell is sufficient, internal signals prepare the cell for an upcoming division.

This process begins with the duplication of the DNA in the **S (DNA synthesis) phase**. During the 7–8-hour-long S phase, the long double helices of DNA in the chromosomes unwind at various locations along their length, and two new DNA strands form as exact replicas of the original DNA strands (Fig. 1.21). Formation of the new DNA strands is coordinated by the enzyme **DNA polymerase**, which fits individual bases onto those exposed by the separation of the

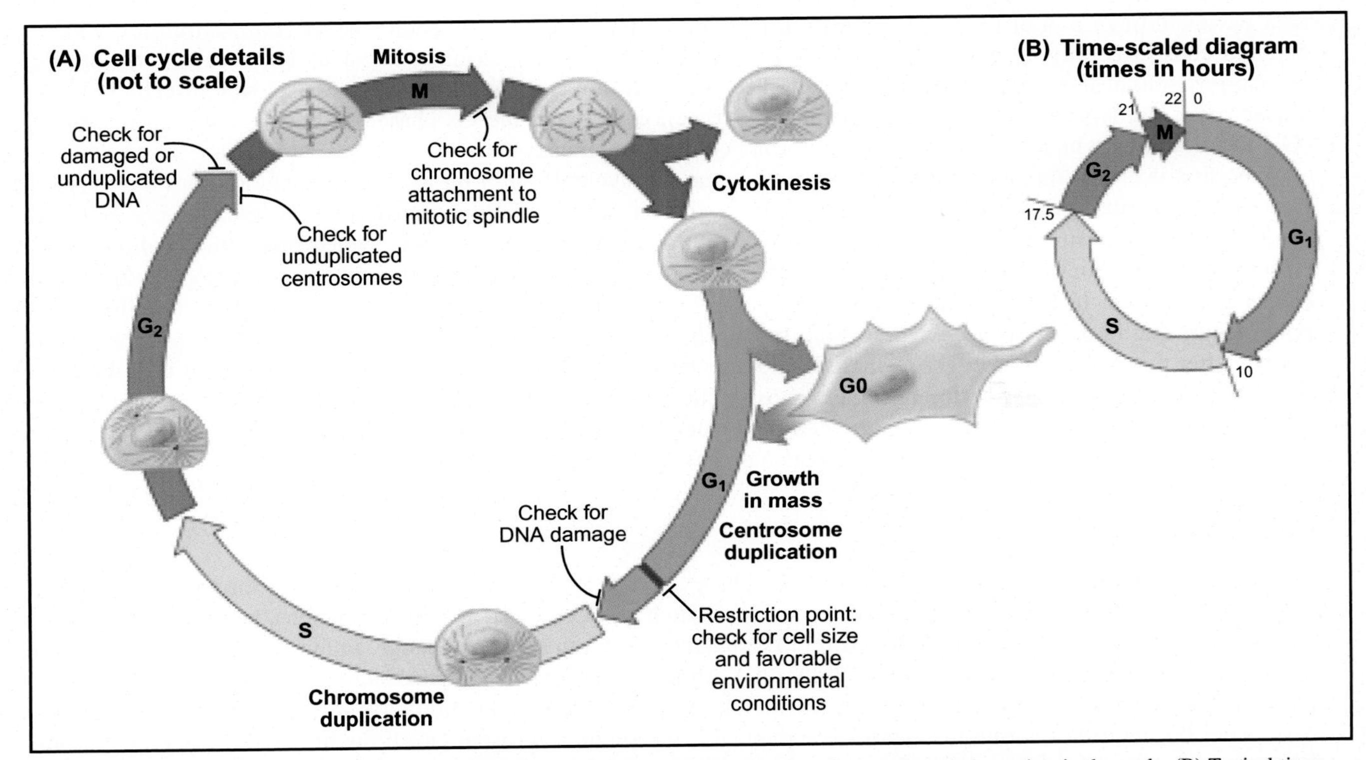

FIGURE 1.20 The mammalian cell cycle. (A) Diagram of the overall cell cycle, showing check and restriction points in the cycle. (B) Typical times for various phases in the cell cycle. *From Pollard and Earnshaw (2004), with permission.*

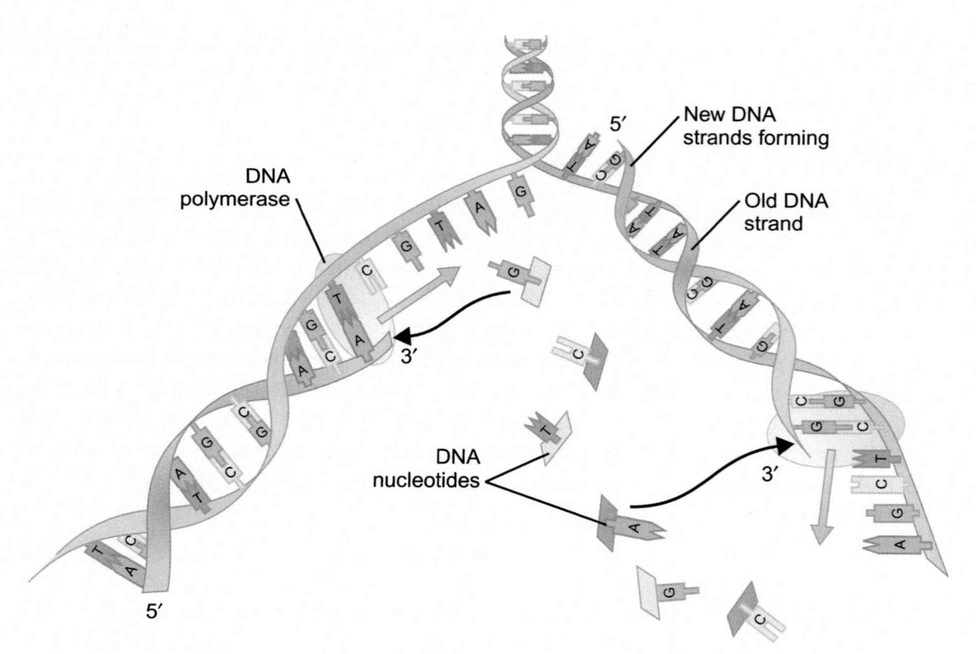

FIGURE 1.21 DNA replication before mitosis. The original double helix separates after the hydrogen bonds that hold the nucleotides together are broken. Then, with the help of DNA polymerase, free nucleotides pair with complementary nucleotides in the separated DNA strands to form new DNA strands. *From Jorde et al. (2010), with permission.*

DNA strands and then connects them to the ends of the newly forming DNA strands. Through the efforts of DNA polymerase, approximately 100 base pairs per second are added to the replicating DNA strands at each site of DNA polymerase action. Without the multiple sites of DNA synthesis, the entire process of replication would take weeks if it were done in a purely linear fashion.

By the time DNA replication is completed, each of the original chromosomes has duplicated. The two halves (chromatids) of the duplicated chromosomes lie side by side and are connected at one point, called the **centromere**. The cell then passes through a brief ($\sim$4 h) **G_2 phase** in preparation for actual cell division (the **M [mitosis]phase**), which in cultured cells lasts only an hour. Many cells leave the cell cycle after a terminal division. These cells carry out their normal physiological functions and then usually die. Such cells are said to be in the **G_0 phase**. Not all cells in G_0 die. Under some circumstances they can reenter the active cell cycle from the G_0 phase. Most cells enter an intermitotic (**G_1**) phase, during which they carry out their normal cellular functions.

Precision is one of the most important elements in the cell cycle. If all of the steps in preparation for mitosis are not perfectly set in place, cell division will be abnormal, and either cell death or transformation into a tumor cell could result. In order to ensure fidelity in the overall division process, several checkpoints, called **restriction points**, exist at critical periods within the cell cycle (see Fig. 1.20). If preparations do not pass muster, the process stops until the deficiency is corrected. If correction does not occur, passage through the rest of the cell cycle is aborted. The first checkpoint occurs late in the G_1 phase where overall monitoring for both cell size and environmental conditions takes place. Shortly after that and just before the S phase is another check for DNA damage. The next critical period is at the end of G_2, just as the cell is preparing to enter into mitosis. This checkpoint monitors for damaged DNA or for unduplicated DNA or chromosomes. If all goes well, the cell then enters mitosis, but a final check occurs midway through that process to ensure that the chromosomes will segregate properly to the two daughter cells.

Cell Division

Cell division (**mitosis**) is broken up into several phases, which can be readily distinguished through a microscope (Fig. 1.22). In **prophase**, the individual chromosomes, which are spread out within the nucleus, begin to condense and can be recognized as interwoven strands within the nucleus. The other major change within the nucleus is the breakdown and disappearance of the nucleolus. Within the cytoplasm, much of the cytoskeleton becomes reorganized and is often broken down into component parts. At this point, the centrosome begins to play a very prominent role in the process. After the centrioles within the centrosome duplicate, the single centrosome itself doubles and the two centrosomes begin to move apart toward opposite ends of the cell. As they are doing this, new microtubules sprout from each centrosome as clusters of star-shaped rays, called **asters**.

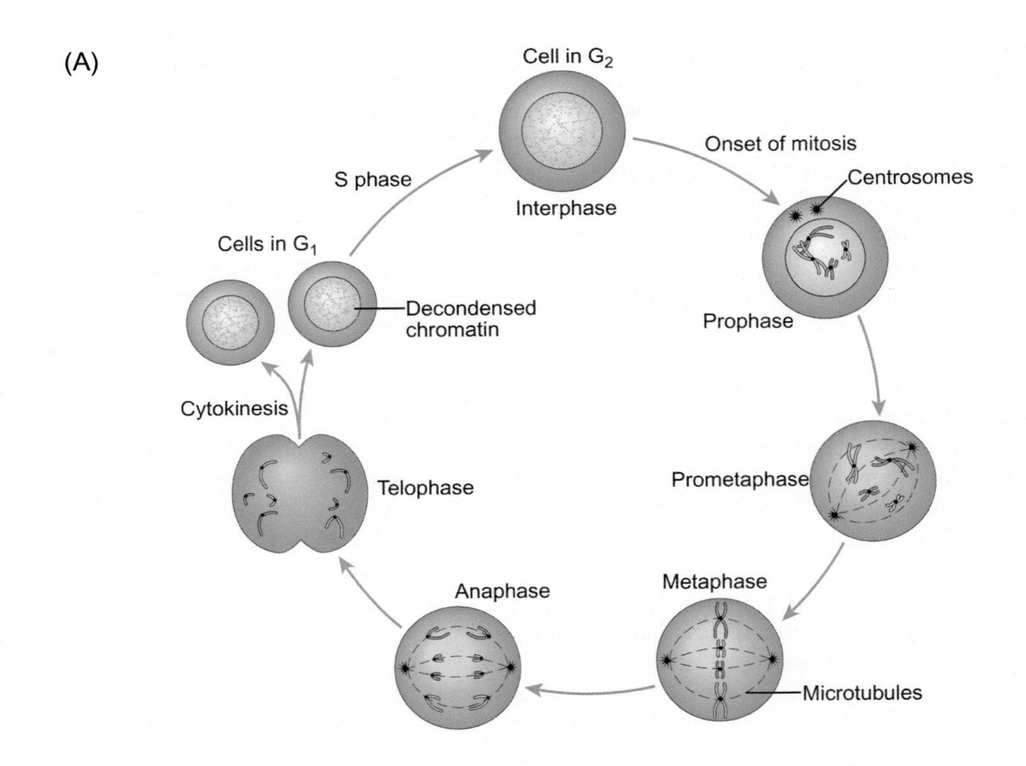

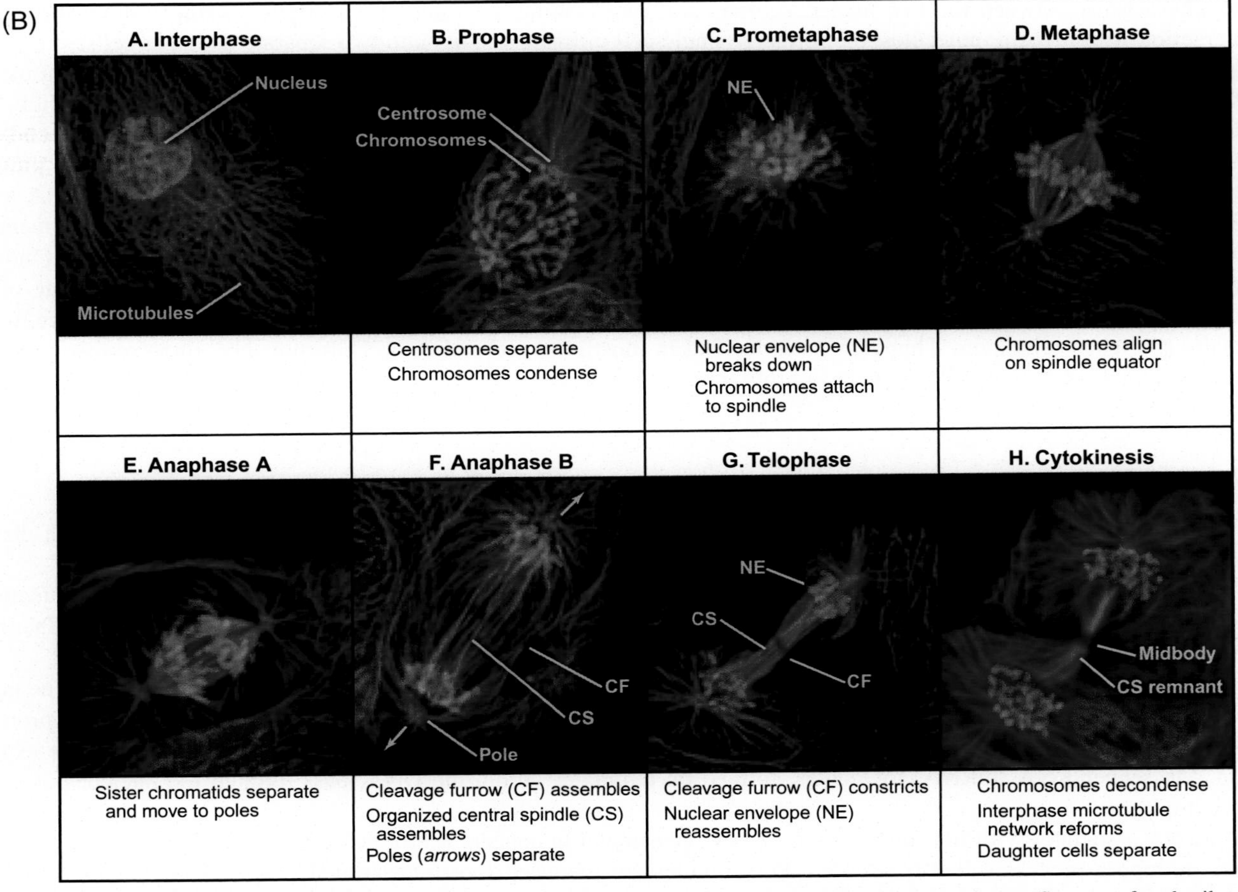

FIGURE 1.22 (A) Schematic representation of the mitotic cycle. Only two pairs of chromosomes are shown. See text for details. (B) Photomicrographs of stages of mitosis. Chromosomes are stained blue, and microtubules of the spindle apparatus are red. *(A) From Nussbaum et al. (2016), with permission. (B) From Pollard and Earnshaw (2004), with permission.*

With the dissolution of the nuclear envelope, mitosis enters metaphase. The chromosomes condense even further, and on either side of the centromere, aggregates of proteins form **kinetochores**, attachment sites for microtubules extending from the centrioles. Metaphase is the period when the mitotic spindle gets organized. In essence, the spindle consists of microtubules growing out from the centromeres on either side of the cell toward the condensed chromosomes in the middle. When they arrive, their ends attach to the newly formed kinetochores (up to 20 microtubules per kinetochore) on the chromosomes to form the mitotic spindle. Microtubules from one centrosome attach to one chromatid, and microtubules from the second centrosome attach to the other sister chromatid. With a lot of jiggling and adjusting, the condensed chromosomes become situated much like a disk oriented perpendicularly to the long axis of the mitotic spindle in what is sometimes called the **metaphase plate**. A last checkpoint ensures that all of the chromosomes are properly aligned along the metaphase plate. If their orientation is appropriate, then the chromosomes are allowed to begin to separate.

Anaphase is the name given to the period when the paired chromosomes begin to move apart (see Fig. 1.22E,F). This cannot happen until a protein, called **cohesin**, that binds the two sister chromatids together at the centromere is degraded. Then through a combination of shortening of the microtubules and the action of motor proteins, the chromatids from each chromosome begin to separate from one another and move up their respective halves of the mitotic spindle toward the centrosomes. Separation of the chromosomes from each other is also facilitated by overall lengthening of the mitotic spindle through the elongation of sets of spindle microtubules that were never connected to the chromosomes. The attachment of aster microtubules to the cytoplasmic cortex on opposite sides of the cell provides an anchor that allows force to be exerted on the separating chromosomes.

While the chromosomes are separating during late anaphase, poorly defined signals emanating from the mitotic spindle prepare the cytoplasm for **cytokinesis**, the actual division of the entire cell. Cytokinesis occurs during the last phase of the mitotic cycle, **telophase**. During telophase, a new nuclear envelope begins to form around the now completely separated sets of chromosomes. The chromosomes themselves begin to decondense and a nucleolus reforms within the nucleus. At the same time, a ring of actin filaments in conjunction with a special form of myosin, begins to contract, forming a waist between the two nuclei. As the actin ring continues to constrict in a purse-string fashion through its interaction with myosin molecules, the formerly single cell ultimately splits into two separate daughter cells.

Most cells do not have an unlimited capacity to replicate. A typical human cell can divide 40–60 times, but usually no more than that. One of the main reasons for this limit (often called the **Hayflick limit** after Leonard Hayflick, who discerned it) is the **telomere**—a nucleotide sequence of CCCTAA that is repeated nearly 1000 times at the end of a DNA strand. There are several reasons why telomeres may exist. They range from stabilizing the chromosome, preventing inappropriate DNA repair at the normal end of a chromosome, to permitting proper replication of the DNA at the end of the chromosome. These reasons aside, each time a cell divides, it loses up to 100 bases from the telomeres of each chromosome. After a number of divisions, the length of the telomere segment becomes so short that normal mitotic mechanisms are interfered with, and the cell stops dividing. Telomere shortening is the basis for one of the major theories of aging. Certain cells, such as gametes, which must replicate often, have an enzyme, **telomerase**, which repairs shortened telomeres. This enzyme is lacking in most other normal cells. Unfortunately, some cancer cells also possess telomerase, which allows them to proliferate indefinitely.

SUMMARY

A cell is the fundamental unit of life. Most cells consist of a nucleus and cytoplasm. Cells are covered by a plasma membrane. The cytoplasm contains many organelles.

The plasma membrane consists of a lipid bilayer, studded by a large variety of proteins. Among many functions, these proteins can also serve as receptors, enzymes, channel proteins, pumps, and adhesion molecules. Channels, pumps, and carriers are all involved in the movement of molecules across cellular membranes.

The nucleus contains the cell's genetic material, which is embedded within the 23 pairs of chromosomes. The genes are encoded within the DNA of the chromosomes. The chromosomes themselves are associated with histone proteins, which help to organize the configuration of the chromosomes and have important functions in regulating gene expression. Messenger RNA (mRNA) molecules, formed on templates within the DNA, pass through the nuclear pores into the cytoplasm.

mRNA associates with ribosomes within the cytoplasm and in concert with transfer RNA (tRNA) produces polypeptides through a process called translation. Polypeptides are modified (posttranslational modification) in both the RER and the Golgi apparatus before being released as functional proteins.

Mitochondria serve as the powerhouses of the cell and are heavily engaged in energy production, whereas lysosomes contain many degradative enzymes that break down nonfunctional or foreign molecules. Peroxisomes and proteasomes also play a degradative role in cellular functions.

Cells contain a cytoskeleton consisting of microtubules, microfilaments, and intermediate filaments. In addition to playing a structural role, microtubules are also involved in transportation in some cells. Microfilaments, which are composed principally of actin proteins, provide mechanical stability to cells and also play important roles in changing the shape of cells. Intermediate filaments are tough and help to maintain structural stability of cells.

Cilia and flagellae are found on the surfaces of many types of cells. These structures beat regularly and are involved in either moving cells or in moving materials over the surfaces of cells.

Transport of materials across cellular membranes is accomplished by endocytosis, exocytosis, or transcytosis. In endocytosis, the plasma membrane envelops a foreign molecule or material, bringing it into the cell. In exocytosis, membrane-bound vesicles within the cytoplasm join with the plasma membrane, open up and discharge their contents outside the cell. Transcytosis is a process by which materials are taken into a cell, surrounded by a membrane and then transported through the cell and released from the other side.

Most cells divide, and before division they go through a cell cycle in which DNA duplication first occurs. This is followed by chromosomal duplication. Actual cell division (mitosis) consists of several phases. In prophase, the chromosomes condense and become recognizable under a microscope. This is followed by dissolution of the nuclear envelope and the formation of a mitotic spindle and movement of the chromosomes to a middle position along spindle fibers during metaphase. Individual chromosomes then move in opposite directions away from the middle metaphase plate during anaphase. Actual cell division (cytokinesis) occurs during telophase. This consists of the formation of a nuclear envelope around the chromosomes of each future daughter cell, followed by constriction of the original cell to form two separate daughter cells.

Chapter 2

Tissues

Few cells function as independent units. Most are associated with other cells in the form of tissues. Tissues are aggregations of cells that work together to accomplish some major biological function. Although occasionally homogeneous, most tissues are composed of several types of cells, which serve different aspects of the overall function of the tissue. Details on how these cells work together to serve the overall functions of many tissues are covered in subsequent chapters. The main goal of this chapter is to outline in general terms how the main types of tissues in the body are organized and how their structural organization helps them to fulfill their main functions. Traditionally, tissues have been subdivided into four main types—epithelia, connective tissues, muscle, and nervous tissue.

EPITHELIA

At the broadest level, **epithelia** fulfill two main functions. One is to cover surfaces (e.g., the epidermis of the skin) or line tubes (e.g., the inner lining of the digestive system). The second main function is to make substances to be excreted.

Both the synthesis of substances and their subsequent excretion are accomplished by glandular epithelial cells. Such cells may be embedded within lining epithelia and function as individual cellular units (e.g., mucus-secreting cells in the epithelial lining of the colon) or collectively as epithelial components of large multicellular glands, such as the pancreas or liver.

Epithelia fulfill a wide variety of functions within the body (Table 2.1). It is not uncommon for a single epithelium to play a variety of roles. Many epithelia, such as the cells lining kidney tubules, are involved in both secretion and absorption. Other internal epithelia that line body cavities also produce secretions for lubrication.

Epithelial Cells

A typical **epithelial cell** in a single-layered epithelium has an apical surface, a basal surface, and lateral surfaces (Fig. 2.1). Each surface fulfills a different function within the epithelium. The apical surface is specialized for exchange functions involving the passage of materials into or out from the cell. In single-layered epithelia and in the bottom layer of multilayered epithelia, the basal surface is tightly connected to a **basal lamina (basement membrane)**.

The basal lamina is a thin carpet-like, noncellular layer that underlies epithelia and that surrounds muscle and peripheral nerve cells like a sheath. Typically, cells secrete their own basal lamina. The basal lamina consists principally of loose mats of two main proteins—**laminin** and **type IV collagen** (see Table 2.4). Epithelial cells, as well as muscle cells, are tightly connected to their respective basal laminae by a class of plasma membrane proteins, called **integrins** (Box 2.1).

The configuration of the apical surface of an epithelial cell varies greatly, depending upon its function. Cells involved in secretion or absorption have apical specializations that greatly increase the surface area of the apical plasma membrane. Most commonly, these are **microvilli**—tiny fingerlike structures that project into the lumen of a tube (Fig. 2.3). Epithelial cells lining the small intestine have up to 1000 microvilli per cell. Epithelial cells involved in moving mucus contain hundreds of beating cilia on their apical surfaces. Olfactory epithelial cells have modified cilia on their apical surfaces. These serve as odorant receptors. In contrast, epithelial cells subject to abrasion have few or no apical specializations. The entire outer surface of the epidermis consists of several layers of tightly connected dead cells.

The Human Body. DOI: https://doi.org/10.1016/B978-0-12-804254-0.00002-8

TABLE 2.1 Some Functions of Epithelial Cells

Protection and covering	Epidermis, oral mucosa
Secretion	Glandular epithelia, kidney tubules
Absorption	Gut epithelium, kidney tubules
Lining function	Peritoneal membrane, pleura
Barrier function	Renal tubular epithelium
Producing other cell types	Seminiferous epithelium
Receiving sensory signals	Taste buds, olfactory epithelium
Moving mucous sheets	Respiratory epithelium
Excretion	Kidney tubules

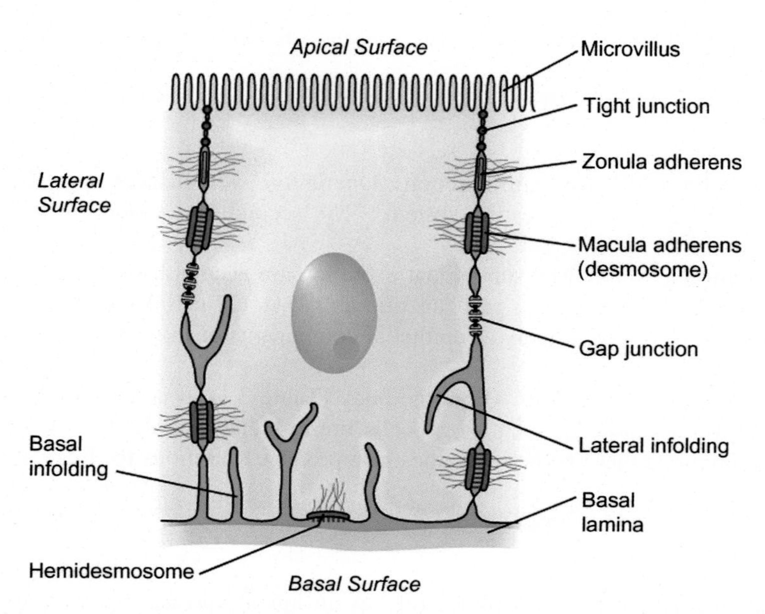

FIGURE 2.1 Diagram of a generalized epithelial cell, showing various surface specializations and intercellular connections.

BOX 2.1 Types and properties of integrins

An **integrin** molecule consists of two subunits—α and β. As the result of alternative splicing of their mRNAs (see p. 11), there are 18 types of α chains and 8 types of β chains, making a very large number of possible combinations of the two subunits. Each combination has slightly different binding characteristics to other molecules. Integrin subunits are shaped like pipes, with the narrow end of the stem embedded in the plasma membrane and the expanded bowl end bonding with molecules of the extracellular matrix (Fig. 2.2). The binding is highly specific for a sequence of three amino acids—RGD[1]—on the matrix molecules. For example, $\alpha_6\beta_1$ integrin has a specific binding site for a portion of the laminin molecule. This accounts for much of the strength of the connection between an epithelium and its underlying basal lamina. On the intracellular side, linker molecules connect the embedded part of the integrin subunits to the cytoskeleton. In addition to protecting the integrity of the epithelium, this provides a mechanism whereby gross forces in the body can be translated through the basal lamina to an epithelium. These forces can stimulate a number of intracellular processes from gene expression to changes in electrical potential. In skeletal muscle fibers, integrin connections with the basal lamina represent one of the main ways by which contractile forces generated by

(Continued)

BOX 2.1 (Continued)

muscle fibers can be translated into gross movements. During embryonic development, the movements of cells from one place to another depend to a great degree on the presence of a laminin substrate to which the migrating cells make temporary attachments by way of integrins.

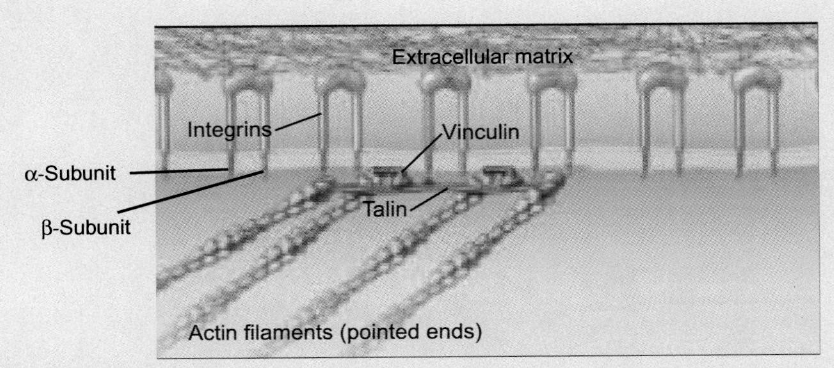

FIGURE 2.2 Integrins connect extracellular matrix molecules (top) with intracellular cytoskeletal elements, for example, actin filaments, through connecting molecules, such as vinculin and talin. *From Pollard and Earnshaw (2004), with permission.*

1. Like the nucleotides of DNA and RNA, individual amino acids in proteins are given shorthand names. In this case, R = arginine, G = glycine, and D = aspartic acid.

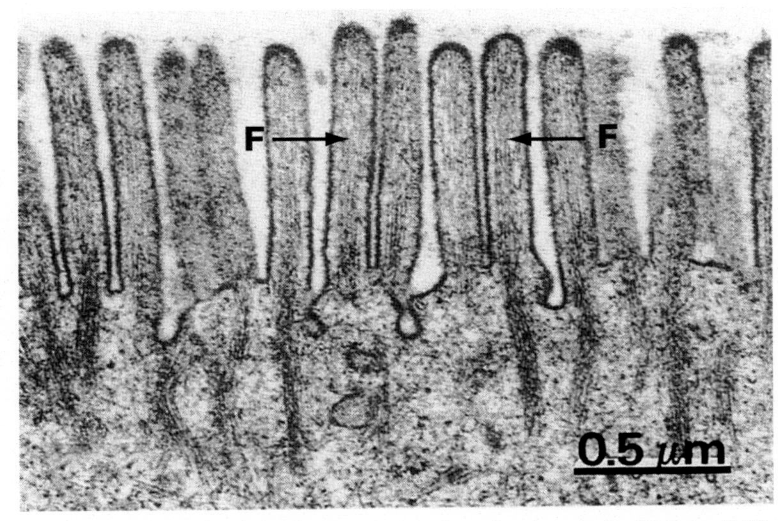

FIGURE 2.3 Electron micrograph of microvilli protruding from the apical surface of an intestinal epithelial cell. The core of a microvillus contains actin filaments, which provide stability. *From Young et al. (2006), with permission.*

Two main structural and functional features characterize the lateral surfaces. One is intercellular connections that bind together neighboring epithelial cells. The other is folds of the lateral plasma membrane. Near their apical end, epithelial cells are bound together by an assortment of intercellular junctions that tightly link one epithelial cell to its neighbors (Fig. 2.4; Table 2.2). Closest to the apical surface is a **tight junction (zonula occludens)**, which encircles the cell like a girdle. The zonula occludens also serves as an intracellular boundary separating the biochemically distinct

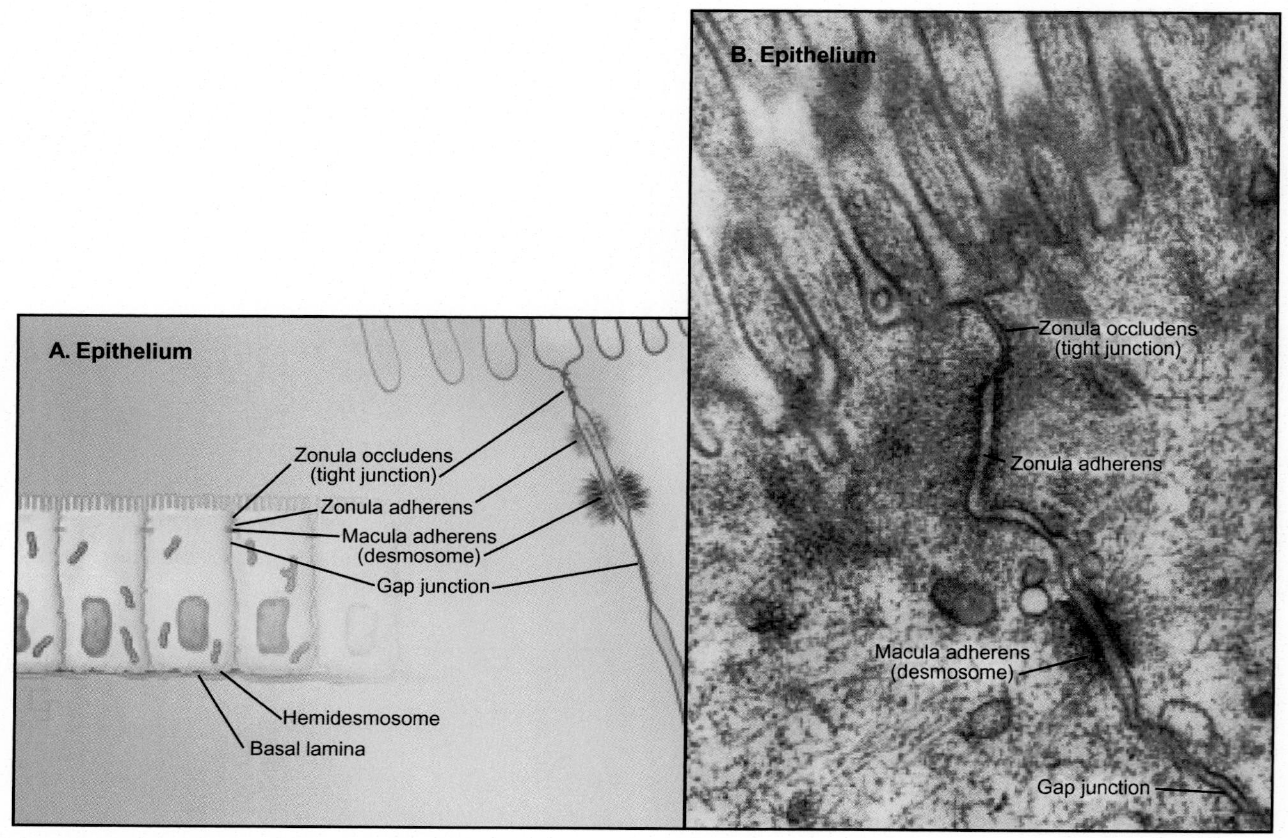

FIGURE 2.4 Lateral connections between epithelial cells. The electron micrograph shows a typical junctional complex, with a beltlike zonula occludens and zonula adherens, each of which surrounds the entire cell and provides tight junctions. Beneath that is a macula adherens, which acts as a biological spot weld causing firm attachments between the two cells. *From Pollard and Earnshaw (2004), with permission.*

TABLE 2.2 Types of Epithelial Intercellular and Cell-Matrix Junctions

Type of Junction	Function
Tight junction	Barrier between inside and outside of epithelium
Zonula adherens	Broad intercellular adhesion
Desmosome	Focal intercellular adhesion
Gap junction	Intercellular communication
Hemidesmosome	Adhesion to basal lamina
Focal contact	Adhesion to basal lamina

apical from the basolateral surface of the cell. Through adhesive proteins, called **occludins** and **claudins**, a tight junction both provides mechanical stability and forms a barrier that prevents the passage of water, ions, and other solutes through the intercellular spaces. The nature of the barrier varies from one type of epithelium to another. Another zonular junction, the **adherens junction**, also encircles the epithelial cell. Projecting from the plasma membrane in this region are densely packed protein molecules, called **cadherins**, that tightly bind to cadherins from the neighboring cell through the mediation of Ca^{++} to form a stable mechanical connection. Within the cell, a meshwork of cortical **actin microfilaments** is connected to the cadherins through linker proteins to add further strength to the junction. The third type of intercellular junction, the **desmosome**, comes in the form of a spot weld that also uses the interactions of cadherin molecules to bind two cells together. Like the zonula adherens, much of the strength of a desmosome lies in the connection between intracellular filaments and the molecules at the plasma membrane at the site of the connection. Desmosomes are especially important in attaching epidermal cells to their neighbors. Through these connections, epidermal cells are able to resist most abrasive forces. People with the genetic disease **pemphigus** have defective desmosomes, and the disease is characterized by the excessive shedding of epidermal cells. Desmosomes linking cells are also important in maintaining the physical strength of heart muscle and the tissues of the cervix.

A different class of intercellular junction is a **gap junction** (see Fig. 2.1), which serves as a channel for electrical communication between two epithelial cells. A gap junction looks like a tiny spot filled with barrel-like channels, called **connexons**. In two adjacent cells, the spots line up, and within them the connexons also line up, forming open channels that extend from one cell to the other. These channels are gated, in the sense that when open, they allow the passage of ions and other small molecules from cell to cell. Gap junctions also exist between heart muscle cells. Rapidly transmitted electrical signals passing directly from one heart muscle cell to the next permit the heart to contract in a coordinated manner.

Some epithelial cells that are heavily involved in fluid exchange, for example many of those that line kidney tubules (see p. 365), have broad membrane folds near the basal part of the cell, through which fluids pass from inside the cells to the outside. Such folds increase the overall surface area available for exchange.

Types of Epithelia

Epithelial cells come in a variety of sizes and shapes (Fig. 2.5). As a rule of thumb, epithelial cells exposed to mechanical forces are thin and scale-like (**squamous** cells). At the other extreme are tall **columnar** cells, which are typically found lining large tubes. When such an epithelium is designed to move a layer of mucus or some other materials, the apical surfaces of these cells are covered with cilia, which beat in unison to propel the material across its surface. Other columnar epithelial cells, called **goblet cells**, manufacture and secrete copious amounts of the mucus that covers the epithelium. Columnar cells are almost always found as single-layered (simple) epithelia. One variant, found in the nose and trachea, appears to be double-layered, but all of the cells are in contact with the underlying basal lamina. Such an epithelium is called **pseudostratified**. **Stratified columnar epithelia** are found only in a few places in the body, such as the male urethra and large ducts of some glands.

Squamous cells fall into two general categories. Those that cover the outer surface of the skin or that line the mouth, pharynx, and esophagus are multilayered or stratified. These cells are subject to drying or abrasion and have strong

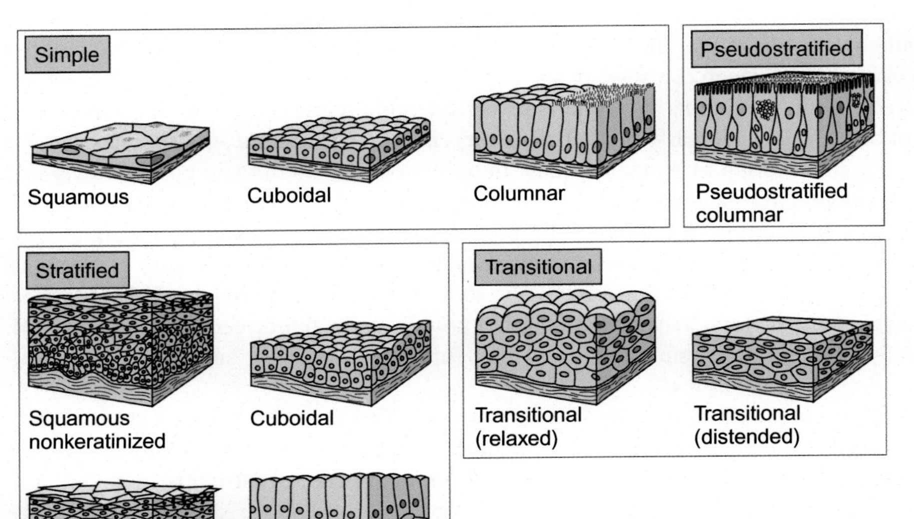

FIGURE 2.5 Major types of epithelia. *From Gartner and Hiatt (2011), with permission.*

lateral connections in the form of tight junctions that keep the epithelium from shedding excessively. The outer surface of stratified squamous epithelia exposed to the air or mechanical abrasion consists of several layers of dead cells. Such epithelia are called cornified (from the Latin word meaning horn), because the keratinized dead epithelial cells are somewhat hard in texture.

Simple single-layered squamous epithelia can be found in situations where transport of liquid or gas across the epithelium is of paramount importance. The thin epithelial linings of capillaries or the air sacs (alveoli) of the lungs reflect this function. These cells are so thin that they resemble little more than fine lines when viewed in cross-section by a light microscope.

Between squamous and columnar types of epithelial cells are the **cuboidal epithelia**. Cuboidal epithelial cells are often found in areas of high metabolic activity, whether the synthesis of secretory products or the exchange of small molecules. Especially in glands, the shapes of individual epithelial cells deviate from strict cuboidal forms to pyramidal or even trapezoidal configurations. Depending upon the specifics of their cellular activity, the disposition, and numbers of their organelles reflect that activity.

Epithelia make up both the secretory and duct systems of glands. The epithelial part of a gland is known as the **parenchyma**, and the supporting connective tissue elements are the **stroma**. There are two broad types of glands—endocrine and exocrine. **Endocrine glands** deposit their secretions within the blood stream and do not have duct systems. In **exocrine glands**, the secretory components are small grapelike clusters of epithelial cells called **acini**. The secretions from the glandular epithelial cells of the acini pass into a system of converging ducts of increasingly larger diameter, which collect the secretions and pass them to some other area of the body.

Several modes of exocrine secretion have been described. The most common is **merocrine**, in which the secretory product is deposited into the ducts by exocytosis from the glandular epithelial cells. A good example is the exocrine component of the pancreas. Individual pancreatic epithelial cells secrete digestive enzymes into the pancreatic ducts in an inactive form. These enzymes are then carried through the duct system to the duodenum, where they become activated and break down food materials. A second, more extreme, mode of secretion is **holocrine**. Seen in the epithelium of sebaceous glands, the entire epithelial cell breaks down and is secreted as an oily substance that moisturizes the skin. The third type of secretion is called **apocrine**, in which the tops of epithelial cells are said to break off along with an intracellular secretion. A classical example of this type is the mammary epithelium as it secretes milk. Recent research, however, suggests that this may actually be an extreme form of merocrine secretion.

Some epithelia are so specialized that they defy easy classification. Examples include the lining of seminiferous tubules, which produces spermatozoa, or the varieties of epithelial cells found within the inner ear. Another unusual variant is the epithelium lining the urinary bladder, called a **transitional epithelium**. In the empty bladder, it contains dome-shaped columnar cells that, when stretched as the bladder fills, become more squamous-like in character (see Fig. 2.5).

One uniform characteristic of epithelial tissues is that they are not directly vascularized. The basal lamina underlying the epithelium acts as a barrier to vascular ingrowth. Because of this, epithelia have to be thin enough to obtain oxygen and nutrients by diffusion. The avascularity of normal epithelia is of considerable relevance in the development of **carcinomas** (tumors derived from epithelial cells). An early epithelial tumor, called a **carcinoma in situ**, consists of a malignant transformation of a single or a small group of epithelial cells. Initially, a carcinoma in situ grows slowly and is very small. At some point in its development, the tumor cells begin producing enzymes, such as collagenase IV, which digest the underlying basal lamina. In addition, the tumor cells produce fewer cadherins (thus weakening their connections to other epithelial cells) and increase production of integrins, which promote attachments with elements of the extracellular matrix of the underlying connective tissue matrix. Finally, they stimulate the ingrowth of blood vessels by their production and secretion of angiogenic factors. When the small, but now invasive tumor becomes vascularized, its rate of growth increases dramatically. In addition, some tumor cells may enter the circulation and spread (**metastasize**) to other regions of the body.

Many epithelia do not consist of uniform populations of cells. The cells in stratified epithelia begin as **stem cells** in the basal layer. These cells divide and produce daughter cells that become more specialized over time as they are pushed toward the surface by more newly formed cells until ultimately they are shed from the surface of the epithelium.

Other cells of an entirely different type are found in some epithelia. In the epidermis of the skin are found large numbers of pigment cells and immune cells that have an entirely different embryological origin from that of the epidermis itself. Epithelia lining the gut and the respiratory tree contain isolated cells or clumps of cells that produce a variety of local hormones.

TABLE 2.3 Some Functions of Connective Tissues

Support	Skeleton
Protection of soft tissues	Skull
Transmission of mechanical stress	Fascia
Storage	White fat
Heat production	Brown fat
Producing blood cells	Bone marrow
Packing and insulation	Fascia, fat
Connecting organs to skeleton	Ligaments, tendons
Transport	Blood
Immune protection	Lymphoid tissues
Tissue repair, scarring	Fibroelastic connective tissue
Lubrication	Joint cavities, synovial membrane
Cellular filtration	Reticular connective tissue, lymphoid tissue

CONNECTIVE TISSUES

As the name implies, a prime function of the **connective tissues** is to connect. This is accomplished by the production and secretion of copious amounts of molecules that embed the connective tissue cells in an extracellular matrix. Most connective tissues serve to support or protect soft tissues within the body or to transmit mechanical forces,[2] but connective tissues also have other functions, as well (Table 2.3).

Types of Connective Tissue

There are many types of connective tissue, and each is characterized by a different dominant type of cell (Fig. 2.6). All of these cells arise in the embryo from a generalized connective tissue precursor cell called a **mesenchymal cell**. Within the embryo, a variety of signaling events from other tissues, called **inductions**, steer the development of mesenchymal cells down specific pathways of specialization (**differentiation**). For example, in the early embryonic head, inductive signals from scalp epidermis stimulate underlying mesenchymal cells to become osteoblasts (bone precursors), whereas signals from the anterior end of the gut induce nearby mesenchymal cells to form the cartilaginous precursors (chondroblasts) of the deep bones of the skull.

Fibroelastic Connective Tissue

The prototypical cell in connective tissue is the **fibroblast**. This term actually includes a wide variety of cells with a similar appearance, but different specific functions. A fibroblast is a spindle-shaped cell that is designed to produce proteins and carbohydrates for export (Fig. 2.7). To fulfill these functions, it contains an active rough endoplasmic reticulum (RER) and Golgi apparatus. The cytoskeleton of a fibroblast is very sensitive to its mechanical environment and is organized according to the mechanical stresses placed upon the cell and its surrounding matrix (Box 2.2).

Fibroblasts do not produce a basal lamina, but they are completely surrounded by the extracellular matrix that they secrete (see Fig. 2.7). The extracellular matrix consists of three major components—fibers, ground substance, and fluid.

2. An interesting example of the mechanical functions of connective tissues is the **nuchal ligament**, a band of connective tissue that runs along the midline of the neck from the very back of the skull to the dorsal spine of the lowest (7th) cervical vertebra. The nuchal ligament functions as a stabilizer of the head while walking or running. As you take a step, the falling arm and shoulder on the same side as the footfall pulls gently on the nuchal ligament and prevents the head from bobbing forward.

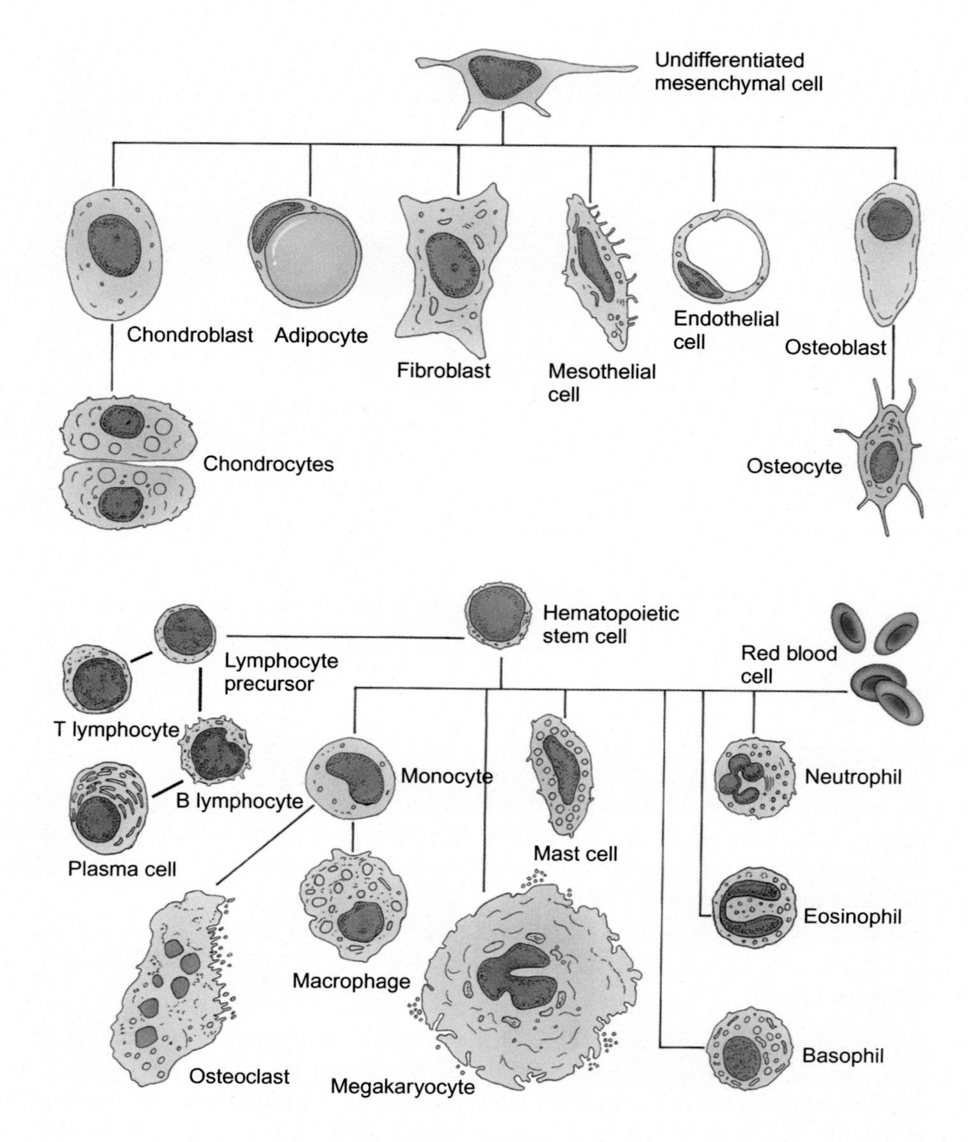

FIGURE 2.6 Major types of connective tissue and blood cells. Their pathways of development from an undifferentiated mesenchymal cell or a hematopoietic stem cell are indicated by the solid lines. *From Gartner and Hiatt (2011), with permission.*

The amount, qualities and proportions of each reflect the particular environment surrounding the cell that produces the matrix.

Collagen Fibers

Connective tissue fibers come in three forms—collagen, elastic, and reticular. **Collagen** is the most abundant protein in the body, constituting about 25% of all bodily proteins. Collagen fibrils (see Box 2.3), of which there are at least 28 types (Table 2.4), are specialized for strength, and some types (**type I**) have a higher tensile strength than steel. Collagen fibers form and are arranged according to lines of stress, which are reflected in both their density and orientation. Different types of collagen serve different functions. On one end of the spectrum are the ropes of type I collagen that make up the bulk of a tendon, where the collagen fibers are so abundant that the fibroblasts that produced them are barely seen as thin elongated nuclei aligned alongside thick bundles of collagen fibers. Collagen fibers also play an important role in supporting the walls of internal organs, but they are not as regularly oriented as those of tendons because the mechanical forces involved in supporting many organs are more diffuse than those of a tendon or ligament.

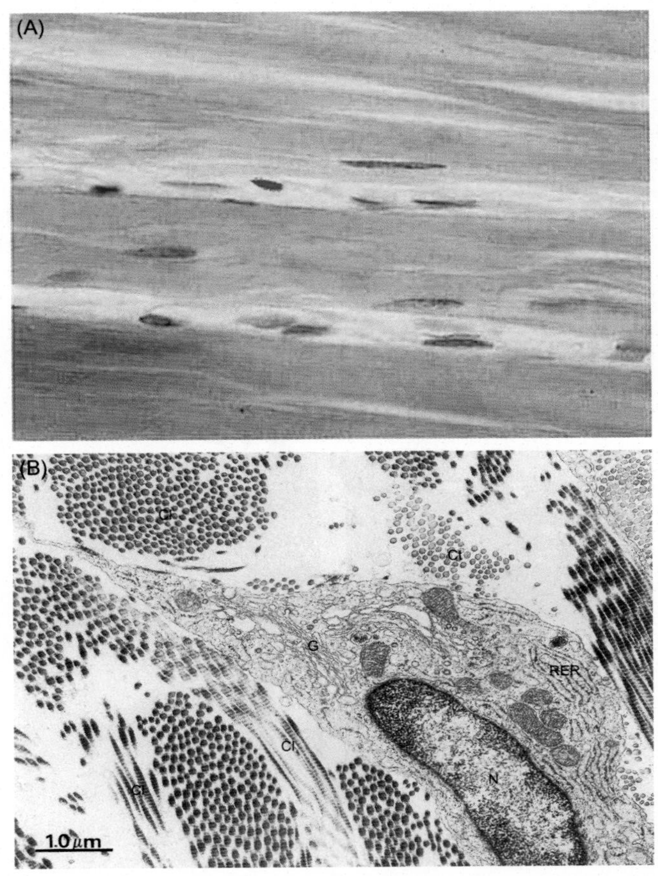

FIGURE 2.7 Fibroblasts. (A) Light micrograph of a tendon. The highly elongated purple nuclei represent the fibroblasts. The pink bands are bundles of collagen fibers, which are products of the fibroblasts. (B) Electron micrograph of a fibroblast. The cytoplasm contains the organelles needed for the synthesis and export of an extracellular product, in this case, collagen fibrils. *N*, nucleus; *RER*, rough endoplasmic reticulum; *G*, Golgi apparatus; *C*, collagen fibrils. *From (A) Erlandsen and Magney (1992), with permission. Young et al. (2006), with permission.*

BOX 2.2 Fibroblasts and connective tissue mechanics

Even under resting conditions, most connective tissues and many parts of the body are under some degree of tension. For example, if part of a blood vessel or a peripheral nerve is isolated from the body, it will shorten by almost 25% from its original length. When an individual is growing, the tensions produced by the elongating skeleton stimulate the fibroblasts in connective tissue to produce more collagen molecules to keep up with the lengthening bones. Under this circumstance, the fibroblasts are able to translate signals from their mechanical environment into gene expression and protein synthesis. When a tissue is stable, the fibroblasts are relatively quiescent even though the overall connective tissue matrix may be under tension. They become "stress-shielded" from the larger forces acting upon the connective tissue as a whole by the cross-linked local extracellular matrix in which they reside. We are beginning to understand how normal fibroblasts sense and respond to their local mechanical environment, but a form of fibroblast that is found in healing wounds and in certain pathologies has given important clues about these processes.

The **myofibroblast** is a specialized form of fibroblast that, as the name implies, has the ability to contract. Although they can arise from several cellular sources, most of them come from cells that look like ordinary fibroblasts. In the dermis of a freshly healing skin wound, the fibroblastic precursor cells are attracted to the wound area by **cytokines** (stimulatory signaling molecules) released from inflammatory cells. These fibroblasts produce matrix molecules, but because the area of the wound is unstable, they are exposed to direct mechanical tension. In response to the tension, the fibroblasts begin to form internal stress fibers composed of cytoplasmic actin strands (Fig. 2.8). They also produce the extracellular matrix molecule, **fibronectin**, as well as integrins, which reside in the plasma membrane in areas called focal adhesion sites. Integrins connect internal stress fibers with external fibronectin molecules which, in turn, bind to collagen fibers. These cells are called **proto-myofibroblasts** and are

(Continued)

BOX 2.2 (Continued)

intermediate stages on the way to becoming myofibroblasts. Their full conversion requires continued episodes of tension and the presence of a growth factor, **transforming growth factor-β** (TGF-β). In response, the cells produce more stress fibers and fibronectin and larger focal adhesion sites. Importantly, they also produce highly aligned **α-smooth muscle actin** fibrils, which allows them to undergo sustained contractions (Fig. 2.9).

Within the healing wound, the mature myofibroblasts contract, pulling bundles of collagen fibers inward. At the same time, enzymes begin to break down parts of these collagen fibers, while new cross-links among collagen fibers conform to the constricting wound area. Ultimately, a small open wound shrinks noticeably (Fig. 2.10), and dense scar tissue typically forms. When the scar becomes stable, the mechanical load on the myofibroblasts is released, and through an incompletely understood mechanism, the myofibroblasts begin to die, leaving a scar containing relatively few cells, but abundant dense collagenous matrix. After severe burns, excessive constriction of a scar can cause wound contractures that can be quite disfiguring. Some people are genetically predisposed to form excessive scar tissue (**keloids**) after simple skin wounds (Fig. 2.11).

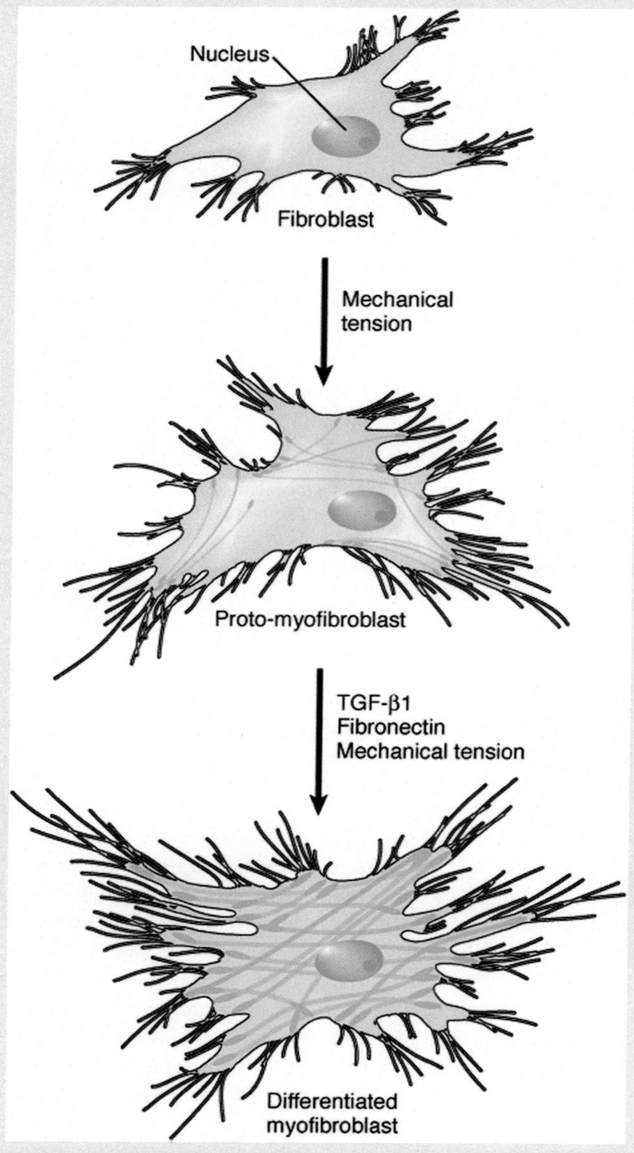

FIGURE 2.8 Stages in the differentiation of a myofibroblast in a healing skin wound. Mechanical tension stimulates fibroblasts to form proto-myofibroblasts, containing actin stress fibers that form complexes with extracellular connective tissue fibers. The growth factor, transforming growth factor β1 (TGF β1), along with the extracellular connecting molecule fibronectin, promotes the continuing development of the proto-myofibroblasts into mature myofibroblasts, which can contract with considerable force.

(*Continued*)

BOX 2.2 (Continued)

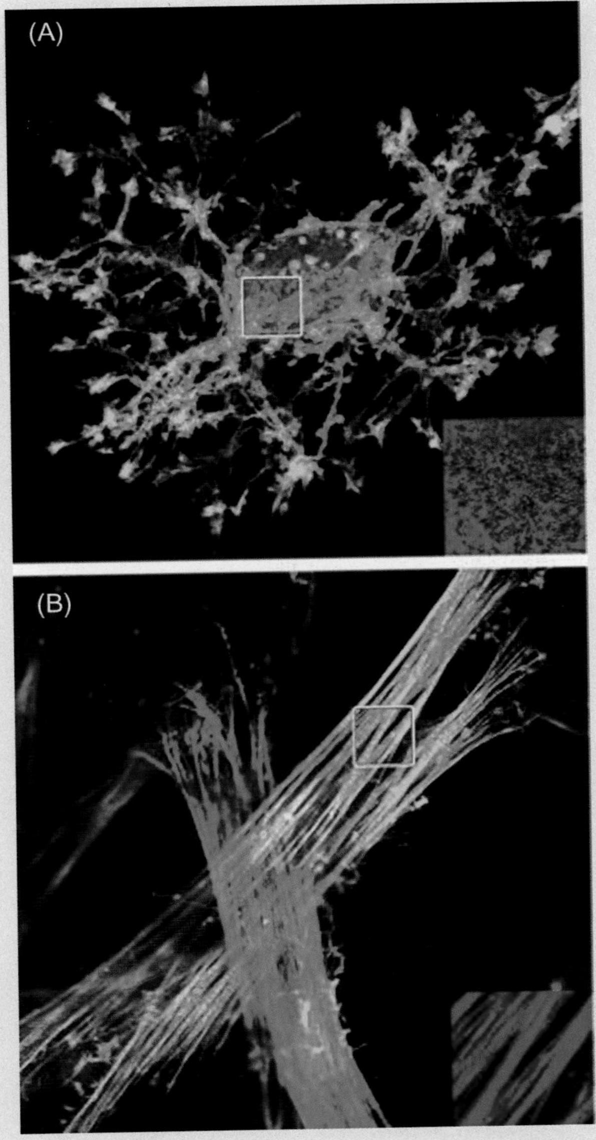

FIGURE 2.9 Photomicrographs of cultured myofibroblasts. (A) Grown under conditions of little tension. (B) Grown under tension. Those grown under tension show prominent well-aligned stress fibers (green). Nuclei are blue. *From Hinz (2007), with permission.*

(Continued)

BOX 2.2 (Continued)

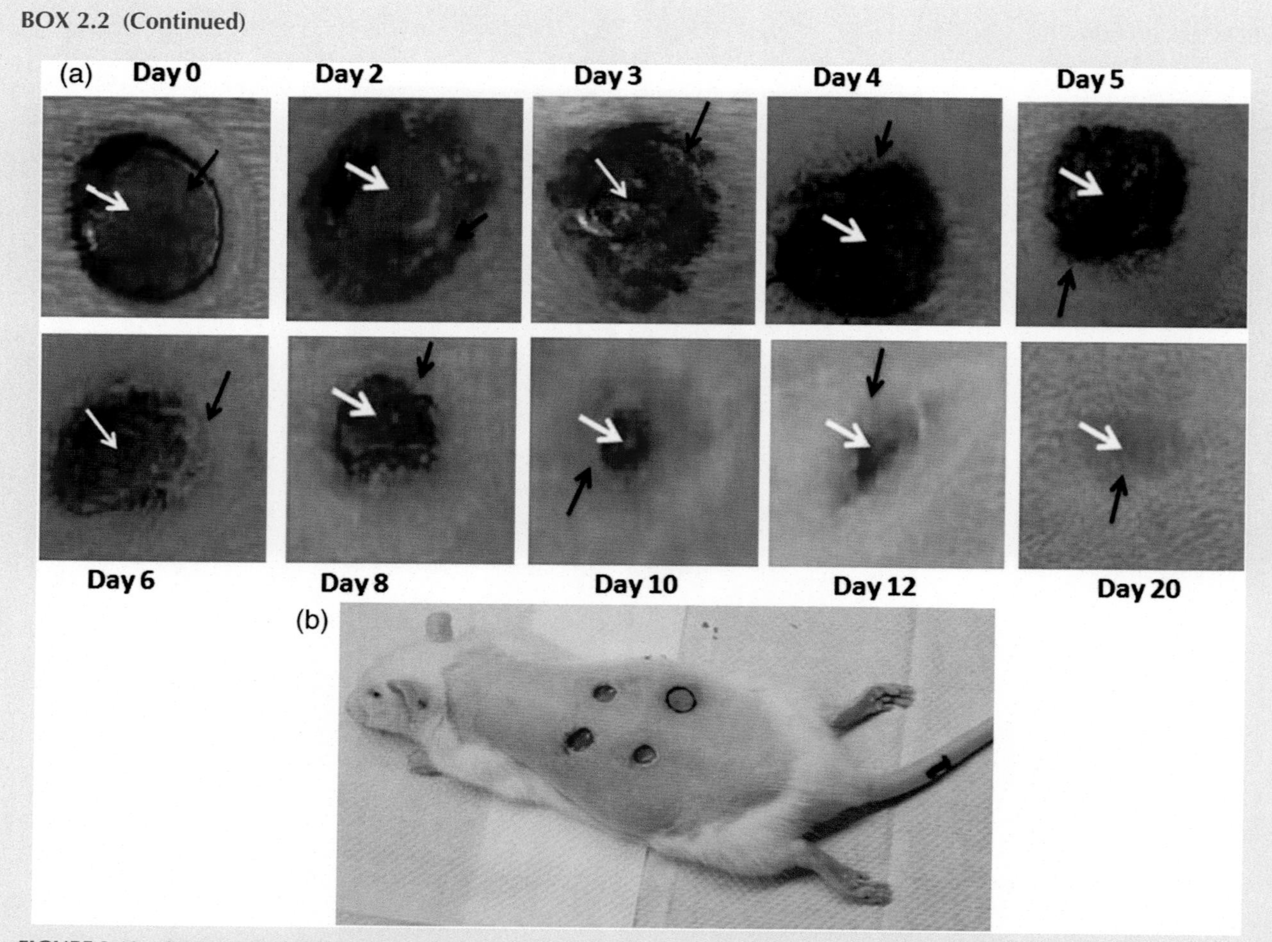

FIGURE 2.10 Contraction of a skin wound in a rat over 20 days. *From Internet http//:biomedicaloptics.spiedigitallibrary.org/data/Journals/BI...*

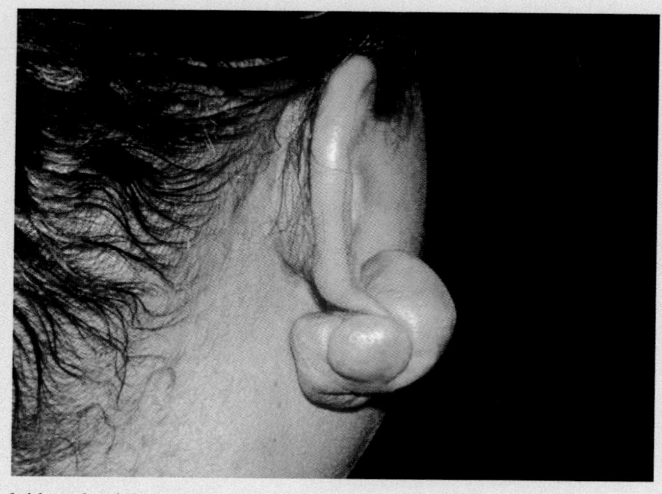

FIGURE 2.11 A multilobular keloid on the right earlobe. *From Taylor SC et al. (2011).*

BOX 2.3 Collagen synthesis

The structure and synthesis **of type I collagen**, which is the dominant form of collagen found in the human body, provides a good general example of how a protein is made and how it contributes to a structure that can be seen with the naked eye. As an example, we will use a tendon of the biceps muscle. The tendon that one can feel beneath the skin is composed of many fibrous bundles, each one of which consists of a batch of tightly packed type I collagen fibers (Fig. 2.12). Like a sewing thread, each collagen fiber consists of many thin fibrils, which are themselves formed from large numbers of tightly interwoven molecular units, called **tropocollagen**. Tropocollagen is a triple helix consisting of three interwoven peptide chains.

The formation of tropocollagen begins in the nucleus of a fibroblast, when a messenger RNA (mRNA) molecule is formed on a stretch of DNA (a collagen gene) through the process of transcription. The mRNA leaves the nucleus and becomes associated with the ribosomal units that stud the membranes of the RER. Through the process of translation, polypeptide chains form on the mRNA template and are released into the RER (Fig. 2.13). After processing, which includes the attachment of some sugar molecules to specific amino acids, the polypeptide chains become organized into triple helical structures, called **procollagen** molecules. The procollagen molecules are then shipped to the Golgi apparatus, the standard pathway for proteins designed for export. Within the Golgi apparatus, additional carbohydrate units in the form of disaccharides are added to the procollagen molecules. At this point, the procollagen is packaged into secretory vesicles within the Golgi complex. The procollagen molecules are then transferred to the outside of the fibroblast through the process of exocytosis. As they leave the cell, some amino acids are cleaved from the ends of the procollagen molecules, resulting in a remodeled protein called **tropocollagen**. Through more cross-linking, the tropocollagen molecules then become linked in a highly orderly manner to form a collagen fibril.

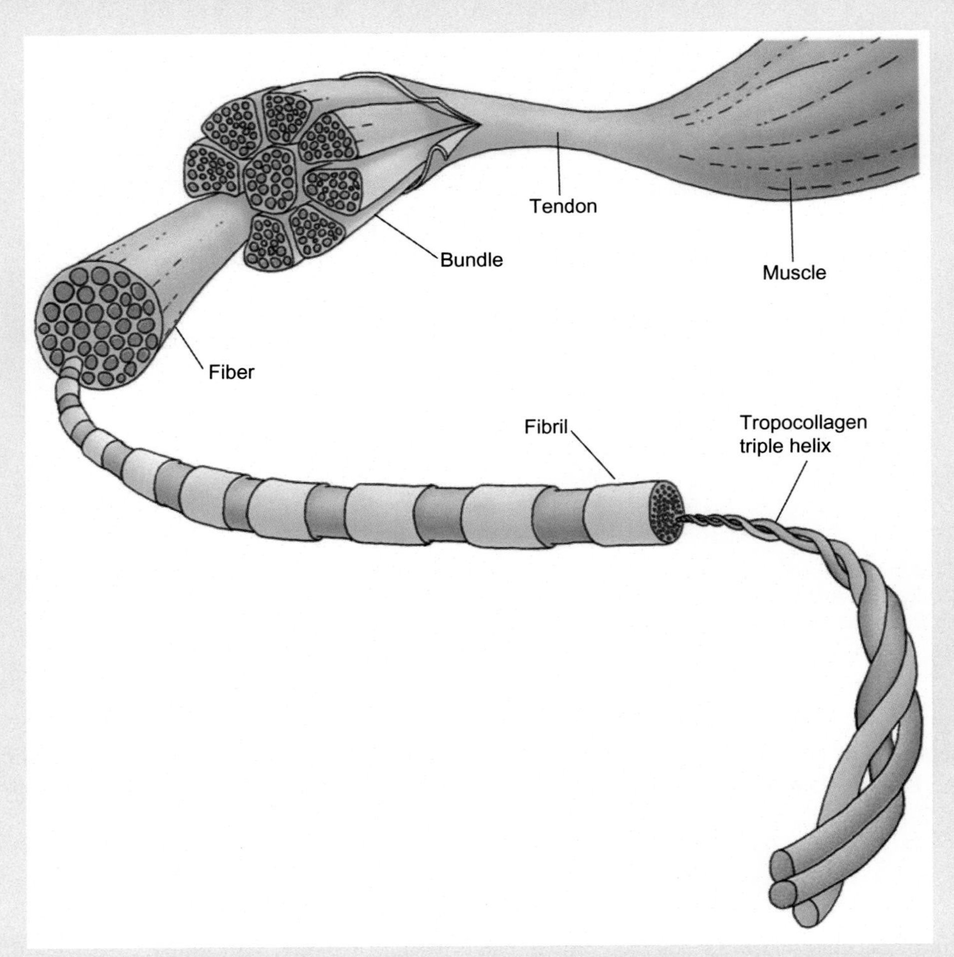

FIGURE 2.12 The structural basis of a collagen fiber. Collagen fibers consist of bundles of smaller collagen fibrils, which in turn are aggregates of individual collagen molecules. A tropocollagen molecule is a triple helical structure made up of three chains. After they are secreted from the fibroblast, tropocollagen molecules self-assemble into collagen fibrils. *From Gartner and Hiatt (2011), with permission.*

(Continued)

BOX 2.3 (Continued)

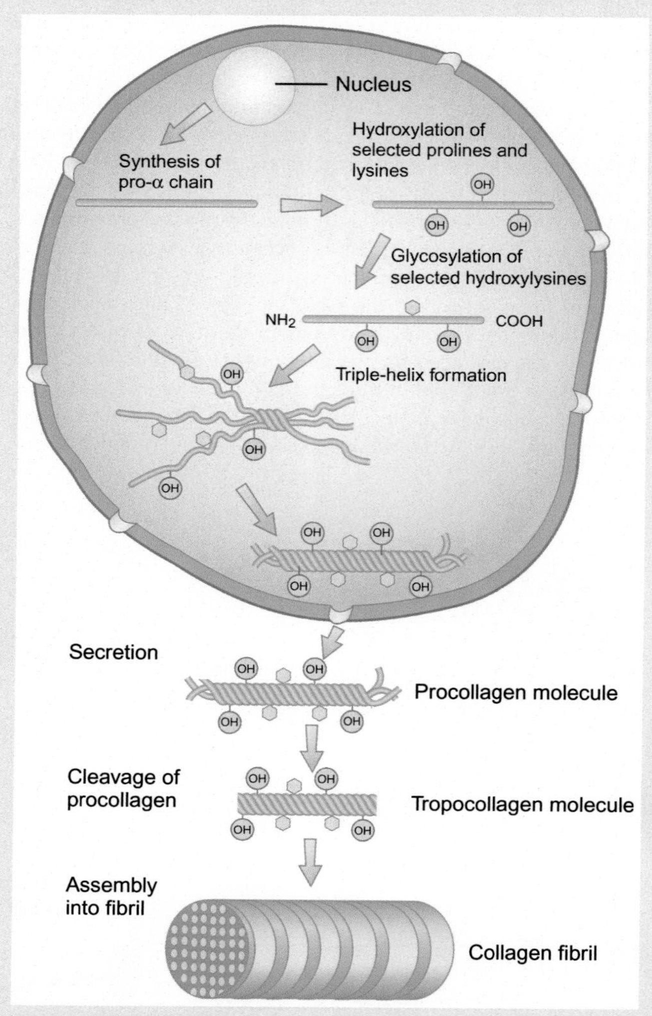

FIGURE 2.13 The process of collagen synthesis, starting with the synthesis of the first peptide chain (pro-α chain), to intracellular modifications of the molecule, followed by its secretion from the cell and further extracellular modification before its ultimate assembly as part of a collagen fibril. *From Jorde et al. (2010), with permission.*

A number of mutations of collagen genes produce demonstrable pathology. One of the more unusual collagen deficiencies results in the **Ehlers–Danlos syndrome**, which is characterized by the hyperextensibility of joints and hyperelasticity of skin (Fig. 2.14).

Elastic Fibers

As their name implies, **elastic fibers** act like rubber bands within a tissue. Especially in tissues that are subject to mechanical deformation, elastic fibers serve to bring them back to their original shape. Good examples are the dermis of the skin, the lungs, and large arteries, all of which have elastic fibers interwoven among collagen fibers. Arteries stretch and then contract in response to the blood pressure generated by each heartbeat, and over the course of a human life, they may go through as many as three billion cycles of expansion and recoil. Elastic fibers are made mainly by embryonic and juvenile fibroblasts. Very few are produced in adulthood. As a result, aging is accompanied by a steady

TABLE 2.4 Most Important Types of Collagen

Collagen type	Characteristics and where found
Type I	Coarse fibers, most connective tissues, especially skin, tendons, ligaments, and bone
Type II	Characteristic of cartilage matrix
Type III	Fine fibers, found in loose connective tissue, often associated with type I collagen
Type IV	Characteristic of basement membranes
Type X	Characteristic of hypertrophic cartilage

FIGURE 2.14 Extreme hyperelasticity of the skin in Ehlers–Danlos syndrome, characterized by defects in types I and III collagen. *From Turnpenny and Ellard (2012), with permission.*

reduction in the prominence of elastic fibers throughout the body. The wrinkles characteristic of aging skin are the result of diminished function of the elastic component of the dermal connective tissue.

Reticular Fibers

Reticular fibers, which are found in many locations within the body, are composed of **type III collagen**. Within connective tissues, reticular fibers appear as thin strands, often seemingly isolated, rather than aggregated into tight bundles. They are especially prominent in the loose connective tissue that supports delicate organs, such as lymphoid tissue or fat. Their normal function is supposed to be support, but remarkably little is yet known about reticular fibers.

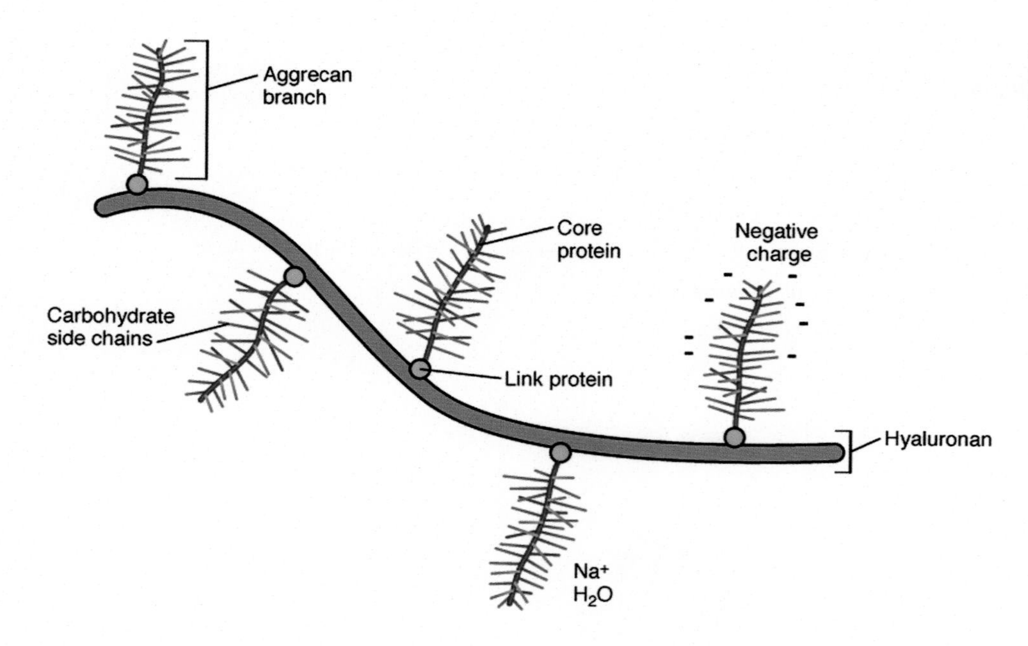

FIGURE 2.15 The structure of a large proteoglycan molecule found in connective tissue.

Ground Substance

The nonfibrous component of connective tissue ranges from gelatinous to rock hard. The so-called **ground substance** consists largely of huge carbohydrate molecules, which act as microsponges to bind water molecules. They also bind growth factors and release them when it is necessary to initiate or modify growth and developmental processes. In embryos, the nonfibrous matrix is almost watery. As an individual matures and ages, the water content of the connective tissues becomes progressively reduced. The main water-binding molecules in connective tissue are called **hyaluronan** and **proteoglycans** (Fig. 2.15), which are extremely large polymers of linked disaccharide units, whose size is variable depending upon the type of tissue. In addition to serving as important space-fillers and shock absorbers, the ground substance acts as a lubricant in joint cavities and restricts the movements of bacteria within connective tissues. Certain pathogenic bacteria (*Streptococcus*, *Pneumococcus*) produce the enzyme **hyaluronidase**, which allows them to penetrate tissues by breaking down hyaluronan in the extracellular matrix.

The matrix of connective tissues used for mechanical support is of an entirely different character. In bone, inorganic crystals of calcium phosphate in the form of the molecule **hydroxyapatite** give the bone its brittle character, whereas the collagen fibers embedded within the bony matrix supply the strength to resist breaking. Collagen accounts for one-third of the dry weight of bone; bone minerals make up the other two thirds.

Another variant is **loose (areolar) connective tissue**. This is a diffuse form of connective tissue found in delicate tissues, such as the mesenteries. Such tissue contains moderate amounts of cells, fibers, and ground substance and its arrangement is characteristic of connective tissue in the absence of highly directed mechanical forces.

Cartilage

Cartilage employs a different strategy to provide strength and support. Cartilage cells (**chondrocytes**) synthesize a special type of collagen (**type II**) that is dispersed in mats that combine with proteoglycans and other carbohydrates to form a firm matrix that is somewhat resilient. Cartilage is an avascular tissue. As a result, cartilage heals very poorly after it has been injured.

Hyaline cartilage, the most common variety in the body, is especially prominent in the embryonic and growing skeleton. As the name implies, it has a somewhat glassy appearance and has an almost slippery feel when fresh. In growing skeletal elements, **chondrocytes** (mature cartilage cells) develop from elongated fibroblast-like **chondroblasts** in the surrounding **perichondrium** (Fig. 2.16). Chondroblasts actively produce and surround themselves with matrix components. As they do so, the cells round up and become chondrocytes. Embryonic cartilage grows by the addition of new chondroblasts from the perichondrium, as well as by the division of chondrocytes already surrounded by cartilage matrix. Progeny of the latter divisions are organized as cell aggregations, called **isogenous groups**, within the cartilage matrix (see Fig. 2.16).

FIGURE 2.16 Hyaline cartilage. Large chondrocytes embedded in a hyaline matrix are situated below a fibrous appearing perichondrium. *From Erlandsen and Magney (1992), with permission.*

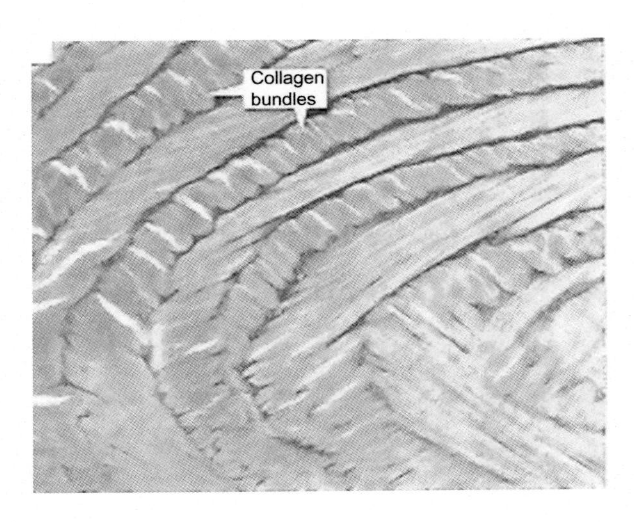

FIGURE 2.17 Orthogonally arranged collagen fibers in an intervertebral disk. This arrangement provides both strength and resistance to shear forces while allowing flexibility. *From Kerr (2010), with permission.*

Other specific functional needs of cartilage are served by special types of extracellular matrices. For example, the cartilage in the ear (**elastic cartilage**) contains a high proportion of elastic fibers that enable the ear to pop back into place after it has been tweaked. **Fibrocartilage**, as in the intervertebral disks, consists of densely packed layers of fibrous connective tissue containing only a few chondrocytes that are surrounded by thin capsules of cartilage matrix. This arrangement allows the disks to withstand the great pressures exerted in the spinal column by the moving body. In many tissues that are characterized by layers of collagen, the fibers in the adjacent layers are usually oriented orthogonally to each other (at 90-degree angles) as a means of increasing strength (Fig. 2.17). This is the case in intervertebral disks.

Bone

As noted above, a major characteristic of bone is the large proportion of type I collagen fibers and inorganic crystals of hydroxyapatite within the matrix. Bone is a well-vascularized tissue that is also metabolically active. It is the main reservoir of calcium in the body. Adult bone is organized in layers, with loose rows of cells (**osteocytes**) alternating with rows of hard bony matrix. In order to obtain adequate oxygen and nutrition, osteocytes must not be located more than a couple hundred micrometers from the nearest capillaries (Fig. 2.18). As is seen in Chapter 4, the internal structure of bone constantly undergoes remodeling, usually responding to its local mechanical environment.

Adipose Tissue

Fat or **adipose tissue**, is a highly specialized and still poorly understood type of connective tissue. For many years, fat cells (**adipocytes**) were thought to be merely fibroblasts that became filled with globules of lipid. Now it is understood that adipocytes represent a special lineage of connective tissue cells. Even more recently, adipose tissue has been recognized as a major repository of adult stem cells.

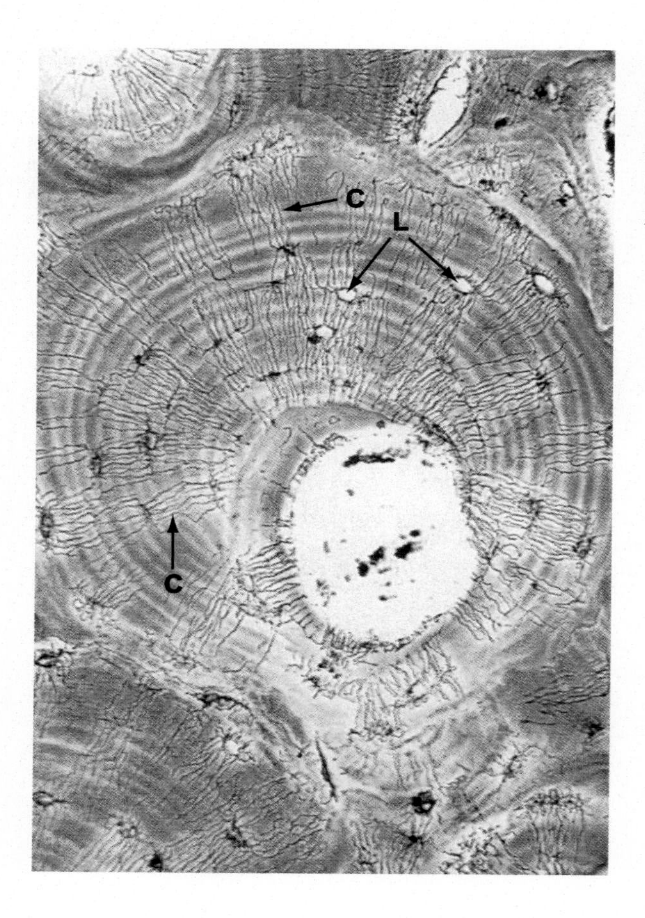

FIGURE 2.18 Cross-section through an osteone (Haversian system), showing concentric circular lamellae of bony tissue and lacunae (*L*) situated between lamellae. In this ground bone preparation, the osteocytes, which occupy the lacunae, are not present. The spidery canaliculae (*C*) contain interconnecting processes of the osteocytes The large empty space is the Haversian canal, which contains a blood vessel and nerve fibers. *From Young et al. (2006), with permission.*

There are three types of fat in the body. The most recognizable, **white fat**, is the dominant form present in the adult body and something most people would like to have less of. A second type, called **brown fat**, is seen most prominently in human fetuses and newborns, although its presence has recently been confirmed in adults. Brown fat accounts for up to 5% of the weight of a newborn infant. Deposits of brown fat are concentrated most prominently in the shoulder region of the back and neck (Fig. 2.19). A newly recognized variety of adipose tissue is **beige fat**. Whereas white fat serves a physiological storage function, brown fat and beige fat are much more active metabolically and are involved in temperature control.

White fat is the body's largest reservoir of energy. In addition, white fat is now known to produce a variety of biologically active substances and has even been classified as an endocrine organ. Healthy amounts of fat in adult males are 15%–18% of total body weight, and in females, 20%–22%. The distribution of white fat differs between males and females (Fig. 2.20). The location of white fat is important metabolically. White fat is present in two major anatomical compartments—subcutaneous and intraabdominal—each of which is served by a different blood supply. In most people, about 90% of their fat is subcutaneous; the other 10% is intraabdominal. **Intraabdominal fat** is drained by the hepatic portal venous system (see p. 295) so that metabolic products (free fatty acids) from the breakdown of fat molecules (triglycerides) flow directly to the liver for further processing. In contrast, **subcutaneous fat** supplies metabolites to skeletal muscle during exercise or fasting. For a variety of reasons, intraabdominal fat poses a greater danger to one's overall health, especially, cardiovascular disease and diabetes, than does subcutaneous fat. Estrogen tends to cause fat to be deposited around the hips and legs in females, whereas testosterone effects lead to the deposition of fat in the abdominal area (both subcutaneous and intraabdominal) in males. After menopause, the reduction in estrogen levels in females results in a greater tendency toward increasing the proportion of intraabdominal fat.

White fat cells (adipocytes) contain a single large globule of fat in the form of triglycerides that occupies 80%–90% of the total cell volume (Fig. 2.21). The yellowish color of white fat is due to the presence of carotene pigments. Adipocytes arise from the proliferation of precursor cells, called **preadipocytes**. Preadipocytes persist in adult adipose tissue and can differentiate into mature adipocytes if the need arises. Major functions of adipocytes are lipid storage and breakdown, which are subject to multiple regulatory factors, especially the action of insulin, which stimulates fat

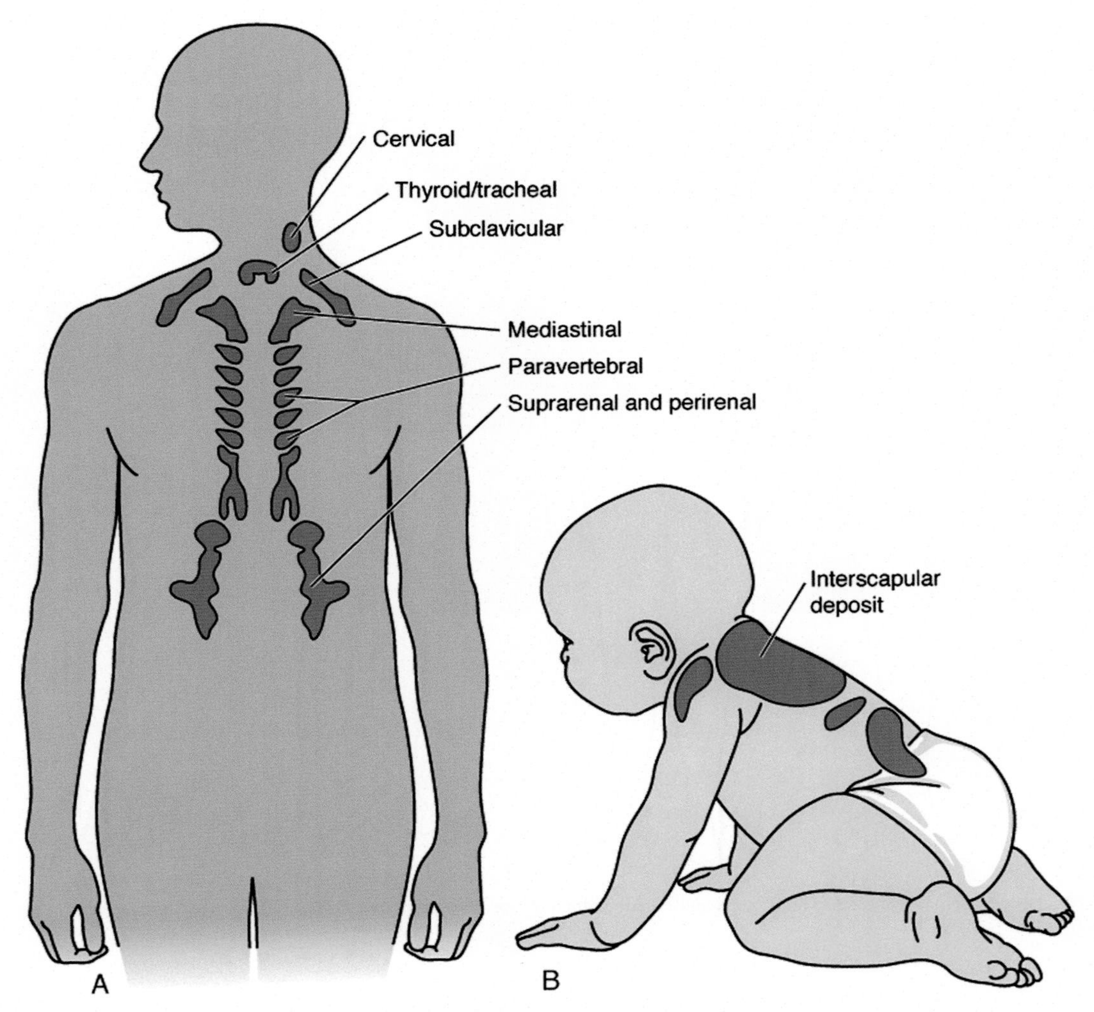

FIGURE 2.19 Distribution of brown fat deposits in adult and infant human. *Redrawn from Diaz et al. (2014), with permission.*

formation (**lipogenesis**). The storage of lipids in fat is actually a protective mechanism, because the deposition of fat in other types of tissues is associated with various pathologies. Despite the storage function, there is a constant turnover of triglycerides within fat cells. Another important function of white fat is serving as a shock absorber, where fat is deposited in areas like the palms of the hand or soles of the feet, around the kidneys and behind the eyeballs.

The major hormonal product of white fat is **leptin**, sometimes known as a satiety hormone. Leptin is secreted into the abundant blood supply of white fat, and when its effect makes its way to the hypothalamus (see p. 248), it reduces the desire for food intake. A converse effect is the body's response to a reduction in leptin, which stimulates a strong desire to eat, causing some researchers to view its primary function (in its absence) as a starvation hormone. Another hormone derived from adipose tissue is **adiponectin**, which may play a role in increasing insulin resistance by cells and therefore leading to type 2 diabetes (see p. 268).

Brown fat is an important component of the thermoregulatory apparatus of the newborn. Newborn infants cannot generate heat by muscular shivering. Rather, heat is produced from the metabolism of brown fat by a mechanism called **nonshivering thermogenesis**. Many hibernating mammals also generate heat through the activity of brown fat. Brown fat cells are multilocular (many small lipid droplets). They contain many more mitochondria than do white fat cells and are surrounded by a richer capillary supply, both of which make them highly metabolically active cells. Their metabolism is extremely energy-expensive as they burn through nutrients. Brown fat cells get their color by pigmented molecules located in the mitochondria. In addition, the mitochondria contain a protein, called **thermogenin** (or technically UCP1[uncoupling protein 1]), which is involved in the production of metabolic heat. It does this by uncoupling heat

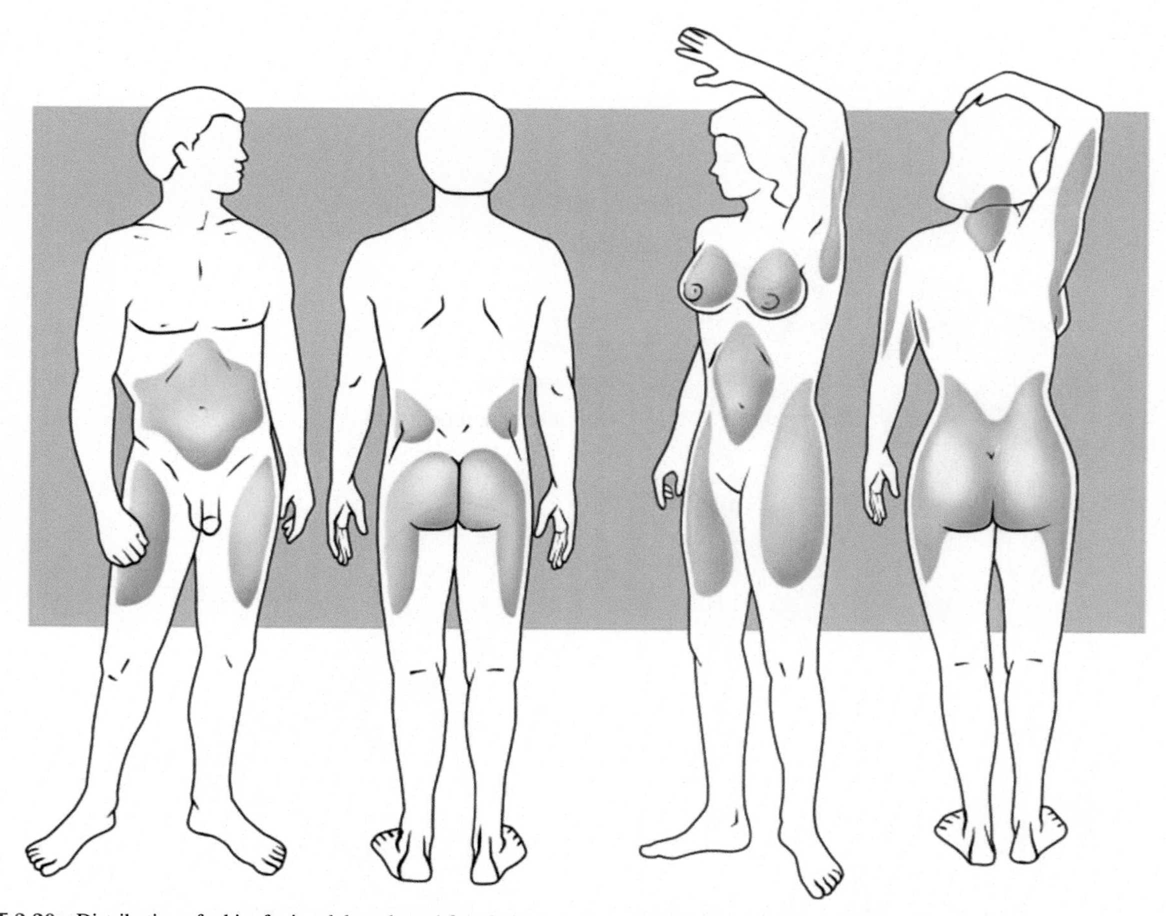

FIGURE 2.20 Distribution of white fat in adult male and female human. *From Thibodeau and Patton (2007), with permission.*

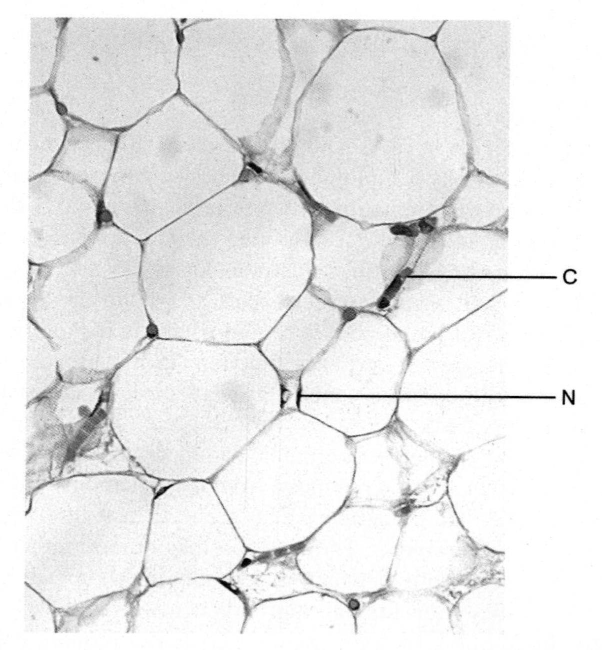

FIGURE 2.21 Microscopic section through white fat. The large empty spaces in each cell contained fat in life, but the fat dissolved during preparation of the microscope slide. *N*, nuclei of fat cells; *C*, capillaries in close association with fat cells. *From Stevens and Lowe (2005), with permission.*

production from ATP (energy) formation during intramitochondrial metabolism. Surprisingly, recent research suggests that instead of originating from the same embryonic stem cells that give rise to white fat cells, brown fat cells, and muscle cells share a similar embryonic origin.

The most recently recognized member of the adipose tissue family is beige fat. Not as dark brown as brown fat, beige fat cells are also multilocular and are also involved in thermogenesis. Beige fat is formed in response to exposure to cold temperatures and is presently thought to arise from either a direct conversion from white fat cells or from preadipocytes that have been steered down a different developmental pathway from the white fat cell lineage. A major stimulus in the formation and function of beige fat is the action of the sympathetic nerves (see p. 140), which richly innervate adipose tissue. Increased exposure to cold increases sympathetic activity, which stimulates both lipid breakdown and the formation of new beige fat cells. The high-energy consumption of both brown and beige fat is being actively explored by pharmaceutical companies on the lookout for physiological means of stimulating weight loss.

Blood

An outlier in the connective tissue family is **blood** (see Chapter 10 for details). Within the circulatory system proper, blood consists entirely of cells carried along by liquid **blood plasma**. However, blood cells arise in **bone marrow**, which has much more of the characteristics of a typical connective tissue. The cells found within the circulating blood could be seen as cellular products of a specialized connective tissue, called the marrow.

Other Components of Connective Tissue

Connective tissue does not consist only of the components described above. With the exception of cartilage, all connective tissues are vascularized to varying degrees. Some, such as the dermis of the face and scalp, are highly vascularized, which explains why wounds to this area bleed so profusely. At the other extreme are ligaments and tendons. The many months required to heal a strained ligament or tendon reflect their poor degree of vascularization.

With the exception of cartilage and blood, connective tissues are usually well innervated, with the degree and pattern of innervation depending upon the type of connective tissue. Bone, for example, contains many sensory nerve endings, which explains why fractures are painful.

Many cells of the immune system leave the blood and reside in the connective tissues. A prime example is the **macrophage**, a cell type derived from circulating monocytes in the blood. Macrophages represent a major line of immune defense (see Chapter 8), and they also play an important role in the removal of damaged tissue. In both of these roles, they are major facilitators of tissue regenerative responses after injury. Other immune cells in connective tissue, such as mast cells, are also blood-derived and play important roles in acute immune reactions, especially in response to allergens.

MUSCLE

A common feature of all muscle tissue is its ability to contract. Three types of muscle—skeletal, cardiac, and smooth—are present in the body (Fig. 2.22). Each type of muscle cell has its own unique characteristics. Skeletal and cardiac muscle fibers are called striated because the highly regular organization of the contractile filaments within them show up as transverse stripes at the light microscopic level. Although smooth muscle cells contain very similar contractile proteins, these proteins are not so regularly organized, and thus smooth muscle appears to be nonstriated. Skeletal muscle is voluntary, meaning that these muscles contract upon conscious demand. Smooth and cardiac muscles, on the other hand, are involuntary in the sense that their contractions are not subject to conscious control.

Skeletal Muscle

Skeletal muscle cells occur in the form of multinucleated fibers that can be up to several centimeters long. In the embryo, a skeletal muscle fiber begins as a single cell, called a **myoblast**. Individual myoblasts begin to fuse with one another, forming elongated cells, called **myotubes**, in which dozens of nuclei are lined up in a central row (Fig. 2.23). The nuclei are large, somewhat elongated and contain a large nucleolus—a characteristic of a cell that is producing large amounts of RNA and proteins. These nuclei and the RNAs that they produce stimulate the synthesis of long contractile proteins that become organized into functional arrays, called **sarcomeres**, located around the periphery of the myotube. Over time, longitudinal series of sarcomeres develop the ability to cause the developing myotube to contract

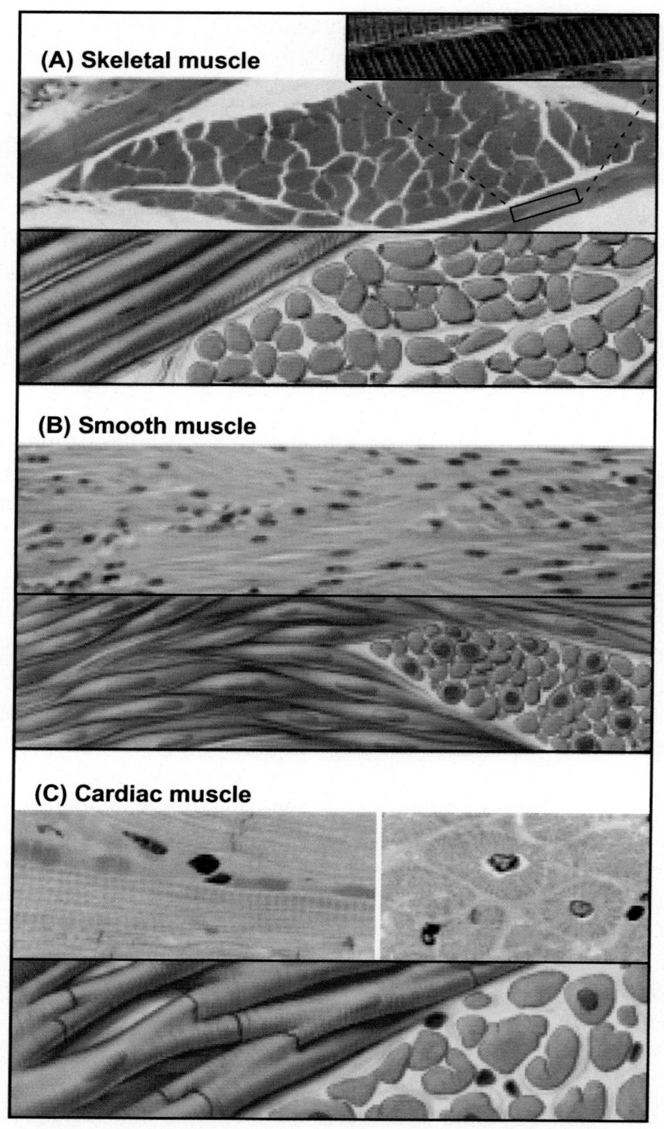

FIGURE 2.22 Drawings and photomicrographs of the three types of muscles. *From Pollard and Earnshaw (2004), with permission.*

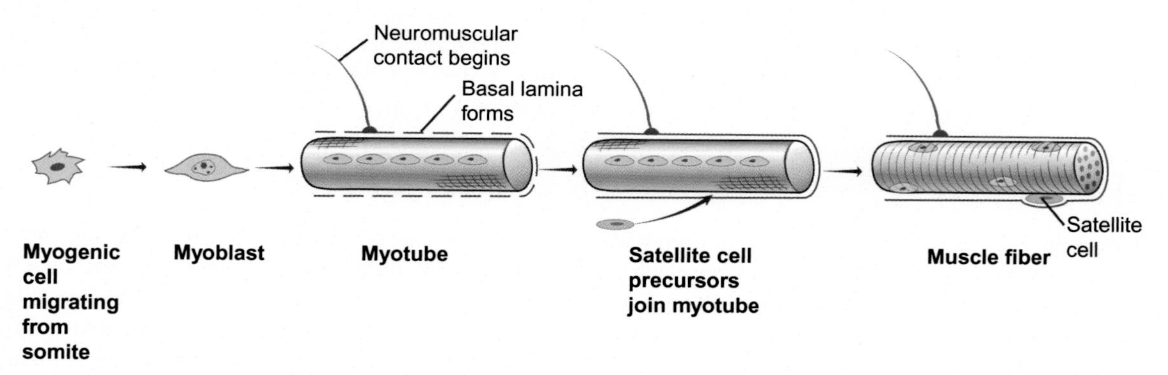

FIGURE 2.23 Embryonic development of a skeletal muscle fiber from a myogenic cell. Many myoblasts fuse to form a single myotube, which then matures to become a muscle fiber. *From Carlson (2007), with permission.*

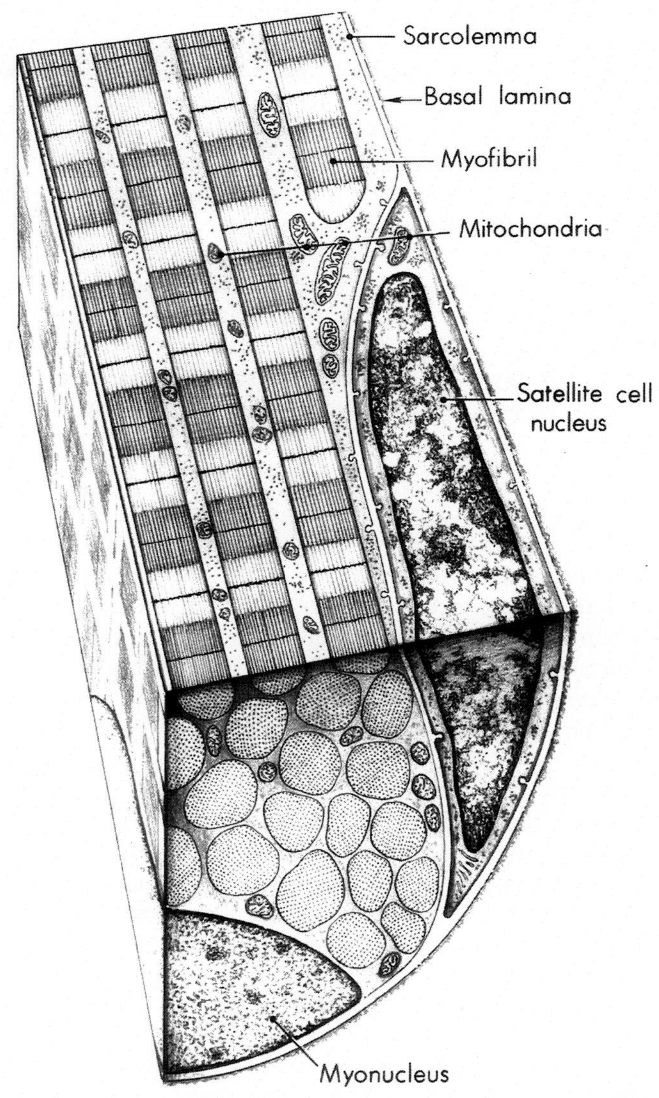

FIGURE 2.24 Drawing of a satellite cell of a skeletal muscle fiber. Mononuclear satellite cells are situated between a muscle fiber and its basal lamina. They serve as adult stem cells for the growth and regeneration of the muscle fiber.

weakly. As the myotube fills with newly synthesized contractile proteins, the myonuclei move to the periphery of what is now called a muscle fiber.

Skeletal muscle fibers make and secrete basal lamina material, which completely surrounds the muscle fibers and mediates the transmission of contractions of the muscle fibers to the surrounding connective tissue. Situated between a muscle fiber and its basal lamina is an undistinguished looking mononuclear cell, called a **satellite cell**, that is distinguishable from myonuclei only under the electron microscope or with special stains (Fig. 2.24). Satellite cells serve as stem cells for skeletal muscle fibers and are necessary for both their growth and regeneration. Multinucleated cells (**syncytia**) do not divide, and few muscle fibers form *de novo* after birth. Therefore for a muscle to grow after birth, the muscle fibers must add new nuclei derived from dividing satellite cells in order to increase in both length and girth.

Details of both the fine structure and contractile physiology of skeletal muscle fibers are presented in Chapter 5. Skeletal muscle fibers are not all homogeneous in either structure or function. Functionally, the speed of their contraction allows them to be divided into two main categories— fast and slow. Based on their need for oxygen while contracting, two types of fast muscle fibers have been described. **Fast glycolytic fibers**, which get their energy by the anaerobic breakdown (through a pathway not requiring oxygen) of glycogen, are those used in heavy lifting or in sprinting. **Fast oxidative muscle fibers** contain more mitochondria than fast glycolytic fibers and come into play where repeated contractions over a longer period of time is required. Fast oxidative muscle fibers are prominent in individuals trained for repetitive exercise, such as long-distance running. All muscle fibers are surrounded by loose networks of capillaries, but because of the need for continuous metabolic exchange, the capillary network surrounding fast oxidative muscle fibers is more extensive than that surrounding fast glycolytic muscle fibers.

Slow muscle fibers are thinner than fast muscle fibers. They contain more mitochondria and are enwrapped by denser networks of capillaries. These muscle fibers are not as strong as fast muscle fibers, but they can contract for much longer periods of time and are therefore endurance fibers. Slow muscle fibers predominate in postural muscles, a good example being the soleus muscle in the calf.

Some animals have fast and slow muscles—for example, dark meat (slow) and white meat (fast) in chickens. The dark color is largely due to the presence of an oxygen-binding protein, **myoglobin**, which is found in slow muscle fibers. Muscles of American jackrabbits, which specialize in long-distance running, contain a high proportion of fast oxidative and slow muscle fibers, whereas cottontail rabbits (hares in Europe), which run in short fast bursts, are endowed with large numbers of fast glycolytic muscle fibers. Muscles of humans and dogs, on the other hand, contain mixtures of fast and slow muscle fibers, although their proportions vary from one muscle to another, depending upon their functional requirements.

Cardiac Muscle

Like skeletal muscle, **cardiac muscle** appears cross-striated when the cells are viewed under a microscope, and the fundamental nature of the contractile machinery is very similar. Beyond that, however, differences outweigh similarities. In contrast to the long fibers containing hundreds of nuclei in skeletal muscle, human cardiac muscle cells (**cardiomyocytes**) normally have only one nucleus (see Fig. 2.22). Cardiac muscle cells begin beating in the human embryo 22–23 days after fertilization, and they must beat continuously throughout life. To do this requires an oxidative metabolism and a good blood supply. To keep up with their energy requirements, cardiomyocytes are richly supplied with mitochondria.

In contrast to skeletal muscle fibers, most cardiomyocytes are not directly innervated. Yet, in order for the heart to function effectively, they must beat in unison. This is accomplished by the presence of gap junctions, which rapidly transmit the signal to contract from the lateral wall of one cardiac muscle cell to the next. Mechanical stability of the cells comprising the heart is attained through arrays of intercellular connectors, such as **desmosomes**, which link one cell to its neighbor in an end-to-end fashion (Fig. 2.25). The junctions between the ends of two cardiac muscle cells look like dark lines in a light microscope and are called **intercalated disks**.

Cardiac muscle does not contain satellite cells, and under normal circumstances, cardiomyocytes do not divide by mitosis after birth. The heart grows by increasing the size of the individual cardiomyocytes (**hypertrophy**). In order to

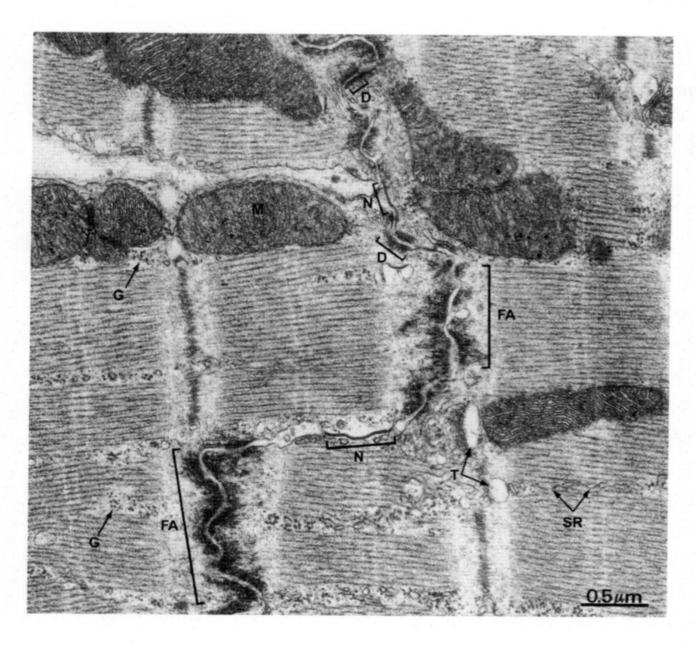

FIGURE 2.25 Electron micrograph of a cardiac muscle cell, showing the intercalated disk that connects two muscle cells. *D*, desmosome; *FA*, fascia adherens, forming tight connections between two cells; *G*, glycogen granules, providing energy stores; *N*, nexus (gap junctions), sites of low electrical resistance where signals pass from one cell to the next; *SR*, sarcoplasmic reticulum; *T*, T tubules. *From Young et al. (2006), with permission.*

produce and maintain the increased amount of cytoplasm, the DNA in the nucleus replicates, so that many mature cardiomyocytes contain from two to four or more times the normal amount of DNA. This condition is called **polyploidy**. Polyploidy allows the production of sufficient RNA to accommodate the need for maintaining the increased amounts of contractile proteins present in mature cardiac muscle cells.

Because of its constant beating function, cardiac muscle must be well supplied with oxygen, which gets to the cells through a rich supply of capillaries that lie adjacent to the individual cardiomyocytes. These capillaries represent the terminal portion of the coronary arterial circulation—all-important in maintaining the health of the heart.

Smooth Muscle

Smooth muscle differs considerably from skeletal and cardiac (striated) muscle. Smooth muscle cells are spindle-shaped and normally contain a single nucleus (see Fig. 2.22). As their name implies, smooth muscle cells do not have a cross-striated appearance when viewed under a microscope, and for many years how they were able to contract was not entirely clear. Now it is known that smooth muscle contains the same major contractile proteins (actin and myosin, see p. 113) found in skeletal and cardiac muscle, but that their arrangement within the cell is quite different. One significant property of smooth muscle is its ability to maintain a contraction for long periods with much less energy consumption than is the case for skeletal muscle. This represents a considerable advantage for certain bodily functions, such as maintaining vascular tone or uterine contractions during lengthy childbirth.

Smooth muscle is seen principally in the walls of tubes, and it functions either to propel material through the tubes (e.g., the digestive system) or to control the diameter of the tube and thereby flow through the tube (e.g., in blood vessels). At one extreme, contractions of the hypertrophied smooth muscle cells constituting the wall of the pregnant uterus push the fetus through the birth canal.

To accomplish these functions, coordinated contractions are often necessary. In some instances (for instance, the iris of the eye or the muscles that raise hairs), autonomic nerves or chemical signals provide the initial stimulus, but commonly intercellular signals transmitted through gap junctions enable a wave of contraction to continue. This is the case in the intestines and in reproductive tissues, such as the uterus or the uterine tubes.

NERVOUS TISSUE

The dominant feature of nervous tissue is the neuron, but the estimated number of neurons in the human brain (86–100 billion) may be actually outnumbered by other cells, which fall into the general category of glial cells (neuroglia). The nervous system as a whole is commonly subdivided into the central and peripheral divisions, and the PNS is broken down further into motor, sensory, and autonomic components. Details of these subdivisions are covered in Chapter 6.

Neurons

Neurons are often very large cells, with a prominent cell body and long cell processes that in some cases can approach a meter in length (Fig. 2.26). These cell processes are of two kinds, axons and dendrites, and their function is to carry signals or molecules along their length. **Axons** transmit electrical signals (action potentials) or transport molecules from the cell body of the neuron to nerve terminals at the end of the axon. **Dendrites** do the reverse. Their terminal branches receive peripheral signals and direct them back to the cell body of the neuron.

The cell body of a neuron is the main synthetic area of the cell, and it is well supplied with the organelles required for such activities, for example, RER, ribosomes, and Golgi apparatus. Axons can possess up to 1000 times more cytoplasm than the cell body, and they contain large numbers of well-ordered microtubules and microfilaments, which are used for both support and as transport mechanisms. These cytoskeletal structures are composed of protein subunits, and the subunits of the microtubules are polarized so that they can direct transport activities down the axon. Such transport is necessary to maintain both the structural integrity of the axon and to allow it to fulfill its normal physiological functions. **Axoplasmic transport** was first demonstrated when peripheral nerves were constricted, and a bulge appeared in the axon between the cell body and the site of the constriction. The bulge indicated that fluid was damming up. The interior of dendrites is more similar to that of the cell body. They contain some mRNA, ribosomes, and Golgi vesicles, as well as some microtubular and microfilamentous material. The latter two, however, are not as regularly organized as are those in the axons.

Three modes of axoplasmic transport have been identified—fast anterograde, slow anterograde, and fast retrograde. **Fast anterograde transport** is a means of moving membrane-bound structures, such as mitochondria and vesicles

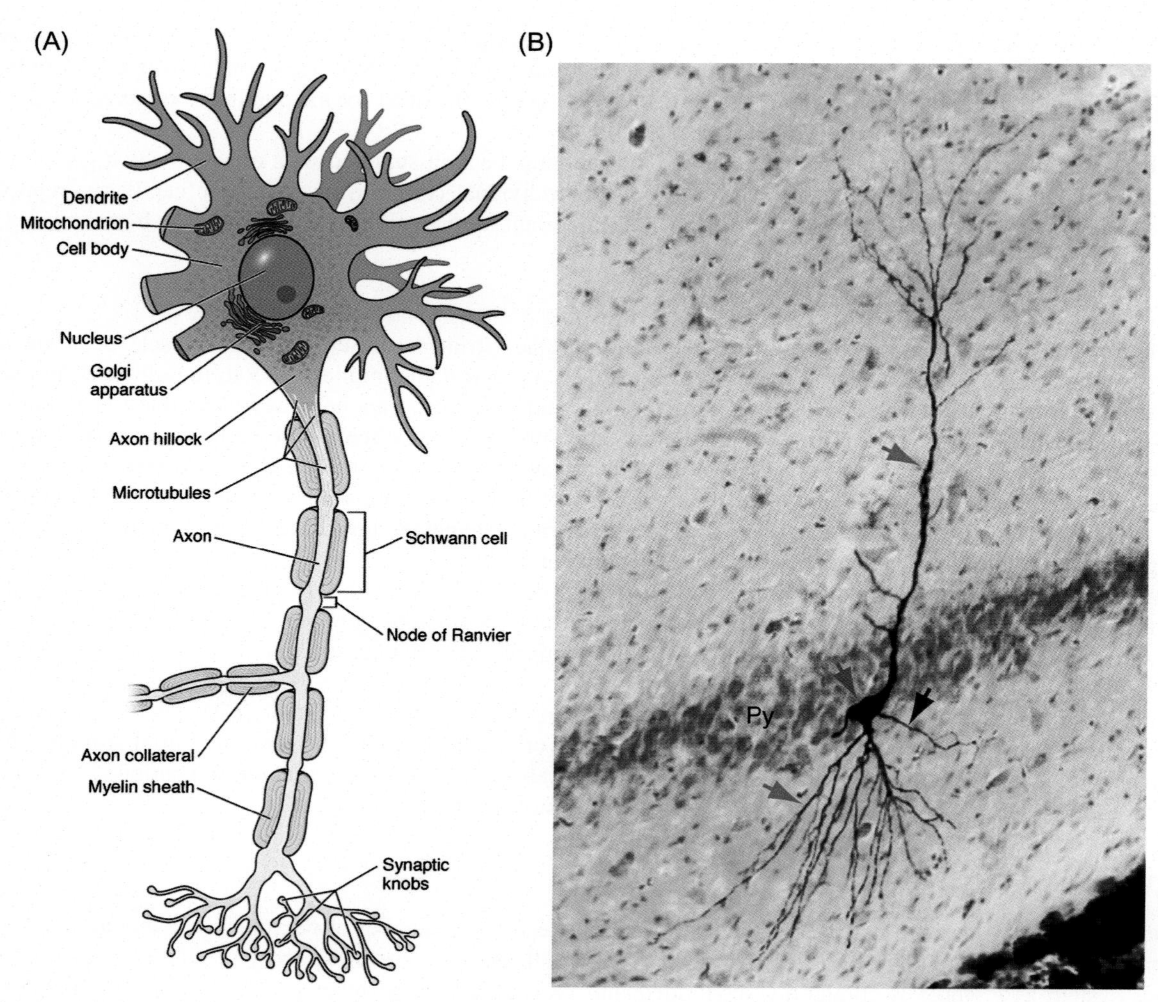

FIGURE 2.26 (A) Drawing of a generic nerve cell (neuron), (B) Photomicrograph of a single neuron (black) in the hippocampus of a rat. *Blue arrow*, cell body; *red arrows*, dendrites; *black arrow*, small axon. *(B) From Nolte (2009), with permission.*

containing functionally important molecules, toward the end of the axon. This is accomplished with the help of **kinesin**, a protein linking transported items to the microtubules, which serve as a track. Fast anterograde transport proceeds at a rate of up to 400 mm/day. In contrast, **slow anterograde transport** carries subunits of microtubules and microfilaments toward the distal end of the axon at rates of only 0.2–8 mm/day. Paradoxically, **fast retrograde transport**, assisted by the protein **dynein** (which is also involved in ciliary motion), moves materials up the microtubules toward the cell body of the neuron at a rate of 200–300 mm/day. Retrograde transport may be the mechanism by which growth factors required for the wellbeing and survival of neurons are carried to them from peripheral regions.

Nerve processes come in an amazing variety of sizes and shapes, especially in the central nervous system (CNS) (brain and spinal cord) (Fig. 2.27). Each of these is related to some functional quality. For example, very simple bipolar neurons contain two short processes which connect to other neurons in the immediate area, and they are often part of a chain that transmits sensory impulses to central processing centers. On a much grander scale are motor neurons, which have a number of short dendritic processes that convey information to a large cell body, as well as a very long axon that is the structural basis for sending a signal to some distant part (e.g., a neuron stimulating part of a toe muscle to contract). Other neurons in the brain receive input from numerous dendrites and are called stellate cells because of their resemblance to stars. These neurons are designed to receive and integrate signals from numerous other neurons in the brain.

Interneuronal connections are accomplished through **synapses** of various types. These synaptic connections, which number 10^{15} in the human brain, are the basis for the incredibly complex integrative functions of the entire CNS. The dendrites of many types of neurons in the CNS are studded with knob-like projections called **dendritic spines**

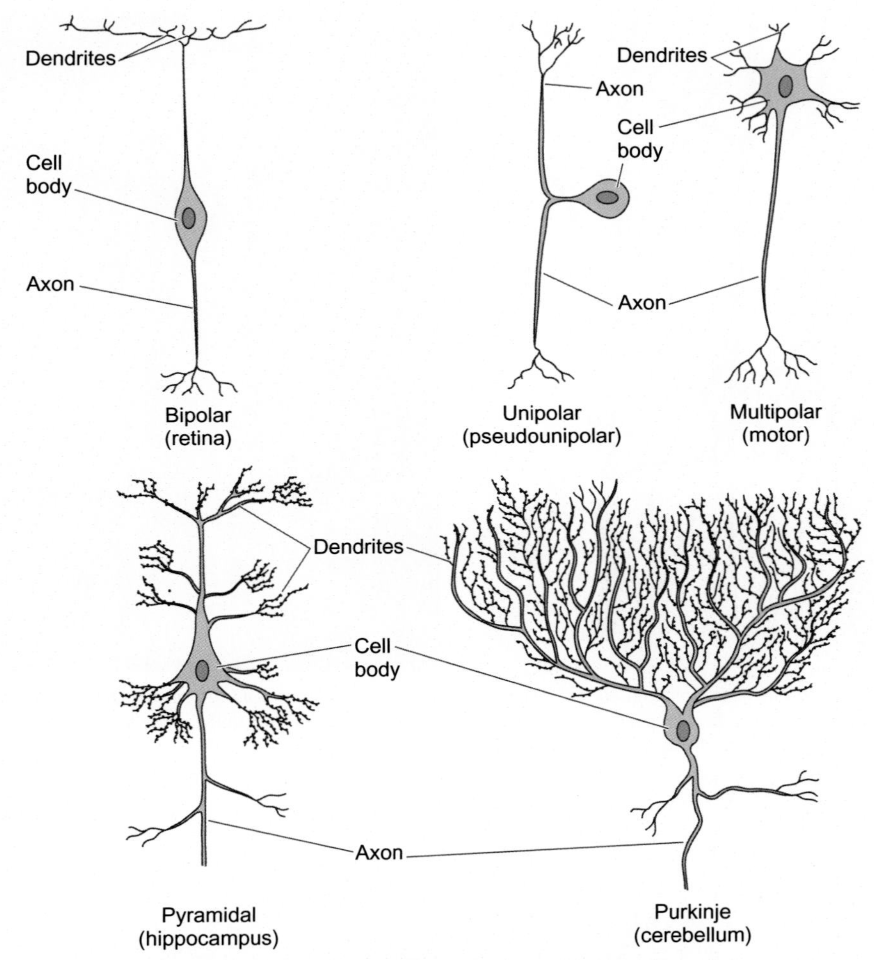

FIGURE 2.27 Some of the many different types of neurons. *From Gartner and Hiatt (2011), with permission.*

(Fig. 2.28). Each dendritic spine receives input from other neurons in the brain. Given that a single stellate cell can have up to several thousand connections with other neurons, the number of possible combinations among neurons is staggering.

The very end of an axon or its branches is a specialized region called the **presynaptic terminal**. Here electrical signals going down the axon (see p. 62) become converted to a chemical signal by causing the release of packets of chemicals called **neurotransmitters**, which are stored in the presynaptic terminals.

A neuron sends its signal to the next element of the chain by a chemical neurotransmitter passing through a narrow synaptic cleft between the nerve terminal and the **postsynaptic membrane** of another neuron or a muscle fiber. The neurotransmitter, of which there is a variety of types, is bound to a chemical receptor molecule on the postsynaptic membrane and then typically stimulates the postsynaptic cell to generate another electrical action potential that continues the signal.

Glial Cells

Glial cells have long remained the most enigmatic component of the nervous system (Fig. 2.29). By default, they have been defined as any kind of nervous tissue cell that is not a neuron. Their total numbers at least equal the number of neurons within the CNS,[3] but the proportions depend upon the part of the brain. For example, in the cerebral cortex, glial cells outnumber neurons by almost 3:1, whereas in the cerebellum the ratio is reversed.

3. Estimates of the proportion of glial cells in relation to neurons in the human brain vary wildly, from less than 1:1 to 50:1.

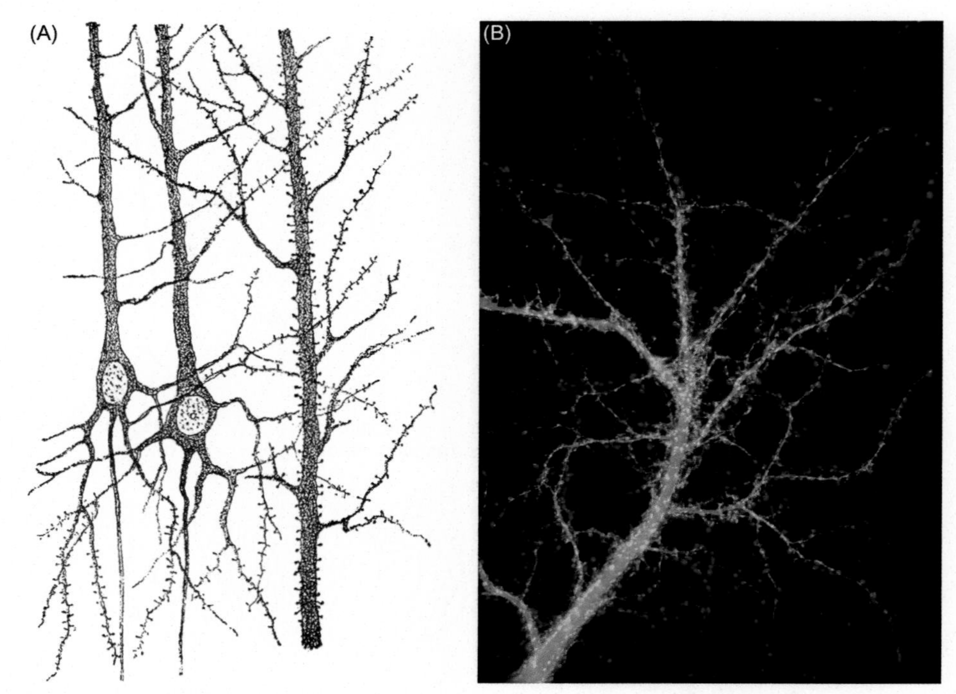

FIGURE 2.28 Synapses on neurons of the cerebral cortex. (A) dendrites studded with tiny dendritic spines. (B) Fluorescence micrograph of a dendrite from a hippocampal neuron. Green, distribution of MAP2 (a microtubule-associated protein in the dendrite); Red, a synaptic protein, showing the locations of a large number of synapses on the dendrite. *From Nolte (2009), with permission.*

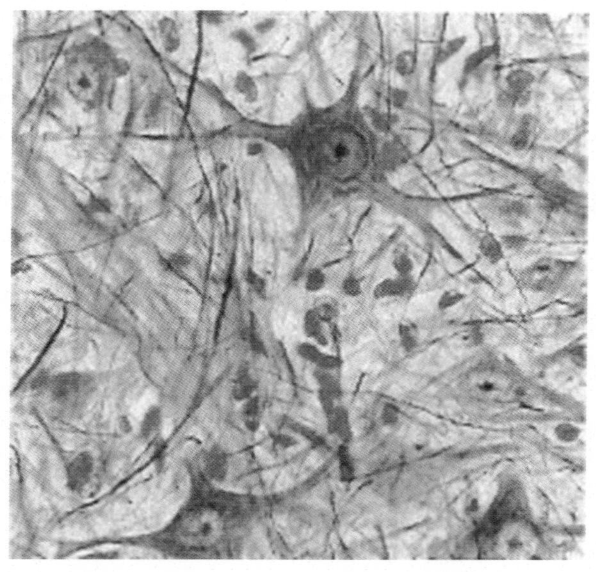

FIGURE 2.29 Photomicrograph of neurons (large brown structures) and glial cells (smaller blue structures). *From Waugh and Grant (2014), with permission.*

Commonly, six types of glial cells have been defined. Four types of them are located in the brain and spinal cord, and two are found in the PNS (Table 2.5). **Astrocytes** and **oligodendrocytes** are closely associated with neurons and arise in the embryo from the same precursor cells as do neurons. Tiny **microglial cells** are derived from a different embryonic stem cell lineage and serve as scavengers of dead cellular material within the brain. The fourth glial cell type, **ependymal cells**, is arranged like an epithelium and lines the central canal of the brain and spinal cord.

TABLE 2.5 Types of Glial Cells

Cell Type	Function
Within Central Nervous System	
Oligodendrocyte	Myelination within CNS
Astrocyte	Creation of blood–brain barrier
	Mechanical support of neurons
	Stimulate synapse formation
	Modulate neuronal signaling
	Metabolize nutrients for neurons
	Secrete growth factors
	Form scars after damage to CNS
Microglia	Remove debris from damaged or dead cells
	Produce signaling molecules (cytokines)
Ependymal cells	Secrete and absorb cerebrospinal fluid
	Stem cell function
Within Peripheral Nervous System	
Schwann cells	Myelination within PNS
	Production of growth factors
Satellite cells	Surround, nourish, and support peripheral neurons

Ependymal cells produce cerebrospinal fluid, but in recent years the ependymal epithelium has also been found to be associated with neuronal stem cells. Within the cerebral cortex, 76% of the glial cells consist of oligodendrocytes, 17% astrocytes, and 7% microglia. The glial cells of the PNS are called **Schwann cells** and **satellite cells**. Schwann cells play the same functional role in the PNS as oligodendrocytes do in the CNS, namely providing an insulating cover of myelin over axons and dendrites. Satellite cells surround the cell bodies of peripheral neurons; their function remains obscure.

Astrocytes, which have many processes occurring off the cell body, come in two main shapes. In gray matter, the processes of what is called protoplasmic astrocytes are relatively fleshy, and they contact both synapses and blood vessels. The astrocytes of white matter have much thinner processes, which contact both blood vessels and nodes of Ranvier. Many functions are attributed to astrocytes. A major one is creating the blood–brain barrier, which prevents many cells and large molecules from reaching the nervous tissues from the blood (Fig. 2.30). Astrocyte processes around synapses promote their initial development and modulate their ongoing function by controlling levels of K^+ and some neurotransmitters. After injury to the brain or spinal cord, astrocytes form scar-like masses that inhibit the regeneration of severed axons and dendrites.

In order to function properly, within both the CNS and PNS, nerve processes are surrounded by a lipid-rich cellular insulation, called **myelin**. The Schwann cells in the PNS and the oligodendrocytes in the CNS wrap around the axons and dendrites in a manner that in cross-section very much resembles layers of electrical tape wound around a wire (Fig. 2.31). Spaces between these cells (**nodes of Ranvier**) and also the insulation that the Schwann cells provide constitute an important structural basis for the transmission of an electrical signal (action potential) down a nerve process.

In peripheral nerves, the cell bodies of the Schwann cells lie outside the multiple layers of myelin sheathing. These are collectively covered by a basal lamina of their own making and a thin layer of collagen fibers. This layer is called the **endoneurium** (Fig. 2.32), and constitutes the first layer of the fasciae that provide mechanical strength to the actual nerve. Surrounding bundles of axons, each covered with its own endoneurium, is a thicker and tougher layer of connective tissue, called the **perineurium**. Finally, ensheathing many bundles of nerve fibers is a still thicker **epineurium**, which surrounds the entire nerve. These coverings are of great importance to nerve surgeons, who suture the severed

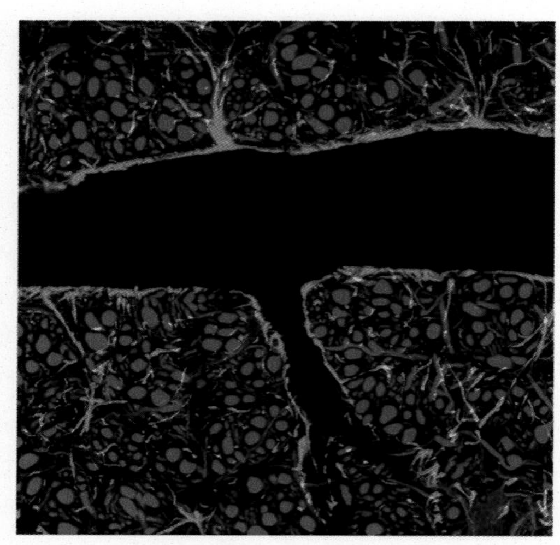

FIGURE 2.30 Fluorescence micrograph, showing the blood–brain barrier. Red, neurons; green, astrocytes, which form a tight barrier between the blood (black space) and the brain tissue. *From Underwood (2015), with permission.*

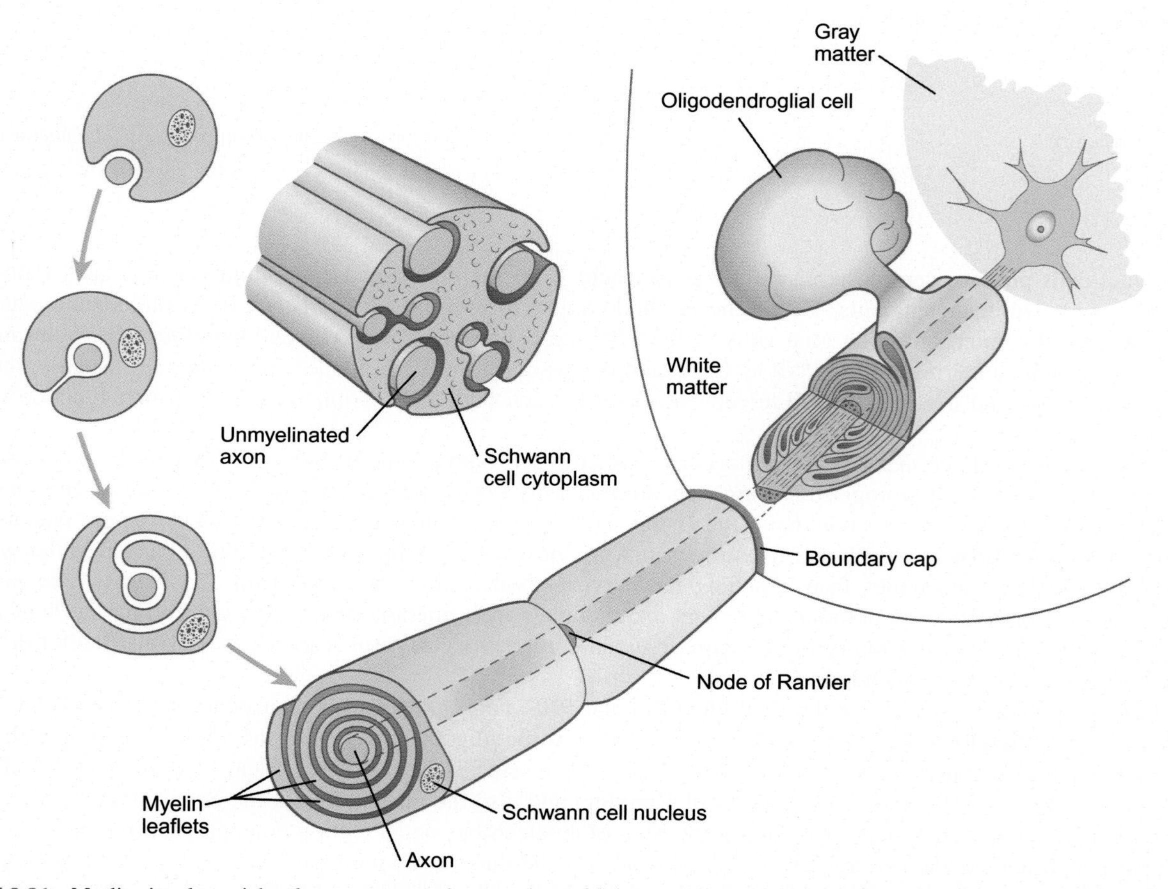

FIGURE 2.31 Myelination the peripheral nervous system by wrappings of Schwann cells and in the central nervous system by similar wrappings of oligodendroglial cells. *From Carlson (2014), with permission.*

ends of perineurium or epineurium together to provide appropriate channels for regenerating nerve fibers after trauma to a nerve.

How a Neuron Conducts Signals

One of the main activities of a neuron is to conduct signals from one end of the cell to the other and then transmit the signals to other cells (neurons or end organs, such as muscle fibers). This is accomplished mainly through electrical signals that flow from the tips of the dendrites to the neuronal cell body and then down to the terminal processes of the axons (see Fig. 2.26). These electrical signals are based principally on the flow of charged ions across the plasma membrane of the neuron.

The starting point is what is called the **resting membrane potential** (RMP) of a neuron. If one measures the electrical charge (voltage) across the plasma membrane of neurons (and other cells), the inside of the cell is negatively charged (commonly by ~40–90 mV) with respect to the immediate outside of the cell (see Box 2.4). This condition is

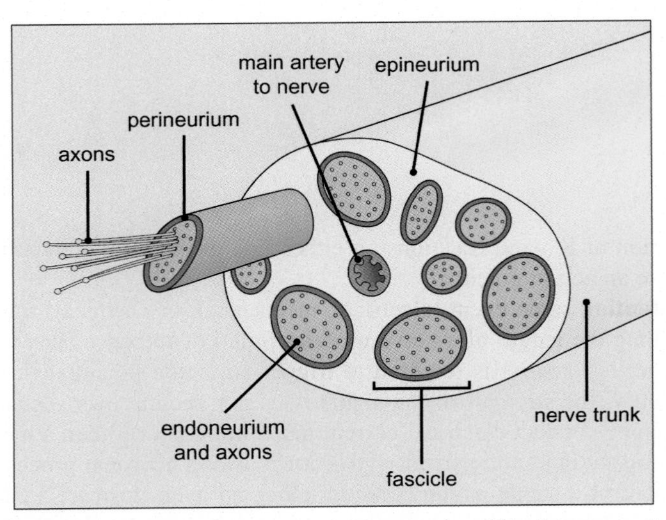

FIGURE 2.32 Diagram showing the relationships among the connective tissue sheaths within the cross-section of a peripheral nerve fiber. *From Stevens and Lowe (2005), with permission.*

BOX 2.4 Resting Membrane Potential

A RMP is not unique to neurons, but it is a property of essentially all animal cells, ranging from ~ −5 mV in red blood cells to −95 mV in skeletal muscle fibers. The RMP of a typical neuron is about −65 mV, with the interior of the cell negative in charge to the outside. The basis for the RMP is the unequal distribution of ions, in particular K^+ and Na^+, on either side of the plasma membrane, with a K^+ concentration of ~140 mmol/l inside and 5 mmol/l outside the cell. Conversely, the concentration of Na^+ inside the neuron is ~14 mmol/l and that on the outside is ~145 mmol/l. These static figures mask the fact that there is a constant flow of ions across the plasma membrane. Several factors modulate the flow of ions, both positive and negative—(1) diffusion through ion channels according to concentration gradients (ions flow from areas of high to low concentrations), (2) properties of specific ion transporter molecules located in the plasma membrane, and (3) attraction of oppositely charged ions to one another.

An important player in this mix is the **sodium-potassium pump** (Fig. 2.33). Proteins in these pumps contain receptor sites, which bind two K^+ ions on the outside of the cell and three Na^+ ions on the inside. The inner portion of the same receptor protein functions as an ATPase. The ATPase function becomes activated when the receptor sites are fully bound with K^+ and Na^+. The ATPase activity then cleaves a molecule of ATP into ADP plus phosphate, with the liberation of sufficient energy by the breakdown of the ATP to stimulate the influx of the two potassium ions into the cell and the efflux of the three sodium ions to

(Continued)

BOX 2.4 (Continued)

the outside. In conjunction with passive leak channels, which allow the passage of both sodium and potassium, the ionic balance that maintains the RMP is maintained.

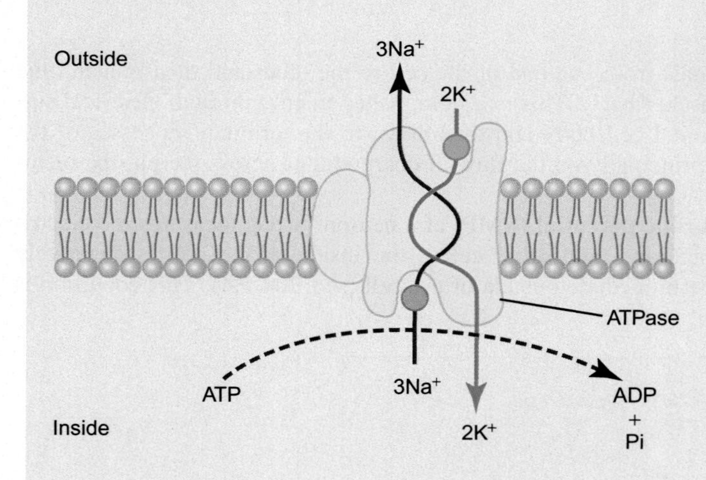

FIGURE 2.33 Functions of the sodium-potassium pump in a nerve cell. As Na^+ leaves and K^+ enters the cell, ATP is converted to ADP, with the release of energy. *From Guyton and Hall (2016), with permission.*

based on an unequal distribution of K^+ and Na^+ ions on either side of the plasma membrane, and it is maintained until the neuron is stimulated to fire an action potential.

To produce an **action potential**, some event (electrical, mechanical, or chemical) must build up an electrical potential that can be propagated along the length of a nerve process. Small disturbances can result in the appearance of local potentials, but a larger change in potential is required to trigger an action potential. In a nerve, an action potential is self-propagating and importantly, the strength of the signal does not decline over distance, as is the case with a local current. Neuronal processes alone conduct electrical current much more poorly than a metal wire. Action potentials represent a means for quickly propagating an electrical signal down a long neuronal process. They do not accomplish this through the seamless movement of a single action potential along an axon. Instead, a single action potential stimulates the creation of another action potential farther along the axon in a manner similar to the falling of a line of upturned dominoes, with each domino representing a single action potential (Box 2.5).

Once the impulse reaches the nerve terminals at the end of the axon, it reaches a synapse (20–24 nm space) that connects the nerve terminal to either another nerve process or a muscle fiber. Although some synapses are electrical and join axons and dendrites of two neurons through gap junctions, most are **chemical synapses** where the electrical impulse going down the axon is converted to a chemical signal. The chemical signals are called transmitters, of which there are over 100 types, and they are received by a postsynaptic apparatus to be reconverted to an electrical signal. For perspective, a large spinal neuron may have 10,000 synapses impinging upon it from other neurons. At the other end, terminal branches from a single motor axon can innervate from 3–6 muscle fibers in the case of extraocular muscles to well over 1,000 muscle fibers in some of the large leg muscles.

In a chemical synapse, the first step is the synthesis of the transmitter in the **presynaptic terminal** and its storage in **synaptic vesicles** (Fig. 2.37). When an action potential reaches the presynaptic terminal, depolarization of the membrane causes the opening of gated Ca^{++} channels and the influx of extracellular Ca^{++} into the terminal. This causes the synaptic vesicles to fuse with the plasma membrane of the terminal and to release their contents of transmitter into the synaptic cleft through the process of exocytosis. The transmitter molecules then diffuse toward the postsynaptic membrane (the other side of the synapse) where they are bound to receptor molecules. At this point, the receptors, which are an integral part of ion channels, open and allow an influx of ions through the channels into the postsynaptic cell, whether it be another nerve or a muscle fiber. This ion flow changes the electrical excitability of the postsynaptic cell, which then reacts in an excitatory or an inhibitory manner, for instance, stimulating the contraction of a muscle fiber.

BOX 2.5 Propagation of an Action Potential

A typical action potential is illustrated in Fig. 2.34. It begins when a local stimulus results in an influx of Na^+ into the cell. This causes the transmembrane potential to rise from negative toward positive. If the rise exceeds a threshold value (~ −55 mV), then voltage-regulated gates open and more Na^+ rushes into the cell than K^+ is removed. Within a few milliseconds the membrane potential quickly depolarizes and rises to a positive level (~ +30−50 mV). As the potential rises above zero, the sodium gates rapidly begin to close, while at the same time, potassium gates more slowly open, causing K^+ to exit the cell in response to the influx of Na^+ into the cell.

The action potential peaks when the cytoplasm of the neuron contains the maximum number of sodium ions, and an equivalent amount of potassium ions has not yet left. The K^+ gates remain open, and as potassium ions continue to leave the cell, the membrane potential quickly declines from a positive value to a negative one and even overshoots the mark by a bit, resulting in a brief period of **hyperpolarization**. Then the membrane potential returns to its normal resting value with the distribution of Na^+ and K^+ to their normal preaction potential values.

A single action potential alone is not sufficient to propagate an electrical signal, since it remains local. It is important to understand how the impulse is able to travel along the length of a neuronal process. A few facts are useful as background. First, without action potentials, the strength of a depolarization event becomes attenuated as it moves down a neural process, whereas one mediated by action potentials preserves its full strength throughout its entire transit. Second, passive current flow is propagated more rapidly in axons with a large, rather than a small diameter due to the lesser internal resistance of a large-diameter fiber. Third, the speed of propagation of action potentials in a myelinated nerve fiber is considerably faster than that in an unmyelinated fiber.

Let us first examine the propagation of an action potential down the length of an unmyelinated nerve fiber (Fig. 2.35). When the action potential is generated at a specific site, it generates a passive depolarizing current that travels down the axon. This current decreases in intensity over distance, but nearby it is sufficient to cause the opening of Na^+ channels down the axon; thus stimulating the initiation of a new action potential of the same strength. This process repeats itself until it reaches the end of the axon. Further down the axon from the new action potential, the current is not sufficient to trigger an action potential. On the back side of the action potential, the membrane is not capable of generating a new action potential as long as the Na^+ gates are open and are in the process of closing. This is known as the **refractory period**, and because of it, the action potential cannot progress in a backward direction. Thus ensures that it will only be propagated in a forward direction. Once the second action potential has been elicited, the overall process is repeated as the signal proceeds down the axon and triggers the third action potential. With a process like that outlined above, an unmyelinated fiber can conduct an impulse at a rate of 0.5−10 m/s, but the conduction velocity in a myelinated fiber can reach up to 150 m/s. What accounts for the difference?

Much of the length of a myelinated fiber is covered by thick layers of insulation in the form of Schwann cell or oligodendrocyte wrappings (Fig. 2.36). At the **nodes of Ranvier**, situated between two insulation wrappings, the plasma membrane contains numerous voltage-gated Na^+ channels whereas few or none of these channels exist between nodes. Therefore action potentials can only be generated at the nodes. The myelin wrappings also ensure that the passive local current does not escape through the internode membranes. The local current moves at a greater speed than the time taken to generate an action potential, so by reducing the frequency of action potentials to only the areas of the nodes, the compressed (nonleaky) local currents move the overall impulse from a point A to a point B (one node to the next) much more rapidly than would be the case for an unmyelinated fiber, which would require the formation of several action potentials for the impulse to reach the same point. Because current only flows across the plasma membrane at the nodes, this mode is called **saltatory conduction** (jumping from node to node).

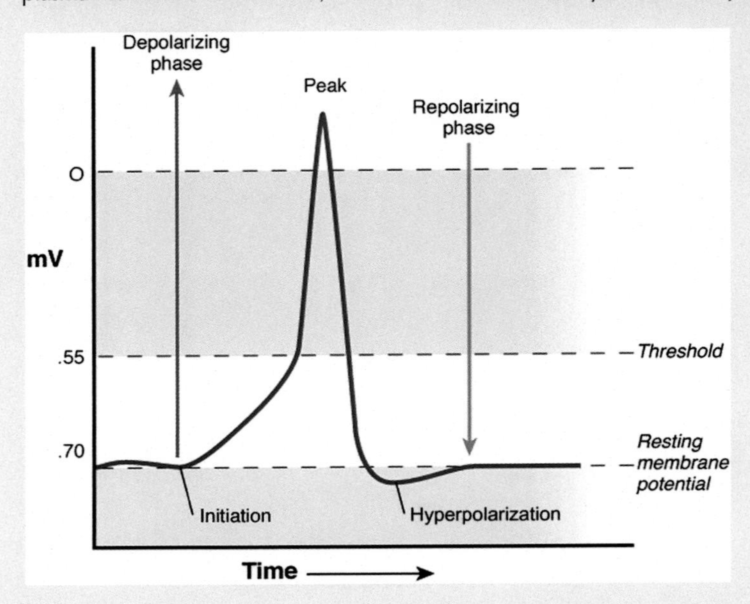

FIGURE 2.34 Diagram of an action potential along the axon of a nerve. *Redrawn from Purves et al. (2008), with permission.*

(Continued)

BOX 2.5 (Continued)

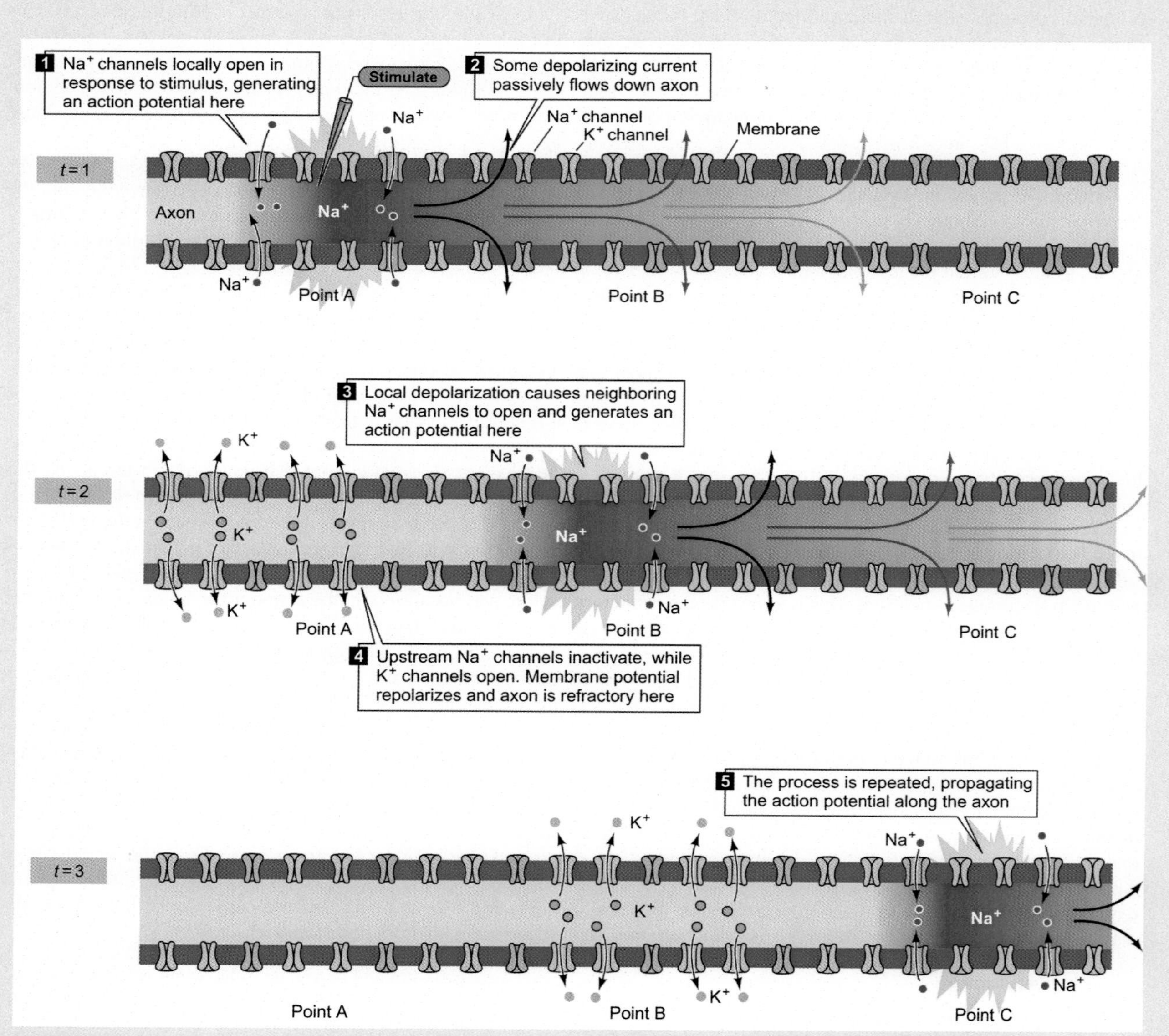

FIGURE 2.35 Conduction of an action potential down an unmyelinated axon. It begins with depolarization of the axonal membrane at point A at the top ($t = 1$). Passive current flow depolarizes the membrane until the opening of new Na^+ channels at point B generates an action potential at that site ($t = 2$). This process is repeated ($t = 3$), as the action potential is propagated at point C. *From Purves et al. (2008), with permission.*

(Continued)

BOX 2.5 (Continued)

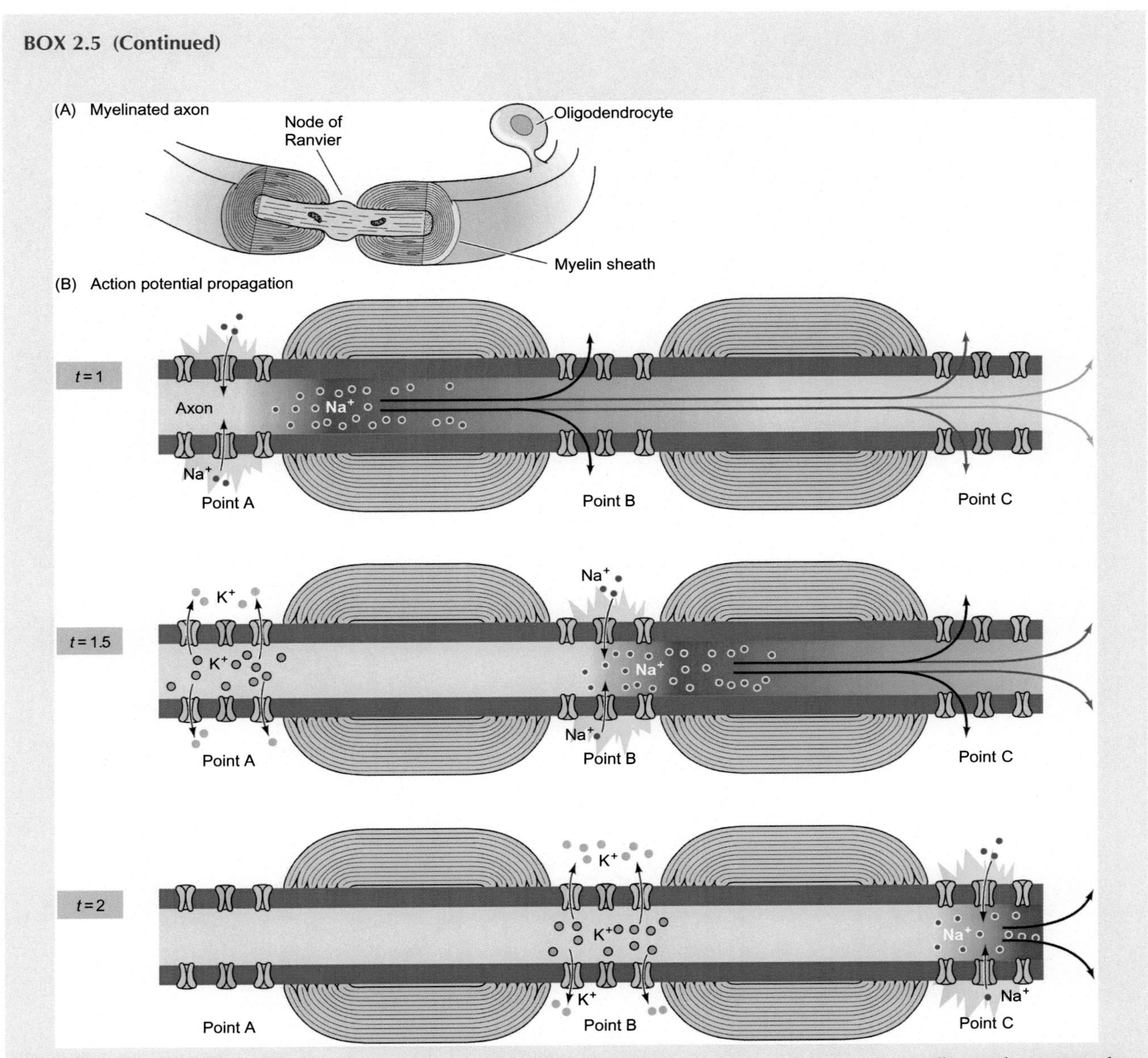

FIGURE 2.36 Propagation of an action potential down a myelinated axon by means of saltatory conduction. The myelin covering prevents the leakage of current through the axonal membrane, but at the nodes of Ranvier (see Part A), the lack of myelin allows the activation of sodium channels and the propagation of the action potential from one node to the next one down the axon. This mode of propagation of the action potential greatly speeds up the overall rate of transmission. *From Purves et al. (2008), with permission.*

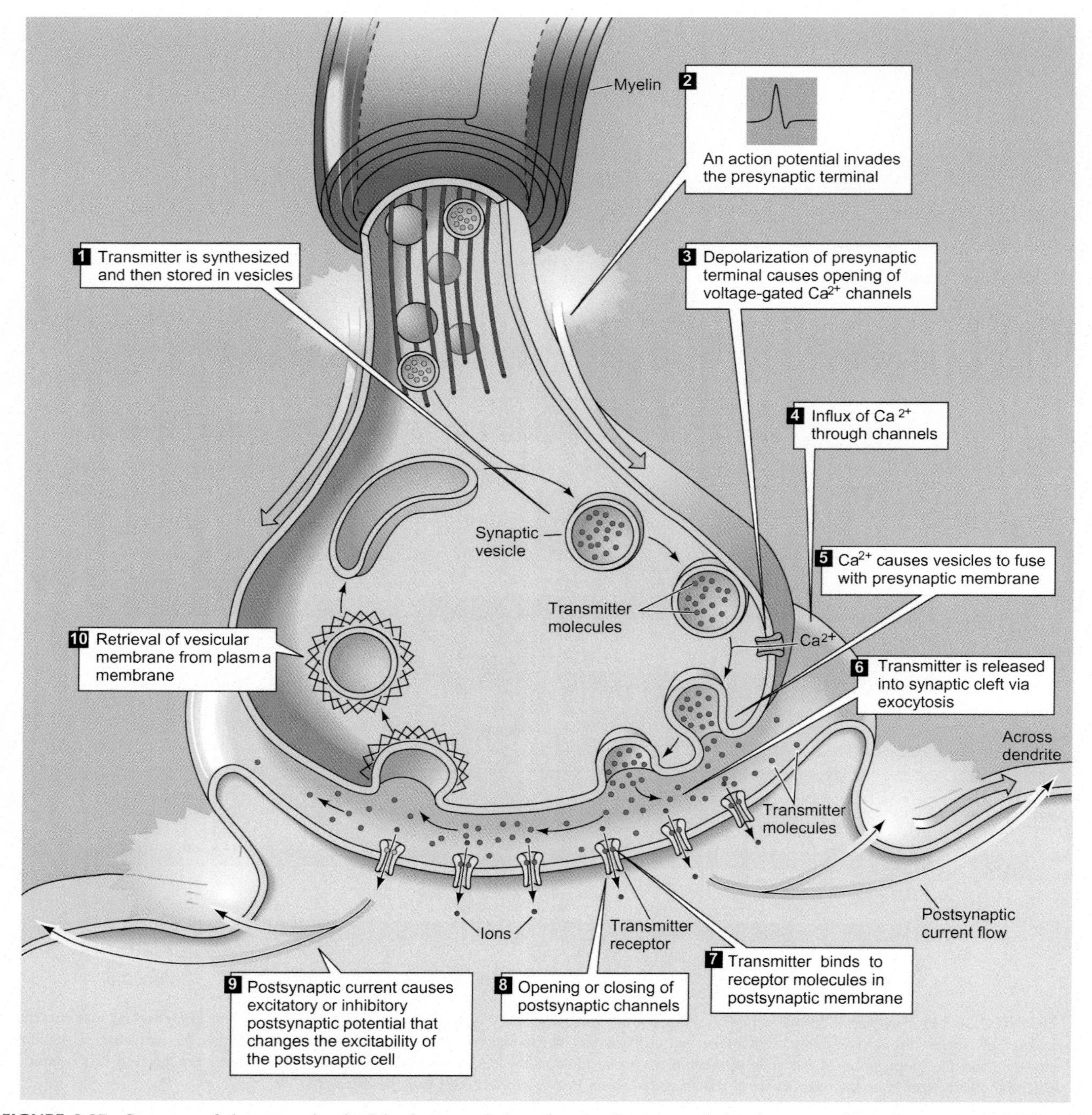

FIGURE 2.37 Summary of the events involved in the transmission of a signal across a chemical synapse. *From Purves et al. (2008), with permission.*

SUMMARY

Tissues are aggregations of cells that work together to accomplish some major biological function. Tissues have traditionally been divided into four types—epithelia, connective tissues, muscle, and nervous tissue.

Two major functions of epithelia are covering surfaces or lining tubes and making substances to be excreted. A typical epithelial cell has an apical surface, a basal surface, and lateral surfaces. The basal surface is tightly connected to an underlying basal lamina. Lateral surfaces of epithelial cells are bound together by a variety of intercellular junctions and adhesion molecules. Gap junctions allow the passage of electrical signals between cells.

Epithelial cells come in a variety of sizes and shapes, ranging from thin scale-like squamous cells to tall columnar cells. Between these are cuboidal cells that are often found in areas of high metabolic activity. Epithelia can be single- or multiple-layered.

Glands are made up of epithelial cells. Glands that secrete their products into ducts are called exocrine glands, whereas glands that secrete their products directly into the blood stream are called endocrine glands. Glandular epithelial cells secrete by one of three modes—merocrine, in which the secretory product is released by exocytosis; holocrine, in which the entire cell breaks down and is secreted; and apocrine, in which the tops of the epithelial cells break off along with an intracellular secretion.

Connective tissues consist of cells that are embedded within their own secretions—the extracellular matrix. Connective tissue types include fibroelastic, adipose, cartilage, bone, and blood.

The prototypical connective tissue cell is the fibroblast, a cell that produces proteins and carbohydrates for export. These substances produce the extracellular matrix, which consists of fibers, ground substances, and fluid. Connective tissue fibers come in three forms—collagen, elastic, and reticular. The consistency of ground substance can range from gelatinous to rock hard, and the water content varies accordingly.

Cartilage consists of chondrocytes that are embedded in a firm, but resilient matrix of their own making. In addition to the hyaline cartilage of joint surfaces, elastic cartilage, such as in the ear, contains a large proportion of elastic fibers. In contrast, fibrocartilage, such as in the intervertebral disks, contains layers of dense, highly oriented collagen fibers.

There are three types of adipose tissue—white, beige, and brown fat. In addition to storing lipids for energy, white fat cells produce a variety of active substances, some of which have an endocrine function. Brown fat, mostly found in fetuses and newborns allows them to generate heat metabolically. Beige fat is also involved in thermogenesis and can appear later in postnatal life.

Blood is a highly specialized connective tissue. Red blood cells carry oxygen, whereas white blood cells are heavily involved in immune functions and defenses against pathogens.

Three types of muscles—skeletal, cardiac, and smooth—are present in the body. Skeletal and cardiac muscle are cross-striated because of the regular arrangement of their contractile proteins.

Skeletal muscle consists of long multinucleated fibers, which can be divided into two main categories—fast and slow—based on their speed of contraction. Slow muscle fibers are dark in color due to their content of myoglobin, an oxygen-binding protein. Each skeletal muscle fiber is connected to a motor nerve terminal.

Cardiac muscle consists of mononuclear cells that are tightly connected by intercellular junctions. Gap junctions permit contractile signals to pass from one cardiomyocyte to another. Cardiac muscle cells have an intrinsic beat, and most are not directly innervated.

Smooth muscle cells are spindle-shaped. Their contractile proteins are not arranged in a regular order. As a result, smooth muscle cells are not cross-striated. Smooth muscle is seen principally in the walls of tubes, where they act to propel material through the tubes or to control their diameter.

Nervous tissue consists of neurons and supporting cells. The supporting cells in the central nervous system (CNS) are glial cells, whereas those in the peripheral nervous system (PNS) are Schwann cells.

Neurons are often very large cells, with a prominent cell body and long cell processes that can be up to a meter in length. Axons carry signals away from the nerve cell body whereas dendrites carry signals toward the nerve cell body.

Molecules are transported along neuronal processes through arrays of microtubules. In contrast, neuronal signals are rapidly carried along the plasma membranes of these processes by means of action potentials.

Neurons connect with one another or with skeletal muscle fibers through specialized endings, called synapses. Action potentials stimulate the release of neurotransmitters across synapses. Action potentials are based on the flow of Na^+ and K^+ ions across the plasma membranes of the neuronal processes, and once begun, they are self-perpetuating until they reach the nerve terminal.

Major types of glial cells in the brain are astrocytes, oligodendrocytes, microglia, and ependymal cells. Astrocytes form the basis of the blood–brain barrier and have many other functions, as well. Ependymal cells produce cerebrospinal fluid. Many nerve processes within both the CNS and PNS are covered with myelin, a fatty covering consisting of cytoplasmic folds of oligodendrocytes in the CNS and of Schwann cells in the PNS.

Chapter 3

Skin

The skin is a most remarkable organ, with probably more kinds of functions than any other organ in the body. It represents the body's face to the outside world and in many respects it is a mirror of our emotions and overall state of health. One of the largest organs in the body, it weighs about 4 kg in adults and has a surface area of 1½–2 m^2. One has only to compare the skin of the scalp with that of the sole of the foot to recognize that the skin is not homogeneous throughout the body, but its structure varies from place to place depending on the functions that a particular region requires.

STRUCTURE OF SKIN

The skin is made up of three layers, each of which contains multiple components (Fig. 3.1). An **epidermis**, composed of stratified squamous epithelial cells called **keratinocytes**, covers the entire surface of the skin. These cells begin as stem cells located at the bottom of the epidermis (Fig. 3.2). Like many kinds of stem cells, individual stem cells divide into two daughter cells, one of which retains its stem cell properties. The other daughter cell moves on and farther up within the epidermis and begins several cycles of cell division, resulting in a much larger population of epidermal cells that become more specialized with each division. Specialization in the epidermis consists largely of producing **keratin** proteins, which provide strength to the epidermal cells. As the epidermal cells approach the surface, they cease dividing and undergo the final stages of specialization before they die and become the most superficial layers of the skin.

The epidermis is underlain by a tough **dermis**, whose mechanical characteristics are determined principally by the interwoven collagen bundles and elastic fibers that it contains. The dermis consists of a **papillary layer**, arranged as dermal pegs and located just beneath the epidermis, and a deeper **reticular layer** (see Fig. 3.1). The collagen and elastic fibers in the papillary layer are distributed as a fine meshwork, whereas those in the reticular layer are coarse and serve as the main source of tensile strength of the skin. Beneath the dermis is the **hypodermis**, which in most regions contains adipose tissue interspersed with looser bundles of collagenous connective tissue. The distribution of subcutaneous fat differs between the sexes with selective accumulation in the abdomen in males and around the hips in females (see Fig. 2.20). Epidermally-derived skin appendages, as different as hairs, nails, sweat glands, and mammary glands play important functional roles. A number of different types of adult stem cells are associated with both hair follicles and subcutaneous fat. Table 3.1 lists some of the most important functions served by the skin.

MECHANICAL PROPERTIES OF SKIN

From a mechanical standpoint the general function of the skin is to keep the insides in and the outside out of the body, but the skin has to fulfill a number of less apparent mechanical functions, as well. It needs sufficient strength to resist rupturing from mechanical stress, but it must also permit movement around joints. It needs to respond to simple sustained pressure, as well as to repeated bouts of mechanical pressure and removal of that pressure, often associated with movement.

All three components of the skin are in some way involved in mechanical functions. The epidermis per se has little mechanical strength, but its continuity must be maintained when the skin is subjected to deformation or abrasion. To a considerable extent, this is accomplished by the network of **hemidesmosomes** (see Fig. 2.1) and associated intermediate filaments that anchor the deepest layer of the epidermis to its basement membrane. Strengthening is furthered by epidermal downgrowths (**rete ridges**) that form an irregular epidermal–dermal interface (Fig. 3.3). Within the epidermis itself, intercellular junctions (**desmosomes**) and a variety of adhesive proteins between keratinocytes ensure

The Human Body. DOI: https://doi.org/10.1016/B978-0-12-804254-0.00003-X

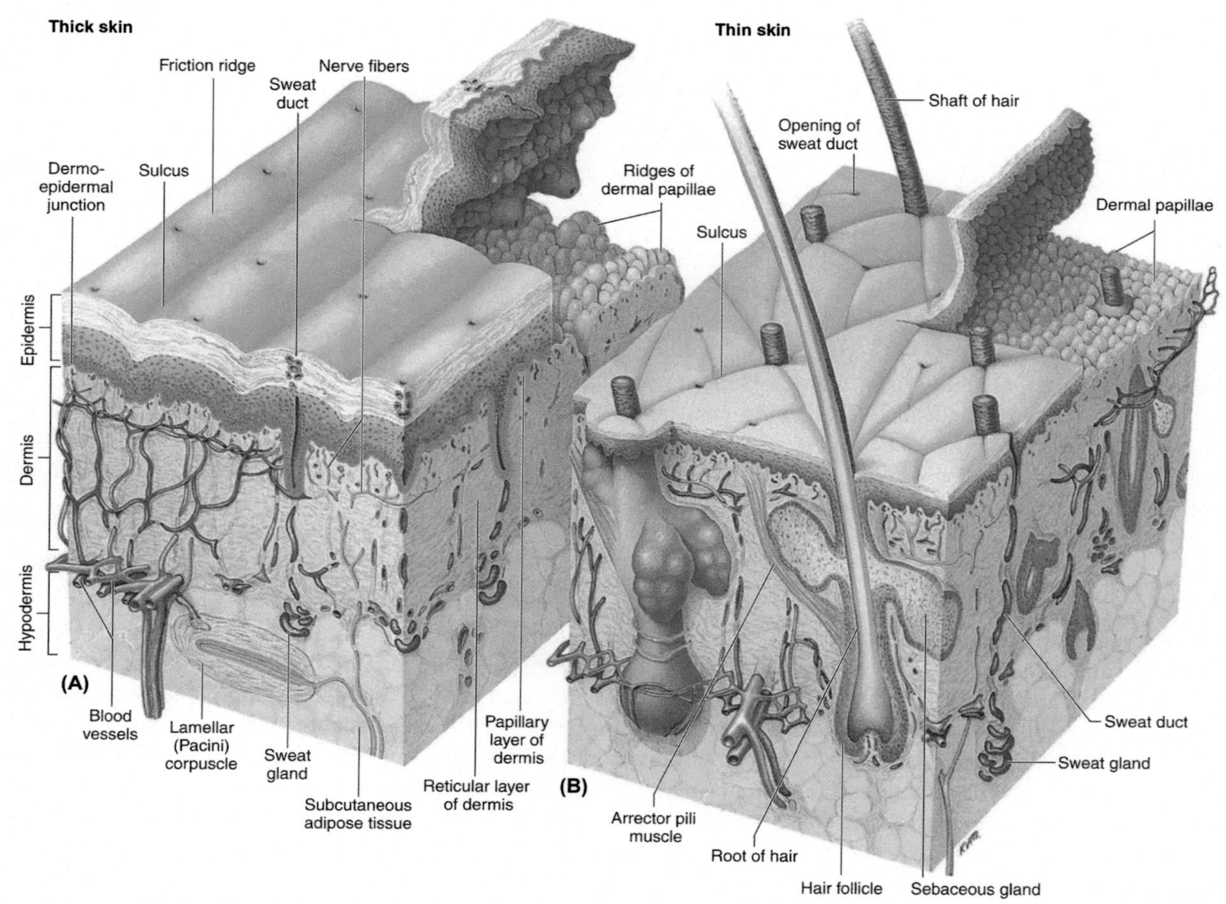

FIGURE 3.1 Structures of thick and thin skin. (A) Thick skin found on the palms of the hands and soles of the feet. Note the relative thickness of the epidermis and the absence of hairs. (B) Thin skin covering most of the body. In addition to the thin epidermis, note the presence of hairs and associated sebaceous glands. *From Thibodeau and Patton (2007), with permission.*

stability. The importance of the epidermal–dermal interface is highlighted by the genetic condition, **epidermolysis bullosa**, which is caused by mutations in junctional proteins that are involved in attachment of the epidermis to the underlying basal lamina. In people afflicted with this condition the epidermis cannot tolerate friction and it undergoes blistering and shredding in response to mechanical friction.

Hard epidermal thickenings (**calluses**) form when the skin, especially on the hands or feet, is subjected to repeated friction or pressure. In addition to preventing abrasion of the epidermis, another function of a callus is to prevent deep tissue damage by spreading an external force over a larger area of skin. The epidermis responds to a mechanical stimulus by increasing proliferation of the cells in its lower layers. The resulting cells undergo increased cornification and become more strongly adherent to one another than do normal epidermal cells. The combination of increased proliferation, increased adhesiveness, and thicker cells due to greater intracellular deposition of keratin proteins within them results in a dramatically thicker layer of epidermis (a callus) over the affected areas.

The dermis is the principal determinant of overall mechanical properties of the skin. Fundamental mechanical properties of the dermis are high tensile strength and the ability to undergo extensive elastic deformations after the application of low force. In essence the dermis (and the skin overall) must be able to stretch over joints when a joint is moved and then return to its original form upon completion of the motion. It also has to be able to resist major pulling or indentation forces without rupturing. This is a reflection of the presence and orientation of bundles of mainly type I collagen fibers, with strands of elastic fibers interdigitated among them. In a nonstretched condition the collagen bundles are oriented in a seemingly random fashion. As skin is first stretched under low tension, both the collagen and elastic fibers begin to show signs of parallel orientation. When greater tension is applied, however, the majority of the collagen fibers become parallel and the skin actually stiffens because the collagen fibers behave almost like mini tendons at this point. When the

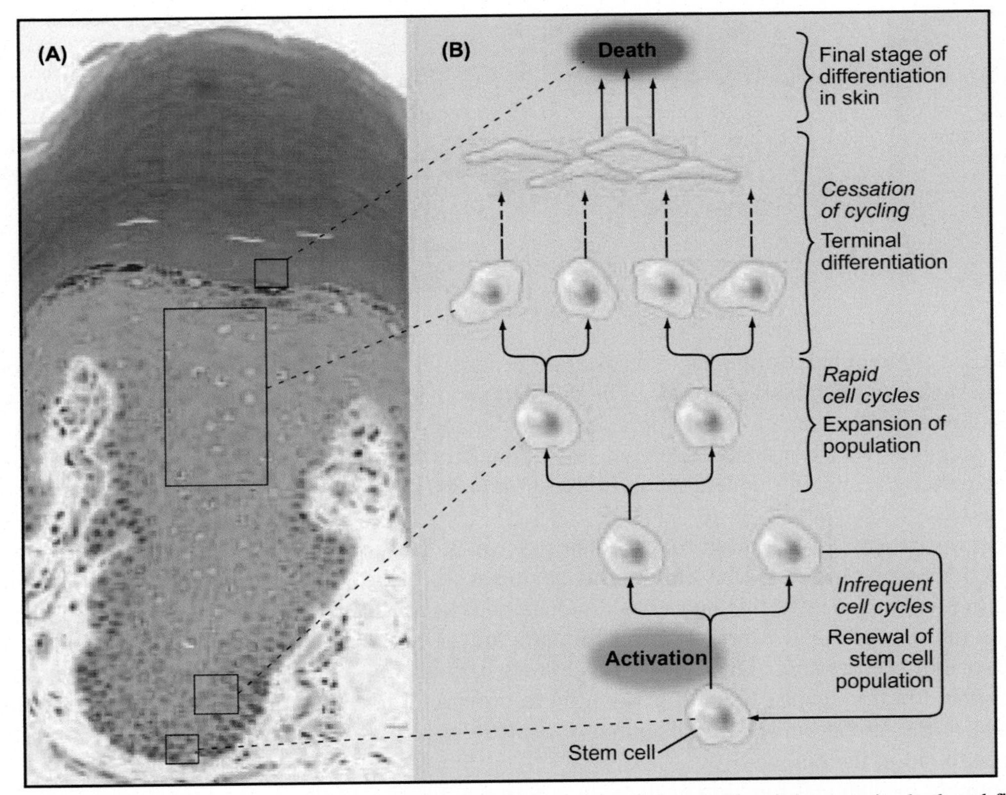

FIGURE 3.2 Summary of the life history of an epidermal cell, starting with a stem cell deep in the epidermis to its death and final shedding. (A) Photomicrograph. (B) Scheme of cell cycles and differentiation. *From Pollard and Earnshaw (2004), with permission.*

TABLE 3.1 Important Functions Served by the Skin
Mechanical protection combined with flexibility
Thermal regulation
Sensory functions—Pain, touch, vibration, temperature, pressure
Protection from UV rays
Synthesis of vitamin D
Immune protection
Fluid retention
Excretion of salts
Emotional signaling
Sexual attraction
Milk production
Self-repair

tensile strength of the collagen fibers is exceeded, the skin tears. The skin easily returns to its original condition after removal of the stress, largely due to the recoil of the elastic fibers. Surprisingly little is yet known about the orientation of collagen and elastic fibers in normal unstressed skin except that even at rest there seems to be some degree of directionality in their orientation. Prolonged mild tension results in growth of the skin. This is the basis of skin expansion

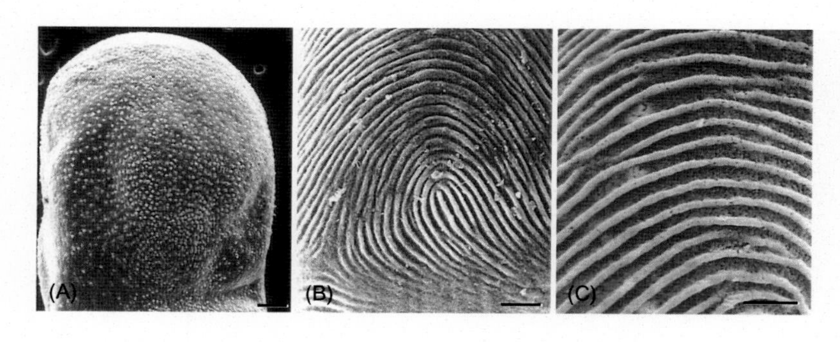

FIGURE 3.3 Scanning electron micrographs of digital palmar skin from a 14-week human fetus, showing (A) the surface of the fingertip, (B) the surface of the dermis with primary dermal ridges, and (C) the lower surface of the epidermis with corresponding ridges. These ridges, which appear early in the embryo, are the basis for fingerprints. *From Misumi et al., 1991, with permission.*

techniques that are commonly used in plastic surgical procedures. Excessive stretching results in permanent traces on the skin, such as the **striae** (stretch marks) produced on the abdominal skin of many women during pregnancy.

When pressure is applied, such as sitting in a chair or standing, the skin undergoes deformation and becomes thinner in the area of pressure. To a considerable extent this is due to the forcing out of fluid from the gel-like ground substance, such as proteoglycans, in the dermis, as well as compression of blood vessels. During a normal recovery phase, the fluid returns.

The hypodermis consists principally of fatty connective tissue and in some areas, thin sheets of muscle. Many mammals possess a thin layer of muscle, the **panniculus carnosus**, that allows them to wrinkle their skin throughout much of their body. In humans much of this layer has disappeared, but evolutionary remnants persist as the platysma muscle in the neck, the muscles of facial expression, the palmaris brevis muscle of the hand and the dartos muscle, which produces contractions of the scrotum. The hypodermis allows movement of the skin over the underlying tissues and also acts as a mechanical buffer against trauma. In addition to providing insulation and serving as a metabolic reservoir, the fatty component of the hypodermis is a rich source of adult stem cells, which have the capacity to transformed into different cell types found in the human body.

A less prominent mechanical function of the skin is found at the tips of the digits. Fingerprints (see Fig. 3.3) and toe prints are based on dermal ridges that form unique patterns for each individual and have served as the traditional means of identifying individuals in forensic investigations. Fingerprints serve to increase friction of the digits, and evolutionarily may have assisted arboreal primates in climbing. In fact the undersides of the prehensile tails of certain New World monkeys also have epidermal ridges. Recent research suggests that the wrinkles that appear on the fingertips after extended soaking in water serve to increase friction and allow better gripping of smooth objects under water. The epidermal ridges in fingerprints are richly supplied with eccrine sweat glands, sensory nerve fiber endings, and blood vessels. On the opposite side of the digits are the nails, which serve important tool-like functions.

SKIN AS A SENSE ORGAN

The human skin is an exquisitely sensitive sense organ that is supplied with a large variety of types of nerve endings serving an equally large number of sensory functions. In keeping with other regional specializations of the skin, certain sensory modalities are better represented in some regions than in others. The large surface area of the skin also permits the spatial integration of many environmental stimuli. The types and locations of the sensory receptors in the skin are summarized in Fig. 3.4 and Table 3.2.

Nonencapsulated Nerve Endings

Free Nerve Endings

The simplest sensory receptors in the skin are the free nerve endings located in the epidermis or the epithelium of the cornea. The skin is organized into a mosaic of territories of free nerve endings, each of which is most sensitive to either heat, cold, touch, or pain. Each territory exhibits a single response, regardless of the stimulus. For example, a territory supplied by a "pain" nerve ending can also respond to a cold stimulus, but it sends a pain signal to the brain. However, the intensity of the cold stimulus needs to be relatively greater than a pain stimulus in order to generate a pain signal in that territory.

Different free nerve endings respond to different stimuli, and the characteristics of their nerve fibers vary, although all are relatively small in diameter. Acute pain is mediated principally by type A∂ nerve fibers, which have a moderate

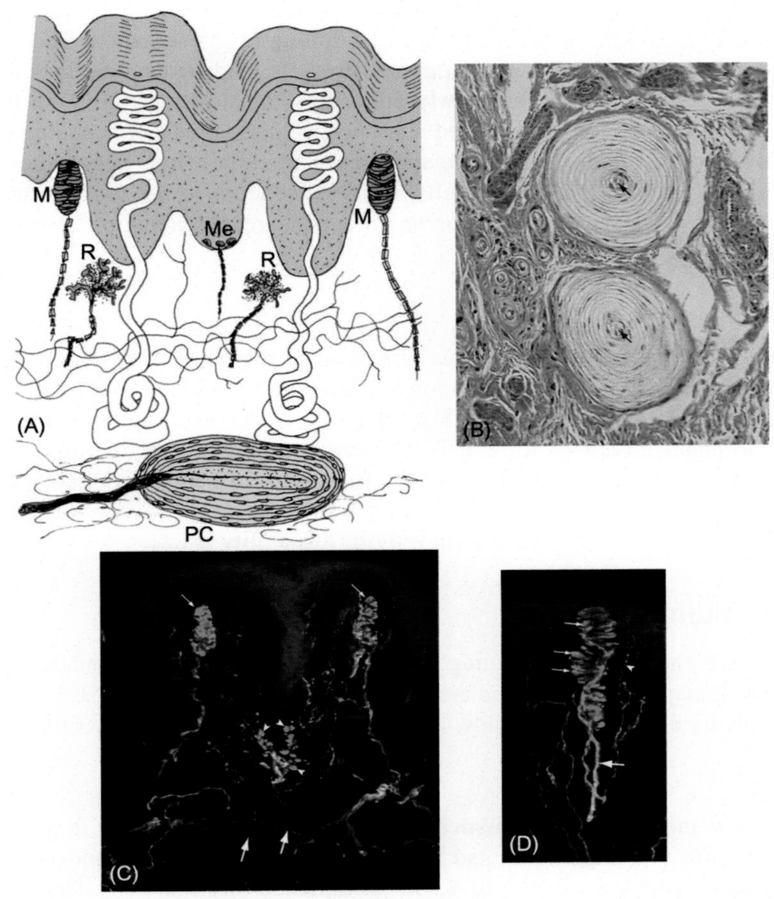

FIGURE 3.4 Sensory receptors in the human skin. (A) Schematic diagram. *M*, Meissner corpuscle; *Me*, Merkel cell; *PC*, Pacinian corpuscle; *R*, ending of Ruffini; (B) Photomicrograph of Pacinian corpuscles. Arrows point to the mechanosensitive nerve endings. (C) Low power photomicrograph of skin showing green Meissner's corpuscles (thin arrows) and Merkel endings (arrowheads). (D) Higher power view of Meissner corpuscle (small arrows). The large arrow points to a myelinated nerve fiber supplying the corpuscle. *From Nolte (2009), with permission.*

TABLE 3.2 Sensory Receptors Found in the Skin

Type	Function	Location
Free nerve endings	Pain, temperature, mechanical stimuli	Epidermis
Peritrichial nerve ending	Detect hair movement	Base of hair follicle
Meissner corpuscle	Touch	Outer dermis
Merkel cell	High-resolution touch	Basal layer of epidermis
Ruffini end organ	Stretch, heat	Outer dermis
Pacinian corpuscle	Deep pressure, vibration	Hypodermis
End bulbs (Krause)	Touch	Mucous membranes

conduction velocity (4–30 m/s) and are **fast-adapting**. The latter means that relatively quickly after the initial stimulus the frequency of firing of the nerve fiber is reduced causing a dulling of the sensation. This is what occurs after a pin-prick. In contrast reception of other stimuli, such as intense heat, is mediated by endings of very thin unmyelinated nerves, which conduct slowly (1–2 m/s) and are **slowly adapting**. The slow adaptation results in continued discharge of the action potentials of the nerves and a prolonged sensation of pain. Free nerve endings can also register generic touch or pressure, such as cotton lightly pressed against the skin. Some free nerve endings respond to chemical mediators released by injured or virally infected cells. These mediators lower the threshold for pain, resulting in **hyperalgesia**—e.g., the increased pain sensitivity of an inflamed area or a region affected by shingles (a herpes infection of nerve roots).

Peritrichial Nerve Endings

Peritrichial nerve endings are free nerve endings that wrap around the bases of hair follicles. A single sensory nerve fiber sends terminal branches to many follicles, and an individual follicle is innervated by branches from 2 to 20 different nerve fibers. These branches terminate in a highly organized collar that encircles the hair below the sebaceous glands in primates. This arrangement allows for exquisitely sensitive mapping of light touch throughout the regions of hairy skin. An extreme form of this is seen in the **vibrissae** (whiskers) of many mammals (cats, mice, and even seals). In fact harbor seals can track the path of a herring swimming in turbid water and catch it by following the turbulence of the water left behind by the fish. If the vibrissae are shaved off, this ability is lost.

Encapsulated Nerve Endings

Other cutaneous receptors are encapsulated, meaning that their nerve fibers are surrounded by some form of cellular/connective tissue covering. Encapsulated receptors are not found in hairy skin. Some are fast adapting and lose sensation quickly after the initial stimulus. Others are slow adapting and continuously register the presence of the stimulus.

Meissner Corpuscles

Finely discriminating touch is modulated by **Meissner corpuscles** located principally in papillary ridges of the dermis at the tips of digits and the lips (see Figs. 3.4D and 3.5). These detect light touch and vibrations less than 50 Hz. A Meissner corpuscle is barrel shaped and consists of a core of flattened disc-like cells derived from Schwann cells that are closely associated with a sensory nerve fiber. Each core cell is attached to a tiny bundle of collagen fibers that penetrate the connective tissue capsule of the Meissner corpuscle and attach it to one of the basal epidermal cells. Pressure on the epidermis is transmitted to the Meissner corpuscle through the bundles of collagen fibers, which then deform the capsule. The deformation (compression) generates action potentials and fires the sensory nerve, but because it is a fast-adapting nerve, the tactile sense only persists when the object is moved along the skin. At the fingertips, deformation of the papillary ridges in the fingerprint area (digital pad) magnifies the stimulus, rendering them even more sensitive. The sensitivity of Meissner corpuscles is considerably accentuated in people with long-term blindness, and it facilitates their reading by Braille. There is evidence that in such individuals the tactile signals spread into the visual cortex of the brain, thus providing a form of tactile patterning not available to most sighted people.

Merkel Cells

The sensation of continuous touch is mediated by **Merkel cells** located within the lower layers of the epidermis (see Fig. 3.4A). Merkel cells are not encapsulated and are found throughout the skin, both hairy and glabrous. They are extremely sensitive to tissue displacement of as little as 1 μm and to low frequency vibration in the range of 5–15 Hz.

Merkel cells are embedded in the basal layer of the epidermis and are tightly connected to surrounding epidermal cells by hemidesmosomes. The lower side of a Merkel cell is apposed to the flattened terminal of a sensory nerve fiber. When a Merkel cell is compressed, it releases neuropeptides that cross the space to the nerve terminal and stimulate the firing of the nerve fiber.

The nerve fibers terminating on Merkel cells are slowly adapting and will continue to fire as long as 30 minutes after the initial stimulus if that stimulus is maintained. Both Merkel cells and Meissner corpuscles are distributed into receptive fields. A **receptive field** is an area served by one receptor. Within a single receptive field, two points of pressure are only felt as one. The smaller and closer together two receptive fields are to one another, the better the two-point discrimination. Not surprisingly, two-point discrimination is best in the fingertips, the lips, and other highly sensitive parts of the body.

(A) RECEPTIVE FIELD OF PACINI'S CORPUSCLES

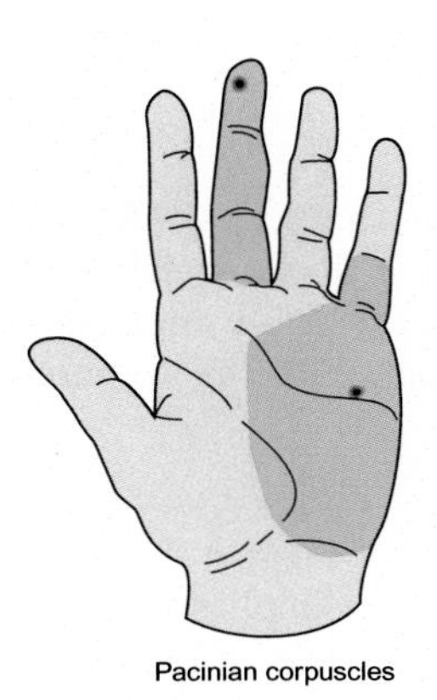

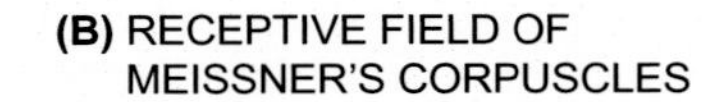

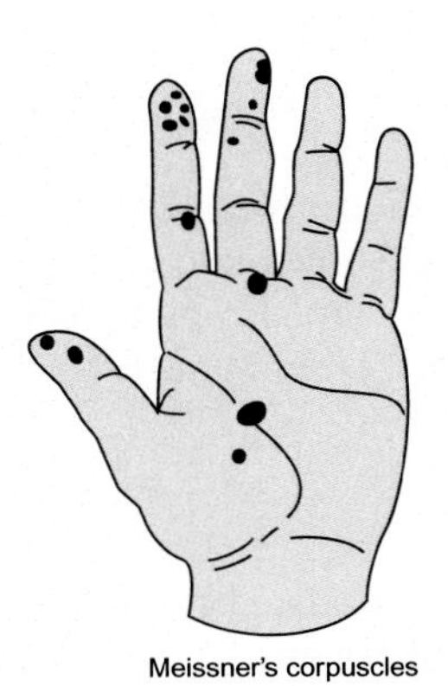

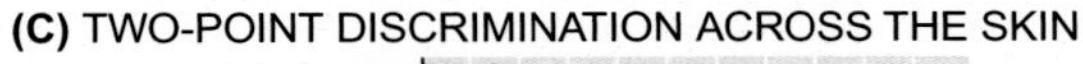

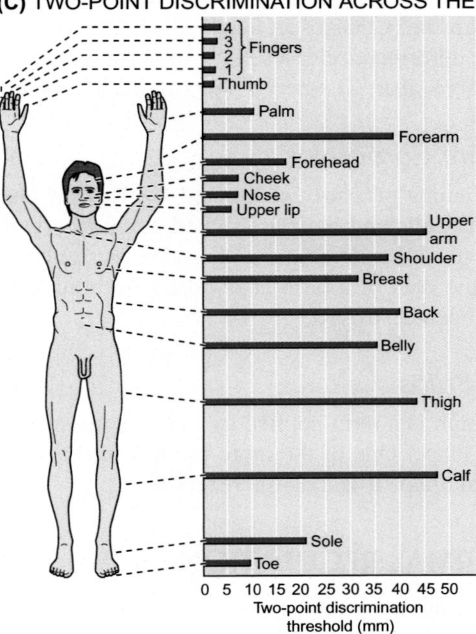

(D)

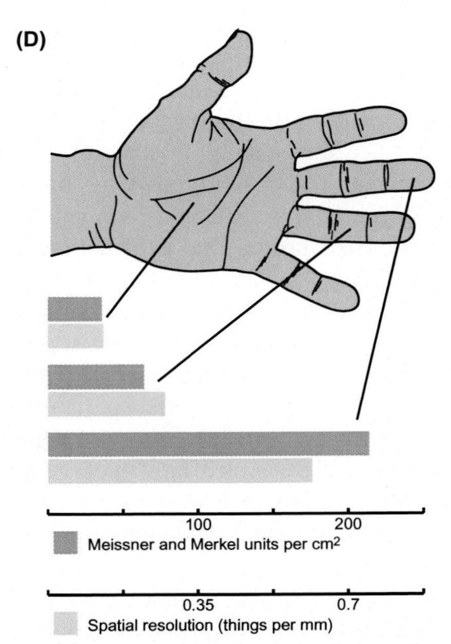

FIGURE 3.5 Receptive fields and spatial discrimination of skin mechanoreceptors. (A) The two black dots represent the areas of maximal sensitivity of each of two Meissner's corpuscles. The surrounding blue areas represent the entire receptive field for that corpuscle, i.e., area where the corpuscle will respond to a sufficiently strong stimulus. (B) Maximum receptive fields for a variety of Meissner's corpuscles are mapped out. (C) A map showing the two-point discrimination distances at various parts of the body. Two-point discrimination is the smallest distance at which two pinpricks can be distinguished as being separate. This distance is smallest in the fingertips and toes, and largest in the trunk and legs. (D) Correlation between spatial resolution and the number of cutaneous receptors in different areas of the human hand. *From Nolte (2009), with permission. (A–C) From Boron and Boulpaep (2012), with permission.*

Pacinian Corpuscles

Like touch, the sensation of pressure is mediated by two types of receptors. Deep touch is received by large (up to 3–4 mm) **Pacinian corpuscles** located deep in the hypodermis, as well as internal tissues, such as the mesentery and periosteum (see Fig. 3.4A,B). They are also sensitive to vibration, with a maximum sensitivity of 250 Hz. These corpuscles, consist of a central fluid-filled cavity containing a single sensory nerve fiber surrounded by as many as 60 concentric layers of flattened Schwann cells arranged in onion skin fashion and finally enwrapped by a connective tissue capsule. Changes in mechanical pressure are transmitted through the lamellae to the tip of the neuron, which upon being deformed, releases Na^+. In sufficient amounts the Na^+ release stimulates the generation of an action potential in the nerve fiber. The nerve fibers in Pacinian corpuscles are rapidly adapting, which allows them to detect vibrations.

Ruffini Corpuscles

Continuous pressure is detected by **Ruffini corpuscles**, which are located deep in the dermis (see Fig. 3.4A). Like Merkel cells and their associated nerve fibers, Ruffini corpuscles feed into slowly adapting nerve fibers, which allows them to register continuous pressure or stretch of the skin. They are also purported to sense heat, which would explain why the pain of a burn is so persistent after the initial stimulus. Deep burns are not painful because Ruffini corpuscles have been destroyed.

End Bulbs

End bulb is a term applied to a poorly investigated class of encapsulated corpuscle that, as the name implies, has a bulbous shape. One of the more prominent types, the **end bulb of Krause**, is found in mucous membranes of the lips and genital areas. It is a tactile receptor.

THERMAL REGULATION

One of the most important functions of the skin is thermoregulation. This involves both cooling and insulation against cold. Insulation is accomplished at two main levels. Protection from severe cold is provided by the fatty layer of the hypodermis. An extreme example of this is the thick layer of blubber that allows large marine mammals to survive in icy waters. The other mechanism of insulation involves hair or feathers. Both of these produce air-filled spaces that trap body heat or keep out the heat of the sun. In extremely cold temperatures fluffing the hairs or feathers increases the depth of the air spaces and provides extra warmth. Some hairs, for example those of deer, are hollow, which adds an extra dimension of insulation. This property is why trout fishermen use deer hair for making floating dry flies. Presumably because of their evolutionary origin in the tropics, humans have lost most of their hairy covering, but this refers to the size of individual hairs rather than their number. The appearance of goose bumps[1] when the skin gets cold is a reminder of our evolutionary past. Interestingly humans have retained hair on their heads. One explanation proposed for this is that head hairs insulate the brain from excessive heat on hot sunny days.

The vascular network of the dermis and hypodermis plays an important role in thermoregulation by the skin. At a gross anatomical level the vasculature of the skin is organized into discrete territories (**angiosomes**), which are supplied by arteries rising from similar territories in deeper tissues, especially muscles and the deep fascia. Knowledge of these territories forms the basis for the successful transplantation of large flaps of skin and/or muscle to other parts of the body because survival of the transplants can be assured by anastomosing the root vessel of a skin territory to a local vessel at the site of the transplant. Once at the level of the skin, the blood vessels are organized into three general layers, one at the base of the hypodermis, another at the junction of the hypodermis and reticular layer of the dermis, and a meshwork of finer vessels at the junction of the papillary and reticular layers of the dermis. The uppermost layer is of particular importance in temperature regulation.

Heat loss is reduced by constriction of the arterioles that form a dense network supplying the capillaries within the papillary layer of the dermis, especially in the extremities. The presence of **arteriovenous anastomoses** (arterioles joining venules in the absence of an intervening capillary bed) provides another mechanism for shutting down the blood flow to superficial capillaries (see p. 293). The vasoconstrictive response is mediated by the autonomic nervous system, which richly supplies these vessels. It redirects blood flow and heat to the core of the body. The concurrent constriction of dermal blood vessels and piloerection in a cold environment are both coordinated by the autonomic nervous system.

Heat loss from the body is accomplished by means of conduction, convection, radiation, and evaporation (Fig. 3.6). **Conduction** involves the transfer of heat from exposed areas of skin directly to cooler objects that are in contact with the skin. **Convection** occurs when heat is carried from the body by air currents, whereas **radiation** is the transfer of heat to or from the body through the air. These phenomena can occur when the heat of the body is greater than that of its surroundings. If the surrounding air is warmer than the body, the only effective means of losing heat is through **evaporation**.

---

1. The erection of hairs is made possible by the contraction of tiny smooth muscles (arrector pili muscle, see Fig. 3.1) that are attached to each hair. Piloerection is stimulated by sympathetic nerve endings attached to the arrector pili muscles.

FIGURE 3.6 Mechanisms of heat loss from the body. *From Waugh and Grant (2014), with permission.*

Cooling of the body by evaporation depends principally upon the secretion of a watery liquid over the surface of the skin by eccrine sweat glands and the subsequent evaporation of that liquid. The evaporative cooling is transmitted to the blood in the superficial capillaries of the skin, and the cooled blood contributes toward reducing the core temperature of the body.

Eccrine sweat glands are derivatives of the fetal epidermis that grow into the dermis or even hypodermal tissues of the palms and soles during the fourth month of pregnancy and throughout the remainder of the skin after the fifth month. The human skin contains up to 4 million sweat glands, which are distributed throughout the body surface (150–300/cm^2) but are most concentrated in the palms and soles. Beyond that, their density is greatest in the forehead, followed by the arms, the trunk and then the legs. Unlike most other glands of the skin, eccrine sweat glands are independent of hairs. Sweat glands can secrete large amounts of liquid—up to 3½ L/h in extreme cases—during intense exercise under warm conditions. The liquid is hypoosmotic to blood plasma, but contains a variety of ions, among them Na^+, K^+, and Cl^-. People acclimatized to increased sweating in hot weather are protected against excessive Na^+ loss by having a lower concentration of Na^+ in their sweat.

Eccrine sweat glands are relatively simple coiled structures at their base, with a straight duct rising toward the surface of the skin (Fig. 3.7). Liquid and ion secretion is accomplished by **clear cells** located in the basal layer of the coils. These cells contain numerous mitochondria and have prominent basal infoldings, a characteristic of other cells that produce watery secretions. Apically located dark cells within the coils produce glycoproteins that are added to the sweat. Surrounding the coils are **myoepithelial cells**, the contraction of which helps to move the secretions through the duct of the gland. Within the ducts an ion channel regulator (cystic fibrosis transmembrane conductance regulator—CFTR) is involved in the resorption of Na^+ and especially Cl^- ions from the initial secretion in order to prevent significant ion loss from the body during sweating. This protein is defective in patients with **cystic fibrosis**. As a result these

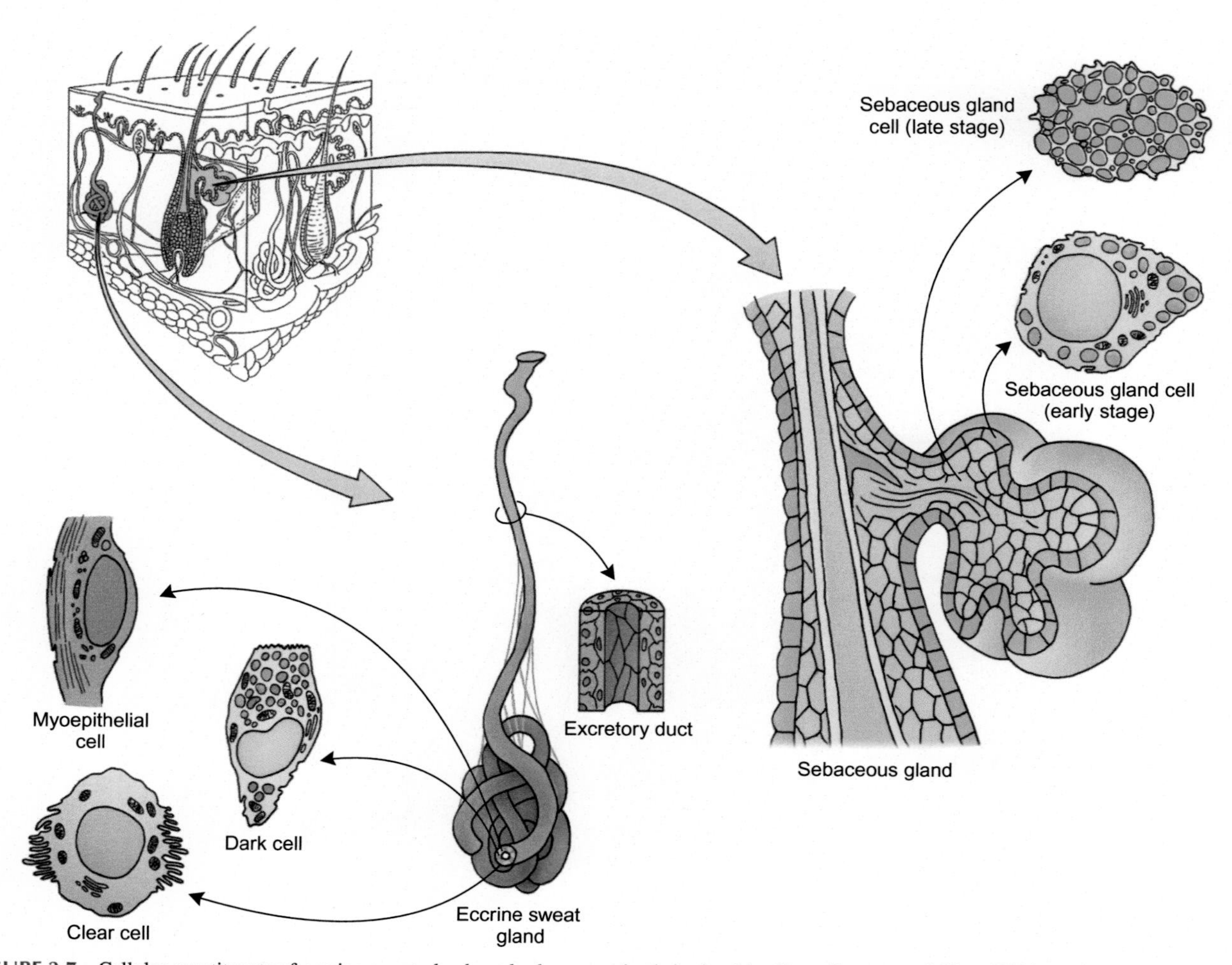

FIGURE 3.7 Cellular constituents of eccrine sweat glands and sebaceous glands in the skin. *From Gartner and Hiatt (2011), with permission.*

individuals lose excessive amounts of Cl^- in their sweat. Sweat glands are innervated by cholinergic sympathetic nerve fibers (see Chapter 6) (in the embryo, these nerve fibers switch from being noradrenergic to cholinergic), and sweating is initiated in the hypothalamus in response to increased core body temperature.

Superficial vasodilatation is another important mechanism for removing excessive body heat to the outside. Flushing of the skin, especially of the face, and the increased prominence of superficial veins in the hands attest to this response. Much of the pink color of a sunburn is due to vasodilatation.

Vasodilatation is accomplished by either passive or active means, both involving the autonomic nervous system. Different types of skin utilize different mechanisms. Nonhairy (**glabrous**) skin utilizes only passive means of vasodilatation through withdrawing stimuli from the sympathetic nerves. (Sympathetic activity normally constricts superficial blood vessels.) On the other hand, vasodilatation in hairy (nonglabrous) skin involves both the same passive mechanisms and also active means, which are not yet fully understood, but likely involve the parasympathetic nervous system.

EMOTIONS AND THE SKIN

The skin is one of the most important means of consciously or unconsciously expressing our emotions. At a gross level, emotions can be expressed by deformations of the facial skin by the actions of the muscles of facial expression (see Fig. 12.4). In many mammals, aggression is signaled by piloerection. Similarly, erection of hair or feathers can be a defense mechanism to make a potential prey object seem to be bigger. Other highly visible emotional responses seen through the skin are the flush of anger or the blush of embarrassment, both of which are reflections of the state of the underlying vasculature of the skin. The sympathetic-based "flight or fight" response, mediated by norepinephrine, is manifest in the skin by superficial vasoconstriction, sweating, and piloerection.

A somewhat more subtle manifestation of emotion through skin responses is emotional sweating, which is also modulated by the sympathetic nervous system. Perhaps as an evolutionary remnant, the eccrine sweat glands of the palms and soles respond primarily to emotional, rather than to thermal stimuli. The eccrine glands of the axilla, the groin and face respond to both thermal and emotional stimuli, whereas those covering the remainder of the skin are mostly unresponsive to emotion. Emotional sweating of the palms is the basis for one component of polygraph tests, which have been purported to detect individuals giving false answers to questions.

Sweating of the palms and soles decreases slippage and is considered to be of survival value in arboreal primates. The stimulation of palmar sweating through stimulation of the sympathetic nervous system is a fear mechanism that might have helped monkeys escape predators in the trees by allowing them to have a better grip on the branches. Some anthropologists feel that emotional sweating of human palms and soles is a residue of our evolutionary past.

In addition to eccrine sweat glands the human skin also contains **apocrine sweat glands**, which are associated with hairs and are concentrated in the axillae, the groin, and the perineum, but are also found in the areolae of the breasts and by the alae of the nose. The **ceruminous (wax) glands** of the external ear canals and the ducts of mammary glands are modified apocrine glands. Apocrine glands only become functional after puberty, and their function is strongly influenced by sex hormones. Although apocrine glands can produce sweat, they also produce other secretions, the functions of which remain poorly understood in humans. Apocrine glands are evolutionary remnants of scent glands, and there is evidence in humans that their secretions may serve a pheromonal function in sexual attraction. Their secretion, which passes through the hair follicle, is normally odorless but bacterial decomposition can result in an unpleasant odor. The letdown of milk in lactating women is an emotional response to the suckling by the infant and is mediated through the hypothalamus (see p. 394).

Emotional stress, whether acute or chronic, is known to have a significant impact on skin functions. Emotional stress effects are mediated from the brain by either neural (autonomic nervous system) or hormonal (mainly from the adrenal cortex [see p. 255]) means. These effects include impaired resistance to infection, impaired wound healing, and altered local immune and barrier functions of the skin.

PIGMENTATION AND PROTECTION FROM ULTRAVIOLET RADIATION

One of the greatest dangers to terrestrial animals is **ultraviolet radiation** from sunlight. For humans, who have lost their protective covering of hair, pigmentation of the epidermis is the first line of defense against UV rays (Fig. 3.8). The evolutionary biology of human skin pigmentation is interesting. Most anthropologists believe that as modern humans evolved in the tropics and lost their hairy covering, increased pigmentation in the epidermis was the mechanism that protected the cells in their skin from UV ray–induced damage to the DNA. Protection from UV rays, however, must be balanced with the need for sunlight to stimulate the synthesis of vitamin D in the skin. As early humans migrated out from the tropics into more northern or southern latitudes, a lessened need for UV protection was also connected with a greater need for exposure of the skin for the production of vitamin D. These evolutionary pressures resulted in progressively decreasing amounts of skin pigmentation, especially as humans moved into more northerly regions. Groups, such as the Samis who live in Lapland, have extremely fair skin, reflecting their need for exposure to the relatively small amount of available sunlight. Even the tanning process, which is most evident in moderately pigmented people living in middle latitudes, reflects a balance between the competing needs for UV protection and vitamin D synthesis. In the summertime, when direct exposure to the sun is greatest, the increased pigmentation from tanning adds extra UV protection, and the rapid fading of the tan in the fall gives cells of the skin greater access to sunlight so that they can continue to produce vitamin D.

Skin pigmentation is based on the presence of **melanin** granules, which are present in **melanocytes** (see Fig. 3.8), immigrant cells in the epidermis that are derived from a special embryonic tissue, called neural crest. Melanin is produced in melanocytes from the oxidation of the amino acid **tyrosine** through a DOPA (dihydroxyphenylalanine) intermediate before becoming melanin. Within the melanocytes, packets of melanin, called **melanosomes**, are distributed along long dendritic processes that interdigitate among epidermal keratinocytes. Many of these melanosome packets are transferred to the keratinocytes themselves via exocytosis from the dendritic processes, followed by endocytosis by the keratinocytes. **Albinism** can result from the mutation of any of a number of genes that control the synthesis of melanin. Albinos typically have a normal number of melanocytes, but they are just not pigmented. A reddish melanin variant, **pheomelanin**, is found in red-haired Europeans.

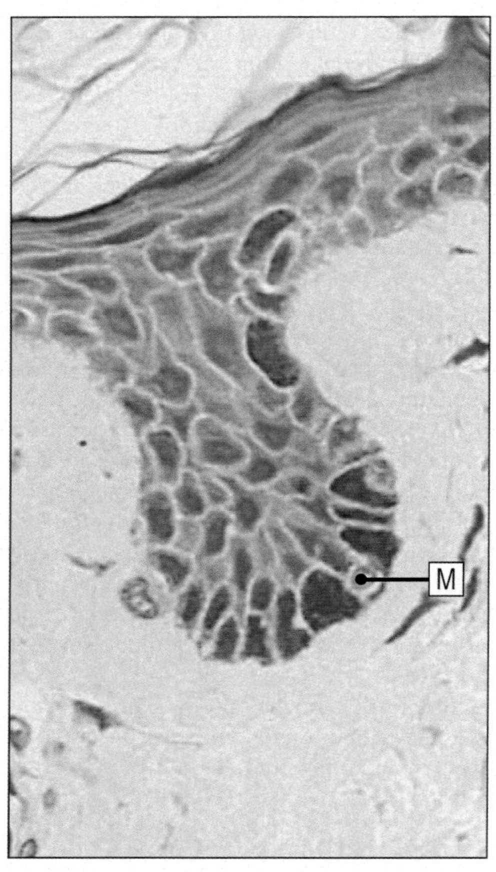

FIGURE 3.8 Cellular basis for skin pigmentation. Darkly pigmented skin has a layer of densely pigmented melanocytes (M) in the basal layer of the epidermis. *From Stevens and Lowe (2005), with permission.*

Vitamin D is produced photochemically in the lower layers of the epidermis from 7-dehydrocholesterol. A feedback loop prevents excessive synthesis of the compound. Inadequate production of vitamin D by the skin can be compensated for by oral intake. This is necessary for heavily pigmented individuals living near the poles. In addition to low levels of sunlight, the use of sunscreens inhibits the production of vitamin D within the epidermis.

Pigmentation is not evenly distributed within the skin. In most quadripedal animals, pigmentation is greatest on the dorsal skin and lightest on the belly. Humans retain some traces of this pattern, and the most lightly pigmented skin is found on the inner surface of the upper arm. **Freckles**, which are most common in lightly pigmented individuals, result from the uneven distribution of melanocytes throughout the epidermis. They darken with increased exposure to sunlight. Another example of differential expression of pigmentation is the "mask of pregnancy," darkening of the skin of the cheeks and upper face as pregnancy progresses. This is an example of **melasma**, which is also expressed in the darkening of the nipples and areolae during pregnancy. These changes in skin pigmentation are related to levels of reproductive hormones (estrogens and progesterone), and can to a certain extent occur in women who have been long-term users of oral contraceptives.

SKIN AS A BARRIER

Another important function of the skin is that of a barrier. The role of pigmentation as a barrier against the effects of UV rays has already been discussed. The cornified epidermis and its underlying basement membrane serve as an important barrier against water loss or entry of excessive fluid into the body. Contributing to the barrier function of the skin are the secretions of the many sebaceous glands. **Sebaceous glands** (see Fig. 3.7), which are usually closely associated with hair follicles, produce an oily secretion by a holocrine mechanism involving destruction and expulsion of the cells that produce the secretion (**sebum**). The importance of the barrier function of the skin is seen in patients with extensive

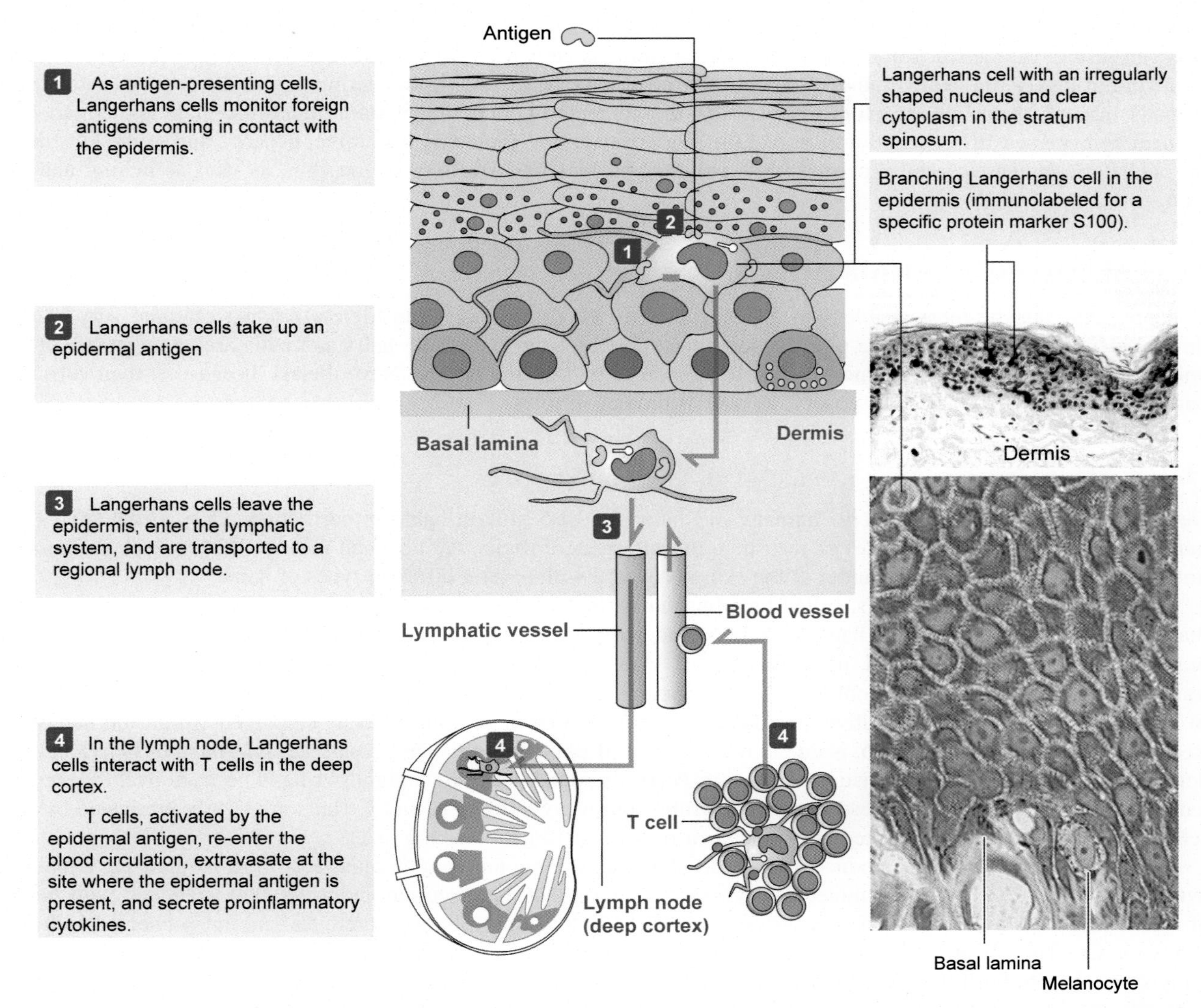

FIGURE 3.9 Immune mechanisms in the skin, starting with detection of an antigen by Langerhans cells in the epidermis and their transportation via regional lymphatics to local lymph nodes for further immune processing. *From Kierszenbaum and Tres (2012), with permission.*

severe burns. One of the main threats to life is excessive fluid loss through the burned skin. In addition to serving as a fluid barrier the epidermis also restricts the passage of many types of chemical compounds across the skin. In order to administer a number of drugs through the skin, they must be combined with DMSO (dimethyl sulfoxide), which greatly increases the permeability of the epidermis.

The skin also serves as an immunological barrier (Fig. 3.9). Prominent within the epidermis is a large population of immigrant **Langerhans cells** (a subset of dendritic cells—see Chapter 8), which represent the peripheral outposts of the immune system in the skin. Also present in the deeper layers of the skin are ~20 billion T lymphocytes, which are important components of the immune system. Derived from monocyte precursors in the blood, Langerhans cells are found among the keratinocytes and send out processes (dendrites) that interdigitate among the epidermal cells, including those lining the hair follicles and sweat glands. These cells monitor for and take up foreign antigens, especially those derived from bacteria living on the skin. When they have become activated by exposure to an antigen, they produce inflammatory cytokines, such as IL-1 (interleukin-1), which serves as a line of antimicrobial defense by priming the immune system for an attack on that antigen. In addition, they leave the epidermis and through lymphatic channels located within the dermis, they are transported to local lymph nodes, where they interact with T cells and instill in them the memory of that particular antigen. This activates an adaptive immune response to that antigen. The T cells are transported to the site of antigen entry via the blood and leave the blood vessels, at which point they begin to secrete their own inflammatory cytokines directed against the foreign antigens. In the face of prolonged antigenic irritation, **dermatitis** can result. The destruction of the system of Langerhans cells in burned skin is one reason why infection, in addition to fluid loss, is the major cause of death in burn patients. Mast cells in the dermis respond to certain antigens and irritants through their IgE receptors by releasing the contents of their numerous cytoplasmic granules as part of an immediate hypersensitivity reaction (see p. 236). These secretions, including histamine, heparin, and several chemotactic mediators, increase vascular permeability, resulting in localized swelling of the skin, as well as itching and even pain.

SKIN APPENDAGES—HAIR AND NAILS

Hair and nails are the most visible skin appendages, but what one sees is the lifeless product of some very complex developmental processes that take place beneath the surface of the skin. Both hairs and nails are aggregations of dead epidermal cells that have been almost completely replaced by keratin proteins. Nevertheless, because of their subsurface connections, they play important sensory, as well as protective roles.

Hair

Despite being called naked apes, we humans still have almost 5 million hairs on our bodies. Only a few areas of the human body (palms of hands, soles of feet, bottoms and sides of digits, the lips, and parts of the external genitalia) are not covered with hairs. The remainder of the skin is covered with several different types of hairs.

Most prominent are scalp hairs, called **terminal hairs**, that number about 100,000. Why the scalp is the only region that has retained a thick coat of hair is not fully understood, but major functions attributed to it are insulation of the brain from excessive heat or cold, protection from UV rays, and even mechanical protection from some forms of blunt trauma. Like the epidermis, hair color is due to pigments trapped in the hair cells. The spectrum from blond to brown to black hair is based on the relative abundance of melanin present in pigment cells contained within the hair shafts. Red hair, along with freckled skin, is due to the presence of **pheomelanin**. The cross-sectional shape of individual hairs determines the physical characteristics of the hair. Perfectly round hairs are straight. As hairs go from oval to more flattened in cross-section, they progress along a spectrum from wavy to curly to kinky. Hairs are lightly lubricated by small sebaceous glands, which are located along the shaft of the hair follicle (see Fig. 3.1).

The hairs over most of our bodies are called **vellus hairs**. How and why humans lost most of their fur remains an evolutionary mystery, but recent molecular evidence suggests that this occurred about 240,000 years ago with the loss of a gene involved in keratin formation. Vellus hairs are short (<2 mm), light colored, and are not connected with sebaceous glands. They begin to develop during the late fetal period, replacing **lanugo hair**—a type of hair present on fetuses only before birth. In contrast to terminal hairs, vellus hairs grow very little, once formed. Under the influence of androgens (present in both sexes) at puberty, some vellus hairs become transformed into the much coarser terminal hairs, such as those in the axilla, the pubic area and the male beard. In male pattern balding, terminal hairs are replaced by vellus hairs.

Each hair is the product of a beehive of cellular activity located in small hair follicles deeply embedded in the skin. The essence of a **hair follicle** is a deep cylindrical epidermal downgrowth that at its tip surrounds a small papilla of dermal connective tissue, blood vessels, and nerves. Signals from the connective tissue cells of the papilla to the overlying epidermal cells stimulate the process of hair growth. A hair itself is a shaft of heavily cornified (keratinized) epidermal cells that are produced deep within the hair follicle and push up the growing hair from the bottom (for details, see Box 3.1). Hairs are hard because of a specialized type of keratin protein that occupies much of the hair cells in which they were made.

Individual hairs have a life cycle consisting of three phases. The first phase (**anagen**) is one of active hair growth, during which cellular proliferative activity at the base of the hair follicle contributes new cells to the base of the hair, consequently pushing the hair up through the skin at a rate of about 1 cm every 28 days. Scalp hairs have a long period of anagen growth, lasting from 2–7 years. Hairs in different parts of the body have different lengths for the anagen phase. Anagen is followed by a brief (2–3 weeks) **catagen** phase during which the communication between the papilla and the overlying epidermal cells of the follicle is effectively cut off. The result of this rupture in communication is the transition of the root of the hair to a **club hair**, which has a more tenuous connection to the skin than does a hair in the anagen phase. The final phase in the hair cycle is **telogen**, sometimes called the resting phase. During this phase, which can last for several months in scalp hair, individual club hairs are pushed out as a newly forming hair begins to take shape. As many as 50–100 scalp hairs are shed per day. Shedding is accompanied by the renewal of activity of the hair

BOX 3.1 The Hair Follicle

Each hair follicle is a complex mini-organ that includes not only the hair and the hair-making apparatus, but also an associated sebaceous gland and a small slip of smooth muscle fibers called the **arrector pili muscle** (Fig. 3.10). In addition, a bulge of epithelial cells along the hair follicle is a repository of stem cells that are involved in the generation of a hair and serve as a source of new epidermal cells in the event of an extensive skin wound.

The base of a hair follicle is slightly enlarged to form the **hair bulb**. Penetrating the bottom of the hair bulb is the dermal papilla and its associated blood vessels and nerves. At the root of a hair, epidermal cells actively proliferate to contribute to the precursors of the medulla and outer cortex of the hair. The papilla contributes no cells to the hair, but is importantly involved in signal-calling (induction) and nutrition of the forming hair. Melanin granules from pigment cells in the epidermis become embedded within the cortex and are the basis for the pigmentation of the hair. As epidermal cells at the base of the hair become pushed up by more newly added epidermal cells, they synthesize great amounts of hard keratin, which greatly contribute to the strength of the hair. The sebaceous glands, which empty their oily secretions into the hair follicle through a holocrine mechanism (dead cells constitute the secretion), contribute a thin coating of sebum around the hair.

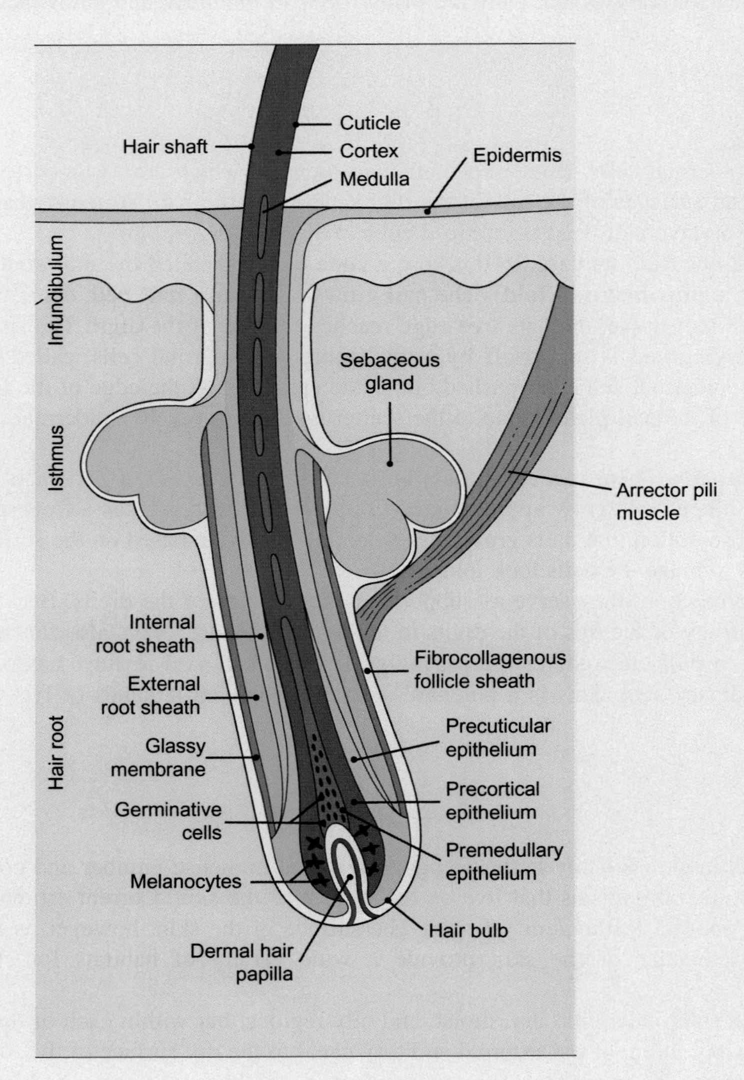

FIGURE 3.10 Diagram showing the main features of a hair follicle. Hair growth takes place at the base of the follicle, where the connective tissue papilla interacts with epidermal cells of the root sheath. The forming hair is then pushed up from its base. The cortex and medulla of the hair contain keratin proteins, which are the basis for the firmness of the hair. *From Stevens and Lowe (2005), with permission.*

follicle. Functional contact between the cells of the papilla and the follicular epithelium (epidermal cells) begins anew and a new hair begins to develop.

The initiation of new hair growth requires the contribution of new epidermal cells derived from stem cells located near the base of the hair follicle. These stem cells are able to sense the balance between multiple stimulatory and inhibitory molecular influences at the base of the follicle. The relative strength of the stimulatory influences determines the configuration of the hair. A reduction in stimulatory signals is one reason why hairs become thinner and less robust as a person ages.

Pattern baldness is a condition that affects 70%–80% of males and up to 40% of females. In males, pattern baldness typically begins at the temples or the top (vertex) of the head, whereas in females, baldness consists of a more diffuse loss of hair all over the top of the head.[2] Baldness has a strong genetic component, and androgenic hormones are heavily involved in bringing about hair loss. The situation is complex because what can promote hair loss in the scalp may actually stimulate hair growth on the face and other parts of the body. In addition to genetically-based baldness, many other factors, such as diet, drugs, stress, radiation, and a host of other factors can also stimulate hair loss, although not of the pattern baldness type. At this point what we know is that there are many forms of baldness, and many factors are involved.

Nails

Like hairs, **nails** consist of highly keratinized epidermal cells. Both fingernails and toenails, which are assumed to be evolutionary derivatives of claws, cover the dorsal surfaces of the tips of the digits. The flexible hard structure that is readily seen is the **nail plate**, consisting of up to 50 layers of dead keratinized epidermal cells (Fig. 3.11).

A nail grows much like a hair. It gets pushed out from its base, in this case a zone of cells, called the **nail root** or matrix, that are located beneath a fold of skin (the proximal **nail fold**). The nail grows out over a **nail bed**, consisting of a richly vascularized[3] and innervated region of soft tissue, until its free edge reaches the end of the digit. The space between the end of the nail plate and the underlying bed is sealed off by a thickening of epidermal cells, called the **hyponychium**. The cuticle is a thin rim of dead epidermal cells that pushed out as the epidermis at the edge of the fold matures. The white crescent (**lunula**) at the base of the nail plate is due to the scattering of light off the epidermal and connective tissue cells in the distal matrix.

Fingernails grow almost 4 times as fast as toenails. There is a correlation between the growth rate of nails and the length of the digital segments that they occupy. Fingernails grow approximately 3 mm/month, and it takes a fingernail between 3 and 6 months to regrow completely. The notion that nails grow after death is an illusion based on the shrinkage of the soft tissues around them, which seems to make the nails look longer.

Nails have several functions. In addition to protection, they serve as support for the soft tips of the digits. Because of their firm support, they also increase the sensitivity of the tips of the digits to touch and pressure. Nails also function as tools, as is evident to anyone who has tried to pull out a sliver. Surprisingly, despite their hard texture, nails are more permeable to water and solutes than is ordinary skin. This is a potential source of problem if hands or feet are immersed in noxious fluids.

MICROBIOLOGY OF THE SKIN

One of the most fascinating recent developments in skin biology is the recognition of the immense number and complexity of the communities of bacteria and other microorganisms that live on the surface of the skin. Current estimates place the average number of surface bacteria at about 1 million/cm^2 of skin. The surface of the skin, however, is not homogeneous, and the different regional characteristics of the skin provide a wide variety of habitats for skin microbiota.

At a broad level the surface of the skin can be subdivided into dry, moist, and oily regions, but within each of these are many microenvironments, ranging from the waxy lining of the external auditory canal to the dry surface of the forearm. Each of these microenvironments of the skin has its own aggregation of resident, nonpathogenic species of bacteria, fungi, and viruses. These organisms derive their nutrition from peptides released from dying keratinocytes or from

2. Baldness is actually not complete hair loss, because the bald scalp is still covered with vellus hairs instead of coarser and longer terminal hairs.
3. The nail plate is translucent, and its light pink color is due to the dense capillary network underlying it. If you press a finger on a nail the nail plate becomes white due to the compression of the capillaries.

(A)

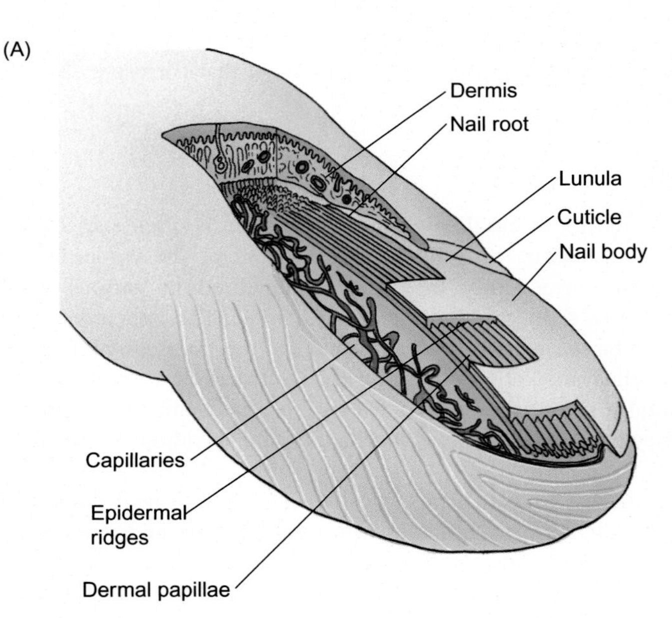

(B)

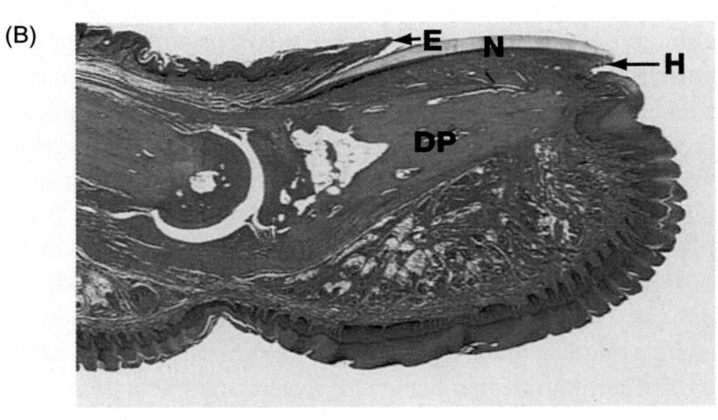

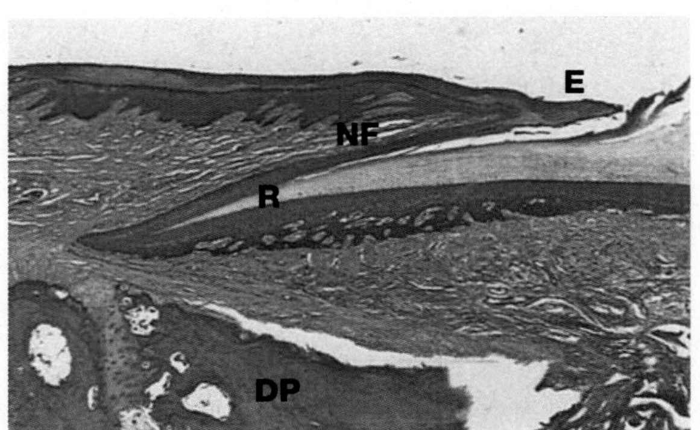

FIGURE 3.11 (A) Diagram of the structure of a fingernail. (B) Histological section through the fingertip of a monkey, showing its relationship to other tissues of the digit. *DP*, distal phalanx; *E*, eponychium; *H*, hyponychium; *N*, nail. *(A) From Gartner and Hiatt (2011), (B) From Young et al. (2006), with permission.*

products of skin secretions, often lipids and their breakdown products, but some bacteria can even use the urea present in sweat as a nutrient.

Just as the regional characteristics of the skin change over one's lifetime the types of microorganisms that inhabit the surfaces of these regions change accordingly. A particularly prominent period of change is puberty, when significant qualitative changes in skin bacteria occur. Teenage acne is a highly visible consequence of these changes.

The normal body maintains a commensal or even synergistic relationship with its skin bacterial communities, and these bacteria work with the body's immunological defenses to produce an environment that is hostile to pathogenic bacteria. These defenses include secretions by fat cells and epidermal cells that are toxic to specific kinds of bacteria. By interacting with the antigen-presenting Langerhans cells in the epidermis and the memory T cells, they stimulate the creation of an environment capable of producing inflammatory cytokines and complement, which act to deter pathogenic bacteria. Each skin microenvironment produces its own immunological microenvironment that corresponds to the characteristics of the microbial communities inhabiting its surface.

MAINTAINING THE INTEGRITY OF THE SKIN

Because the skin represents the interface between the body and its external environment, the skin is often subjected to mechanical trauma or other environmental insults. On a daily basis, cells in the outer layers of the epidermis are removed by various levels of mechanical trauma, and not infrequently the skin is punctured in various ways. Maintenance of the integrity of the epidermis is the result of a tightly orchestrated set of molecular interactions that control the balance between proliferation of stem cells in the basal layers of the epidermis and the progressive differentiation of their daughter cells. As cornified surface cells are shed, they are replaced by new cells arising from the deeper layers, which produce the appropriate combinations of keratin proteins to form a waterproof, mechanically tough surface layer. The time between generation of a new keratinocyte in the basal layer and its ultimate shedding is commonly 8–10 weeks in a normal adult.

When the skin is cut or punctured, a complex set of restorative mechanisms is called into play to heal the wound (Fig. 3.12). A typical skin wound heals in four main stages—hemostasis, inflammation, proliferation, and remodeling. The first stage consists of the formation of a blood clot of fibrin and platelets. Beneath the clot the release of inflammatory factors from the platelets results in the accumulation of leukocytes in the dermal tissue beneath the clot (second stage). Within hours the basal epidermal cells become mobilized and begin to spread out beneath the clot to cover the wound and restore continuity of the epidermal layer. In the wounded dermis below the epidermis, secretory products of macrophages stimulate blood vessels to sprout in the area of the wound and dermal cells to proliferate. Important among these cells are the **myofibroblasts**, a type of cell that has characteristics of both fibroblasts and smooth muscle cells (see Box 2.2). After epidermal continuity has been restored beneath the clot, fibroblasts and myofibroblasts in the dermis begin to produce new fibers—mostly collagen, but some elastic—to restore the dermis. A prominent reaction in the healing of a skin wound is contraction of the wound and remodeling of the dermal connective tissue (Fig. 3.13). This is accomplished in large part through the actions of the myofibroblasts and is evidenced by the reduction in the area of the original wound surface as the original clot (scab) is sloughed. Ultimately the continuity of the skin, both epidermis and dermis, in the area of the wound is restored, but typically the strength of the dermis is not as great as it was before the wound occurred. The tensile strength of a healed skin wound is 40% of normal by 1 month and 70% at the end of the first year. In a large wound, especially if significant inflammation has occurred, a persistent scar marks the location of the wound.

AGING AND THE SKIN

Like every other tissue in the body the skin is affected by aging changes, but because we can see the skin, these changes become painfully obvious. Aging changes are many, and they form the basis for an immense industry that produces various creams and potions designed to forestall or reduce the effects of aging on the skin.

Aging of the skin occurs in two ways. One is intrinsic **biological aging**. The other is **photoaging**—the result of exposure of skin to UV rays. Often what is seen in aging skin is due to a combination of both mechanisms. Biologically aged skin is typically pale, smooth, and finely wrinkled, whereas photoaged skin is coarsely wrinkled with uneven pigmentation.

One of the most prominent manifestations of aging in the skin is the appearance of wrinkles (Fig. 3.14). Wrinkles are the result of many age-related changes in the skin, but one of the most important is the deterioration of both elastic and collagen fibers, as well as a reduction in the amount of hyaluronic acid within the ground substance. When that happens, deformed skin does not return to its original condition. Over time in areas of repeated motions, static wrinkles mark the tracks of habitual movements. Photoaging changes, in particular, appear to be due to the photochemical generation of reactive oxygen species, which through several pathways increase the activity of proteolytic enzymes, especially matrix metalloproteinases. These break down the dermal matrix and interfere with its repair. In adults the amount of dermal collagen decreases by $\sim 1\%$ per year.

Many other cellular and structural components of the skin also decrease or deteriorate during the aging process. The epidermis becomes thinner due in large part to a decrease in proliferation of cells in the basal layer. This probably also accounts for the decreased growth and greater fragility of nails with age. Within the epidermis the numbers of both melanocytes and Langerhans cells decrease. The reduction in melanocytes often leads to irregular pigmentation and the

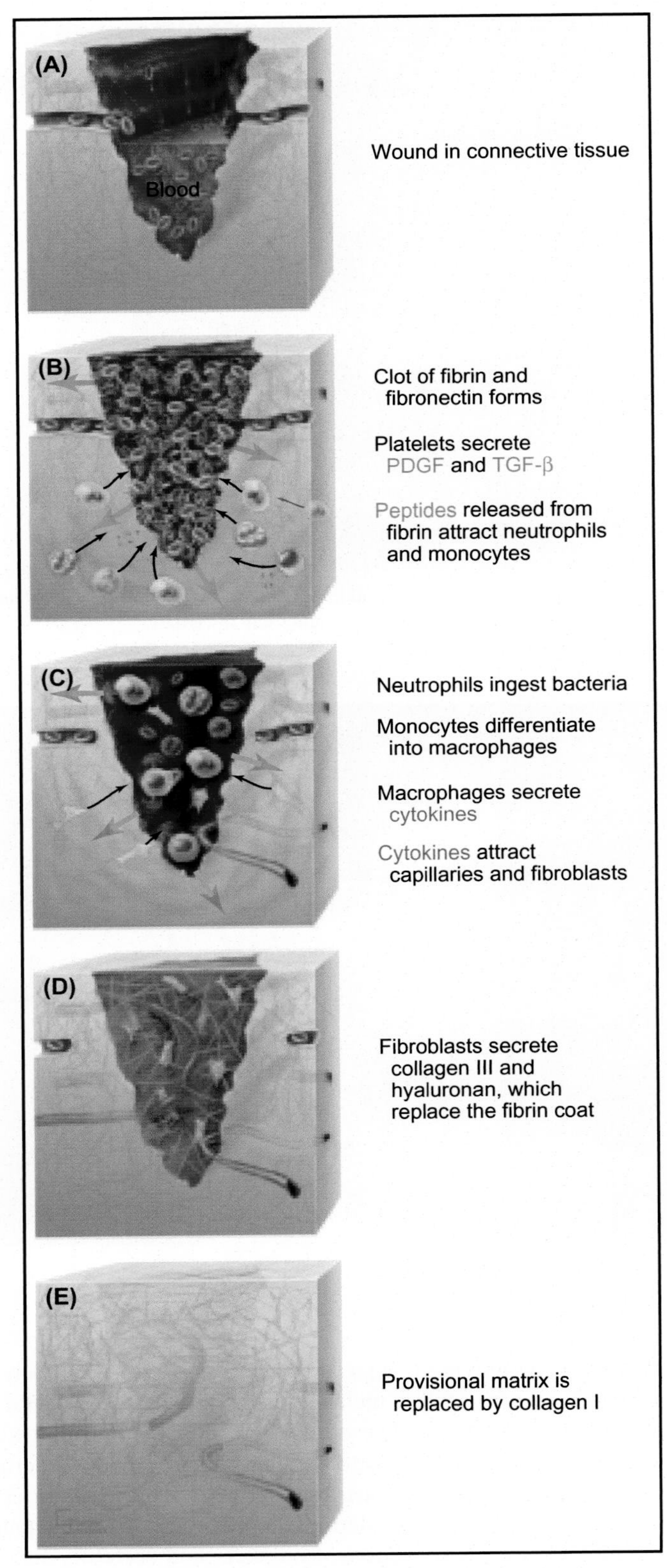

FIGURE 3.12 Healing of a deep skin wound. *From Pollard and Earnshaw (2004), with permission.*

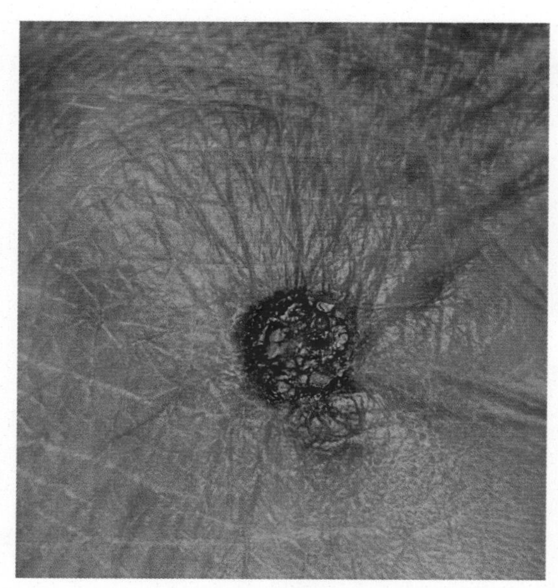

FIGURE 3.13 A healing skin wound on the hand. Note how the wrinkles in the skin extend out from the wound like rays of the sun. This is due to contraction of the original wound by myofibroblasts and pulling the surrounding skin inward.

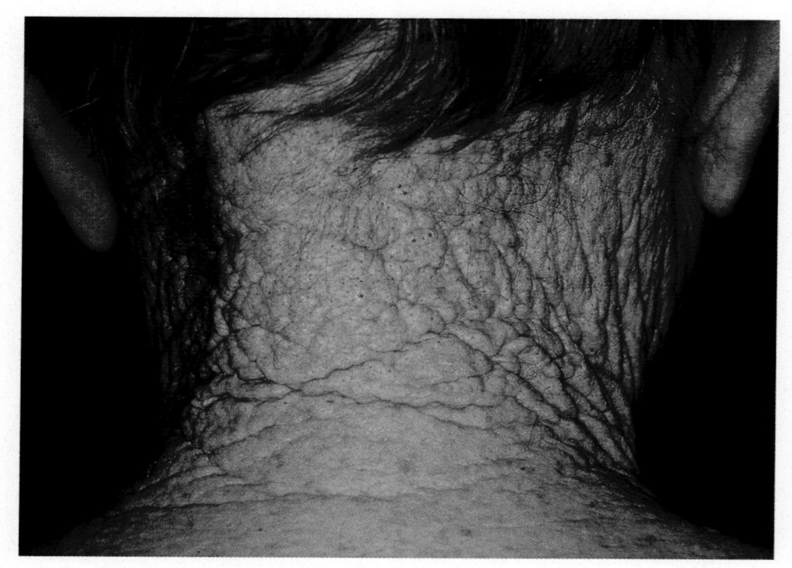

FIGURE 3.14 Wrinkled skin. *From Bolognia et al. (2012), with permission.*

appearance of "age spots" in the skin. The graying of hair is due to losses in the numbers of functional melanocytes at the bases of hair follicles. Hair becomes white when melanocytes in the hairs are lacking. The white color is due to light refracting from air spaces within the medulla. The decline in Langerhans cells results in compromised immune functions, which can be manifest by reduced responses to foreign antigens. The epidermal–dermal interface flattens with age, resulting in a greater susceptibility to blister formation and shearing-type injuries of the skin.

Within the dermis a reduction in size and number of dermal blood vessels results in paler skin, but more importantly can lead to disturbances in thermoregulation. A reduction of numbers and secretory function of both sweat and sebaceous glands contributes to the dryness that is often characteristic of aged skin. The number of encapsulated nerve endings in the skin also decreases, although apparently the number of free nerve endings remains relatively constant. The former is the basis for lessened tactile discrimination, and the latter accounts for the preservation of the sense of pain.

SUMMARY

Skin is made up of three components—the epidermis, the dermis, and the hypodermis. The epidermis is multilayered and is firmly connected to the dermis through a variety of connections. The dermis consists of collagenous connective tissue and is underlain by a fatty hypodermis.

The skin is well designed to resist mechanical forces. The epidermis is strengthened by intercellular junctions, as well as by hemidesmosomes that connect the epidermis to its basement membrane. Rete ridges connect epidermis to dermis. The dermis is characterized by both high tensile strength and elasticity, enabling it to return to its original shape after deformation.

The skin serves as a major sense organ with many types of nerve endings subserving different sensory functions. Free nerve endings are pain receptors, whereas peritrichial nerve endings on hairs detect light touch. Encapsulated nerve endings detect superficial and deep touch, heat and cold.

Thermal regulation is accomplished largely through the constriction or relaxation of vascular plexuses within the dermis. Erection of hairs helps to conserve heat. Evaporation of fluid secreted by sweat glands acts as a cooling mechanism.

The skin is an important means of expressing emotions. This can be accomplished by deformation of facial skin, erection of hairs, or changes in the superficial vasculature. Emotional sweating is an unconscious reaction.

Pigmentation in the form of melanin granules in basal epidermal cells protects the body from ultraviolet (UV) radiation. The relationship between amount of pigmentation, exposure to the sun, and vitamin D production is an important one.

The skin also acts as a barrier against radiations, pathogens, or mechanical influences.

Major skin appendages are hairs and nails. These are both aggregations of dead epidermal cells.

Humans have large numbers of hairs over most of the body. These range from the coarse hairs of the scalp, the axillae and pubic areas to fine vellus hairs that cover much of the skin. Individual hairs go through a hair cycle consisting of three stages—anagen, characterized by active hair growth; catagen, during which growth ceases; and telogen, the final phase during which the hair becomes eventually pushed out.

Nails consist of highly keratinized epidermal cells that take the form of plates. A nail grows from its base, the nail root, and is pushed out from there.

The skin is host to a complex microbiome, with different regions of the skin containing different types of bacteria. The types of bacteria change over the course of one's lifetime. Normal skin bacteria even serve as part of the body's defense against pathogens.

The skin is capable of a remarkable degree of healing after being wounded. The epidermis first seals off the wound, and beneath it dermal fibroblasts secrete new matrix to restore strength.

The skin shows characteristic aging effects. These can be due to chronological aging or photoaging. Wrinkling is largely due to deterioration of the elastic fibers, as well as the breakdown of collagen.

Chapter 4

The Skeleton

The skeleton is commonly viewed as a collection of dry bones strung up on a stand in a classroom. Rather than being the static framework upon which the rest of the body is draped, the skeleton is teeming with cellular and molecular activity. Bones of the skeleton are mechanically connected to each other through ligaments, and to muscles and fascia through tendons and direct insertions of collagen fibers into the substance of the bone. The skeleton not only plays a biomechanical role, but also increasingly it is becoming recognized as influencing many aspects of the overall metabolism of the body. For example, it is the body's most important store of calcium and phosphate. From the early embryo until the time of death, the skeleton is constantly changing in response to its mechanical and hormonal environments, and it is becoming increasingly clear that skeletal tissue influences and is influenced by many other tissues of the body in addition to its mechanical role. After a brief treatment of the evolution of the skeleton, this chapter will follow the life history of the skeleton from its first appearance in the embryo until old age.

EVOLUTIONARY ORIGINS OF THE SKELETON

Both bone and an internal skeleton are unique to the vertebrates. Many invertebrates, even some three-foot long earthworms, have no skeleton and are able to get along quite well without one. Other invertebrates, such as insects and molluscs, use a hard outer covering (**exoskeleton**) for protection and mechanical support. Invertebrate exoskeletons are composed principally of calcium carbonate. There are two main problems with this exoskeletal strategy. One is that, aside from the shells of molluscs, other exoskeletons do not permit growth, and in order to grow, an insect, for example, must undergo a series of molts in order to form new slightly larger exoskeletons to fit the larger soft tissue body. Another problem is that calcium carbonate is relatively unstable, meaning that its calcium and carbonate components readily pass from the exoskeleton into solution.

Vertebrates adopted a different strategy for building a skeleton. The earliest vertebrates (e.g., lancelets), had already evolved a rather stiff rod of nonskeletal tissue, called the **notochord** (hence the name **chordates** for our phylum), that runs the length of the body for longitudinal support. A small basket of cartilage that supports the gill area is one of the earliest traces of vertebrate skeletal structures.

Early jawless fishes developed tiny thorn-like structures (**odontodes**) of bone in the dermis of their skin. Some of the odontodes are thought to have evolved into teeth. Others expanded into dermal plates of bone, which still functioned like an exoskeleton. These dermal plates provided the evolutionary basis for the formation of many flat bones (such as those of the cranium) that are prominent in all higher vertebrates. The next major advance was the development of a bony internal skeleton (**endoskeleton**), which could grow with the animal without the necessity of molting. Our endoskeleton—the long bones of the limbs and the vertebral column—provides internal attachments for muscles and allows both strength in movements and the ability to grow to great size in some species.

The dermal plates in early fishes formed by the direct conversion of connective tissue cells into bone—a mode of bone formation called **intramembranous bone formation** (see p. 91). In a later evolutionary development, a new mode of bone formation (**endochondral bone formation**—see p. 94) involves the replacement of cartilaginous skeletal elements by bone. Most of the endoskeleton forms by endochondral bone formation. The original notochord is still prominent in early embryos, but in higher vertebrates it virtually disappears during later embryonic development and is only represented by fragments in the adult vertebral bodies and in the intervertebral disks.

The skeletons of humans and most higher vertebrates still retain three fundamental forms of skeletal tissues. Teeth are used for predation and eating; dermal bones (e.g., the cranium), for protection; and endochondral bones (limbs and vertebral column), for support and locomotion.

The Human Body. DOI: https://doi.org/10.1016/B978-0-12-804254-0.00004-1

Vertebrate bone utilizes calcium phosphate as the major inorganic constituent of bone. This material has considerable advantages over the calcium carbonate used by invertebrates. A main advantage is its greater metabolic stability. Even though there is a constant turnover of calcium and phosphate ions in the vertebrate skeleton, it is nowhere as great as that for calcium carbonate, especially under conditions, such as hard exercise, when the pH of body fluids may temporarily fall from 7.41 to 7.15. With such a drop in pH, much more calcium carbonate than calcium phosphate is released from skeletal tissues. Thus, the presence of calcium phosphate increases the stability of the vertebrate skeleton.

EMBRYONIC ORIGINS OF THE SKELETON

The skeleton begins in the early embryo with aggregations of a type of connective tissue called **mesenchyme** (Fig. 4.1). Mesenchymal cells are precursors to a number of types of tissues, specifically cartilage, bone, connective tissue, muscle, blood, and adipose tissue.

Two major conditions are required to initiate the formation of a part of the skeleton from a mass of mesenchymal cells. One is a stimulus for a group of cells to develop (**differentiate**) into either cartilage or bone cells. The other is for the cells comprising the future skeletal elements to receive or express information that will guide the development of their unique shape and location. This latter process remains one of the least understood aspects of development. A rule of thumb is that the general form of a skeletal element is determined through the expression of an embryonic pattern-forming mechanism, but that the final form of a bone is strongly influenced by the mechanical environment in which it resides.

A good example is the development of an individual vertebra in the spinal column. Each of the 24 vertebrae, plus sacrum and coccyx, has a different readily recognizable shape. At a fundamental level, the characteristic of each of these is determined by the pattern of expression of a group of powerful genes, called ***Hox* genes**. The existence of these genes was first recognized many years ago in fruit flies (*Drosophila*). Flies, like all other insects (and humans), are highly segmented, and each segment is different. Genetic experiments involving some highly disruptive mutations, for example, resulting in legs forming in the place of antennae, allowed geneticists to recognize a series of genes that control the identity of each segment of the fly's body. Almost exact copies of these genes are found in humans, and they, too, control the form of our bodily segments. Although at first glance, our bodies do not appear to be segmented, consider the rib cage. Each set of ribs and the vertebra to which it is attached represents a body segment. Many other parts of the human body are also segmented in the early embryo, but by adulthood, these segments are not often conspicuous.

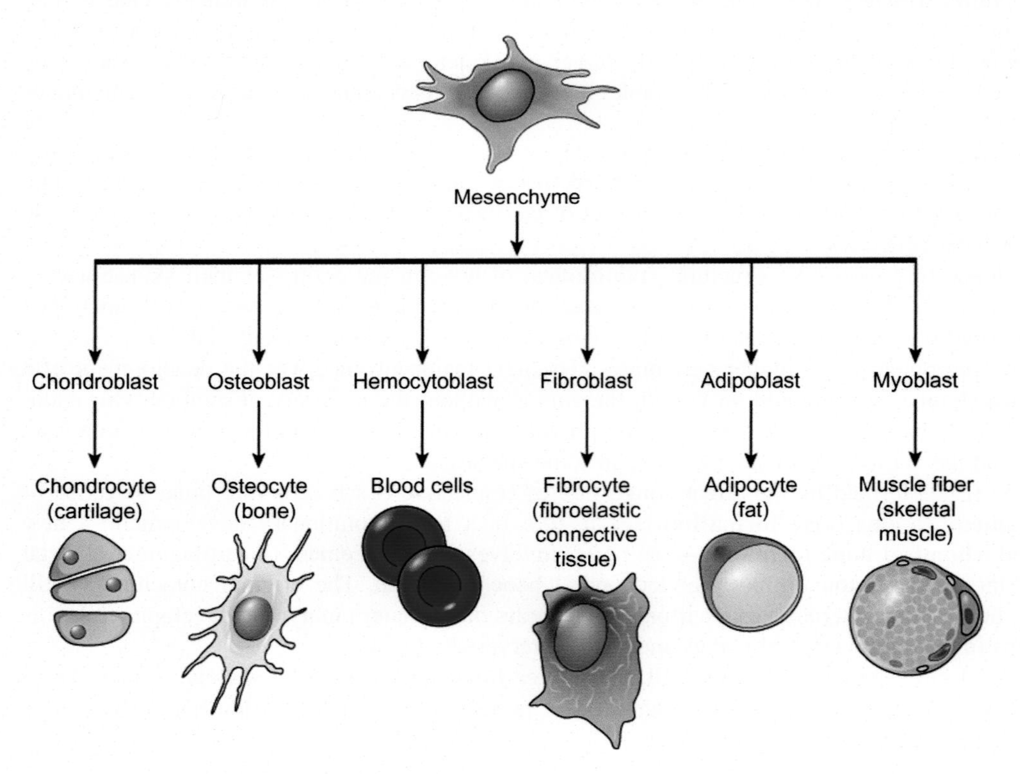

FIGURE 4.1 Mesenchyme and its cellular derivatives.

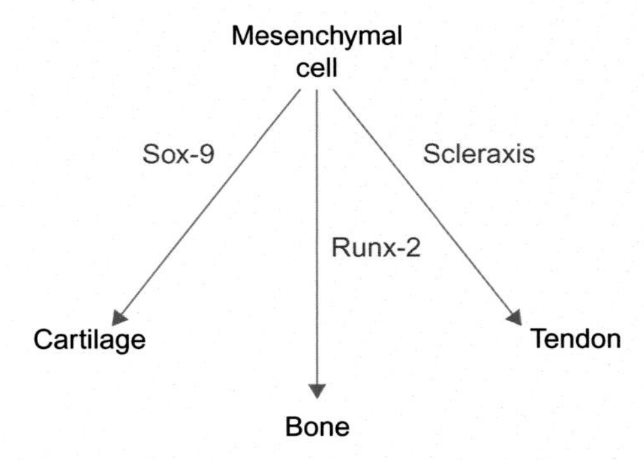

FIGURE 4.2 Transcription factors involved in mesenchymal cell differentiation.

Occasionally a person is found to possess an extra thoracic vertebra and to be missing a lumbar vertebra, as one example. We now know that this is due to a genetic mutation that alters the balance of pattern-forming information inherent in the array of *Hox* genes and that various mutations can cause what are called level shifts in the characters of the body segments, such as the example involving *Drosophila* given earlier. This is just one example of how genetic information controls basic patterning of many parts of the body. We are still just scratching the surface regarding our understanding of embryonic patterning mechanisms.

More is understood about what determines whether a mesenchymal cell will develop into bone or cartilage. In an area where a skeletal element will form, the mesenchymal cells move closer together, with less intervening extracellular matrix. These cells have the potential to form cartilage, bone, or tendon. The developmental pathway that they take is determined by the action of transcription factors, like the products of the *Hox* genes mentioned earlier. **Transcription factors** act on genes and unlock specific developmental programs that guide the differentiation of the mesenchymal cells. For example, the factor **Sox-9** steers mesenchymal cells down a cartilage-forming pathway, whereas exposure to **Runx-*2*** results in the formation of bone, and **scleraxis** controls the formation of tendon (Fig. 4.2).

The embryo uses several strategies in forming the skeleton, which is often subdivided into three main components—the **axial skeleton**, the **appendicular skeleton**, and the **cranial bones**. The axial skeleton, which extends from the base of the brain, through the vertebrae and ends with the tip of the coccyx, is highly segmented and under the developmental control of the *Hox* genes. These bones begin as cartilaginous models of the individual bones and later become true bone through endochondral ossification. The appendicular skeleton consists of the bones of the arms and legs, along with the pelvic and pectoral girdles that connect them to the rest of the body. The cranial bones cover the brain and facial structures and differ from the others in that most take shape through intramembranous bone formation.

CARTILAGE

Under the influence of the transcription factor Sox-9, cells destined to form cartilage begin to produce the set of molecules unique to the extracellular matrix of cartilage. Chief among these is a special type of collagen (**type II collagen**) that serves as a major binding agent for the other molecules of the matrix. The other major matrix component is an enormous molecule consisting of many molecules of a proteoglycan called **aggrecan** attached to a backbone of hyaluronic acid (see Fig. 2.15). This complex polymer, which can be up to 4–5 μm long, is highly negatively charged. The negative charge attracts Na^+, which, in turn, attracts and binds to H_2O. The combination of the strength of the collagen fibers and the space-filling and water-binding properties of the proteoglycan molecules gives cartilage its unique biomechanical properties. An adhesive protein, called **chondronectin**, helps to bind the chondrocytes to molecules of the extracellular matrix just as fibronectin does for fibrous connective tissue. The collagen fibers impart tensile strength and the proteoglycan molecules, resiliency to the matrix. This enables the cartilage in joints, like the knee, to tolerate both the weight and mechanical stresses placed upon them.

In the embryo, the densely packed cartilage-forming cells (**chondroblasts**) begin secreting matrix molecules, which surround the cells with an ever thickening gel-like capsule. This pushes the individual cells farther apart within the initial nodule of young cartilage. Growth of the initial cartilaginous nodule occurs in two ways. Cartilage cells already surrounded by matrix (now called **chondrocytes**) are still able to divide, and they expand the size of the nodule from the inside by a process called **interstitial growth**. Daughter cells from such divisions are located close to one another

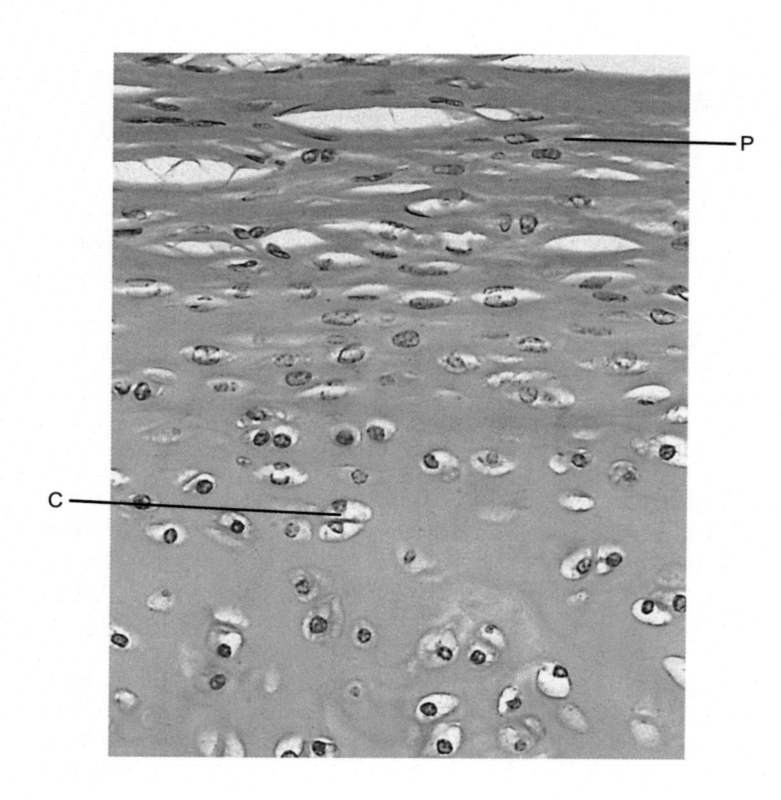

FIGURE 4.3 Hyaline cartilage. P—perichondrium; C—cluster of chondrocytes (isogenous group). *From Stevens and Lowe (2005), with permission.*

and form recognizable nests of cells (Fig. 4.3). Such a group of cells, along with the matrix immediately surrounding it, is called a **chondrone**.

The other mechanism of cartilaginous growth involves laying down new cartilage on the outside edge of the nodule. This is called **appositional growth**. Fibroblast-like cells immediately surrounding the nodule (the **perichondrium**) have the capacity to form cartilage, which is added to the existing cartilaginous surface in successive layers. A unique feature of cartilage is that it does not contain blood vessels, because the chondrocytes secrete molecules that inhibit the penetration of blood vessels into the cartilaginous matrix. Fortunately, chondrocytes are adapted to functioning best under low oxygen pressures. Cartilage also does not contain nerves. Without either blood vessels or nerves, cartilage remains one of the structurally most simple tissues.

The initial shape of the purely cartilaginous model of an embryonic bone, for example, the humerus in the upper arm, is largely genetically determined, and it approximates the shape of the adult bone. As the embryo grows, the size of the bone becomes too large for adequate nutrition of an entirely cartilaginous structure because of its lack of a direct blood supply. At some point in development (specific to every bone), the skeletal element undergoes a transition from a purely cartilaginous structure to one in which parts become replaced with true bone through endochondral ossification.

Especially in the major joints, like the knee, cartilage can be damaged through direct trauma (tears) or degeneration (arthritis). Unfortunately, cartilage has one of the poorest self-repair potentials of any tissue in the body. The reasons are not fully understood, but among the contributing factors are its lack of vascularization and the separation of chondrocytes from one another by the surrounding matrix. Many types of procedures have been devised to promote the repair of hyaline cartilage, but to date none has been totally successful. A common strategy, especially in the knee, has been to drill small holes through the damaged region of cartilage into the underlying bone in order to allow stem cells from the bone marrow and blood vessels to enter the hole. Sometimes additional cartilage or stem cells have also been placed into the hole. Some degree of repair often occurs, but the tissue formed is typically fibrocartilage, not hyaline cartilage. Unfortunately, fibrocartilage does not have the mechanical properties, especially the resiliency, of hyaline cartilage, and over time the fibrocartilage degenerates. More recently researchers have taken healthy chondrocytes or stem cells from the patient and grown them in cell culture. These cells, plus artificial matrix material, have been placed into areas of damage. There have been reports of some cartilage regeneration, but unfortunately such cartilage takes a very long time to mature. Even at 2 years, maturation of repaired cartilage is not complete. This is more than twice the time required for the maturation of healing muscle or bone tissue.

BONE

Direct (Intramembranous) Bone Formation in the Embryo

In the skull and a few other places in the body, bone forms directly from mesenchyme, rather than replacing a cartilaginous model that formed first. This process is called intramembranous bone formation, and it begins with an inductive signal from a nearby embryonic epithelium. Responding to the inductive signal, the mesenchymal recipients of the signal form a cellular condensation (as is the case for cartilage) and produce the transcription factor Runx-2, which steers these cells directly into a bone-forming mode. Through a couple of intermediate steps, the mesenchymal cells differentiate into **osteoblasts**—cells that actively produce the matrix of bone tissue. Osteoblasts are extremely active in the production of proteins and carbohydrates designed for export, and their cellular organelles reflect this activity. They have abundant rough endoplasmic reticulum and a prominent Golgi apparatus, along with many mitochondria to supply energy. Overall, the osteoblast has the form of a cuboidal epithelial cell. The organic matrix secreted by the osteoblasts is called **osteoid**. Osteoid consists of a high concentration of type I collagen and smaller amounts of types II and III collagen (collagen constitutes almost 90% of its dry weight), along with some ground substance consisting of several major glycoproteins and some growth factors, in particular one called **bone morphogenetic protein (BMP)**. Osteoid is nonmineralized matrix, and in humans mineralization does not begin until almost 2 weeks after the secretion of osteoid.

Once mineralization (calcification) begins, the process is quite rapid, often occurring within hours. Through mechanisms that are yet incompletely understood, crystals of calcium phosphate, called **hydroxyapatite**, form in depressions in the collagen molecules. This greatly increases the stiffness of the osteoid and essentially squeezes out some of the noncollagenous components of the osteoid matrix to form areas of true bone. In this type of bone, called **woven bone**, neither the cells nor the collagen fibers are oriented in any particular way. Woven bone, which is more cellular and weaker than standard adult bone, is characteristic of bone that is first formed in the embryo and also in a healing fracture.

Woven embryonic bone begins as tiny slivers or **spicules** of osteoid. This matrix completely surrounds some of the osteoblasts, which then drastically reduce their production of additional matrix and transition into osteocytes. Although in humans the lifespan of an osteoblast may extend to 8 weeks, only a small percentage ($\sim$20%) of osteoblasts become osteocytes. Most of the remainder die, but a few remain on some surfaces of the bony spicule.

The transition from osteoblast to osteocyte involves at least two stages—an intermediate **osteoid osteocyte** and a mature osteocyte (Fig. 4.4). During the transition from osteoblast to osteoid osteocyte, the cell body loses its cuboidal shape and almost 70% of its volume, but begins to produce elongated cytoplasmic processes directed toward the region of osteoid where mineralization is occurring. Elongation of the osteocytic cell processes is facilitated by the production of enzymes that locally degrade the osteoid matrix in advance of the elongating cytoplasmic processes. As the osteoid osteocytes within the osteoid mature further, they influence the osteoid matrix, causing it to undergo mineralization. Even though surrounded by matrix, osteocytes maintain connections with neighboring osteoblasts and osteocytes by 40–60 long cytoplasmic processes, which lie in tiny fluid-filled **canaliculi** that radiate out from the osteocytes much

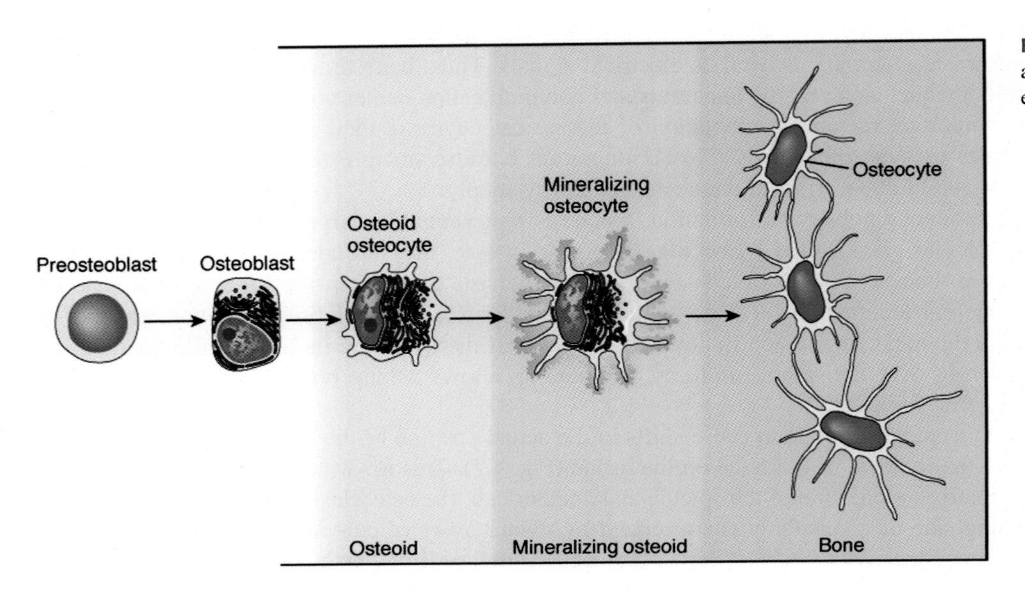

FIGURE 4.4 The transition from an osteoblast to a mature osteocyte embedded in bone matrix.

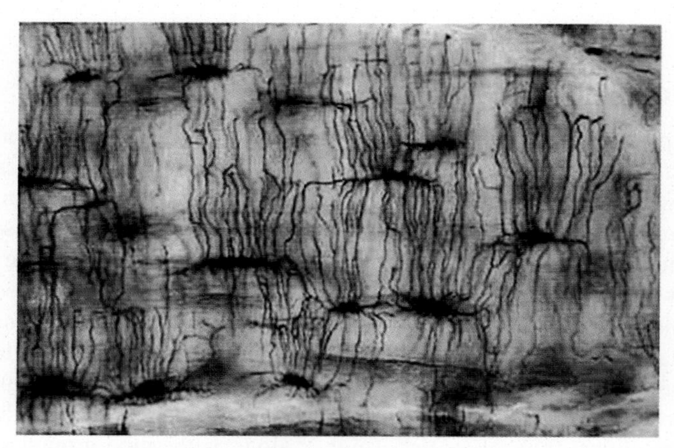

FIGURE 4.5 Longitudinal section through ground compact bone. The dark lacunae, normally occupied by osteocytes, give off many thin canaliculae, which contain interconnecting osteocytic processes. *From Erlandsen and Magney (1992), with permission.*

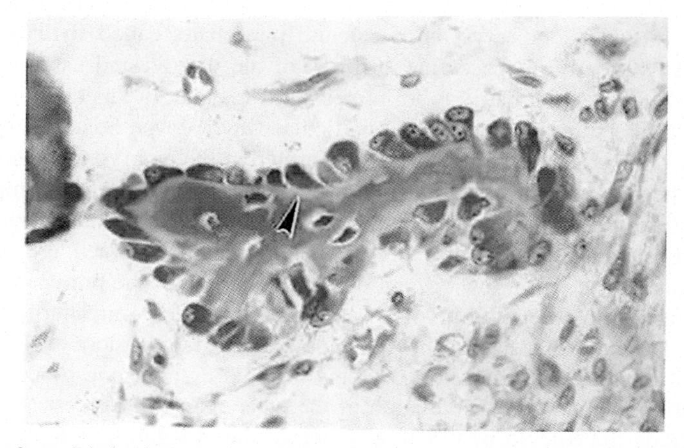

FIGURE 4.6 A bony spicule newly formed during intramembranous bone formation. Beneath a row of osteoblasts is a thin light pink layer of osteoid (tip of *arrowhead*). Over time, the osteoblasts will become surrounded by bone matrix and become osteocytes. *From Erlandsen and Magney (1992), with permission.*

like legs of a spider (Fig. 4.5). The osteocytic processes communicate with one another by gap junctions, which allow the exchange of nutrients and waste materials, as well as electrical signals. Thus, bone tissue is in reality an immense network of interconnected cells that act as a sensing apparatus and communication center within what is often assumed to be a rigid static bone. An important metabolic adaptation of mature osteocytes is their ability to function at lower oxygen tensions than osteoblasts or most other cells. This is important because many osteocytes in mature bone are situated a considerable distance (up to 200 μm) from the nearest capillary supply.

During early embryonic intramembranous bone formation, a bone is represented by large numbers of isolated bony spicules (Fig. 4.6). As bony spicules (also called **trabeculae**) increase in size, they are already subject to genetic or mechanical influences that cause them to be remodeled—a theme that persists throughout the lifetime of a bone (see p. 101). Remodeling of the internal architecture of a bone can accomplish a number of functions. For bones of the cranium, remodeling represents the means by which the overall diameter of the skull can increase while continuing to provide a hard protective covering over the brain. Similarly, remodeling allows a long bone to increase in diameter while maintaining a hollow marrow cavity.

Remodeling the internal architecture of a bone to cause shifts in the actual position of the bone requires the formation of new bone on the outside and the removal of old bone on the inside (Fig. 4.7). This involves the coordinated actions of osteoblasts forming new bone matrix on one side of the spicule and another cell, the osteoclast, which removes bone from the other side. The bone-forming side of a spicule is characterized by regular rows of cuboidal osteoblasts, whereas the opposite surface is lined by flattened inactive osteoblast derivatives, called **surface lining cells**.

(A)

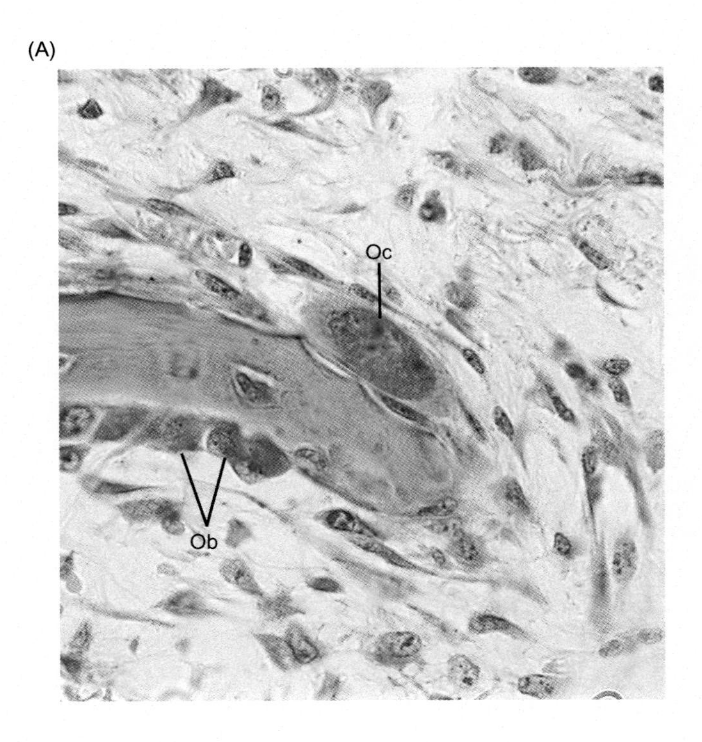

(B)

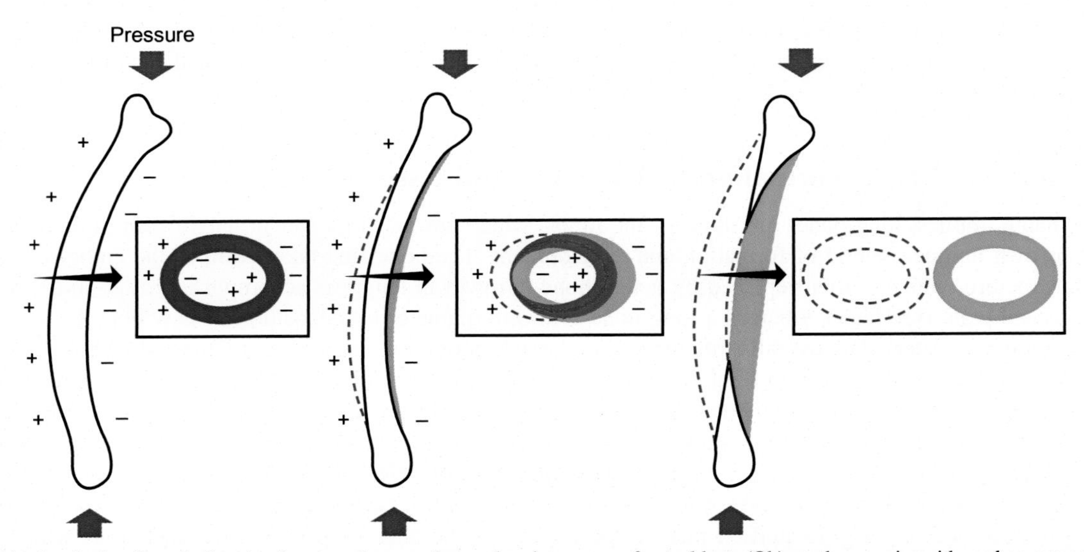

FIGURE 4.7 (A) A spicule of newly forming intramembranous bone, showing a row of osteoblasts (Ob) on the growing side and an osteoclast (Oc) on the resorbing side. (B) Diagrams illustrating the remodeling of a bent bone in relation to its mechanical environment. + and −, net electrical charge; blue, position of original bone; red, locations of newly added bone. *From (A) Gartner and Hiatt (2007) and (B) Carlson (2007), with permission.*

The **osteoclast** is a large multinucleated cell that arises from the fusion of several monocytes from the peripheral blood. As its name implies, an osteoclast is a cell that destroys existing bone matrix. After activation by local growth factors, it attaches to a bony surface like a suction cup, the rim of which forms a seal through binding between integrins on its plasma membrane and several matrix proteins of the bone (Fig. 4.8). The space between the osteoclast and the bone is a closed microenvironment. Within that space, hydrochloric acid secreted by the osteoclast activates enzymes that dissolve the apatite crystals of the bone matrix, and proteolytic enzymes digest the collagen fibers. The highly

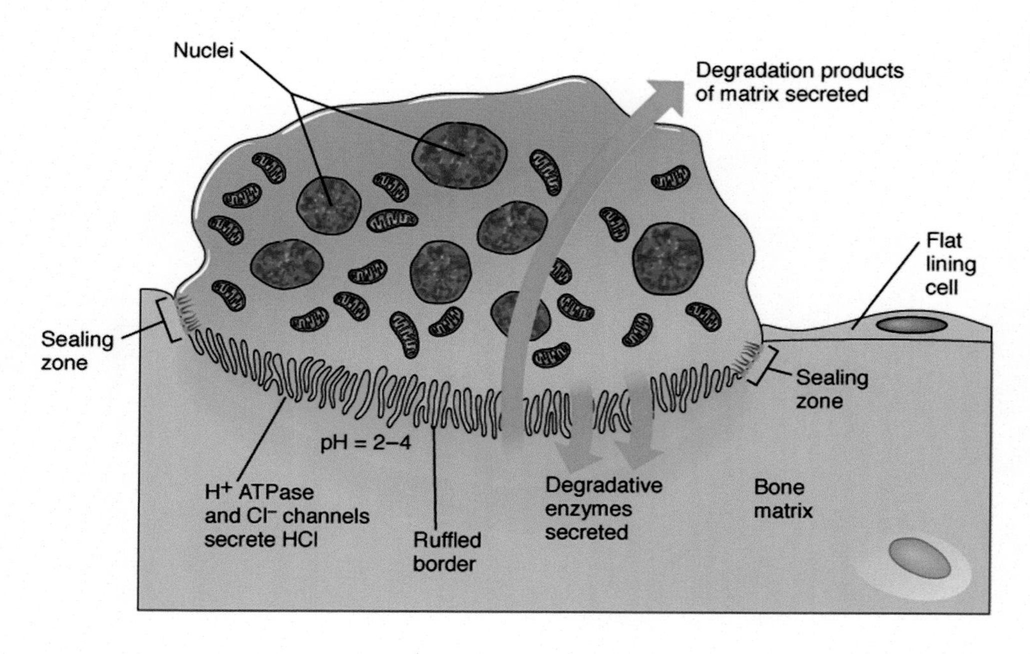

FIGURE 4.8 Osteoclast eroding bone matrix.

acidic environment (pH 2–4) is generated through the actions of hydrogen pumps and chloride channels within the osteoclast. As is typical of cells that produce a very acidic environment (see p. 338), the inner plasma membrane of an osteoclast is highly convoluted to increase its surface area. Both inorganic and organic degradation products of bone matrix pass through the osteoclasts and are extruded from the free surface of the osteoclast into tissue fluids for re-use.

The theme of addition of new bone on one side parallel to bone resorption on the other is a common one wherever bone is growing or remodeling in response to local or systemic environmental signals. The local signals may be mechanical or hormonal. As more is learned about them, we realize that the signals and factors modulating them are extremely complex.

Long Bones and Their Formation by Endochondral Ossification

As their name implies, long bones are those of the limbs, which are much greater in length than in diameter. Long bones have their own unique method of formation and development. The bone first takes form in the embryo as a purely cartilaginous structure, which is then replaced by actual bone. Growth in length is accomplished by a unique structure called the epiphyseal plate (see later), whereas a long bone grows in diameter by deposition of new bone by cells derived from the periosteum (see later). The overall structure and organization of a mature long bone are outlined in Box 4.1.

The first cartilaginous models of long bones in the embryo measure in millimeters. In the adult, the same bone could approach a half meter in length, in the case of a very tall person. In order to accomplish this remarkable feat of growth in a rigid structure, major changes must take place in the bone. A fundamental change is the replacement of the embryonic cartilage by bone through the process of **endochondral ossification**. This process does not necessarily involve the direct conversion of cartilage into bone, but rather the replacement of cartilage by bone (but see footnote on p. 97). Yet, despite the presence of a rigid calcified matrix, the bone must be able to grow in all dimensions without becoming a solid rod, which would be both much heavier and much weaker than a tubular bone.

The transformation of the cartilaginous rod into its bony successor begins midway along the shaft, where the inner cells of the perichondrium develop the capacity to produce osteoblasts instead of chondroblasts (Fig. 4.12). At this point the perichondrium has become a **periosteum**. At the same time the cartilage in the mid-shaft is undergoing significant changes. The chondrocytes hypertrophy and begin producing **type X collagen**, which predisposes the matrix to calcification.[1] As the matrix calcifies, it becomes less permeable to the passage of oxygen and nutrients. As a consequence, many of the chondrocytes enclosed by the calcified matrix die. Products of the dying chondrocytes are **angiogenic**

1. Calcification refers to the deposition of calcium salts in a soft tissue. Ossification refers to the formation of true bone, with osteocytes embedded within a calcified osteoid matrix. Under pathologic conditions, calcification can occur in a number of normally soft tissues, such as the walls of blood vessels, heart valves, brain and breast tissue.

BOX 4.1 The Structure of a Long Bone

A mature long bone contains a large number of parts (Fig. 4.9). From each end, it can be subdivided into a broad terminal **epiphysis**, a long narrow **diaphysis** in the middle, and a short tapering **metaphysis** linking the two. The ends of the bone are capped with a layer of hyaline cartilage that has remained cartilaginous since embryonic times. This cartilage represents the joint surface (see p. 108) and is naked cartilage with no fibrous perichondrial covering. The epiphysis and most of the metaphysis consist of spongy bone, with red hematopoietic marrow filling in the space among the spicules.

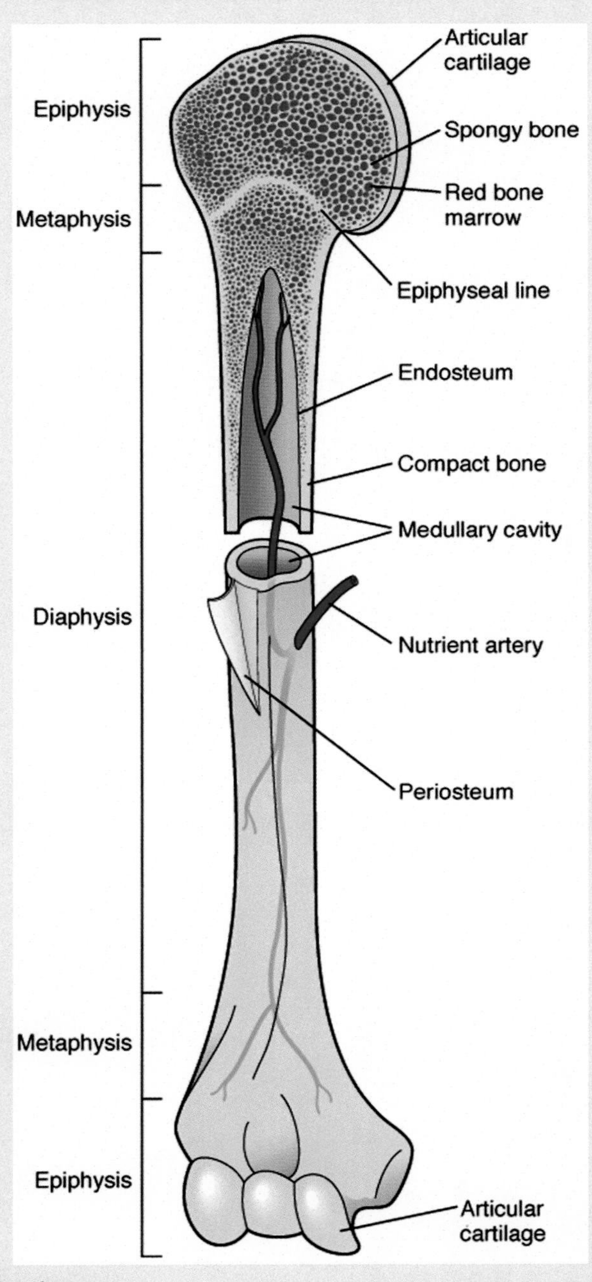

FIGURE 4.9 Overall structure of a long bone.

The shaft of a long bone is dominated by a thick cortex of compact bone, with its internal infrastructure of large numbers of multigenerational osteones (Fig. 4.10). Surrounding the cortex is the periosteum, consisting of an outer layer of dense fibrous connective tissue and a thinner inner layer of osteogenic cells. The outer layer is strongly connected to the surrounding fascial tissues, forming a close connection between the skeleton and the other movable tissues of the body. These connections are made even firmer by **Sharpey's fibers**, bundles of collagen fibers leading from the periosteum into the bone, where they are strongly embedded within the matrix of the bone (Fig. 4.11).

(*Continued*)

BOX 4.1 (Continued)

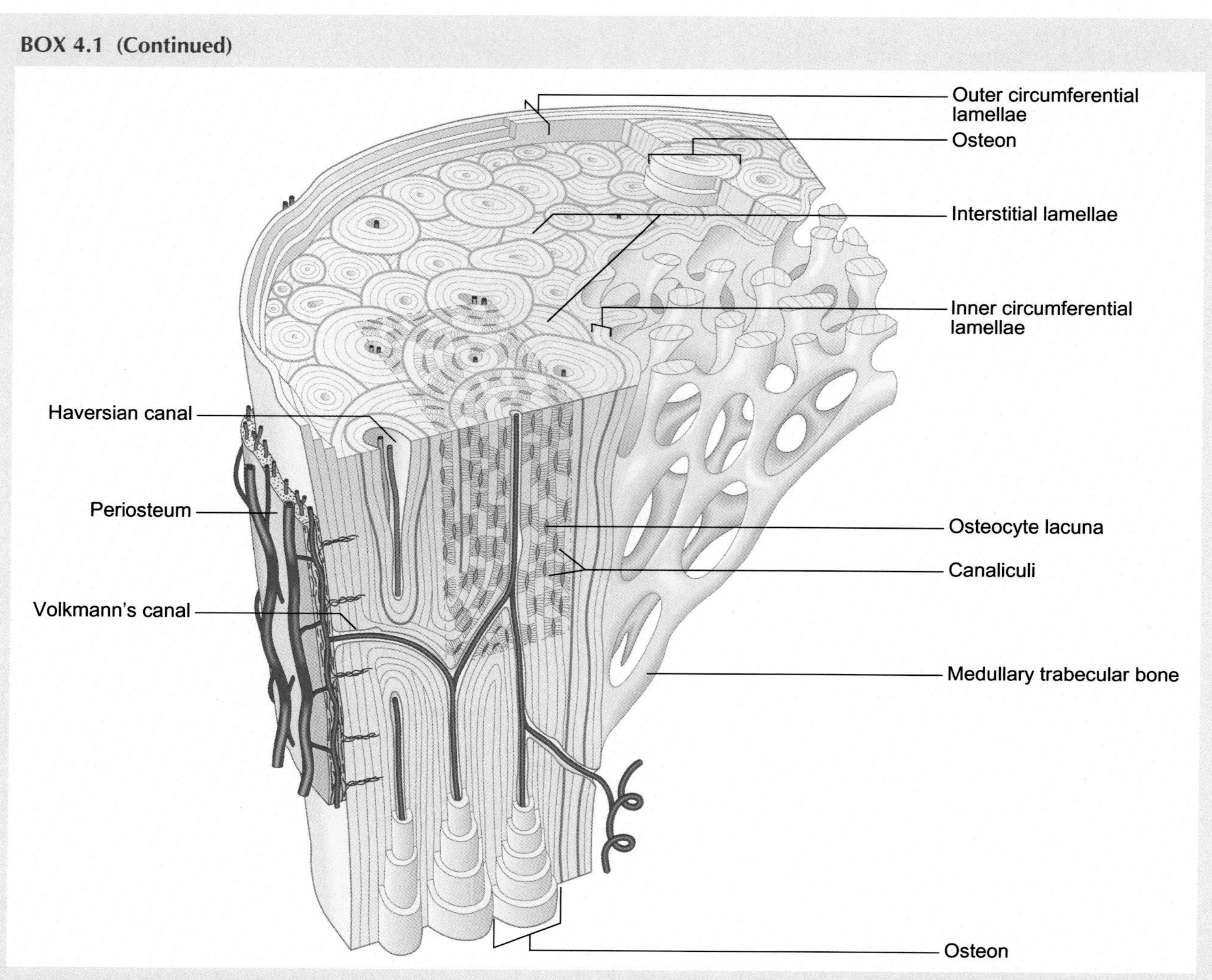

FIGURE 4.10 Main structural features of a mature long bone, with compact bone on the outside and spongy (cancellous) bone toward the middle. *From Drake et al. (2005), with permission.*

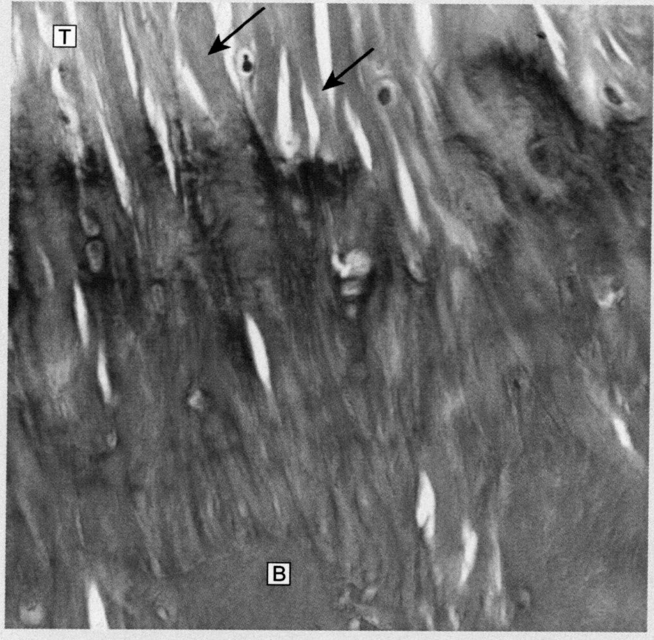

FIGURE 4.11 The site of tendon insertion into a bone. Collagen fibers (arrow) from the tendon (T) insert into the bone (B), at which point they are known as Sharpey's fibers. *From Stevens and Lowe (2005), with permission.*

(*Continued*)

BOX 4.1 (Continued)

The central region of the bone (the medulla) is occupied by fatty marrow, which is bounded by a thin layer of cells, the **endosteum**, that lines the inner surface of the cortex. In addition to fat cells and some mesenchymal stem cells, the marrow contains blood vessels and sympathetic nerve fibers. Most bones have a main **nutrient artery** and vein that penetrate the mid-shaft region through what is called a **nutrient foramen**, but blood vessels also enter the bone in other areas, especially on either side of the epiphyseal line. Smaller blood vessels enter the marrow cavity of a bone through transverse canals that cut through various regions of the cortex. The locations of larger blood vessels and these canals, called Volkmann's canals (see Fig. 4.10), are of considerable importance, especially when a bone is fractured (see p. 104), because regions of a fractured bone that are cut off from their blood supply die.

The dominant component of bone is its extracellular matrix. The matrix has two components—the organic and the inorganic matrix. In a dried bone, the inorganic matrix is most prominent, but the remains of the organic matrix are embedded within it. In the absence of an organic matrix, bone would maintain its shape, but would be easily susceptible to fracturing or crumbling when under pressure. In contrast, the organic component of the matrix can be demonstrated by soaking a bone in a weak acid, which removes the inorganic component of the matrix. After such treatment, a thin bone like the fibula can be bent and literally tied into a knot. It is the combination of the inorganic and organic matrices that gives bone its remarkable mechanical properties.

The gross outlines of a bone are anything but smooth. In addition to the flared epiphyseal ends, the surface of a long bone is punctuated by various bumps (**tubercles**) and ridges. These represent sites of attachment of muscles and ligaments, as well as sheets of fascia. Although their initial location may be determined genetically, they represent areas where the tension of a tissue pulling on a bone causes the deposition of bone or fibrocartilage at the interface. As a general rule, the stronger the tensile force, the larger the bump or ridge.

(stimulate vascular ingrowth), and blood vessels begin to grow from the new periosteum toward the calcified cartilage. After the blood vessels grow inward, some of the calcium salts of the calcified matrix dissolve. More importantly, **chondroclasts**—the functional equivalent of osteoclasts—attach to and digest the now defunct matrix, creating spaces within the formerly solid cartilaginous matrix.

Osteogenic (bone-forming) stem cells accompany the ingrowing blood vessels, and settle down on the exposed surfaces of the cartilaginous matrix. With that as a nidus, they become transformed into osteoblasts[2] and begin laying down a layer of osteoid matrix in the same manner that was described earlier for intramembranous bone formation. As the osteoblasts become surrounded by osteoid, they undergo the transition into osteocytes embedded in a true bony matrix, which initially surrounds a tiny remnant of calcified cartilage.

Meanwhile, and even preceding the above, osteogenic cells in the innermost layer of the periosteum begin to form tiny bony spicules on the surface of the bone by means of intramembranous bone formation. These bony spicules differ from the ones in the interior in that they do not form on top of cartilaginous remnants. The periosteally derived bone forms a collar around the mid-shaft of the long bone.

At this point, all of the newly forming bone has the configuration of spongy bone. Spongy, or trabecular, bone consists of irregular spicules or shelves of bone that surround soft tissue—the bone marrow (Figs. 4.13 and 4.16). Stem cells of the marrow come in with the blood vessels and create the hematopoietic (blood-forming) system in what is called the red marrow. The counterpart to spongy bone is compact bone, which constitutes most of the shaft of an adult long bone. Its structure and mode of formation is discussed on p. 100.

Once the process is established, the calcification of central cartilage, its erosion, and eventual replacement by spongy bone proceeds toward either end of the bone. At genetically predetermined times, unique to each bone, a similar, but more limited process of matrix calcification and replacement by spongy bone in what is called a **secondary center of ossification** takes place at each end of the bone (see Fig. 4.12). This leaves at either end of the bone a thin layer of cartilage that is of critical importance for further growth and development of the bone. It is called the **epiphyseal plate**, and the cellular dynamics of this region are responsible for the growth in length of a bone.

The Epiphyseal Plate and Growth in Length of a Bone

The power to increase the length of a bone is concentrated in the cartilaginous **epiphyseal plates** located near each end of the bone. These plates are situated between the shaft of the bone and the secondary ossification centers within the

2. Recent studies have shown that some of the chondrocytes that are embedded in calcified cartilage can survive and undergo a direct transformation into osteoblasts.

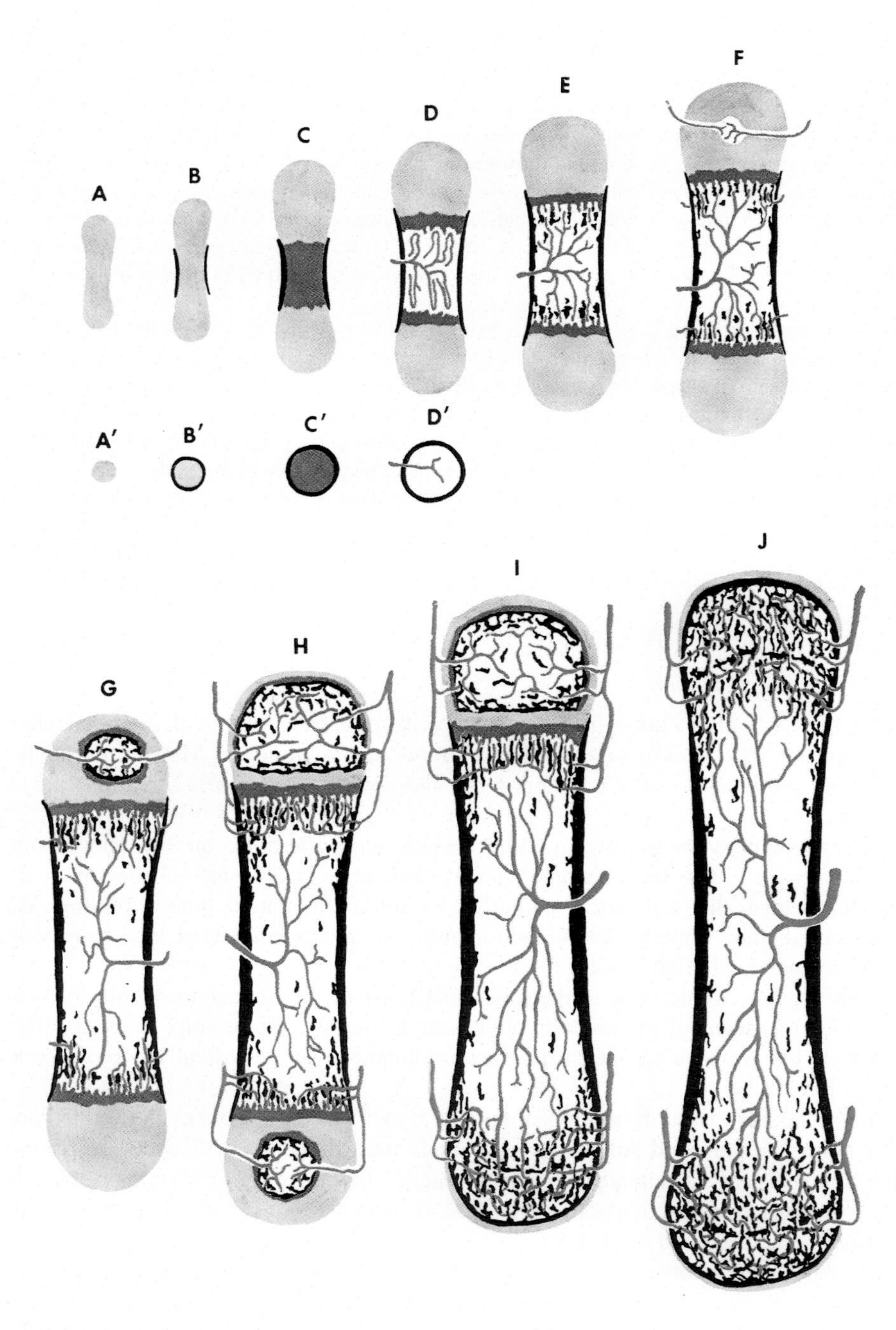

FIGURE 4.12 Sequence of events in endochondral bone formation. (A) A cartilage (blue) model of a long bone in the early embryo. (A′–D′) Cross-sections through the middle of the bone. (B) A periosteal collar begins to form around the bone. (C) Cartilage in the middle of the bone becomes hypertrophied and calcified (purple). (D) Blood vessels invade the area of calcified cartilage, bringing in osteoblasts that lay down spicules of new bone (black) in the primary site of ossification (D and E). (F) Invasion of the end (epiphysis) of the bone by blood vessels. (G) Formation of a center of secondary osteogenesis within the epiphysis. (H) Formation of an epiphyseal plate between the primary and secondary sites of ossification. (I) Growth in length of the bone through the activities of the epiphyseal plate. (J) Closure of the epiphyseal plate (note the absence of blue and purple) at the end of bone growth. Along the shaft (diaphysis) of the bone, bony spicules are replaced by compact bone (dense black areas). *From Bloom and Fawcett (1975), with permission.*

epiphyses. The general functional principle is expansion of the cartilage toward the epiphysis, with simultaneous removal of cartilage and its replacement by bone on the diaphyseal side, all while remaining at about the same thickness. In terms of absolute measurements, the epiphyseal plate moves away from the midpoint of the bone as the bone lengthens. A complex set of cellular phenomena, involving several zones of activity, is required to accomplish this.

Starting from the epiphyseal end of the plate and working inward toward the shaft, the first zone is a layer of resting cartilage (Fig. 4.14). Immediately beneath is a region of chondrocyte proliferation. The resulting daughter cells line up into columns of flattened cells, much like a stack of pancakes. These cells hypertrophy, changing the shape of the flattened cells to an almost cuboidal configuration. This change of shape expands the epiphyseal plate toward the end of the bone, thereby lengthening the bone. Then the hypertrophied chondrocytes secrete type X collagen into the cartilaginous matrix, and the matrix begins to calcify. With the calcification of the matrix, most of the chondrocytes die. The dead cells and parts of the matrix are removed, leaving behind some vertical columns of calcified matrix. Blood vessels

(A)

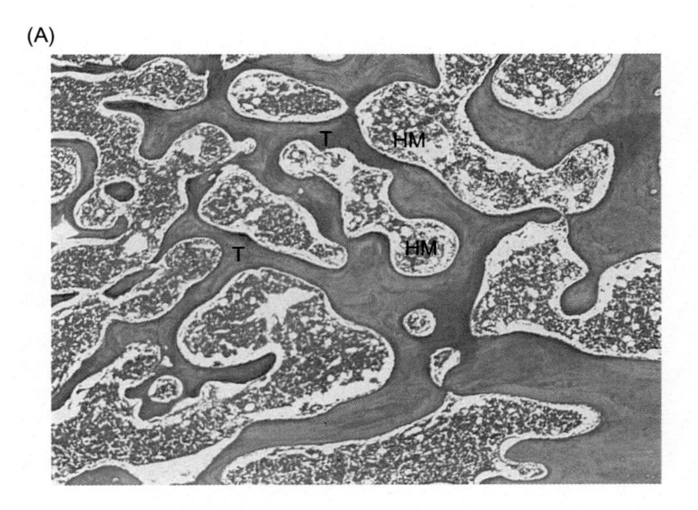

(B)

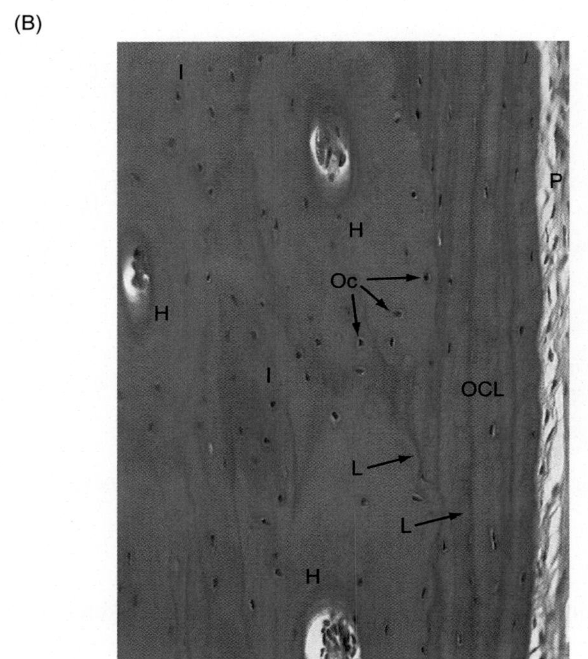

FIGURE 4.13 (A) Spongy bone at the end of a long bone. HM, hematopoietic marrow; T, trabeculae. (B) Compact bone along the edge of a long bone. Longitudinal section. H, Haversian system; I, interstitial lamellae; L, dark (purple) cement lines between adjacent lamellae; Oc, osteocytes, recognizable by their dark purple nuclei; OCL, outer circumferential lamellae deposited at the edge of the bone. *From Young et al. (2006), with permission.*

and other red marrow constituents fill in the spaces between the columns of remaining calcified cartilage matrix, and osteogenic cells accompanying them line up on the columns and begin to deposit a layer of osteoid upon the cartilage remnants. At the same time, chondrocytic proliferation on the epiphyseal side of the plate continues, adding to the total mass of viable cartilage.

Under the influence of growth hormone, the process of forming new cartilage on one side of the epiphyseal plate and its removal and replacement by new spongy bone on the other continues as long as the bone is growing. As an individual reaches the end of the overall growth period, caused by increased concentrations of sex steroid hormones during puberty, the cartilage of the epiphyseal plate diminishes and is finally replaced by bone to form the **epiphyseal line**. Called **closure of the epiphyseal plate**, this marks the end of growth of that particular bone. Estrogen in both sexes is critical for proper closure of the epiphyseal plate and the cessation of growth. In its absence, growth continues into adulthood. The two epiphyseal plates within a single bone may close at different times, as do those of different bones.

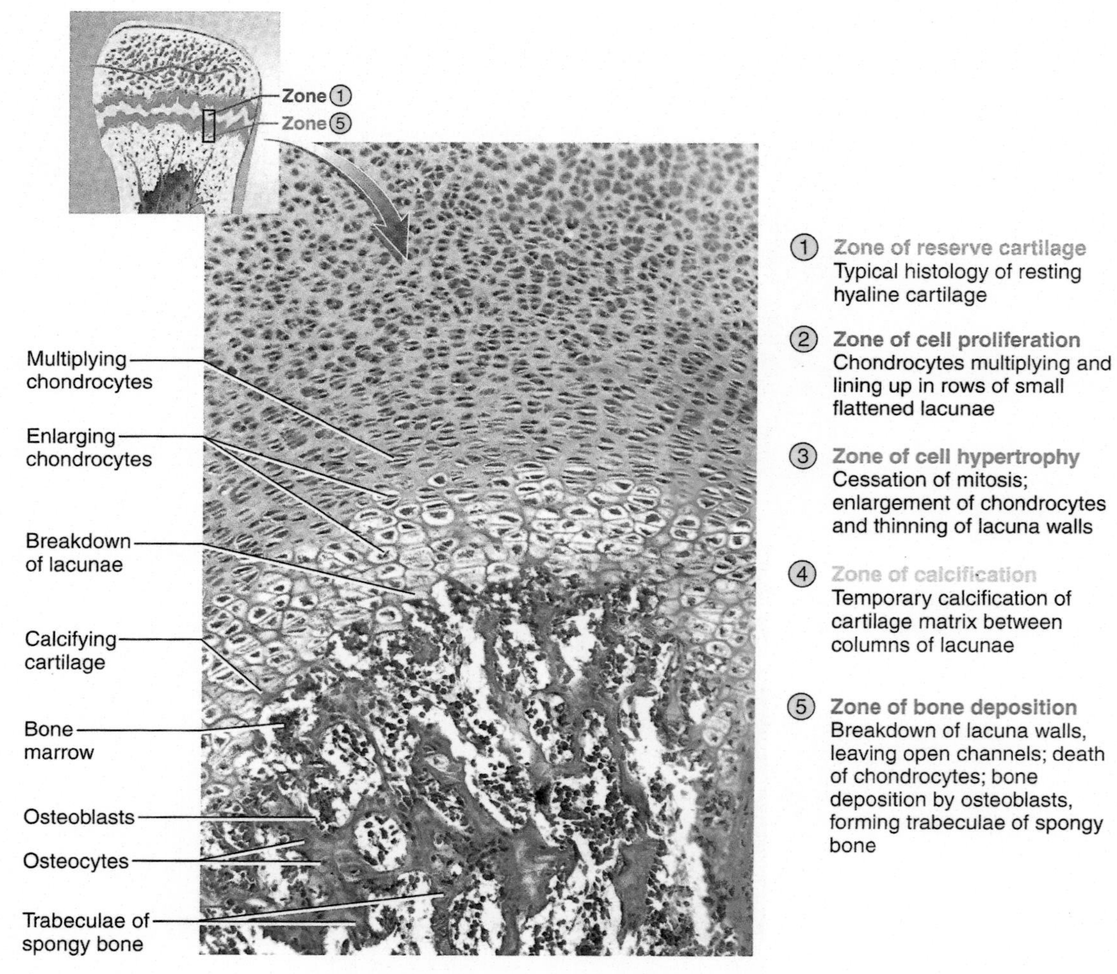

FIGURE 4.14 Structure and dynamics of an epiphyseal plate in a growing long bone. *From Saladin (2007), with permission.*

Recognition of these sequences allows anthropologists to accurately age the skeleton of an individual who has died in the late teenage years or early 20s.

Growth in Diameter and the Formation of Compact Bone

Growth in diameter of a long bone follows the general principle of adding bone on the outside while removing bone from the inside (see Fig. 4.7B), but as this is occurring, the substance of the growing bone becomes transformed from spongy to compact bone. To understand this, we turn back to the early embryo, when the cartilaginous model first becomes replaced by true bone.

The first bony tissue to appear is deposits of woven or nonlamellar bone beneath the periosteum at the midpoint of the bone. This bone first appears as spicules, which in aggregate are organized as spongy bone. Over time, the spongy bone becomes transformed into compact bone by a characteristic microscopic process. The first step is the creation of hollow cylinders among the spicules (Fig. 4.15). The cylinders are formed through the erosive actions of osteoclasts on the matrix of the bony spicules. The orientation of a cylinder is determined by the course of a blood vessel, which provides the precursor cells of the osteoclasts and then occupies the central core of the cylinder.

Following the creation of a cylinder, osteoblasts begin to line up along its inner wall and lay down a layer of bone in the usual way. Then additional layers of bone are laid down from outside in until the innermost layer lies almost adjacent to the blood vessel at the center of the cylinder. By this time the formerly hollow cylinder is completely filled in by layers of bone that in cross-section resemble a target. Each of the layers is called a **lamella**, and the entire

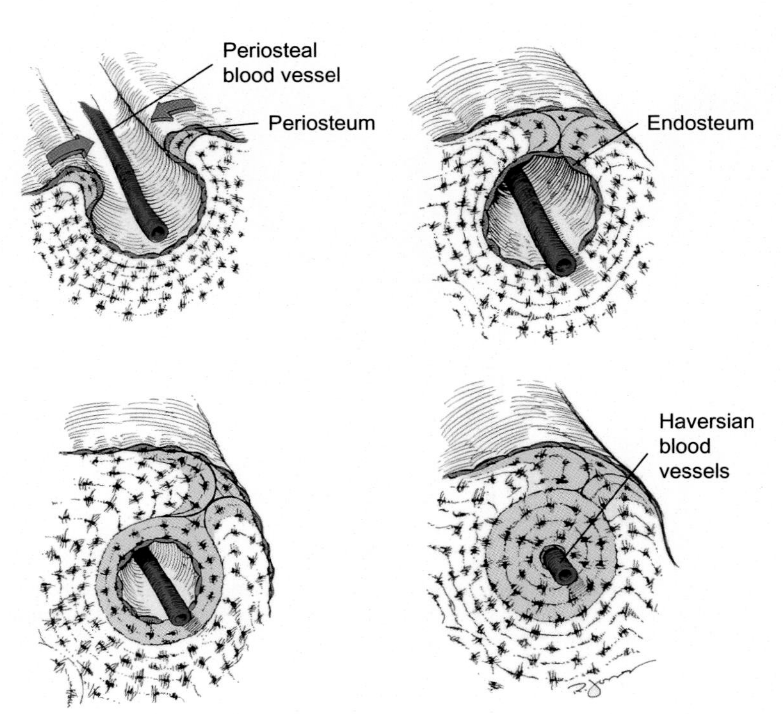

FIGURE 4.15 Formation of an osteone by laying down concentric layers of bony lamellae from the outside in. *From Thibodeau and Patton (2007), with permission.*

bony cylinder is called an **osteone** (see Fig. 4.15). When many osteones have formed, the outer (cortical) area of the bone contains much more hard than soft material. Therefore it is called **compact bone**. The innermost core of the osteone, containing a blood vessel and a nerve fiber, is called a **Haversian canal**. The inside (medulla) of the shaft of a long bone is hollow and is filled by a marrow that is fatty, rather than hematopoietic in character. Such marrow is often called **yellow marrow**, in contrast to the hematopoietic marrow in spongy bone, which is called red marrow.

Further Modeling of a Long Bone

If one looks at the outlines of a typical long bone, it is apparent that the ends are wider than the shaft. As the bone grows in both length and girth, it is apparent that some remodeling must take place in the area of the epiphyseal plate in order to maintain its shape. The width of the epiphyseal plate is increased through the addition of chondrocytes beneath the perichondrium at the edges of the plate. A narrow transitional region between the epiphyseal plate and the diaphysis (shaft) is called the **metaphysis** (see Fig. 4.9). In this area, the removal of newly formed bone just beneath the epiphyseal plate molds the taper of the metaphysis from the larger width of the epiphyseal plate to the narrower diameter of the diaphysis.

Reorganization (Remodeling) of Compact Bone

Despite the seemingly high degree of organization and appearance of stability, compact bone and its constituent osteones are in a constant, but sometimes slow state of flux. In response to growth pressures and changing mechanical circumstances, existing osteones get removed, and others are formed (see below for greater detail on underlying mechanisms). Remodeling to form a new osteone begins with the removal of existing bone by osteoclastic action, resulting in the formation of a new hollow cylinder, which normally is not confined to a single osteone. Instead, the orientation of a new osteone corresponds to the new biomechanical environment of that part of the bone. In general, the process of creating a new hollow cylinder and its filling in by concentric layers of fresh bone follows the pattern seen in the formation of all individual osteones. After several generations of osteone remodeling, one can recognize the succession of osteone formation by examination of the superimposition of parts of newer osteones over older ones (see Fig. 4.10).

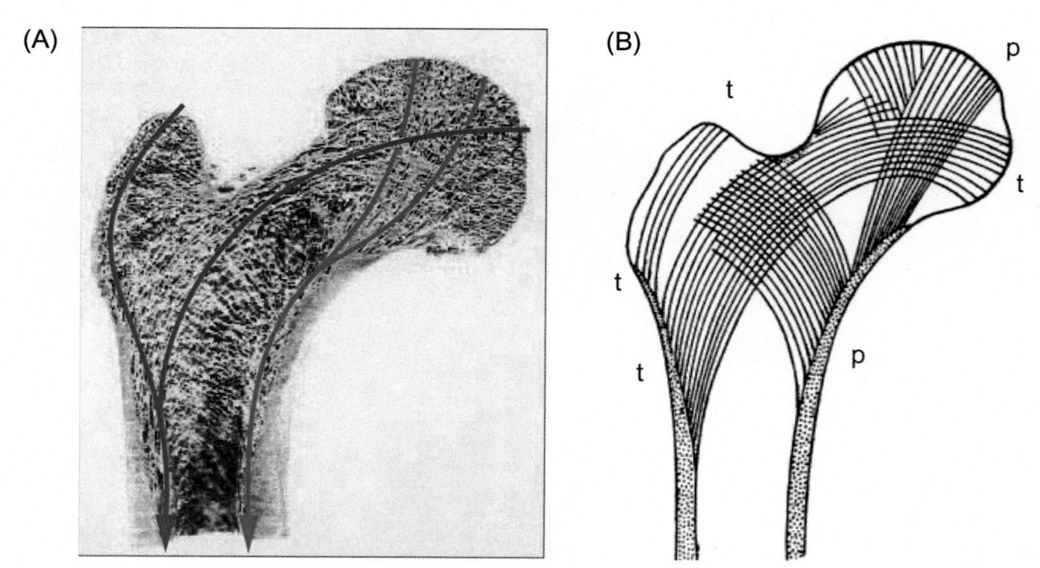

FIGURE 4.16 (A) Head and neck of a human femur, showing areas of compact and cancellous bone. Note the linear arrangement of cancellous bone spicules in the head and neck of the femur. (B) Schematic drawing of the head and neck of the femur, with lines of bony spicules (trabeculae) oriented according to pressure (p) and tension (t) points along the bone. Compare with the orientation of the trabeculae in part A. *From (A) Saladin (2007) (B) Hall (2005), with permission.*

Mechanical Stimuli, Remodeling, and the Internal Architecture of a Bone

The structure of bone is exquisitely sensitive to its mechanical and hormonal environment. Mechanical-loading results in the increased deposition of bone. For example, the humerus in the dominant arm of professional tennis players often contains 50% more bone than does that of the nondominant arm. At the other extreme, bone loss follows an almost straight line decline under conditions of weightlessness. One of the most important issues facing astronauts is maintaining the mass of their skeleton while up in space. In the early days of space flight astronauts who had been in space for several months were placed on stretchers immediately after return to earth to prevent their bones from fracturing from the gravitational pull of earth.

All bones show evidence of their response to their mechanical environment. The relationship between tension and the size of bony tubercles to which tendons are attached has already been discussed. Internally, the tissue architecture of a bone also betrays the mechanical forces imposed upon it. The classical example is the head and neck of the femur. Seen in section (Fig. 4.16), the trabeculae in the femoral neck follow remarkably closely the lines of tension and compression found in the arm of an industrial crane, an observation made as early as the 19th century. The configuration of the internal architecture of both spongy and compact bone reflects its current biomechanical environment, but should that environment change, the bony architecture correspondingly changes. Such changes in internal architecture are pronounced in the case of bones around the foot and heel of individuals taking up classical ballet, but even minor variations in mechanical-loading result in adaptive responses. All bones are different, and not all bones respond to stimuli in exactly the same way. Compare the bones of the cranium and those of the limbs. The cranium is designed to protect the brain and is subject to far fewer mechanical stressors than are bones of the limbs. Nevertheless, skull bones also respond to mechanical stimuli. For example, behind each of my ears are pronounced grooves in the skull caused by the pressures of the temple pieces of my eyeglasses.

Because of the nature of bone matrix, adaptive changes are rarely instantaneous, but are measured in days and weeks. On a long-term basis, the entire skeleton is renewed over a period of about 10 years. Nevertheless, the process of adaptation begins within minutes of a mechanical stimulus. Changing the internal architecture of a mature bone in response to a stimulus is called **remodeling**. The fundamental basis for remodeling is the recognition of a mechanical or hormonal stimulus and the translation of that stimulus into cellular activity. This section will concentrate on the response to mechanical stimuli.

Fundamental to understanding bone remodeling is recognition of the way in which a bone is put together. Central to this are the osteocytes. From the time when they become entombed in osteoid matrix, osteocytes are connected to one another, as well as to osteoblasts through their ~50 elongated cytoplasmic processes that are connected to those of neighboring cells by gap junctions. This forms a large interconnected cellular meshwork that remarkably resembles Indra's net of Indian mythology (Fig. 4.17). In both cases a push or pull on one node, whether an osteocyte in bone or a jewel in Indra's net affects many others through the threadlike interconnections.

In bone, the osteocytes are situated in small lacunae within the bone matrix, and their cytoplasmic processes occupy **canaliculi**—thin spider leg-like tunnels in the bone matrix. Between the cells and the bony matrix is a thin layer of fluid. Adhesion molecules connect some regions of the cells to the matrix.

External mechanical forces create stresses on a bone that can cause compression or tension on parts of the bone. Since bone is a rigid material, one of the major problems is translation of minor displacement of the bone itself (as little as 0.1% displacement) to a mechanical force that can be recognized by the cells of the osteocytic meshwork. The entire osteocytic meshwork acts as a three-dimensional strain sensor. One adaptation of many long bones lies in their gross form. Many show a slight curvature (or pronounced curvature in many four-footed animals), which by itself increases the level of strain for a given load. It is now recognized that even minor dynamic mechanical forces, such as repeated standing on one's toes, can produce an adaptive response, whereas steady force is less likely to do so. Although many details remain uncertain, a generally accepted model of the mechanisms underlying remodeling is taking shape.

The cellular response to a mechanical stimulus begins with the exposure of an osteocyte to shear stress resulting from flow of the interstitial fluid within the canaliculi in response to the mechanical deformation.[3] The fluid flow and resulting shear stress differs throughout the cellular meshwork depending upon the site and strength of the mechanical force applied to the bone. What part of the cell actually receives and responds to the fluid flow remains uncertain, but there are three main options. One is displacement of integrins and other molecules that attach the cells to the matrix. Another is pressure-sensitive ion channels. A third is the displacement of **primary cilia**, single cilia that are present on many cells of the body.

Detection of shear stress elicits two principal types of response. One is an almost immediate release of local (paracrine) chemical messengers, in particular nitric oxide gas and a prostaglandin (prostaglandin E_2). The other response remains within the cell. Displacement of integrins on the cell surface is translated to the cytoskeleton in the interior of

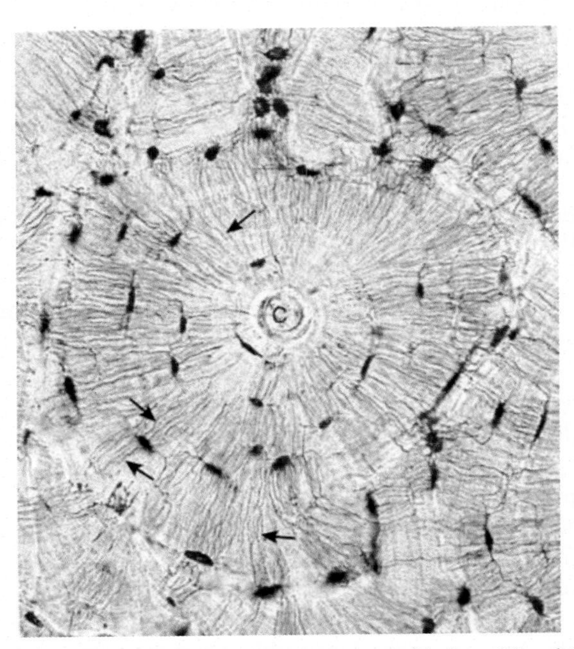

FIGURE 4.17 Section of ground bone, showing the relationship between osteocytes (black ovals) and the canaliculae (arrows) connecting them. C, Haversian canal. *From Gartner and Hiatt (1997), by permission.*

3. For a number of years, many bone biologists felt that bioelectric signals generated by the deformation of bone represented the initial signal for remodeling. Bone is piezoelectric—meaning that, when a substance is mechanically deformed, it becomes electrically polarized. A net negative charge builds up on the side of the bone subject to compression, and a net positive charge is found on the side exposed to tension (see Fig. 4.7). As a generalization, there is a net deposition of bone on a negatively charged surface and net resorption of bone on the electropositive surface. This knowledge led to the development of a number of electrical devices for the healing of nonunion fractures. Although such devices are still commonly used today, their mechanism of action remains uncertain, but it does not seem to correspond to the processes that occur during the normal remodeling of bone.

the cell and is ultimately transmitted to the nucleus, where it affects gene expression. This results in the synthesis of molecules that directly affect the cells involved in bone remodeling. At a cellular level, this means the activation of osteoclasts for the removal of existing bone and osteoblasts for producing new bone matrix.

Activation of osteoclasts occurs in response to signals from both osteocytes and osteoblasts, and possibly the smooth bone-lining cells, as well. In addition to molecular signals from living osteocytes that have been subjected to fluid pulses, degeneration products of dead osteocytes that have died in response to microfractures in deformed bone are suspected to stimulate osteoclastic activity. Influences, both stimulatory and inhibitory, of all cell types are closely interlinked. The recruitment of new osteoblasts from progenitor populations and their activation is a response to stimulatory signals from both osteoclasts and smooth bone lining cells and the reduction in inhibitory influences from osteocytes. Osteoblasts are preferentially attracted to areas of recent matrix erosion, where they begin laying down new osteoid that conforms to the new functional needs.

The site-appropriate removal old bone and deposition of new bone result in the gradual shifting of trabeculae of spongy bone to locations and orientations that are more favorable to the new mechanical microenvironment. Within compact bone, the removal of old and the formation of new osteones follow the same principles. When a bone is mechanically loaded, the net result is the deposition of more bony material, whereas in an unloaded environment, net resorption leads to less dense and usually weakened bone.

Fracture Healing

Despite the overall strength and the mechanical buffering capacity of the body, bones do break. Bones can and do break in a variety of ways, but for the sake of simplicity, the repair of only one type of fracture—a clean transverse break—will be discussed here.

A simple fracture is usually accompanied by immediate intense pain because of the rich innervation of the bone and its surrounding periosteum. This is followed by bleeding from blood vessels that were ruptured by the fracture. The extravasated blood coagulates into a **hematoma** (blood clot) at the site of the fracture (Fig. 4.18A). One of the earliest consequences of the fracture is the interruption of the local blood flow to parts of the bone supplied by the ruptured vessel. This can range from a small local area to something as large as the entire head of the femur, in the case of certain fractures of the femoral neck. When deprived of a local blood supply, few cells can survive for more than 2–3 hours. In bone, this results in the death of the osteocytes in the region supplied by the ruptured vessels, and irreversible **necrosis** (cell death) of that part of the bone (Fig. 4.18B).

Through mechanisms yet incompletely understood, the clotted blood at the fracture site releases cytokines that activate the formerly dormant cells of the periosteum and endosteum and cause many of these cells to divide. Mesenchymal stem cells resident in the marrow may also be called into action. These activated cells begin to produce matrix molecules, the character of which depends upon the local vascular environment. In a highly vascular environment, the activated cells begin to lay down woven bone very similar in character to that formed in the early embryo. If local revascularization has been less efficient and the oxygen tensions are lower, the activated cells first produce cartilage, which later becomes replaced by bone (Fig. 4.18C). Regardless of its character, the region of new skeletal tissue formation is called a **fracture callus**. When seen on X-rays, a callus looks like a broad cuff encircling the fracture site. Because of the thickness of the periosteum relative to that of the endosteum, a callus is more highly developed on the outer than the inner surface of a bone.

Over a few weeks, the new skeletal tissue formed in the callus begins to undergo remodeling in response to the mechanical environment at the fracture site. The amount of local remodeling that can occur to restore a nonsplinted displaced fracture can be astounding (Fig. 4.19).

In close to 10% of all fractures, good healing does not take place at the fracture site, resulting in a false joint (**pseudarthrosis**). A common site for this to occur is the lower tibia. Over the years, many techniques have been devised to stimulate the healing of nonunited fractures. Many of these take advantage of mechanisms that are operative in the normal growth and remodeling of a bone. One involves the application of the growth factor, **bone morphogenetic protein (BMP)**, which is normally bound to the organic matrix of the bone. BMP stimulates the recruitment and activation of osteogenic cells, which are stimulated to lay down new bone matrix material. Another popular method for treating nonhealing fractures is electrical stimulation. A number of electrical healing devices, operating on several different principles, have been used by orthopedic surgeons on hundreds of thousands of patients since the 1970s. Despite success rates for both BMP and electrical healing methods that are typically about 80%, some patients respond better to one versus the other form of treatment.

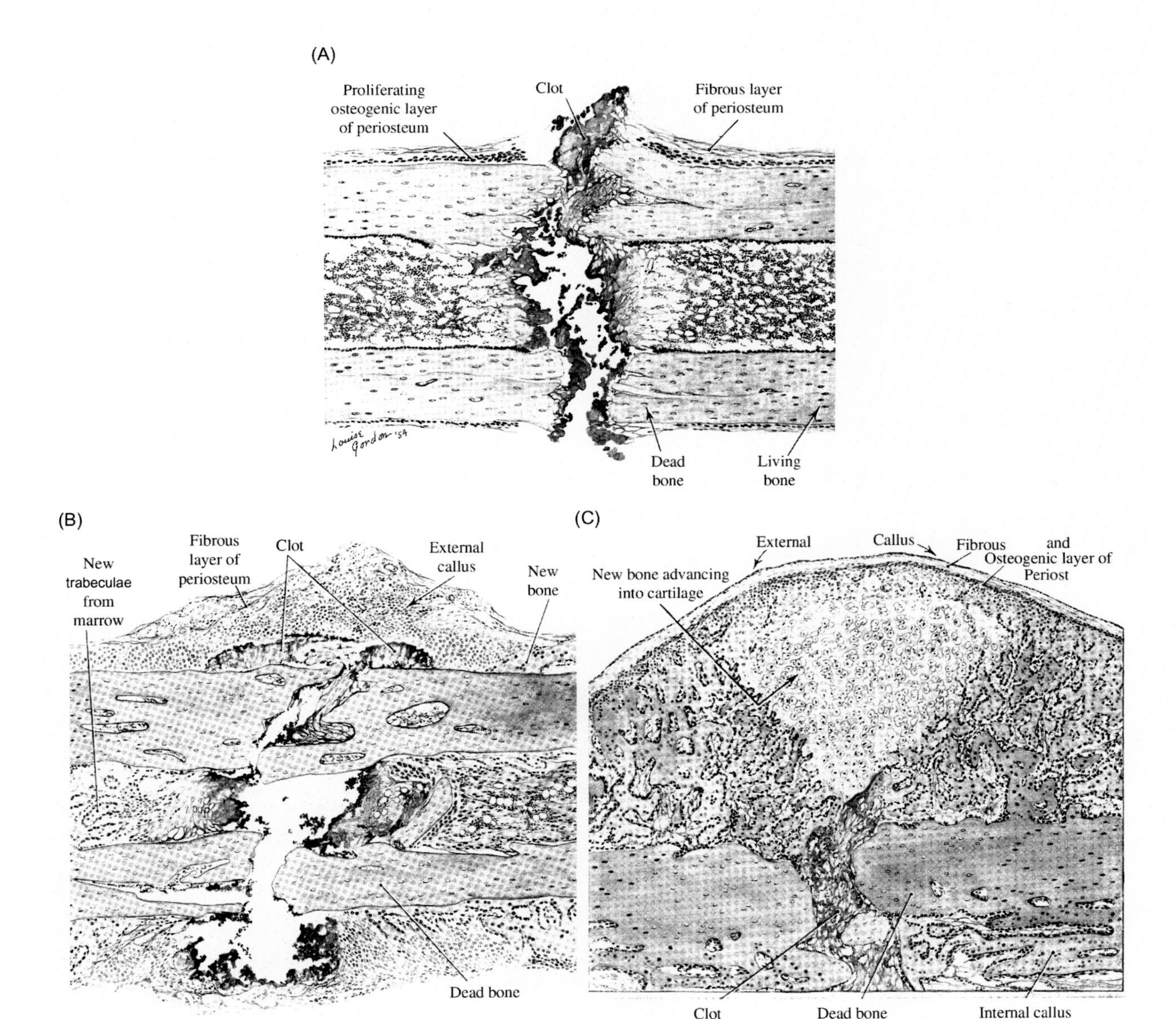

FIGURE 4.18 (A–C) Major stages in the healing of a complete fracture of a rabbit's rib. *From Bourne (1956), with permission.*

An interesting technique for lengthening bones involving principles of normal fracture healing was devised by a Siberian surgeon, Gavriil Ilizarov in the 1950s. Now called **distraction osteogenesis**, it consists of making a transverse cut through a bone and then allowing early healing and callus formation to take place. Then an external apparatus is applied outside the bone that can gradually move the two cut ends of the bone apart at a rate of as much as 1–2 mm/day (Fig. 4.20). This lengthens the callus until the overall length of the bone has reached its desired length. At that point the callus is allowed to harden and heal. This technique has proven to be very effective in the treatment of shortened segments of limbs, and increases in length of over 3 cm have been achieved.

Endocrines and Bone

Bone represents the largest store of calcium in the body, and it has long been known that circulating hormones modulate the transfer of Ca^{++} between bone and the body (see Fig. 9.21). Only much more recently as it been recognized that other aspects of bone are also affected by circulating factors and that bone itself is source of circulating factors that affect other organs in the body. Ironically, some hormonal influences on bone are quite paradoxical; direct effects on bone may stimulate deposition of bone, whereas indirect actions through other organs may have the opposite effect.

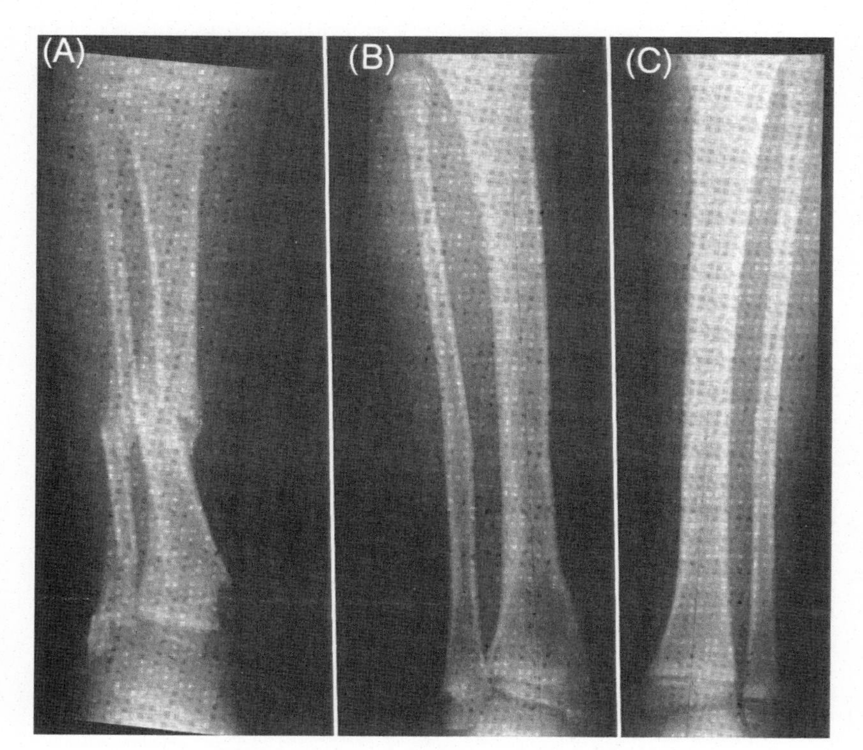

FIGURE 4.19 X-ray images of healing a severe leg fracture in a young patient. (A) Early stages of healing, with callus material filling in the gaps between the broken ends. (B) Two years after injury, the fractures had healed and remodeling had straightened out the contours of both fractured bones. (C) Normal bones in the contralateral leg. *From Frost (1973), with permission.*

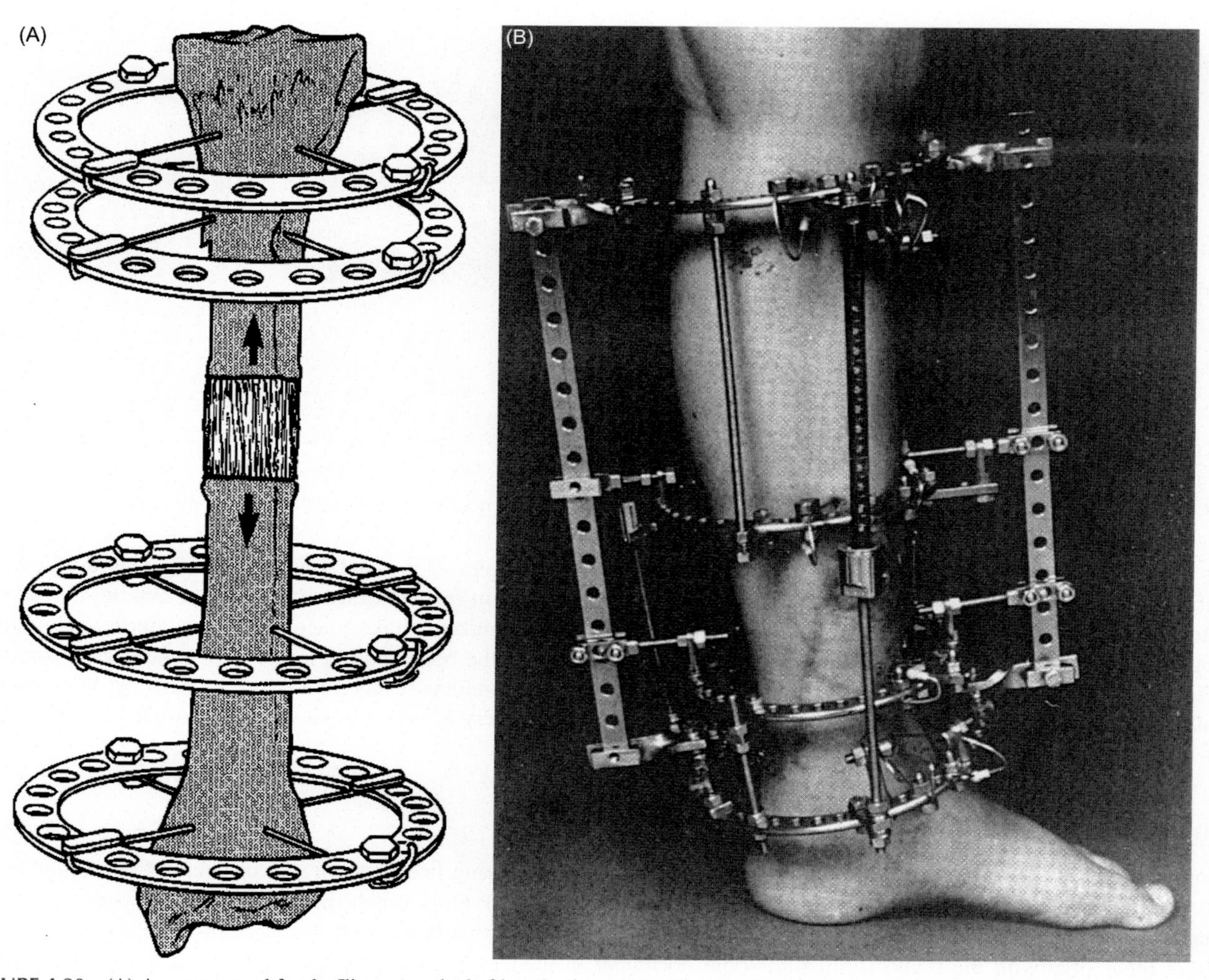

FIGURE 4.20 (A) Apparatus used for the Ilizarov method of lengthening a bone. The circular rings (A) are fixed to the bisected bone, and the metal rods are used to gradually force apart the open ends of the fractured bone. As the ends of the bone are spread apart, callus material fills in the gap and ultimately ossifies, thereby lengthening the bone. (B) The apparatus in place on a patient. *From Browner et al. (2003), with permission.*

The classic hormone affecting Ca^{++} metabolism is **parathyroid hormone**, which increases circulating Ca^{++} levels by promoting the resorption of bone through the indirect stimulation of osteoclastic activity via a direct influence on both osteoblasts and osteoclast precursor cells. These cells then stimulate the resorption of bone by releasing a cytokine [interleukin-6 (IL-6)], which stimulates osteoclastic activity. The counterpart to PTH is **calcitonin**, a hormone secreted by C cells embedded within the thyroid gland. Calcitonin acts directly on osteoclasts to reduce their rate of bone resorption and slow the rate of bone turnover. Through its actions, it reduces the level of Ca^{++} in the blood.

Vitamin D has long been associated with bone because a deficiency of this vitamin results in **rickets**, a softening of the skeleton through a lack of mineralization—in effect a persistence of an osteoid matrix. Administration of vitamin D acts on the small intestine and kidneys to promote the absorption of Ca^{++} and make it available through the blood for mineralization of the osteoid. Yet the direct effect of vitamin D on bone is to mobilize Ca^{++} from the bone matrix, although its indirect effects through the kidneys and intestine overshadow its direct effects on bone itself.

Fat-derived **leptins** (see p. 45) exert a major indirect effect on bone through the central nervous system. Working through the hypothalamus in the brain, they influence the activity of the sympathetic nervous system (see p. 140) and therefore the sympathetic nerve fibers found in bone. Sympathetic signals received by osteoblasts result in the loss of mass in spongy bone, but the maintenance of mass of dense cortical bone and the deposition of subperiosteal bone on the periphery. The net result is a widening of the diameter of long bones, enabling them to support the increased weight caused by the fat. Paradoxically, the effect of leptins acting directly on bone cells is to inhibit bone resorption.

Other hormones also act on bone. Testosterone in males and estrogens (estradiol) in females act to maintain skeletal mass. Estrogen favors bone stability, and the rapid reduction of estradiol levels in women after menopause can result in increased bone resorption and decreased formation of new bone, with **osteoporosis** as the result. Surprisingly, adipose tissue produces some estrogen, so fat in older women may play a role in reducing the risk for osteoporosis.

Bone itself is now known to be a source of a hormone, called **osteocalcin**. Produced by osteoblasts, it not only plays a role in regulation of bone mass, but also it is secreted into the blood. It promotes insulin secretion by the pancreas and increases the sensitivity of muscle and the liver to insulin. Osteocalcin also maintains testosterone secretion by the Leydig cells of the testes, although it does not have a parallel effect on the ovaries.

The Aging Skeleton

The aging of the skeleton is reflected in both the increased number of fractures in older individuals, especially women, and characteristic posture seen in some older people. These are both consequences of significant internal changes in the aging skeleton. As a general rule, the balance between the deposition and removal of bone changes, with removal being the dominant process. This change is particularly abrupt and prominent in postmenopausal women. The decline in estrogen levels removes a significant positive influence for new bone formation, and the imbalance between reduced bone formation and greater bone removal can lead to **osteoporosis**. Males, however, are not exempt from this condition, but the more gradual decline in testosterone levels results in less abrupt changes in the bones. Osteoporosis has a strong genetic component, with over 50% of the risk related to heredity factors.

Spongy bone is especially strongly affected by aging changes, and after age 75 years relatively little spongy bone remains in the skeleton. The reduction in spongy bone results in the extension of the central marrow cavity toward each end of a long bone. This significantly weakens the neck (metaphyseal region) and accounts in large part for the high incidence of fractures through the necks of bones seen in the elderly. The bent back (**kyphosis** or dowager's hump) sometimes seen in elderly individuals is commonly due to accumulated microcompression fractures of vertebrae, which ultimately result in deformation of individual vertebrae. Most aspects of strength of long bones (tensile, compression, torsional, and bending strength) decrease from 15% to 40% with aging. With the weakening of bones, the character of fractures differs between young and old people. Fractures of young bones are often of the greenstick variety, with a significant longitudinal component, whereas fractures of old bones are more often purely transverse.

With the aging of the population, significant effort has been directed toward identifying means of maintaining bone structure and strength. In addition to a number of pharmacological strategies, one of the most effective is physical exercise including some weight bearing. Based on the biology of bone, dynamic exercise, not necessarily of long duration, is a reasonable choice.

JOINTS

Joints are present any time two bones come together. Many types of joints exist, ranging from the bony sutures (**synostoses**) that fuse cranial bones together to synovial joints (**synarthroses**) that are found between the ends of adjacent long bones. Joints that are composed of bone or dense fibrous tissue allow only limited movement, whereas those of the vertebral column, with intervertebral disks intervening between adjacent vertebrae, allow somewhat more movement. Synovial joints are designed to allow the freest movement, but the exact nature of the allowed movements depends upon the gross configuration of the joint. In contrast to the single movements of the interphalangeal joints of the digits, the shoulder joint (glenohumeral joint) permits a multiplicity of movements.

A synovial joint consists of plates of articular cartilage, roughly 2 mm thick, that cap the ends of the two bones. Surrounding the area of bone abutment is a joint capsule, consisting of a synovial membrane that is continuous with, but does not cover the articular cartilages, and a fibrous capsule overlying that (Fig. 4.21). Associated with the joint are a number of other structures, including ligaments that connect bone to bone, tendons that attach muscles to the bones, and bursae, fluid-filled sacs that buffer mechanical influences on the joints. Larger joints also contain cartilaginous articular disks or menisci.

The main function of a joint is to allow movement, and the main function of the tissues immediately surrounding the joint cavity is to make the movement efficient and pain-free over the course of a lifetime. The articular cartilage is hyaline cartilage that is designed to provide a smooth gliding surface and to distribute the weight over as great a surface area as possible. In order to withstand the millions of movements over a lifetime, articular cartilage must be strong and also resistant to wear. Strength is provided by the orientation of the type II collagen fibrils within the matrix. At the surface of the cartilage, the collagen fibers are oriented parallel to the surface, and the chondrocytes are flattened. Farther down from the surface, the fibers are more randomly oriented, and close to the region where the articular cartilage interfaces with actual bone, the collagen fibers are oriented perpendicularly to the joint surface, but parallel with the long axis of the bone.

Critical to maintaining the integrity of the joint cartilages is lubrication of the articular surfaces. This is the function of the **synovial fluid**, which is present in remarkably small amounts (~5–6 mL in the knee joint). Synovial fluid is a complex mixture of components that come from different sources (see Fig. 4.21). A key contributor is the **synovial membrane**, a thin layer of fibrous connective tissue lining those parts of the joint capsule not occupied by joint

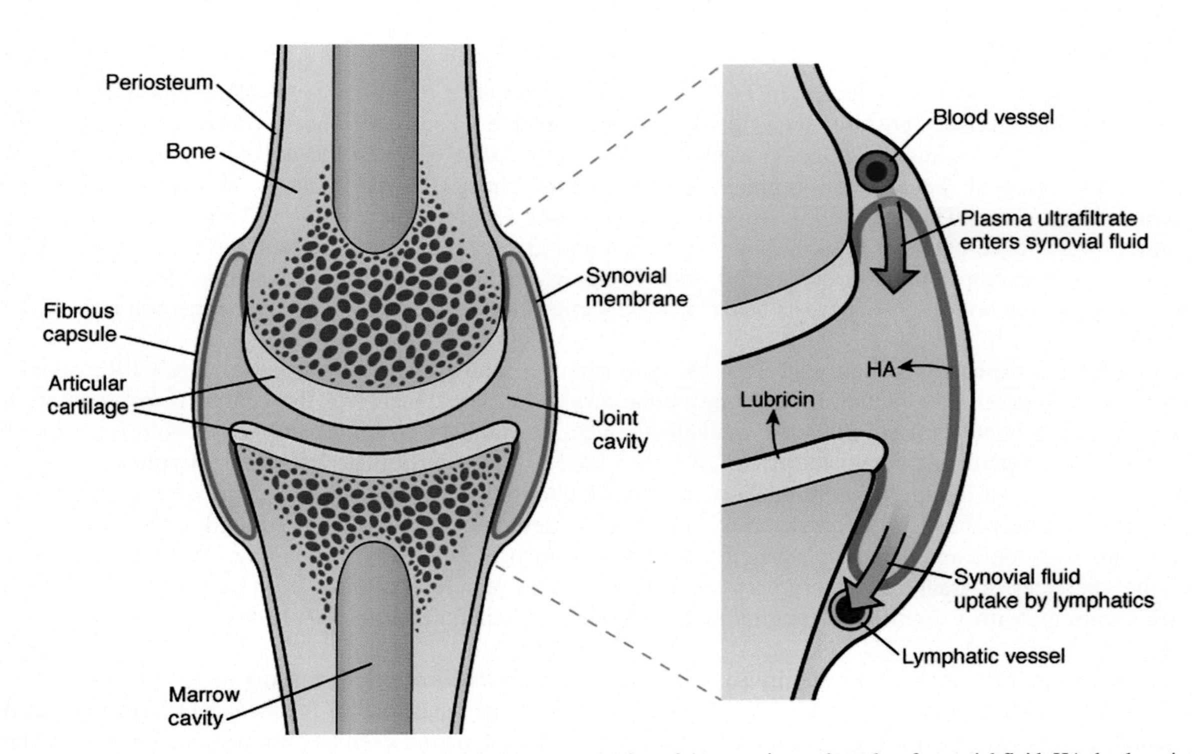

FIGURE 4.21 A simple synovial joint. The inset provides a basic representation of the secretion and uptake of synovial fluid. HA, hyaluronic acid.

cartilage. The main cellular components are fibroblast-like **synoviocytes** and macrophages. The synovial membrane also contains blood and lymphatic vessels and nerve fibers.

Much of synovial fluid is an ultrafiltrate of blood plasma, containing most of the small proteins, for example, albumin, but not the larger blood proteins. In response to stretch, the synoviocytes synthesize and secrete hyaluronic acid, which is a major contributor to the viscosity of synovial fluid. When subjected to dynamic shear stresses, superficial chondrocytes of the articular cartilage secrete a glycoprotein, **lubricin**, which contributes to the slipperiness of the fluid. An additional component, phospholipid molecules, acts as an interface between synovial fluid and the surface of the articular cartilage. Synovial fluid contains few cells, but it contains some enzymes in latent form. Most important are **matrix metalloproteinases**, which if activated, begin to degrade the articular (joint) matrix and can lead to the development of joint pathology. Synovial fluid also contains a mix of growth factors and cytokines. To complete the dynamic equilibrium, synovial fluid leaves the joint cavity via lymphatics within the synovial membrane.

When pressure is placed upon a joint, fluid and wastes are squeezed out of the matrix of the articular cartilage, and at rest, fluids enter. Because hyaline cartilage matrix is not rigid, its shape adapts to some degree to the pressures generated by the movements or load-bearing. The overall shape of the normal joint contributes greatly to distributing pressure and weight. The synovial fluid itself dissipates some of the energy absorbed by the joint.

Osteoarthritis

Osteoarthritis is a condition that affects more than 80% people older than 70 years of age. It begins with degenerative changes in the articular cartilage in response to two important risk factors—abnormal loading on normal cartilage or normal loading on abnormal cartilage. Age alone and heredity factors represent other risk factors.

Whether some time after an acute injury or as a reaction to repeated minor levels of trauma, the nature of the articular cartilage changes. The articular surface softens and its tensile strength decreases. These changes are in response to degradation of the collagen and aggrecan proteoglycan molecules within the matrix subsequent to the production of a variety of cytokines (especially IL-1), which activate enzymes that degrade major matrix components. Cytokines also promote inflammatory responses, especially of the synovial membrane. Inflammation of the synovial membrane stimulates the firing of sensory nerve fibers within the membrane, resulting in the sensation of pain in the joint. Along with changes in the articular cartilage, the underlying bone is affected. The subchondral plate of bone underlying the cartilage becomes thicker and may develop cysts. These changes make the overall mechanical properties of the articular cartilage less favorable. In addition, new bony outgrowths, called **osteophytes**, form along the edges of the affected joint. Ultimately the articular cartilage degenerates to the point where the underlying bone instead of cartilage is present at the articular surface. Much research is currently being directed at devising methods to stimulate the repair of cartilage, but to date progress has been slow.

Rheumatoid arthritis is a much more severe condition than is osteoarthritis. It typically appears earlier in life and most commonly occurs in women. The pathology and mechanisms underlying the disease are quite different from those in osteoarthritis, as well. It is an autoimmune disease that principally attacks the soft tissue of the joint capsule, causing chronic inflammation, most commonly in the joints of the hands and feet. Rheumatoid arthritis has a strong genetic component, which is a risk factor for 50% of the cases. Among environmental factors, smoking is a significant risk factor. Most treatment is directed at reducing the inflammatory response.

SUMMARY

The skeleton begins in the embryo as aggregations of mesenchymal cells that soon begin to produce cartilage models of individual bones. Over time, most of the cartilage is replaced by true bony tissue.

As differentiating mesenchymal cells begin to produce a cartilaginous matrix, one of its most characteristic elements is type II collagen. Cartilage can grow by interstitial growth (the division of chondrocytes already embedded in matrix) or by appositional growth (addition of chondrocytes along the edge).

Bone can be formed by two means in the embryo. One is intramembranous bone formation, by which mesenchymal cells differentiate directly into bone. Intramembranous bone formation occurs mostly in the skull. The other is endochondral bone formation, which consists of bone gradually replacing the original cartilage in a long bone.

The differentiation of bone consists of the transformation of mesenchymal cells into osteoblasts, which produce an organic matrix called osteoid. Once the osteoblasts become fully embedded in the osteoid matrix, they are called osteocytes. Under the influence of bone morphogenetic protein (BMP), the osteoid becomes calcified to form a true bone matrix. Even in mature bone, osteocytes are connected with one another through long cellular processes that occupy canaliculi within the bone matrix.

Bone is constantly being remodeled. In addition to new bone formation by osteoblasts, bone is removed by multinucleated osteoclasts. Osteoclasts, which are derived from the blood, adhere to bone matrix and then secrete enzymes that dissolve the matrix.

Long bones of the limbs develop through endochondral ossification. At some point, the cartilage in an embryonic long bone becomes calcified, and blood vessels penetrate the calcified cartilage. Bone-forming cells, entering along with the blood vessels, begin to lay down layers of bone matrix onto the remnants of the calcified cartilage. Over time, the conversion of cartilage to bone progresses toward the ends of the bone until finally only a thin disk of cartilage remains. This is called the epiphyseal plate and represents the mechanism that produces longitudinal growth of the bone. Bones grow in diameter as cells of the periosteum become converted into osteoblasts, which lay down new layers of bone on the outside of the bone.

The shafts of long bones consist of compact bone, which consists of many cylindrical osteones. At the center of an osteone is a Haversian canal, which is occupied by nerve fibers and a blood vessel. Compact bone responds to mechanical stress, and the osteones undergo remodeling in response to the stresses. Old osteones are removed, and new ones, corresponding to new lines of stress, form. The center of a long bone is occupied by fatty (yellow) marrow. The ends of a long bone consist of spongy bone containing red (blood-forming) marrow.

After a fracture, bones heal through a mechanism that forms a callus of bone-forming cells at the site of the fracture. Over time, the new bone at the site of the fracture becomes remodeled in a manner corresponding to the mechanical environment of that bone.

Bone is the largest store of calcium in the body. Parathyroid hormone (PTH) promotes resorption of bone, whereas calcitonin reduces the rate of bone resorption. Vitamin D is necessary for healthy bone formation. Other hormones, such as testosterone and estrogens, also affect bone, especially in growing individuals.

During aging, a decline in estrogen in women can result in osteoporosis due to a shift in the balance between bone deposition and resorption. Over time, there is a decline in the amount of spongy bone in the skeleton.

Joints are present any time two bones come together. Synostoses, seen in the skull, fuse bones together. Synarthroses (synovial joints) are the moveable joints found between the ends of two long bones. Synovial joints are lined by a synovial membrane that produces the lubricating synovial fluid.

Chapter 5

The Muscular System

The basic mechanical function of a muscle is to contract and generate force so that something can be moved. This simple sentence covers an amazing dynamic range. The essence of muscle contraction takes place at the molecular level, where two types of filamentous protein molecules slide past one another. This action is then almost instantaneously translated up layers of complexity from the molecule, to the cell, to the tissue, to the organ, to the body part, and finally to the body as a whole.

Probably nowhere is the overall integrative function of the body better displayed than the contraction of a muscle. Starting with a thought in the brain, the signal to contract is sent via peripheral motor nerves to usually more than one muscle. Actual contraction of the muscle(s) then acts through connective tissue to pull on a tendon, but often layers of fascia, as well. The principal recipient of the muscular pull is one or more bones, which then move some part of the body. While (and after) this is occurring, the blood vessels within the muscle react, usually by increasing their blood flow. The heart follows suit by increasing its beat in parallel to the increased rate of breathing to accommodate the body's need for more oxygen. At an unseen level, carbohydrate metabolism within both the muscle and other areas of the body increases in order to supply energy. These are just some of the responses of the body to a seemingly simple action stimulated by a thought.

In this chapter, the story of muscle begins with a brief structural dissection of a muscle from the gross to the molecular level. It then transitions to an analysis of muscle function beginning with molecular events and continuing up the levels of organization to whole body movements. This background of normal muscle structure and function is then used as the basis for a discussion of muscle growth, adaptation, repair, and aging.

THE OVERALL ORGANIZATION OF A MUSCLE

Most of the more than 600 muscles in the human body are built along the same lines. They have tendinous origins and insertions that attach to bones at either end and between them a belly of thousands of individual muscle fibers (Fig. 5.1). An important, but often neglected component of a muscle is the sheets of connective tissue that line various parts of the muscle. The outside of the entire muscle is covered with a thin translucent sheet of connective tissue called the **epimysium**. Within the substance of the muscle are a number of compartments of muscle fibers held together by a thinner sheet of connective tissue called the **perimysium**. At the cellular level, each muscle fiber is enveloped by a wispy **endomysium**. Each of these connective tissue layers is interconnected to the next, and the entire muscle has connections with both its tendons, as well as neighboring fascial sheaths.

Muscles have a major and sometimes several minor territories where a substantial artery and nerve branch enter the muscle and a major vein exits. Within the muscle, the larger blood vessels branch repeatedly until each muscle fiber is surrounded by a network of capillaries. The nerve supplying the muscle sends individual nerve fibers or their terminal branches to each muscle fiber. Parallel to the blood vasculature is a network of lymphatic vessels that arise as blind pouches close to the terminal venous branches. Like the veins, smaller lymphatic vessels collect into larger ones that are situated alongside the arterial branches. Because the small lymphatic vessels do not have muscular walls, flow of lymph relies upon pulsations of the arteries and contractions of the muscle itself to propel lymph flow.

Muscle fibers themselves are elongated multinuclear structures that arise in the embryo from the fusion of many mononucleated cells called **myoblasts** (see Fig. 2.23). In normal muscle, the nuclei (**myonuclei**) are located at the periphery of the muscle fiber. Each muscle fiber is surrounded by a basal lamina that is connected to the endomysium. Between the muscle fiber and its basal lamina are scattered mononuclear **satellite cells** (see Fig. 2.24) that serve as stem cells for muscle growth, adaptation, and regeneration. Human muscle fibers range from a few mm to several cm in length.

Skeletal muscle is also called cross-striated muscle because at the light microscopic level a muscle fiber is seen to have alternating light- and dark-staining bands that run perpendicular to the length of the fiber (Fig. 5.2). Closer examination at a somewhat higher power shows a very narrow dark-staining band running through each light-staining band.

The Human Body. DOI: https://doi.org/10.1016/B978-0-12-804254-0.00005-3

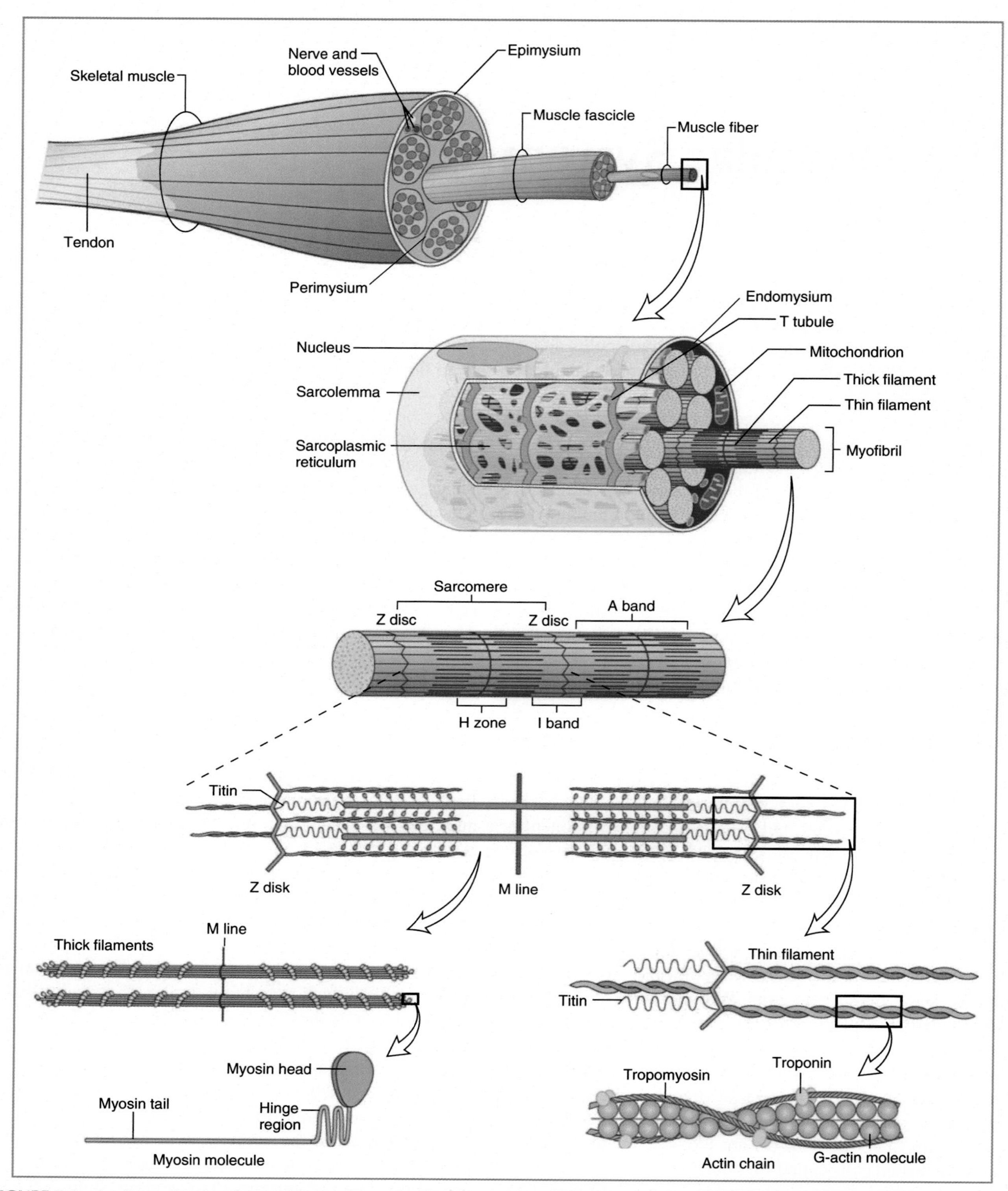

FIGURE 5.1 A scheme showing the breakdown of an entire skeletal muscle into successively smaller units, ending with the actual contractile proteins. *From Carroll (2007), with permission.*

These thin lines are called Z-bands. A segment bounded by two Z-bands is called a **sarcomere**, and it represents the fundamental contractile unit of muscle (see Fig. 5.2). Further dissection of a sarcomere, which is 2.3 μm in length, can only be done with the aid of an electron microscope. When seen in cross section, a muscle fiber is seen to consist of as many as a thousand **myofibrils**—bundles of contractile proteins—arranged in parallel. The bundles of myofibrils connect to specialized Z-line proteins at either end of a sarcomere (Fig. 5.3).

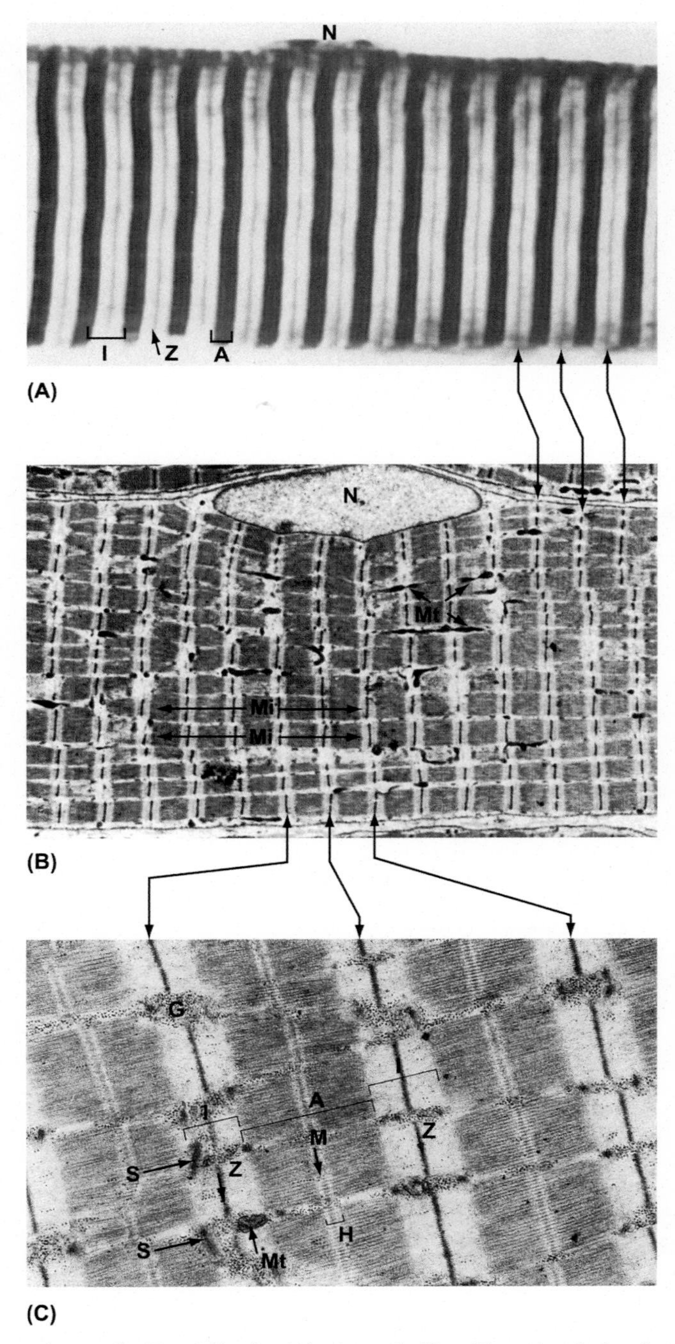

FIGURE 5.2 Light (A) and electron micrographs (B and C) of a skeletal muscle fiber, illustrating the banding patterns (A, H, I, M, Z) that reflect the high degree of organization of the contractile proteins. Two Z-bands are the boundaries of a sarcomere, the fundamental contractile unit of a muscle fiber. *G*, Glycogen deposits; *N*, nucleus; *Mt*, mitochondria; *S*, sarcoplasmic reticulum. *From Young et al. (2006), with permission.*

The most prominent visible components of a sarcomere are thick filaments alternating with thin filaments (see Fig. 5.3). **Thick filaments** are aggregates of pipe-shaped **myosin** proteins, with the bowls of the pipes (the myosin heads) protruding from the bundle. These protrusions are of vital importance in generating the contraction of a muscle fiber (see p. 118). **Thin filaments** consist of two intertwined chains of filamentous **actin** molecules. As is true for the actin cytoskeleton of other cells (see p. 17), filamentous actin is a polymer of G-actin molecules. Situated in the parallel grooves between the actin filaments are elongated chains of **tropomyosin**, a regulator of muscle contraction (Fig. 5.4). Attached to these chains at periodic intervals are globular aggregates of **troponin** molecules, which play a primary regulatory role.

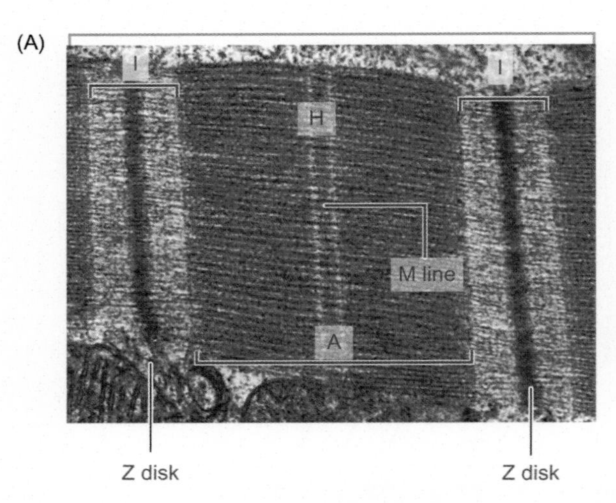

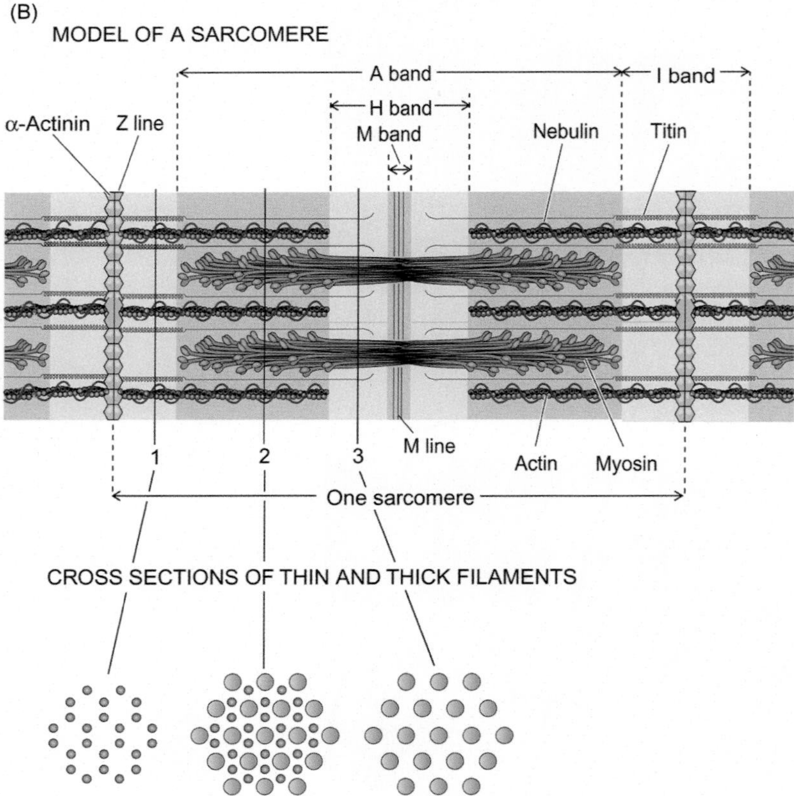

FIGURE 5.3 The structure of a sarcomere. (A) Electron micrograph. (B) Molecular model of a sarcomere. *(A) From Kierszenbaum and Tres (2012), with permission. (B) From Boron and Boulpaep (2012), with permission.*

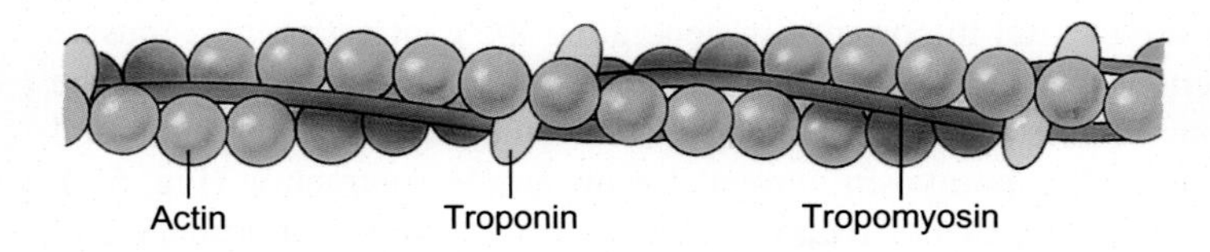

FIGURE 5.4 Structure of a thin filament, showing the relationship between tropomyosin and troponin to the actin chains. *From Thibodeau and Patton (2007), with permission.*

Within the length of a sarcomere, the thick filaments occupy a central position, with thin filaments interdigitated partway at either end of the band of thick filaments. Both thick and thin filaments show a high degree of registration, which is why the cross-striations are so regular and highly visible. The Z-line is the area where thin actin filaments from adjacent sarcomeres come together in a zigzag fashion in conjunction with another large protein called **α-actinin** (see Fig. 5.3). This molecular linkage allows the contractions of individual sarcomeres to occur in a continuous fashion. In addition neighboring sarcomeres are linked side-by-side through lateral connections at the Z-lines. Such linkage keeps the sarcomeres in register.

Surrounding the myofibrils is an irregular network of smooth endoplasmic reticulum, called the **sarcoplasmic reticulum** in muscle (Fig. 5.5). The sarcoplasmic reticulum stores Ca^{++}, which plays a critical role in the initiation of muscle contraction. Scattered through the interspaces among the myofibrils are rows of mitochondria, which provide much of the energy used in muscle contraction. As will be seen later, the density of mitochondria corresponds closely with the functional nature of the muscle fiber. A final membranous structure of great functional importance is the **T-tubule**

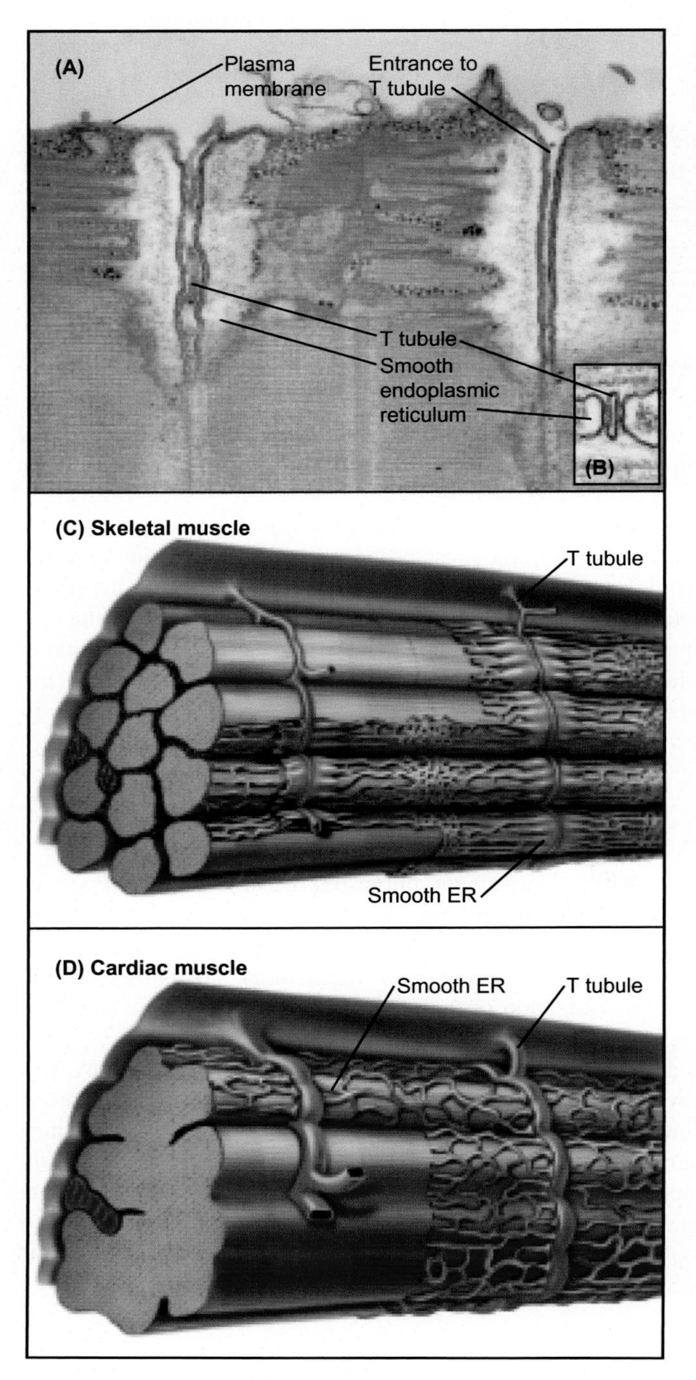

FIGURE 5.5 Membrane specializations of striated muscle fibers. (A) Electron micrograph of T-tubules penetrating into the muscle fiber from the plasma membrane (sarcolemma). (B) The close association between T-tubules and blind endings (cysternae) of the smooth endoplasmic reticulum (sarcoplasmic reticulum) is called a triad and is the site at which contractile stimulus carried by the action potential on the plasma membrane is transferred to the interior of the muscle fiber. (C and D) Drawings of the three-dimensional arrangement of the T-tubules and sarcoplasmic reticulum in skeletal and cardiac muscle fibers. *From Pollard and Earnshaw (2004), with permission.*

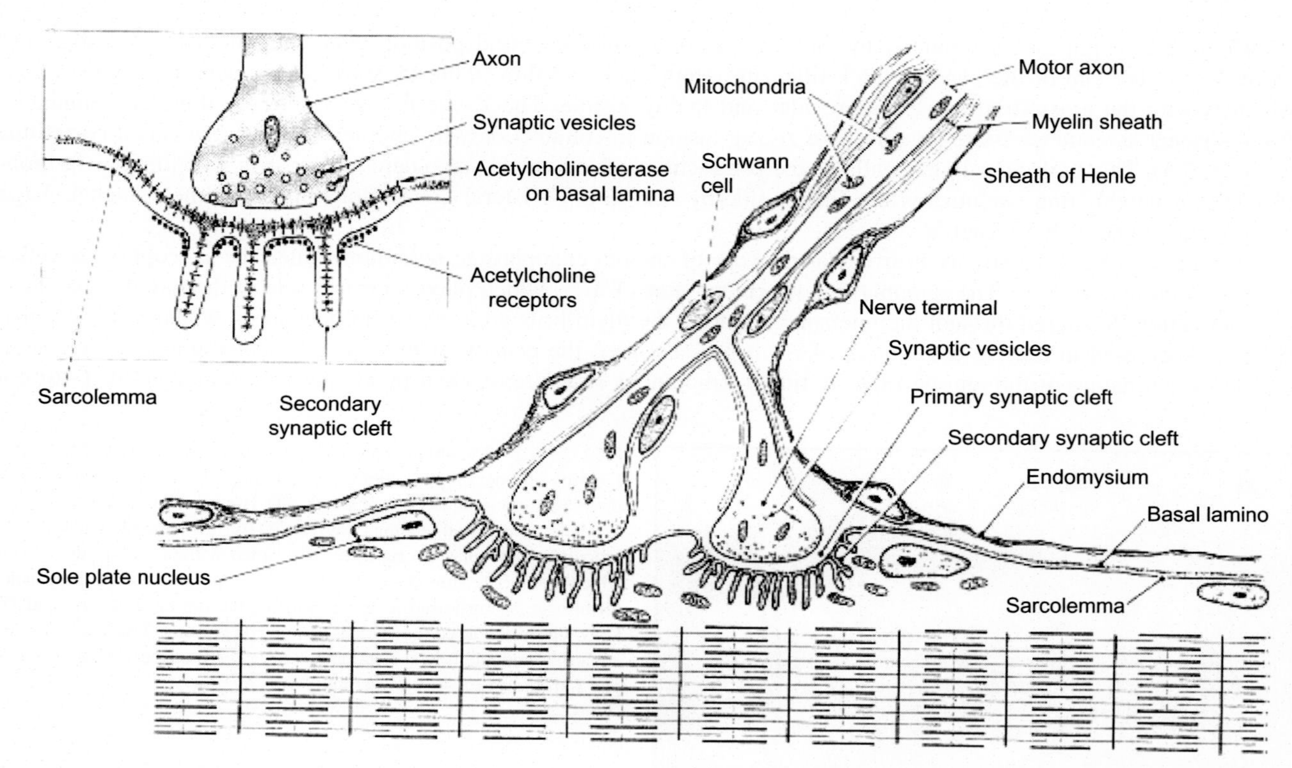

FIGURE 5.6 Drawing of a neuromuscular junction. The inset shows details of the nerve terminal where actual transmission of the neural stimulus occurs.

(transverse tubule), which is an extension of the plasma membrane and dives deep within the muscle fiber at regular intervals along the sarcomeres. Terminal channels of sarcoplasmic reticulum, called **cysternae**, run parallel to each side of the T-tubules to form triads (see Fig. 5.5).

A final single structure present on every skeletal muscle fiber is the **neuromuscular junction** (Fig. 5.6). A modified synapse, this is the spot where a motor nerve terminal makes contact with the muscle fiber and where the action potential of the nerve becomes transferred to the muscle fiber as a stimulus to contract. Within the rounded boutons on the tips of the branches of the nerve terminal are many synaptic vesicles containing the neurotransmitter **acetylcholine**. At the synaptic cleft, the nerve terminal and the muscle fiber are separated by the basal lamina of the muscle fiber. The region of the muscle fiber underlying the nerve terminal is a specialized region sometimes called the motor end plate. In this region, the plasma membrane of the muscle fiber (**sarcolemma**) is thrown up into a number of folds, which increase its surface area. Embedded in the postsynaptic folds of the sarcolemma are large numbers of acetylcholine receptors.

EVOLUTION OF MUSCLE

One of the chief characteristics of animals is movement, and without movement, either internal or external, almost no animal would be able to survive. It is likely that the common unicellular ancestor of all animals already possessed the fundamental molecular units required for movement. A common core of contractile proteins, consisting of actin, myosin, and some associated regulatory proteins, such as tropomyosin, is conserved in all **metazoans**. The ability of individual cells in the human body (e.g., macrophages) to move is a reflection of the persistence of these ancient means of producing movement.

Some of the most primitive animals, such as sponges, do not have identifiable striated muscle fibers, but they, too, possess contractile proteins in cells that regulate water flow through the pores in their body wall. Jellyfish and some of their close relatives, which diverged from the main line of metazoan evolution hundreds of millions of years ago, contain both skeletal and smooth muscle fibers. Although the presence of muscle in jellyfish has suggested a common origin for all muscle cells in the animal kingdom, molecular studies have shown that muscle cells have likely evolved independently a number of times. Although striated muscle fibers may look similar under a microscope, there is considerable variation in how the components of the Z-line and the molecular elements that regulate the interactions between actin and myosin are organized. Nevertheless, among the vertebrates a common basis for muscle contraction is present.

MUSCLE CONTRACTION

The Stimulus to Contract

The direct stimulus for muscle contraction begins with the motor nerve. As the action potential reaches the nerve terminal, voltage-gated Ca^{++} channels open, and Ca^{++} enters the **boutons** (rounded ends) of the nerve terminals (Fig. 5.7). The increase in concentration of Ca^{++} within the boutons stimulates the synaptic vesicles to fuse with the plasma membrane and to expel their content of **acetylcholine** into the synaptic space. The acetylcholine then diffuses across the synaptic space and binds to the acetylcholine receptors located on the folds of the postsynaptic membrane of the muscle fiber. The **acetylcholine receptor** is a complex of five proteins that together form an ion channel. When two molecules of acetylcholine bind to the receptor complex, the channel opens and both Na^+ and K^+ pass through the channel into the muscle fiber. In the meantime, any remaining unbound acetylcholine molecules in the synaptic cleft are quickly destroyed by **acetylcholinesterase** that had been bound to the region of the basal lamina within the synaptic cleft. Such a cleanup action is important to allow simple, rapid, and repeated contractile stimuli to reach the muscle fiber.

With the passage of Na^+ and K^+ through the acetylcholine receptor channels, the stimulus to contract has passed from the nerve to the muscle fiber. When the influx of these ions in the motor end plate region raises the local membrane potential past a critical point, it triggers a second action potential—this time along the surface of the muscle fiber.

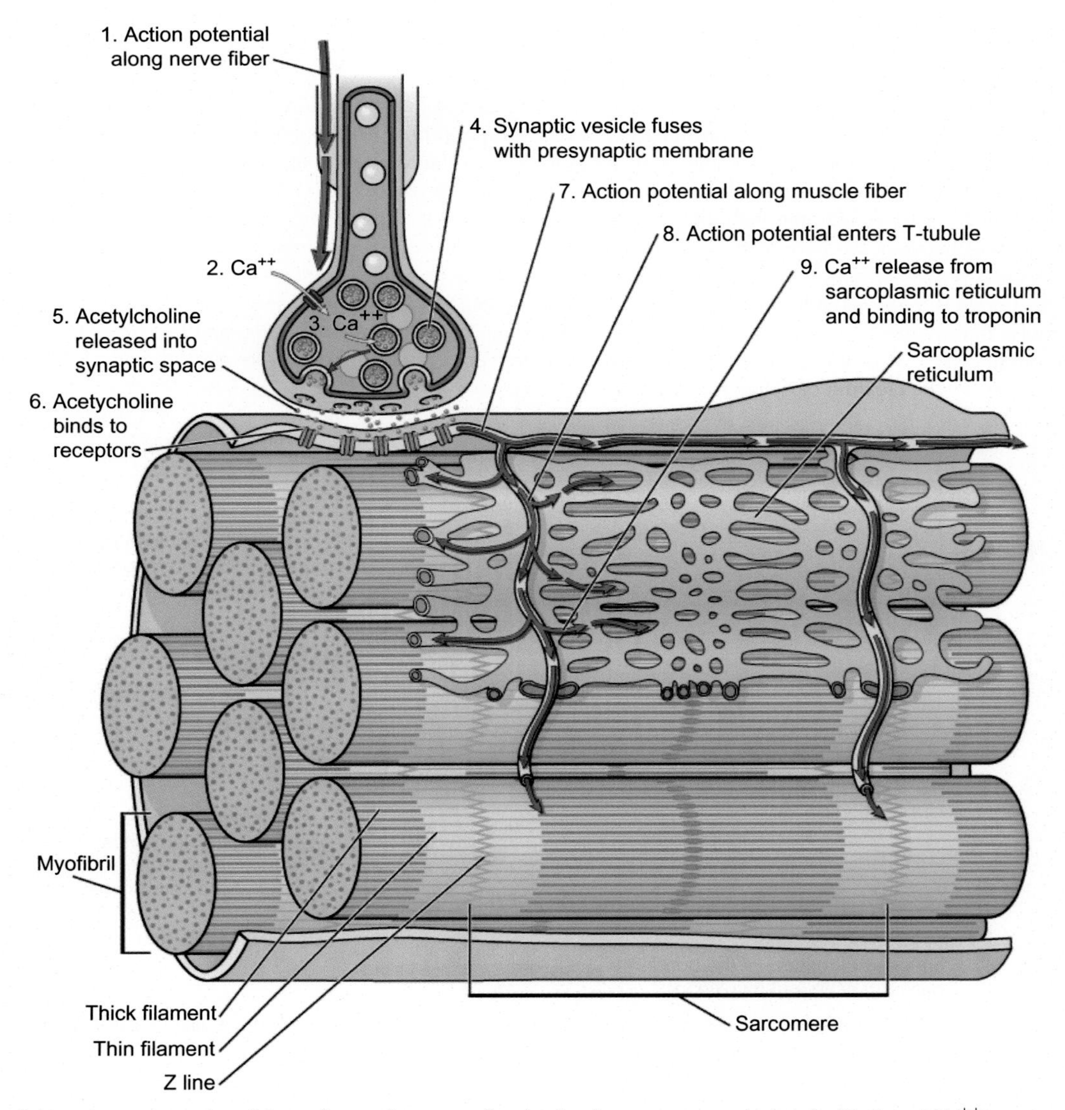

FIGURE 5.7 Schematic representation of the pathway of a contractile stimulus from a nerve terminal to the binding of Ca^{++} to troponin on the thin filaments.

Muscle fibers are very large cells, both in length and in diameter. Therefore the rapid transmission of the contractile stimulus in the form of an action potential is necessary. The function of the T-tubules (see Fig. 5.5) is to allow the action potential to spread quickly into the interior of the muscle fiber, as well as along its length.

At the interior of the T-tubule system, the stimulus to contract is translated from an electric to a chemical one in a process called **excitation–contraction coupling**. The action potential that passes along the membrane of the T-tubule opens Ca^{++} channels that are connected to Ca^{++} channels within the membranes of the cysternae of the sarcoplasmic reticulum (see Fig. 5.7). This opens Ca^{++} release channels in the sarcoplasmic reticulum with the subsequent release of Ca^{++} into the cytoplasm of the muscle fiber. There it binds to troponin, and the next phase of muscle contraction begins.

The Molecular Basis of Muscle Contraction

The molecular essence of muscle contraction is the sliding of thin filaments past thick filaments, resulting in the shortening of the muscle fiber. This is accomplished through a phenomenon called **cross-bridging**—the formation of temporary molecular connections between the thick and thin filaments—and a molecular mechanism that shortens the sarcomeres by a rowing-like mechanism that propels the thin actin filaments along the thick myosin filaments. Individual cross-bridges are formed and then broken as a myosin head connects with the next available binding site along the thin filament, thus pulling the thin filament along the length of the thick filament (for details, see Box 5.1).

BOX 5.1 The Molecular Basis of Muscle Contraction

The first stage in muscle contraction is the formation of the cross-bridges, and it requires a deeper understanding of how thick and thin filaments are put together. The protruding heads of the myosin proteins that make up the thick filaments contain two important binding sites—one for actin and the other for ATP. Within the thin filaments, the tropomyosin chains (see Fig. 5.4) block the binding sites for myosin that are located on the actin chains. The binding of Ca^{++} to one of the molecules in the troponin complex initiates a series of conformational events that results in the shifting of the tropomyosin chains to a position deeper within the grooves between the two intertwined actin chains. This shift exposes the myosin-binding sites on the actin and sets the stage for cross-bridge formation between thick and thin filaments (actin and myosin).

The cross-bridge is a reversible bond between the head of a myosin molecule and a binding site on the actin filament. Actual movement of a thin filament along a thick filament is accomplished by the repeated binding and release of myosin heads from their actin binding sites. This process is energy intensive and involves a power stroke by the myosin heads, followed by release and recovery, before the creation of a new binding event and repetition of the sequence (Fig. 5.8).

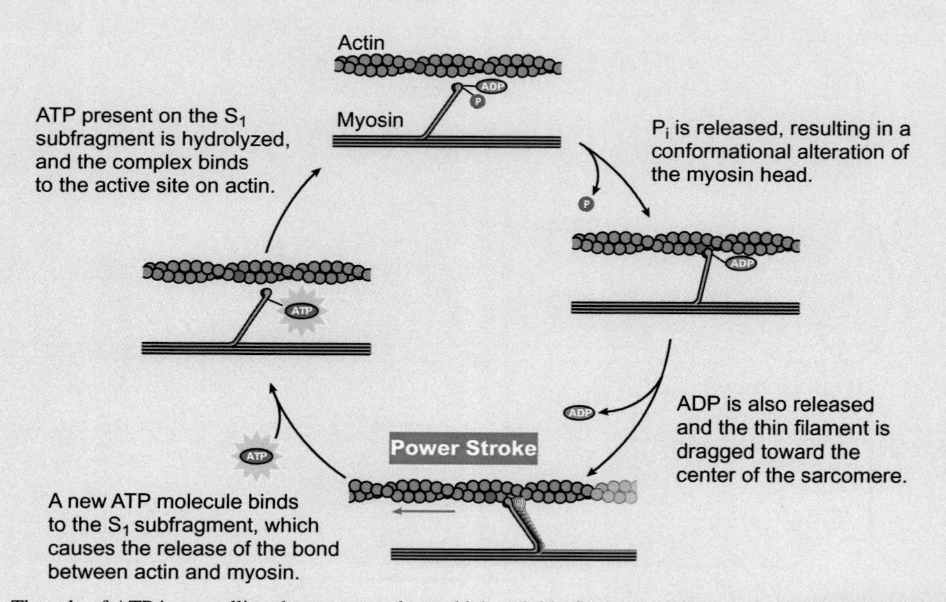

FIGURE 5.8 The role of ATP in propelling the power stroke as thick and thin filaments slide past one another during muscle fiber contraction. *From Gartner, and Hiatt, (2011), with permission.*

(Continued)

BOX 5.1 (Continued)

The cross-bridge cycle begins when a myosin head (which protrudes 90 degrees from the long axis of the thick filament) with bound ADP plus an inorganic phosphate group (P_i) attaches to a myosin-binding site on the actin filament. Release of P_i from the complex triggers the power stroke, during which the myosin head bends from 90 to 45 degrees. The bending of the myosin heads moves the actin filament 10–12 nm along the myosin toward the center of the shortening sarcomere, and this is the action that generates the force in muscle contraction (Fig. 5.9). ADP is then dissociated from the myosin molecule, and the myosin head (the cross-bridge) remains firmly attached to actin. This condition persists until a new molecule of ATP binds to the myosin head, causing it to be released from the actin filament.[1] Following the release of the myosin head from actin, hydrolysis of the ATP to ADP + P_i causes the myosin head to become re-cocked from the 45- to the 90-degree position, and the cross-bridge cycle begins again. During the re-cocking phase, the cross-bridges are not in contact with the actin filament, and the muscle is in the relaxed state.

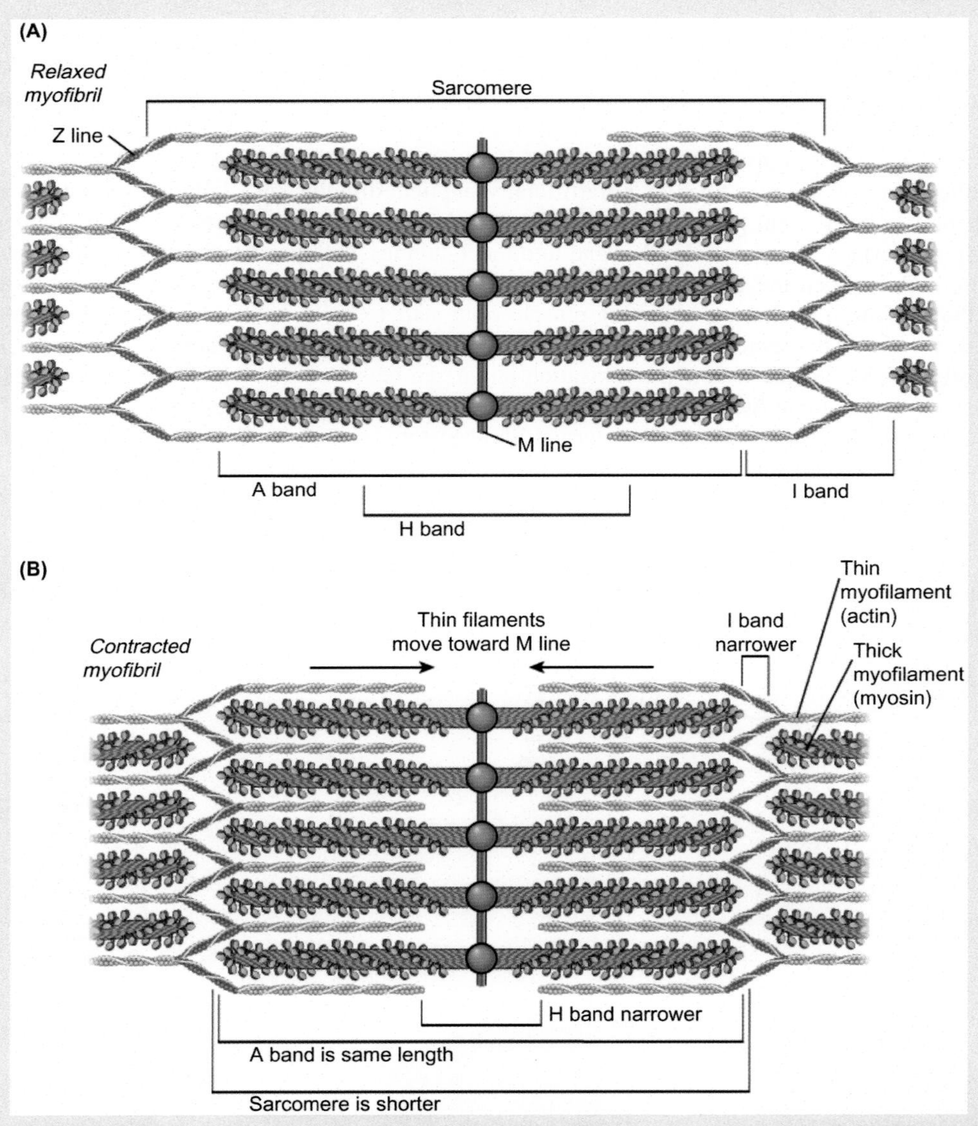

FIGURE 5.9 The sliding filament model of muscle contraction. (A) Relationship between thick and thin filaments in a relaxed muscle fiber. (B) The same, in a contracted muscle fiber.

Muscle contraction is highly energy consuming. Each cycle of cross-bridge formation and dissociation requires one molecule of ATP per myosin head. In order to maintain the state of contraction, the supply of ATP must be rapidly replenished. One mechanism is the transfer of a high-energy P_i group from the molecule **phosphocreatine** to ADP through the actions of an enzyme creatine kinase, thereby generating a new ATP molecule. Another more abundant energy source within muscle is stored **glycogen,** a carbohydrate polymer. During its enzymatic breakdown, additional ATP molecules are created.

1. In the absence of ATP, the cross-bridges remain firmly in place and the muscle stiffens. The condition of *rigor mortis* after death is due to the exhaustion of ATP and the continued maintenance of myosin cross-bridges tightly bound to actin.

As long as stimulation from the motor nerve causes action potentials to sweep across the sarcolemma of the muscle fiber, contraction will continue, but upon cessation of that stimulus, contraction stops. In order for contraction to cease and for relaxation to follow, Ca^{++} must be removed from the sarcoplasm. This is accomplished by a calcium pump that transports Ca^{++} into the sarcoplasmic reticulum, where it is bound to a molecule appropriately named **calsequestrin**. Some Ca^{++} is also removed from the muscle fiber through calcium pumps in the sarcolemma, but this contributes in only a small way to removal of calcium. If this mechanism played a larger role, too much Ca^{++} would be removed from the muscle fiber, leaving not enough to activate a complete contractile response.

The Cellular Basis of Muscle Contraction

The net result of the molecular sliding described earlier is contraction of the muscle fiber and generation of force. In aggregate, the contractions of many individual muscle fibers cause the muscle as a whole to contract. Many of the fundamental principles and phenomena of whole muscle contraction can be seen at the level of a single muscle fiber.

One of the simplest reactions of a muscle fiber to a brief stimulus is a **twitch**. The tension generated in a simple twitch follows a characteristic curve, which has three components (Fig. 5.10). The time between the initial stimulus, whether it be the firing of a nerve fiber or an artificially applied stimulus in a laboratory, and the first measurable increase in tension is a brief latent period of a few milliseconds. This period represents the time required for the stimulus to pass along the surface of the muscle fiber and into the T-tubules, and then for the release of Ca^{++} from the sarcoplasmic reticulum. Only when the Ca^{++} frees the myosin-binding sites on the actin filaments and cross-bridges begin to form, does the tension of the now contracting muscle fiber start to build up. The contraction phase, represented by increasing tension, reflects the increasing number of cross-bridges being formed. Then as the number of cross-bridges starts to diminish, the muscle fiber enters the relaxation phase and tension is reduced to resting levels. Depending upon the type of muscle fiber, a single twitch is completed in ~ 100 ms.

Simple twitches are relatively uncommon in normal life. Most muscles are called upon to undergo sustained contractions, which physiologists call tetanus or **tetanic contractions** (see Fig. 5.10). A tetanic contraction is the result of repeated stimuli at such short intervals that the muscle fiber (or muscle) doesn't have time to fully relax before it is called upon to contract again. Regular contractions of muscle are tetanic in nature.

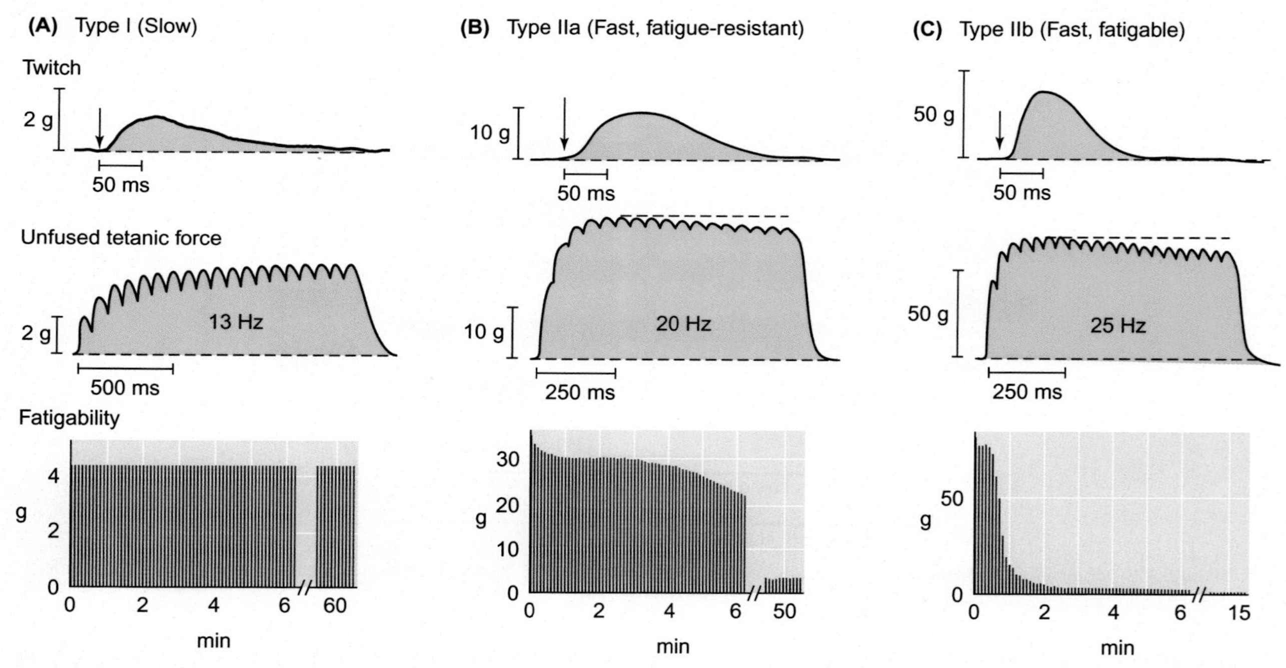

FIGURE 5.10 Contractile characteristics of three main types of skeletal muscle fibers. *X* axis—time; *Y* axis—contractile force. (A) Slow (type I); (B) fast, fatigue resistant (type IIa); (C) fast, fatigable (type IIb). Top row—characteristics of a single twitch. Slow muscle fibers generate much less tension, but over a longer period than do the other two types of fibers. Middle row—slow muscle fibers take longer to fuse individual twitches into a tetanic contraction, which is also relatively weak compared to that of fast muscle fibers. Bottom row—fatigability. Although relatively weak, slow muscles can contract at a maximal level for prolonged periods of time, whereas the strong contractions of fast muscle fibers fade quickly because of fatigue. *From Boron and Boulpaep (2012), with permission.*

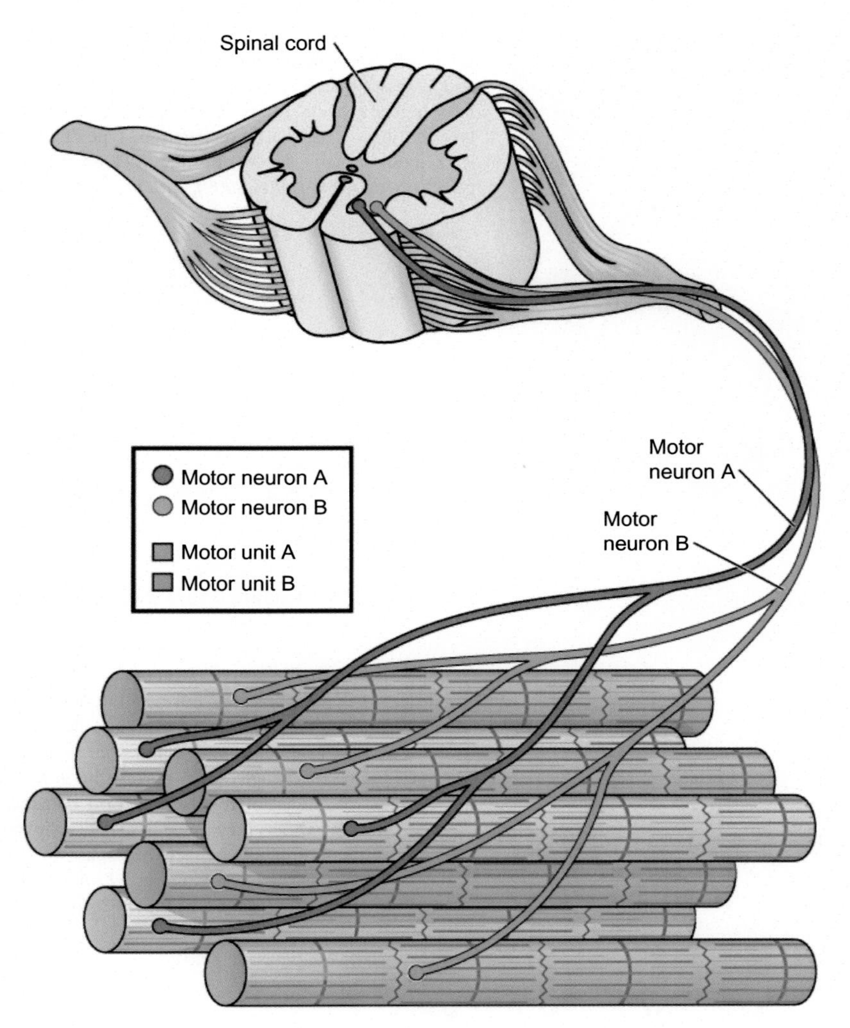

FIGURE 5.11 Diagrammatic representation of two motor units in a skeletal muscle.

Almost never does a single muscle fiber contract in the living body. A single motor nerve fiber sends branches to a number of muscle fibers, and when that nerve fiber fires, all of the muscle fibers to which it is connected contract simultaneously. The combination of a motor nerve fiber and the muscle fibers that it supplies is called a **motor unit** (Fig. 5.11), and within a muscle a motor unit represents the fundamental contractile unit.

Motor units are of different sizes, and their size reflects the function of the muscle. Muscles for which fine motor control is important have small motor units, whereas muscles that generate large amounts of power have large ones. For example, motor units in small extraocular muscles may consist of only four muscle fibers connected to the same nerve fiber. On the other hand, single motor units in some of the larger leg muscles may include over 1000 muscle fibers. The muscle fibers within motor units do not occupy adjacent spaces. Rather, the muscle fibers of various motor units are highly interdigitated. As many as 50 motor units can contribute to a block of 100 muscle fibers.

Not all motor units are of the same functional type. Based on their speed of contraction and susceptibility to fatigue, three different varieties have been characterized. These varieties correspond to the functional type of muscle fibers within the respective motor units. All muscle fibers within a single motor unit are of the same functional type.

Slow motor units (type I) and the muscle fibers that comprise them contract relatively slowly, do not generate a high degree of force, and are highly resistant to fatigue. Correlated with this is the frequency of firing of the motor nerve fibers supplying these muscle fibers (10–50 Hz). In comparison with fast muscle fibers (see later), **type I (slow twitch) muscle fibers** (Fig. 5.12) have a smaller cross-sectional area, contain more mitochondria, and are able to sustain their function by gaining energy through the aerobic metabolism of sugars and lipids, largely because the capillary networks surrounding the muscle fibers are quite dense. Their speed of contraction is correlated with the type of myosin that they contain. Type I muscle fibers contain the oxygen-binding protein **myoglobin**, which gives the muscle fibers a reddish color and is the basis for the dark color of slow muscles (e.g., leg muscles in chickens).

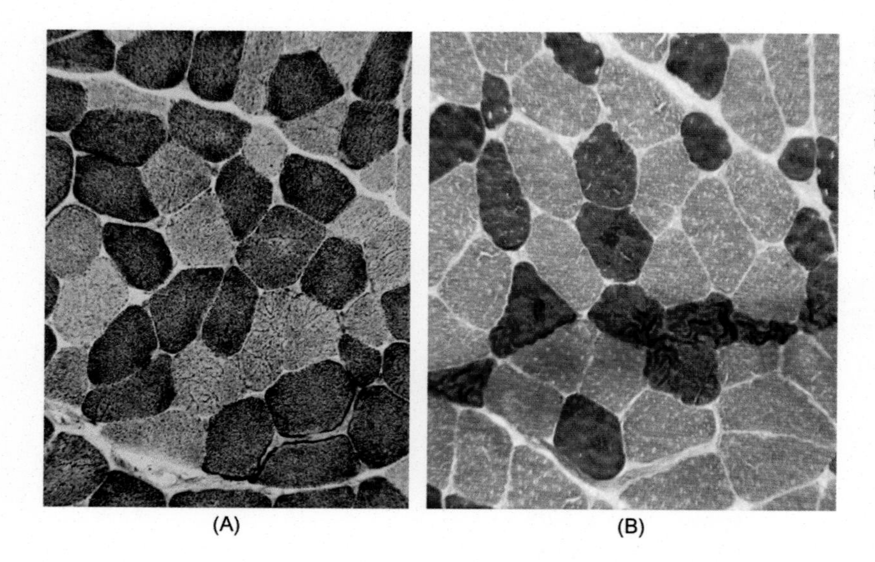

FIGURE 5.12 Histochemical preparations of skeletal muscle, showing differences between type I (slow) and II (fast) muscle fibers. (A) Type I fibers (dark) have higher concentrations of oxidative enzymes than type II fibers. (B) In this preparation, type II fibers stain more strongly for myofibrillar ATPase than do type I fibers. *From Nolte (2009), with permission.*

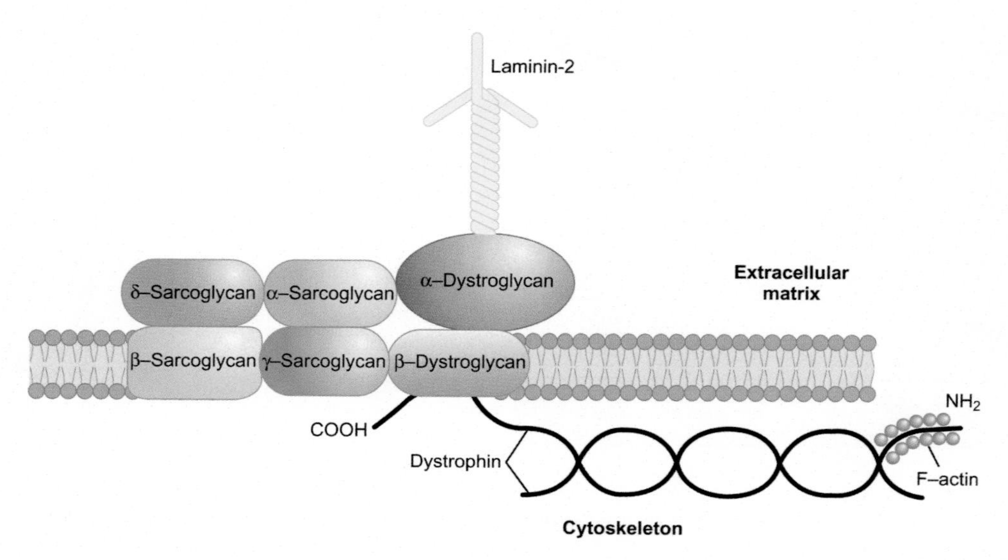

FIGURE 5.13 Molecular attachments between the actin filaments within a muscle fiber and the extracellular matrix (laminin). The chief intracellular linker is the huge protein dystrophin, which is defective in boys with Duchenne muscular dystrophy and makes the muscle fiber more susceptible to damage. Other membrane-associated linker proteins connect dystrophin to the extracellular matrix. *From Jorde et al. (2010), with permission.*

Fast motor units and muscle fibers (**type II**) contain a fast type myosin, and are larger in diameter than slow muscle fibers. As a result, each muscle fiber generates more force when contracting than does a slow muscle fiber. The motor nerves that supply them fire at a rate of 30 to more than 100 Hz. There are two types of fast muscle fibers and motor units—**fast fatigue-resistant** (type IIA) and **fast fatigable** (type IIX). The former contain fast myosin, as well as many mitochondria, and are also surrounded by a prominent capillary network. They are also able to utilize aerobic metabolic pathways for energy during contraction, making them resistant to fatigue. Fast fatigable are the largest and strongest muscle fibers, but because of a relatively small number of mitochondria and a reduced capillary network, they are not able to produce sufficient energy through aerobic means. Instead, they must rely upon the anaerobic breakdown of carbohydrates, e.g., glycogen, with which they are well supplied, to produce energy. The speed of a twitch contraction is several times faster than that of a slow muscle fiber.

Muscle fibers do not contract in a vacuum. Their contractions can only be effective if their contractile force is transmitted from the muscle fiber itself to the surrounding connective tissue. This is accomplished through a complex of proteins that link the sliding actin filaments first to the sarcolemma and then to the basal lamina overlying the muscle fiber. A major intracellular protein in this complex is **dystrophin**, a very large molecule that is important in maintaining the integrity of the sarcolemma during contraction. Through several other linker proteins, intracellular molecules, such as dystrophin, are connected to components of the basal lamina surrounding the muscle fiber, especially **laminin** (Fig. 5.13). Other molecular links connect the basal lamina to the endomysial connective tissue.

Within the muscle fiber is **titan**, the largest known protein in the human body. Titan, which is present as filaments 1.2 μm long, runs parallel to the thick and thin filaments from the Z-line to the center of the sarcomere. In addition to keeping the thick and thin filaments in line, titan is elastic and is largely responsible for the passive resistance to stretch in a relaxed muscle.

Contraction of an Entire Muscle

The contraction of an entire muscle is the result of interplay among many individual components. The prime movers are the muscle fibers/motor units themselves, but without the layers of connective tissue overlying them, the contractions of muscle fibers would be ineffective. Muscles also contain sensory components, which play a regulatory role in whole muscle contraction. These sensors are muscle spindles and Golgi tendon organs. The vasculature plays an ancillary role by supporting sustained or repeated muscle contractions.

Motor Units

Whole muscle contractions are graded, rather than all-or-none actions. For the smallest movements, only a few motor units (see Fig. 5.11) may be placed into use. As the need for power increases, additional motor units are pressed into service. Almost never do all motor units contract at the same time. A standard pattern of recruitment of individual motor units accounts for much of the smoothness in a muscle contraction. Recall that human muscles contain a mix of fast and slow muscle fibers. When a stimulus to contract arrives from the brain, the first motor neurons to react are small ones that supply slow motor units of muscle fibers. This provides for a level of relatively weak, but sustained contraction. As the intensity of the stimulus from the brain increases, larger motor neurons begin to fire. These supply fast motor units, which provide greater power, but are less able to sustain a contraction for a long period. Motor units are recruited at staggered intervals, another feature that results in a smooth contraction of the muscle. During relaxation, motor units cease firing in reverse order. Large fatigable units become quiescent first, followed by progressively smaller ones until the last remaining ones are the slow motor units.

For less than maximal contractions, a given level of force is accomplished in two ways. One is the alternation of different motor units which, in aggregate, produce that amount of force. The other is changes in the firing rate of whatever motor units are active at a given time. All of these modulations are influenced by concurrent sensory signals coming to the brain from the muscle spindles and Golgi tendon organs.

The Sensory Apparatus of a Muscle

Feedback from muscle to brain is called **proprioception** (sense of place), and the two principal sensors in muscle have very different functions. **Golgi tendon organs** are nerve endings located in the tendon near the myotendinous junction at each end of the muscle (Fig. 5.14). The organ consists of strands of collagen emanating from 10–20 representative motor units and is enwrapped by nerve fibers. A fibrous capsule encloses the organ. Their function is to measure the amount of force generated by the muscle through sensing the degree of tension within the tendon. This is accomplished by compression of the nerve endings when the muscle contracts. This activates stretch-sensitive cation (positively charged ions) channels resulting in depolarization of the nerve endings, which then send a message (an action potential) to the spinal cord for further processing higher in the central nervous system. Golgi tendon organs sense and signal force generated by the muscle throughout its entire physiological range.

Muscle spindles play an entirely different role. They measure the length and rate of stretch of a muscle. In contrast to Golgi tendon organs, muscle spindles are located at irregular sites throughout the belly of the muscle. Golgi tendon organs are situated in series with the collagen fibers of the tendon, whereas muscle spindles function in parallel with other muscle fibers within the muscle. Although very small, muscle spindles have a complex structure (see Fig. 5.14). A muscle spindle contains 3–12 very small **intrafusal muscle fibers**, which are designated as **nuclear bag** or **nuclear chain** fibers. (Ordinary muscle fibers are called **extrafusal fibers**.) Nuclear bag fibers contain a clump of nuclei in the central region of the fiber, and in nuclear chain fibers, the nuclei are spread out in a single longitudinal row in the central region. Intrafusal muscle fibers contain contractile proteins at either end of the fiber, but not in the middle nucleated region. The contractile ends of the intrafusal fibers are innervated by terminals of special motor nerve fibers. Endings of two types of sensory nerve fibers wrap around the central regions of both types of fibers.

Nuclear chain and one type of nuclear bag fiber are called static sensors, and another type of nuclear bag fiber is a dynamic sense receptor. Because muscle spindles are arranged in parallel with extrafusal muscle fibers, it is necessary to maintain a certain degree of tension on the central areas of the intrafusal fiber regardless of the state of contraction

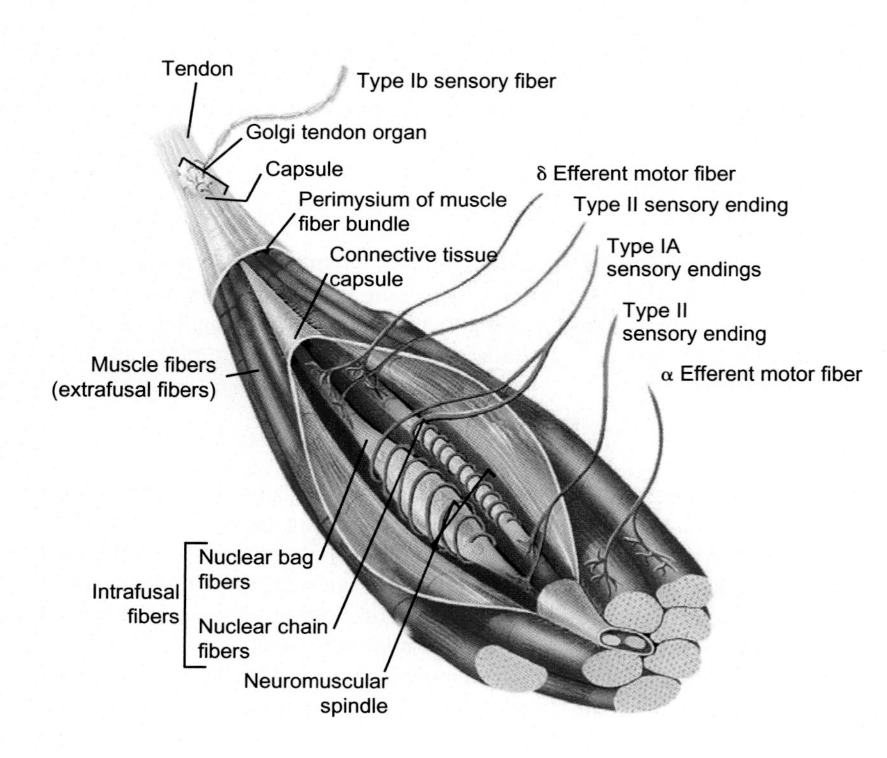

FIGURE 5.14 Diagrammatic representation of a muscle spindle and Golgi tendon organ in a skeletal muscle. *From Thibodeau and Patton (2007), with permission.*

of the muscle as a whole. Maintaining such tension is the function of the contractile ends of the intrafusal fibers. Thus, whenever the muscle is stretched, the degree of stretch is registered by the sensory nerve fibers innervating the central regions. When they sense stretch, stretch-sensitive ion channels are activated and Na^+ rushes in. This raises the resting membrane potential of the sensory nerve terminals and consequently, the likelihood that they will send an action potential signal to centers in the central nervous system for processing in much the same manner as occurs in the nerves innervating the Golgi tendon organs.

The density of muscle spindles in a given muscle depends greatly upon its function. A muscle that has very sensitive functions is much more plentifully endowed with spindles than one used for more gross movements. For example, extraocular muscles have ~36 spindles per gram of tissue, whereas the massive gluteus maximus muscle contains less than one spindle per gram.

Proprioception is not confined to muscles themselves. The connective tissues around them contain many sensory receptors, as well. Ligaments contain at least four types of mechanoreceptors, and within fascia, **Ruffini endings** (see p. 69) detect tension and shear. Deep **Pacinian corpuscles** detect pressure. In sum, the brain continually receives signals from a very large number of sources that contribute to one's sense of overall position throughout the body.

The Connective Tissue in a Muscle

A typical muscle contains hundreds of thousands to millions of muscle fibers, each of which contracts as an individual unit. Without a way to focus the contractions of these individual muscle fibers, it would be impossible to generate the effective contraction of an entire muscle. The task of focusing these individual contractions falls to the connective tissue components of the muscle. The most important concept is that connective tissues really do connect and that all levels of organization above that of the muscle fibers are interconnected through fibrous connective tissue links.

As noted earlier, muscle fibers are connected to the endomysial connective tissue through links to the basal lamina (see Fig. 5.13). The delicate endomysium forms an integrated meshwork linking many muscle fibers into a smoothly functioning unit. This endomysial network smoothly blends into a more robust layer of connective tissue called the perimysium (see Fig. 5.1), which is able to aggregate the forces generated by the muscle fibers within it. The muscle fibers within a perimysial sheath constitute a **muscle fasicle**. The perimysial connective tissue converges onto the tendons at either end of the muscle and also connects with the epimysium, the layer of connective tissue that surrounds the entire muscle. Connective tissue has quite different mechanical properties from muscle fibers, the latter being elastic and the former being plastic. Whereas muscle fibers repeatedly stretch and contract, the fibrous connective tissue within a muscle does not have those properties. Connective tissue responds to prolonged stretch by lengthening, and it has a tendency to shorten if not periodically stretched. This is why stretching is important to prevent hamstring shortening and subsequent injuries in runners.

A very important component of a muscle is the tendon, through which the force generated by the muscle fibers becomes concentrated and transmitted to the bone to which the muscle is attached. The bulk of a tendon consists of robust parallel bundles of type I collagen fibers with scattered rows of fibrocytes distributed among them. The vascularization of a tendon is poor. Of vital importance to a tendon is its attachment to the skeleton on one end and the belly of the muscle on the other.

At the skeletal end, the collagen fibers of the tendon meld seamlessly with the collagen of the periosteum, and through Sharpey's fibers the tendon connections penetrate into the matrix of the bone itself (see Fig. 4.11). The muscle end of the tendon (**myotendinous junction**) is much more specialized. Its prime functional requirement is to firmly join cells specialized for producing force (the muscle fibers) to collagen fibers specialized for transmitting the sum of the forces generated by the individual muscle fibers. The connections at myotendinous junctions are so strong that a muscle tear almost never occurs in that region. Instead, tears occur in the tendon proper or the belly of the muscle.

The terminal end of the muscle fiber is shaped to maximize its surface area for attachments. Instead of a smooth ending, numerous finger-like processes project from the tendinous end of the muscle fiber. Inside the muscle fiber, thin actin filaments project from the last Z-line toward the membrane covering the projections. These are connected to the extracellular matrix through two types of molecular complexes. One involves the large intracellular protein dystrophin and its associated linking proteins that link the actin thin filaments to laminin, the extracellular matrix protein present in the basal lamina, and that also extends beyond the basal lamina into the collagen fibers of the tendon. The other molecular connector occurs via the membrane-associated integrin proteins. These also bind to laminin molecules. Each individual myotendinous junction contributes to the total force that passes through a tendon. In contrast to the way in which they are often portrayed, tendons often extend like a partially opened fan for a considerable distance along the surface of a muscle, thus providing a broad surface for attachments of individual muscle fibers.

Muscles do not always terminate in tendons. In some muscles, the muscle fibers terminate directly on the bone, and the epimysial connective tissue merges with the collagen fibers of the periosteum. This is the case for the intercostal muscles, where the muscle fibers run from one rib to the next. Another type of connection is called an **aponeurosis**, in which the muscle fibers of a broad flat muscle terminate in an extensive sheet of dense connective tissue. The oblique muscles of the abdomen insert into a sheet of fibrous connective tissue that covers the anterior surface of the abdomen. Other muscles terminate in the dermis of the skin. This is how many superficial muscles of the face are able to alter one's facial expression.

Muscle Architecture in Relation to Function

Muscles come in a variety of shapes, all designed to maximize the efficiency of certain mechanical functions. Not only the external shape, but also the internal architecture of a muscle has a profound effect on the amount and direction of tension developed by the contracting muscle fibers. Some of the most important shapes and architectures are illustrated in Fig. 5.15.

Two examples will illustrate the importance of internal architecture in relation to muscle function. In the thigh, the hamstring and other flexor muscles have fibers that run parallel to the long axis of the muscle. Because these muscles are so long, individual muscle fibers don't extend the entire length of the muscle. Rather, several muscle fibers are arranged in series and are functionally connected through the connective tissue that invests them. Such muscles are said to have a long excursion (% of shortening), but they are only moderately strong in terms of the whole muscle force that they can generate. In contrast, those muscles on the extensor side of the thigh, e.g., the quads (quadriceps femoris), are **pennate muscles**. The fibers of these muscles are shorter and are situated at an angle to the long axis of the muscle. The overall excursion of these muscles is less because of the pennation, which results in a lateral, as well as longitudinal component to the shortening process (Fig. 5.16). On the other hand, contraction of these muscles generates more whole muscle force, because of what is called the physiological cross-sectional area, which is measured perpendicular to the axis of the muscle fibers, not to the axis of the whole muscle. Thus, if two muscles were of equal cross-sectional area, a pennate muscle would be stronger than a parallel fibered one because of its larger physiological cross-sectional area. In the case of human thigh muscles these architectural adaptations fit the functions nicely. The quads provide the force needed for tasks such as climbing stairs, whereas the hamstrings allow the degree of shortening needed for running. An extreme example of the importance of internal architecture is the soleus muscle, the main postural muscle in

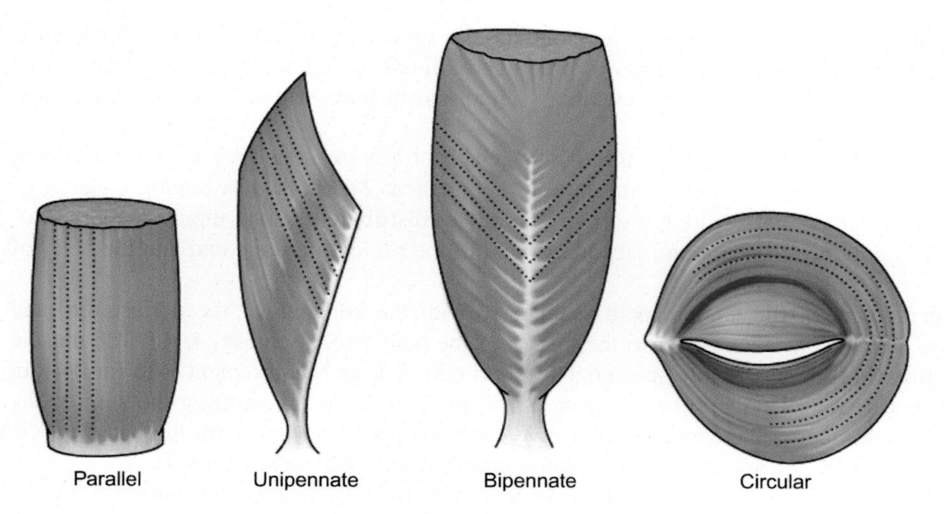

FIGURE 5.15 Fundamental types of internal architecture in skeletal muscles.

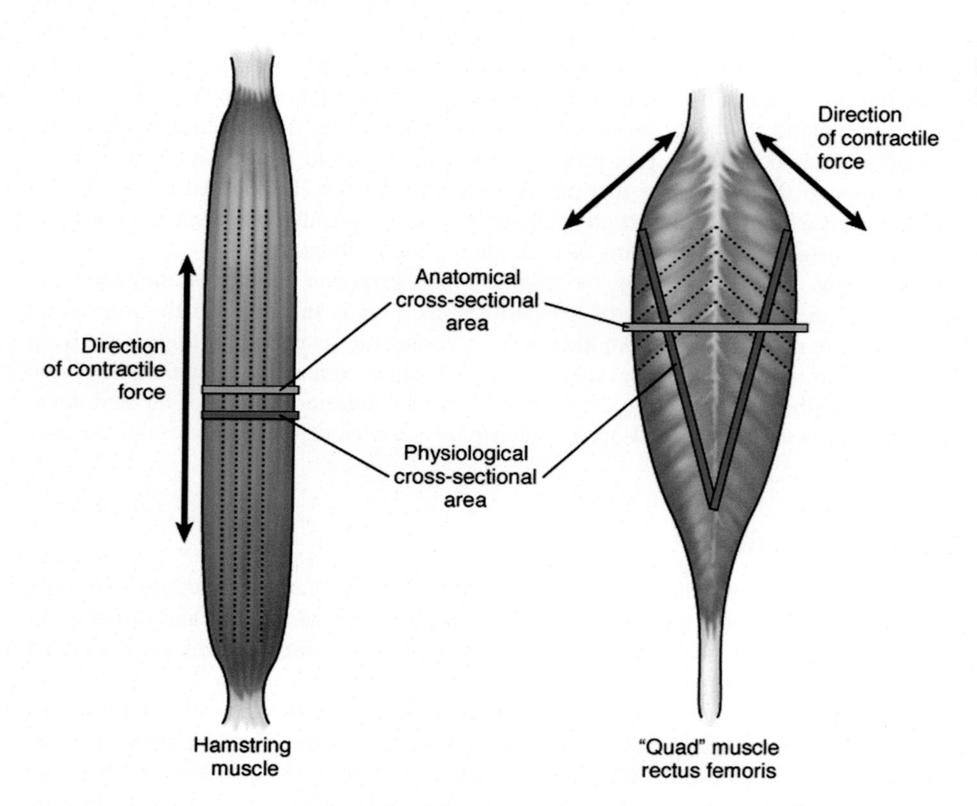

FIGURE 5.16 Functional architecture of flexor (hamstring) versus extensor (quad) muscles in the human thigh.

the calf. It has a highly pennate organization and is able to exert a great deal of force for its overall size. It also has a high percentage of slow muscle fibers, which allows it to function on an almost continuous basis without fatiguing.

One of the more unusual architectural adaptations of muscle is seen in the intrinsic muscles of the tongue (see Fig. 12.12), the trunk of elephants, and the tentacles of an octopus. In these, muscle fibers are organized in small groups that are situated at right angles to one another in each of the three dimensions. This arrangement allows the multiple movements, including extension, of which each of these structures is capable.

Another aspect of internal architecture involves the muscle fibers themselves. At equilibrium every muscle fiber has a fixed number of sarcomeres, which allows the muscle to contract at maximal efficiency. If a muscle is kept in a shortened position for an extended period of time, the muscle fibers lose sarcomeres to adjust for the new shortened length. Conversely, if a muscle is maintained in a stretched position, new sarcomeres are added. The surrounding connective tissue similarly adapts to the changes in muscle length.

Muscle Contractions and the Whole Body

Covering the spectrum of gross muscle movements and their effects on the body could fill a large book, but some important general principles can be extracted from the larger body of knowledge. Most important is that muscles do not contract in isolation. Whenever a muscle contracts, it affects many other parts of the body, both directly and indirectly. It has been common to look at both the anatomy and function of muscles as individual units, but a fuller understanding of muscle function can be gained by looking at their function in an integrated manner.

Although the most prominent connections between muscles and the rest of the body are through tendons and aponeuroses, lateral connections with neighboring muscles or other structures through the surrounding fascia play a significant role in coordinated movements. Along the main force-producing axis, connections between the epimysium and ligaments or fascial sheets often link distant parts of the body together. An important recognition of these functional connections is embedded in the *Anatomy Trains* concept that points out functional muscle-fascia connections throughout the entire body (Fig. 5.17). Such connections not only explain many normal bodily movements or postures, but they also provide important clues for discovering the basis for many chronic pains of the musculoskeletal system.

Many muscles of the extremities can be classified as one-joint or multijoint (usually two) muscles. Multijoint muscles, such as the biceps muscles of both the arm and leg, are typically long and superficial. They can often affect the movement of the joints that they cover in a variety of ways. In contrast, beneath these superficial multijoint muscles is a layer of deep single joint muscles (Fig. 5.18). These more obscure muscles only move a single joint and often serve a postural function rather than being primary actors in large movements.

Where a muscle attaches to a bone is very important in transmitting the contractile force generated by a muscle. The biceps brachii of the upper arm inserts on the radius (as well as a fascial band) of the forearm in the manner of a **type III lever** (Fig. 5.19). The farther from the elbow that its distal tendon inserts onto the radius, the more efficient is the ability of the biceps to flex the arm. Another type of lever (**type II**) is seen at the attachment of the Achilles tendon to the calcaneous (heel bone), where the gastrocnemius muscle is lifting the weight of the body on the same side of the fulcrum (the ball of the foot). The simple **type I lever** arrangement is seen in only a few places in the body, but the muscles that extend or flex the skull pull on it opposite the fulcrum (the atlanto-occipital joint) in much the same manner as a seesaw functions.

Specific bones improve the leverage of the muscles attached to them by improving the mechanical advantage of the lever system. An excellent example is the overall lever function of the calcaneous (heel) bone (see Fig. 5.19B). Another is the patella (kneecap), which greatly improves the overall efficiency of the quadriceps muscles.

For any given gross movement, different muscles or groups of muscles play different roles. For example, when flexing the forearm, the contracting biceps brachii is the prime mover or agonist. At the same time, the triceps brachii on the other side of the arm relaxes proportionally to allow for a smooth motion. Other muscles—the brachialis in this case—contract with the prime mover and serve to steady the overall movement. Still other muscles may act as fixators. In this example, many of the muscles of the scapula contract to stabilize the origin of the prime mover (biceps). All of these coordinated muscular movements require a great deal of integrated signaling within the central nervous system. Certain neurological diseases in which such central integration is defective result in grossly distorted movements.

Muscle Metabolism During Contraction

The driving factor in muscle metabolism is the need to produce energy to support muscular contractions. ATP provides the energy by cleaving its third phosphate group with the concomitant release of energy to form ADP (adenosine diphosphate). A second molecule with high-energy bonds, **creatine phosphate**, is also present in muscle fibers.

When a muscle fiber contracts, its intrinsic supply of ATP is only sufficient to maintain the contraction for a few seconds. Thereafter, energy comes from the transfer of the phosphate group from creatine phosphate to the ADP molecule. This provides enough energy for another 15 seconds of contraction. After that, energy must come from the breakdown of carbohydrates and lipids—first from within the muscle fiber and then from elsewhere in the body.

The main intrinsic source of metabolic energy in a muscle fiber comes from the breakdown of **glycogen**, a carbohydrate polymer that yields the simple sugar glucose when it is enzymatically broken down into individual subunits. In the absence of oxygen (**anaerobic glycolysis**), a molecule of glucose is broken down into **lactic acid**, with the generation of two ATP molecules during the process (Fig. 5.20). This process is energetically inefficient and can sustain moderate physical activity for only about a minute. Nevertheless, anaerobic glycolysis provides ~100 times more immediate energy than can be provided by the immediate breakdown of ATP and creatine phosphate. Beyond that, O_2 must be injected into the system in order to allow sustained muscular activity through a process called oxidative

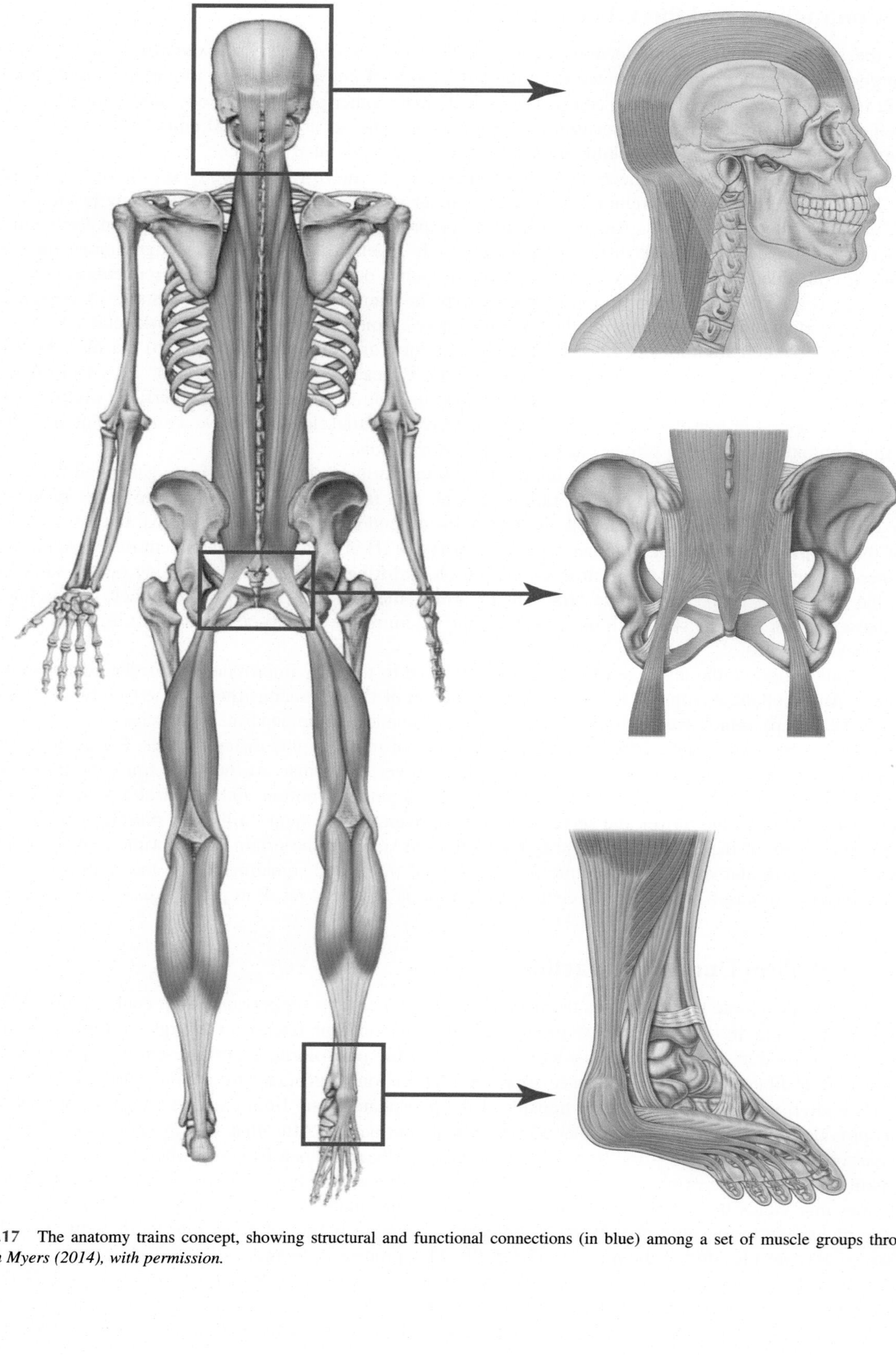

FIGURE 5.17 The anatomy trains concept, showing structural and functional connections (in blue) among a set of muscle groups throughout the body. *From Myers (2014), with permission.*

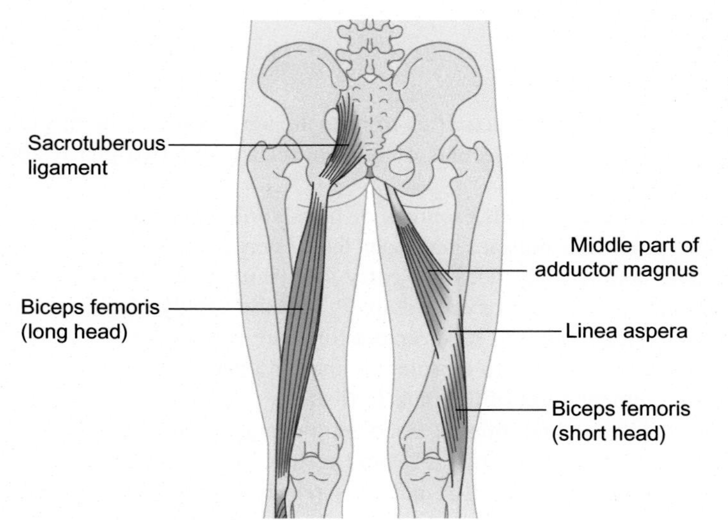

FIGURE 5.18 Two joint (e.g., biceps femoris long head) versus single-joint (e.g., biceps femoris short head and adductor magnus) muscles. Two joint muscles play an important power function, whereas single-joint muscles are more important in adjusting posture. *From Myers (2014), with permission.*

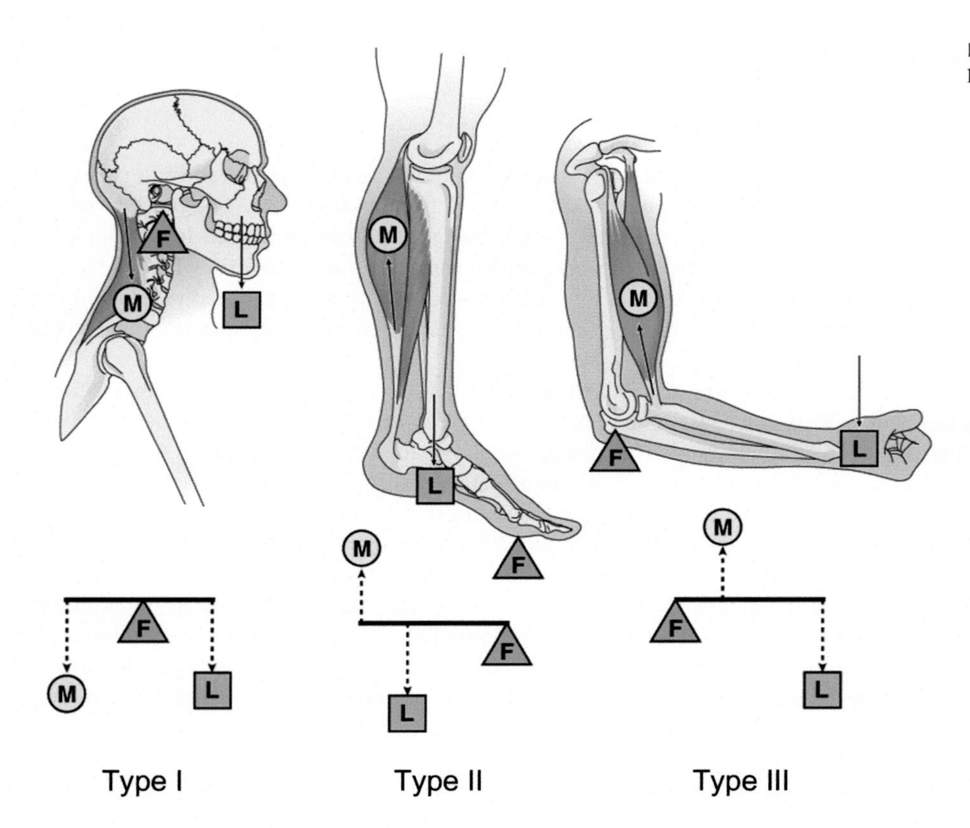

FIGURE 5.19 Muscles functioning as levers. *F*, Fulcrum; *L*, load; *M*, muscle.

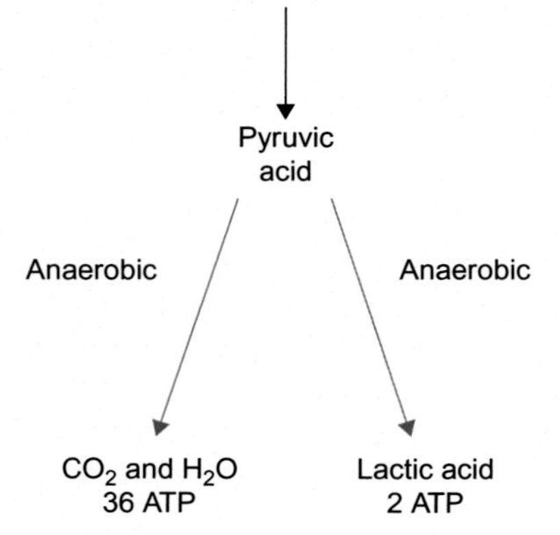

FIGURE 5.20 Abbreviated scheme of ATP production in muscle through aerobic and anaerobic pathways.

metabolism or aerobic respiration. In the presence of O_2 and through the activities of mitochondria, a molecule of glucose gets broken down into CO_2 and H_2O, with the generation of 36 molecules of ATP instead of the two that result from anaerobic glycolysis.

Aerobic respiration can sustain muscular activity from intrinsic muscular sources of metabolites until the supply of intramuscular glycogen and some lipids runs out. At that point, the muscles must get their energy sources from elsewhere in the body, especially the liver, via the blood. The resting body does not carry sufficient O_2 to sustain major muscle activity, and the heart and lungs require 1–2 minutes to increase oxygen delivery and the supply of glucose to meet the needs of the muscles. The body's response to the exercise-induced requirement for more oxygen involves three levels: (1) more intake of air (O_2) by the lungs, (2) faster delivery of O_2 to muscles by the circulatory system, and (3) increased extraction of O_2 from the red blood cells into the mitochondria of the muscle fiber.

There are advantages and disadvantages to the various ways in which muscle fibers gain energy. Anaerobic glycolysis occurs rapidly and doesn't require oxygen, but it produces relatively little ATP and results in the accumulation of lactic acid in the muscle. Aerobic respiration produces a great deal of ATP, but it is relatively slow in onset and requires the presence of oxygen. Adaptations of muscle fibers (see later) allow muscles to make the most efficient use of each of these methods of energy production. Nevertheless, if one measures the overall efficiency of muscle (mechanical work/metabolic cost), muscle operates at only ~20% efficiency.

Muscle Fatigue and Recovery

A simple definition of fatigue is a decline in muscular performance during continuous activity, followed by a period of recovery. Beyond this definition, fatigue is a complex phenomenon. Causes of fatigue also vary. Fatigue caused by high-frequency, high-intensity contractions does not involve the same factors as fatigue from prolonged low-frequency, moderate-intensity contractions. High-frequency-induced fatigue occurs rapidly and has a fast recovery, whereas fatigue induced by low-frequency contractions occurs slowly, but recovery from this type of fatigue is also slow. In addition, fatigue can be due to factors within the central nervous system (central fatigue) or within the muscles themselves (peripheral fatigue).

Central fatigue is well recognized, but less well understood. When an individual begins an intensive training regime, a measurable improvement in function and reduction in fatigue may occur within days, well before the muscles involved have had time to adapt to the new demands placed upon them. This phenomenon is called **central training**. Although the higher level processes underlying central fatigue remain obscure, they can affect the firing rate of motor neurons.

Causes of **peripheral fatigue** are now known to be more complex than simply a build-up of lactic acid within the muscle fibers. One factor is simply depletion of biochemical fuel, such as ATP or glycogen. Other factors are abnormal distributions of intracellular ions. During intense exercise, both Na^+ and K^+ accumulate within the muscle fiber, making it difficult to generate and propagate action potentials. Accumulation of lactic acid and a reduction in intracellular pH play a role in the reduction of Ca^{++} release from the sarcoplasmic reticulum. This affects cross-bridge formation by reducing the effects of Ca^{++} on troponin (see p. 118). At a gross level, when a muscle exceeds 50% of its maximum contractile force, the circulation within it collapses. This greatly reduces a major potential source of both oxygen and substrates for energy and explains why maximal contractions of muscles cannot be maintained for long periods of time because they must rely upon anaerobic glycolysis for their energy.

Recovery of a muscle from fatigue is accompanied by heavy breathing after the cessation of exercise. During this period, the body is replacing its oxygen reserves that were depleted during the initial moments after the start of the exercise period. At the same time, the muscle fibers are replenishing their intrinsic stores of ATP and creatine phosphate, which were used up before oxidative metabolism kicked in. Finally, most of the lactic acid that had built up during the exercise and has entered the bloodstream becomes oxidized in the liver to another acid—pyruvic acid—which can be utilized as a substrate in aerobic metabolism.

GROWTH AND ADAPTATION OF MUSCLE

Like the skeleton, the muscular system is highly dynamic, and individual muscles or even parts of muscles are responsive to body growth or changes in the functional demands placed upon them. Fundamental to growth is lengthening the muscle to keep up with growth of the skeleton. Once an individual has stopped growing, muscles readily adapt to their functional environment. When additional demands are placed upon muscles, they become either bigger and stronger or develop greater endurance. When fewer demands are placed upon muscles (e.g., during bed rest or under conditions of microgravity in space), they adapt by becoming smaller (atrophy). Muscle atrophy becomes even more pronounced when their innervation is interrupted. Similarly, the loss of muscle tissue accompanies the aging process.

Muscle Growth

By the time of birth, human muscles already contain almost the full adult complement of muscle fibers. This means that as the body grows, the existing muscle fibers will need to increase their length in proportion to the growth of the rest of the body; they also need to grow in diameter in order to provide the strength needed to move the ever enlarging mass of the body.

The ability of an existing muscle fiber to accomplish these tasks is limited by the number of myonuclei that it contains. Each myonucleus serves a limited domain of a muscle fiber in terms of the amount of protein that it can produce. Beyond the collective capacity of all the existing nuclear domains within a muscle fiber, a muscle fiber must add new myonuclei in order to expand its overall synthetic capacities. As is the rule for nuclei within a **syncytium** (multinucleated cell), myonuclei do not multiply by mitotic division. Instead, addition of new myonuclei is accomplished by the fusion of progeny of satellite cells (see p. 48) to the muscle fiber and their nuclei becoming functional myonuclei. Each new nucleus added to the muscle fiber adds to its capacity to synthesize the subcellular machinery needed to function at a higher level.

Growth in length of a muscle fiber is accomplished by the addition of new sarcomeres (contractile units). The muscle fiber does this in response to the increasing tension caused by the lengthening skeletal parts or soft tissues to which it is connected. Growth in diameter occurs by the formation of new myofibrils in parallel to existing ones. Because the strength of a muscle (or muscle fiber) is proportional to the total cross-sectional area of the contractile material, the formation of new myofibrils keeps up with the increasing requirement for strength in a growing muscle. Under normal circumstances, new muscle fibers are rarely formed in human muscle.

Traditionally, muscle growth was assumed to be controlled by the functional demands placed upon the muscle fibers. This idea, however, was upset by analysis of the nature of a condition that produced animals with a massive musculature, e.g., double-muscled cattle (Fig. 5.21). These animals did not develop their tremendous muscles through exercise. Rather, they were found to have a mutation of a gene that controls the formation of a protein, called **myostatin**. Myostatin effectively puts the brakes onto skeletal muscle development, and in its absence the growth of muscles is unchecked.

Training Effects on Muscle

Training involves the entire body, but muscle is the focal point. Numerous training protocols have been devised for specific purposes, and an important generalization is that training for one type of task usually does not carry over for another except in the most general terms. For purposes of this section, the effects of two generic types of training will be covered—strength training and endurance training.

FIGURE 5.21 Double-muscled bull.

Muscle strength results from a combination of three factors—physiological strength of the muscle itself (size, cross-sectional area, contractile proteins), neurological strength (central conditioning, motor unit recruitment, etc.), and mechanical strength (lever characteristics, joint properties, muscle architecture). Especially because of its mechanical strength, the masseter muscle (which closes the jaw) exerts the greatest force—the equivalent of bench-pressing almost 1000 pounds.

Strength training utilizes increasing resistance against contracting muscles with the aim of increasing muscle mass, strength, and anaerobic endurance. Although humans do not all show the same degree of responsiveness to strength training, an increase in cross-sectional area of individual muscle fibers of ~25% is considered to be a reasonable expectation. Any improvement in muscle strength over the first few days after the start of training is almost certainly due to neural conditioning, rather than to an effect on the muscle tissue itself. One of the results of strength training is the increased utilization of motor units during muscle contractions and better coordination of individual motor units as a muscle is contracting.

The response of muscle to **strength training** occurs in two major phases. The first, which is measured in days, involves the muscle contractile proteins. In the first hours after the start of an exercise program, the breakdown of existing proteins is dominant. This is followed by a phase of increased protein synthesis that occurs, at most, over a few days. The increase in amount of protein synthesis is limited by the number of nuclei within the muscle fiber and the capabilities of the existing protein-synthetic machinery within the cell. The most critical of the proteins are the myosins. **Myosins**, in particular the part of the molecule called the heavy chain (MyHC), come in three varieties—one slow (MyHC I) and two fast (MyHC IIA and MyHC IIX). Depending upon the exact nature of the strength training, there can be some interconversion between the two fast myosins, but normally fast myosins do not replace slow ones, or vice versa.

The second phase of muscular response to strength training occurs at the cellular level over days to a few weeks. This involves the satellite cells, which become activated by resistance training and begin to divide. Some of the daughter cells of satellite cells become incorporated into the muscle fibers as new myonuclei, which then begin to produce contractile proteins through their own synthetic machinery. The more satellite cells incorporated into a muscle fiber, the more myonuclei and the more contractile proteins can be synthesized. Since the muscle fibers normally don't increase in length, the increased amount of new muscle protein results in hypertrophied muscle fibers and larger muscles. The changes described earlier are mainly seen in fast muscle fibers, which undergo double the degree of hypertrophy seen in slow muscle fibers within the same muscle. An ancillary change in strength-trained muscle is increased glycogen storage within the muscle fibers. This provides a greater store of energy for short-term powerful contractions. Measurable hypertrophy in strength-trained muscles can be seen in 4–6 weeks. Testosterone is heavily involved in muscle hypertrophy, which is why weight training produces larger muscles in men than in women.

Endurance training follows quite different protocols from those used in strength training, and the results of endurance training are also quite different. Muscle fiber hypertrophy is minimal, but significant changes occur in other aspects of the muscle (Fig. 5.22). Beginning with a significant increase in capillary density and oxygen uptake and up to a doubling of mitochondrial number, the training effects are directed at increasing the oxidative metabolic functions of the muscle. These increases occur over a period of several months before leveling off. Unfortunately, after the cessation of training, it takes only a few weeks for the muscles to regress to pretraining levels of these parameters.

A day or two after beginning some types of training, people may experience soreness of certain muscles. Although the cause is not definitively known, this phenomenon is currently thought to be caused by small muscle fiber tears that occur when a muscle is contracting while it is being stretched (**eccentric contractions**).[2] Walking or running downhill causes eccentric contractions. After a short period of training, the connective tissue within the muscles adapts, and such injury does not recur. The damaged muscle fibers then regenerate (see later).

Muscle Atrophy

An often unrecognized adaptation of skeletal muscle to functional demands is a reduction of muscle size when its functional needs are reduced. As seen in Fig. 5.22, detraining after a period of intensive training results in the muscles' soon returning to their untrained condition. At the other extreme, an individual placed in bed rest for an extended period of time loses muscle mass because of the inactivity of many of the muscles, especially postural muscles. A particularly vexing problem for astronauts spending extended periods in space is the progressive loss of both bone and muscle mass

2. Contractions while a muscle is shortening are called **concentric contractions**.

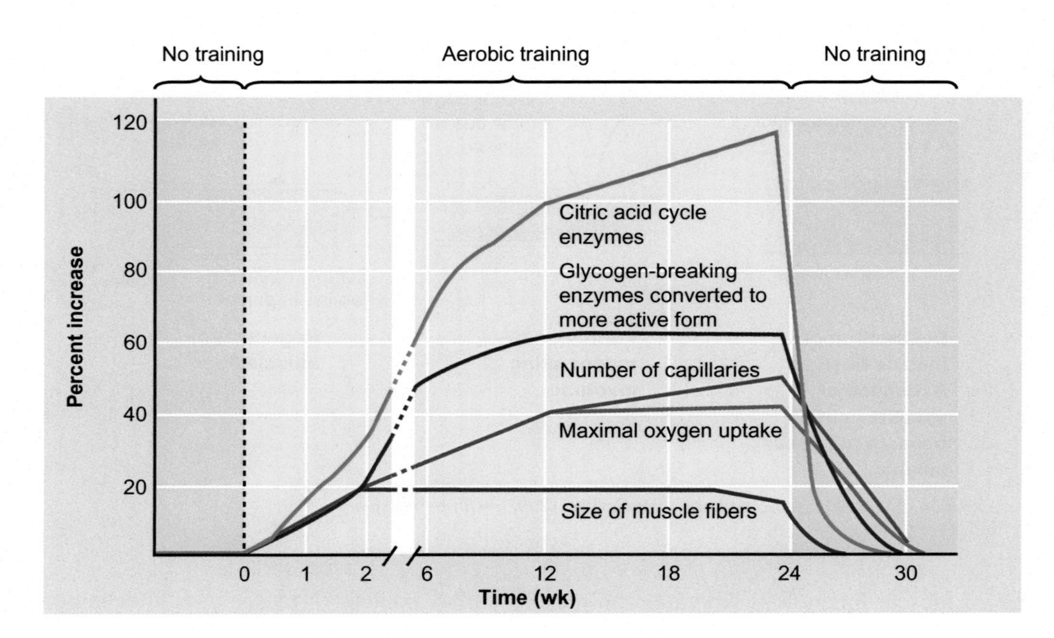

FIGURE 5.22 Effect of aerobic training and detraining on a number of properties of skeletal muscles. *From Thibodeau and Patton (2007), with permission.*

(up to 30% in muscle) while in space. Considerable effort has been undertaken by the various space agencies to devise resistance exercise protocols that would slow down microgravity-induced muscle and bone loss.

One of the most profound types of **muscle atrophy** occurs after the motor innervation to a muscle has been interrupted through injury to the nerve. Following denervation, a muscle rapidly loses functional mass, meaning a massive reduction in its contractile proteins. Over many weeks and months, new bundles of collagen fibers begin to form within the endomysial connective tissue as the muscle fibers concurrently diminish in size. For many years the nature of the neural influence on muscle fibers was not understood. A popular early hypothesis that motor nerves provided a chemical trophic factor that maintains the integrity of muscle fibers was supplanted by the recognition that the action potentials resulting from intact innervation are sufficient to maintain muscle mass. Techniques are presently being developed for maintaining the functional mass of denervated muscles through electrical stimulation.

Muscle Regeneration

When a muscle is injured, the damaged muscle fibers attempt to repair themselves through regeneration. The muscle fibers themselves have little capacity for self-repair. Instead, the damaged muscle fibers become invaded by large numbers of macrophages, descended from monocytes in the peripheral blood (the same types of cells that produce osteoclasts, see p. 93, Fig. 5.23). The macrophages remove the damaged components of the muscle fibers through a process of phagocytosis. While doing this, they also secrete cytokines that activate the satellite cells residing beneath the basal laminae of the damaged muscle fibers. The satellite cells proliferate and ultimately fuse with one another to form newly regenerated muscle fibers in a manner that closely parallels the development of muscle fibers in the embryo. In humans, muscle regeneration is not always successful. In cases of massive muscle trauma or when the local blood supply is inadequate, damaged muscle tends to heal by scarring, rather than by true muscle fiber regeneration.

Muscle fiber degeneration and regeneration plays an important role in the pathology of most **muscular dystrophies**. One of the most prominent muscular dystrophies is the Duchenne variety, which for genetic reasons affects mainly young boys. **Duchenne muscular dystrophy** is caused by mutations in the gene coding for the huge molecule dystrophin, which is one of the proteins that stabilizes the sarcolemma of muscle fibers (see p. 122). When the dystrophin protein is defective, the sarcolemma has a tendency to tear upon exertion. This sets into motion the sequence of muscle fiber degeneration and regeneration described earlier. Over the course of years, muscle fibers of boys with muscular dystrophy have undergone many cycles of muscle fiber breakdown and repair. Ultimately, the repair system seems to become exhausted, and the damaged muscle fibers are no longer able to regenerate, a condition that ultimately leads to death. Attempts to facilitate muscle fiber regeneration by introducing healthy satellite cells into dystrophic muscles

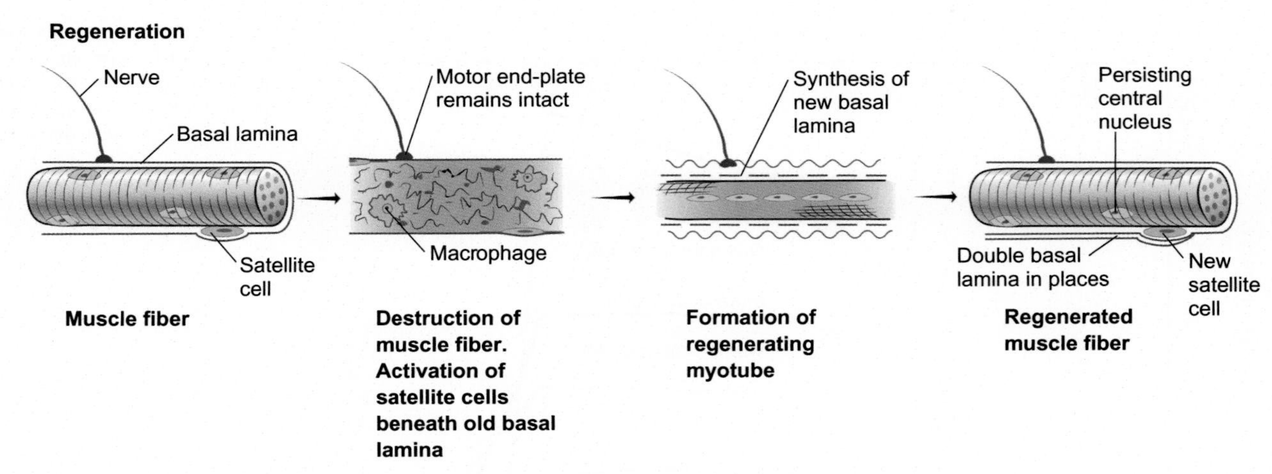

FIGURE 5.23 Major steps in the regeneration of a skeletal muscle fiber. *From Carlson (2007), with permission.*

have resulted in improvements of muscle function in laboratory animals, but for a number of reasons, such techniques have been much less successful in humans.

The Aging of Muscle

As in many other systems in the body, aging is no friend of the muscular system. After middle age, muscle mass and function begins a slow and inexorable period of decline that accelerates with increasing age (Fig. 5.24). In some individuals, the decline is sufficiently significant that the term **sarcopenia** (deficiency of flesh) has been applied to the condition. Fortunately, through increased strength training, the pattern of loss can be partially reversed, but for any given level of activity or training, the age-related slope of muscle decline remains roughly the same.

Much of the age-related decline of muscle is secondary to age changes in the peripheral nervous system. As the body ages, individual neurons innervating motor units die, leaving all of the muscle fibers attached to them in a denervated condition. Neurons innervating fast motor units are most likely to die. These denervated fast muscle fibers have two probable fates. The first is to remain denervated and undergo profound denervation atrophy, without contributing to the strength of the muscle. A second fate is that some of these denervated muscle fibers become reinnervated by sprouts growing from branches of nearby motor nerve fibers. Although some of the sprouts could come from fast nerve fibers, they are more likely to arise from slow nerve fibers. This converts the formerly fast muscle fibers into slow muscle fibers.[3] The net result is a reduction in the absolute number and proportion of fast muscle fibers and a relative increase in slow muscle fibers.

In practical terms, this means that as one ages, one's muscles lose power because of the smaller number of fast muscle fibers. On the other hand, the aerobic capacity of a muscle is more likely to be preserved because of the increased number of slow muscle fibers. This change (along with changes in lung capacity and cardiac function) is reflected in winning times of foot races for various age groups, all of which decline with age (Fig. 5.25).

The loss of motor units with aging can be significant. Up to the age of 50, the number of muscle fibers and motor units remains relatively constant. After that age, the number of motor units supplying a muscle declines steeply, with the number of muscle fibers declining somewhat less steeply. One study showed that in a group of subjects in their mid-80s, the number of innervated motor units of a muscle in the foot declined by more than 85%. The loss of large numbers of functional motor units not only weakens a muscle, but it reduces the ability of that muscle to contract smoothly. This is likely a contributing factor to the loss of balance that often plagues the elderly.

3. Experiments conducted in the 1960s showed that one can change the functional characteristics of a muscle by changing its innervation. When a fast muscle is connected to a slow nerve, it begins to function as a slow muscle, and when a slow muscle is reconnected to a fast nerve, it takes on the characteristics of a fast muscle.

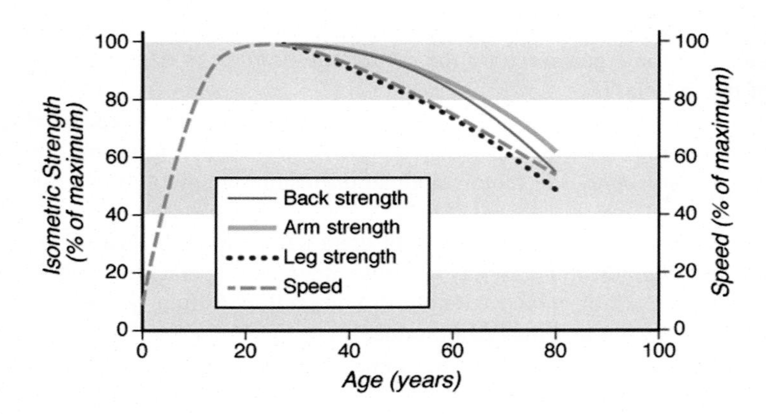

FIGURE 5.24 Graph illustrating changes in muscle power throughout the human life span. Data on strength are based on a 30-year-old base of 100%. The speed curve is presented as percentage of maximum speed.

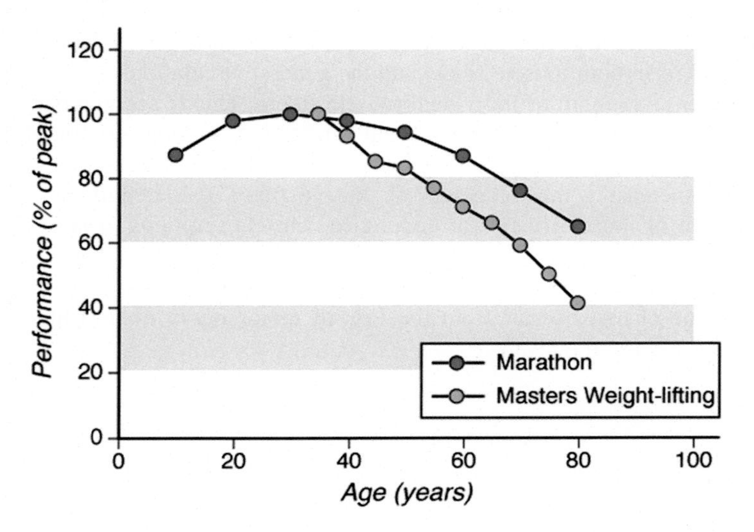

FIGURE 5.25 Age-related declines in performance of master athletes in the marathon and in weight-lifting (clean and jerk).

SUMMARY

Most muscles consist of a belly of skeletal muscle fibers, with tendons at each end. Individual muscle fibers are surrounded by a thin endomysium. Bundles of muscle fibers are bound together by a more robust perimysium, and an entire muscle is covered by an epimysium.

A single skeletal muscle fiber is composed of a series of sarcomeres (bundles of contractile proteins) packed beneath the sarcolemma. Each muscle fiber contains many myonuclei situated just beneath the sarcolemma. A mesh-like sarcoplasmic reticulum envelops bundles of contractile proteins, called myofibrils. A muscle fiber is surrounded by a basal lamina, and between the muscle fiber and its basal lamina are satellite cells, which act as stem cells for allowing muscle growth and regeneration.

The principal contractile units of muscle fibers are thick filaments, composed of myosin, and thin filaments, composed of actin. Associated with these are regulatory proteins, such as troponin and tropomyosin.

Contraction of a muscle fiber begins with the release of acetylcholine from the motor nerve terminal. This stimulates an action potential that spreads over the sarcolemma and enters the interior of the muscle fiber through T-tubules. The electrical signal is translated into a chemical signal in the sarcoplasmic reticulum, with the release of Ca^{++}.

The essence of muscle contraction is the sliding of thin filaments past thick filaments after the release of Ca^{++} from the sarcoplasmic reticulum. This is accomplished by the energy-requiring formation of cross-bridges between actin and myosin and a shift in position of the cross-bridges.

Individual muscle fibers contract as simple twitches or, more usually, through tetanic contractions, which represent the summation of many twitches. One motor nerve fiber supplies many muscle fibers. This constitutes a motor unit. Depending upon the type of nerve and the type of myosin within the muscle fiber, motor units are slow (type I) or fast (type II) contracting. Slow motor units are fatigue resistant. Fast motor units are either fatigable or fatigue resistant.

The contractile units of a muscle fiber are connected to the sarcolemma and then to the endomysium through a variety of proteins, such as dystrophin. Through this means the force generated by the contractile proteins is transmitted to the connective tissue and ultimately to the tendons of the muscle.

Muscle contraction is highly modulated at both the motor and sensory ends. Modulation of contraction is organized through the central nervous system. A muscle contains muscle spindles and Golgi tendon organs, which represent the sensory apparatus of a muscle. Coordination between sensory input and motor modulation results in smoothly coordinated muscle contractions.

The internal architecture of a muscle reflects its function. Muscles that have long contractions have muscle fibers that run parallel to the length of the muscle. For greater strength, muscles have various degrees of pennation, which increase the physiological cross section of the muscles and result in greater strength of contraction. Similarly, the way muscles connect to the skeleton determines the amount of leverage that they can produce.

During contraction, muscle fibers use a great deal of energy. The main intrinsic source of energy is glycogen, which is broken down by anaerobic glycolysis to generate ATP (adenosine triphosphate). Longer-term muscle contractions depend upon aerobic breakdown of carbohydrate, which depends upon the circulation to provide both oxygen and glucose.

Muscle fatigue can arise both centrally and peripherally. Causes of both are more complex than was earlier thought. Recovery from fatigue requires the removal of lactic acid from both muscle fibers and the general circulation.

Muscles grow mainly through the increase in diameter and length of individual muscle fibers. This is accomplished through the incorporation of satellite cells into the muscle fibers and the subsequent production of more contractile material. Myostatin regulates the growth of muscles.

Training increases the effectiveness of muscles by increasing the diameters of muscle fibers for strength or by increasing the capillary supply and mitochondrial content of muscle fibers for endurance. Muscle responds to lack of use by undergoing atrophy.

Injured muscle is capable of regeneration. Regenerating muscle fibers originate from satellite cells.

Muscle mass is lost with increasing age. Many of the changes result from the loss of motor nerve fibers (motor units), with atrophy or loss of the denervated muscle fibers being the result.

Chapter 6

The Nervous System

If any component of the body could take credit as its master controller, it would be the nervous system. Its functions are sufficiently diverse and complex that entire textbooks and courses are devoted to it. Structurally, the brain and spinal cord, at first glance, seem to be almost featureless, with the consistency of a thick pudding. Appearances, however, are deceptive. When analyzed with the appropriate tools, the nervous system consists of an incredibly complex network of cellular pathways and molecular domains. Although neuroscientists have outlined the main features of the nervous system, some of the largest questions, such as what is consciousness or the details of memory, still elude contemporary science.

The functions of the nervous system are myriad. At the most simple level, it receives signals and sends out messages, but the essence of the nervous system is what goes on in between. Sensory outposts of the nervous system pick up features of the external environment and transmit them to the brain as electrical signals running through sensory nerve fibers. These nerve fibers are components of peripheral nerves, which course throughout the body and finally enter the central nervous system (CNS—either the spinal cord or the brain), where they make connections (**synapses**) with other nerve fibers that are completely embedded within the CNS (Fig. 6.1). A large variety of **tracts** (nerve fibers of similar functionality) within the CNS carry the sensory messages to various parts of the brain, although other connections of the same tracts are purely local and serve immediate reflexes. A good example of a local reflex is the act of immediately drawing one's finger away from a hot object.

Within the brain, sensory input is organized, integrated and coordinated with many centers in a manner that can result in output ranging from muscular movements to changes in mood. The complexity of these internal connections is staggering, with some neurons connecting to as many as several thousand other neurons. At some point, the brain sends out messages that leave the CNS via motor nerve fibers or components of the autonomic nervous system. Motor nerve fibers terminate on groups of muscle fibers (motor units, see p. 121), and their aggregate effect results in coordinated muscular movements. Autonomic signals, which influence the functions of most internal organs, are more subtle and less immediate in their effect. Some functions generated by the CNS, such as the control of breathing rate, are so ingrained that we do not notice them.

The cells of the nervous system also produce an astounding number of molecules. In addition to the housekeeping molecules that all cells need to function, neurons produce over 100 neurotransmitters that carry messages from one neuron to the next. Increasingly, the nervous system has been found to produce hormones that exert either local or systemic effects.

This chapter is intended to give an overview of the structure and functions of the nervous system by presenting specific examples of neural functions rather than attempting anything resembling complete coverage. Readers wishing to delve more deeply into the nervous system should consult the many excellent neuroscience and neuroanatomy books.

EVOLUTION OF THE NERVOUS SYSTEM

Communication from one part of the body to another arose among some early animals even before the development of nerves or nervous systems. In fact, even among single-celled bacteria and protozoans sensory and motor functions are mediated by intracellular organelles. The most primitive multicellular animals, such as sponges, do not have recognizable nerves, but yet they are able to detect and respond to some stimuli by slow movements. The motive force for some of these reactions in primitive animals is an action potential-like phenomenon that is driven by Ca^{++} rather than by Na^{+} as in higher animals. Despite the absence of nerve cells and any forms of synaptic connections, sponges seem to contain most of the genes required for higher nervous function, but the actions of these genes have not been coordinated in a manner leading to synaptic function.

The Human Body. DOI: https://doi.org/10.1016/B978-0-12-804254-0.00006-5

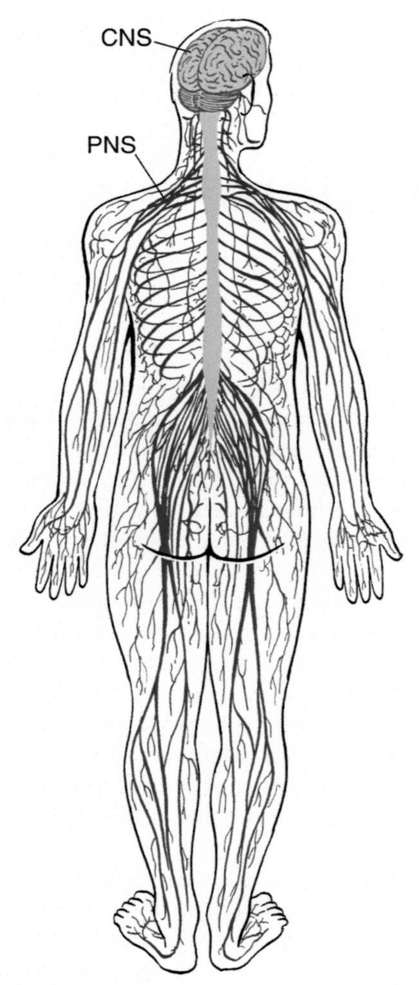

FIGURE 6.1 Location of the central nervous system (CNS), and the distribution of nerves of the peripheral nervous system (PNS) throughout the body. *From Nolte (2009), with permission.*

The first actual nerves, in the form of diffuse nets, are found in coelenterates, such as *Hydra* or jellyfish. Their function is recognized by the graceful rhythmic contractions of the "umbrella" portion of a jellyfish as it slowly swims through the oceanic waters. Nervous systems with an organization somewhat like ours arose in the Precambrian period somewhat over 550 million years ago when animals with bilateral symmetry appeared on the scene. The fundamental organization of the nervous system consists of a long nerve cord(s) with collections of nerve cell bodies (ganglia) serving each bodily segment and an aggregation of ganglia around the oral region. The latter ultimately evolved into the brain.

Many invertebrates (**Protostomes**—worms, molluscs, and arthropods) have a ventral nerve cord, whereas the spinal cord in chordates (**Deuterostomes**) is located in the dorsal side of the body. Once the brain of vertebrates evolved, its fundamental characteristics have changed remarkably little over the past half billion years. Different regions have developed to serve various sensory functions, such as vision, hearing, and tactile, mechanical and electromagnetic senses. Birds and mammals have evolved increasingly complex cerebral cortices for greater integration of the sensory input to the brain. High-level cortical processing and cognition in some mammals have been among the most recent evolutionary trends.

OVERALL STRUCTURE OF THE NERVOUS SYSTEM

The broadest subdivision of the nervous system is that between the **central (CNS)** and **peripheral (PNS) nervous systems** (Fig. 6.2). The CNS includes the brain and spinal cord. The PNS (essentially what we think of as nerves) includes three subdivisions—the somatic, autonomic, and enteric.

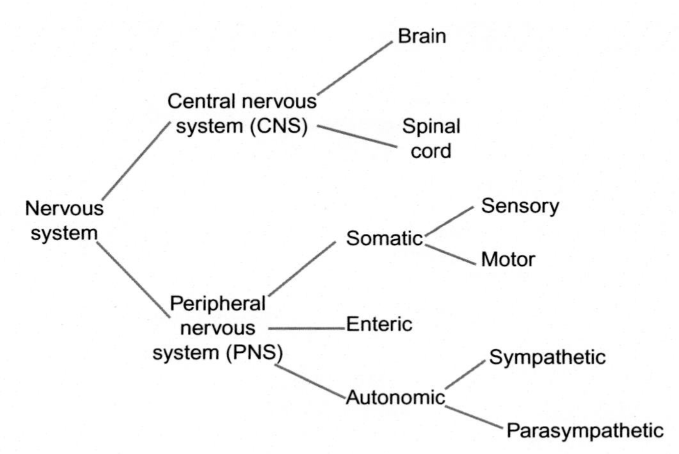

FIGURE 6.2 Overall organization of the nervous system.

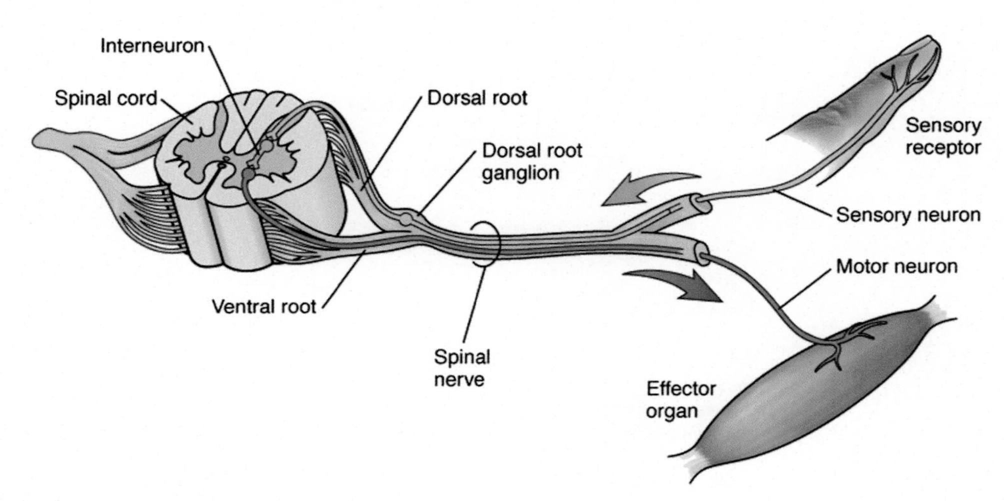

FIGURE 6.3 Components of a simple reflex arc.

Peripheral Nervous System

Somatic nerves are the most conspicuous components of the PNS. These are often large nerves, white in color, which run down the limbs and anywhere in the body where there are muscles. A typical peripheral somatic nerve contains both sensory and motor fibers and likely, some autonomic nerve fibers, as well. The classical simple version of the organization of a peripheral nerve is a sensory nerve fiber with a long **dendrite** originating far in the periphery and its cell body situated in a **sensory ganglion** close to the spinal cord. From the cell body in the ganglion, a short **axon** extends into the spinal cord and connects with several other neuronal processes. The simplest are local **interneurons** that make immediate connections with the cell bodies of motor neurons, which are embedded within the gray matter at the same level of the spinal cord. The motor neurons send out axons that terminate on muscle fibers. These three cells form the basis for a **local reflex**, which is rapidly acting (Fig. 6.3). The sensory axons also connect with neurons that run up the spinal cord and bring the same sensory impulse into the brain for higher level processing.

The **autonomic nervous system** follows much less well-defined pathways in much of the body, but within the trunk, it can be identified grossly as a system of ganglia and bundles of interconnecting nerve fibers (Fig. 6.4). Autonomic nerves consist of two neurons arranged in series. The first neurons, called **preganglionic**, are situated within the spinal cord and send out their axons, which are myelinated, into the periphery. The second neurons, whose processes are non-myelinated, are called **postganglionic**. These terminate in a large number of internal structures. In contrast to somatic nerves, autonomic nerves are relatively slow-acting.

The two main divisions of the autonomic nervous system—the **sympathetic** and **parasympathetic**—often, but not always, have counteracting effects on the body's physiology. These two divisions arise in distinctly different locations (see Fig. 6.4). Sympathetic nerves come from the spinal cord at levels from the thoracic to the mid-lumbar area,

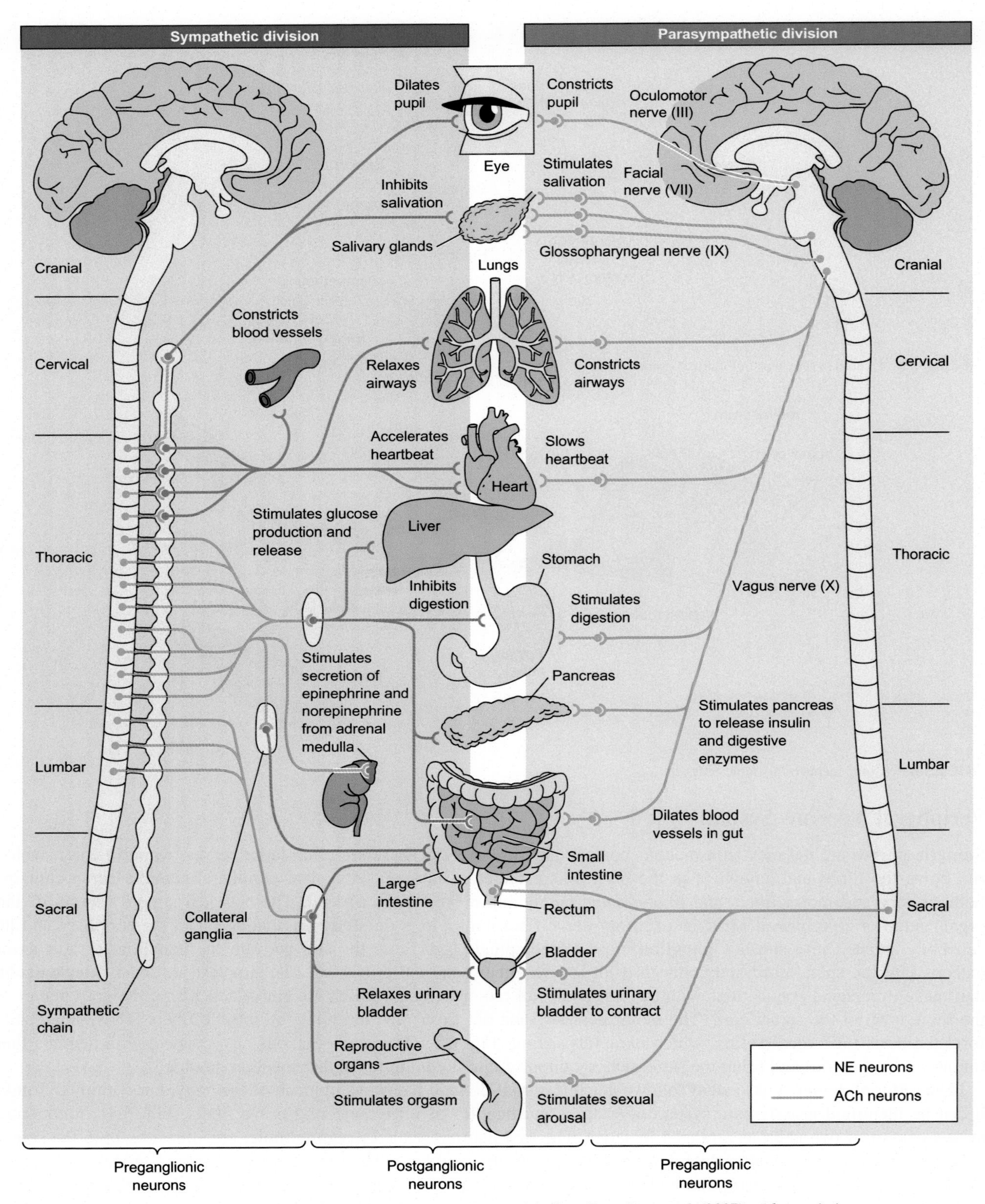

FIGURE 6.4 The autonomic nervous system. *NE*, norepinephrine; *ACh*, acetylcholine. *From Bear et al. (2007), with permission.*

whereas parasympathetic nerves arise in two disparate locations—in conjunction with some of the cranial nerves and in the sacral spinal cord.

In addition to different sites of origin, other significant differences characterize the two types of autonomic nerves. In sympathetic nerves, the myelinated preganglionic neurons synapse with the nonmyelinated postganglionic neurons in well-defined ganglia close to the spinal cord. One set, the **sympathetic chain ganglia**, consists of a string of interconnected ganglia running along each side of the vertebrae; hence, the alternate name **paravertebral**. Another set consists of several nonconnected ganglia (**collateral ganglia**) which are located farther from the vertebrae and are associated with the abdominal aorta. Because of their location close to the spinal cord, the postganglionic nerve fibers are often quite long in order to reach their end organs. The ganglia of parasympathetic nerves, on the other hand, are smaller and are located in the organs which they innervate.

These differences account for some important functional properties of the autonomic nerves. Because of their nonmyelinated postganglionic segment, autonomic nerves conduct impulses more slowly than do somatic nerves. Even the structural differences within the autonomic nerves have significant functional consequences. Sympathetic nerves tend to serve more generalized physiological responses because of the diffuse distribution of their fibers. Due to the considerable length of their nonmyelinated postganglionic segments, the sympathetic response is relatively slow. Parasympathetic nerves are focused more tightly on individual organs, and because of their long myelinated preganglionic segment and very short postganglionic segment, they conduct impulses more quickly than do sympathetic nerves. Thus, the parasympathetic response of a structure, such as constricting the pupil of the eye by the muscles of the iris, is a more rapid response than the widening of the pupil, which is part of a more generalized "fight-or-flight" sympathetic response. The differential response to sympathetic versus parasympathetic stimulation of the same organ is largely due to the transmitter emitted by the terminals of the postganglionic axons. Although within their ganglia both types of nerves use acetylcholine as the transmitter, the terminal postganglionic branches of sympathetic nerve fibers secrete mainly **norepinephrine**, whereas parasympathetic nerve fibers secrete **acetylcholine**.

The third component of the peripheral nervous system (PNS), the **enteric nervous system**, is organized and acts much like the CNS. In fact, it is now often called the second brain. Located within the walls of the gut, it consists of over 100 million neurons with many local interconnections. Many of these neurons are not directly connected with the CNS, and if the neural connections between the gut and the CNS are severed, the gut is still able to operate with a remarkable degree of autonomy thanks to the organization of the enteric nervous system.

Central Nervous System

The brain and spinal cord have a common embryological origin, and even in the adult many parts of the brain still bear traces of their commonality with the spinal cord. The human brain, however, has evolved in directions that have added layers of structural and functional complexity over the basic organization of the CNS.

Spinal Cord

The spinal cord extends within the protective cover of the vertebrae from the base of the skull down to the end of the first lumbar (L1) vertebra (Fig. 6.5). Like the brain, it is encased by three layers of soft tissue (Fig. 6.6A). The outer layer, the **dura mater**, is a tough protective membrane consisting principally of collagen fibers. The dura mater is innervated by sensory nerve fibers and can sense pain. No tissues beneath the dura are pain-sensitive. Immediately beneath that is a fine meshwork of cobweb-like fibers, called the **arachnoid layer**. Directly adherent to the spinal cord and brain is a transparent vascularized **pia mater** layer. The space between the spinal cord and the dura is filled with **cerebrospinal fluid (CSF)** (Box 6.1), which among other things, acts as a mechanical buffer for the soft tissues of the brain and spinal cord.

The spinal cord itself is grossly relatively featureless except for the pairs of spinal nerves that exit from the cord with every body segment, for example, at each vertebra. Although the spinal cord per se ends at the L1 level, a thin cord, called the **filum terminale** (see Fig. 6.5) extends down from this point and connects it to the end of the sacral vertebrae. Projecting past the terminal end of the spinal cord is a large group of spinal nerves, called the **cauda equina** (horse's tail). These nerves originate within the spinal cord, but individual nerves forming the cauda exit the vertebral column between each pair of lumbar and sacral/coccygeal vertebra.

The disparity between the length of the spinal cord and the vertebral canal is due to differential growth between the embryonic spinal cord and the vertebral column, which elongates more than the spinal cord as the fetus grows. Thanks to this disparity, the space below the end of the spinal cord is available for lumbar punctures for obtaining samples of CSF without the clinician's being in danger of injuring the spinal cord itself during the procedure.

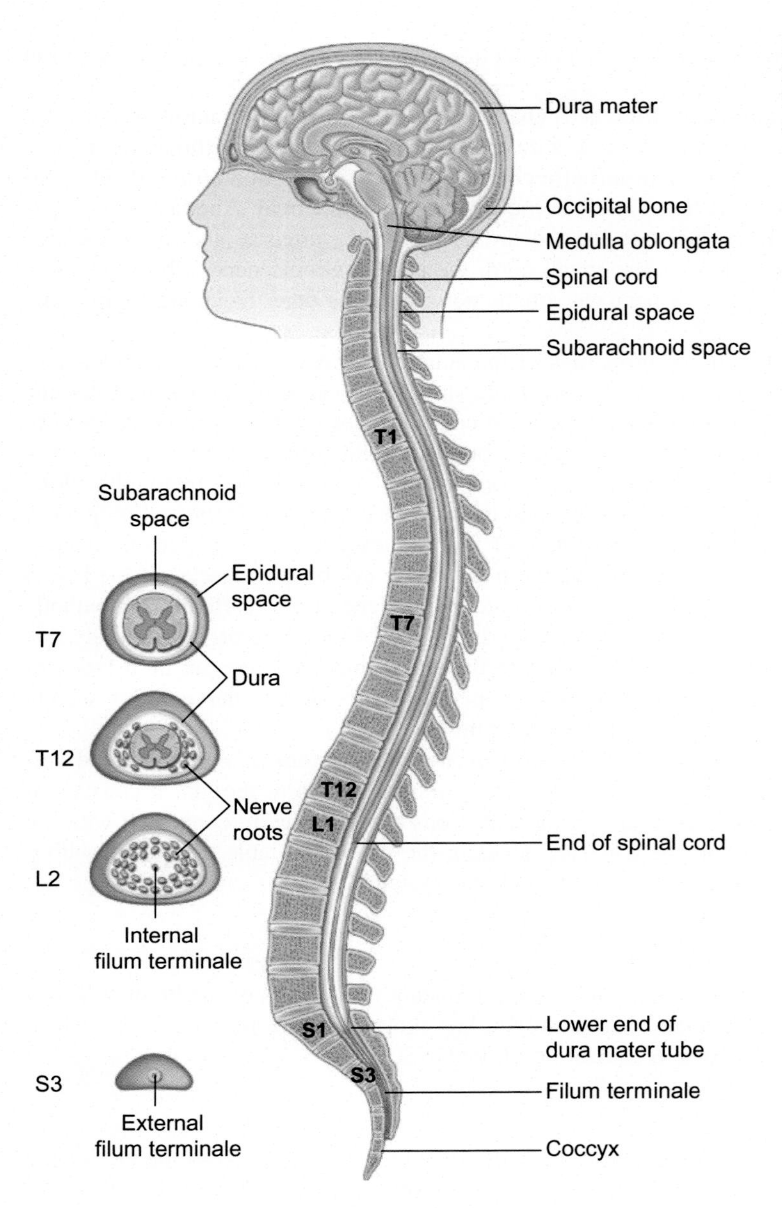

FIGURE 6.5 Overall structure of the central nervous system. Note the termination of the spinal cord at the level of L-1. *L*, lumbar; *S*, sacral; *T*, thoracic vertebrae. *From Waugh and Grant (2014), with permission.*

The internal organization of the spinal cord is relatively constant over its length (see Fig. 6.6A). In the very center is a small **central canal** through which CSF circulates. This canal is lined by a thin epithelium, called the **ependyma**, which has increasingly been recognized as a source of stem cells. Surrounding the central canal is a core of **gray matter**—butterfly-shaped in cross-section (see Fig. 6.6A). Gray matter is gray because the nerve processes within it are nonmyelinated. The gray matter also contains the cell bodies of neurons. The remainder of the spinal cord consists of **white matter**—groups of myelinated nerve processes that appear white because of their fatty coverings. The axons and dendrites within the white matter are grouped into bundles, called tracts, each of which serves a specific function (Fig. 6.6B).

Nerves leaving the spinal cord at all levels along the vertebral column serve a wide variety of functions that are associated with specific segments of the cord (Fig. 6.8). The nerves supplying the arms and legs are much more robust than those that supply the thorax or abdomen, and these differences are reflected in the spinal cord. Elegant experiments in chick embryos performed many years ago showed the relationship between the size of the peripheral load and the size of the nerves emanating from the spinal cord at that level. Grafting extra limb buds to the flank in chick embryos resulted in a considerable enlargement of the nerves that supplied the new limb and a corresponding enlargement of the gray matter in the area of the spinal cord from which the motor nerve fibers originated.

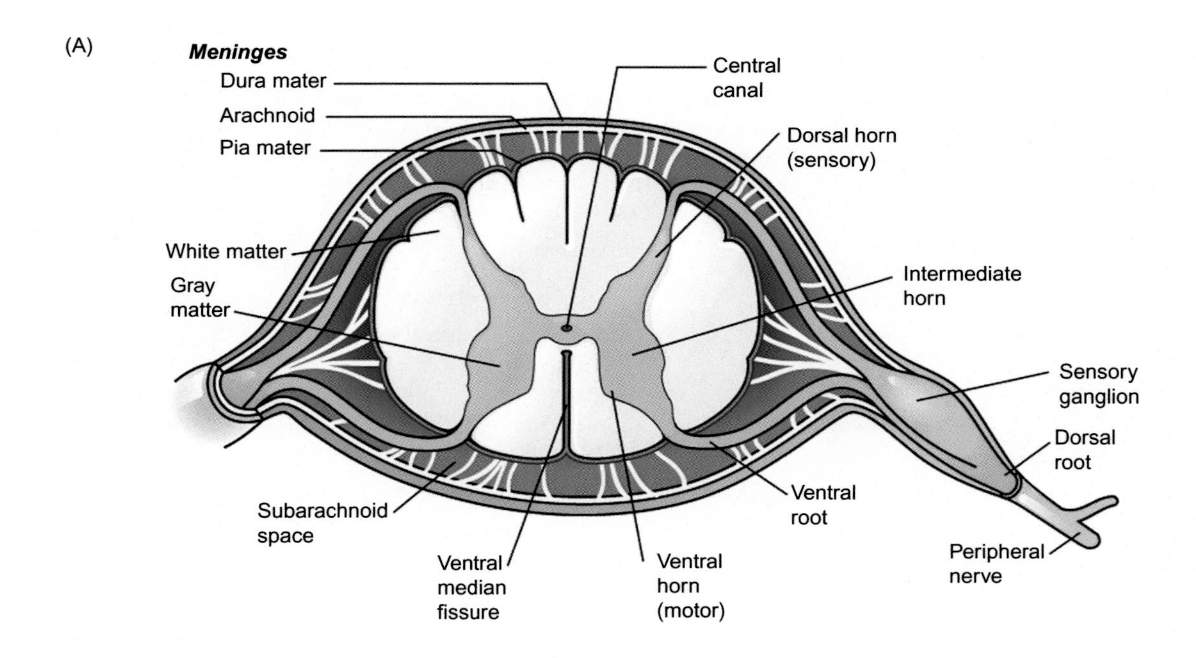

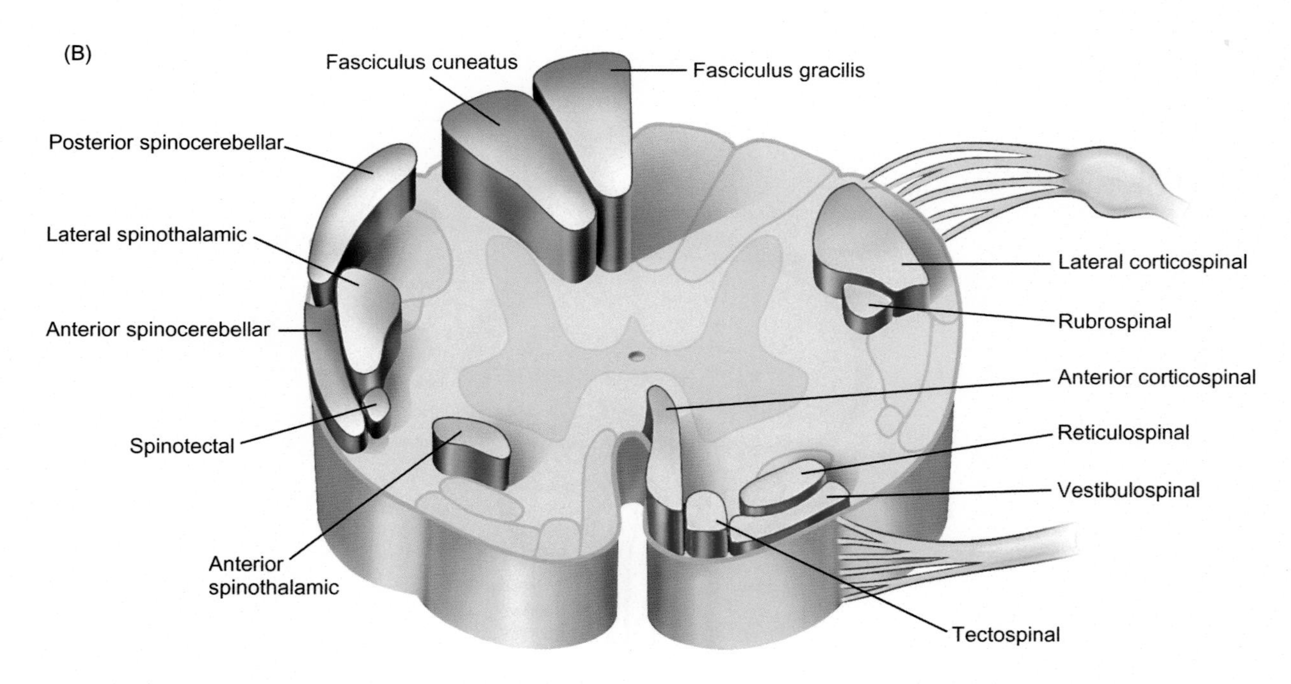

FIGURE 6.6 (A) Cross-sectional anatomy of the spinal cord and meninges. (B) Cross-section of the spinal cord, showing the locations of the major tracts. *Red*, motor tracts; *blue*, sensory tracts. *From Thibodeau and Patton (2007), with permission.*

Brain

The gross structure of the brain is reasonably straightforward, but this masks a bewilderingly complex internal structure and functionality. In the embryo, the brain initially arises as a simple tube in much the same manner as the spinal cord, but soon various parts of the developing brain go their own way, in manners unique to their specific regions (Fig. 6.9). From a basic tube, the brain first becomes subdivided into three regions and shortly thereafter, into five (Fig. 6.10). Starting as early as the sixth week of embryonic development, many of the adult regions of the brain are recognizable derivatives from the basic five subdivisions.

BOX 6.1 Cerebrospinal fluid and the ventricular system of the central nervous system

The entire CNS is bathed in CSF, a filtrate of blood plasma. The main functions of CSF are mechanical. Because the brain and spinal cord have essentially the same specific gravity as CSF, they float in a bath that both buffers them against mechanical trauma and reduces tension on nerves and blood vessels connected with the CNS. In addition, some metabolites from brain tissue are removed via CSF. CSF provides a stable ionic environment for the cells and tissues of the CNS.

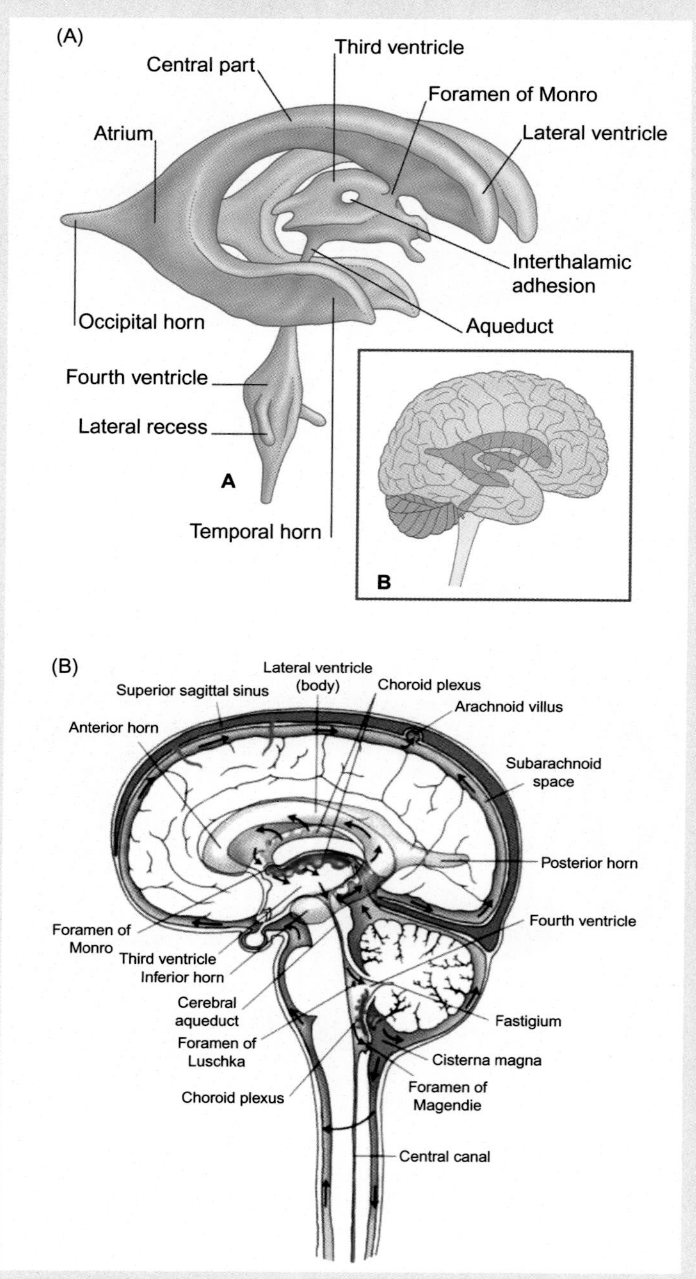

FIGURE 6.7 The ventricular system of the brain. (A) Anatomy. (B) Flow of cerebrospinal fluid (*black arrows*). *From (A) Fitzgerald et al. (2012) and (B) Siegel and Sapru (2011), with permission.*

Overall, CSF is formed in specialized areas, the highly vascularized **choroid plexuses**, within the ventricular system of the brain. Some, however, is formed within the brain tissue itself. Within the brain, the original central canal of the embryonic CNS has expanded to an interconnected group of four ventricles (Fig. 6.7A). The first two large ventricles, called **lateral ventricles,**

(*Continued*)

BOX 6.1 (Continued)

are enclosed by the cerebral hemispheres. Through short channels, called **foramina of Monro** (interventricular foramina), these are connected to a median third ventricle, which is situated at the base of the brain and is largely enclosed laterally and ventrally by the thalamus and hypothalamus. Through another channel, the **aqueduct of Sylvius** (cerebral aqueduct) the third ventricle is connected to the fourth ventricle, which underlies the cerebellum and occupies the central region of the upper medulla. The fourth ventricle is continuous with the central canal of the spinal cord.

Most CSF is formed by choroid plexuses located within the lateral ventricles. The first step in its formation is rough filtration by the capillaries within the choroid plexus to prevent the exit of blood cells. Blood proteins are filtered out by tight junctions between the epithelial cells that constitute the lining of the choroid plexus. Beyond that, active membrane gates modulate the concentrations of inorganic ions within the CSF. Some glucose is present in CSF.

Once formed, CSF then flows into the third ventricle and through the aqueduct of Sylvius into the fourth ventricle, where another area of choroid plexus secretes additional CSF into the mix. Within the fourth ventricle, CSF can also leave the ventricular system through the **foramina of Magendie and Luschka**. It then circulates around the subdural space that surrounds the brain and spinal cord (see Fig. 6.7B). In order to maintain an equilibrium, CSF is removed from the subdural space via **arachnoid villi**, which project into venous sinuses within the dural layer covering the brain. Although the total volume of CSF ranges from 90 to 140 ml, about 500 ml is produced and recycled each day.

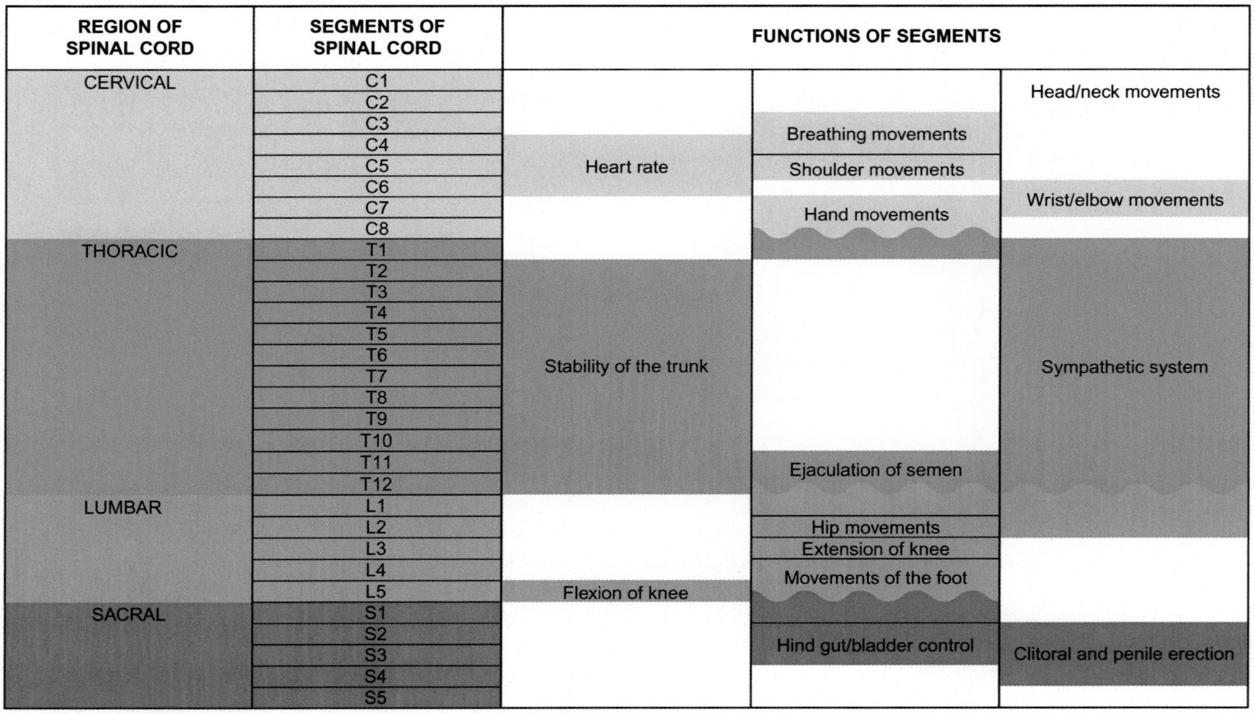

REGION OF SPINAL CORD	SEGMENTS OF SPINAL CORD	FUNCTIONS OF SEGMENTS		
CERVICAL	C1			
	C2			Head/neck movements
	C3		Breathing movements	
	C4			
	C5	Heart rate	Shoulder movements	
	C6			
	C7		Hand movements	Wrist/elbow movements
	C8			
THORACIC	T1			
	T2			
	T3			
	T4			
	T5			
	T6	Stability of the trunk		Sympathetic system
	T7			
	T8			
	T9			
	T10			
	T11		Ejaculation of semen	
	T12			
LUMBAR	L1			
	L2		Hip movements	
	L3		Extension of knee	
	L4		Movements of the foot	
	L5	Flexion of knee		
SACRAL	S1			
	S2		Hind gut/bladder control	Clitoral and penile erection
	S3			
	S4			
	S5			

FIGURE 6.8 Functions of the segments of the spinal cord. *From England and Wakely (2006), with permission.*

Brainstem

From both a structural and functional standpoint, the **brainstem** is a significant component of the brain (Fig. 6.11). The brainstem, which includes major derivatives of the embryonic mesencephalon, metencephalon, and myelencephalon, bears considerable resemblance to the spinal cord, and many of the functions of the brainstem are fundamental to life. The structures found in the human brainstem constitute much of the CNS in earlier vertebrates, and the brainstem is considered to be an evolutionarily primitive part of the brain. Some neuroscientists also include part of the diencephalon, called the thalamus, as part of the brainstem.

Specific structures aside, three broad categories of function characterize the brainstem as a whole. First, it serves as a conduit, through which almost all of the major tracts of nerve processes pass from the spinal cord to the brain or vice

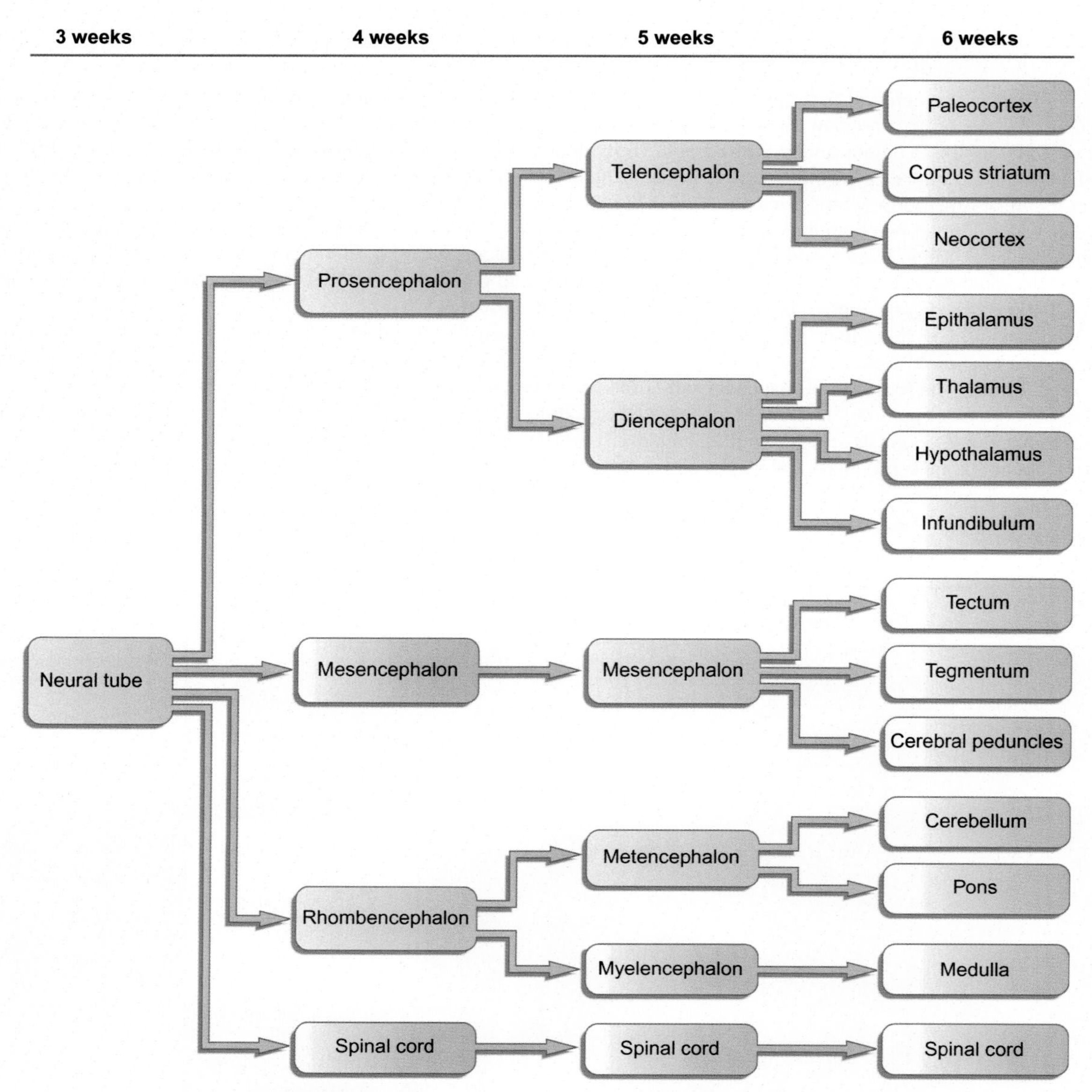

FIGURE 6.9 Increasing levels of complexity in the developing human brain. *From Carlson (2014), with permission.*

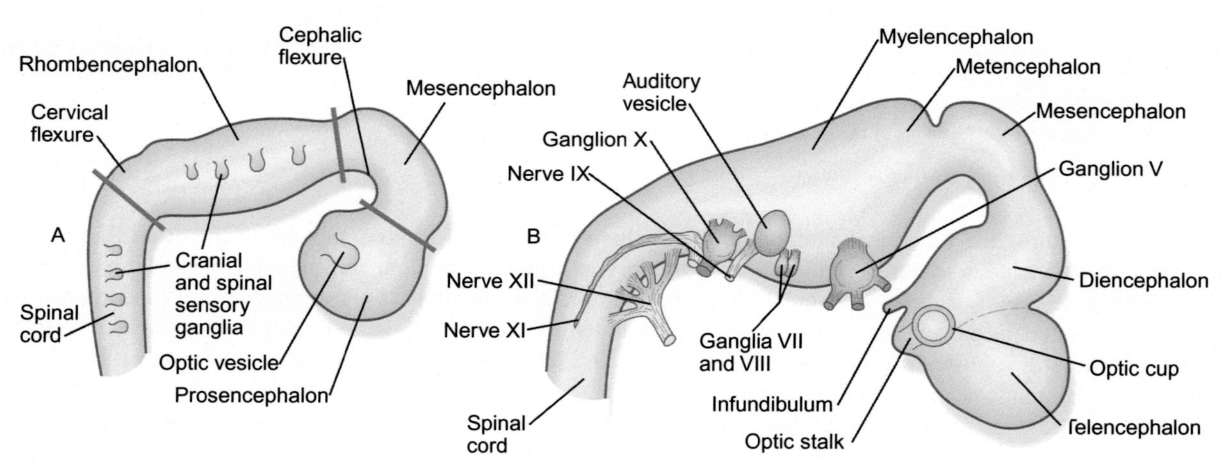

FIGURE 6.10 The basic anatomy of the three- and five-part brain in the human embryo. *From Carlson (2014), with permission.*

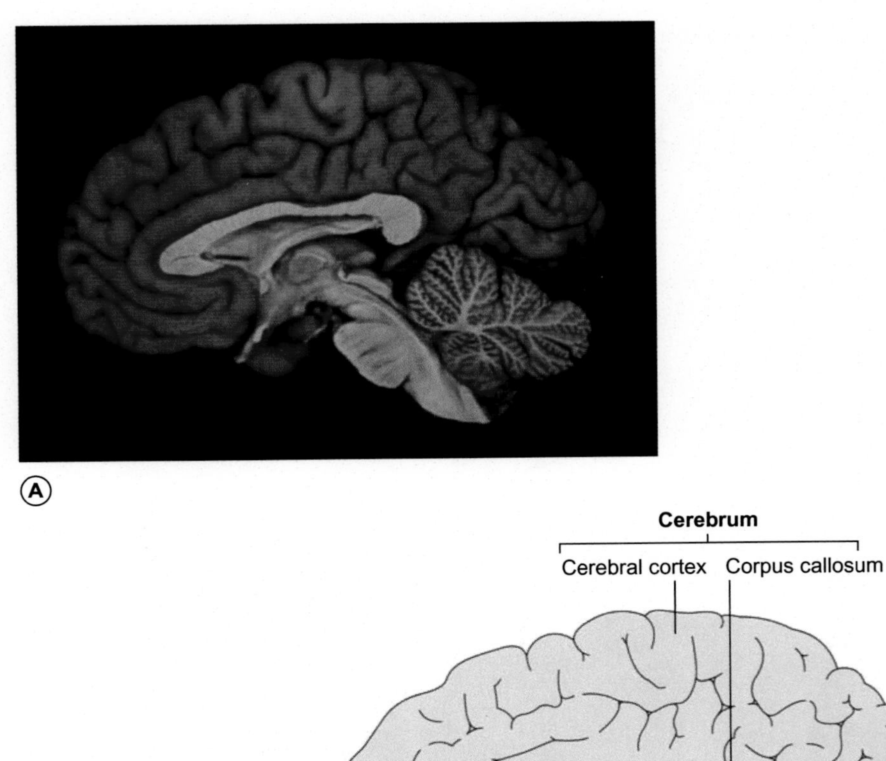

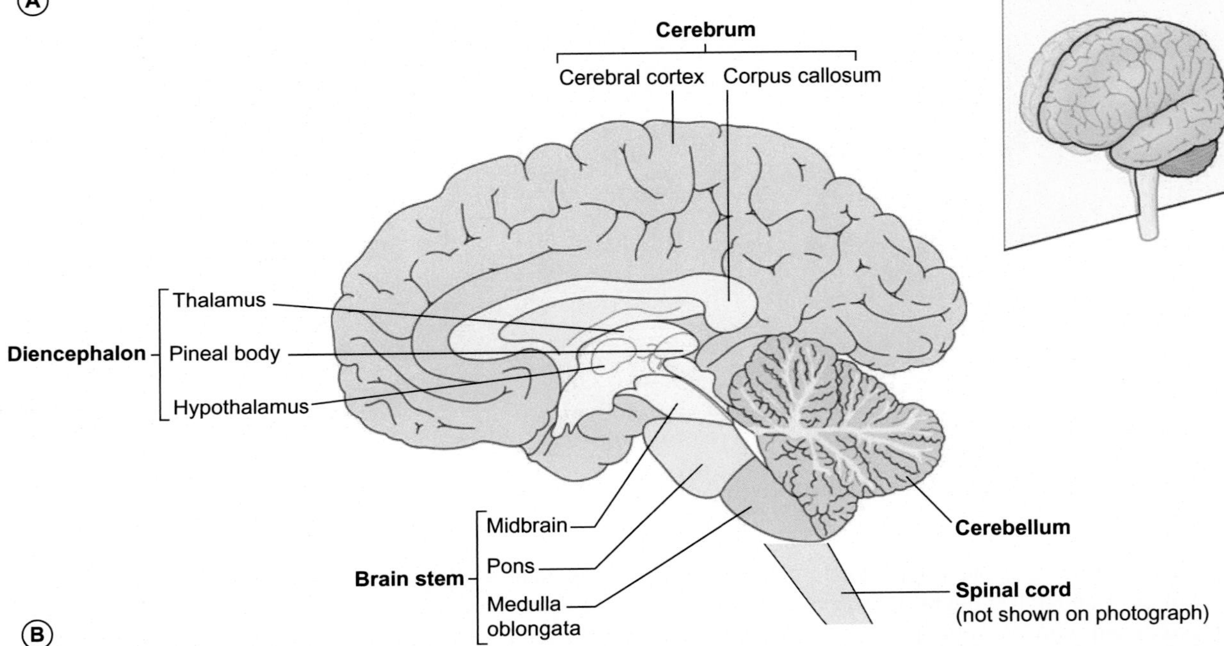

FIGURE 6.11 (A) Midsagittal section through the human brain. (B) Drawing, indicating the major components. *From Waugh and Grant (2014), with permission.*

versa. Much of the bulk of the brainstem is occupied by these tracts. A second feature of the brainstem is that most of the cranial nerves enter into and exit from it (Fig. 6.12; Table 6.1). With some modifications, the presence of segmental cranial nerves coming from the brainstem roughly parallels the arrangement of spinal nerves in relation to the spinal cord. The third functional feature is the presence of integrative centers that are critical for life. These integrative centers take the form of brainstem nuclei, which are groups of neurons embedded within the various regions of the brainstem that collect and process sensory information from a variety of sources and then send out motor signals. Most of these integrative centers receive information gathered by peripheral receptors or generated by higher brain centers and process that information in a way that leads to a specific response, such as changes in heart or breathing rate or blood pressure. Some of the integration and subsequent responses are purely reflex in nature. These reflex centers bypass higher coordinating centers in the brain and, as a result, produce a more rapid response, such as a sneeze or a cough.

In the roof of the midbrain are two pairs of hillocks, called the superior and inferior colliculi. As is covered in greater detail in Chapter 7, the **superior colliculi** are major visual coordinating centers, whereas the **inferior colliculi** serve important auditory processing functions.

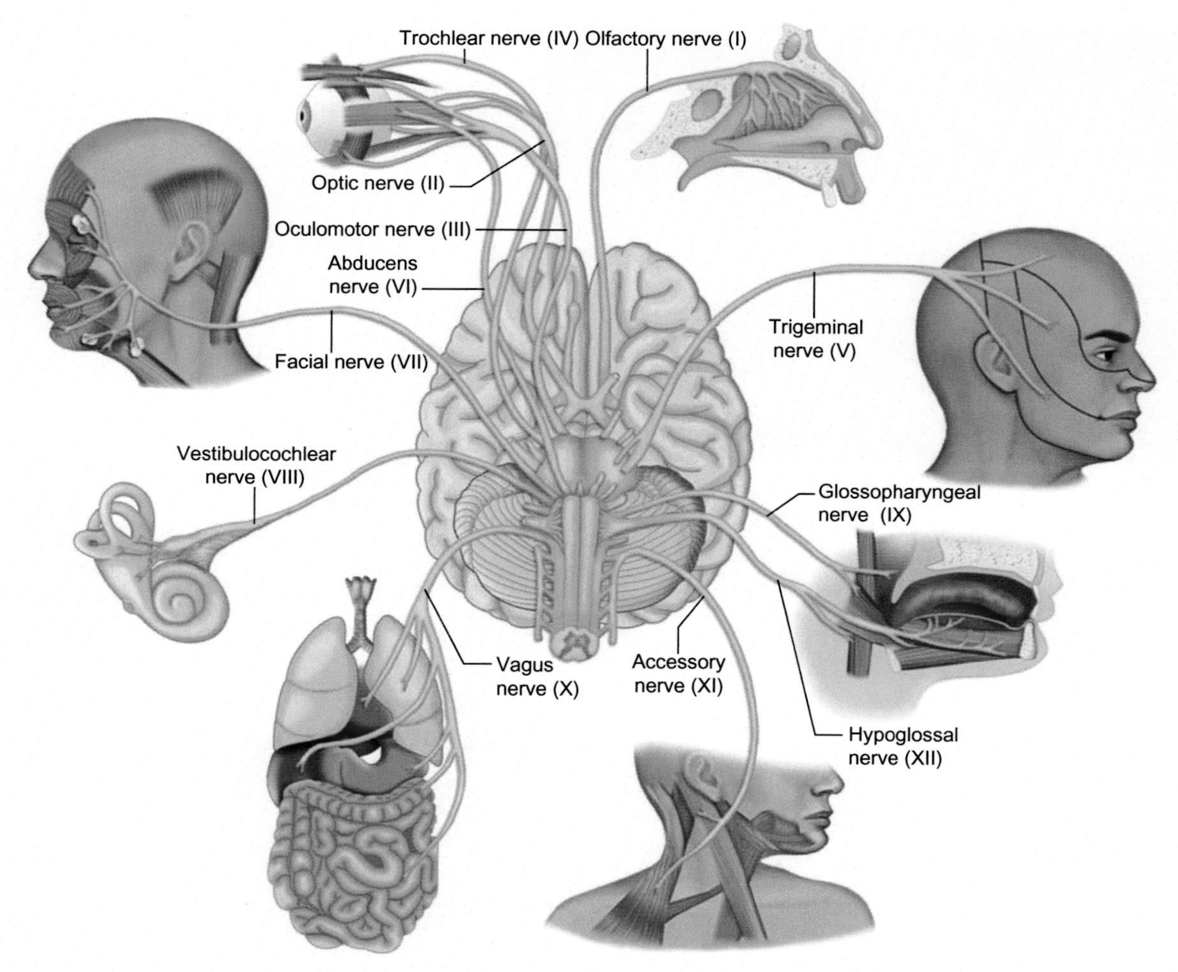

FIGURE 6.12 Anatomical distribution of the cranial nerves. *From Thibodeau and Patton (2007), with permission.*

A vitally important, but poorly anatomically defined component of the brainstem is the **reticular formation**, an aggregation of about 100 neural networks and **nuclei**[1] that are distributed throughout the brainstem. Many of the functions associated with the reticular formation operate at a subconscious level, and many of these modulate major influences that pass in both directions between the brain and spinal cord. It provides an overlay to motor control by participating in the maintenance of balance, posture, and muscle tone. The cardiac and vasomotor control centers of the medulla are components of the reticular formation. It plays a complex role in the modulation of pain, because some pain fibers pass through it on their way to the cerebral cortex. One analgesic pathway can block some pain signals from reaching the cerebrum. In still poorly understood ways, the reticular formation is heavily involved in the generation and maintenance of states of alertness and sleep. In addition, **habituation** (the ability to ignore repetitive sensory signals while being responsive to others, for example, a baby's cry in a noisy environment) is an important function of the reticular formation.

Cerebellum

The **cerebellum** is an outgrowth from the brainstem that plays a vital role in modulating movements. In humans, it is tucked beneath the posterior part of the cerebrum, and when sectioned, it looks something like a tightly folded cauliflower (see Fig. 6.11). The human cerebellum consists of two main lateral lobes, with a thin wormlike connection,

1. In the context of neuroanatomy, a nucleus is an aggregation of neuronal cell bodies within the gray matter of the central nervous system. Typically, the neurons constituting an individual nucleus subserve a specific function(s).

TABLE 6.1 Functions of the Cranial Nerves

Nerve		Functions
I	Olfactory	Olfaction
II	Optic	Vision
III	Oculomotor	Motor to several extraocular muscles
		Parasympathetic constriction of pupils
IV	Trochlear	Motor to extraocular muscle (superior oblique)
V	Trigeminal	Motor to muscles of mastication and others
		Sensory from skin of head and neck, tongue, meninges, sinuses, tympanic membrane
VI	Abducens	Motor to extraocular muscle (lateral rectus)
VII	Facial	Motor to muscles of facial expression and others
		Parasympathetic supply to most glands of head
		Sensory from area around ear
		Taste—anterior 2/3 of tongue
VIII	Vestibulocochlear	Hearing, equilibrium
IX	Glossopharyngeal	Motor to stylopharyngeus muscle
		Sensory from posterior 1/3 of tongue, inner tympanic membrane
		Sensory from carotid body
		Taste from posterior 1/3 of tongue
X	Vagus	Motor to muscles of pharynx and larynx
		Sensory from pharynx, larynx, and viscera
		Parasympathetic supply to pharynx, larynx, thoracic, and abdominal viscera, gut
XI	Spinal accessory	Motor to sternocleidomastoid and trapezius muscles
XII	Hypoglossal	Motor to muscles of tongue

appropriately called the **vermis**, connecting them in the midline. Tucked in alongside, and including part of the vermis, is the evolutionary ancient **flocculonodular lobe**, which in the human is heavily involved in balance.

The cerebellum is characterized by a number of unusual features. Although accounting for only about 10% of the volume of the brain, it yet contains more than 50% of the total number of neurons in the brain. Like in the cerebrum, the gray matter of the cerebellum—the cortex, containing the neurons—is located on the outside, and the white matter, constituting myelinated nerve processes, on the inside. The cerebellar cortex itself is subdivided into three layers. The innermost layer, called the **granular layer**, is composed of huge numbers of very small neurons—the **granule cells**. The middle layer consists of a single sheet of huge **Purkinje neurons** (Fig. 6.13). These cells have a single axon that penetrates into the interior white matter of the cerebellum and an enormous arborization of dendrites that extend into the outer **molecular layer**, which consists of a maze of neural processes interconnected by synapses.

The interior of the cerebellum contains an assortment of nuclei that receive input from other parts of the brain or the cerebellum itself and relay neural signals to other areas. Three paired bundles of nerve fibers, called **cerebellar peduncles**, carry neural signals to or from the cerebellum. The **inferior cerebellar peduncles** contain several types of fibers with distinct functions. An important one is a tract that carries sensory (**proprioceptive**) information from muscles and joints via the spinal cord and medulla. Another brings to the cerebellum proprioceptive and balance information from the head and the vestibular apparatus of the ear (see p. 205). In addition, it contains fibers that regulate both cerebral and cerebellar functions. The **middle cerebellar peduncles** carry commands for movements generated in the cerebrum into the cerebellum for further modulation. The **superior cerebellar peduncles** carry mainly messages from the cerebellum to the cerebrum via the thalamus (a major relay station in the brain).

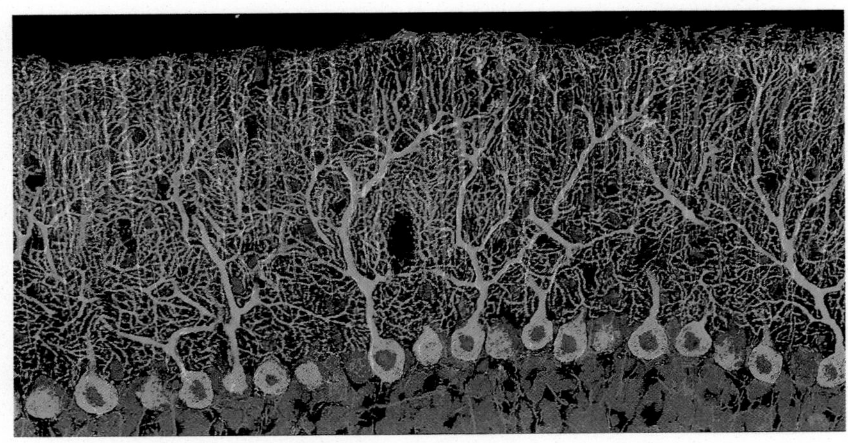

FIGURE 6.13 Purkinje neurons (green) in the cerebellar cortex. *From Kerr (2010), with permission.*

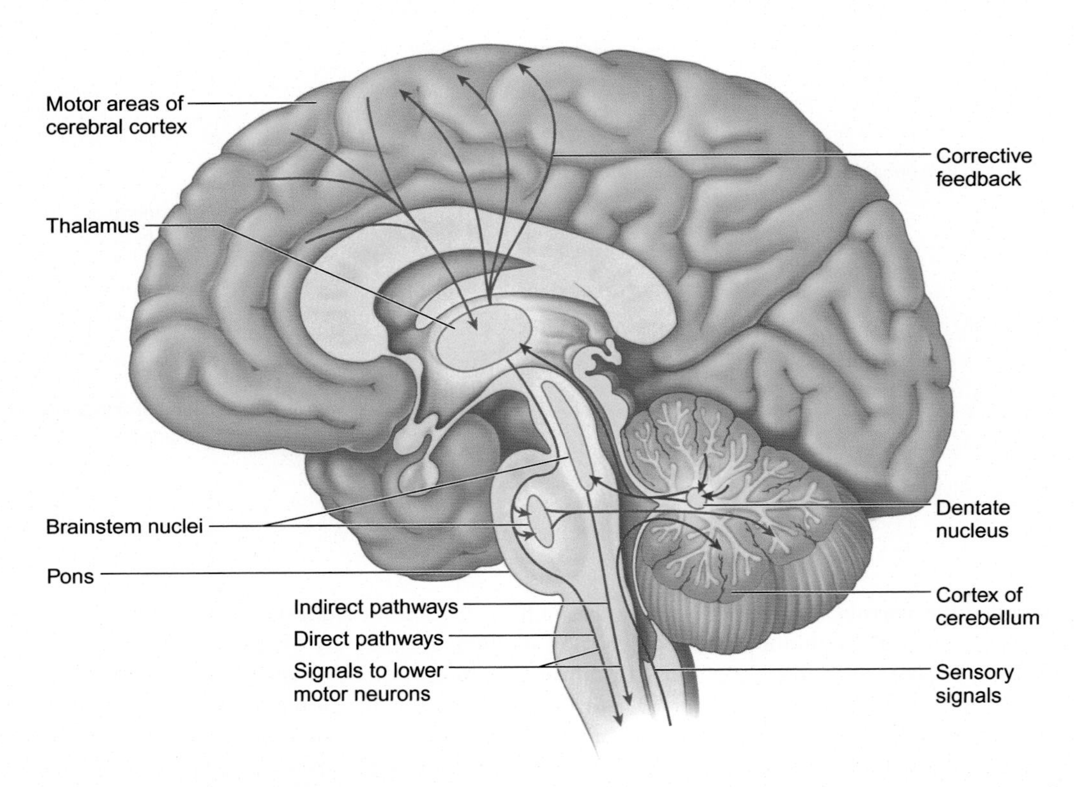

FIGURE 6.14 Major coordinating pathways from the cerebellum. *Blue arrows—* sensory/afferent pathways; *red arrows*—motor/efferent pathways. *From Thibodeau and Patton (2007), with permission.*

One of the main cerebellar functions is the modulation of conscious skilled motor movements through the activities of the lateral lobes. The basic movements are mainly initiated within the cerebrum and are then relayed to the cerebellum through tracts of fibers that pass through a variety of nuclei located within the brainstem and the cerebellum itself (Fig. 6.14). Other sensory afferent input from proprioceptors located within muscles and joints (e.g., muscle spindles and Golgi tendon organs, see p. 124) enters the cerebellum from the spinal cord or nerves in the head. Integration of this information is done through the participation of intrinsic cerebellar nuclei and tightly organized connections between several types of intracerebellar nerve fibers that converge upon the Purkinje cells through many synaptic connections within the outer molecular layer of the cerebellar cortex. Some of these broad connections with dendrites of many Purkinje cells account for smooth actions of many groups of axial and trunk muscles. Motor learning, the improvement of quality of specific actions with repeated practice (e.g., shooting basketball free throws), is an important function of the cerebellum. In the function of maintaining balance and posture, the cerebellum works closely with the vestibular system, as it does in coordinating many complex movements. The cerebellum is becoming increasingly recognized as a participant in cognition and functions, such as language.

Traditionally, much of our knowledge of cerebellar function came from studying patients with specific lesions of the cerebellum. Most of these lesions were characterized by movement disorders. Common symptoms involve jerky or noncoordinated movements. Other problems associated with cerebellar lesions are **intention tremors**— tremors increasing in intensity when approaching a target—and difficulties in performing rapidly alternating movements (**dysdiadochokinesia**). Motor learning is also compromised in certain cerebellar lesions. Newer functional imaging studies have shown more participation of the cerebellum in cognitive tasks and affective states than was earlier imagined.

Diencephalon

Situated just above the brainstem and mostly surrounded by the cerebrum, the **diencephalon** (see Fig. 6.11) is a key part of the brain through which almost all messages to and many from the cerebrum and the rest of the body pass. In addition, the diencephalon serves some functions that are unique to that area. The two main adult components of the diencephalon are the thalamus and hypothalamus. Associated with them is the **epithalamus**, which contains the pineal gland, and the **optic chiasma**, the crossroads between the two optic nerves—the only cranial nerve connected with the diencephalon. Central to the diencephalon is a fluid-filled space, the third ventricle of the brain (see Fig. 6.7). The thalamus forms most of the lateral walls of the third ventricle, and the hypothalamus constitutes the floor.

Thalamus. The **thalamus** is commonly known as the gateway to the cerebral cortex because many of the sensations received from the body are directed to some of the more than 20 nuclei located within the thalamus. From these nuclei, new nerve fiber tracts arise and are distributed to a wide variety of anatomical and functional regions within the cerebral cortex. Through the processing and redistribution of the sensory signals, the thalamus serves as a critical relay station within the CNS. In addition to processing sensory information, the thalamus is also an important processing center for motor stimuli emanating from the cerebellum or other deep areas of the brain and directed toward the cerebrum.

The thalamic nuclei have been grouped into several types, which relate to their ultimate function. Those nuclei that process most standard somatic sensory stimuli receive input from well-defined tracts running up the spinal cord or from the cranial nerves,[2] and their output is directed toward equally well-defined functional regions within the cerebral cortex (see Fig. 6.14). For this reason, they are often referred to as **specific nuclei**. Other thalamic nuclei receive input from a variety of deep brain centers, the cerebral cortex itself and some other thalamic nuclei. Called **association nuclei**, these also project to highly specific areas of the cerebral cortex. The third category of thalamic nuclei, called **nonspecific nuclei**, project outgoing fibers to wide regions of the cerebral cortex rather than to specific regions. They do, however, project to some specific and association nuclei within the thalamus.

In addition to more straightforward relay functions, such as those described earlier, the thalamus is also involved in more complex integrative functions, such as learning, memory, and emotions. For these, nuclei within the thalamus are components of neuronal circuits involving regions of the cerebrum, such as the limbic system and many areas of the cerebral cortex, and the reticular formation, which is centered in the midbrain. Given the immense number of possibilities for different combinations of connections of nerve fibers entering and leaving the thalamus, it is easy to understand why the thalamus is such an important integrative center within the brain.

Hypothalamus. Situated beneath the thalamus, as its name implies, the **hypothalamus** occupies only a small part of the brain (see Fig. 6.11), but its influence on the entire body is immense. Whereas through its aggregation of relay centers the thalamus is involved in variety of higher level functions, the hypothalamus serves as the initiator and coordinator of an important set of more elemental functions. Although many important hypothalamic functions are generated principally from within the hypothalamus itself, almost all depend upon input coming from other parts of the nervous system or the bloodstream.

Like the thalamus, the hypothalamus is dominated by a group of nuclei that coordinate the main activities of this region of the brain. These nuclei receive major afferent input from parts of the brain (limbic system) and brainstem (reticular formation) that are commonly associated with subconscious activities. In turn, they also send out efferent projections to both higher (e.g., to the thalamus and the cerebral cortex) and lower (midbrain and medulla) brain centers. A unique feature of the hypothalamus is its association with and control over functions of the pituitary gland (Box 6.2).

2. Major sensory modalities that are relayed through the thalamus include pain, temperature, hearing, vision, taste, balance, and smell.

BOX 6.2 The hypothalamus and the pituitary gland

In both structure and function, the hypothalamus and the pituitary gland are intimately linked (Fig. 6.15). The **pituitary gland** (**hypophysis**) consists of two very different components, and their relationship to the hypothalamus is equally different. The anterior pituitary (**adenohypophysis**) is a classic endocrine gland, with different types of cells producing different hormones (see p. 250). The posterior pituitary (**neurohypophysis**), on the other hand, is a neuroendocrine organ in which the hormones are directly secreted by neurons.

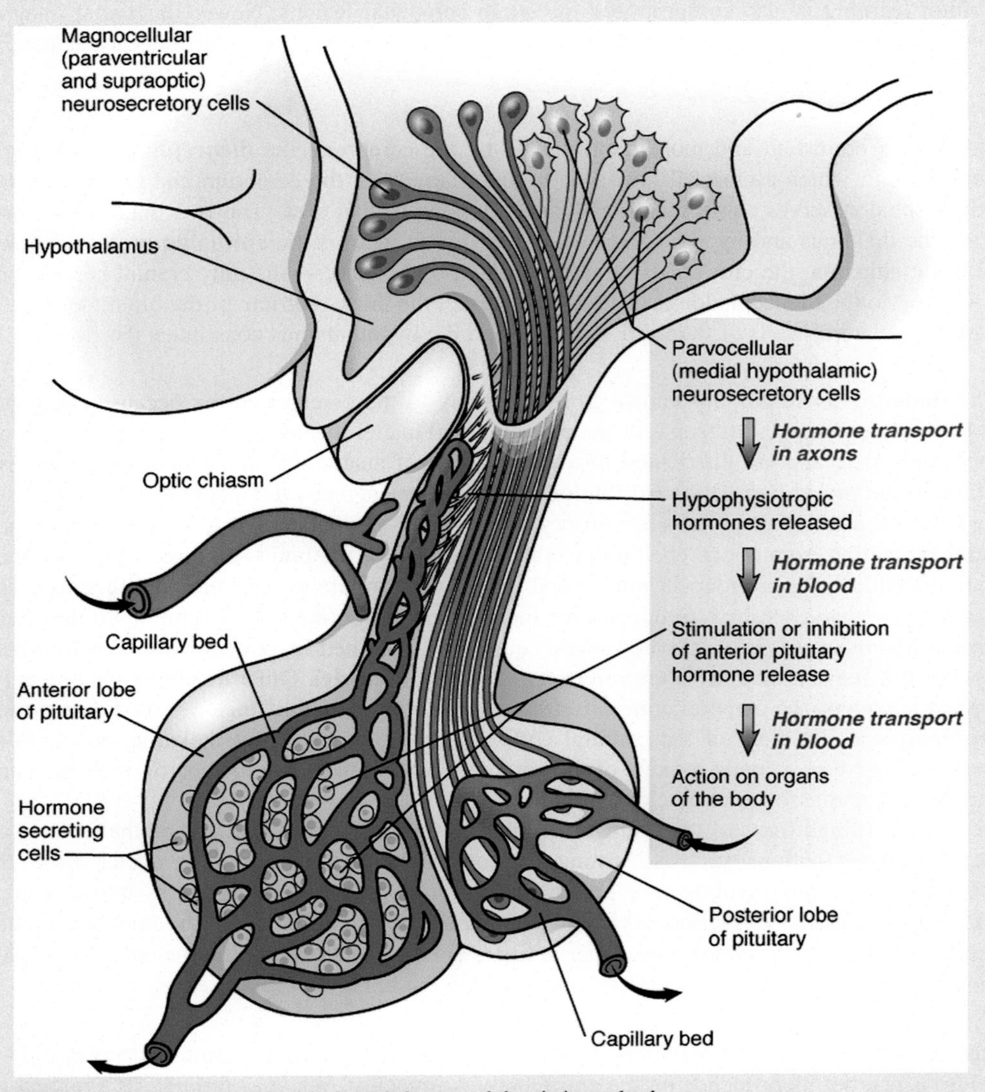

FIGURE 6.15 Relationships between the hypothalamus and the pituitary gland.

Hormones are released by cells of the anterior pituitary in response to **releasing hormones** produced by neurons in various locations within the hypothalamus. The releasing hormones, each of which targets a specific hormone-secreting cell type in the anterior pituitary, are secreted from nerve terminals into lumens of the meshwork of tiny venous vessels that constitute the **hypothalamohypophyseal portal system**.[3] The releasing hormones are then carried down to a capillary bed that surrounds the endocrine-secreting cells of the anterior pituitary. They then leave the portal circulation and stimulate the release of the primary hormones that are synthesized in the anterior pituitary (Table 6.2).

(Continued)

BOX 6.2 (Continued)

TABLE 6.2 Hypophyseal Releasing Hormones and Their Targets in The Pituitary

Releasing Hormone	Pituitary Cell Target	Pituitary Hormone
GHRH	Somatotroph	Growth hormone (GH)
GnRH	Gonadotroph	Follicle-stimulating hormone (FSH)
GnRH	Gonadotroph	Luteinizing hormone (LH)
TRH	Thyrotroph	Thyroid-stimulating hormone (TSH)
CRH	Corticotroph	Adrenocorticotropic hormone (ACTH)
Dopamine[a]	Lactotroph	Prolactin (PRL)

Abbreviations: GHRH, growth hormone-releasing hormone; GnRH, gonadotropin-releasing hormone; TRH, thyrotropin-releasing hormone; CRH, corticotropin-releasing hormone.
[a]*Prolactin is unusual in that it does not require a releasing hormone. Control of its release is, instead, accomplished through an inhibitor, namely dopamine, from the hypothalamus.*

In contrast to the anterior pituitary, the posterior pituitary is directly supplied by nerve fibers extending from large neurons located within two nuclei of the hypothalamus, the **paraventricular** and **supraoptic nuclei** (see Fig. 6.15). These neurons produce two hormones, antidiuretic hormone (ADH), sometimes called vasopressin, and oxytocin. These hormones are released within the posterior pituitary and are taken up by a local capillary bed that brings them into the general circulation.

Antidiuretic hormone acts directly on the collecting ducts of the kidney (see p. 367) to increase water absorption. Its release by the hypothalamic neurons responds to the sensing of the osmolarity of the blood by cells of the supraoptic and paraventricular nuclei. Recognition of changes in osmolarity is then translated to the neurons that synthesize and secrete ADH. When the body becomes dehydrated, the reduced amount of water within the blood concentrates the solutes, thereby increasing its osmolarity. This results in the release of ADH. In contrast, ADH secretion is reduced under conditions of overhydration.

Oxytocin is produced and released by neurons located within the supraoptic and paraventricular nuclei of the hypothalamus. Most of the oxytocin is secreted into the bloodstream, but in order to bypass the blood–brain barrier some oxytocin is secreted directly into the parenchyma (soft tissue) of the brain, where it can act on other neurons. Classically, oxytocin was considered to have two main functions—inducing milk letdown and stimulating the contractions of uterine smooth muscle cells, but in recent years multiple quite different functions have been attributed to this hormone.

Oxytocin's role in milk letdown begins with the stimulus of suckling at the nipple. This stimulus is carried to the spinal cord through spinal nerves and is relayed to the hypothalamus. It then causes the oxytocin-producing neurons to fire bursts of action potentials, resulting in the secretion of oxytocin from the nerve terminals and into the bloodstream. When circulating oxytocin reaches the mammary gland, it acts on the **myoepithelial cells** situated alongside the mammary ducts (see Fig. 14.21). The myoepithelial cells contract, squeezing the milk into the ducts, which also widen under the influence of oxytocin.

The other traditional role ascribed to oxytocin is its ability to stimulate contraction of the smooth muscle cells of the uterine wall. During early labor, it stimulates dilatation of the cervix, and later it causes strong contractions as the fetus is being expelled from the uterus. After birth, oxytocin stimulates strong contraction of the uterine wall in the area of the former placental attachment, thus reducing postpartum hemorrhage. In cases of excessive hemorrhage after delivery of the placenta, oxytocin is often delivered as a drug to reduce bleeding. Even during the first week after birth, nursing women may feel mildly painful contractions of the uterus as the oxytocin released during the nursing process continues to stimulate contraction of the uterine smooth muscle.

More recent research has ascribed a quite different set of new functions for oxytocin. Sometimes called the love hormone or contentment hormone in the popular press, oxytocin now appears to play a role in a variety of interpersonal situations. It is released through positive social reactions and is said to increase trust and empathy as it facilitates bonding, especially within individuals of a group, but not among those not belonging to the group. It is also said to bring about feelings of contentment after sexual arousal.

3. A portal system is a group of veins that begins from a capillary network and ends in another capillary network without first going to the heart. The largest portal system in the human is the hepatic portal system (see p. 295), which begins as capillary networks within the wall of the gut and proceeds through progressively larger veins that again terminate in a capillary network within the liver. In the case of the hypothalamohypophyseal portal system, a normal capillary network in the hypothalamus drains into small veins that pass down the stalk of the pituitary gland before breaking up into another capillary network within the body of the anterior pituitary into which pituitary hormones are secreted. Because of the short biological half-life of the releasing hormones, the portal system is an efficient way to direct them from the hypothalamus to the anterior pituitary before they become degraded.

The **pineal gland**, situated on the posterior wall of the third ventricle, also has important connections with the hypothalamus.

Functionally, the hypothalamus is a central player in the integrative functions of the nervous system. At the periphery, many of these functions are carried out by the autonomic nervous system. Within the CNS, the hypothalamus is a collecting point for many sets of behaviors that take their origins in the limbic system (see p. 156) or the reticular formation.

Temperature regulation well illustrates the different levels at which the hypothalamus can influence important physiological and behavioral functions. Within the hypothalamus itself is a group of neurons that are sensitive to changes in blood temperature. Impulses generated by these neurons are propagated to autonomic nerves surrounding blood vessels, which expand when the blood temperature is too high and constrict when it is too cold. At a more chronic level, the hypothalamus releases thyrotropin-releasing hormone, which in the anterior pituitary releases thyroid-stimulating hormone, a hormone that increases the overall metabolic rate of the body in a cold environment. At a more immediate time frame, signals from the hypothalamus induce shivering when cold and panting when warm.

Although the main center for cardiovascular control resides in the medulla, signals from both the hypothalamus and limbic system influence the cardiovascular control centers within the brainstem. Acting through sympathetic nerves, they increase the heart rate; conversely, through parasympathetic stimulation, they slow the heart rate.

The hypothalamus is a major center controlling feeding behavior. Much of our knowledge concerning this role comes from experiments in which regions of the hypothalamus were stimulated or lesioned. It is now clear that the lateral hypothalamus contains a **feeding center** and that the ventromedial region contains a **satiety center**. These regions respond to various environmental signals, often hormonal, which are generated by eating or by hunger. A good example is **leptin**, the protein secreted by adipocytes (see p. 45). This compound binds to a specific nucleus within the hypothalamus (arcuate nucleus) and leads to a reduction in the desire to eat. Sensory stimuli, taste and olfaction, reach the hypothalamus through the limbic system, specifically the amygdala (see p. 158) and increase the desire to eat. Similarly, hypothalamic signals stimulate the desire to drink, and under conditions of dehydration, water retention is promoted through the action of antidiuretic hormone, secreted by hypothalamic neurons terminating in the posterior pituitary.

Experiments conducted principally on cats have shown the hypothalamus to be the dominant center coordinating aggressive and defensive rage behavior. Much of the conscious or subconscious basis for rage behavior comes from higher brain centers, such as the cerebral cortex and limbic system, but these are filtered through hypothalamic nuclei that integrate these signals and turn them into defined actions.

Acting mainly through its hormonal influences on the pituitary gland, the hypothalamus also influences sexual behavior in both males and females. In order to do this, the hypothalamic neurons must gauge the internal hormonal environment through specific receptors on their surfaces and respond to that information by modulating the synthesis and release of the hormones that control the hormonal output of the pituitary gland. External environmental signals, such as changing light levels, odors, and temperature, are often processed through the hypothalamus and play a major role in the reproductive cycles of many animals and a lesser role in humans.

Many biological rhythms, including sleep, are also coordinated through the hypothalamus. Some of these rhythms are generated through the activities of clock genes that regulate transcriptional activity in a highly ordered manner. Many other rhythms are generated through the perception of light cycles by the retina of the eye. A good example of **daily (circadian) rhythms** in the human is the interaction among the retina, the hypothalamus and the pineal gland in controlling the secretion of the **melatonin**, a hormone derived from serotonin and heavily involved in sleep cycles (Fig. 6.16). The 24-hour cycle of light and darkness perceived by the retina is translated to the **suprachiasmatic nucleus** of the hypothalamus, which coordinates the internal rhythmic activity. Neural output from this nucleus extends to other hypothalamic nuclei (especially the **paraventricular nucleus**) and from there to the spinal cord. Sympathetic nerve fibers from the spinal cord extend to the pineal gland,[4] where the rhythmic synthesis and secretion of melatonin into the blood occurs. Melatonin secretion occurs as darkness falls, and it is an important mediator of sleep rhythms. Secretion of melatonin during daylight is reduced by inhibitory signals emanating from the suprachiasmatic nucleus to the paraventricular nucleus. These signals reduce the stimulation of the sympathetic nervous system through the paraventricular nucleus, with the result that melatonin synthesis in the pineal gland is reduced. On a long-term basis, seasonal changes in the length of daylight and darkness are also registered by the suprachiasmatic nucleus, which also regularly resets the circadian clock.

4. The pineal gland, a pine cone-shaped structure about the size of a grain of rice and located in the midline just behind the thalamus, is one of the few unpaired structures in the brain. Called the seat of the soul by Descartes, it is heavily vascularized as befits a secretory gland.

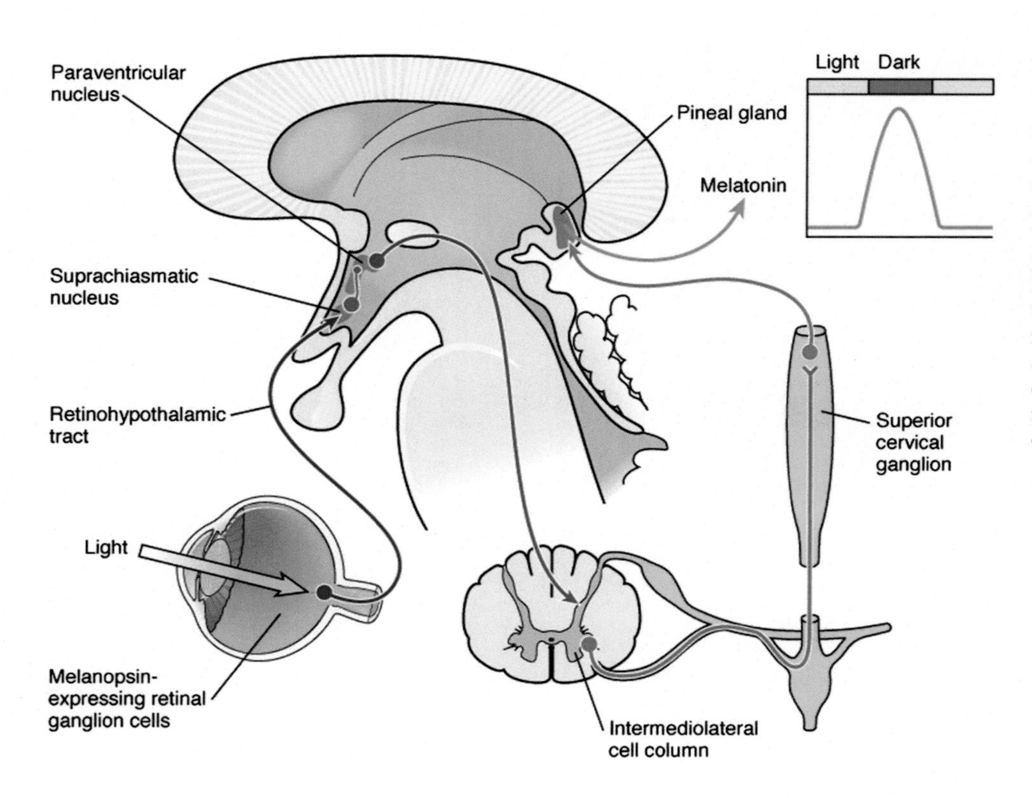

FIGURE 6.16 Neuronal circuits involved in the production and secretion of melatonin. Light-hitting photosensitive ganglion cells in the retina stimulates a signal going to the circadian pacemaker (suprachiasmatic nucleus). Then an inhibitory signal to the paraventricular nucleus of the hypothalamus reduces the sympathetic output to the pineal gland, resulting in decreased synthesis and release of melatonin. In the dark, this inhibition is reduced, and melatonin synthesis and release occurs.

Telencephalon

The **telencephalon** is the terminal part of the central nervous system and is dominated by the cerebral hemispheres, which represent the culmination of a long series of evolutionary developments in the CNS of vertebrates. The essence of the evolution of the vertebrate brain is the retention of critical regions and functions throughout phylogeny, with the addition of new structures and functions as needed by the evolving species. All vertebrates contain the same subdivisions of the CNS, but as they have evolved into more complex forms, their brains have become correspondingly more complex. The structure and functions of the brainstem (called the **archipallium** [Fig. 6.17], according to one classification), the most primitive part of the brain, have changed relatively little over millions of years. Even the fundamental organization of the cerebellum has remained remarkably stable (it is only absent in hagfish and lampreys), although its relative size and complexity have increased considerably in the higher vertebrates. The brainstem controls basic bodily functions, such as breathing, heart rate, temperature, and balance. Another primitive part of the brain is the area concerned with olfaction. This covers a number of anatomical regions, but its fundamental organization is remarkably well preserved.

Structures in the midbrain and forebrain relating to special senses—olfaction, vision, and hearing—vary among species depending upon how important they are to the animals. For example, the visual system and its representation in the brain is very prominent in birds, whereas the olfactory system is less so. With the rise of mammals, the prominence of the visual system has declined, possibly because the early mammals were largely nocturnal, but in most mammals, the olfactory system has assumed increased importance and size. Especially among fishes, electromagnetic sensation has evolved independently several times.

Throughout vertebrate phylogeny, the hypothalamic area has been a vital region that controls many autonomic and hormonal functions, but later in evolution, the appearance of a set of structures, commonly known as the **limbic system**, provided the basis for an increasingly complex mechanism of regulating and modulating the functions of the hypothalamus. In higher mammals, the limbic system (**paleopallium**) has become a primary region of the brain in which emotions are consolidated. The limbic system probably arose in the prehistoric ancestors of reptiles and birds, but it has attained its greatest degree of complexity in mammals. In keeping with their phylogenetic antiquity, the olfactory and limbic systems are closely connected, both anatomically and functionally.

The disproportionate rise in complexity of the cerebral **cortex (neocortex)** and associated deep structures (collectively known as the **neopallium**) in several groups of mammals (primates, cetaceans, and elephants) has resulted in the

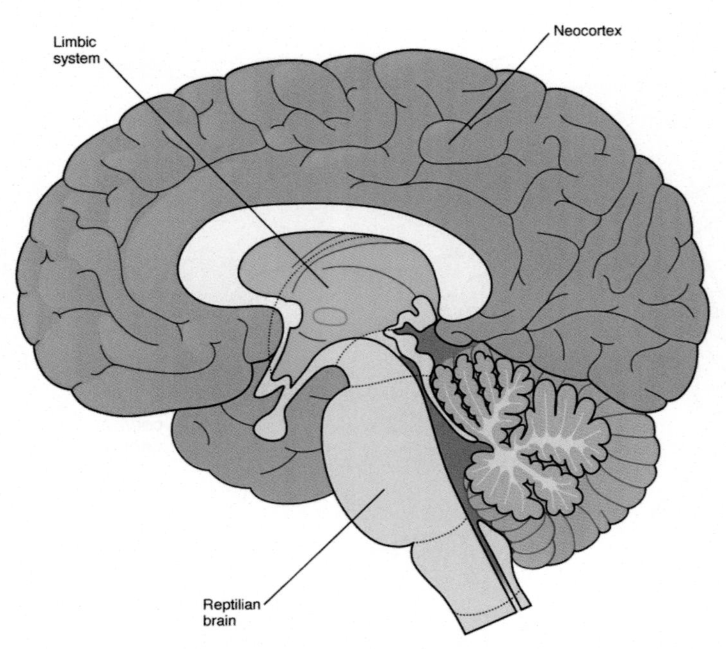

FIGURE 6.17 The three major evolutionary divisions of the brain. Gray, the reptilian brain (archipallium); pink, the limbic system (paleopallium); purple, the neocortex (neopallium).

human brain as we know it. The increase in overall mass of the cerebral cortex has been associated with the appearance of large numbers of neural coordinating centers (nuclei) deep within the cerebrum. This is a general evolutionary trend and is also evident in the number of deep nuclei also present within the enlarged mammalian cerebellum. The extensive folding of the cerebral cortex is a function of its relative size. Mammals with a smaller brain have a smooth cortical surface, whereas folds and crevasses dominate the surface topography of the brains of humans and dolphins. The greatly increased amount of gray matter made possible by the extensive folding of the cerebral cortex has allowed the development of the incredibly complex neural interconnections that form the structural basis for higher level intellectual processes.

Olfactory Region. Perhaps the most ancient component of the forebrain, the **olfactory system** overlies the olfactory region of the nose and extends back along the base of the prefrontal cortex until it reaches components of the limbic system (Fig. 6.18). This configuration directly connects the sense of smell to the region of the brain that is the seat of our emotions.

The **olfactory neurons**, located in the lining of the upper chamber of the nose (see p. 310), send axons (as many as three million per side) through numerous holes in the cribriform plate of the ethmoid bone, which separates the nasal chambers from the brain. Collectively, this diffuse stream of axons constitutes the **olfactory nerve** (cranial nerve I). Groups of axons, bound by Schwann cells into bundles called fila, enter the olfactory bulb and synapse with roughly 50,000 neurons (mitral cells), which in turn, are concentrated into ~2000 **glomeruli**. Within the glomeruli, other cells modulate the olfactory signals. There is evidence that individual glomeruli respond to different odorants. From the glomeruli, many axons extend directly into the amygdala and associated limbic structures (see below), where they interact with other neural processes that collectively influence much of our emotional makeup.

Limbic System. The **limbic system** consists of an aggregation of structures that operate at an almost purely subconscious level. Although they can exert a profound influence upon our emotions and sense of wellbeing, many limbic system effects modulate the even more elemental activities of the hypothalamus or spread to the cerebral cortex, where they add subtle overtones to conscious actions. Increasingly, the limbic system has been found to play a vital role in spatial learning and memory—especially short-term memory.

Centered at the seat of the cerebrum, the core of the limbic system includes the hippocampal formation, the septal area, and the amygdala, as well as the closely associated prefrontal cortex and the cingulate gyrus (Fig. 6.19). All of these structures receive input from many other parts of the brain—both lower, from the brainstem, and higher, from

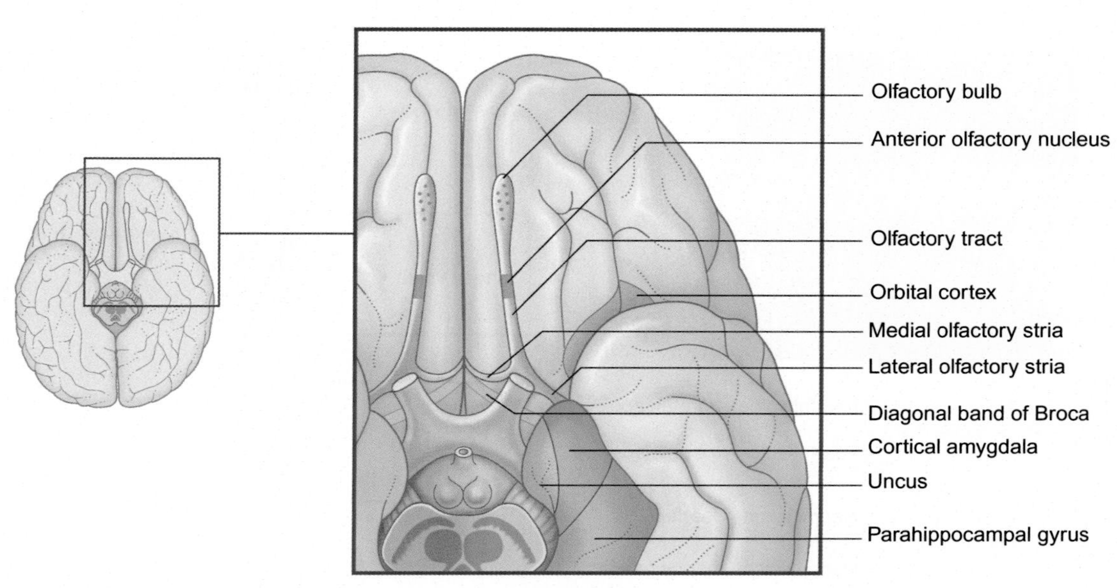

FIGURE 6.18 Ventral view of the brain, highlighting the olfactory areas. *From Fitzgerald et al. (2012), with permission.*

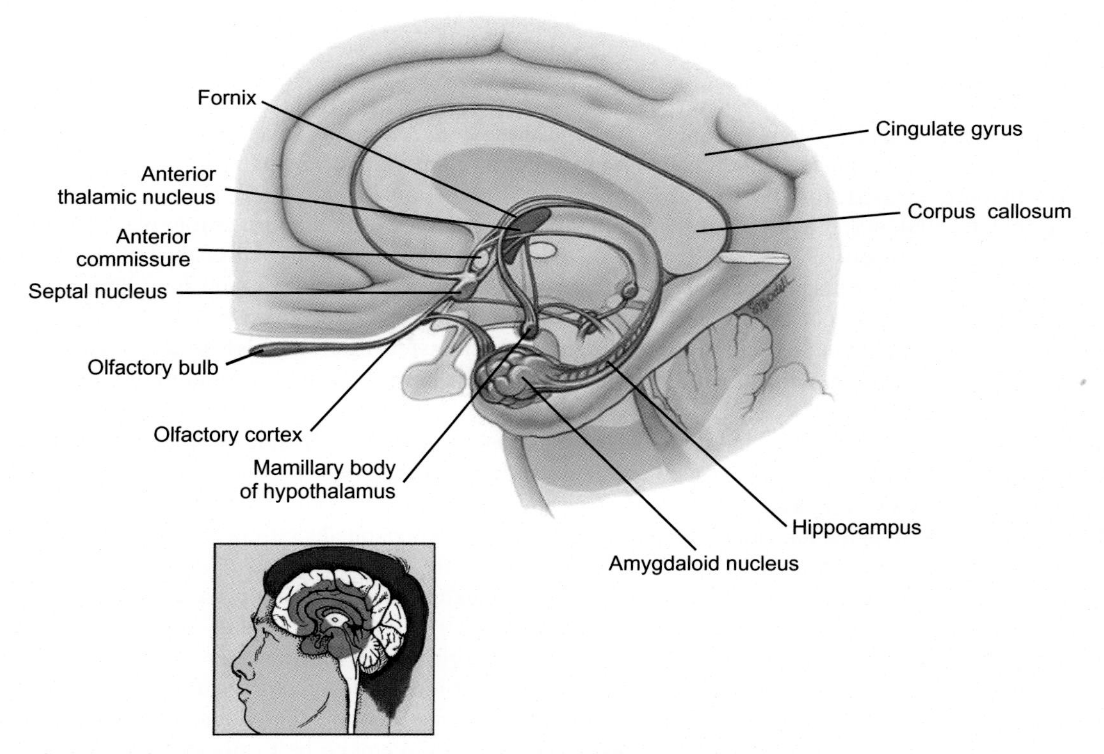

FIGURE 6.19 Structures of the limbic system. *From Thibodeau and Patton (2007), with permission.*

many regions of the cerebral cortex. Likewise, they send out efferent fibers to an equally large number of cortical destinations, but much of the output of the limbic system is directed toward the hypothalamus.

One of the most important functions of the **hippocampus** is the modulation of functions carried out by the hypothalamus. Because of its extensive connections with other parts of the brain, the hippocampus incorporates both subconscious and conscious input to its modulatory activities. Much of the hippocampal output to the hypothalamus is directed through tracts passing through the relatively inconspicuous **septal area** (see Fig. 6.19), which acts as a relay station

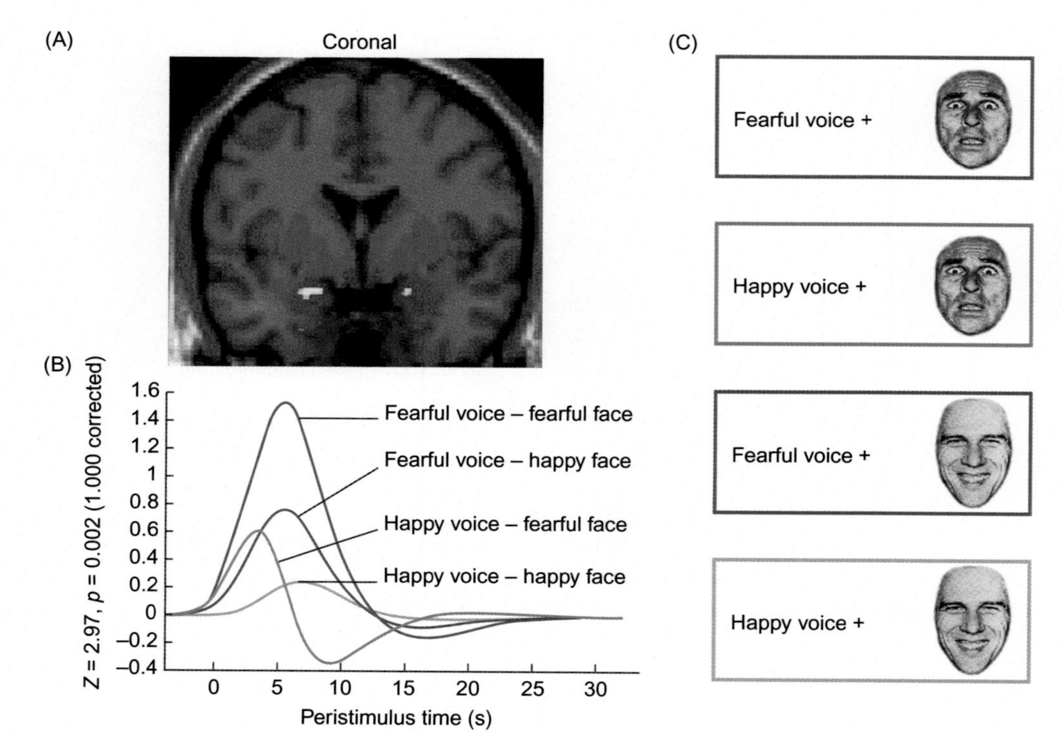

FIGURE 6.20 (A) fMRI image of the brain, showing activation of the amygdala (yellow) after a volunteer is exposed to a fearful face and voice. (B) and (C) Graph showing the intensity of amygdala responses to combinations of happy and fearful faces and voices. *From Fitzgerald et al. (2012), with permission.*

between the hippocampus and the hypothalamus. Through these connections, both autonomic and endocrine activities of the hypothalamus are fine-tuned. Another important function of the hippocampus is to temper the aggression and rage behaviors that are released from the hypothalamus.

Both experimental studies and analysis of human pathology have shown the important role played by the hippocampal formation in certain aspects of memory. If parts of the hippocampus are destroyed or dysfunctional, patients have trouble remembering events or material presented within recent minutes, even though their long-term memories remain intact. This can include reading comprehension, because of an inability to remember previously-read lines of text. The hippocampus may contribute to the overall process of memory by transferring from short- to long-term memory centers. Research studies have shown that animals with hippocampal lesions have difficulty negotiating mazes and other activities involving spatial memory.

The **amygdala**, a small group of nuclei at the end of the hippocampus, is well situated to exert a profound influence on many of the autonomic and visceral responses mediated by the hypothalamus. It receives major input from the olfactory system (as well as taste), the brainstem (reticular system) and many areas of the cerebral cortex. Its neural output is heavily focused on the hypothalamus and the part of the midbrain (periaqueductal gray) that, with the hypothalamus, exerts a powerful control over autonomic functions.

The amygdala plays a particularly important role in fear and anxiety reactions. If a person is presented with a fearful image or sound, activity of the amygdala greatly increases (Fig. 6.20). Different parts of the amygdala can either dampen or intensify rage or aggressive reactions in humans. Abnormal rage reactions are seen in certain types of epilepsy, and surgical lesions in parts of the amygdala can reduce their intensity. According to much current opinion, anxiety and phobias in many adults may be based on fearful situations that the individuals encountered during early childhood. Although these situations cannot be recalled, their memory seems to be firmly embedded in the inner workings of the amygdala. Because of its connections with the hypothalamus, emotional reactions generated in the amygdala can also be translated into changes in blood pressure, heart rate, pupillary dilation, breathing rate, and other autonomic functions controlled by the hypothalamus. Even eating, drinking, and some endocrine functions have been shown to be strongly influenced by the amygdala.

The **prefrontal cortex**, a region of cerebral cortex abutting the septal area (see Fig. 6.32), has increasingly been shown to be a major link between the limbic system and the integrative centers of the entire cerebral cortex. This region has become more prominent as the mammalian brain has enlarged during evolution. It also has indirect connections with the hypothalamus, allowing its cortical input to be translated into the modulation of visceral functions.

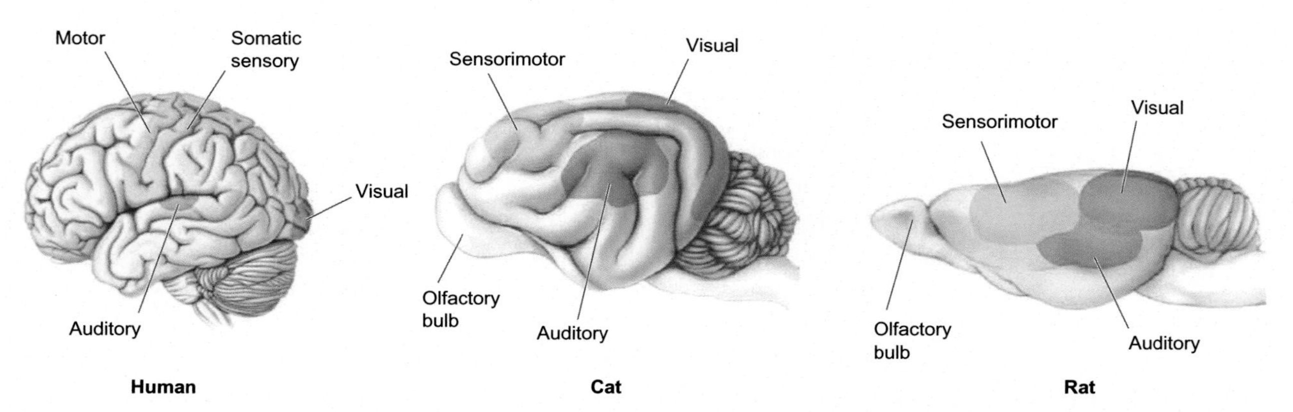

FIGURE 6.21 Relative locations and prominence of cortical sensory areas in rat, cat, and human brains. *From Bear et al. (2007), with permission.*

Cerebral Cortex. The rise in complexity of the **cerebrum** has been one of the cardinal features of mammalian evolution. As mammals became larger and more behaviorally complex, the cerebral cortex has greatly expanded. The cortex of a monkey is 100 times greater than that of a rat, and that of a human, 10 times larger than that of a monkey. Along with the increase in the absolute size of the cerebral cortex is a remarkable increase in the complexity of its organization. The surface contours of a rat's cortex are smooth, but as brains got larger they developed convolutions (**gyri** [folds], **sulci** [grooves]), which enabled more efficient packing of the expanded surface area of the cortex (Fig. 6.21). Such folding is a classical biological strategy for increasing surface area within a confined space (e.g., folding of the surface of the gut, p. 340). Another pronounced change accompanying the evolutionary development of the brain is the proportion devoted to defined sensory and motor functions. Most of the brain of the rat is occupied by areas dedicated to sensory or motor functions (see Fig. 6.21). As the brain has expanded, the proportion of such areas has been reduced, and areas involved in processing sensory information or integrating cerebral functions (**association areas**) have greatly increased to the extent that such regions occupy the majority of the human cerebral cortex (**neocortex**). Internally, as is also the case with the cerebellum, an increase in the number of deep nuclei that coordinate cerebral functions is a concomitant feature of its greater size.

Characteristic of both the cerebral and cerebellar cortex is the location of the gray matter (which contains the neurons) on the outside of the white matter (which consists of tracts of white myelinated nerve fibers). The shift in gray matter to the outside occurs in the embryo, when the precursors of cerebral neurons migrate along special glial cells (**radial glial cells**) from the interior of the neural tube to the exterior. This occurs in successive waves (interestingly, from the inside to the outside layers) until ultimately six layers of cells and neuronal processes have been laid down to form the layers of the neocortex.

The basic six-layered arrangement is remarkably constant throughout the 2.4 mm thick neocortex, but phylogenetically older parts of the brain (e.g., parts of the limbic and olfactory systems) have fewer (between 3 and 5) layers. Nevertheless, subtle structural specifics of the layers vary among many parts of the brain. A century ago, Brodmann described up to 50 different microscopic architectures for different regions of the brain and assumed that these structural differences were the basis for different functions. Subsequent physiological research has borne this out.

Layer I (the outermost) of the cortex consists principally of fibers from deeper neurons and is an area where cross-connections are made (Table 6.3). Layers III and V contain prominent medium to large neurons, called **pyramidal cells** because of their shape. Many of these large cells send out axons that extend long distances from the cell body, and many enter the spinal cord. Pyramidal neurons, which can be either excitatory or inhibitory, integrate the activities of the neurons that are connected to them by synapses. In contrast, layers II and IV are characterized by the presence of large numbers of smaller granule and stellate cells that are involved in more local connections. Layer VI consists of a mixture of cell types. Each layer has characteristic inputs and outputs (Fig. 6.22).

In addition to the six horizontal layers, the neocortex is subdivided into as many as 100 million **cell columns** (sometimes called modules). Although they are poorly defined structurally, neurophysiological studies have shown that within a vertical area 50–100 μm in diameter and up to a few hundred neurons deep, all cells within the column respond to the same type of sensory modality, but not to others. Because of this property, cell columns are considered to be microscopic functional units of the cortex.

Through observations of patients with a variety of cerebral lesions and, later, physiological experiments, it has long been known that different functions are associated with different parts of the brain. More recent studies, using the tools

TABLE 6.3 Layers of the Cerebral Cortex

Layer	Name	Connections
I	Molecular	Receiving area for nonspecific afferent fibers from the thalamus and monoaminergic neurons from brainstem
II	Outer granular	Receiving area for cortical afferent fibers (callosal and association)
III	Outer pyramidal	Mainly efferent—callosal and cortical association fibers
IV	Inner granular	Receiving area for afferents from thalamus
V	Inner pyramidal	Mainly efferent—to neostriatum, brainstem, and spinal cord
VI	Multiform	Efferents to thalamus and short intracolumnar projections

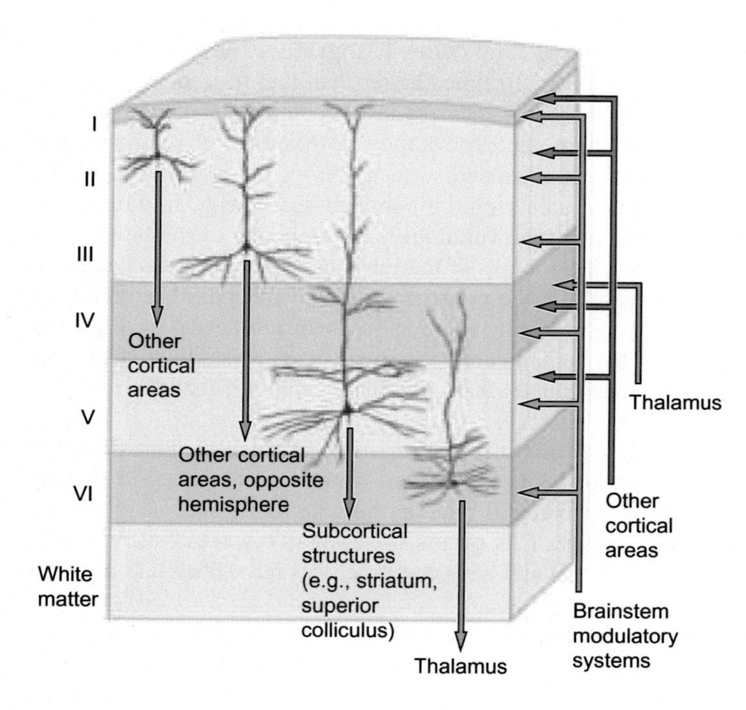

FIGURE 6.22 Inputs to and outputs from the cortical layers. Overall, layers 1–4 project to other cortical areas, and layers 5 and 6 to subcortical areas. *From Purves et al. (2008), with permission.*

of **positron emission tomography** (PET) and **functional magnetic resonance imaging** (fMRI), have considerably sharpened the localization of many physiological functions within the brain. fMRI, in particular, provides fine-level imaging through detecting changes in blood flow in active parts of the brain. These are then represented as colored images showing "hotspots" for a particular function (Fig. 6.23).

The most dramatic localizations of function are seen in the primary somatic sensory and motor areas of the brain. Located in the **precentral gyrus** (motor) and the **postcentral gyrus** (sensory), these areas encompass seemingly distorted maps (**somatotopic maps**) of the major areas of the human body (Fig. 6.24). In reality, these maps reflect the density of innervation of the areas of the body (**homunculus**) represented in these gyri. Other major functional areas of the brain are represented in Fig. 6.25. Details of the optic and auditory centers of the brain are presented in Chapter 7.

The various functional areas of the cerebral cortex do not operate independently. Although the motor and sensory functions of the pre and postcentral gyri are highly prominent, in each case cortical regions associated with them are of vital importance in modulating the perception of sensory stimuli or the character of motor responses. For example, output from the primary motor cortex (precentral gyrus) is directed principally toward controlling contractions of the limb musculature. However, inputs from a variety of sources influence these signals (Fig. 6.26). Two other areas of the frontal cortex, the supplementary motor cortex, and the premotor area, receive considerable sensory input from the posterior

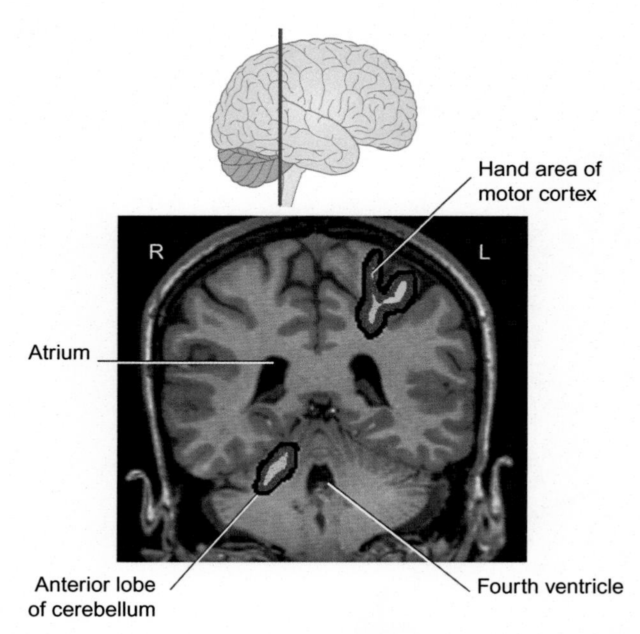

FIGURE 6.23 Sample fMRI image from the brain, in this case, showing responses to repetitive movements of the fingers of the right hand. *From Fitzgerald et al. (2012), with permission.*

parietal cortex. This input, itself, represents the amalgamation of a variety of sensory stimuli, with a strong visual component, which is processed in the posterior parietal lobe. This sensory information is used by the premotor cortex in assembling the sequence of actions required for complex motor tasks, such as tying one's shoelaces. The sum of this processed information is then passed onto the precentral gyrus, which is the primary driver of the motor action. Fiber projections from the premotor area even extend down to the level of the spinal cord for further modulation of motor actions. Many other levels of control operate in conjunction with one another to allow even repetitive motions like ordinary walking to take place (Fig. 6.27).

At a higher level are the areas of **association cortex** (Fig. 6.28), which represent the parts of the cortex that support higher intellectual functions, such as thought and abstract reasoning. These functions fall under the general rubric of cognition. Many neuroscientists believe that association areas occupy the majority of the human cerebral cortex. Commonly, association cortical areas are located adjacent to regions that have been shown to have specific sensory or motor functions. They are characterized by numerous interconnecting fiber tracts.

Beneath the expanded human cerebral cortex lies another layer of structures that have evolved to coordinate the activities of the cortex. These constitute the **basal ganglia**, a group of nuclei that partially surround the thalamus and largely communicate to the cortex through thalamic connections. They receive input from cortical association areas, the limbic system, and the eye fields. Thus they are able to add a significant cognitive overlay onto motor actions. Various components of the basal ganglia generate both excitatory and inhibitory influences on higher cortical functions. For this reason, pathology of the basal ganglia involves both hypokinetic and hyperkinetic disorders. A good example of a **hypokinetic disorder** is **Parkinson's disease**, which is due to a loss of dopamine-secreting neurons from the **substantia nigra** (a component of the basal ganglia). At the other extreme are **Huntington's chorea** and **hemiballism**, both of which are characterized by violent uncontrolled movements of the limb musculature. Both are due to degeneration, either genetically based or due to a stroke, of components of the basal ganglia that would normally exert an inhibitory effect on specific cortical functions.

An important property of the brain is its functional asymmetry. Although many brain functions are shared roughly equally between the two sides of the brain, others are not. A number of cortical asymmetries are due to the crossing of tracts of nerve fibers from one side to the other. These, however, are balanced by crossed tracts going to the other hemisphere. Other functions are truly asymmetrically represented in one side of the cerebral cortex or the other (Table 6.4, Fig. 6.29).

The differing functions of the two cerebral hemispheres were outlined through research on the split brain phenomenon. Normally, the left and right sides of the brain are interconnected through several massive tracts of nerve fibers that cross the midline. The largest of these connecting links is the **corpus callosum**, located above the thalamus (see Fig. 6.11).

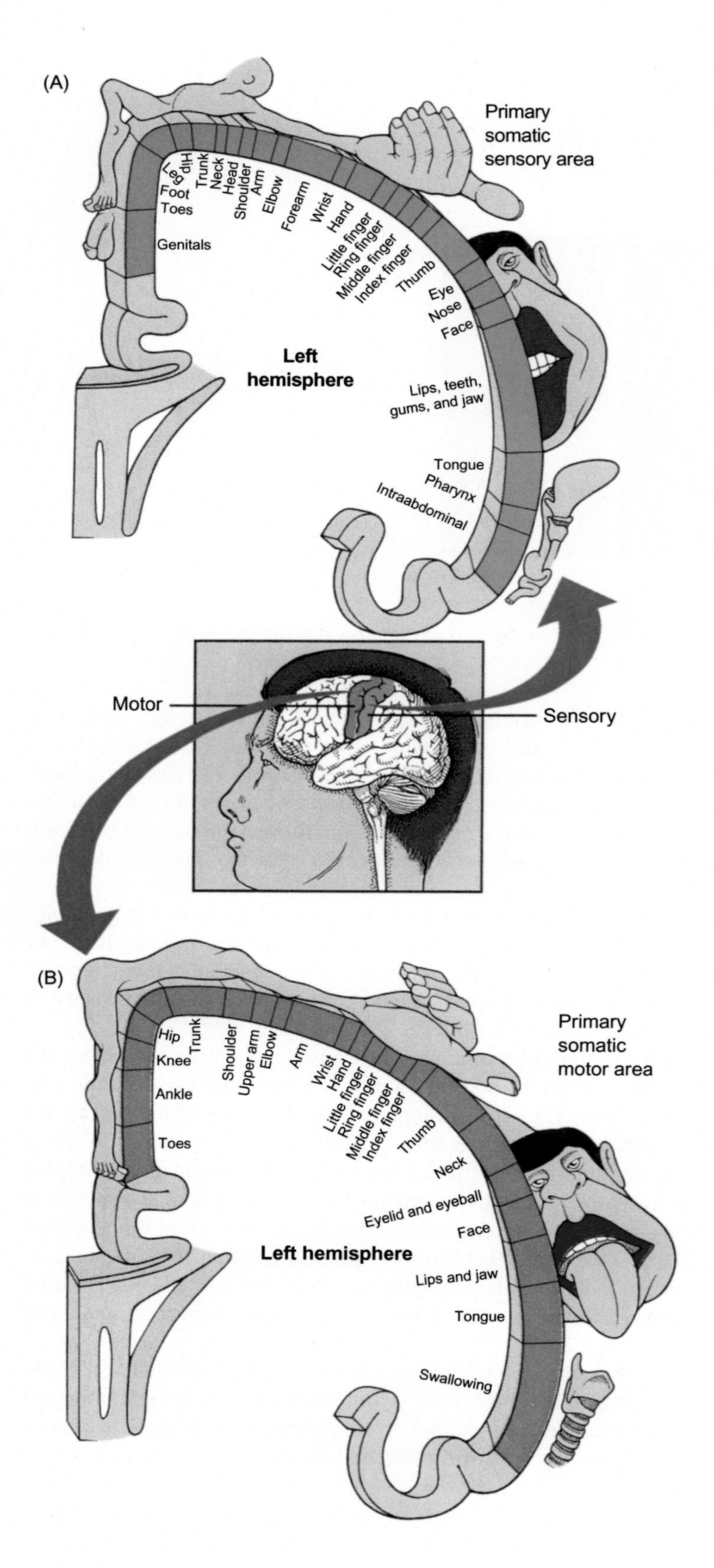

FIGURE 6.24 Primary sensory (A) and motor (B) areas of the cerebral cortex, showing the relative prominence of projections from the most sensitive regions of the body. The face and fingers are particularly well represented. *From Thibodeau and Patton (2007), with permission.*

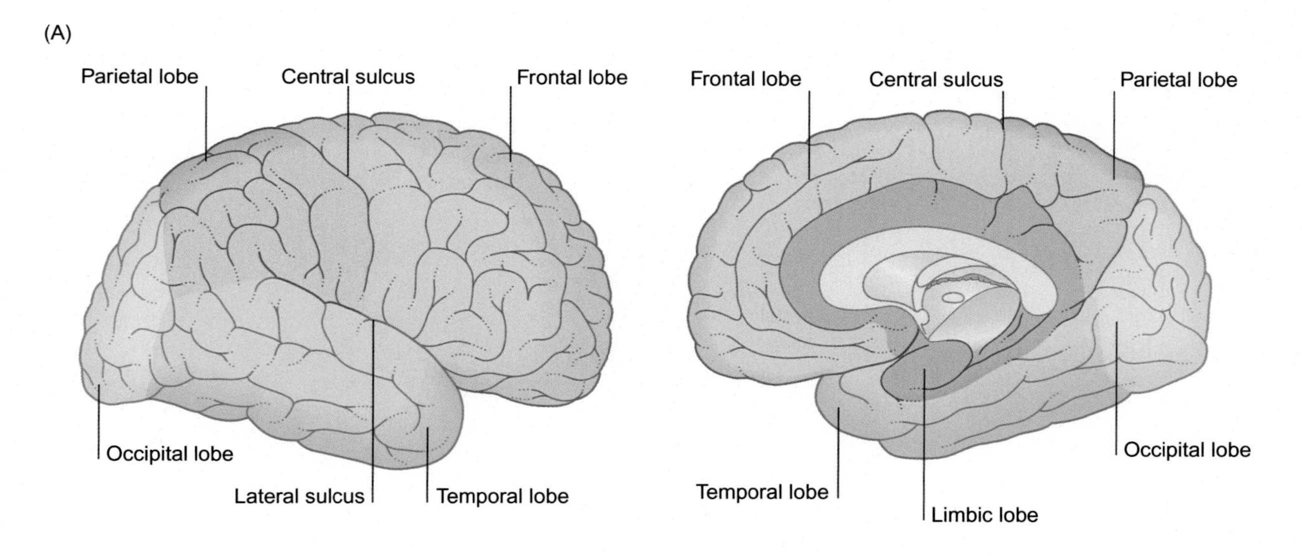

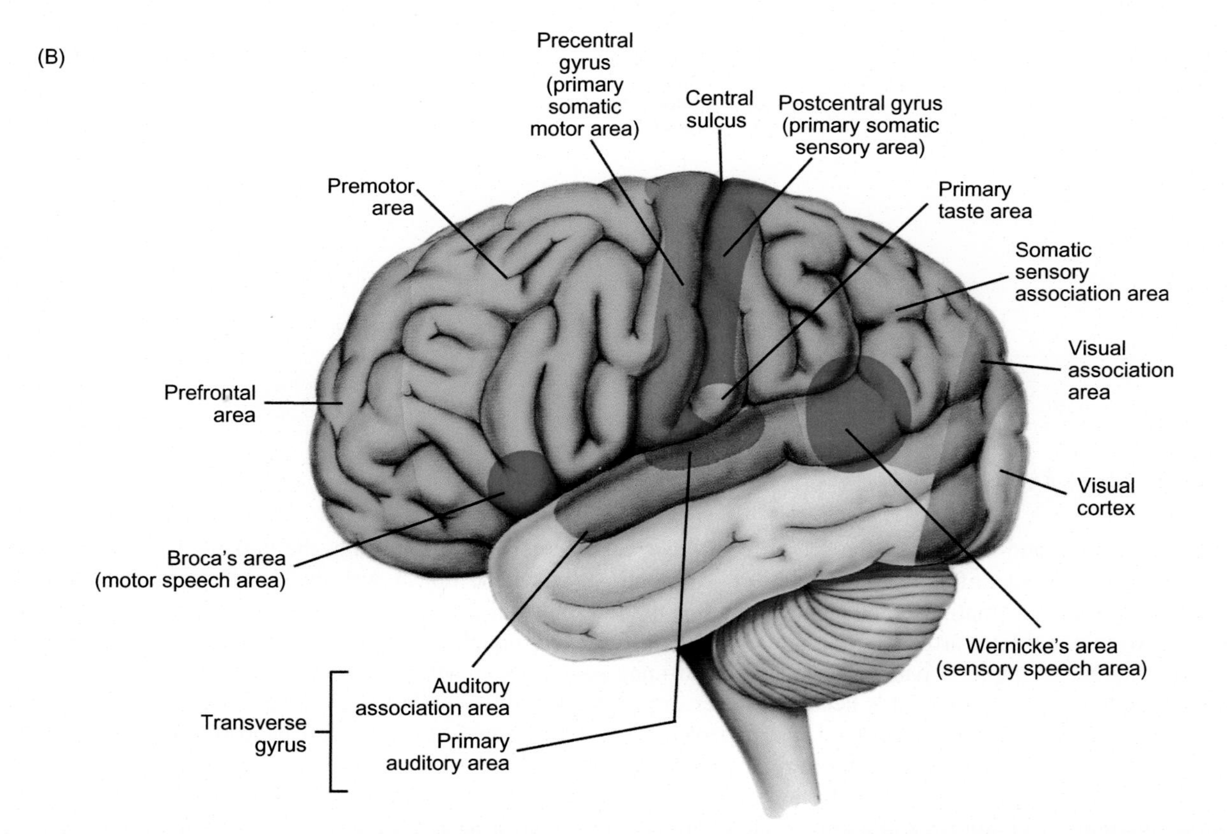

FIGURE 6.25 (A) Anatomical regions of the cerebrum. *Left*, lateral surface; *right*, medial surface. (B) Functional areas of the cerebral cortex. Lateral view. *From (A) Fitzgerald et al. (2012) and (B) Thibodeau and Patton (2007), with permission.*

As many as 200 million nerve fibers cross the corpus callosum. If the corpus callosum has been severed (sometimes to prevent the spread of epilepsy), there are surprisingly few behavioral consequences under normal living conditions, but experiments conducted on such individuals in the 1960s showed some striking effects.

In one experiment, patients were given a familiar object in their right hand without seeing it. When asked to name the object, they had no trouble. When given the same object in the left hand, however, they were unable to name it. The reason is that sensation surrounding the object in the right hand reaches the left brain through pathways other than

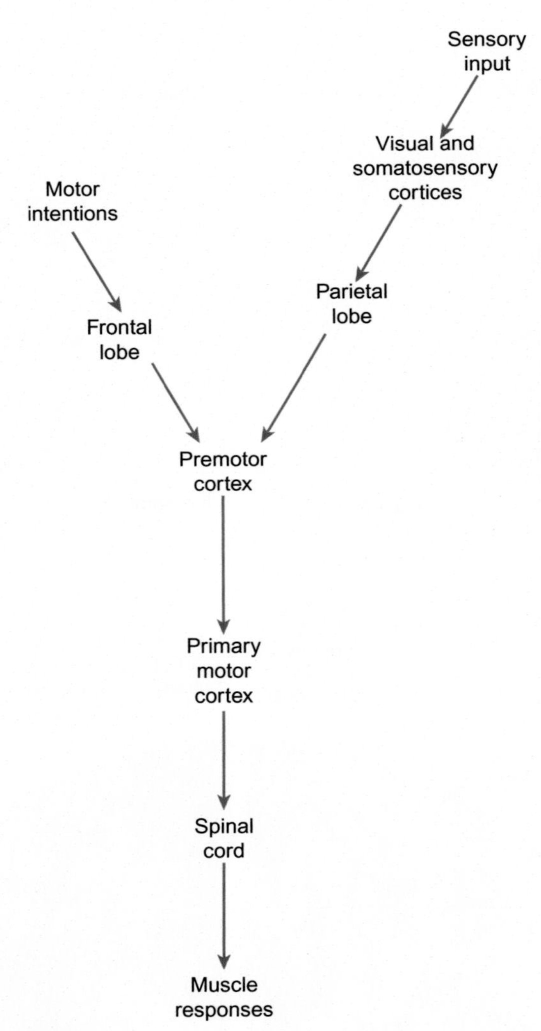

FIGURE 6.26 Scheme of inputs to motor functions.

the corpus callosum, and because of the presence of the speech center in the left brain, the patients were easily able to verbally express their identification of the object. By contrast, when the object was in the left hand, the right brain received the sensory information in the same manner as the left, but in the absence of a speech center in the right brain, the patients were not able to articulate the nature of the object, even though they had a general sense of what they were holding. In another example, a patient was asked to arrange some blocks with his right hand to match a pattern illustrated on a card. He could not do this because the sensory signals went to the left side of the brain, which is not proficient in accomplishing such tasks. Instead, the patient instinctively wanted to use his left hand to accomplish the task, because the left hand was represented in the right side of his brain, which is dominant for spatial reasoning.

The cerebral cortices, especially the frontal and parietal cortex, are heavily involved in higher level functions, such as abstract thought, associations, and learning and memory. The physical basis for many of these higher functions is still not well understood and remains one of the frontiers in neuroscience.

NEURONAL PATHWAYS AND CIRCUITS

Thanks to their elongated axons and dendrites, neurons are able to connect with distant locations. For the most part, neural processes (axons or dendrites) are grouped according to their general functions. These groups of extended neural processes are called **tracts** when located within the CNS. Tracts transmitting sensory information carry signals from the periphery to the brain; conversely, motor tracts carry signals from the brain to the periphery. Textbooks of neuroanatomy devote hundreds of pages to describing details of the many pathways that are used for communication within the

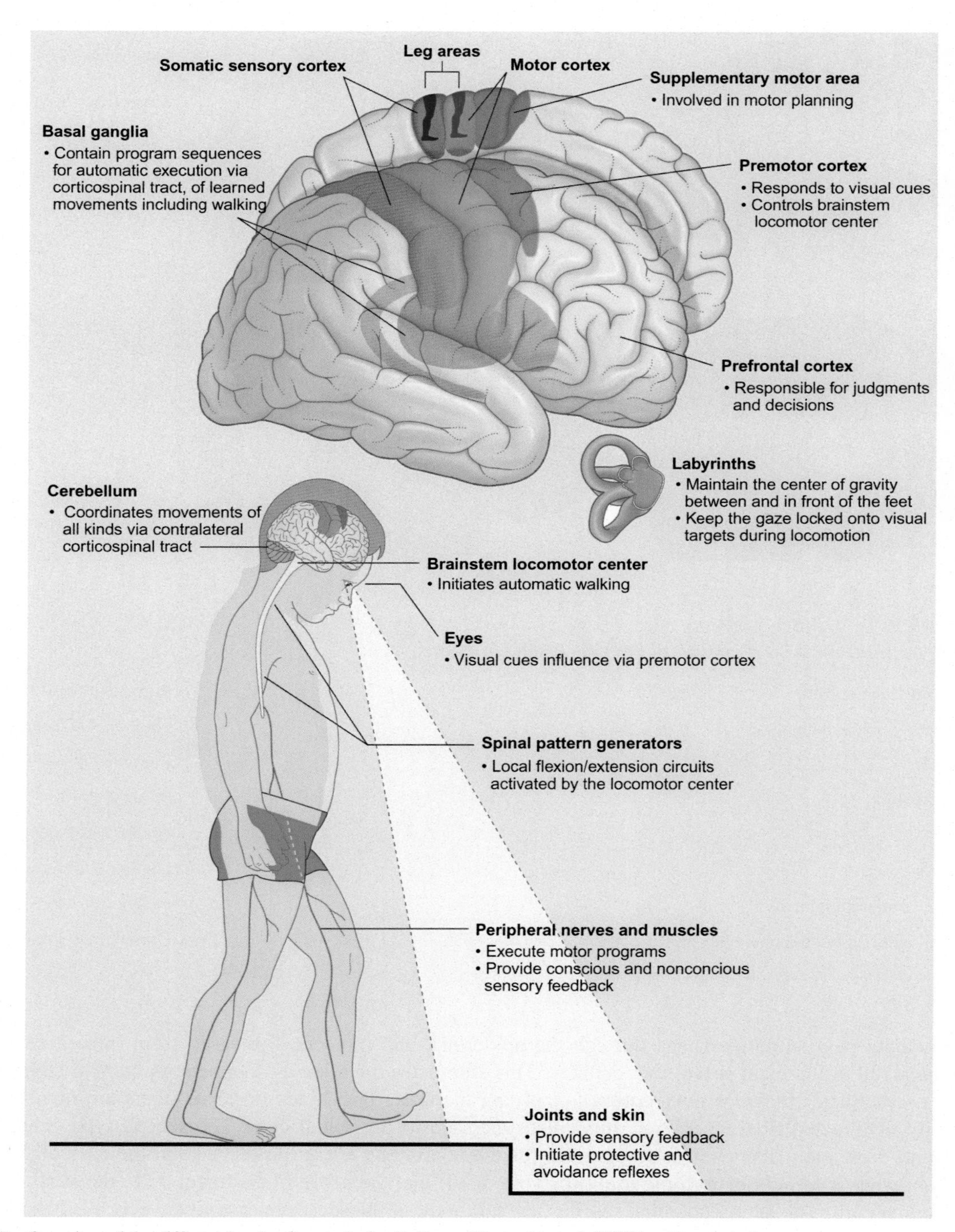

FIGURE 6.27 Overview of the different levels of control of gait. *From Fitzgerald et al. (2012), with permission.*

CNS. This section will provide details of two such pathways—one for sensory information (in this case, pain) and another for somatic motor functions—as an illustration of the types of interconnectivity found within the CNS. Some other pathways, for example, for vision and hearing, are described in later chapters.

Pain Pathway

The response to accidentally placing one's right thumb on a hot stove burner will be used to illustrate the major pain pathways. For almost any painful stimulus, the first reaction is an almost instantaneous withdrawal from the pain-inducing agent, in this case, a hot stove. The immediate reaction is often followed by actions dictated by cognitive decisions.

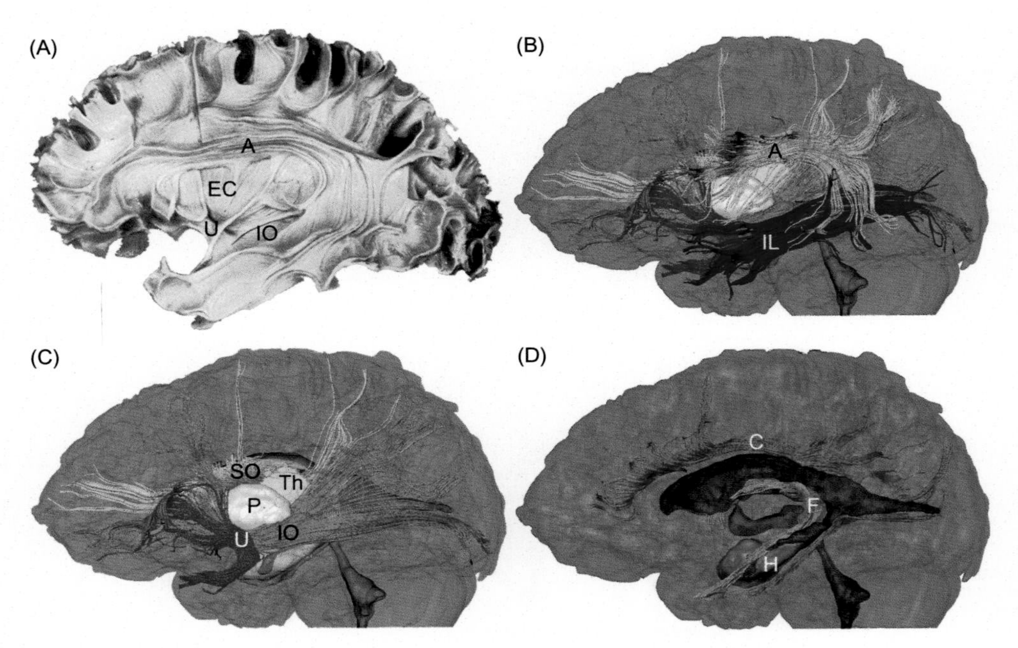

FIGURE 6.28 Long association bundles in the cerebrum as seen from the medial side. (A) Dissection. (B)–(D) Various imaging products. *From Nolte (2009), with permission.*

TABLE 6.4 Asymmetrical Localization of Specific Brain Functions

Left Cerebral Hemisphere	Right Cerebral Hemisphere
Understanding syntactic language	Emotional coloring of language
Motor speech	Rudimentary speech
Mathematical ability	Musical ability
Writing	Spatial reasoning
Symbolic processing	Emotional processing
Responds to written commands	Responds to nonverbal stimuli
Dominant in right-handed people	Dominant in left-handed people

The immediate pain stimulus passes through the epidermis, and molecules released from injured cells reduce the activation threshold of the local nerve fiber endings. This stimulates the release of an action potential along dendrites of the sensory nerve fibers (**thermal nociceptors**) supplying the area. Two kinds of nerve fibers are involved. One type (large, heavily myelinated fibers) sends an immediate message to the spinal cord. The second type (small, unmyelinated) represents slow pain fibers and is responsible for the continuing sensation of pain. In the case of burned thumb, we know from what is called **dermatome mapping** (Fig. 6.30) that the nerve fibers travel up to the sixth cervical nerve toward the spinal cord. The action potential reaches the cell body of the sensory nerve fiber, which is located in a spinal ganglion close to the spinal cord (see Fig. 6.3). The neural signal then passes into the spinal cord through axons of the stimulated nerve fibers. At this point, a local reflex circuit is activated. Because of branching of the entering nerve fibers, this involves several segments of the spinal cord above and below the primary entry point. At the same time, the signal passes from the primary sensory nerve fibers to secondary nerve fibers in tracts that run up the spinal cord.

Fast transmission of pain passes from the primary into secondary neurons. Axons from the secondary neurons pass to the other side (in this case, the left) of the spinal cord and collect into the **lateral spinothalamic tract** (Fig. 6.31, see Fig. 6.6B), which directly passes into the thalamus. From the thalamus, tertiary neurons project into the thumb area (see Fig. 6.24) of the left primary sensory cortex in the postcentral gyrus for immediate localization of the site and intensity of the pain. Some other fibers leaving the thalamus go to the part of the cerebral cortex called the **insula**, which introduces some emotional content to the sensation.

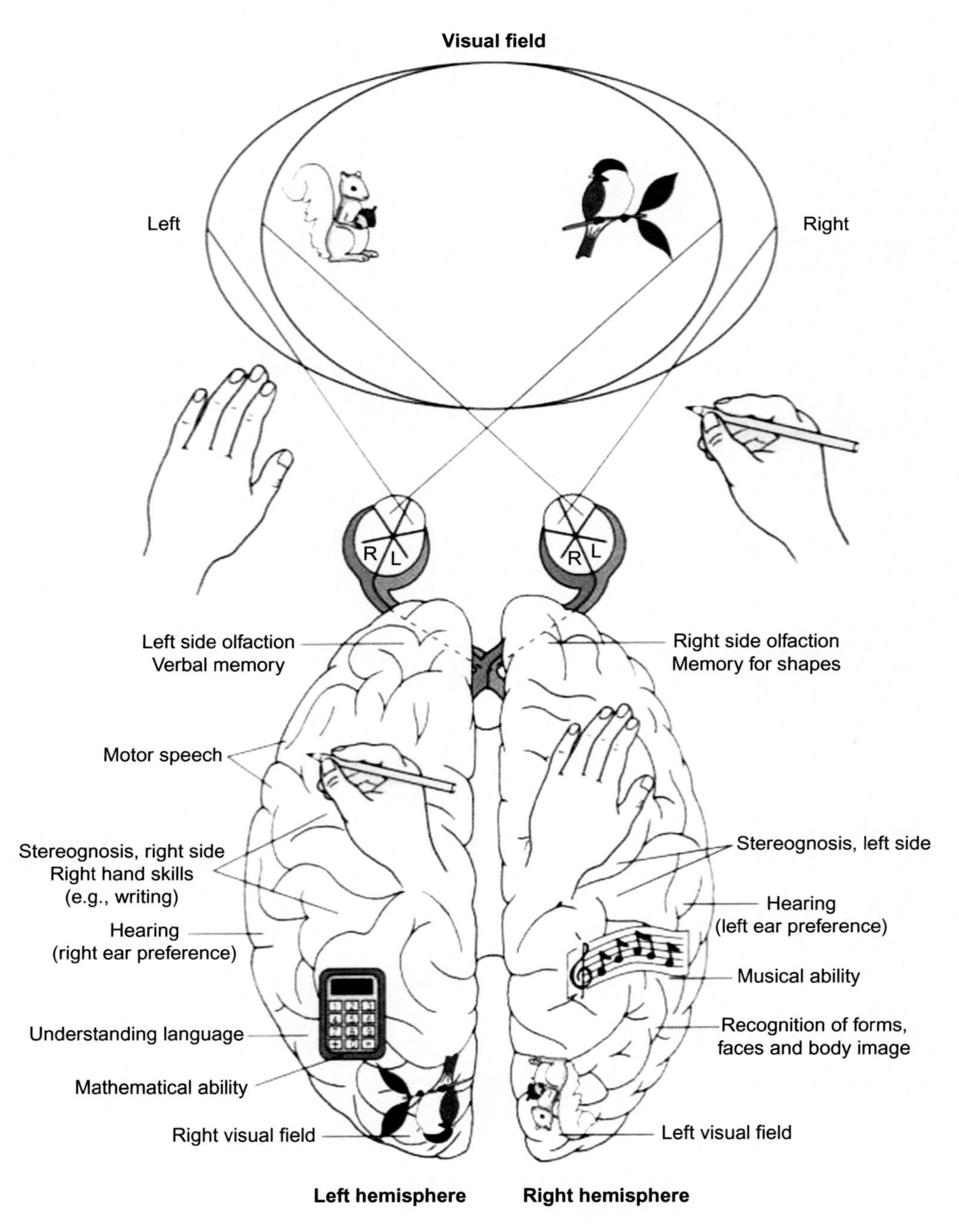

FIGURE 6.29 Lateralization of cortical function. The illustration represents functions dominant in either side of the brain. This information came largely from clinical cases involving patients with split brains due to sectioning of the corpus callosum. *From Siegel and Sapru (2011), with permission.*

Other secondary nerve fibers that connect to the slow unmyelinated primary fibers at the level of entry into the spinal cord send axons that also cross to the other (left) side before entering another tract that run to the reticular formation in the brainstem. Here the pain is processed in a variety of ways, including the possibility of introducing opiates (e.g., **endorphins**) into the circuit in ways that are still not completely understood. Processing slow pain signals in the reticular formation and the periaqueductal gray within the brainstem can activate descending pain pathways that do not reach the cerebral cortex. Through inhibitory mechanisms, these can substantially modify the perception of pain by introducing levels of analgesia into the system. Other nerve fibers pass from the reticular system into the thalamus and from there connect to nerve fibers projecting to the limbic system, specifically the prefrontal cortex and the cingulate gyrus. This pathway introduces fear and other emotional overtones to the entire process of perceiving the pain of the burn. Still other projections lead to optic centers, allowing the person to quickly locate the source of the pain.

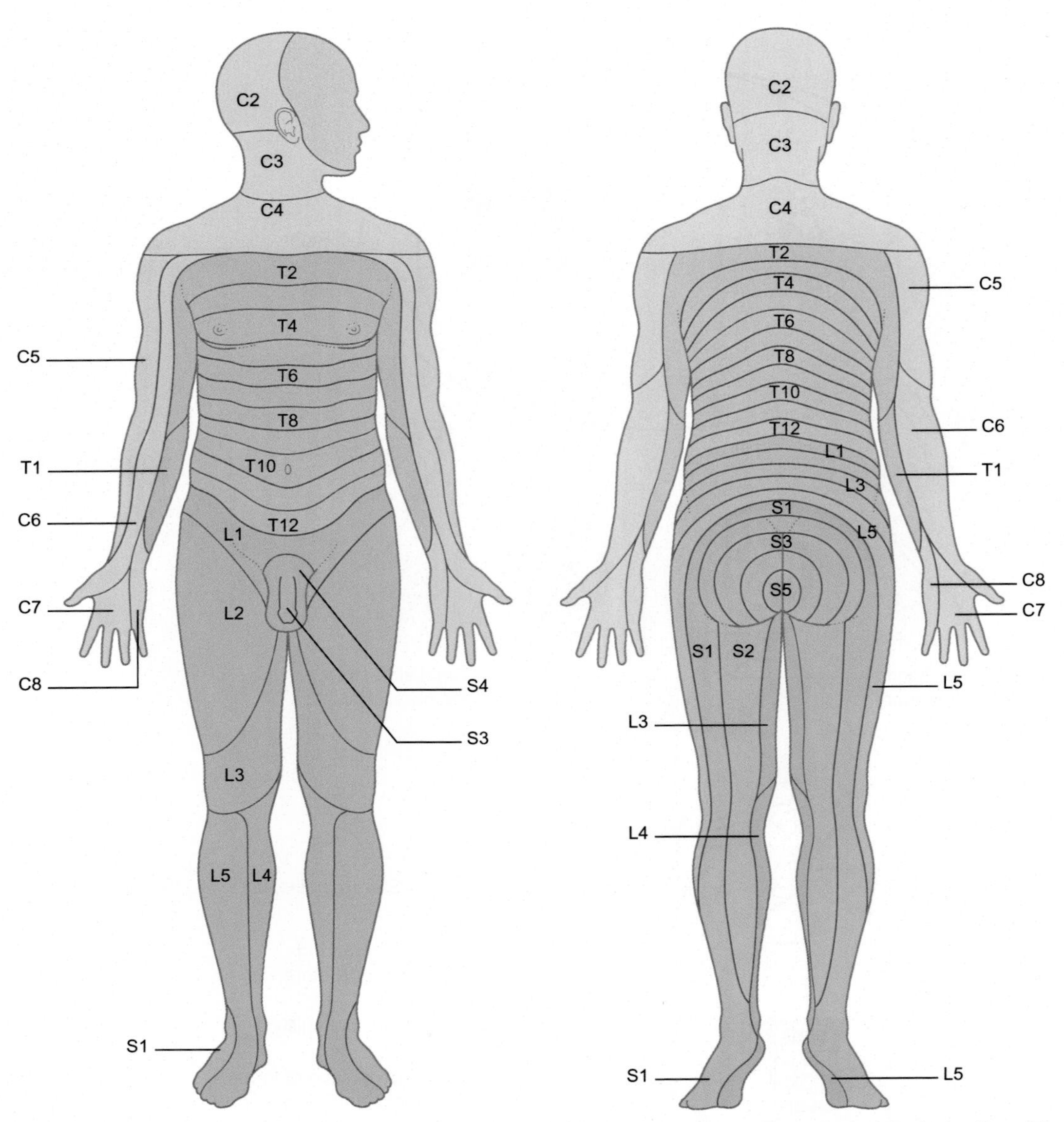

FIGURE 6.30 Dermatome patterns (areas supplied by nerve fibers from specific spinal levels) on the surface of the body. *From Fitzgerald et al. (2012), with permission.*

From these sensory receiving and processing centers, association fibers lead to the motor cortex of the precentral gyrus. The sum of the collective sensory and emotional processing then becomes integrated into a higher level motor response to the burn than the local reflex present within the spinal cord.

Somatic Motor Pathways

The motor response to a burned thumb can be extremely varied. The immediate withdrawal of the thumb from the stove is accomplished principally through the local spinal cord-level reflex pathway illustrated in Fig. 6.3. Subsequent motor actions are much more complex and involve sensory information that has traveled up the spinal cord and into the brainstem and the cerebral cortex. Some of these actions are also almost reflex in nature, such as saying ouch or something stronger in response to the burn. This requires rapid processing of the sensory information and its consolidation in the **speech center** (Broca's area) on the left side of the brain and the attendant mobilization of the facial and laryngeal muscles required to make an intelligible sound.

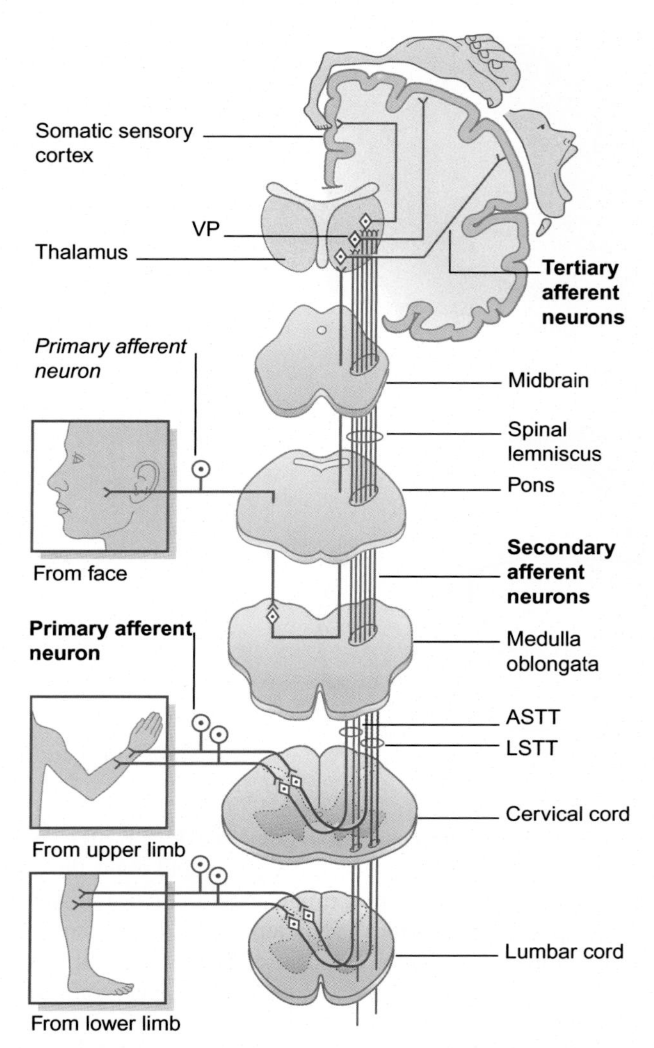

FIGURE 6.31 The spinothalamic pathway, carrying pain and some other sensory stimuli from peripheral receptors to the sensory cortex. *ASTT*, anterior spinothalamic tract; *LSTT*, lateral spinothalamic tract; *VP*, ventral posterior nucleus of the thalamus. *From Fitzgerald et al. (2012), with permission.*

Subsequent motor actions to the burn, such as shaking the hand or dipping the thumb in cold water, often involve intent. Here, the sensory information is processed and integrated through many association areas of the brain. Ultimately the frontal lobe, which receives much of their integrated input, sends commands through the supplementary motor area and the premotor cortex to the primary motor cortex in the precentral gyrus (Fig. 6.32). These commands, combined with other modulating signals from the cerebellum and the basal ganglia, elicit a definitive motor response from the primary motor cortex.

The large motor neurons in the motor cortex send axons down into the white matter of the brain (Fig. 6.33) through a broad fiber-containing area called the **corona radiata**. From there, they pass through a narrower region (**internal capsule**) and enter the midbrain as part of a massive aggregation of fiber tracts, known as the **cerebral peduncles**, which carry many axons from the cerebrum into the brainstem. In the medulla, in an area called the pyramid, over 80% of the motor fibers cross-over to the other side of the brainstem at what is called the **pyramidal decussation**.[5] The axons then

5. Crossing of nerve fibers to the other side in neuroscience is known as **decussation**. Decussation is a widespread phenomenon within the nervous system. It occurs in both sensory and motor tracts within the spinal cord and brainstem, and it is very prominent within the visual system. This phenomenon occurs throughout the vertebrates, although not among the invertebrates. Amazingly we have almost no idea about how this anatomical phenomenon occurred during phylogeny or why it is evolutionarily advantageous. Speculation on the basis for decussation is virtually absent in neuroscience or neuroanatomy books.

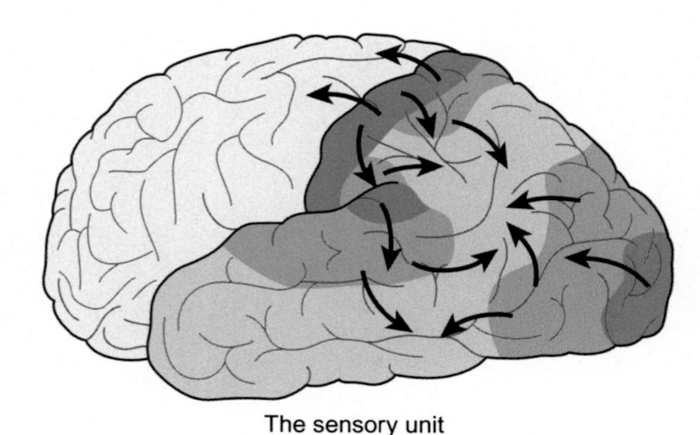

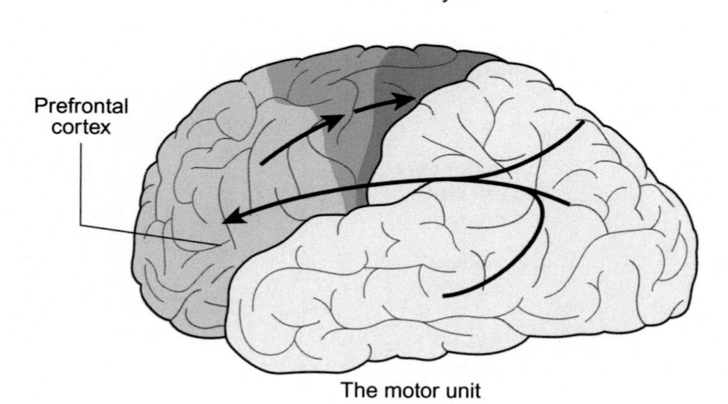

FIGURE 6.32 Pathways of sensory and motor information flow in the brain. The darkest shading indicates primary areas, intermediate, and light shading indicate secondary and tertiary areas. *McBean and van Wijck (2013), with permission.*

descend into the spinal cord as the **lateral corticospinal tracts** (see Fig. 6.6B)—one on each side. Slightly farther down the brainstem another ~10% of the motor fibers also decussate and enter the anterior corticospinal tract. Only a small percentage of the original motor fibers leaving the primary motor cortex remains on the same side to form the lateral corticospinal tract. Once in the spinal cord, various axons terminate at different levels, depending upon their origin within the homunculus in the motor cortex (see Fig. 6.24). Those for the arm and hand terminate and make many synaptic connections in the lower cervical and upper thoracic levels of the spinal cord. Still within the corticospinal tracts, those axons for the leg, foot, and lower body continue down the spinal cord until they reach their appropriate levels.

The high degree of spatial order that characterizes nerve fibers in the motor homunculus in the precentral gyrus continues down the corticospinal tract until the termination of the axons. Here the axons of the **upper motor neurons**, as those that originate within the brain are called, synapse with dendrites of neurons (**lower motor neurons**) located within the anterior horn of gray matter within the spinal cord. The anterior horn neurons are also highly organized spatially, with groups of neurons serving both the flexor and extensor muscles of different areas located close to one another (Fig. 6.34). Many of the connections between the upper and lower motor neurons are mediated by small local internuncial neurons, which can be either excitatory or inhibitory in nature. The inhibitory fibers are particularly important for smooth motions, because before any group of muscles (**agonists**) can effectively contract, their **antagonists** must relax. For delicate movements, such as the movement of a single finger, fine branches of certain of the upper motor neurons functionally innervate and stimulate only small numbers of motor neurons. These and many other feedback and modulatory influences must all operate in concert to produce the coordinated movements that we often take for granted.

NEURAL REGENERATION AND STEM CELLS

For many years, conventional scientific wisdom dictated that the brain is a postmitotic tissue that has no ability to produce new cells or to regenerate either neurons or neuronal processes. In contrast, peripheral nerves have a robust capacity for regeneration of neuronal processes. In humans, the rule of thumb is that fibers in a peripheral nerve can regenerate at a rate of ~1 mm per day. New research conducted in the later years of the 20th century have substantially changed the prospects for some forms of regeneration within the brain and spinal cord.

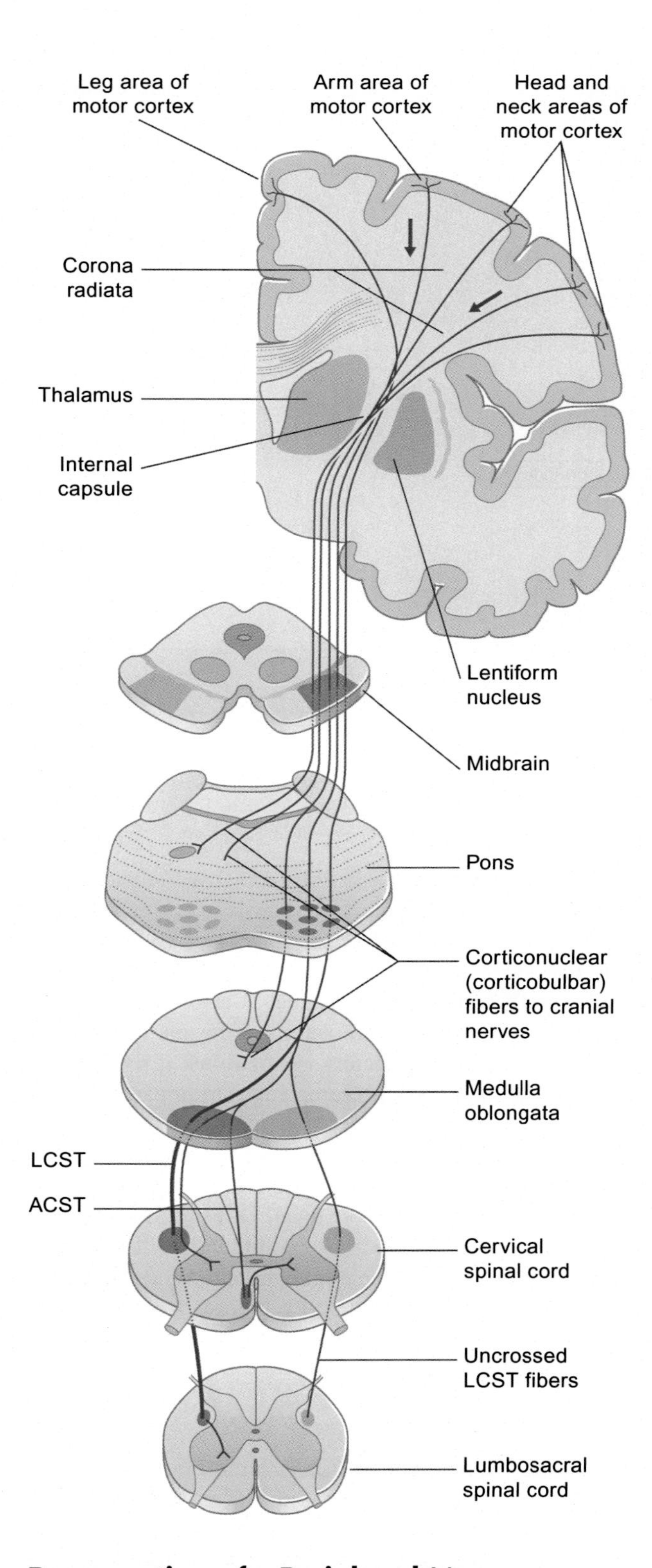

FIGURE 6.33 The pyramidal tract carrying motor signals from the cerebral cortex through the spinal cord to the periphery. *ACST*, anterior corticospinal tract; *LCST*, lateral corticospinal tract. *From Fitzgerald et al. (2012), with permission.*

Regeneration of a Peripheral Nerve

Peripheral nerves can be damaged by compression or by being severed through some sort of trauma. In an area of compression, the axon begins to degenerate, as does the axonal material distal to the lesion. Significant changes occur to the Schwann cells and myelin sheaths that surround individual degenerating axons. Soon after the traumatic event, the cell bodies of the Schwann cells separate from the remainder of their myelin-filled cytoplasm. Then stimulated by the

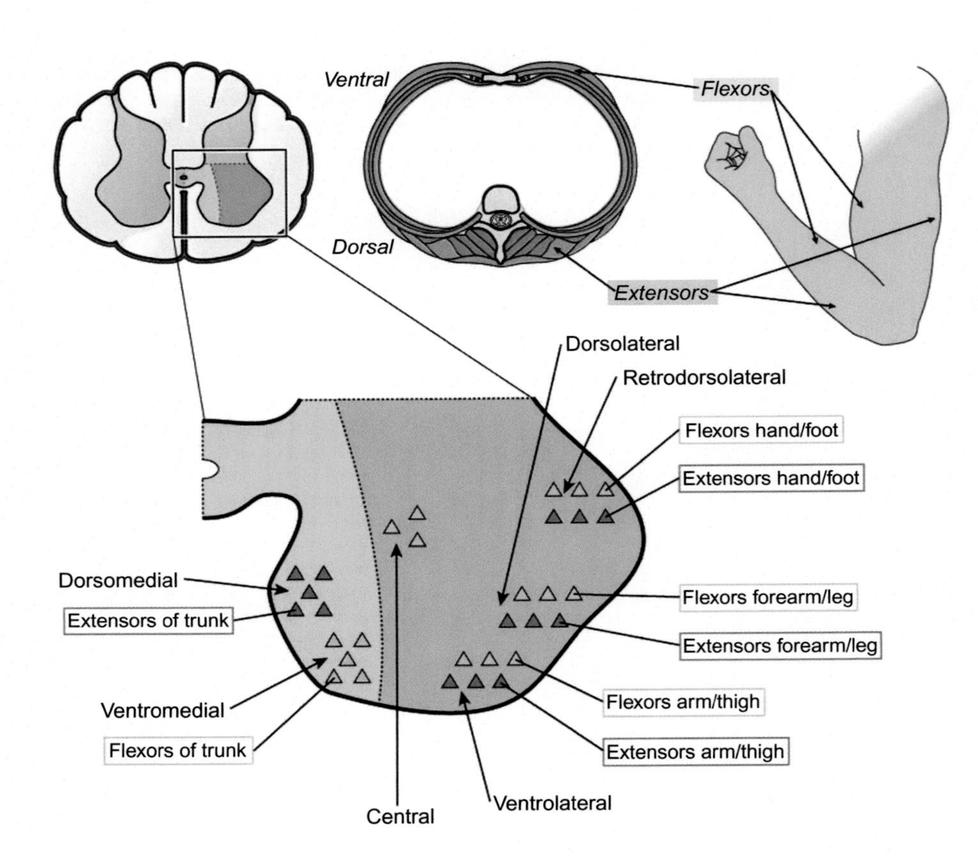

FIGURE 6.34 Location of cell columns in the anterior spinal cord, showing the somatotopic organization of motor axons. *From Fitzgerald et al. (2012), with permission.*

release of local cytokinins, macrophages penetrate the basal lamina that surrounds the axon and its myelin sheath, and they begin to engulf the myelin fragments surrounding the degenerating axon. Myelin removal is a critical phase of peripheral nerve regeneration, because myelin inhibits the regeneration of an axon. While these degenerative events are occurring distal to the lesion, axonal transport (see p. 52) brings to the degenerating end of the axon the materials that it needs to begin elongating. The end of the damaged axon forms a **growth cone**, a structure that is also involved in the outgrowth of nerve fibers in the embryo (Fig. 6.35). Critical to axonal outgrowth are interactions between the axon and Schwann cells, as well as the influence of several growth factors. One of the growth factors is **fibroblast growth factor** (FGF), which is bound to molecules in the basal lamina that surrounds the axon. After trauma and during early regeneration, the FGF is released from the basal lamina and facilitates elongation of the axon.

Even at a rate of 1 mm/day, the dimensions of human peripheral nerves mean that effective regeneration can take many months. A common form of injury is brachial plexus trauma. Because of the length of the arm, complete regeneration can take well over a year. Unfortunately, during the period of regeneration, the most distal part of the nerve commonly undergoes changes that are inimical to successful regeneration. One of the most common problems is the deposition of collagen fibers (a form of scar tissue) in the channels emptied by the degenerated axons. When the regenerating axons reach the collagen deposits, their further progress is likely to be impeded and reinnervation incomplete. For shorter distances, axonal regeneration is often remarkably successful.

Regeneration in the Brain and Spinal Cord

Under normal circumstances, regeneration of damaged axons in the brain and spinal cord is minimal. For many years, the lack of regeneration was attributed to the inability of central axons to grow out after injury or to the blocking of regeneration by a dense scar of glial tissue. An experiment in 1981 provided the basis for a major conceptual breakthrough in the study of regeneration in the CNS. The question was whether the poor regeneration of central axons was due to some intrinsic property of the neurons or whether it was connected to the environment of the CNS. Lesions were made in the CNS of rats, and pieces of peripheral nerve were grafted to the severed central tracts. Surprisingly, the central axons regenerated beautifully into the peripheral nerve graft, but when they got to the end of the graft and were confronted with the environment of the brain, regeneration stopped abruptly. This seemingly simple experiment

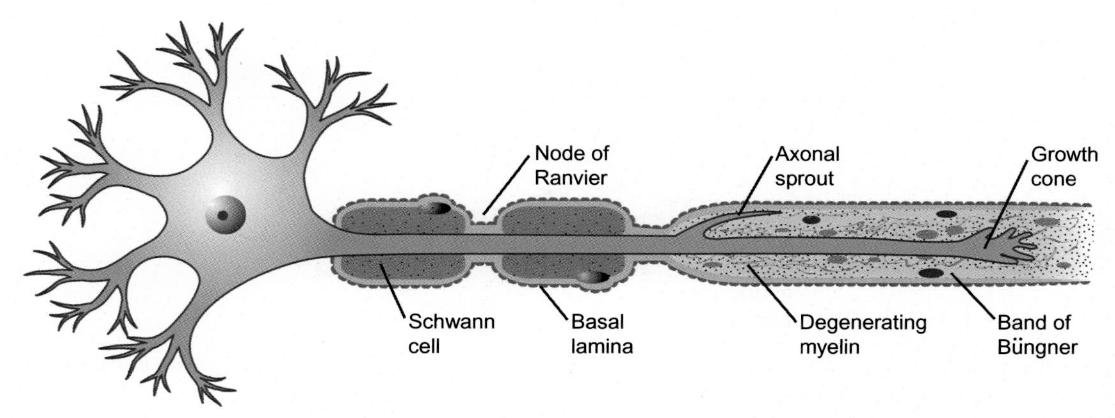

FIGURE 6.35 The regeneration of an axon in a damaged peripheral nerve. The regenerating axon (blue) sends out sprouts and a terminal growth cone into areas associated with the persisting basal lamina and dedifferentiating Schwann cells, which form bands of Büngner. *From Carlson (2007), with permission.*

demonstrated that (1) central axons are capable of robust regeneration after injury and (2) the local environment is a major determining factor in permitting or not permitting regeneration.

Subsequent experiments showed that the persistence of myelin after injury in the CNS is inhibitory. Several inhibitory molecules have been identified. Many fishes and amphibians, which are able to regenerate the spinal cord and parts of the brain, do not have these inhibitory molecules in their CNS. In peripheral nerves, invading macrophages and the local Schwann cells effectively remove the myelin, whereas in the CNS oligodendrocytes, which produce the central myelin, do not possess those properties. In addition, macrophages are much less likely to gain access to a site of injury in the CNS. Classical experiments like those described earlier have provided the impetus for much contemporary research to identify and create local environmental conditions that will support the directed outgrowth of central axons.

The other dogma that was held far too long by the neuroscience community was that new neurons do not form in the CNS, despite evidence to the contrary as early as the mid-1960s. The brain has traditionally been considered to be a postmitotic tissue, and there is little evidence that mature neurons can divide. However, as the result of research conducted in the late 1980s, we now know that the adult CNS, like many other tissues, possesses significant stores of neural stem cells. Neural stem cells have the ability to develop into new neurons and glial cells, although their ability to differentiate into nonneural cells is limited. Significant concentrations of neural stem cells have been found in the subventricular zone,[6] the dentate gyrus of the hippocampus and the striatum of the cerebral cortex.

Under normal circumstances, stem cells from the subventricular zone begin to mature and migrate from there toward the olfactory bulb in what is called the **rostral migratory stream** (Fig. 6.36). There, they differentiate into small interneurons (e.g., granule or glomerular cells). Stem cells in the dentate gyrus also differentiate into granule neurons that are involved in learning and memory. Unfortunately, there is little evidence that neural stem cells normally form neurons with long axonal projections—the type that would be needed to repair significant functional lesions within the brain and spinal cord. A great deal of research and clinical experimentation is presently directed toward implanting stem cells of various sorts into the brain for the treatment of conditions, such as Parkinson's disease, in which relatively small neurons produce vital secretory products. Ideally, implanted stem cells would generate axonal processes that grow out to the appropriate parts of the brain and then establish functional connections.

AGING IN THE NERVOUS SYSTEM

Few bodily systems exhibit aging changes more prominently than the nervous system. Unfortunately, these changes are better described than understood, but this is a reflection of our lack of understanding of the mechanisms underlying

6. The **subventricular zone** is a region of cells immediately underlying the epithelial lining of the central canal (the ependyma) in both the brain and spinal cord. The subventricular stem cells seem to be remnant populations of embryonic neurogenic ependymal cells. These cells survive in local niches that both nourish the stem cell precursors and prevent their premature differentiation into other cell types. The most prominent concentration of subventricular cells is in the inner walls of the lateral ventricles of the brain, but neural stem cells have also been found in the subependymal regions of the midbrain, the cerebellum, and the spinal cord.

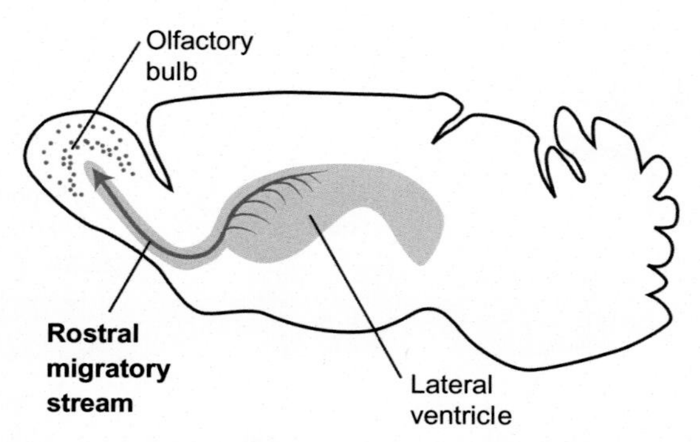

FIGURE 6.36 Representation of the rostral migratory stream, which carries neurogenic cells generated in the subventricular zone along the lateral ventricle of the brain into the region of the olfactory bulb in the rat. *From Carlson (2007), with permission.*

many major functional aspects of the nervous system. One of the difficulties in describing aging in the nervous system is variability among individuals and the difficulty in separating normal aging changes from age-related pathology.

Overall, there is a reduction in weight of the brain and a decrease in the number of neurons in selected areas, even while the number of glial cells often increases. At the cellular level, individual neurons show a loss in the number of dendrites, dendritic spines, and the number of synapses, but these numbers differ from one part of the brain to another. Complicating extrapolations from numerical loss is the issue of redundancy of connections within the brain. It is not known how much actual loss is compensated for by the persistence of redundant connections. At the biochemical level, there is a general decrease in metabolism and a decrease in protein and lipid synthesis.

At a functional level, motor changes are easiest to document. These include slower movements and reflexes, and a progressive loss of balance. Many of these changes can be attributed to the loss of motor neurons and downstream changes in motor units and muscle fibers, as well as to changes in the cerebellum. Certain aspects of memory are often compromised with increased age. These include short-term memory, while at the same time long-term memory tends to be much better preserved. Problems with word-finding (a component of semantic memory), especially of nouns, are commonly experienced as one age. Unfortunately, our understanding of the basis for normal memory is not sufficient to explain the changes associated with aging.

One common age-related pathology—**Parkinson's disease**—can be related to a specific functional deficit, namely the failure of dopamine production (a transmitter) by cells of the substantia nigra in the basal ganglia region of the brain. The other major scourge of aging is **dementia**, exemplified by **Alzheimer's disease**. This condition results in a major loss of cognitive functions. Estimates of frequency in populations over 80 years range from 20% to much higher. Alzheimer's patients have severe gross brain atrophy, including great loss of pyramidal cells in the cerebral cortex. Many surviving neurons, especially in the hippocampus and amygdala, are filled with tangles of microtubular proteins (**tau protein**), and plaques of a deposit called **amyloid** are also prominent in affected areas. Understanding the relationship between observed pathology and the functional changes in the brain in dementia remains a major challenge.

SUMMARY

The nervous system is subdivided into the central and peripheral nervous systems (PNS). The central nervous system (CNS) consists of the brain and spinal cord. The PNS is subdivided into somatic, autonomic, and enteric divisions. Somatic nerves are either motor or sensory, and autonomic nerves are sympathetic or parasympathetic. The enteric nervous system supplies the gut.

The spinal cord, which runs within the vertebral column, is ensheathed by three layers of ependymal tissue—dura mater, the arachnoid layer, and pia mater. The space between the spinal cord and the ependyma is filled with cerebrospinal fluid.

A narrow central canal in the spinal cord is surrounded by a layer of ependymal cells. Outside of that is a butterfly-shaped zone of gray matter, containing the cell bodies of neurons and unmyelinated nerve cell processes. Beyond that is the white matter, which consists of bundles of myelinated nerve fibers called tracts.

The brain consists of a brainstem, which controls many basic bodily functions; the cerebellum, which coordinates movements; the hypothalamus (a physiological coordinating center), and thalamus (a major relay station to the brain) make up the diencephalon; the telencephalon is dominated by the cerebral hemispheres and is the chief integrating center of the brain.

The telencephalon is evolutionarily the most recent component of the brain. The olfactory region processes olfactory signals and distributes them to other areas of the brain, where they are translated into emotions or physiological responses. The limbic system, containing the hippocampus and amygdala, operates at a subconscious level and is involved in learning and memory, as well as fear and anxiety responses. The cerebral cortex is heavily involved in integration of a large variety of neural signals, which are ultimately translated into motor responses and thoughts. The left and right sides of the brain are not functionally equal.

The pain pathway is an example of coordination between initial reception of pain by terminals of sensory nerve fibers in the skin and processing by the brain. These signals are carried to the spinal cord and are then carried up the spinal cord toward the thalamus via the lateral spinothalamic tract. From the thalamus, other nerve fibers carry the signal to the sensory cortex, from which association fibers ultimately converge upon the motor cortex. These signals are translated into a motor response, which results in removing the affected part of the body from the source of pain.

Peripheral nerve fibers regenerate well, whereas those of the brain do not. A severed peripheral nerve fiber forms a growth cone that extends down the nerve at a rate of ~1 mm/day. In the CNS, scarring, as well as degeneration products of myelin inhibit regeneration. Stem cells that can produce new neurons have been identified within areas of the CNS.

Chapter 7

Special Senses—Vision and Hearing

For all practical purposes, the entire outer surface of our body is a sense organ. The skin (see Chapter 3) is studded with special nerve endings that measure our mechanical and thermal environment. The nose (see Chapter 11) is the seat of our olfactory sense, and the tongue (see Chapter 12) senses the chemical and mechanical environment inherent in the food and liquid that we take in. The remaining major senses are concentrated in two very complex sense organs—the eyes and ears and their neural connections.

Both vision and hearing have ancient evolutionary origins. A common theme is that molecules and structures that originally served one function were adapted over time for their new sensory roles. Many of the fundamental patterns of vision, in particular, were set long ago—before and during the Cambrian explosion of diversity more than 500 million years ago. Hearing and ears are unique to vertebrates, but a vibratory sense, which is the essence of hearing, is well developed in many invertebrates. As a general rule, individual sense organs developed before their processing centers appeared in the brain, because without the sense organ, there would be no reason for the corresponding part of the brain to evolve.

In their contemporary form, both the visual and auditory systems illustrate well how structural adaptations of the eye and ear greatly facilitate their sensory functions. Nevertheless, some structural arrangements are not ideal and may reflect evolutionary wrong turns. Yet, the overall efficiency of both the eyes and ears is remarkable.

VISUAL SYSTEM

Evolution of Vision

Vision is one of the most broadly distributed functions throughout the animal kingdom. In its most elemental form, vision consists of light recognition and is accompanied by some form of response. Even some unicellular plants contain an eyespot, which reacts to light by molecular mechanisms remarkably similar to those that occur in the human eye. Much of the transition from simple light receptors to eyes as we know them occurred during the Cambrian explosion of species between 600 and 500 million years ago.

The earliest light sensors were used largely for detection of light levels and for entraining daily (diurnal) and seasonal rhythms. These functions persist today, even in the most complex animals. Eyes that detect visual patterns have evolved a number of times and in a number of different directions. The evolution of vision involved two major steps.

The first was the development of light-sensitive molecules. This seems to have been a single evolutionary step that is fundamental to all animals and even some members of the plant kingdom. A family of light-sensitive molecules, called **opsins**, forms the molecular basis for light reception. Opsins consist of a protein (seven-member transmembrane) molecule that is connected with a vitamin A (retinoic acid) derivative—**retinal**, in the case of humans and many other species. To this date, more than 1000 different opsins have been identified among animal species. When a photon of light hits an opsin molecule, retinal or its equivalent undergoes a **conformational change** (change in shape). This is first translated into a chemical reaction and ultimately into an ionic change through the activity of ion channels. The ionic change leads to an action potential, through which the primary visual signal in humans and most animals is transmitted to the brain.

The second major evolutionary step was the development of light-sensitive cells that house the various opsins. Light-sensitive groups of cells, collectively called **eyespots**, have evolved independently among invertebrates at least 50 times. Such eyespots are capable of measuring light intensity and some directionality, but they cannot form images. Higher level visual receptors that can distinguish directionality and images have appeared in only six phyla, but these groups contain the vast majority of animal species.

The Human Body. DOI: https://doi.org/10.1016/B978-0-12-804254-0.00007-7

Within these phyla, two different strategies were followed. Both cases involved the transformation of the light-sensitive cells to a form that greatly increased the amount of membrane surface that houses the opsin molecules. Three major invertebrate groups—the annelids, molluscs, and arthropods—utilize cells that have transformed large numbers of microvilli into platforms (**rhabdomeres**) containing visual pigments. The chordates, on the other hand, make use of transformed ciliary cells, which contain stacks of platforms with the opsins embedded within their membranes. Each type of cells is associated with its own family of opsins.

Visual acuity requires much more than light-sensitive cells. The ability to discriminate directionality of light arose when the cells containing the visual pigments (the opsins) began to sink beneath the epithelial surface into a pit, or later, a cup-like structure—an arrangement that in a highly modified form persists in the retina of the human eye. Because the ambient light entering a pit or cup hits the light-sensitive cells at different angles, directionality can be determined. This is not possible on a flat optical surface. Even better directional sensing was obtained by reducing the size of the opening to the cup (analogous to the pupil of the human eye), and by this mechanism some poorly resolved images could be obtained.

A next phase in the evolution of eyes was the growth of a layer of transparent cells (the equivalent of the human cornea) over the opening to the optic pit. This, plus the later evolution of a lens, greatly improved the focusing of the eye and created much sharper images. These and other developments resulted in the eye's functioning very much as a photographic camera.

Ultimately, the light collected by the sensitive cells must be translated into some sort of response. This has been accomplished by neural outgrowths from the light-collecting apparatus. In a few primitive animals, like jellyfish, the eyes connect directly with their muscles, but in most species, the information collected by the eyes flows to the brain in the form of neural signals. The number of ways that the information is processed is almost as great as the variety of different types of eyes.

The evolution of the eye has also involved a larger dimension. Many invertebrates, especially insects, have developed complex **compound eyes** with many cellular **facets** that function as independent units, while collectively providing a more coherent overall image. The position of mammalian eyes is of considerable functional significance. Predators have evolved eyes with binocular vision in order to provide a better focus on potential prey, whereas prey animals have eyes located on the sides of the head for better range of vision. A number of cave-dwelling species of fishes and amphibians have surprisingly rapidly lost much of their visual apparatus.

One of the most important evolutionary changes in vertebrates was the development of specialized cells (rods) for night vision, in contrast to the cones, which react to daylight colors. Many subtle subcellular and molecular modifications underlay this change. Most of the major evolutionary steps in the creation of the vertebrate camera eye were being set in place as the protochordates gave rise to the more familiar vertebrate species over 500 million years ago.

The Human Eye as a Camera

The human eye is a complex structure, but the roles of the various structural components can be best understood by making the analogy of the eye to a camera (see Fig. 7.3B) and following the path of a ray of light as it enters the eye and then passes through it until it hits the retina.

Eyelids and Tears

The eyelids and tears, produced by the lacrimal glands, are the rough equivalent to the lens cap and lens cleaner of a camera. The **eyelids** (Fig. 7.1) are thin flaps that help to ensure that the surface of the cornea will be adequately lubricated. When closed, the eyelids protect the eye against mechanical, thermal, or photic damage. The eyelids contain a surprisingly diverse number of tissue components. Their outer surface is covered by skin—hairless except for the eyelashes protruding from the free margin. The eyelashes offer some protection from particulate matter that could injure the cornea. The inner surface of the lids is covered by the **conjunctiva**, a thin epithelium studded with goblet cells, which produce a mucous secretion that plays an important role in the spread of tears (see later). Stretched out within the eyelid is the **tarsal plate**, a layer of dense connective tissue that gives some substance and mechanical stability to the eyelid. Embedded within the tarsal plate are **Meibomian glands**, a row of 20–30 modified sebaceous glands, which produce an oily secretion that is also important in tear dynamics. Connected to the soft tissue of the upper eyelid are the tendinous ends of the **levator palpebrae superioris muscle**, the main muscle that elevates the upper lid. Beneath that is a thin layer of smooth muscle that also plays a role in eyelid dynamics. Finally, descending from the region of the eyebrow are slips from a larger circular muscle, the **orbicularis oculi**, which completely surrounds the eye.

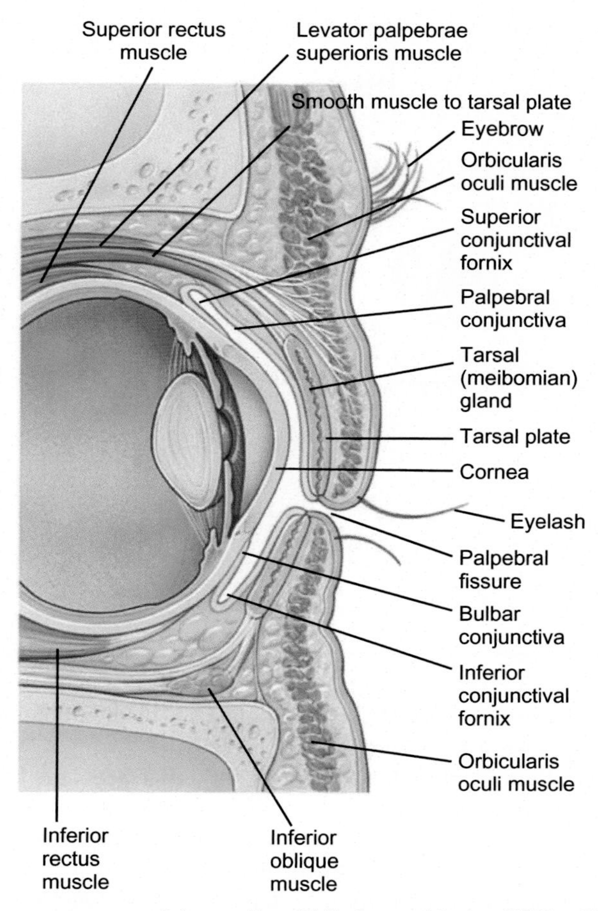

FIGURE 7.1 Eyelids and other accessory structures around the eye. *From Thibodeau and Patton (2007), with permission.*

Most people blink unconsciously every 4–10 seconds in order to maintain an even spread of tears across the cornea. Spontaneous blinking originates in a blinking center within the globus pallidus of the cerebrum and involves relaxation of the levator palpebrae muscle along with activation of the orbicularis oculi muscle. Another form of blinking occurs in response to irritation of the cornea. This corneal reflex is faster than the normal blink and is controlled by a reflex center in the pons. The orbicularis oculi muscle is more heavily involved in a reflex blink.

A major function of blinking is to ensure the even spread of **lacrimal fluid** (tears) across the surface of the eye. The largest component of tears is the secretion of the almond-size **lacrimal gland**, which is situated lateral to and above the eye in a recess of the frontal bone (Fig. 7.2). Tears leave the lacrimal gland through several small ducts and spread over the surface of the cornea from the lateral to the medial corners of the eye, where they collect in a soft tissue mass called the **caruncle** (see Fig. 7.2). From there, they empty into a lacrimal sac and into a nasolacrimal duct that leads to the nose.

The film of lacrimal fluid over the surface of the eye consists of three layers. The innermost layer, the mucus secretion from the goblet cells of the conjunctiva, coats the surface of the cornea and provides a hydrophilic layer that allows an even spread of the lacrimal fluid. The middle aqueous layer consists of the secretion of the lacrimal glands and forms the bulk of the lacrimal fluid. This secretion normally consists of water, electrolytes, antibodies for immune protection from foreign antigens and lysozyme, which breaks down the cell walls of certain pathogenic bacteria. All of these components individually provide some measure of protection to the cornea. In addition, dissolved oxygen in this fluid diffuses into the cornea, which is avascular and is not able to get sufficient oxygen from other sources.[1] The outer layer of

1. One problem with contact lenses is that they reduce or prevent the diffusion of oxygen into the cornea.

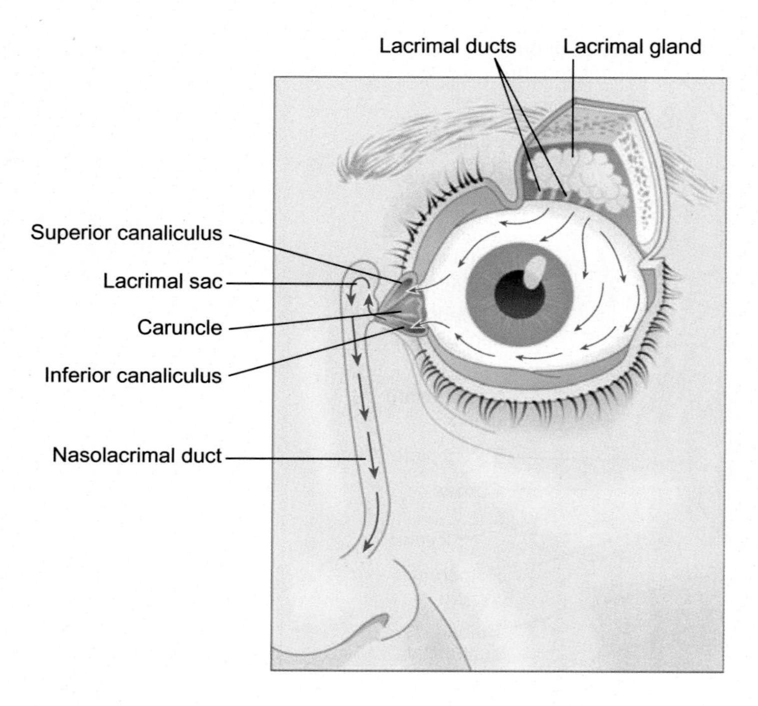

FIGURE 7.2 The lacrimal apparatus. Arrows show the flow of tears. *From Waugh and Grant (2014), with permission.*

lacrimal fluid consists of the oily secretion of the Meibomian glands. This hydrophobic layer envelops the film of lacrimal fluid and prevents the spilling over of the watery middle layer.

Three main types of tears have been identified. Normal tear fluid represents the basal level, and from 0.75 to 1.1 mL of this fluid is secreted by the lacrimal glands per day and drains through the nasolacrimal ducts. In response to irritants, the lacrimal glands produce greater amounts of a second type of tears through a neural reflex pathway that begins with a branch of the fifth cranial (trigeminal) nerve, the main sensory nerve of the face. A third type is emotional tears which, interestingly, have a different composition from normal tears. Emotional tears contain some protein hormones, such as prolactin, adrenocorticotrophic hormone, and leu-enkephalin (an opiate). Emotional tears are known to have behavioral effects on certain laboratory animals and may have some effect on interpersonal interactions in humans, as well. An unusual mammal, the naked mole rat, spreads emotional tears over its body in order to keep other aggressive rats away. According to one study, emotional tears in a woman may reduce sexual arousal in a man. When the fifth cranial nerve is cut, the secretion of reflex tears ceases, but emotional tears can still form.

The Eyeball

The **eyeball** (Fig. 7.3), which is about 25 mm in diameter, is the biological analogue of the camera itself. Its overall structure can be broken down into three layers or tunics—an outer tunic, which is analogous to the camera housing and lens, a middle layer, which is involved in the size of the aperture, and an inner tunic, which would represent the film (Fig. 7.4). The **outer tunic** consists of the sclera and the cornea. Components of the **middle tunic**, called collectively the **uvea**, are the choroid layer, the ciliary body, and the iris. The middle layer is richly vascularized. The **inner tunic** is the retina and its backing, the pigmented epithelium of the retina.

Cornea and Sclera

The outer tunic of the eye is a tough layer of connective tissue consisting of the cornea and the sclera. The **sclera** (white of the eye) comprises five-sixths of the eyeball and is composed mainly of type I collagen fibers, the irregular arrangement of which makes the sclera opaque. In keeping with the eye's origin as an outgrowth of the embryonic brain, it is not surprising that the sclera of the eye is continuous with the dura mater covering the brain. The sclera serves as the attachment site for the six **extraocular muscles**, which move the eyeball (see later). Humans are among the few mammalian species in which the sclera is readily visible. Because of that, it is not difficult to determine the direction of a person's gaze by looking at the iris. According to one hypothesis, this arrangement has evolved as a means of nonverbal social communication.

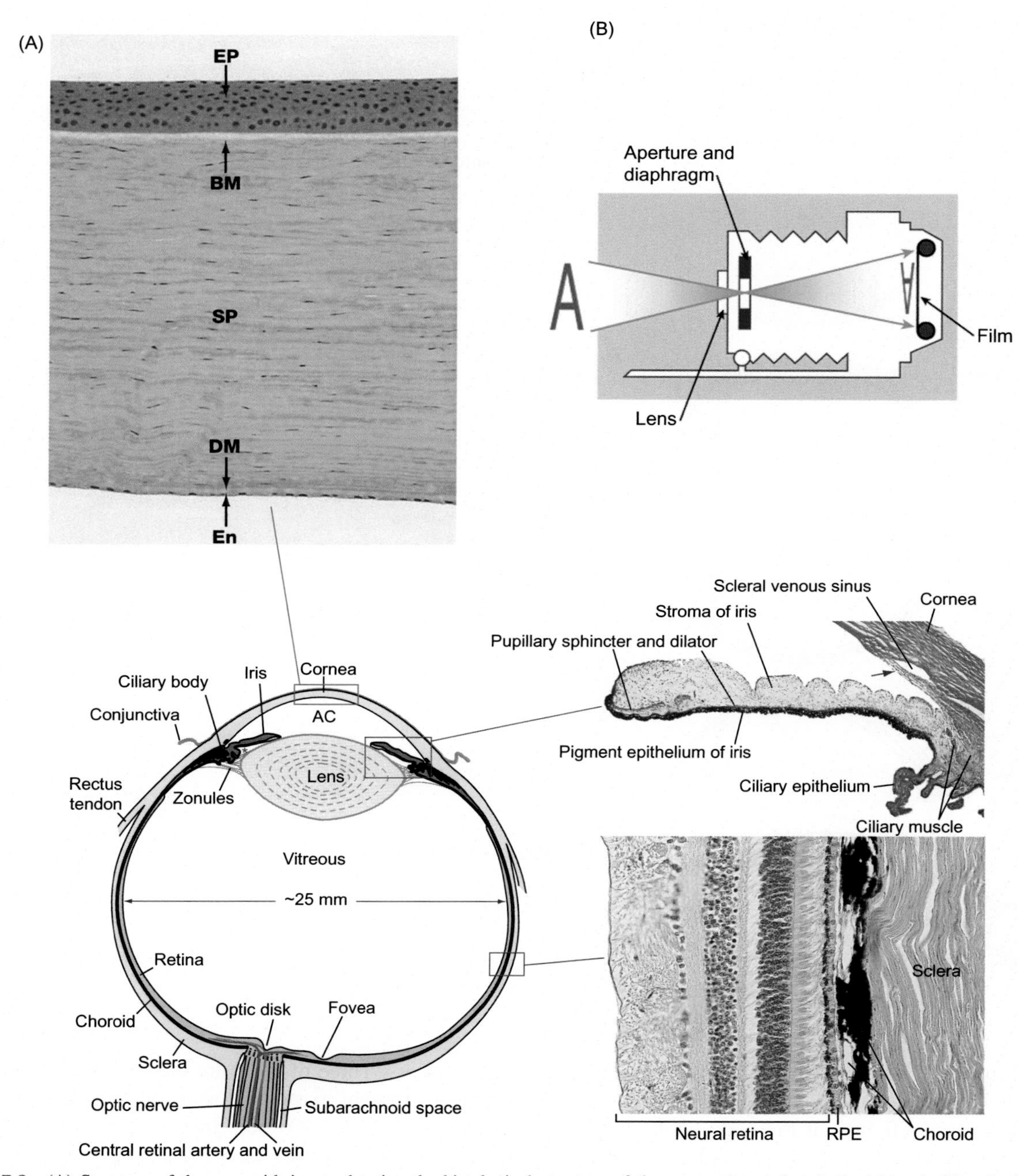

FIGURE 7.3 (A) Structure of the eye, with insets showing the histological structure of the cornea (upper inset), the iris and ciliary body (middle inset) and retina (lower inset). (B) Comparable structure of a film camera, with a lens, aperture, and diaphragm (analogous to iris and ciliary body) and film (analogous to retina). B, Bowman's membrane; DM, Descemet's membrane; En endothelium; EP, epithelium; RPE, retinal pigment epithelium; SP, stroma. *(A) From Nolte (2009); Young et al. (2006), with permission.*

The **cornea**, which is continuous with the sclera, is a significant component of the lens apparatus in the eye. Like the sclera, it is a tough tissue composed principally of type I collagen fibers, but because of their regular arrangement, the cornea is transparent. Because of its curvature, is responsible for about two-thirds of its total focusing power. Light hitting the corneal surface at the air/water interface refracts (bends) toward the lens in much the way a stick protruding from the water seems to change direction once it enters the water.

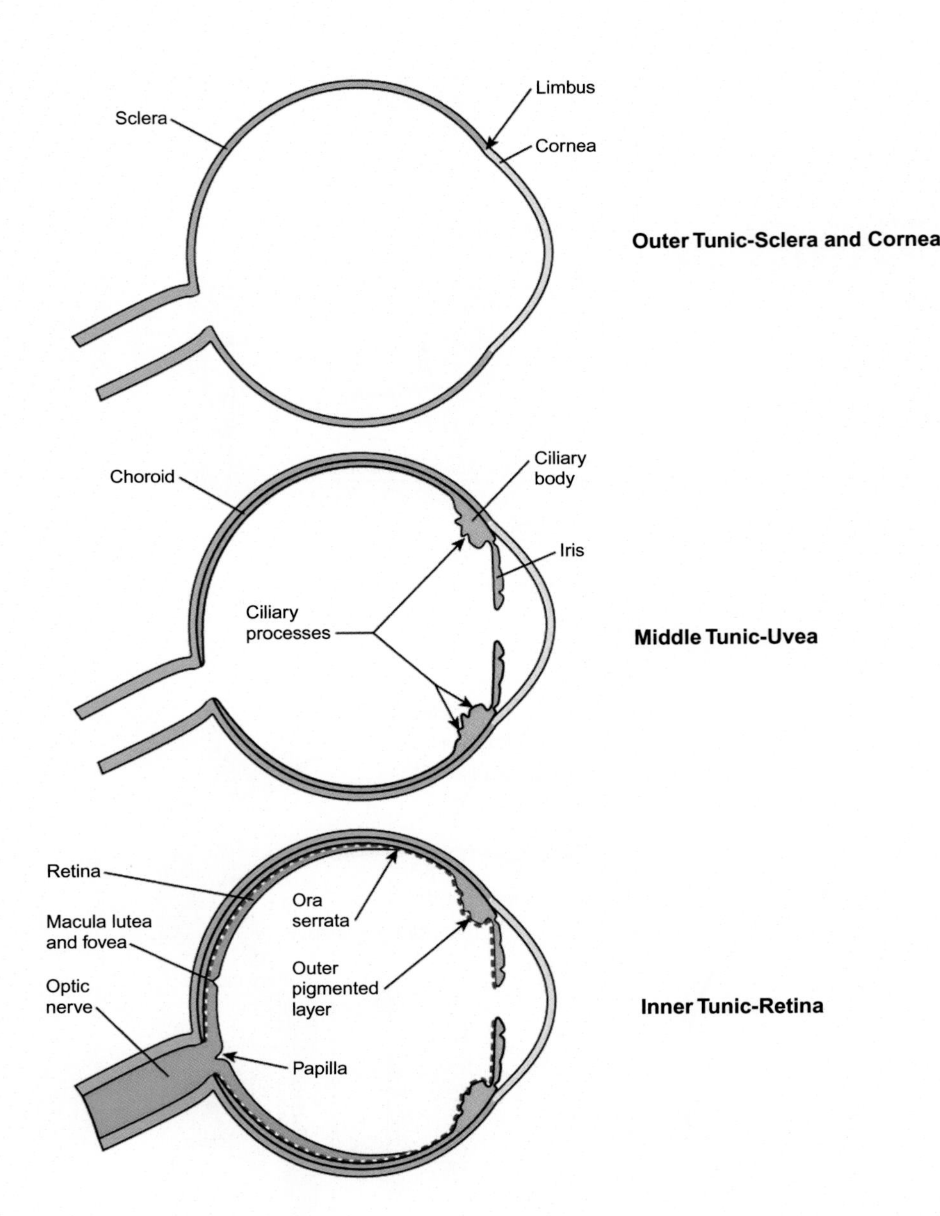

FIGURE 7.4 The three tunics of the eye. *From Kierszenbaum and Tres (2012), with permission.*

The cornea is a complex tissue, consisting of five anatomical layers. Three of these are most important functionally (see Fig. 7.3A). The outer corneal epithelium blocks the passage of foreign matter and water into the eye. It is one of the most sensitive tissues of the body, with a density of pain fiber endings 300–600 times greater than that of skin. Irritation of these nerve endings (part of cranial nerve V) initiates the corneal reflex. Dissolved oxygen from tears enters the cornea through the corneal epithelium.

The middle layer (the stroma) consists of about 200 layers of highly oriented collagen fibers and represents 90% of the thickness of the cornea. The fibers in each layer are oriented orthogonally to those of the next layer (Fig. 7.5) and some combination of the arrangement and spacing of the fibers accounts for the transparency of the cornea. About 78% of the corneal stroma consists of water. Importantly, it is not vascularized in order to maintain its transparency.

Maintenance of the integrity of the cornea is largely a function of the inner corneal endothelium. This single-cell layer is responsible for pumping excess water from the corneal stroma. In keeping with the energetic requirements for such pumping action, the corneal endothelial cells are richly supplied with mitochondria. During embryonic development, thyroid hormone stimulates the endothelial cells to begin pumping water out of the stroma. In contrast to the outer corneal epithelium, the corneal endothelium does not regenerate after damage or cell loss. When endothelial cells are lost from disease or aging, the remaining cells are not capable of removing sufficient quantities of water. As a result the waterlogged corneal stroma becomes cloudy.

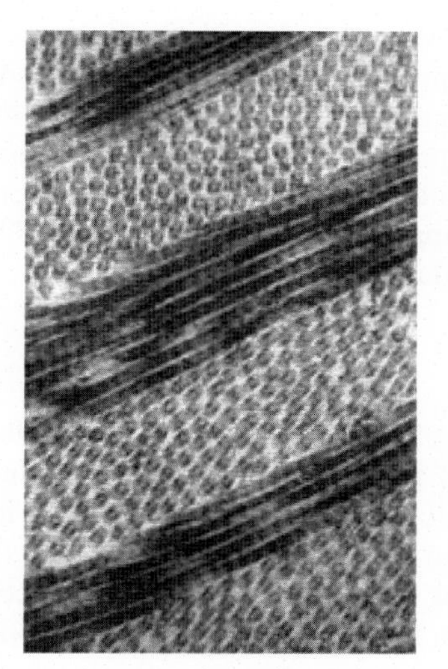

FIGURE 7.5 Electron micrograph of corneal collagen, showing the orthogonal arrangement of the layers of collagen fibers. *From Pollard and Earnshaw (2004), with permission.*

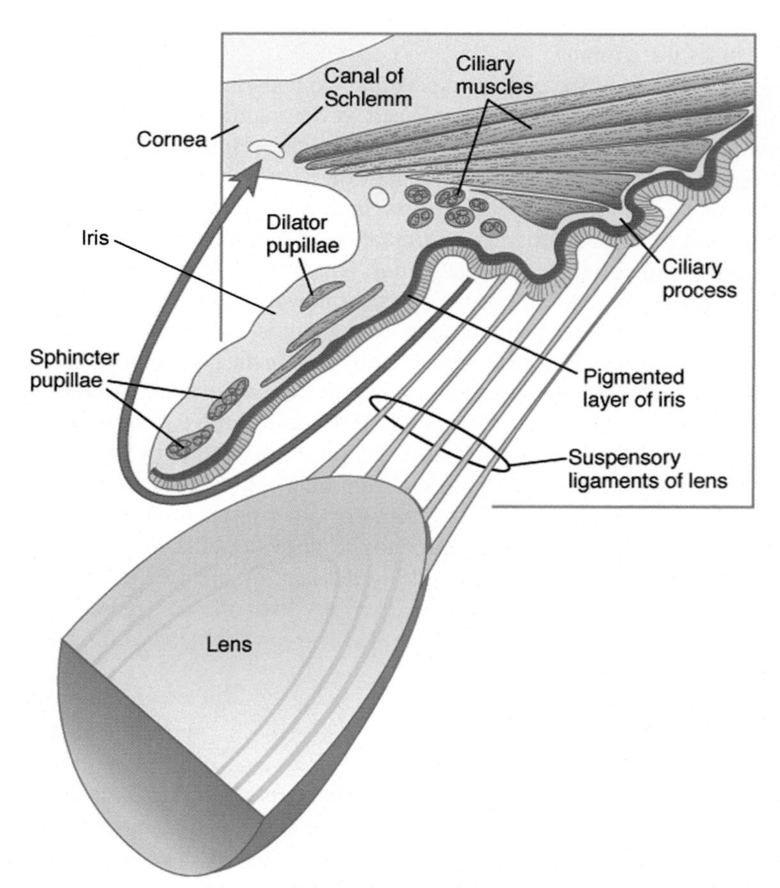

FIGURE 7.6 Details of the iris and ciliary body. The red arrow shows the pathway of circulation of aqueous humor, from its origin behind the iris to its removal through the canal of Schlemm. *From Carlson BM (2014), with permission.*

The endothelial layer of the cornea borders the anterior chamber of the eye (see Fig. 7.3A), which is filled with a fluid called **aqueous humor** through which light rays must pass. Aqueous humor, found in both the anterior and the posterior chambers, is produced by ciliary processes extending out from the ciliary body (Fig. 7.6) as an ultrafiltrate of blood. Large blood proteins are filtered out by connections between ciliary epithelial cells before the fluid reaches the chambers. Aqueous humor consists of 98% water, along with electrolytes, amino acids, glucose, ascorbic acid, immunoglobulins, and small amounts of other components. Some of these components are taken up by the corneal endothelium and nourish the corneal tissues. The immunoglobulins represent a line of resistance to disease.

Aqueous humor is present under slight pressure (15 mm Hg over atmospheric pressure), and this maintains the firmness and curvature of the cornea. There is a delicate balance between its production (from the ciliary body, see Fig. 7.6) and removal (through the **canal of Schlemm**). A disturbance in the equilibrium, due usually to drainage problems, can lead to increased intraocular pressure and **glaucoma**.

Iris and Pupil

Continuing the camera analogy, the **pupil** of the eye, through which light must pass, is the aperture, and the **iris** controls the size of the aperture (like f-stops in a camera). The iris is part of the middle tunic of the eye (uvea, see Fig. 7.4). Its outer edge connects to the ciliary body and sclera, and its inner margin outlines the pupil of the eye. Its cellular structure is well adapted for its functions. The back side of the iris is covered by a two-cell thick epithelium that is heavily laden with melanin pigment granules. This shields the retina from excessive light and ensures that the light reaching the retina has passed through the pupil—important for proper focusing. The bulk of the iris consists of a well-vascularized, loose connective tissue containing two smooth muscles, a constrictor muscle surrounding the margin of the pupil and a dilator muscle consisting of radially oriented strands of smooth muscle-like myoepithelial cells passing through the stroma just in front of the pigmented epithelium. The front surface of the iris—that facing the cornea—is irregular and is made up of a layer of fibroblasts and some pigment cells. Aqueous humor passes through crypts in this layer to nourish the cells of the stroma.

The iris, which is the colored part of the eye, contracts and expands the pupil, depending upon the amount of ambient light. Reducing the diameter of the pupil is accomplished by contractions of the **constrictor muscle**. This is accomplished through the pupillary light reflex in which light hitting the retina sends signals to midbrain centers. These centers then send out efferent signals along parasympathetic nerve fibers traveling along with the oculomotor nerve (cranial nerve III). Dilation of the pupil, in response to low light or a general increase in sympathetic activity (fight or flight reaction), is accomplished through sympathetic stimulation of the dilator muscle.

The basis of both eye color and patterns of pigmentation is surprisingly complex. Genetic studies have shown that many genes are involved in eye pigmentation, but the physical construction of the components of the iris also plays a very important role in eye color. There are many variants of eye color, but brown and blue are the most common, followed by green, hazel, and almost black. Brown is a genetically dominant form, but there are many variations and interactions with other pigmentation genes. Brown eye color is due to the presence of **melanin** granules in cells at the outer border of the iris. The greater the concentration, the deeper the shade of brown. Surprisingly, the dense layer of pigment cells at the back of the iris plays little role in eye color. People with blue eyes have little outer pigment in the iris. Much of the blueness is due to the scattering of light rays from bundles of collagen within the stroma of the iris. Other shades of eye color involve the presence of varying quantities of other pigment molecules (eumelanin, pheomelanin), in addition to melanin. The pink eyes of true albinos are due to the visibility of the blood vessels within the iridial stroma. In addition, the sensitivity to light in many albinos is due to the lack of pigment in the back of the iris. Color patterns in the iris can also be genetically based, but much of pattern involves the configuration of the iridial stroma. All of these variables result in any individual's iris being unique. This serves as the basis for the use of the iris in identifying individuals by customs and law enforcement agents.

Lens and Ciliary Apparatus

The **lens** (see Fig. 7.3A) is the second component of the eye that acts to refract light so that it focuses correctly upon the retina. In humans, the lens accounts for about one-third of the total focusing power of the eye (the cornea, for the remaining two-thirds). In contrast, the much rounder lens of aquatic animals does most of the focusing, because the cornea is flatter and also there is much less refraction of light rays between the surrounding water and the cornea. The lens accommodates to objects at different distances by changing its shape—rounder for nearby objects and flatter for those farther away.

Accommodation of the lens is accomplished by the smooth muscle fibers of the ciliary body and the suspensory ligaments that connect the muscle to the lens. When the ciliary muscle fibers, which are arranged like a sphincter, contract, the suspensory ligaments relax, thus reducing tension on the lens. This allows the lens to round up to a greater degree for focusing on close objects. For distance vision, the ciliary muscles relax, thereby putting tension on the

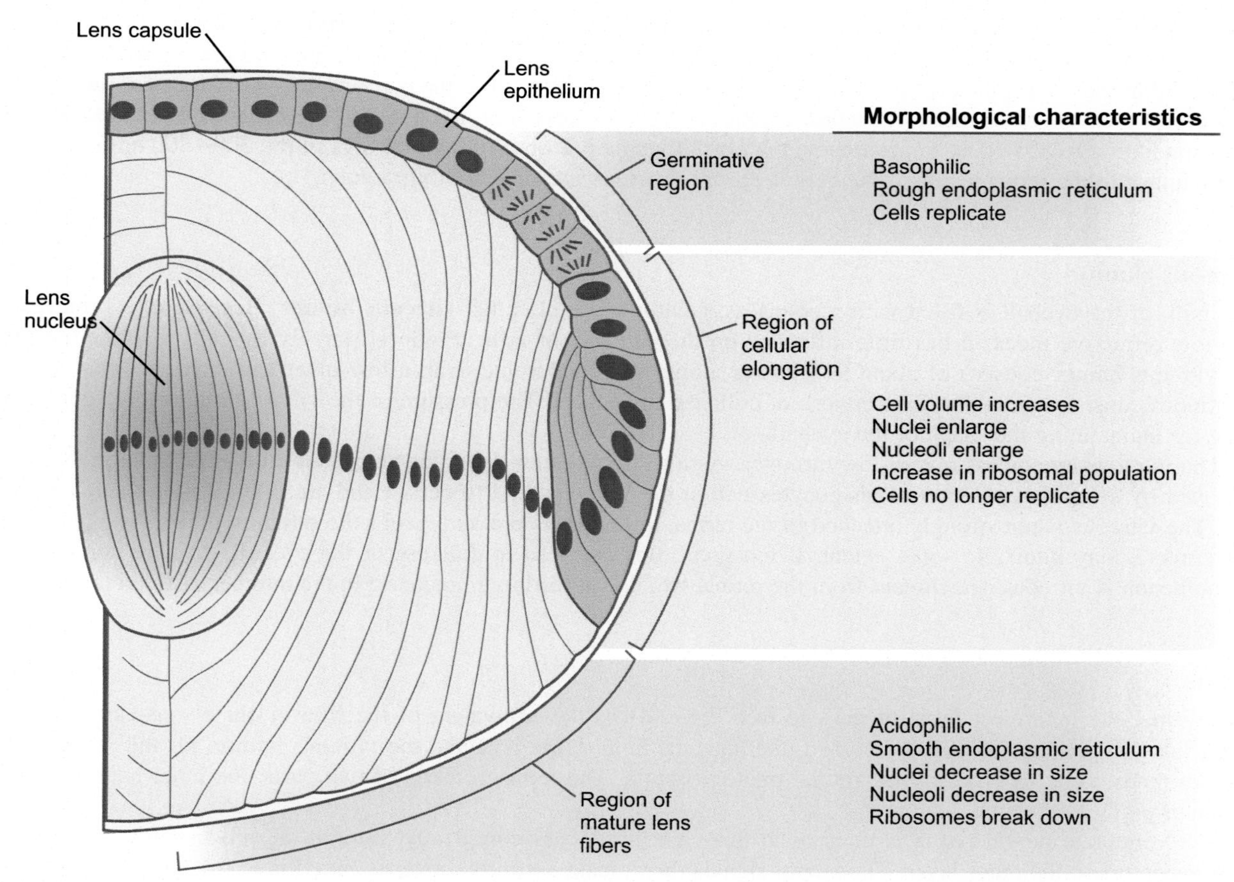

FIGURE 7.7 Organization of the vertebrate lens. As the lens grows, epithelial cells from the germinative region stop dividing, elongate, and differentiate into lens fiber cells that produce lens crystallin proteins. *From Carlson (2014), with permission.*

suspensory ligaments and pulling out on the lens and flattening it. The need for active muscle contraction for close vision may account to some extent for ocular fatigue when reading or doing close work.

The adult human lens is about 10 mm in diameter, and it is located just behind the iris. It is bathed in front by aqueous humor and behind by vitreous humor. The normal lens is transparent and nonvascularized. It is composed of three main regions (Fig. 7.7): (1) a basement membrane, or lens capsule, which completely surrounds the lens; (2) the main body of the lens, consisting of banana-shaped cells, called lens fibers; (3) a single layered cuboidal lens epithelium covering the front of the lens, which is responsible for obtaining the bulk of the needed ions and nutrients for the entire lens from components of the aqueous humor. Ion balance and water content are maintained by active Na^+/K^+-ATPase pumps within the lens epithelium. Most of these materials enter the lens through the polar region (front of the lens) and leave near the equator—the outer rim where the suspensory ligaments attach.

In the adult, the cells constituting the body of the lens (the lens fibers) are organized almost like layers of an onion when viewed in horizontal section (see Fig. 7.7). The center of the lens contains a region of smaller cells, which constitute the lens nucleus. These are the cells that first formed the body of the lens in the embryo. The lens grows by the addition of new lens fibers at the equatorial region. These cells are derived from the lens epithelium. Those epithelial cells close to the equator of the lens begin to elongate, and while doing this they begin to synthesize **crystallin proteins**, which soon fill much of the cell and impart transparency to the cell. Because the elongating cells are intensively producing intracellular proteins, they are richly endowed with ribosomes. As the new lens fibers fill with crystallin proteins, they lose many of their organelles, which is necessary for them to become transparent. The lens grows throughout life and does so by adding new lens fibers from the outside, therefore lens fibers are successively younger as they are situated away from the central lens nucleus.

Members of three classes (α, β, and γ) of crystallins are found in mammalian lens fibers. α-Crystallins evolved from an ancient family of heat shock proteins, which respond to various forms of cellular stress; β- and γ-crystallins have evolved from a common precursor protein that probably originally served an enzymatic purpose. In the lens, these proteins must become tightly packed in the lens fibers so that they are transparent to light and are regularly enough oriented so that they don't distort the light rays. Because the lens fibers lose the capacity to synthesize new proteins when they divest themselves of their organelles, crystallins must be very long-lived since they cannot be replaced within the cell. In addition to providing transparency, the crystallins help absorb ultraviolet rays in the 300–400 nm range to prevent damage to the retina. Cells of the cornea absorb shorter wavelengths of ultraviolet.

Vitreous Humor

The bulk of the eyeball is filled with a colorless, gelatinous fluid called **vitreous humor**. Because of its transparency and low refractive index, it has little influence on the pathway of light. Produced largely by cells of the ciliary body, the vitreous humor consists of about 98% water along with hyaluronic acid, a few other carbohydrates that give it its gelatinous consistency, and a fine network of collagen molecules. The pressure of the vitreous humor is largely responsible for maintaining the shape of the eyeball.

The vitreous humor has a very low turnover. As a result, cellular debris or by-products of inflammation that are not removed by a small population of phagocytes within the vitreous tend to persist and are often visible as floaters in the eye. The vitreous is not strongly attached to the retina, although its pressure keeps the retina in place. As a person ages, the vitreous may liquify to some extent. If too great, this can lead to collapse of the eyeball. A more common aging phenomenon is vitreous detachment from the retina, which can lead to an increase in the number of floaters in the eye.

Retina

The **retina**, the innermost tunic of the eye (see Fig. 7.4), is the equivalent of the film in our eye-as-camera analogy. Light images, passing through the cornea, the aqueous humor, the lens and the vitreous humor, hit the retina in focus and are ready to be received by the retinal photoreceptors. The retina consists of two parts, the inner neural retina and the outer pigmented retina.

The vertebrate neural retina is unusual in that light must pass completely through the retina before it reaches the photoreceptors in the outer layer. The neural signals then must pass in the reverse direction to the inner layer of the retina before they are gathered and flow to the optic nerve, through which they leave the eye. In an amazing series of parallel evolutionary steps, the eyes of cephalopods (octopi and squids) have independently developed an overall form remarkably similar to that of the vertebrate eye, but in their case they did it right. In the cephalopod eye the layers of the retina are reversed, and the photoreceptors are in the inside layer, which makes the first contact with light. The cells that carry the image away from the retina are conveniently located in the outer layer.

The overall organization of the **neural retina** is complex, with as many as 10 anatomical layers having been described. The basic organization, however, is relatively straightforward. In essence, the neural retina consists of three layers of neurons—the outer layer of photoreceptors, a middle layer of bipolar neurons, and an inner layer of ganglion cells that collect visual signals and carry them to the optic nerve via elongated cell processes (Fig. 7.8). Two other types of cells, horizontal cells and amacrine cells, are oriented perpendicularly to the chain of three retinal neurons. They function to connect and integrate various vertical pathways of the three-neuron chain.

One other important cell type in the neural retina is the Müller cell. Müller cells, which extend through almost the entire thickness of the neural retina, serve a number of functions similar to those of glial cells. Their main function is to maintain a stable extracellular environment by regulating the ionic environment, the balance of neurotransmitters, storing energy sources (glycogen), and serving as insulation and mechanical support for neurons.

A second major component of the retina is the **retinal pigment epithelium** (RPE), a separate one-cell thick epithelial layer that in the embryo arises separately from the neural retina (see Fig. 7.3A). The cuboidal cells of the RPE are heavily pigmented. The pigment granules absorb scattered light and by doing so improve the optical qualities of the eye and also protect the sensitive cells of the neural retina from photooxidative stress.

The tightly interlocked cells of the RPE are situated beneath the highly vascular choroid coat (part of the middle tunic of the eye). Because of this location, they serve to transport and regulate the flow of ions, nutrients, metabolites, and water to and from the neural retina. They also play an important role in keeping the eye as an immunologically privileged site. The RPE also plays important supporting roles in the visual process, such as ion transport and the storage, processing and release of retinal (see later). The cells of the RPE also phagocytize discarded membranes from the rod and cone cells.

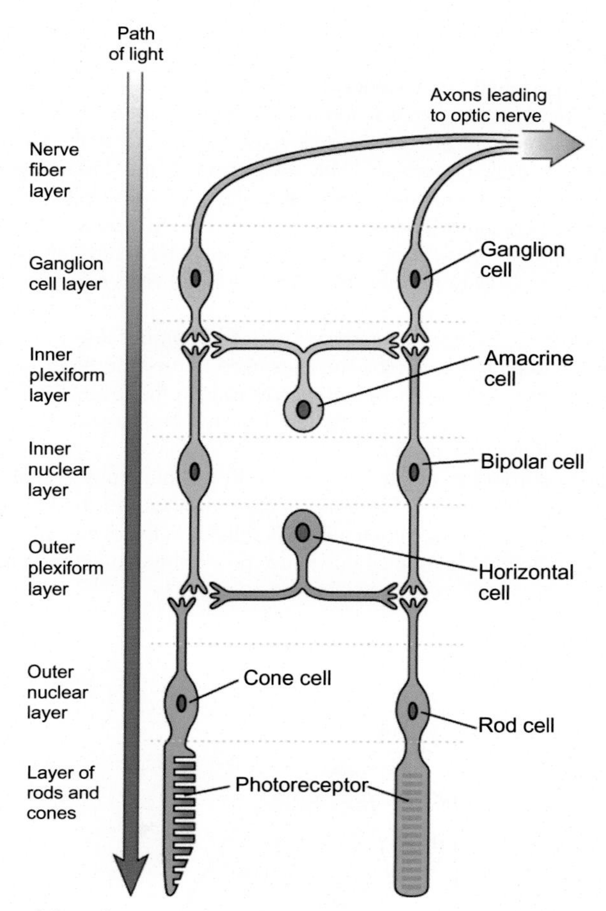

FIGURE 7.8 Basic cellular organization of the retina.

In some salamanders, the entire neural retina can be regenerated from cells of the RPE, although unfortunately this phenomenon does not naturally occur in mammals. Nevertheless, it has been found that adult stem cells capable of differentiating into cells of the neural retina have been found in the mammalian RPE.

Vision

Rods and Cones

The process of vision begins when rays of light hit the photoreceptors within the neural retina. These receptors are highly specialized neurons—the rods and cones. The retina contains about 90–120 million rods and 5–7 million cones. Rods are highly sensitive to light, but do not transmit colors; cones detect the various wavelengths of light but are less sensitive than rods. Because of these properties, rods function optimally in the dark (Box 7.1), whereas cones are designed for color vision in daylight. Within the retina, cones are concentrated in a small area called the **fovea** (see Fig. 7.3A), a region from which rods are excluded.

Many of the characteristics of our vision are related to both the properties of rods and cones and their disposition throughout the retina (Fig. 7.9). The fovea is the area where red and green cones are densely concentrated, and it is the region of the retina most responsible for high resolution vision. Blue cones are scattered outside the fovea. They are much more light sensitive than are red or green cones, a property that to some extent compensates for their small numbers. Rods are concentrated outside the fovea (see Fig. 7.9). A thousand times more sensitive to light than cones, the rods have poorer image resolution, but are better motion sensors. Because of these properties, people focus on the fovea for image resolution, but use more peripheral vision for the detection of motion.

BOX 7.1 Night (Scotopic) Vision

As evening darkness approaches, major functional changes in the retina are reflected in what we see. The amount of incoming light is insufficient to fully stimulate red and green cones, and red colors begin to seem washed out. In contrast, at twilight, green colors often seem relatively brighter, largely because of the high sensitivity of rods to the green wavelengths in low light. In real darkness, all colors except shades of gray disappear. Dark adaptation is not instantaneous, because it takes rods about 30–40 minutes to become totally adapted to darkness. Even a short burst of white light can induce temporary night blindness because of the time necessary for rods to readjust. The rhodopsin (see p. 191) in rods is insensitive to the longer red wavelengths, and because rods don't respond to red light, many instruments used at night have red signals on their screens. This red is picked up by the cones. Each eye responds separately to light. If a light is shined onto one eye after dark, the other remains dark-adapted if it has been shielded from the light.

In the dark, images are less distinct (20/20 vision is reduced to ~20/200), because of the lack of participation by the cones in night vision, but the rods continue to pick up motion. The eyes need to move in order to pick up stationary objects at night. At night, our visual field has a blind spot because of the absence of rods in the fovea. Objects seen with peripheral vision can disappear when the image passes over the fovea. A star (starlight is in the blue-green color spectrum) seen in one's peripheral visual fields may disappear when one looks directly at it.

In many nocturnal animals, ranging from fish to cats, cells in the RPE contain tiny crystalline structures, forming a reflective layer called the **tapetum lucidum**. The tapetum lucidum reflects light rays back to the retina, thus increasing the efficiency of light capture—by as much as a factor of six in cats. When a light is shined on the eye of an animal with a tapetum lucidum after dark, it reflects back a unique color, ranging from orange in walleye pike to blue in cats and deer.

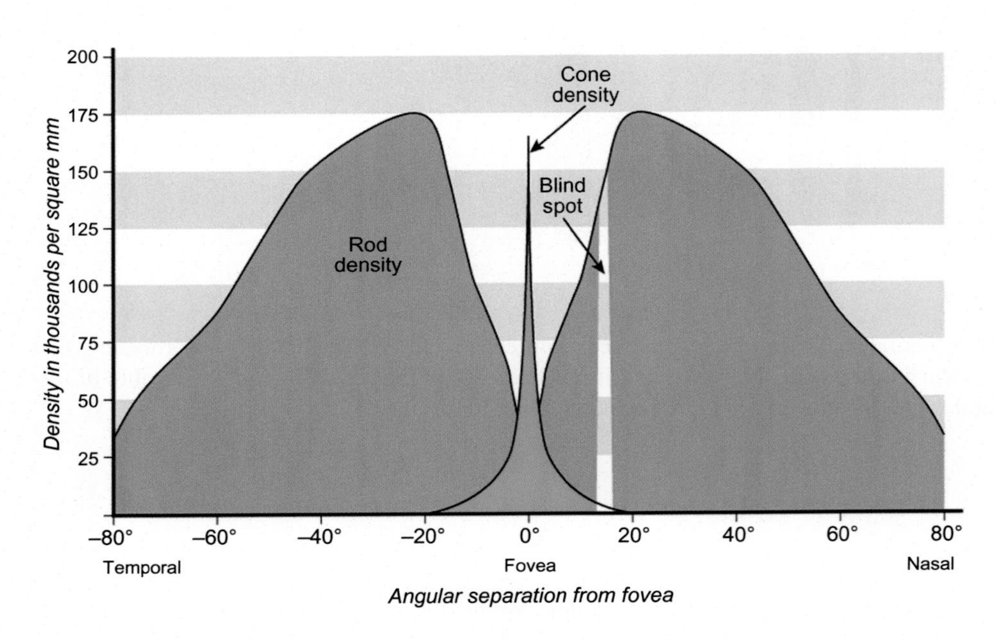

FIGURE 7.9 Density of rods and cones in the human retina. The reference point (0 on the X axis) is the fovea, the small region of greatest visual acuity and the location of most of the cones. The blind spot is the area where the optic nerve fibers leave the retina.

Properties of rods and cones are listed in Table 7.1. The bodies of these cells are embedded in the outer (back) layer of the neural retina, but the light-sensitive parts of these cells protrude from the surface of the neural retina and are partially embedded within the overlying RPE.

Rod and cone cells have three segments (Fig. 7.10). The innermost consists of synaptic terminals along with a short axon that connects them to the cell body. Next is an inner segment, which contains the nucleus, and beyond that, a region of cytoplasm specialized for synthesis—especially of the photopigments. The outer segment is connected to the inner segment by a short stalk containing a modified cilium. Within the outer segments are large numbers of infolded membranes containing embedded photopigments. The outer segments of rods contain stacks of about 1000 free floating membrane disks that have pinched off from the outer cell membrane. In contrast, the disks in cones are infoldings of the outer cell membrane that remain connected to the ciliary stalk. In rods, the youngest disks are closest to the inner segment. Those farthest from the inner segment are the oldest, and after about 12 days they are shed into the cytoplasm

TABLE 7.1 Properties of Rods and Cones

Rods	Cones
More numerous (90–120 million)	Less numerous (5–7 million)
High light sensitivity	Lower light sensitivity
Can respond to single photon	Requires >100 photons for similar response
Slow response to light	Fast response to light
More photopigments	Fewer photopigments
Don't transmit colors	Three types of cones mediate color vision
Function best in night vision	Function best in day vision
Relatively low visual acuity	High visual acuity
Become saturated in daylight	Hard to saturate by light
Absent in fovea	Concentrated in fovea
Concentrated in periphery of retina	Sparse in periphery of retina

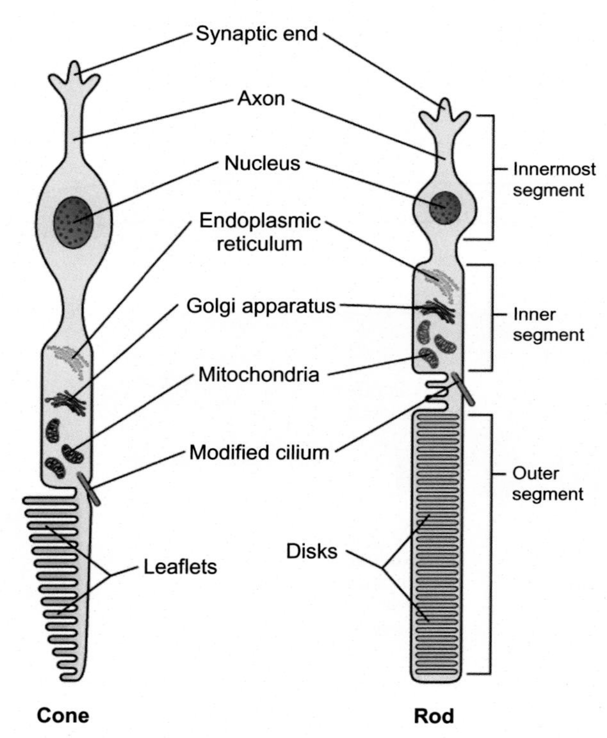

FIGURE 7.10 Cellular structure of rods and cones.

of the retinal pigment epithelial cells, in which the ends of the outer rod segments are embedded. There, the disks are broken down, with some components recycled and others transferred to the choroid circulation for removal.

Phototransduction

Photons of light hitting the disks in the outer segments of either rods or cones begin the molecular process of vision. Only about 10% of the light entering the eye actually contacts the receptors. The remainder is either absorbed by other components of the eye or misses the photoreceptor molecules. The essential photosensitive element in rods and cones is **rhodopsin**, the reddish photopigment that can react to a single photon of light. The essence of phototransduction is a sequence beginning with a photon of light hitting a molecule of rhodopsin. This initiates a chain of molecular events that ends with the depolarization of the plasma membrane of the photoreceptor cell and the transmission of a signal from it to a bipolar neuron in the retina (Box 7.2).

BOX 7.2 The Chain of Events Underlying Phototransduction

Embedded within the disks of rods are approximately 30,000 molecules per μm^2 of rhodopsin. Rhodopsin consists of an opsin peptide combined with the vitamin A derivative, **retinal**. The opsin determines the wavelength of light that is absorbed by retinal. In darkness, retinal exists in the geometrically unstable form of 11-cis-retinal. When a photon of light hits a rhodopsin molecule, the 11-cis-retinal becomes transformed to 11-trans-retinal within 1 ps (Fig. 7.11). The geometric change stimulates an overall change in the shape of the rhodopsin molecule. This change ultimately leads to the separation of retinal from the opsin and results in bleaching of the photopigment to a pale yellow. The liberated opsin stimulates the activity of a protein, called **transducin**. A single rhodopsin molecule is able to activate up to 800 transducin molecules, thus greatly amplifying the light signal. Transducin then activates an enzyme that converts the messenger cGMP (cyclic guanosine monophosphate) into GMP. The absence of cGMP results in the closure of ion channels and the hyperpolarization of the cell membrane. Unusually, hyperpolarization, rather than the formation of an action potential, results in the photoreceptor cells releasing the neurotransmitter at its synaptic end. In the case of both the photoreceptor cells and the bipolar neurons to which they are connected, the axonal length is so short that an action potential is not necessary to carry the signal.

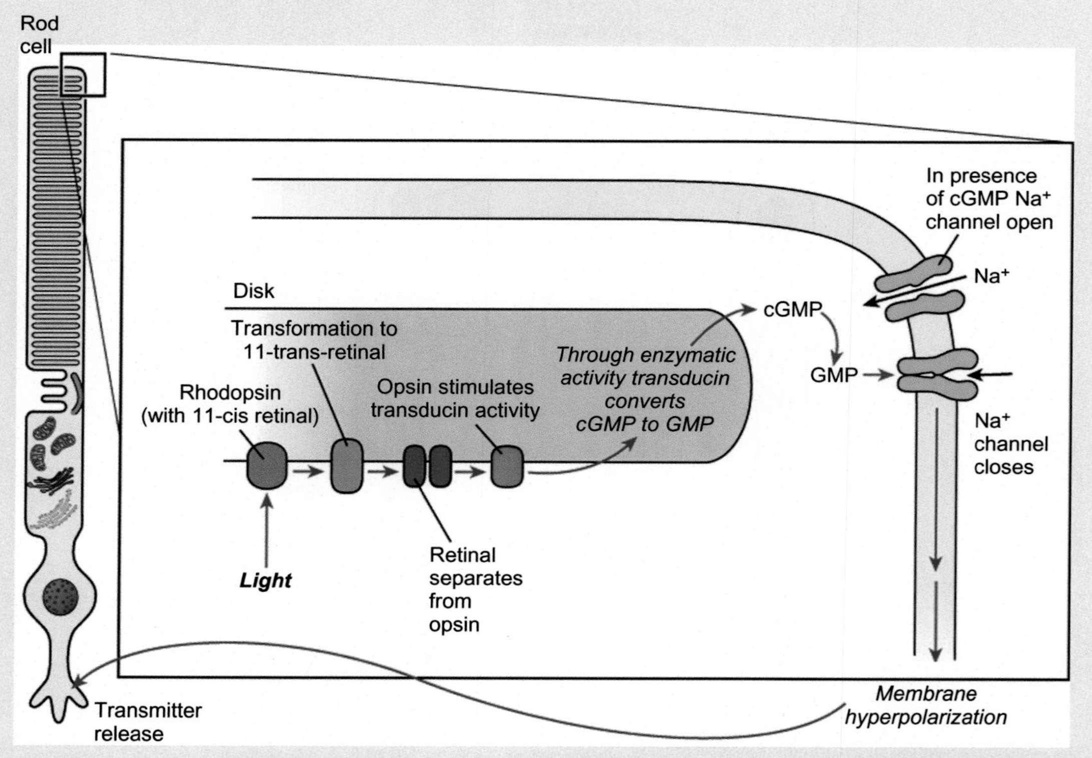

FIGURE 7.11 Phototransduction within a rod cell, showing major molecular events from the time a photon of light hits a disk in the rod to polarization of its cell membrane.

Within a few minutes after light stimulation, the rhodopsin molecule regenerates. Key to this process is the transference of 11-trans-retinal from the photoreceptor cell to the pigment epithelium, where it becomes reconverted to 11-cis-retinal. This compound is then transferred back to the photoreceptor cell, where it then recombines with the opsin to re-form functional rhodopsin.

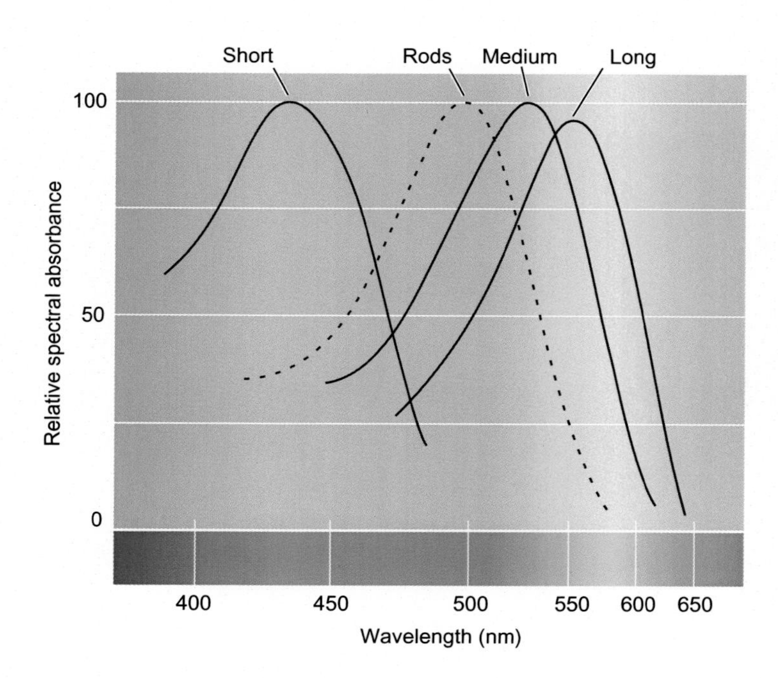

FIGURE 7.12 Absorption spectra of rods and cones. Relative absorbance is a measure of the sensitivity of the individual opsins. Short, blue; medium, green; and long, red cones. *From Purves et al. (2008), with permission.*

Rods contain only the single photopigment, rhodopsin, which has a maximum sensitivity to light waves of about 500 nm. In contrast, there are three different types of cones, each with its particular opsin that is most sensitive to light of a different wavelength. The three human opsins are most sensitive to blue, green, or red light (Fig. 7.12). They can also be stimulated by other wavelengths of light, but with a lesser level of sensitivity. About 64% of cones are most sensitive to red; 32% are most sensitive to green; and only 2% are blue sensitive.

Processing of Visual Signals by the Retina

The direct connection of a light-stimulated photoreceptor cell is the **bipolar neuron**. A receptive field is the area of the retina that stimulates a single bipolar cell after it has been exposed to light. Typically 15–30 rods directly synapse with a single bipolar neuron in the receptive field center, whereas in the fovea of the retina (the area of greatest visual acuity) there is a one-to-one ratio of cones to bipolar cells. A consequence of this arrangement is relatively poor spatial resolution in the dark, when rod function is dominant. A single rod or cone can also synapse with several bipolar cells (Fig. 7.13B). Horizontal cells of the retina, which are located at the photoreceptor/bipolar cell interface, connect more distant photoreceptor cells to a bipolar neuron (Fig. 7.13A). This domain is called the **receptive field surround**, and it represents an integrative mechanism for improving image quality.

The structural and functional relationships between photoreceptors, horizontal cells, and bipolar neurons in the receptive fields are complex, with inhibitory and excitatory processes often occurring in adjoining bipolar cells. The net result of these interactions is the ability to maintain image contrast under a variety of light conditions.

A similar degree of complexity exists at the boundary layer between the bipolar neurons and the retinal ganglion cells. At this juncture another cell type with lateral projections, the **amacrine cell** (see Fig. 7.8), interposes itself between the bipolar cells and the retinal ganglion cells. Like the horizontal cells, the amacrine cells provide a considerable degree of lateral connection and coordination among spatially related visual cellular units (i.e., the photoreceptors and the bipolar cells).

An example will illustrate this point. As mentioned earlier, as many as 30 rods synapse with a single bipolar neuron. The bipolar neurons that service the rods do not synapse directly with retinal ganglion cells. Instead, several rod bipolar cells synapse with a given amacrine cell, resulting in a considerable convergence of individual rod input onto a single amacrine cell. The rod amacrine cells then synapse with cone bipolar cells, which connect to retinal ganglion cells. A beneficial result of this arrangement is that a large number of rods can collect responses from a very weak light source and magnify their input. On the negative side, the fact that input from so many rods converges onto one cell considerably reduces the spatial discrimination by the rods. In contrast, within the fovea, a single cone cell synapses with a single bipolar cell, which synapses with a single midget retinal ganglion cell. This produces excellent spatial resolution.

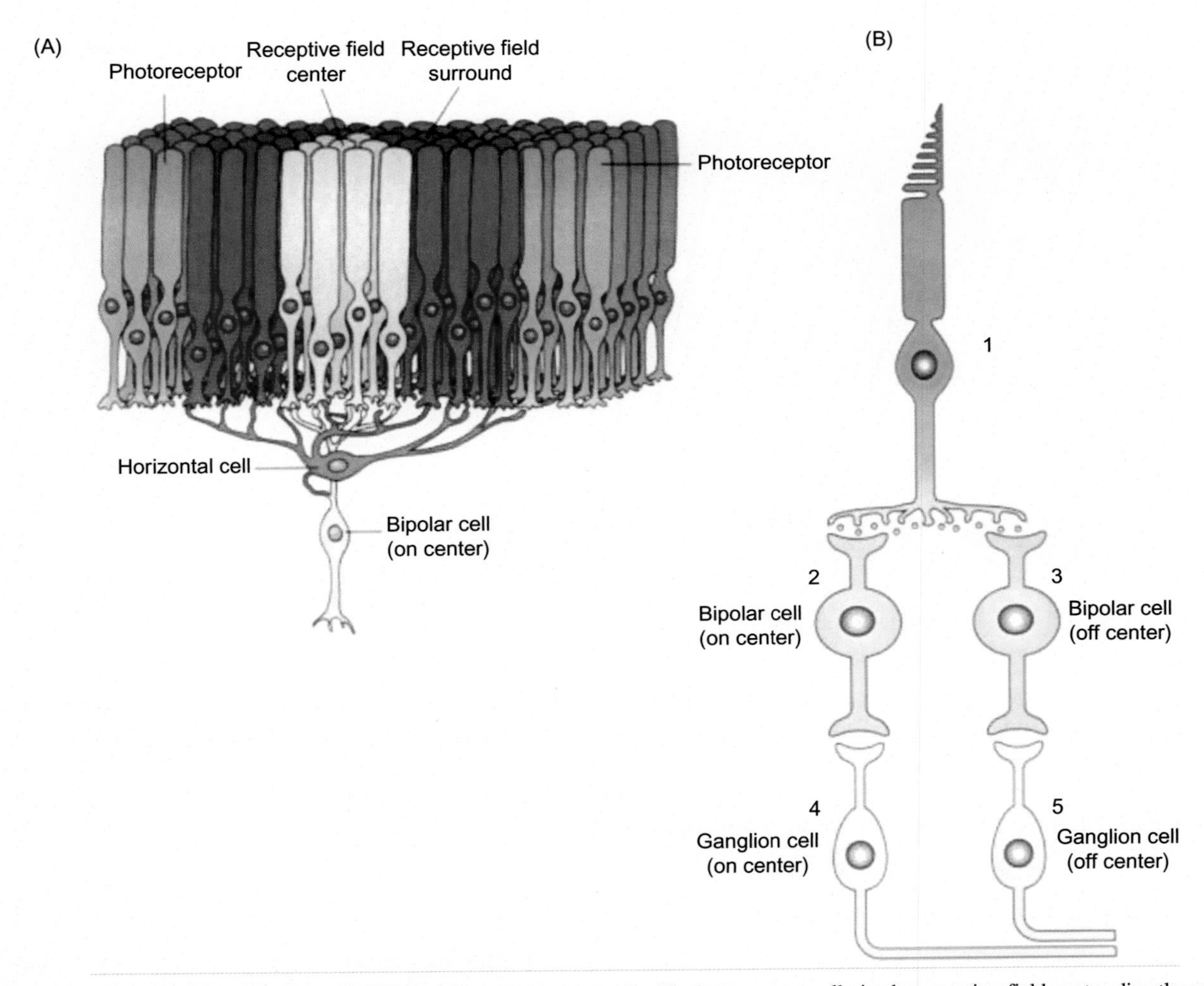

FIGURE 7.13 Receptive field of photoreceptors and their connections. (A) All photoreceptor cells in the receptive field center directly synapse with the underlying bipolar cell. Receptors of the receptive field surround communicate with the bipolar cell through connections with horizontal cells. (B) A single photoreceptor cell may synapse with one or more bipolar cells, one on center and the others, off center (receptive field surround). *From Siegel and Sapru (2011), with permission.*

A recently discovered cell with a different light-sensing function is the **photosensitive retinal ganglion cell**. These cells (1% of all retinal ganglion cells) contain the photopigment **melanopsin**, which is excited by light in the blue range (absorption peak at ~480 nm). They do not register images, but rather measure the intensity of ambient light. A main function appears to be synchronizing circadian rhythms with respect to the 24-hour light–dark cycle, and their input is probably involved in the suppression of melatonin release by the pineal gland (see p. 154). They also contribute to the pupillary light reflex. Their presence explains how animals or humans lacking in rods or cones can still respond to light.

Retinal Output

Both photoreceptor and bipolar neurons have very short axons and consequently don't require action potentials for transmission of impulses. Retinal ganglion cells, on the other hand, send long axons into the central nervous system (CNS) proper and thus operate more like typical neurons. The axons emanating from the retinal ganglion cells converge upon the optic disk (see Fig. 7.3), which represents the starting point of the optic nerve (cranial nerve II).

The **optic nerve** is organized more like a tract in the brain than a peripheral nerve. There is a reason for this. During embryonic development, most of the eye arises as an outpocketing of the diencephalon of the brain, and the optic nerve is the continuous connection between brain and eye. Each eye sends an optic nerve to a region at the base of the diencephalon (the optic chiasm), where they converge and then separate (Fig. 7.14A).

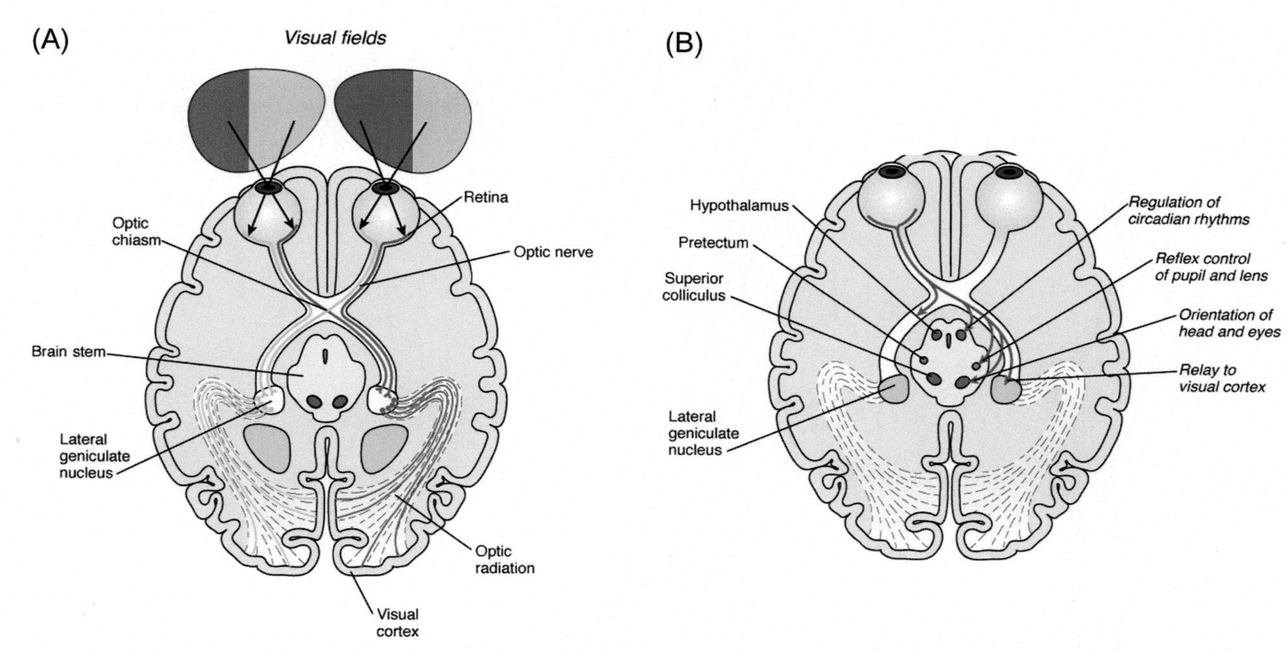

FIGURE 7.14 Visual fields and central pathways. (A) Primary visual pathways. (B) Central connections of visual fields and their functions.

Within the optic nerve are axons that go directly to a variety of locations within the brain. At the **optic chiasm**, axons carrying primary visual signals can take two pathways. Those that originate on the lateral sides of the retina leave the optic chiasm to enter the optic tract on the same side. In contrast, those axons that arise on the medial side of the retina cross over to the optic tract on the other side (see Fig. 7.14A). Retinal axons entering the optic tracts can lead to several functional destinations (Fig. 7.14B). A small number go to the hypothalamus, where they are involved in the regulation of light–dark rhythms. Another group feeds into the midbrain region called the pretectum. These axons are part of a circuit that connects with autonomic nerve fibers to control the pupillary light reflex. Yet another group synapses in the superior colliculus of the midbrain. These are involved in the coordination between head and eye movements and visual targets.

The largest group of visual axons synapses in the **lateral geniculate nucleus** of the thalamus, where they connect with nerve fibers that pass through white matter of the cerebral cortex as the **optic radiations**, (see Fig. 7.14A), which terminate in the visual cortex of the occipital lobe of the cerebrum. From the retina to the visual cortex, the axons maintain a strict topographical order, so that in the visual cortex retinal images are accurately represented. Much processing occurs within the visual cortex, and association projections from there go to many other regions of the cerebral cortex, where further processing and integration into conscious actions take place.

AUDITORY SYSTEM

Human hearing is inextricably connected with the sense of balance. Thus, the term **statoacoustic organ** is commonly applied to the ears and semicircular canals. The sense of hearing was the last major sense that evolved among the vertebrates, and the course of that evolution has been a tortuous one.

Evolution of the Auditory System

Both hearing and balance evolved from a primitive vibration-detection system, the fundamental basis of which has not greatly changed over several hundred million years. These senses both arose from hair cells, which are derived from the precursors of all multicellular animals. The ancestors of us all (**choanoflagellates**) consisted of single cells with a motile **kinocilium** and a surrounding array of microvilli (Fig. 7.15A). The kinocilium created water currents that swept food materials into the microvilli, which brought them into the body of the cell, where they were digested. Similar cells have persisted in all animals and have been adapted for different functions.

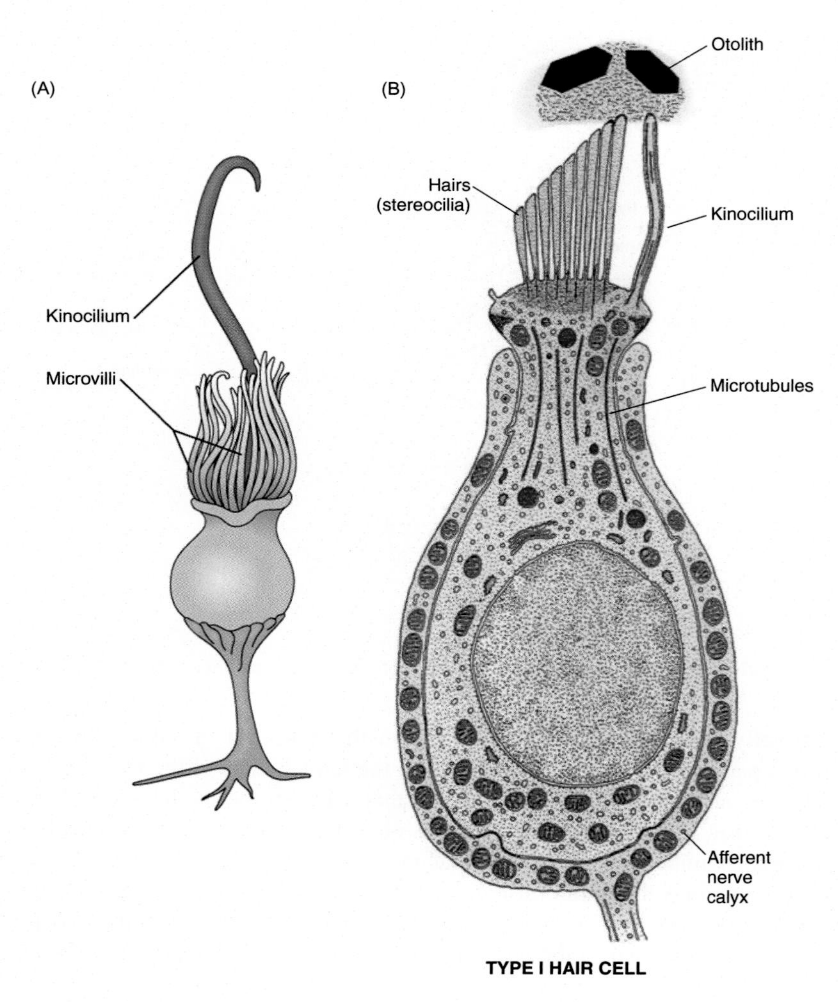

FIGURE 7.15 (A) A choanoflagellate, one of our earliest ancestors, with a central kinocilium and microvilli. (B) A type I hair cell in the semicircular canal system. (B) *From Gartner and Hiatt (2011), with permission.*

In the earliest chordates, **hair cells** were used for two important sensory functions. One, which is unique to aquatic vertebrates, involved the presence of scattered hair cells over the outer epithelium of the skin. These cells monitored water currents (vibrations) in the external environment. The other was an internal structure—a small vesicle filled with fluid and containing a small area of hair cells covered with a gelatinous coat and often a crystalline mass, such as calcium carbonate (Fig. 7.15B). This organ monitored the internal environment of the animal and provided information on its attitude (position) in the water and its own movements. By this time, the original kinocilium had undergone a transformation from a motile organelle to one with a sensory function, and the microvilli had transformed into stereocilia, structures more adapted to sensation. Although the external hair cells did not survive the transition to terrestrial living (aquatic larval salamanders still have them), the internal organ serving equilibrium has persisted in a form remarkably similar to that in the earliest vertebrates.

Hearing in vertebrates has undergone a very complex evolution. The fundamental basis of hearing remains a fluid-filled cavity with one part of the lining consisting of an array of modified hair cells. The basis of hearing is the mechanical effect of waves in the water exciting hair cells that are tuned to a specific frequency. Beyond that, hearing has evolved along separate pathways many times in vertebrate evolution. The mammalian ear consists of three major components—the external ear (**pinna**), the middle ear (often called the tympanic ear), and the inner ear. The inner ear is the evolutionarily stable component, although even that has undergone a number of changes. The origins of both the middle and external ears reflect a fascinating repackaging of components during the course of evolution.

Fishes possess an inner ear, but no middle or external ear. Their body is acoustically transparent, so that an aquatic sound wave can pass right through their body. The reason that a fish can hear is that within its inner ear is a mass of calcium carbonate, called an otolith, that is of a different density from water. The otolith, which functions principally in the sense of balance, is situated over the hair cells of the inner ear. When an aquatic sound wave hits it, it vibrates and stimulates specific hair cells through the gelatinous mass which connects it to the hair cells. This stimulates a hearing response—mainly to low frequencies of 1000 Hz or less.

Hearing underwent a huge transition after the first vertebrates left the water and adopted a terrestrial existence. The transition to terrestrial vertebrates occurred in the Triassic period, between 250 and 200 million years ago. With this, strategies for picking up sound waves changed, although the basic functions of the inner ear did not. The fundamental problem was how to transmit vibrations to the inner ear. Being able to do so was a matter of survival in terms of escaping predators or communication among members of the same species. Some of the earliest low-slung amphibians picked up vibrations from the ground (possibly from the footsteps of predators) by sensing them through the bones of their lower jaw, which rested on the ground. These animals had no middle or external ear, so the vibrations had to be transmitted to the inner ear via other bony structures that could relay the vibrations. The hearing sensation was confined to low frequencies.

A major step in the evolution of the human ear was the origin of the tympanic ear (the equivalent of the middle ear). Although the essence of inner ear function has remained remarkably similar, the means by which sound signals are transmitted to the inner ear have undergone some important transitions. The variable components involve both the middle and external ears. Evolutionary biologists now feel that overall ear structure has evolved independently with the development of each major vertebrate group.

Two developments were critical in the evolution of the mammalian middle ear. One involved bones; the other, a cavity that surrounded them. Of the many evolutionary experiments in the development of hearing, one of the most durable models was the adaptation of a single bone (**hyomandibular**) that originally served as a strut linking the second gill arch in aquatic vertebrates with the skull to one (**columella**) that links the tympanic membrane or its equivalent to the inner ear via the skull bones. This arrangement is still used by present-day amphibians, reptiles, and birds, and in mammals the columella has morphed into the stapes. Continuing the theme of repurposing jaw-related bones to audition, two bones that connected the jaws of fishes to the skull lost their original function when early vertebrates developed a different mode of connection between the jaws and skull. Then these bones shrank in size and evolved into the malleus and incus, which when connected with the stapes, formed the familiar chain of inner ear ossicles seen in humans (see Fig. 7.17).

The other development began with the formation of the **spiracle**, a cavity in the head that arose from the space between the first and second gill arch in fishes and persists in contemporary sharks and rays. As the open space between gill arches in fishes became more sac-like, the opening to the outside shrank and ultimately became covered with a membrane that served as an eardrum. The sac of the spiracle became associated with and surrounded the malleus, incus, and stapes as the middle ear cavity. The human eardrum (**tympanic membrane**), which arose from tissues somewhat deeper within the spiracle, became attached to the malleus, whereas the stapes is directly connected to the inner ear apparatus.

In most nonmammalian vertebrates the middle ear (tympanic) cavity is rather broadly connected with the pharynx through patent canals, with the result that each ear is exposed to sound waves that affect the other. This arrangement allowed these animals to localize low-frequency sound. Higher mammals developed much thinner **Eustachian tubes** that do not allow the free flow of sound waves through their channels. In mammals the localization of sound arose through the development of neural computation within the brain. Coincident with this evolution was the acquisition of the ability to hear high-frequency sounds.

Also important in high-frequency hearing was the development of the external ear—moveable in many mammals. Like most other components of the middle ear, the external ear develops in the embryo from tissue associated with both the first and second gill (branchial) arches.

Accompanying all these changes was the formation of brain nuclei that process the raw auditory input and distribute the modified signals to other integrative centers within the brain. This set the stage for a variety of reflex activities, e.g., the startle reflex.

The Human Ear

After all the evolutionary experimentation, the human (mammalian) ear is a mix of ancient and newer structures that have come together to form a remarkably efficient apparatus for capturing and transmitting sound waves. The ear consists of three tightly linked components—the external, middle, and inner ears (Fig. 7.16).

Actual hearing is accomplished by collecting sound waves in the **external ear** and funneling them through the external ear canal (**external auditory meatus**) until they hit the eardrum (**tympanic membrane**, Fig. 7.17). The vibrations of the tympanic membrane set in motion the three bones of the middle ear. The inner end of this interconnected chain

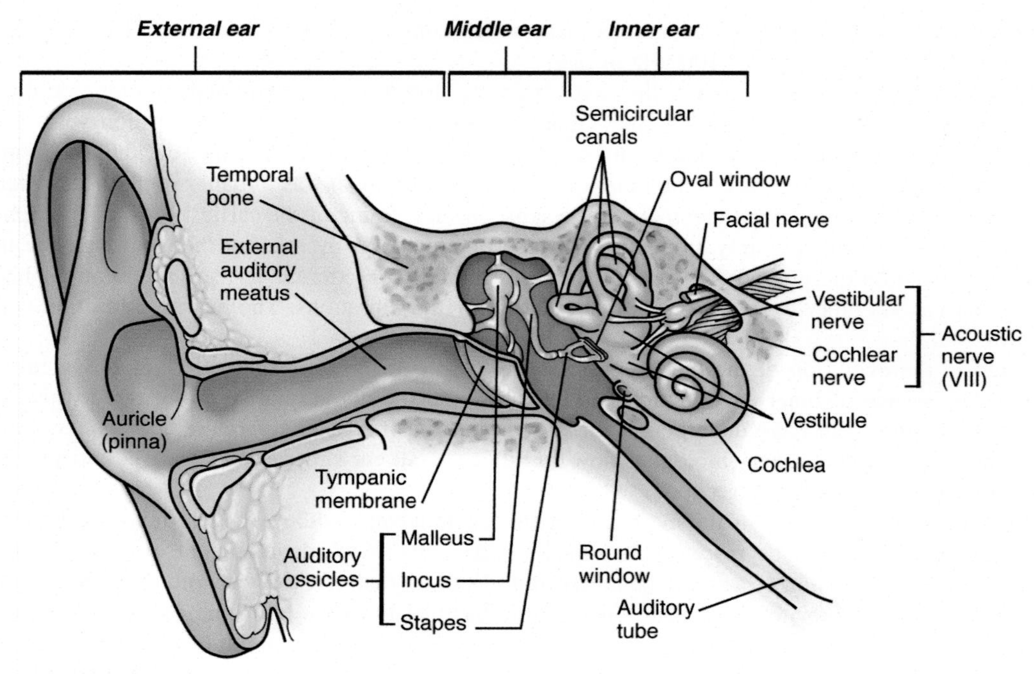

FIGURE 7.16 Overall anatomy of the ear.

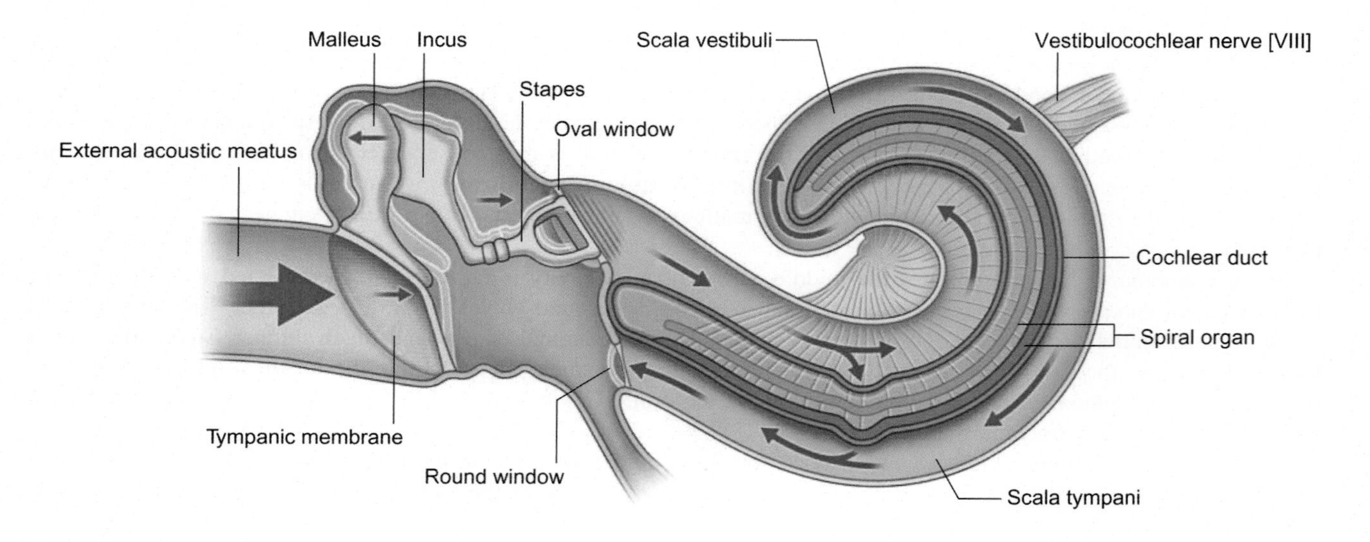

FIGURE 7.17 Transmission of sound (arrows) through the external, middle, and inner ear. *From Drake et al. (2005), with permission.*

of bones abuts another membrane at the oval window, from which the vibrations are carried into the inner ear (cochlea). Within the cochlea, a highly organized set of hair cells responds to the vibrations carried in fluid and transmits the information to the brain as a neural signal. Several acoustic centers within the brain process the raw information and pass it to other regions for integration and motor responses.

External Ear

The external ear (pinna) is a strictly mammalian innovation. Although birds and reptiles and even frogs have a tympanic membrane, it lies flat on the surface of the head. Mammals, by contrast, have prominent external ears that are

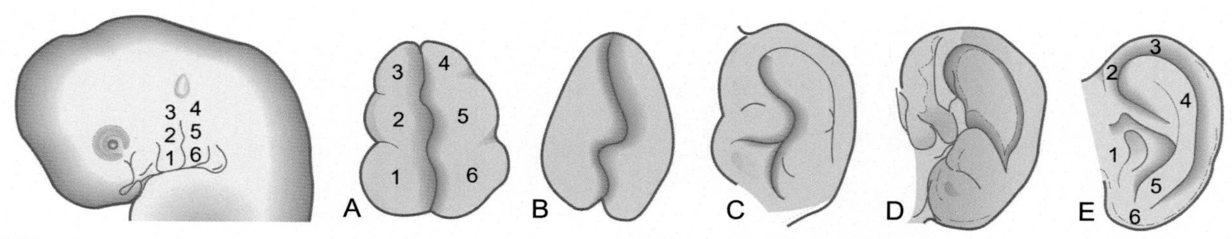

FIGURE 7.18 Embryonic development of the external ear from two pharyngeal arches. Blue, first arch; red, second arch. *From Carlson (2014), with permission.*

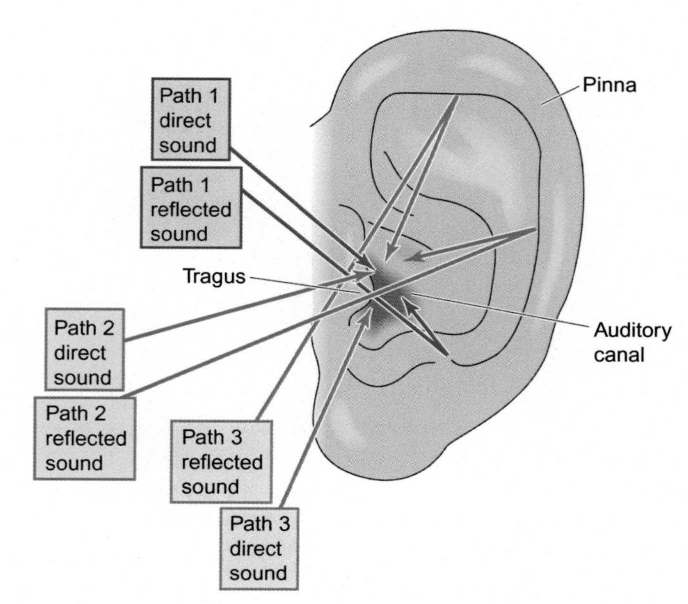

FIGURE 7.19 How the ear detects sound in the vertical plane through either direct entry of sound waves into the auditory canal or by reflection of sound waves into it. *From Boron and Boulpaep (2012), with permission.*

specialized for collecting sound waves. Most mammalian ears are connected to a set of extrinsic muscles that can move the ears to further increase their ability to pick up sound waves. In addition, another six small intrinsic muscles can change the shape of the ear. Human ears have largely lost those capacities, although some people are able to wiggle their ears through contractions of their extrinsic ear muscles.

Human ears come in a wide variety of sizes and shapes. In fact, ear shape has been used as a personal identifier, much as fingerprints and iris patterns. The external ear arises from a series of six bumps on the first two embryonic pharyngeal arches—three on each arch (Fig. 7.18). These sites of origin are consistent with the origins of the components of the middle ear, which also arise from the first and second arches.

The collection of sound waves by the external ear plays a major role in the localization of sound. This function is highly important for survival—both in nature, where it is critical in both finding food and avoiding becoming food, and in civilization, where it used both in communication and avoiding danger from moving objects, like cars. Sound is localized along both vertical and horizontal planes, and each occurs through a different mechanism.

Vertical localization of sound depends upon the shape of the external ear. A given sound can enter the external auditory meatus (see Fig. 7.16) either directly or by reflection of part of the external ear (Fig. 7.19). Reflected sound takes slightly longer to reach the external auditory meatus, and through central analysis of the interference patterns caused by these delays, the brain is able to localize the vertical source of the sound, even with only one ear.

Horizontal localization of sound requires both ears, and it depends upon the frequency of the sound. For frequencies greater than 2000 Hz, the intensity of the sound received by each ear is the determining factor. If the sound is coming from the right side of the head, the left ear is in a sound shadow, resulting in a lesser intensity of the sound (Fig. 7.20A). This difference in intensity is recognized and processed by the brain. For wavelengths less than 2000 Hz, the wavelength is greater than the width of the head. This allows sound waves from the right to reach the left ear with

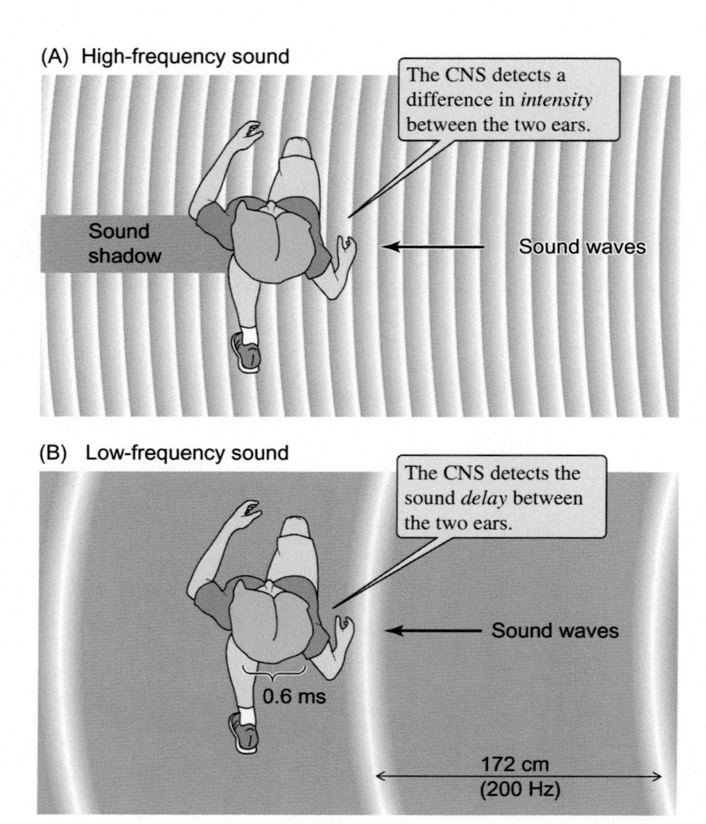

FIGURE 7.20 Sound detection in the horizontal plane. (A) For frequencies between 2 and 20 kHz, horizontal sound localization involves detection of a difference in intensity between the two ears. (B) For low-frequency sound (<2 kHz), the CNS detects the delay between sound reaching one ear and the other. In this example, a 200 Hz sound wave coming from the right would reach the left ear 0.6 ms after it reaches the right. The brain can pick up and process this information. *From Boron and Boulpaep (2012), with permission.*

essentially equal intensity because they diffract around the head, but the sound reaching the left ear is slightly delayed (Fig. 7.20B). In this case, the brain picks up and processes the delay, with the resultant localization of sound.

Once sound waves pass the external ear proper, they enter the external auditory meatus, a tube roughly 30 mm in length. Its surface is lined with skin, with some hairs, and it contains both sebaceous glands and modified apocrine glands. Cerumen (earwax) is a mixture of secretions from both the sebaceous and apocrine glands, along with desquamated epithelial cells, that collects in the ear canals. Cerumen and the hairs protect the auditory meatus against entry by insects and foreign bodies, and it has been found to protect against certain species of pathogenic bacteria. An important auditory function of the external auditory meatus is amplification of sound waves around 3000 Hz, which is in the common range of human speech. The external and middle ears amplify sound by as much as 10–15 dB.

The tympanic membrane lies between the external and middle ear. In mammals, it is largely located within the confines of the temporal bone of the skull. Because of its deep location, it considered not to be homologous with the tympanic membranes of many lower vertebrates. The tympanic membrane is well innervated by several small sensory nerves. For that reason, stretching or rupture of the tympanic membrane is very painful. Its inner surface is the attachment point for the handle of the malleus, one of the inner ear bones.

Among nonhuman mammals, large ears are often adaptations for exceptionally acute hearing. In addition they have sometimes evolved for other functions. For example, the large ears of African elephants are designed to radiate heat as a mechanism for controlling overall body temperature. Similarly, rabbits or hares living in hot deserts have much larger ears than do their Arctic counterparts.

Middle Ear

The middle ear is an air-filled cavity lined by a mucous membrane and occupied by the three middle ear bones (ossicles)—the **malleus** (hammer), **incus** (anvil), and **stapes** (stirrup) (Fig. 7.21). The middle ear cavity is connected to the pharynx by a thin **pharyngotympanic (Eustachian) tube** (see Fig. 7.21), which is normally collapsed, but can equalize

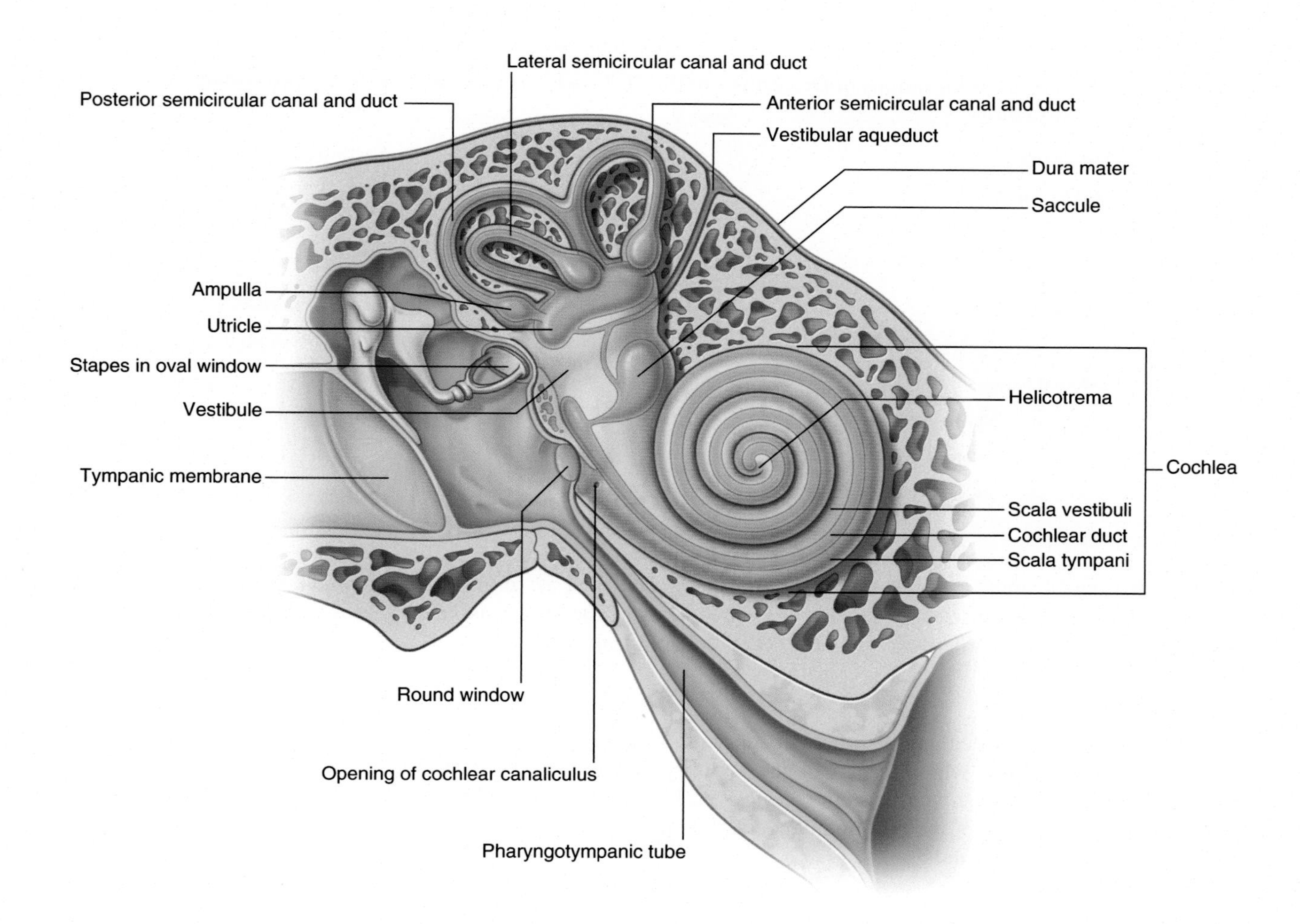

FIGURE 7.21 The anatomy of the middle and inner ear. *From Drake et al. (2005), with permission.*

pressure between the middle ear and the outside air. In addition to equalizing pressure, the Eustachian tube has another important function. The epithelium lining the tube is ciliated, with the beat directed from the middle ear cavity toward the pharynx. This arrangement clears both mucus and pathogens from the middle ear cavity. Interference with the ciliary function can result in increased numbers of middle ear infections. Several small muscles from the upper pharynx attach to the Eustachian tube and dilate it when swallowing. This explains why swallowing equalizes pressure in the middle ear when making a rapid descent in an airplane.

A major function of the middle ear is to convert the low-pressure sound waves in the air to something that can produce waves in the fluid contained in the inner ear. This is accomplished by movements by the chain of three interlinked middle ear ossicles. The chain ends with a firm connection of the stapes at the membrane of the oval window. Several factors contribute to the efficiency of the middle ear as a transmitter of sound waves. One is the much larger diameter of the tympanic membrane in relation to that of the oval window, which helps to concentrate the sound. Another is the lever-like arrangement of the middle ear bones—another amplifying mechanism. The long handle of the malleus is attached to the inner surface of the tympanic membrane, and the stapes is connected to the oval window. These mechanisms amplify the acoustic signal many times by the time the signal enters the fluid of the inner ear.

Two small muscles are located within the middle ear. One, the tensor tympani, is connected to the handle of the malleus; the other, the stapedius, attaches to the stapes. Both of these muscles are activated by a loud noise, and their actions on their respective middle ear bones act as a buffer to prevent noise-induced damage to the middle ear.

The adult human ear reflects its evolutionary and embryological past. Recall that the middle ear bones are derived from bones that originally suspended the jaws of primitive vertebrates (see earlier). In both phylogeny and ontogeny, components of the external and middle ears have undergone major transitions, but still bear evidence of their origins from the first and second pharyngeal (branchial) arches (Table 7.2). In the middle ear, the malleus and incus arise from

TABLE 7.2 Contributions of Derivatives of the First and Second Pharyngeal Arches to Adult Ear Structures

First Arch	Second Arch
Middle Ear	
Malleus	Stapes
Incus	
Tensor tympani muscle	Stapedius muscle
Cranial nerve V (trigeminal)	Cranial nerve VII (facial)
External Ear	
Tragus and anterior ear	Helix
Tympanic Membrane	
Covers slit between first and second arches	

the first pharyngeal arch, as do the tensor tympani muscle and its motor nerve (cranial nerve V) innervating it. Similarly, the stapedius muscle, modulating the movements of the stapes, is innervated by a branch of the facial (VII) nerve, which supplies other derivatives of the second pharyngeal arch.

Inner Ear

The inner ear is a fluid-filled cavity situated deep within the temporal bone of the skull. It consists of two functional components—the **cochlea**, which is part of the auditory system, and the **labyrinth**, which contains the semicircular canals and is part of the vestibular system (see Fig. 7.21). At the interface between the two is a vestibule that occupies the space on the inner side of the partition between the middle and inner ear. To complete the discussion of hearing, the cochlea will be considered first.

Cochlea

Sound waves transmitted from the middle ear ossicles cause the membrane of the **oval window** to vibrate through its connection with the stapes. The vibration of this membrane causes waves in the fluid within the inner ear. As the membrane of the oval window is depressed into the inner ear cavity, the membrane covering the **round window**—another space in the wall between the middle and inner ear—bulges out to compensate. Both of these windows are located close to the base of the cochlea.

The cochlea is the structure in which sound waves are converted to neural signals. As its name implies, the cochlea is a snail-shaped structure that wraps around itself 2.5 times over a conical bony structure called the **modiolus**. Uncoiled, it is only 30–35 mm long. For its small overall volume (about the size of a pea), the cochlea has a very complex structure that is very well adapted for its function. A knowledge of the general structure of the cochlea is essential for understanding its function in hearing.

The essence of cochlear function is best appreciated by looking at an uncoiled representation of the cochlea (Fig. 7.22A). In cross section, the cochlea consists of three fluid-filled compartments (Fig. 7.22B). Two of them—the **scala vestibuli**, which connects with the oval window, and the **scala tympani**, which connects with the round window—are continuous at the apex of the cochlea. Their fluid content, called perilymph, is similar to cerebrospinal fluid, with a low K^+ and a high Na^+ content. The other compartment—the **cochlear duct (scala media)**—is continuous with the vestibular apparatus; together, they and their surrounding membrane are called the **membranous labyrinth** (see Fig. 7.21). The fluid in this compartment (**endolymph**) contains a high concentration of K^+ and a low concentration of Na^+.

The hearing component of the cochlea (spiral organ or **organ of Corti**) is located within the cochlear duct. Its component parts are situated in linear arrays along the length of the cochlea. The base of the organ of Corti is the **basilar membrane**, which plays a fundamental role in the hearing process (Fig. 7.23). The basilar membrane is roughly 5 times wider at the apex than at its base, and the base is about 100 times stiffer than it is at the apex. These properties are critical in sound perception.

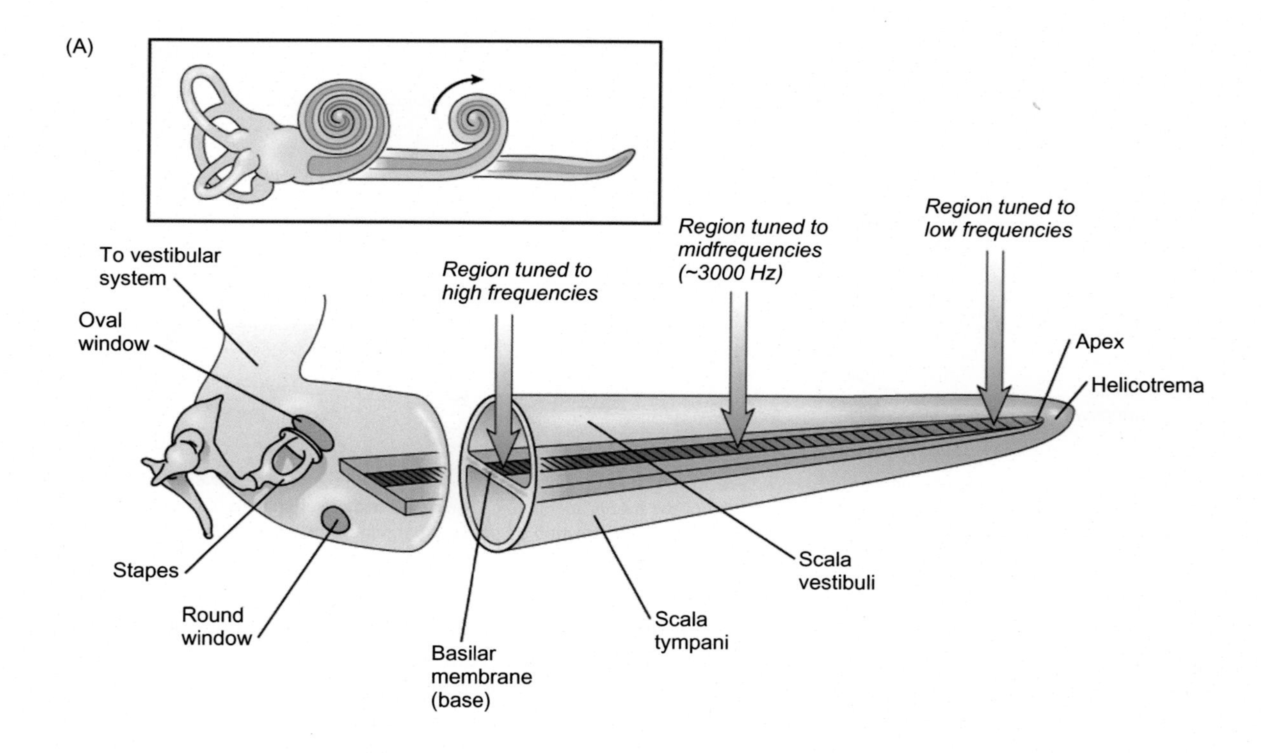

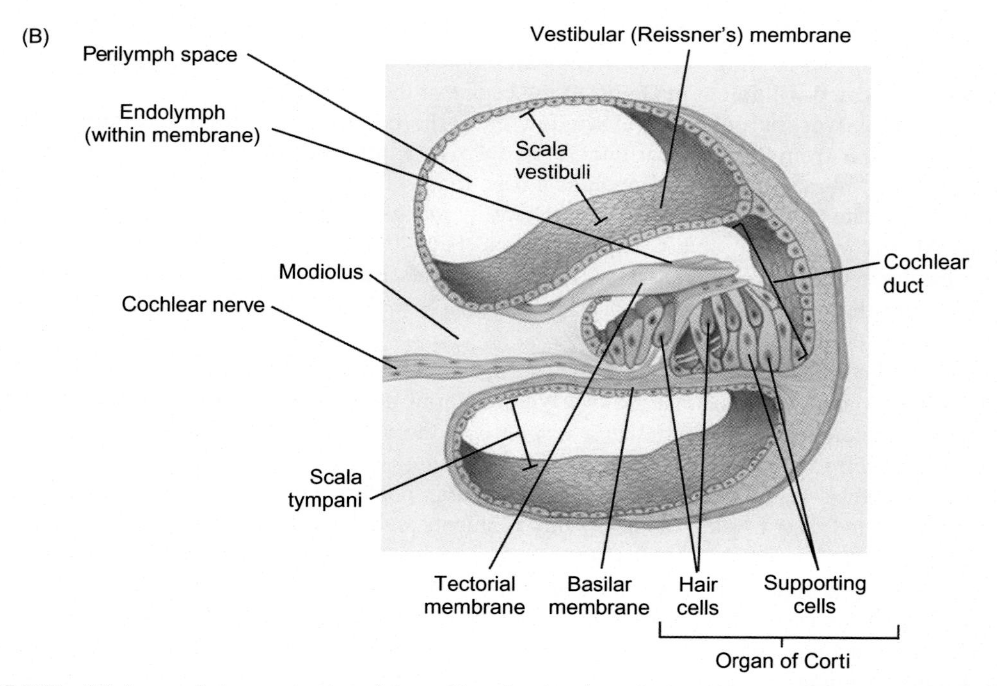

FIGURE 7.22 (A) An uncoiled representation of the cochlea, showing the regions sensitive to different sound frequencies (high frequencies at the base and low frequencies at the apex). (B) A cross section through the bony labyrinth, with its major components. *(B) From Thibodeau and Patton (2007), with permission.*

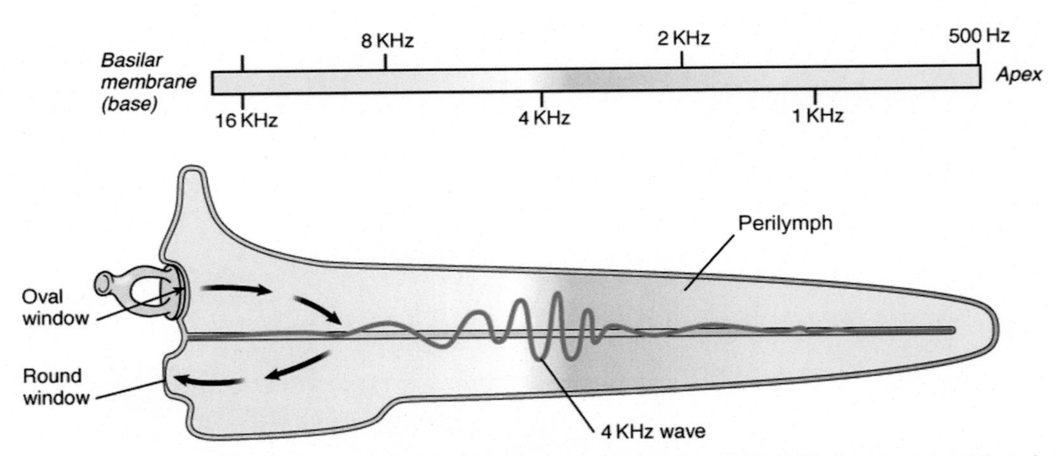

FIGURE 7.23 Resonance of the basilar lamina of the cochlea in response to a sound wave. The red line is an exaggerated version of a traveling wave of 4 kHz, showing the greatest resonance in the middle area of the basilar membrane. The black arrows summarize the response of perilymph to inward pressure on the oval window to an incoming sound wave and its compensation by outward bulging of the round window.

Situated on top of the basilar membrane are two arrays of auditory receptor cells—the inner and outer hair cells. The **inner hair cells** form a single row, consisting of about 3500 cells from the base to the apex of the cochlea. There are three rows of **outer hair cells**, numbering about 15,000 in total. Interspersed among these are supporting cells of various sorts (see Fig. 7.22B). As their name implies, the surfaces of the hair cells are covered with numerous (30 to several hundred) stereocilia (microvilli). In keeping with their long evolutionary history, each of these cells contained a kinocilium (see p. 193), but in humans this is lost early during development. Above the apical surface of the hair cells is a thin reticular lamina, but the stereocilia protrude above this structure. Covering all of this is a **tectorial membrane**, which also plays a fundamental role in hearing. The hair cells are connected by synapses to the dendrites of **bipolar neurons**, whose cell bodies are located within the modiolus at the center of the cochlear spiral.

Sound waves entering the scala vestibuli at the oval window enter the perilymphatic fluid and through deformations of the membrane separating it from the cochlear duct, create waves within the endolymph of the cochlear duct. These waves cause the basilar membrane to vibrate, much as a rope that is flicked. The mechanical properties of the basilar membrane cause it to vibrate at a peak at some point where its intrinsic properties match the frequency of the waves traveling along it. At this point, the wave essentially stops. High-frequency waves find a compatible point close to the base of the basilar membrane, whereas low-frequency waves travel close to the apex before its more floppy properties resonate with these waves (Fig. 7.23). This is the essential mechanism underlying the perception of pitch.

When waves cause the basilar membrane to vibrate greatly in a region corresponding to a particular pitch, the stereocilia of the hair cells in that area brush against the overlying tectorial membrane and become bent. As little as a 0.3 nm lateral movement of the stereocilia can stimulate a response to the softest sounds. Bending of the stereocilia causes a depolarization of a hair cell in a very precise manner.

How bending stereocilia initiates a signal in a hair cell is a remarkable process. On a given hair cell, the stereocilia are arranged in step-like fashion, and their tips are linked by tiny filaments, called **tip links** (Fig. 7.24). At the base of a tip link is a cation channel. When the stereocilia are displaced toward larger ones, the tip links pull on the walls of the ion channels and open them to the passage of K^+ ions. The rush of K^+ into the cell causes a depolarization that opens Ca^{++} channels along the sides of the cell. The entry of Ca^{++} into the cell causes the hair cell to release a neurotransmitter from its base. The neurotransmitter then acts on the dendritic endings of bipolar neurons that are located within the spiral ganglion and stimulates a neural response.

The inner and outer hair cells are differently wired and play different roles in the hearing process. Although the inner hair cells are outnumbered 3 to 1 by the outer hair cells, they are nevertheless connected with more than 95% of the neurons in the spiral ganglion. Each inner hair cell connects to dendrites of about 10 bipolar neurons. By contrast, many outer hair cells synapse with a single bipolar auditory neuron. The inner hair cells are the actual sensory receptors. The outer hair cells play a different role in the hearing process.

Outer hair cells act as sound amplifiers. Along their lateral sides, these cells contain a motor protein that changes the shape of the cell by contracting or relaxing. Because of their connections with the basilar membrane, their

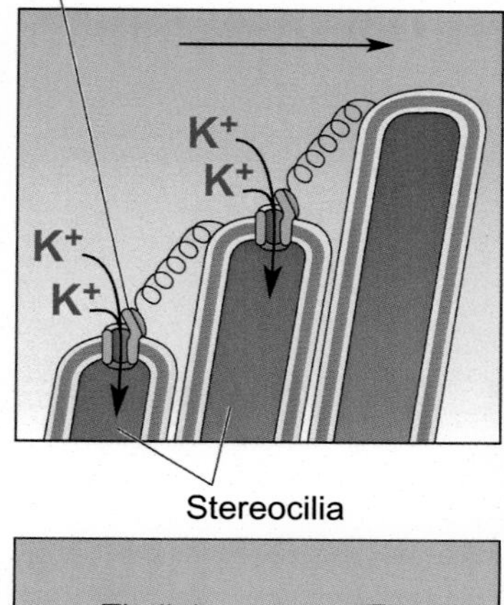

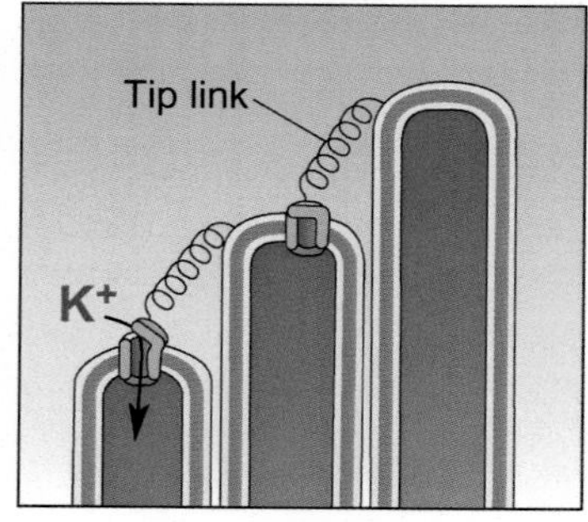

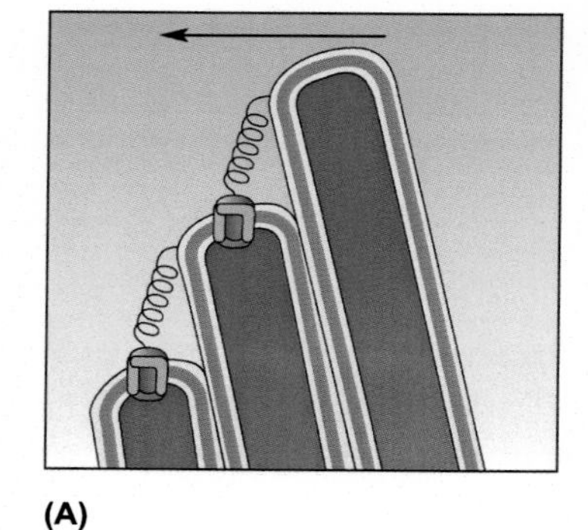

(A)

K+
K+
Endolymph
Reticular lamina
Voltage-gated calcium channel
Depolarization
Ca2+
Inner hair cell
Vesicle filled with excitatory neurotransmitter
Perilymph
Spiral ganglion neurite

(B)

FIGURE 7.24 Signal transduction and depolarization in a hair cell. (A) Mechanical stimulation created by a sound wave opens ion channels at the tips of adjoining stereocilia on the hair cell. (B) The entry of K^+ into the cell depolarizes the hair cell, causing the opening of voltage-gated Ca^{++} channels. This leads to the release of a neurotransmitter from synaptic vesicles and its pickup by terminals of nerve fibers from the spiral ganglion. *From Bear et al. (2007), with permission.*

contractions and relaxations can amplify the wave motions of the basilar membrane, thus intensifying the sound stimulus. Interestingly, this same mechanism operating in reverse can cause sound to be emitted from the ear, by creating wave movements in the cochlear fluids, which then transmit the action through the middle ear ossicles and causes the tympanic membrane to vibrate, thus causing a faint sound.

Central Processing of Hearing

When stimulated by neurotransmitters from the hair cells, the bipolar neurons of the spiral ganglion then send action potentials through their afferent axons into the brain stem. The bundle of axons coming from the spiral ganglion is collectively called the **cochlear** or **auditory nerve**. It soon joins the vestibular nerve, which arises from the region of the semicircular canals (see later), to form the **auditory-vestibular nerve** (cranial nerve VIII). The axons of the cochlear division of this nerve enter the upper medulla, where they synapse with neurons of the upper and lower cochlear nuclei (Fig. 7.25). In a theme that is repeated many times in the CNS, the cochlear axons project onto the cochlear nuclei in a tonotopic manner, meaning that there is a spatial gradient within the nuclei from high to low pitched sounds.

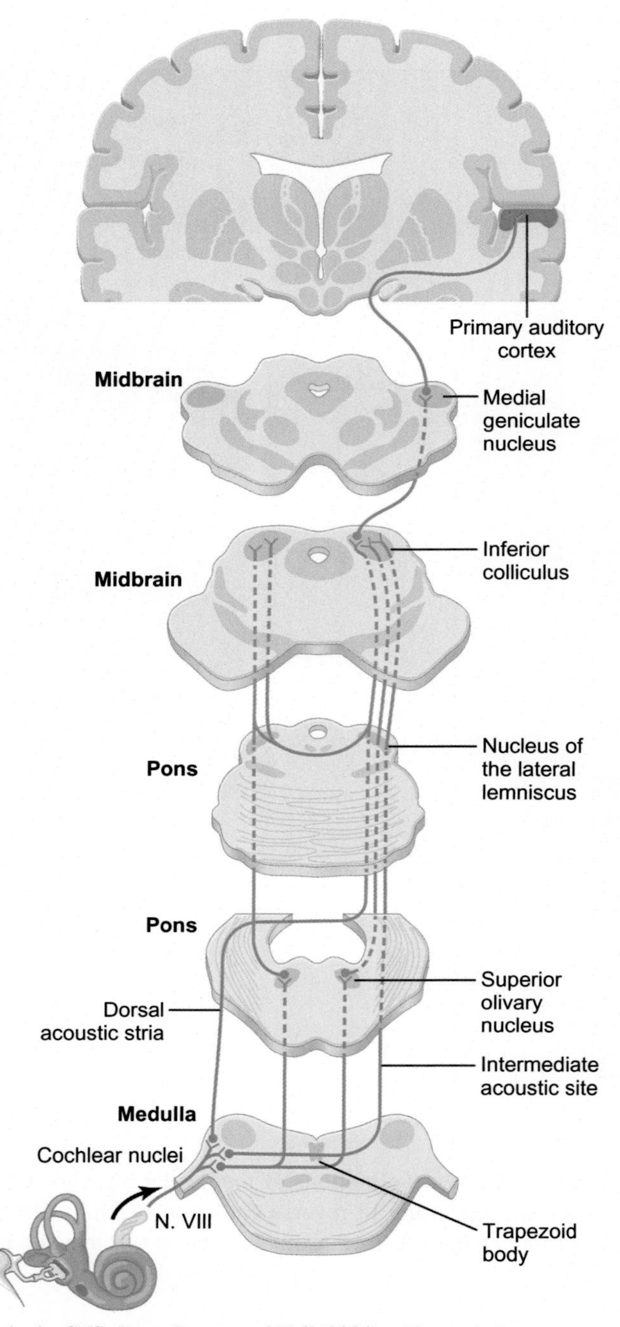

FIGURE 7.25 Major auditory pathways in the CNS. *From Guyton and Hall (2006), with permission.*

Within the CNS, axons from the cochlear nuclei ascend to the **superior olivary complex** in the pons. Input from each ear is processed by both olivary nuclei. Small differences in the timing or intensity of the signals allow the olivary complex to localize the origins of a sound. From the olivary complex, axons go to the inferior colliculus of the midbrain, where again the input is tonotopically organized, with the dorsal portion responding to low-frequency sounds and the ventral portion to high-frequency sounds.

Output from the inferior colliculus then goes to the thalamic gateway region, specifically the medial geniculate nucleus. From there it is distributed to the primary auditory cortex and associated areas within the temporal lobe of the cerebral cortex for higher level processing.

THE VESTIBULAR SYSTEM

All vertebrates, from hagfishes to humans, possess a vestibular system, which functions on the basis of deflection of the microvilli of hair cells in a manner very similar to that of the hair cells of the cochlea. The human vestibular system, like the cochlea, is embedded deep within the temporal bone and includes two main functional components. One, sometimes called the **static labyrinth**, consists of the **saccule** and **utricle** (Fig. 7.26)—structures that respond to static position and linear acceleration of the head. The other, sometimes called the **kinetic labyrinth**, is the group of three semicircular canals, which respond to rotational acceleration. All of these structures are part of the overall membranous labyrinth that also includes the cochlear duct. Like the organ of Corti, the hair cells within the vestibular structures are bathed in endolymph and are separated from the surrounding perilymph by a membranous covering.

The sensory component of both the saccule and utricle is called the **macula** (spot) and consists of a layer of hair cells surrounded by columnar epithelial supporting cells. Projecting from the apical surfaces of the each hair cell are roughly 100 stereocilia and one taller kinocilium organized in stair-step fashion (Fig. 7.27). These structures extend into a gelatinous cap whose surface is studded with tiny calcium carbonate crystals called **otoconia**. The basal surfaces of the hair cells lie next to terminals of dendrites of the vestibular nerve.

When the head is tilted or is subjected to horizontal or vertical acceleration or deceleration, the pressure of the endolymphatic fluid against the gelatinous cap, which is denser because of the otoconia, displaces the stereocilia of the hair cells. Depending upon the direction of displacement, the hair cells either depolarize or hyperpolarize (see p. 202 and Fig. 7.24).

The maculae of the saccule and utricle are organized with a demarcation, called the **striola**, running down the middle. On either side of the striola, the hair cells are oriented in opposite directions. Thus, with any given position or movement, one set of hair cells is depolarized and the other is hyperpolarized. In addition, the maculae are not flat planar surfaces, but rather are slightly curved so that different degrees of tilt of the head can be recognized. The maculae on either side of the head are oriented in mirror images to one another. All of these factors allow complete three-dimensional coverage of head position. Resting hair cells discharge at a rate of about 100 Hz, and for any given position or movement of the head, the discharge rate increases or decreases depending upon how the hair cells are oriented in relation to the tilt of the head.

The three semicircular canals are designed to detect rotation or angular acceleration of the head. They are oriented 90 degrees to one another to cover the three geometric planes, and those on one side of the head are in mirror image orientation to their counterparts on the other side.

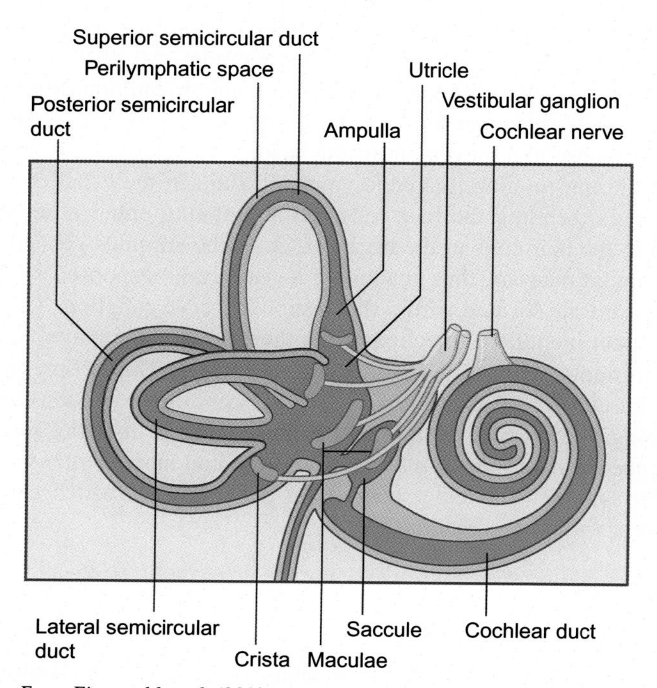

FIGURE 7.26 The vestibular system. *From Fitzgerald et al. (2012), with permission.*

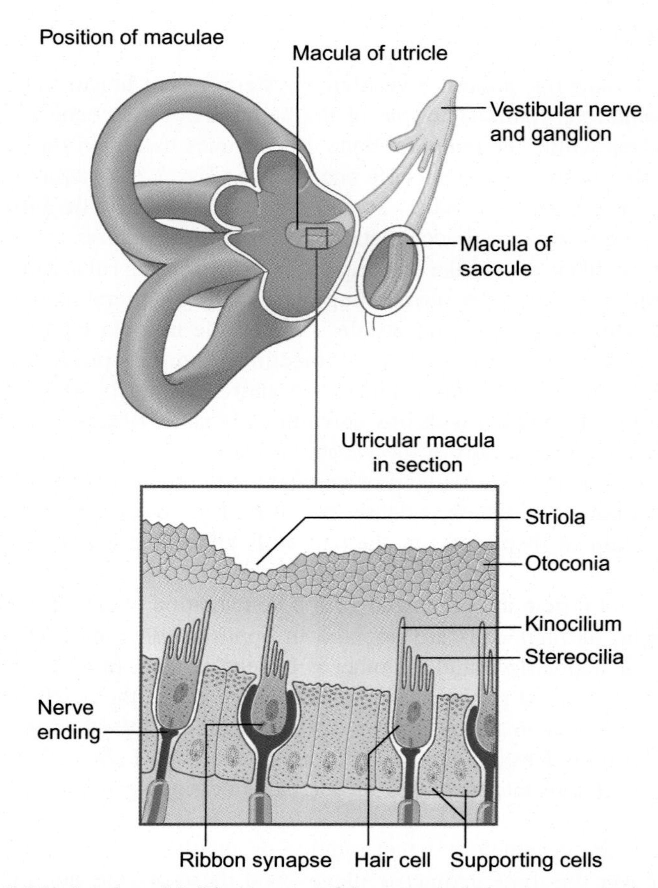

FIGURE 7.27 Structure of the static labyrinth (macula). Displacement of the otoconia during movement stimulates the kinocilium and stereocilia of a hair cell to depolarize (see Fig. 7.24). *From Fitzgerald et al. (2012), with permission.*

One end of a canal is somewhat expanded into an **ampulla**. Within the ampulla is a tuft (**crista**) of hair cells encased in a gelatinous **cupula**, which extends the width of the ampulla. All of the hair cells in a given crista are oriented in the same direction, but those of the different cristae follow the orientation of the individual semicircular canals. The hair cells of comparable canals on either side of the body are of opposite polarity; this allows them to work together to initiate the body's response to a movement of the head.

When the head accelerates during rotation, the endolymphatic fluid in the canal that is oriented along the plane of rotation pushes against the cupula, bending the hair cells and stimulating either a depolarization or hyperpolarization response. In a manner similar to the hair cells in the organ of Corti, the stimulus generated in the hair cells is transmitted to the dendrites of the vestibular neurons, thus generating a true neural response.

Axons from the bipolar neurons are located within the vestibular nerve ganglion. These synapse with neurons of the vestibular nucleus within the upper medulla. Projections from there serve three major types of reflexes. The first is that involved in maintaining equilibrium and gaze when the head is moved. This reflex necessarily includes connections with some of the extraocular muscles. A second reflex involves descending projections to motor nerves and the muscles that they innervate, and it strongly influences posture and balance. A third includes connections with muscle spindles and the cerebellum and influences muscle tone in relation to posture and movement, especially of the head. Other nerve fibers in the vestibular system synapse with other neurons in the thalamus, which carry information on posture and movements to integrative centers within the cerebral cortex.

SUMMARY

Eyes have a long evolutionary history and have evolved many times, often in parallel. Light-sensitive molecules—opsins—form the molecular basis for vision.

The human eye has many parallels with a film camera. Eyelids and tears, produced by the lacrimal glands, are the equivalent of the lens cap. The eyeball itself has three layers or tunics. The outer tunic consists of the cornea and sclera. Components of the middle tunic (uvea) are the choroid layer, the ciliary body, and the iris. The retina and its pigmented epithelium constitute the inner tunic.

The optically clear cornea and lens are responsible for focusing light. The iris and pupil control the size of the aperture (f-stops in a camera). The shape of the lens and its resulting focal distance is controlled by the ciliary apparatus. The lens is composed of crystallin proteins.

The retina is the equivalent of film. The neural retina contains three layers of neurons, which receive the light and then transfer a signal into the optic nerve.

The light receptors are the rods and cones. Rods are highly light sensitive and are critical for night vision. Color is perceived by the cones, which are less sensitive than rods and do not function at night.

Phototransduction begins with photons of light hitting the outer segments of both rods and cones. The photosensitive element in rods is rhodopsin. Three different types of cones contain opsins sensitive to blue, green, or red light. Rhodopsin activates many transducin molecules, thereby amplifying the intensity of the light signal. This is converted to an electrical signal that is then transmitted to bipolar neurons within the retina. These cells connect with terminals of optic nerve fibers. The optic nerve carries the light signal to light-processing regions of the brain.

Hearing and balance evolved from vibration-detecting systems in primitive invertebrates. Both hearing and balance are based on reception by hair cells.

The auditory apparatus consists of the external, middle, and internal ear. Sound entering the external auditory meatus causes the tympanic membrane to vibrate. This vibration is transmitted to a series of three middle ear bones—the malleus, incus, and stapes. The stapes is connected to the oval window. Vibration of the oval window sets up fluid currents in the cochlea. These resonate with hair cells that are attuned to specific frequencies. When activated, the hair cells transmit a signal to fibers of bipolar neurons within the modiolus. From there, the signal passes to fibers of the auditory nerve and ultimately to acoustic centers within the brain.

The sense of balance is served by the vestibular system, which contains three prominent semicircular canals. Displacement of otoconia stimulates hair cells within the vestibular system, which transmit their signal to fibers of the vestibular nerve.

Chapter 8

The Lymphoid System and Immunity

The human body is engaged in a constant struggle to protect itself from foreign invaders, mainly bacteria, viruses, molds, and parasites. This struggle has a long evolutionary history, and even some of the most primitive animals have developed defenses against such invaders. One of the earliest defense strategies was the development of cells that could phagocytize pathogens. The macrophage represents one of the most ancient of these specialized cell types, but even single-celled amebas engage in phagocytosis. Many of these defense strategies depend upon recognition of a pathogen as something foreign, but those strategies are more generic than specific. It was not until the evolution of vertebrates that a highly sophisticated immune strategy, capable of identifying individual foreign antigens and counteracting them with antibodies, developed. A by-product of this strategy is protection from internally generated cancerous cells.

At the level of cellular and molecular detail, the immune system is extremely complex, but the general strategies of immune defense of the body represent excellent examples of the adaptation of structure to functional needs. Most cells and tissues involved in immune defense belong to the lymphoid system, which is designed to patrol and protect all internal and external bodily surfaces. With this set of duties, the lymphoid system is necessarily diffuse, and elements of it are distributed throughout the body (Fig. 8.1).

Immune responses have been subdivided into innate and adaptive varieties. **Innate immunity** represents the body's first line of defense against invading pathogens. If this line of defense is breached, then the **adaptive immune response** comes into play. Both of these responses involve multiple mechanisms, but in broad brushstrokes, the adaptive immune response includes **humoral immunity** (the formation of **antibodies** to specific **antigens**[1] and **cellular immunity** (the use of lymphoid cells themselves as defensive agents). To thwart invasion by pathogens, innate immunity utilizes physical barriers, phagocytosis (engulfing invaders and debris by cells), and more generic chemical defenses.

A hallmark of the immune system is its specificity, as well as its diversity. Immune cells respond to any of the hundreds of thousands of specific foreign antigens that may enter the body. Yet at the same time, they do not react to the large number of antigenic sites present on the cells and tissues of a person's own body. Individual lymphocytes respond specifically to one particular antigen. Despite their sensitivity in responding to foreign antigens, elements of the immune system have built-in safeguards that prevent premature or indiscriminate responses.

This chapter will first introduce the major players of the lymphoid system and their structural organization. Then several examples of immune strategies directed against specific types of invasive pathogens will be described. Another section will cover distinguishing self from nonself and how associated mechanisms play a role in tumor surveillance and the consequent issues of graft rejection.

DEVELOPMENT AND ORGANIZATION OF THE LYMPHOID SYSTEM

Cells of the Lymphoid System

The cells of the lymphoid system arise from hematopoietic (blood-forming) stem cells located within the bone marrow (Fig. 8.2). Those destined to become lymphocytes must spend some time maturing in what are called central (or primary) lymphoid organs. **T cells** (T-lymphocytes) mature in the thymus—the T referring to thymus. **B cells** (B-lymphocytes) remain in the bone marrow to mature, but the B does not refer to bone marrow. Early in the history of modern

1. An antibody is a protein molecule produced by immune cells that can combine with a specific antigen. An antigen is a molecule to which an antibody specifically combines. Most antigens are proteins or peptides, but they can also be complex carbohydrates. Antigens can enter the body from the outside, or they can be produced within the body.

The Human Body. DOI: https://doi.org/10.1016/B978-0-12-804254-0.00008-9

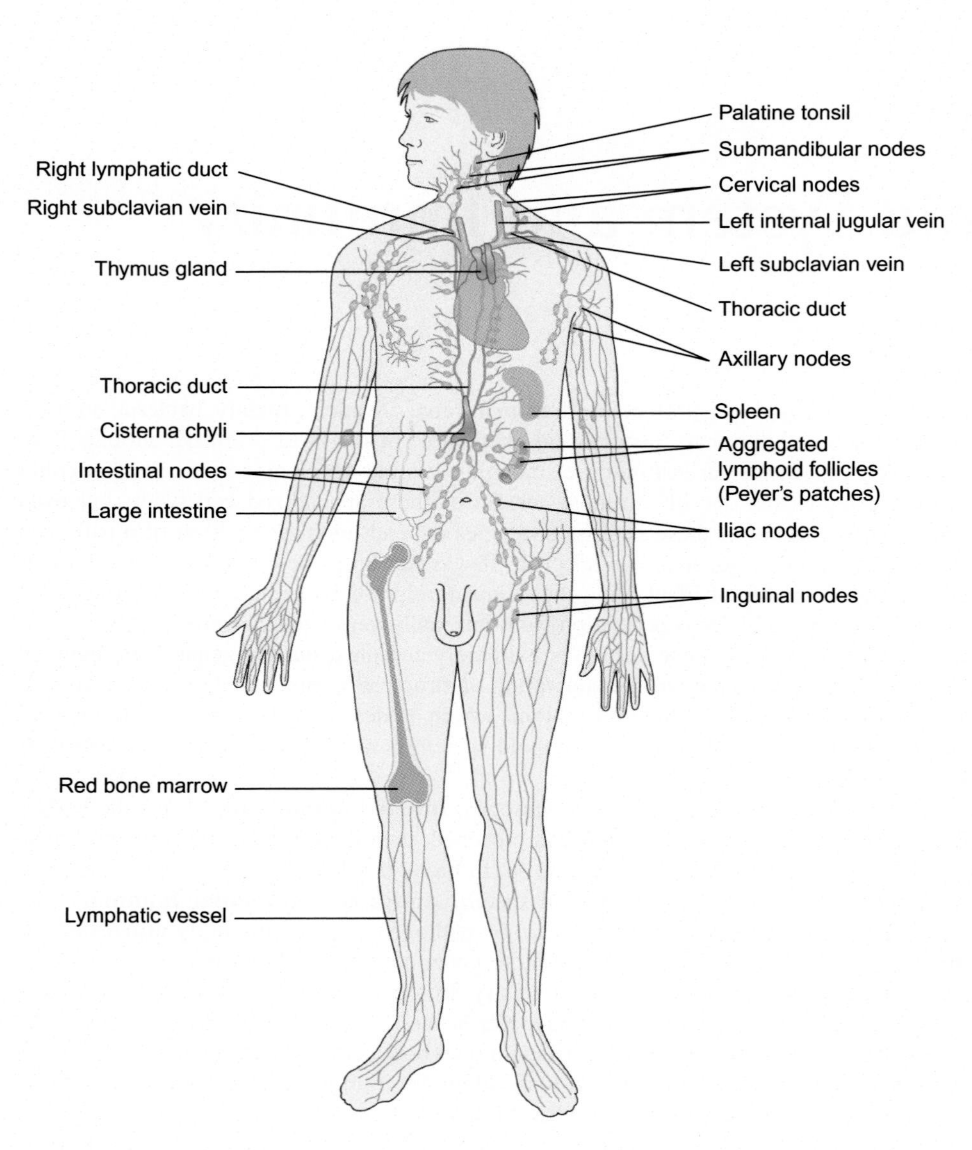

FIGURE 8.1 An anatomical overview of the lymphoid system, including both lymphatic organs and the lymph vascular system. *From Waugh and Grant (2014), with permission.*

immunology, development of immune cells was studied in birds, and in birds B cells mature in a unique lymphoid organ of the gut, called the **bursa of Fabricius**. It is the bursa after which the B in B-lymphocytes is derived.

B-Lymphocytes

B-lymphocytes (B cells) are generated from hematopoietic stem cells and begin to mature in the bone marrow (see Fig. 8.2). They then leave the marrow and enter the blood circulation, which carries many of them to lymph nodes scattered throughout the body or to the spleen. After further processing in the lymph nodes or spleen, B cells are mature, but immunologically naïve, meaning that they have not yet developed the capacity to mount a specific response to antigens. Many B cells die before they ever develop this capacity. Death of cells before attaining their full functional capacity is a common theme throughout the components of the lymphoid system. It is part of a fine-tuning process that, while seemingly wasteful of cells, is actually an important mechanism for maintaining the specificity of the immune system while still protecting the body from being attacked by its own immune system.

Both in the blood and after they enter the lymph nodes or spleen, B-lymphocytes are small cells with a dense nucleus and very little cytoplasm (see Fig. 10.1). When antigenically stimulated, they become activated and produce more cytoplasm. Those B cells destined to produce secreted antibodies differentiate into plasma cells (see below).

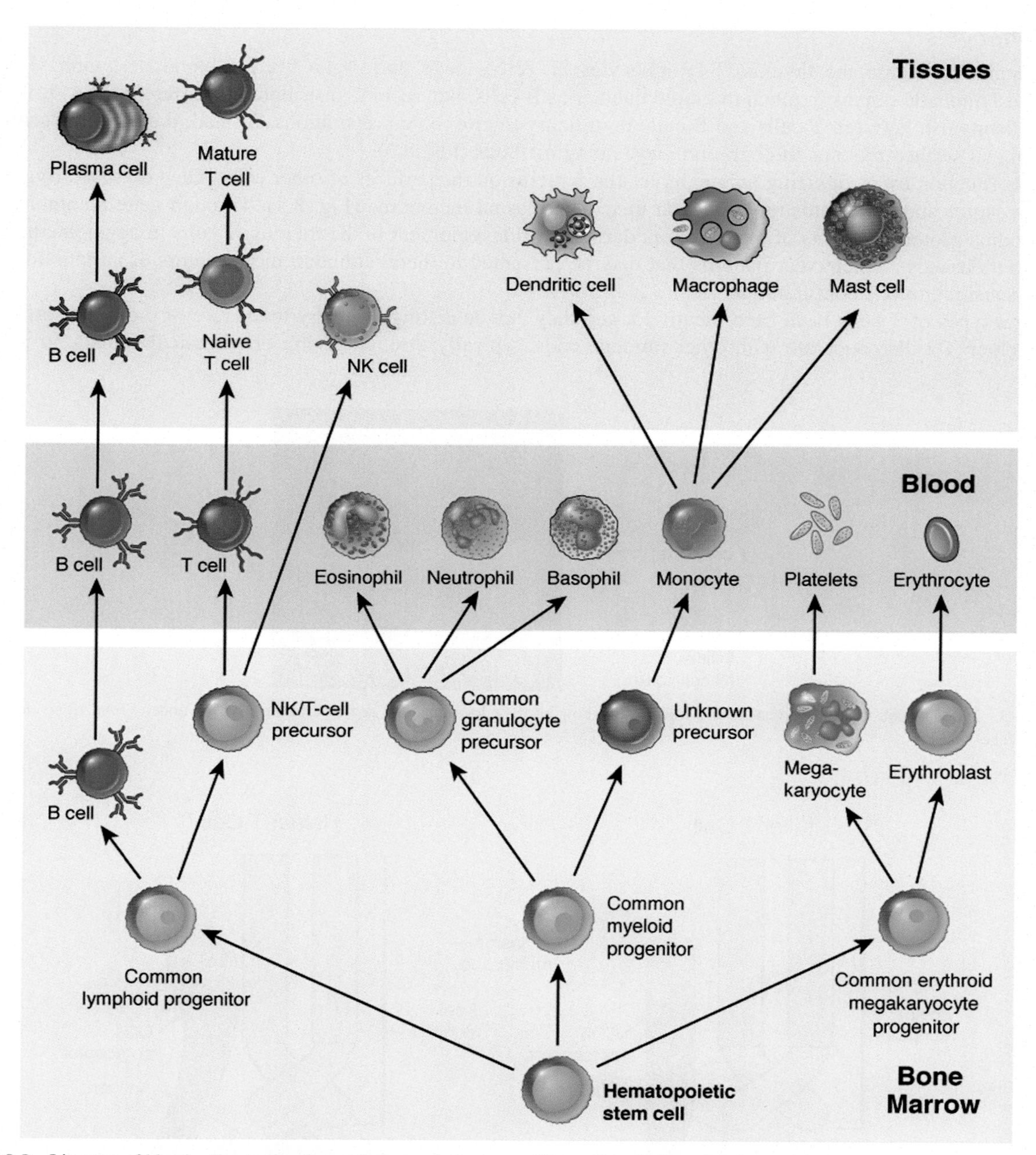

FIGURE 8.2 Lineages of blood cells, starting from a hematopoietic stem cell located in the bone marrow.

A high percentage of the lymphocytes in the body are located in the deeper layers of the skin or the loose connective tissue (lamina propria) that underlies mucosal surfaces—specifically the linings of the digestive, respiratory, and reproductive systems.

Lymphocytes recognize foreign antigens through cell surface receptors. In the case of B cells, the surface receptors are membrane-bound immunoglobulin molecules (see p. 215) that uniquely bind to specific antigens. Once an antigen is bound, an intracellular signaling cascade stimulates a response against that antigen.

Upon antigenic stimulation, many B cells differentiate into **plasma cells**, which are the primary producers of antibodies. Plasma cells are veritable protein (antibody) factories, and their internal structure bears testimony to that function (see Fig. 1.13). They have a huge well-developed rough endoplasmic reticulum and a prominent Golgi apparatus, both of which are designed to produce proteins for export. The antibodies produced by these cells enter the lymph and blood and are distributed throughout the body.

T-Lymphocytes

After maturation within the thymus, T-lymphocytes (T cells) leave that organ via the blood or lymph. They enter peripheral lymphatic organs in much the same manner as B cells, but become distributed to different areas. It is not possible to distinguish between T cells and B cells in ordinary microscopic preparations. Instead, they are characterized by their own cell surface proteins through immunostaining methods (Fig. 8.3).

T cells function by recognizing unique molecular patterns on the surfaces of other cells. They do this through a complex of receptor and coreceptor molecules on their own plasma membrane (Fig. 8.4). Through gene recombinations of the individual protein elements of the receptors during their development in the thymus, T cells in aggregate have sensitivities to thousands of molecular patterns that may be presented to them, although the receptors of an individual T cell are only sensitive to one configuration.

Several types of T cells have been identified, and they can be distinguished by the nature of their cell surface molecules. **Helper T cells** cooperate with other immune cells, typically through giving off chemical signals, in generating

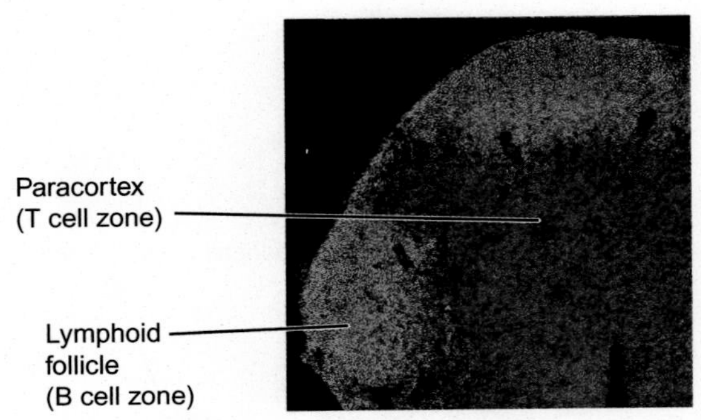

FIGURE 8.3 Immunostaining of B (green) and T (red) cells, showing their locations and segregation in a lymph node. *From Abbas and Lichtman (2009), with permission.*

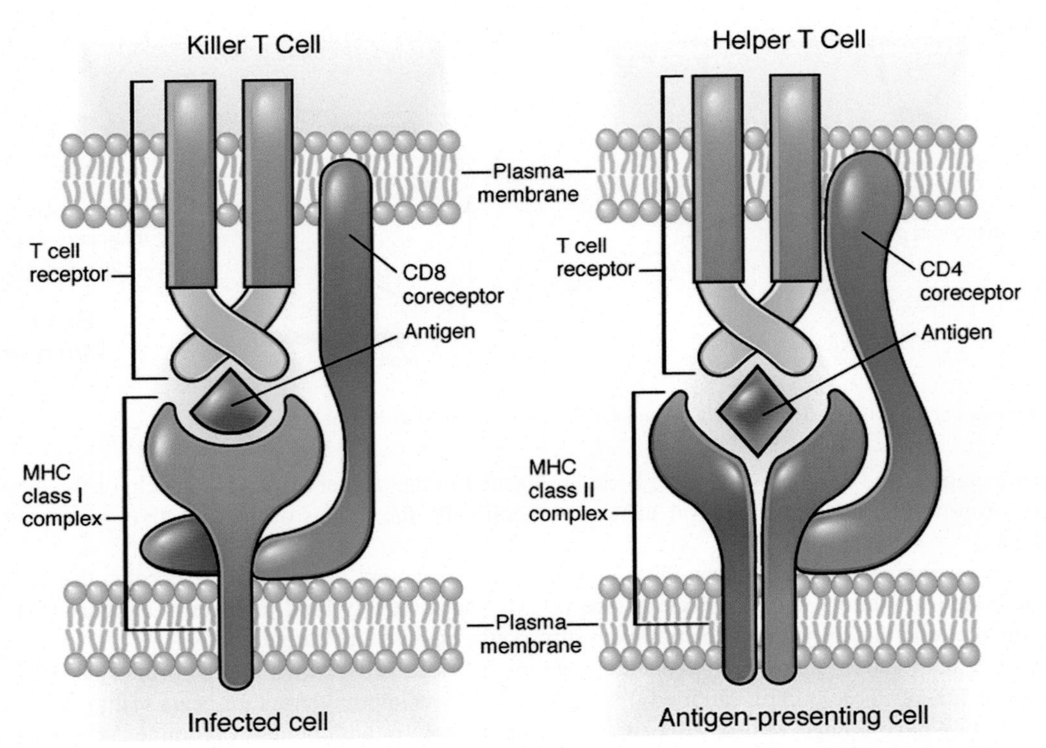

FIGURE 8.4 T cell receptors in killer T cells and helper T cells and their relationship to molecules of the MHC I and II complexes as they bind foreign antigens.

immune responses. The coreceptors of helper T cells are identified as **CD4**—the acronym is given to their specific surface marker. In contrast to the more indirect action of helper T cells, **killer T cells** themselves bind to and kill cells that contain foreign pathogens. The coreceptors of killer T cells are identified as **CD8**. The third type of T cell, about which much less is known, is the **regulatory T cell** which, like helper T cells expresses CD4 on its surface. These T cells place limits on immune reactions that could be damaging to the individual.

Natural Killer cells

A third variety of lymphocyte, in addition to B and T cells, is the **natural killer (NK) cell**. Its origin and lineage are less well understood than those of B and T cells, but they play a vital role in the early lines of defense (innate immunity) against foreign invaders. These cells tend to be concentrated in the liver, spleen, and circulating blood and can be quickly marshaled to sites of local infection.

Macrophages

Another important component of the white blood cells in the circulating blood is the **monocyte**—a cell that looks like a large lymphocyte (see Fig. 10.1). Monocytes leave the blood and settle down in connective tissue as resident **macrophages**. Individually, macrophages are the most potent phagocytic cells in the body. They can phagocytize entire bacteria in areas of infection or large masses of cellular debris in cases of sterile tissue damage. In addition, they secrete a large variety of powerful **cytokines** (signaling molecules), which orchestrate a wide variety of cellular responses to injury. Macrophages also play an important role in presenting foreign antigens to other cellular elements of the immune system for their response. As will be seen below, macrophages may not always be the first responders to a foreign invasion, but they are major players in chronic inflammatory processes. A macrophage can live in tissues for several months.

Neutrophils

Neutrophils (polymorphonuclear leukocytes or PMNs, see Fig. 10.1) are by far the most numerous leukocytes in the blood (40%–75% of all white blood cells). In the immediate aftermath of a pathogenic invasion, they are called from the circulating blood and converge on the site of the invasion. PMNs are not only phagocytic, but also they release from their cytoplasmic granules powerful oxidants that can kill bacteria. On the other hand, these same oxidants can also cause local tissue damage. PMNs have a very short life in tissues (1–2 days). In acute inflammation, PMNs accumulate and die in great numbers. **Pus** is the concentrated residue of huge numbers of dead PMNs. As the acute inflammatory stage subsides, their function is largely taken over by macrophages.

Eosinophils and Basophils

Eosinophils, leukocytes with large granules in their cytoplasm,[2] constitute 1%–3% of the circulating white blood cells. Like monocytes, their main function occurs after they have left the bloodstream. Their cytoplasmic granules contain powerful proteins, such as peroxidase, which can kill invaders, but also have the potential to harm normal cells and tissues. They contain surface receptors with a high affinity for IgE, the major class of antibody involved in warding off parasites or in hypersensitivity reactions. Their normal function from the evolutionary standpoint is in the defense of the body against parasitic worms.

Basophils, like eosinophils, seem to have evolved for the defense against parasites. Present in only minute numbers in the blood (<0.5% of white blood cells), they also contain prominent cytoplasmic granules which, when released in tissues infected by parasites, injure, or kill the invaders.

Mast Cells

Although earlier considered to be descendants of basophils, **mast cells** are now thought to represent a discrete cellular lineage. Generated in the bone marrow, they travel to connective tissues through the blood and undergo their final

2. Eosinophils get their name because their granules stain with the acid dye, eosin, which looks pink under the light microscope (see Fig. 10.1). The granules of basophils, on the other hand, stain with basic dyes, hence their name. The granules of basophils stain a deep purple color. Nucleic acids are also basophilic, and with the standard histological dye hematoxylin and eosin, nuclear chromatin and large concentrations of cytoplasmic RNA take on a purplish cast. In histology, **cytoplasmic basophilia** signifies a cytoplasm that contains a significant number of free ribosomes or a well-developed rough endoplasmic reticulum.

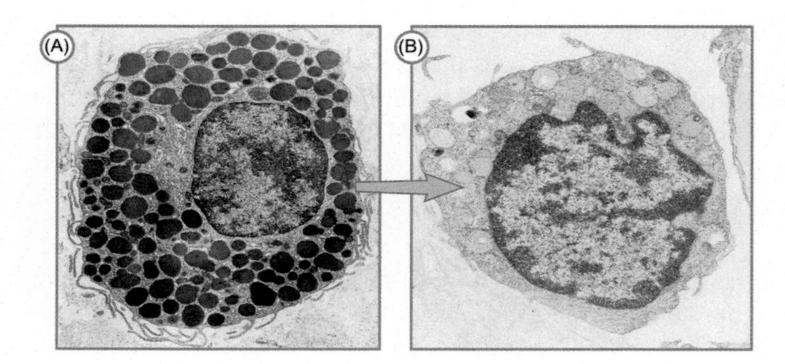

FIGURE 8.5 Electron micrographs of mast cells before and after release of their granules. *From Abbas and Lichtman (2009), with permission.*

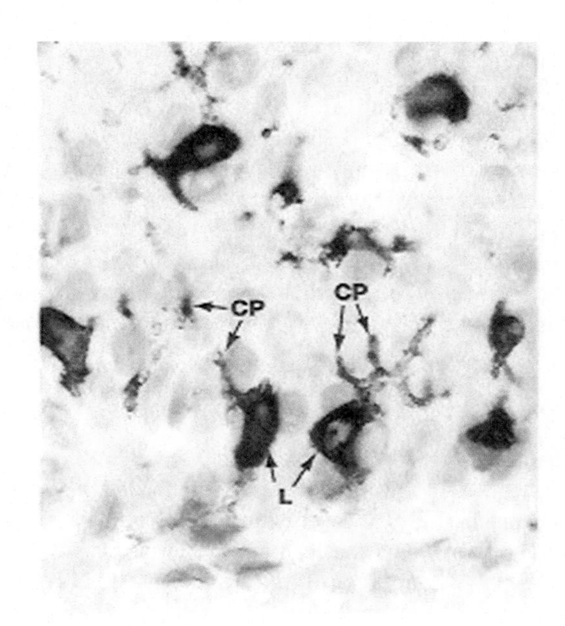

FIGURE 8.6 Dendritic cells (dark cells). *From Young et al. (2006), with permission.*

maturation outside the bloodstream. Like eosinophils and basophils, mast cells have large cytoplasmic granules (Fig. 8.5). Mast cells bind IgE on their surfaces, and when antigenically stimulated, the granules release a large variety of pharmacologically active substances. Most prominent is **histamine**, which causes vasodilation and increased vascular permeability. It also causes smooth muscle contraction and increased mucus production—the basis for asthmatic reactions to allergens taken up by the respiratory system. Stimulated mast cells also secrete **heparin**, which paradoxically, inactivates histamine. Other secretory products stimulate the ingress of neutrophils and eosinophils to the site of pathology; others result in vasoconstriction. When stimulated in excess, mast cell secretions can cause **asthma**.

Dendritic Cells

Dendritic cells were long regarded as obscure residents of a variety of tissues, with little-known function. Now these starfish-shaped cells (Fig. 8.6) are known to play a variety of vital roles in immunity, mainly as antigen-presenting cells (see p. 77). They are components of the innate immune system, but a major function is to stimulate adaptive immune responses by providing information about the type and location of an infection.

Despite a superficial similarity, all dendritic cells are not the same. Two major classes, each derived from different precursor cells, have been identified. One comes from a myeloid[3] precursor cell (see Fig. 8.2), with a monocyte being its immediate precursor. These cells migrate mostly to both the epidermis and dermis of the skin. The other major class,

3. Myeloid is a term referring to a lineage of progenitor cells that gives rise to all white blood cell lineages except lymphocytes (see Fig. 8.2). Lymphocytes arise from a separate lymphoid lineage.

called follicular dendritic cells, which may arise from mesenchymal stromal progenitors, make their way into the paracortical T cell areas of lymph nodes and the periarteriolar sheaths within the spleen (see below). The third type of dendritic cell is found in the medulla of the thymus (see Box 8.2).

The first dendritic cells to be identified historically are called **Langerhans cells** and are located within the epidermis of the skin or the stratified squamous epithelium lining structures such as the esophagus or vagina. These cells are highly phagocytic. They ingest microbes and break them up into small fragments, exposing many microbial antigens in the process. They then carry them to other lymphoid cells (T cells) in lymph nodes or the spleen. Other dendritic cells within the dermis of the skin perform similar functions.

Follicular dendritic cells are resident within lymphatic follicles. Arising from precursor cells in the bone marrow, they do not migrate once they reach a lymph node or the thymus, but form a stable meshwork within a follicle. They do not actively process antigens, but rather present them to B-lymphocytes. Thymic medullary dendritic cells present self-antigens (antigens representative of cells within one's own body) produced within the thymus and present them to maturing thymocytes (T cells) in the thymic medulla.

Molecular Players in Immune Responses

A complete immune response requires the participation of many classes of molecules, which individually play central roles at various stages of the response. This section will serve as a general introduction to many of these important molecular classes. Some will be covered in greater detail when specific types of immune responses are outlined.

Antibodies

Antibodies are Y-shaped molecules composed of four polypeptide chains— two identical **heavy chains** and two identical **light chains** (Fig. 8.7). The light chains are attached to the heavy chains on the outside of the Y, and the bases of the two heavy chains are attached to each other—all by disulfide bonds. The two ends of the Y are variable regions about which more will be said later. The amino acid sequences of these variable regions are the basis for the tremendous diversity of antibodies and are responsible for the recognition of specific antigens. Each complete limb of the Y is called the **Fab region**, and this accounts for the identification of specific antigens. The base of the Y, called the **Fc region**, is the portion that determines its overall biological activity.

There are five major classes of antibodies—each characterized by a different set of heavy chains in the Fc region (Table 8.1). **IgM (immunoglobulin M)** is attached to plasma membranes of mature, but immunologically naïve lymphocytes. Present on the B cell surface as pentamers, IgM serves as a generic antigen receptor. After appropriate stimulation, in the place of IgM, the B cell will secrete more specialized immunoglobulins, for example, IgG, against specific antigens (Box 8.1).

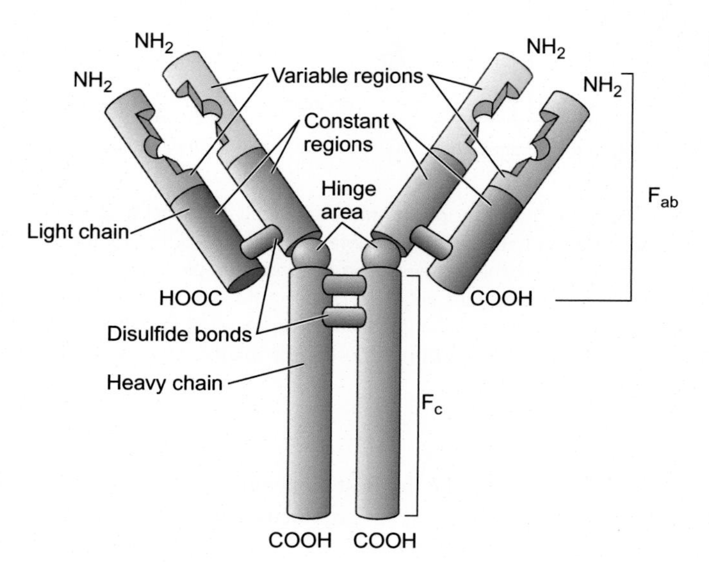

FIGURE 8.7 Representation of an IgG molecule. *From Gartner and Hiatt (2011), with permission.*

TABLE 8.1 Major Classes of Antibodies

Type	Function	Disposition	Form	Half-life (days) in Serum
IgA	Mucosal immunity	Secreted	Monomer, dimer, trimer	6
IgD	Antigen receptor in immature B cells	Membrane-bound	Monomer	Minimal
IgE	Parasite defense Mast cell activation	Secreted	Monomer	2
IgG	Opsonize invaders, complement activation, antibody-mediated T cell cytotoxicity	Secreted	Monomer	23
IgM	Complement activation, antigen receptor in naïve B cells	Membrane-bound	Pentamer	5

BOX 8.1 Maturation of antibody-forming capacity by B-lymphocytes

A naïve B cell expresses surface receptors for IgM, and when activated by an antigen, the first type of antibody produced against that antigen is IgM. IgM is not a highly specific antibody and does not bind tightly, so it is not highly effective against an invading pathogen (Fig. 8.8). But, when a B cell is activated within the germinal center of a lymphoid nodule, it undergoes two fundamental changes that allow it to produce highly effective antibodies against the foreign antigen to which it has been exposed.

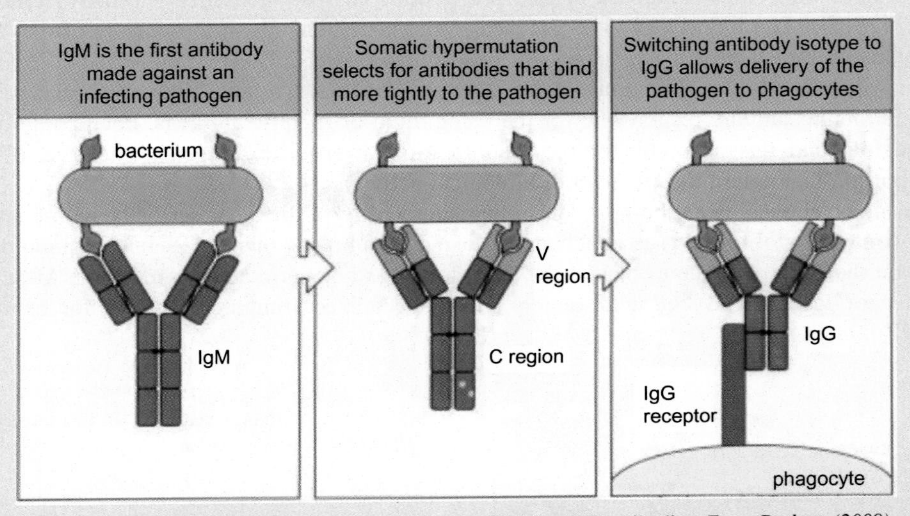

FIGURE 8.8 Somatic hypermutation and the switch from the production of IgM to IgG antibodies. *From Parham (2009), with permission.*

The first is a process called **somatic hypermutation**—a refinement of the capacity of an antibody to react against a specific antigen. This process consists of rearranging genes for both light and heavy chains of the variable (Fab) chains of the antibody molecule to produce proteins that allow the antibody to bind very tightly to the antigen. The second change, called **class-switching**, also involves alteration of the DNA so that the gene for the appropriate Fc class of the antibody molecule (see Table 8-1) is produced. For example, if the Fc component switches from IgM to IgG, the antibody will be secreted into the blood or lymph by plasma cells rather than being bound to the surface of the lymphocytes. Both of these steps involve changes at the gene level of the cells involved and can occur throughout life.

IgG is the most prominent form of secreted antibody and is found in the highest concentration in the blood. B cells—actually their plasma cell descendants—can produce enormous amounts of IgG molecules to bind to any antigen to which the body is exposed. Antibodies produced by B cells to specific antigens are first embedded in the plasma membranes of the B cells by their Fc tail. If a B cell bearing the membrane-bound antibody then comes into contact with the antigen that it recognizes, the B cell becomes activated and produces a clone of plasma cells that produce large amounts of that antibody in a soluble form. **IgA** is associated with the protection of mucosal surfaces and often passes through the epithelial lining of the mucosa to attack foreign pathogens before they actually enter the tissue space of the body. **IgE** is also secreted by B cells and binds to mast cells, causing their degranulation and release of histamine.

Complement

From an evolutionary perspective, the **complement system** represents one of the most ancient defense mechanisms against invading pathogens. It is present in sponges and sea urchins, which appeared on the scene well over a half billion years ago. Like the proteins leading to blood clotting, complement is a system of proteins, produced by the liver, in which breakdown products of one molecule stimulate a reaction in the next step down the cascade. Although complement is principally a component of the innate immune system, certain elements interact with the more recently evolved adaptive immune system. In human embryologic development, complement proteins begin to appear as early as the first trimester of pregnancy.

The components of the complement system are proteins that are associated with cell membranes or are present in the circulating blood. Activation of the complement system can follow any of three initial pathways, but all converge to a final common pathway that leads to the death of infected cells.

There is little relationship between the names given to the three pathways of complement activation and their evolutionary history. The most ancient is called the **alternative pathway** only because it was discovered later than what is called the **classical pathway**. The most abundant complement protein is called C3, and it is rapidly broken down into two subunits. One, called $C3_b$, binds avidly to amino or hydroxyl groups, which are commonly exposed on the surfaces of invading pathogens. If $C3_b$ does not find a binding site in much less than a second, it binds to water and is effectively neutralized.

Binding of $C3_b$ to the surface of a microbe initiates the alternative pathway (Fig. 8.9). This continues as a cascade of other activated complement proteins that ends with the formation of a **membrane attack complex**—a group of complement protein subunits surrounding an open membrane channel formed from several C9 units that leads to the interior of the cell. Water and ions freely enter the channel, resulting in the death of the cell. Fortunately, normal human cells are equipped with mechanisms that prevent the activation of complement on their surface.

The **classical pathway** involves an aggregation of about 30 complement proteins to form a large complex called **C1**. It begins with membrane-bound IgM pentamers bound to an antigen. When two or more C1 complexes are brought together by binding to IgM pentamers on the surface of a cell, the C1 complexes become activated. Through a different activation pathway from that of the alternative pathway, the complement cascade converges to the membrane attack complex and results in the destruction of the invading cell.

The third complement activation pathway is called the **lectin**[4] **pathway**. It begins with the binding of a mannose-binding protein, made in the liver, to mannose groups expressed on the surfaces of many pathogens, ranging from bacteria, to yeasts, to viruses, to parasites. Mannose-binding protein, plus an associated protein, acts to break down C3 into $C3_b$ plus another fragment. Farther down this distinctive cascade, a membrane attack complex is formed, leading to the death of the invader.

In addition to forming membrane attack complexes, components of the complement system fulfill other defense functions. Sometimes $C3_b$ attached to the surface of an invader becomes enzymatically trimmed to form $iC3_b$ units. These are no longer capable of following a pathway leading to the formation of membrane attack complexes. Instead, they coat (opsonize) the invader and provide an attachment surface for macrophages, which then proceed to engulf the invader. The small $C3_a$ and $C5_a$ fragments that are clipped off early in the activation cascades serve as chemoattractants, which both attract neutrophils and macrophages to the site of an infection and activate them.

Major Histocompatibility Complex (MHC)[5]

Like antibodies, the **histocompatibility proteins** are collectively a highly diverse class of molecules. Their basic function is to present antigens so that other cellular components of the immune system can deal with them.

There are two classes, classes I and II, of MHC molecules (see Fig. 8.4). Class I MHC consists of an α chain with its base embedded within the cell membrane. The α chain protein is attached to a β_2-microglobulin beyond the cell membrane.

4. A lectin is a protein that binds to a specific carbohydrate molecule, such as mannose.

5. In humans, the abbreviation HLA (human leukocyte antigen) is commonly used in place of MHC.

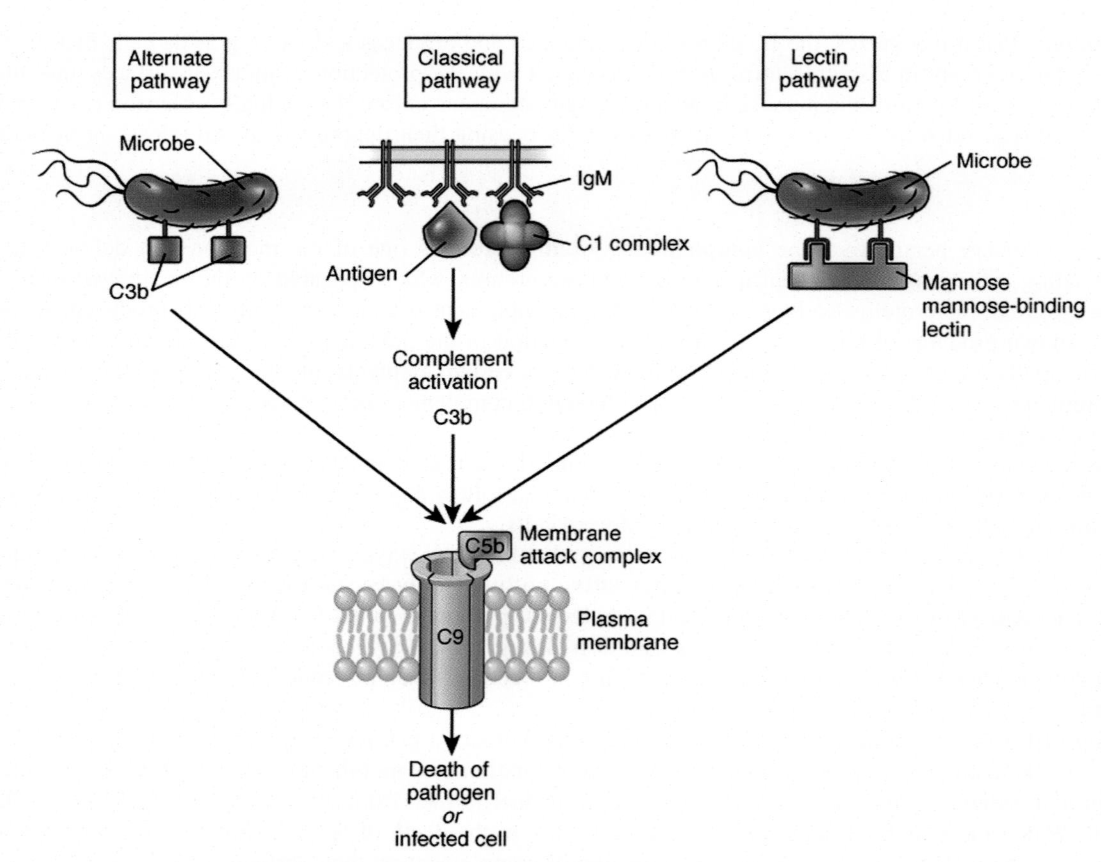

FIGURE 8.9 Simplified schematic representation of the three complement pathways.

Class II MHC is a dimer of an α and a β chain, both of which are also embedded within the cell membrane. Molecules of each class possess a groove that forms complexes with a peptide fragment derived from the breakdown of a foreign protein. The grooves are specific for individual peptides—thus the need for so many varieties of MHC molecules. These antigenic peptide fragments are then presented by the MHC molecule to a cellular component of the immune system.

The two MHC classes have quite different functions. Class I MHCs are expressed on the surface of virtually all nucleated cells in the body. They display their antigen to killer T cells,[6] which determine whether the peptide fragment belongs to or is foreign to the body. If a normal cell is infected by a pathogen, some of the proteins of the pathogen are enzymatically broken down. Within the rough endoplasmic reticulum, peptide fragments (up to 8–9 amino acids) of the proteins are bound into the grooves of the MHC molecule that correspond to those specific peptides. In addition, peptide fragments of normal cells may also form complexes with class I MHC molecules. These MHC molecules, along with the peptides in their grooves, are transported to the surface of the cell via the Golgi apparatus. There they present their peptide antigens to killer T cells. If the peptides are recognized as belonging to the normal body, the cells presenting these self-antigens are spared. On the other hand, if the presented antigen is recognized as foreign by a killer T cell, the infected cell is destroyed. In summary, the class I MHC system monitors the state of health of individual cells throughout the body and is designed to eliminate those that contain foreign pathogens.

In contrast to class I MHCs, class II is designed to monitor the presence of foreign pathogens in the extracellular environment. An antigen-presenting cell, such as a macrophage or dendritic cell, engulfs a pathogen and enzymatically breaks it down to peptide fragments in cytoplasmic phagosomes. Within the cell, these fragments of formerly extracellular molecules (up to 20 amino acids) fit into grooves of the corresponding class II MHC molecules, which are then carried to the surface of the cell for presentation to helper T cells.

6. It is important to recognize that, despite the similarity in their names, NK cells and killer T cells (cytotoxic lymphocytes) are two different varieties of cells. NK cells function principally in innate immune responses, whereas killer T cells belong to the adaptive immune system.

Cytokines

Cytokine is a word commonly used in immunology to describe protein molecules, often produced in small amounts, that exert powerful effects on nearby cells (paracrine effect) or even the cell that produced them (autocrine effect). Many cytokines, called **interleukins** (abbreviated IL-some number, e.g. IL-6), are produced by white blood cells (leukocytes) and act on other leukocytes. In innate immune reactions, macrophages and dendritic cells each produce several types of IL that play roles as diverse as stimulation of cells to leave the blood and migrate to a site of tissue damage or activating the functions of other or the same cells.

Another cytokine, **tumor necrosis factor** (TNF), kills cancerous or virus-infected cells, and it can activate other immune cells. A special family of cytokines is the **interferons** (IFN). These molecules, often secreted by macrophages, can exert powerful antiviral effects.

In the case of infections, cytokines produced by a variety of immune cells provide important signals concerning both the location and type of pathogen. Even at later stages of lymphocyte maturation, cytokines contribute to their functional activation.

Different functional classes of helper T cells secrete different profiles of cytokines. For example, one type of helper T cell (Th1) fights bacterial or viral infections by producing (1) TNF, which activates macrophages and NK cells; (2) IFN-α which helps macrophages to maintain their activity; and (3) IL-2, a growth factor that stimulates the proliferation of killer T cells, NK cells, and Th1 cells themselves. The maturation of a Th1 cell itself is promoted by IL-12, secreted by dendritic cells. Similarly, Th2 cells produce three ILs (IL-4, -5, and -13) that are used to fight against parasites. Th17 cells produce ILs that defend against fungal or yeast infections.

Receptors and Ligands

Most interactions between cells and other cells or their secretions occur at the cell surface. Receptor molecules embedded in the plasma membrane receive signals from other cells by binding to the signaling molecules (called **ligands**). Such binding often changes the configuration of cell surface molecules in a way that sends off a series of signal transduction (see p. 3) events that ultimately affect gene expression within the nucleus. Many of the maturational processes and functional properties of immune cells are initiated by receptors binding to specific ligands. In immunological reactions, most ligands are typically antigens of some sort, with signaling molecules constituting only a small proportion of the types of ligands. These antigens may be present on the surfaces of microbes or they may be isolated molecules, for example, those that cause allergic reactions. Many antigenic ligands are proteins; others are complex carbohydrates. In lymphoid cells, receptors bound to antigens must form close clusters in order to elicit a cellular response.

Lymphoid Tissues and Organs

Cells of the lymphoid system pass through several common stages in their life histories. They are born; they mature; they become activated; and they carry out their designated function before dying. In the embryo, lymphoid cells arise from hematopoietic stem cells located in the liver, spleen, and bone marrow, with the bone marrow assuming prominence as embryonic development proceeds. Members of the lymphocytic lineage arise in the bone marrow and spend the first part of their lives maturing in what are called **central (primary) lymphoid organs**, namely the thymus (for T cells) and the bone marrow (for B cells) (Fig. 8.10). Once they have attained a certain degree of maturity, they leave the central lymphoid organs via the blood as immunologically naïve B and T cells and are seeded into **peripheral (secondary) lymphoid organs**, where they await activation. The most prominent peripheral lymphoid organs are lymph nodes and the spleen, but very large numbers of lymphocytes are distributed into the connective tissue layers immediately underlying the epidermis of the skin and the mucosas of the digestive, respiratory and reproductive tracts. Often these cells are seemingly randomly distributed, but in other cases, they display a significant degree of tissue organization and may even be grossly visible as the tonsils in the pharynx or Peyer's patches in the lining of the intestine (see Chapter 12). Collectively, these tissues are known as **mucosa-associated lymphoid tissues**, or **MALT**.

Thymus

Long a highly enigmatic organ, the **thymus** is now known to be the place where T cells proliferate and mature. Along the way, the vast majority of maturing T cells in the thymus are selected for destruction, mainly because they do not fit the precise requirements of protecting the body from foreign antigens while yet ignoring self-antigens.

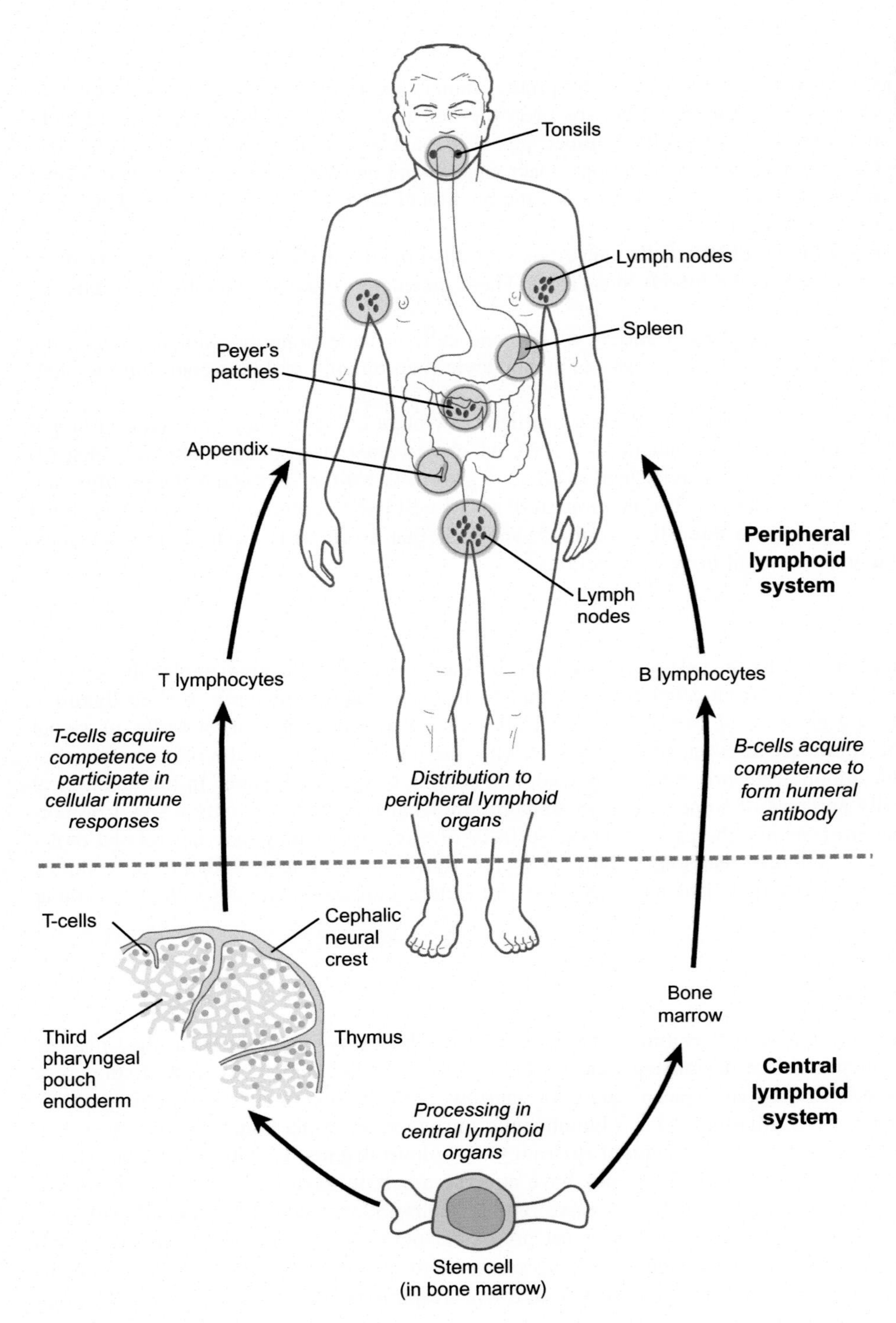

FIGURE 8.10 Embryonic development of the lymphoid system. *From Carlson (2014), with permission.*

Situated just behind the sternum, the thymus is a bilobed structure that is most prominent in young children. Even by the time of puberty, when it reaches its maximum weight of 30–40 gm, the thymus has begun to involute, with its original cells becoming replaced by fat cells. The young thymus consists of a well-defined cortex and medulla (Fig. 8.11). The cortex is densely populated by large numbers of small lymphocytes, whereas the cells of the medulla are larger and, overall, present a less crowded appearance. The cortex of the thymus is covered by a thin capsule of connective tissue, beneath which is a meshwork of cortical epithelial cells that play a major role in the maturation of **thymocytes** (a synonym for thymic lymphocytes) (Box 8.2). The medulla is less packed with thymocytes and contains epithelial cells

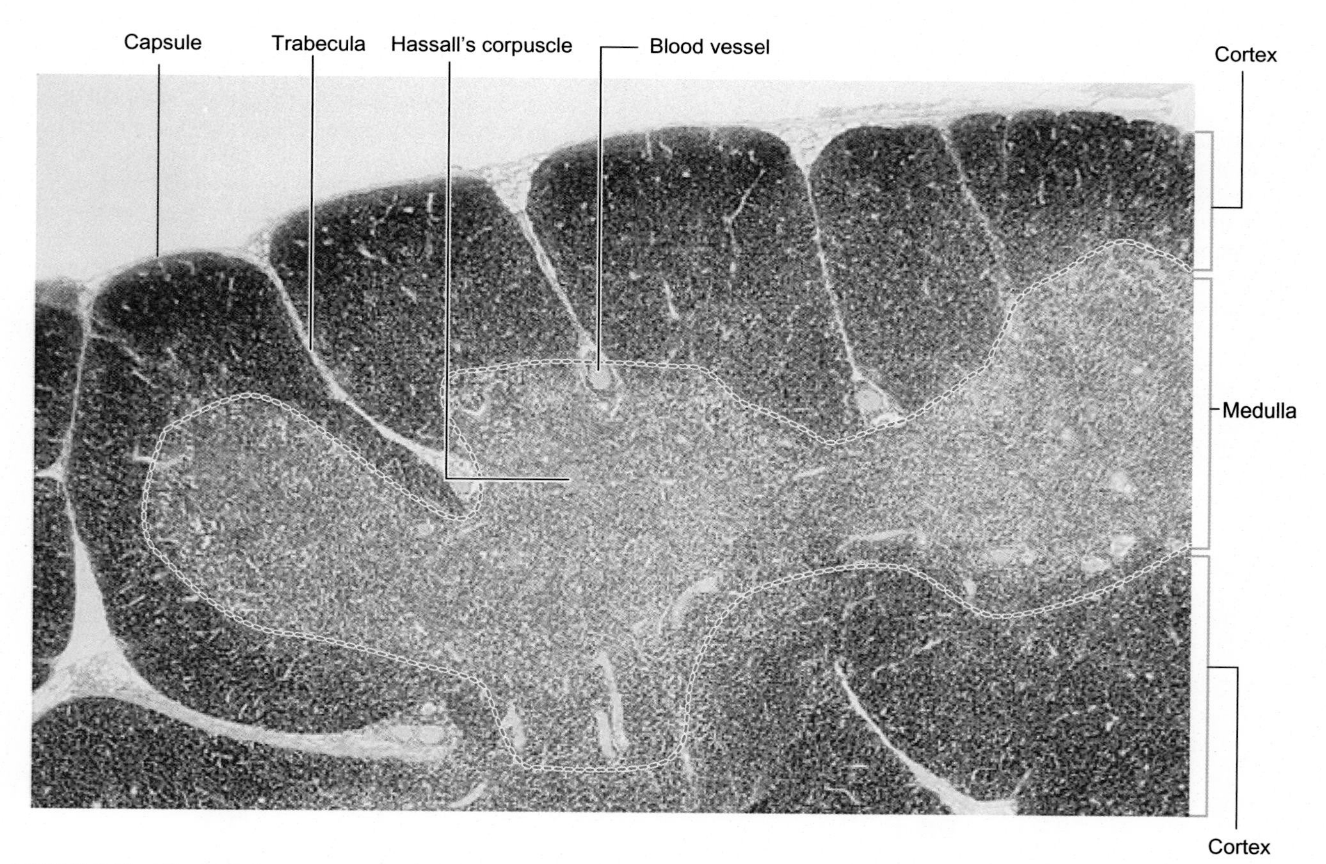

FIGURE 8.11 Histological structure of the thymus, showing a cortex densely populated by developing T cells and a medulla consisting of a greater variety of cells. *From Kierszenbaum and Tres (2012), with permission.*

and medullary dendritic cells, which present self-antigens to nearly mature thymocytes. Within the medulla, some epithelial cells form tightly bound whorls, called **Hassall's corpuscles** (Fig. 8.13). Their function is poorly understood, but they, as do the other stromal cells of the thymus, secrete some cytokines including one that activates thymic dendritic cells. The thymus has a rich vascular supply. It contains efferent, but no afferent lymphatic vessels.

Bone Marrow and B-Lymphocyte Maturation

Bone marrow is the chief incubator of the earliest stages of all cells of the immune system and the blood. Structural details of the bone marrow are given in Chapter 10. Throughout life, the development of B-lymphocytes begins in the bone marrow, where very early in their individual life histories some precursor cells are committed to becoming B cells rather than T cells (Fig. 8.14). Within the bone marrow, these cells are intimately associated with stromal cells, which assist their development through the release of growth factors and other contact-mediated processes. Although the commitment to becoming B cells and early stages in their maturation occurs in the bone marrow, B cells are still functionally immature when they are released from the bone marrow and enter the circulating blood. Details of the early maturational processes of B cells in the bone marrow are presented in Box 8.3.

Lymph Nodes

A typical **lymph node** is a small (a few mm) bean-shaped structure with a hilus in the concave surface through which a small artery and vein enter and leave and from which an efferent lymph vessel exits (Fig. 8.15A). The opposite (convex) surface receives afferent lymphatic vessels. Lymph nodes are distributed as clusters in places (e.g., neck, axilla, and groin) where the afferent vessels can drain defined regions of the body. The lymph nodes essentially strain the incoming lymph to remove and process foreign pathogens or other foreign antigens. In addition to free-floating pathogens, afferent lymph may contain antigen-presenting cells, such as dendritic cells or macrophages.

BOX 8.2 T cell maturation within the thymus

Postnatally, immature T-lymphocytes enter the thymus from their site of origin in the bone marrow. Carried by the blood, they appear to gain access to the thymus at the corticomedullary junction through the walls of high endothelial venules in much the same manner as lymphocytes enter lymph nodes.

Once in the thymus, these lymphocytes undergo a sequence of maturational stages that take them on a well-defined journey from the corticomedullary junction to the outer cortex and then from there back into the medulla before they leave the thymus as mature helper or killer T-lymphocytes. Each stage of the journey involves a topographic site in which the local environment plays an important role in the functional development of the lymphocytes (often called **thymocytes**) while within the thymus. Fig. 8.12 provides a roadmap of the seven stages of their journey.

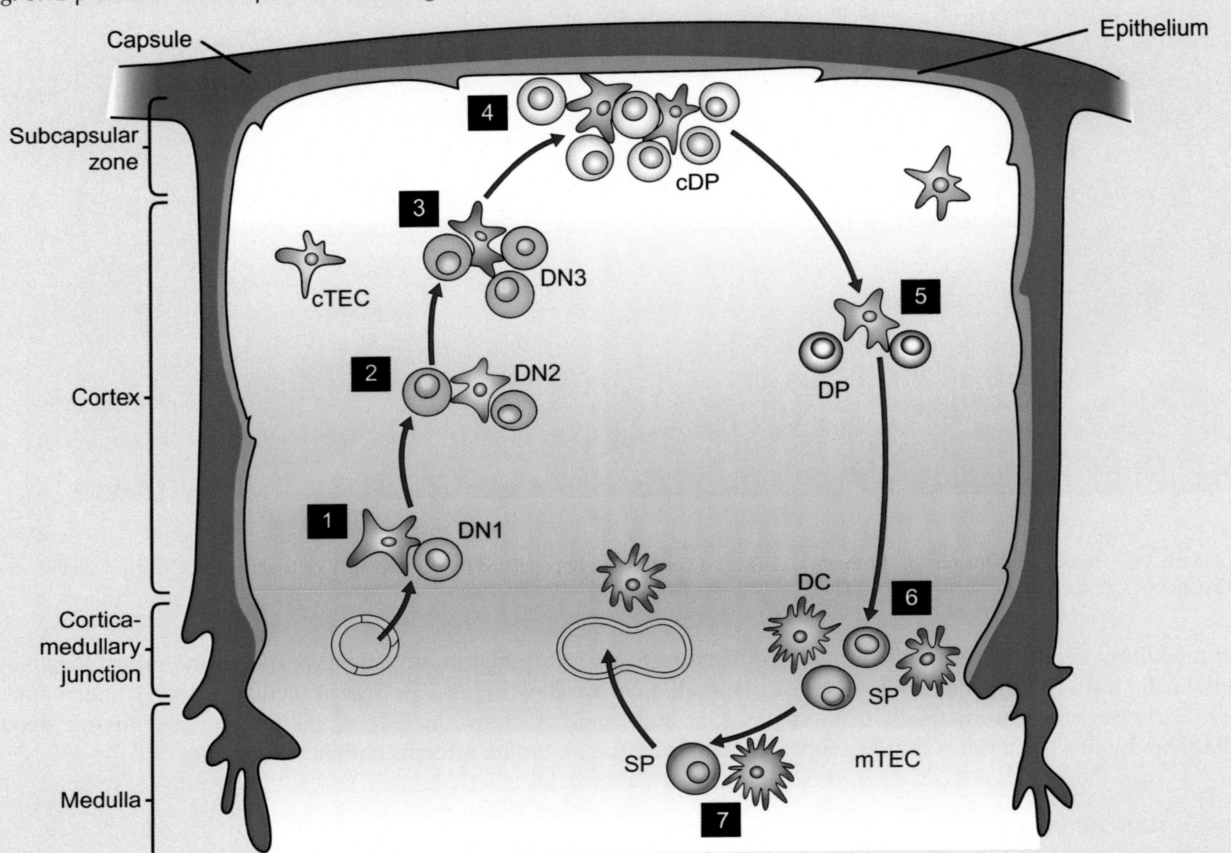

FIGURE 8.12 The pathway of maturation of T cells in the thymus. The numbers refer to zones mentioned in the text. *Adapted from Kerr (2010), with permission.*

Critical to the proliferation and functional differentiation of thymocytes is their relationship with the stromal cells and connective tissue components of the thymus. The thymic cortex is covered by a capsule of dense connective tissue that sends partitions inward to divide the cortex into semi-separate lobules. Within the thymic cortical capsule are two kinds of epithelial cells. One type, sometimes called nurse cells, lies immediately beneath the capsule and stimulates proliferation of immature T cells through the secretion of specific cytokines. The other type of epithelial cell forms a reticular meshwork throughout the cortex. Maturing thymocytes are closely associated with this meshwork. Cortical epithelial cells closely surround the cortical blood vessels. The basal laminae underlying these epithelial cells are closely apposed to the basal laminae surrounding the cortical blood vessels. Together, they form the blood-thymus barrier that effectively protects the developing T cells from exposure to antigens originating outside the thymus.

When T-lymphocyte precursors first enter the thymus (Zone 1), they no longer have the potential to become B-lymphocytes, but they must undergo a long series of developmental events before they become functionally mature T-lymphocytes. After entering the corticomedullary junction, these thymocytes spend about 10 days in Zone 1, during which each cell undergoes about 10 serial divisions produce about 1000 progeny from each entering cell. These newly entered thymocytes lack the surface molecules (CD4 and CD8, see Fig. 8.22) that would identify them as either helper or killer T cells. Because of this, they are called **double negative (DN) cells**.

(Continued)

BOX 8.2 (Continued)

After their tenure in Zone 1, the thymocytes spend the next two days migrating farther along the inner cortex (Zone 2), using processes of the cortical epithelial cells as guides. There, they respond to a variety of incompletely understood molecular cues to become further committed to becoming T cells. These cells continue to divide and are still DN. The process of rearranging the genes coding for subunits of the T receptor proteins begins. The DN cells then move into the outer cortex (Zone 3), where for two more days they continue to proliferate. At this point, they become firmly committed to the T cell lineage, and recombination of the genes encoding the T cell receptor subunits intensifies.

Zone 4, even though passage through it lasts only a day, marks a number of critical turning points in the maturational history of a T cell. This zone is located in the subcapsular region of the thymic cortex. Once the thymocytes get there, most of the T receptor gene arrangements have been completed. The cells that fail to do this successfully die. The remaining cells multiply dramatically and begin to express on their surfaces both the CD4 and CD8 coreceptors (**double-positive**, DP). The polarity of migration also reverses, so that from this stage the thymocytes migrate back through the cortex in the direction of the medulla.

By the time the thymocytes have begun their reverse migration through the cortex in Zone 5, they have lost their proliferative capacity for the remainder of their residence in the thymus. Another change that might explain how they can now migrate in a reverse direction is that because of cell surface changes, these cells no longer need to maintain close contact with cortical epithelial cell processes in order to move. Instead, they now respond to medullary cytokine signals that direct their inward migration.

One of the most important aspects of T cell maturation within the thymus is the selection for survival of thymocytes that are tolerant of self-antigens, but yet recognize foreign antigens. The developing thymocytes must pass both **positive** and **negative selection tests** before they are allowed to leave the thymus. Only about 5% of thymocytes pass both of these tests. The testing begins in the thymic cortex, where the DP thymocytes are exposed to cortical epithelial cells that present a peptide/MHC complex on their surface. Thymocytes that weakly bind to these cells via their own T cell receptors receive a signal that allows them to survive. Those thymocytes that fail to bind are fated to die. This important positive selection event takes place with thymocytes located in Zone 5. As a consequence of their binding, the thymocytes that passed the positive selection test also begin to commit themselves to becoming either helper CD4 or killer CD8 T cells.

Following cytokine cues, the thymocytes next migrate to the outer medulla (Zone 6), where they must pass the next test of negative selection. Importantly, the cellular players with whom they interact directly are dendritic cells, not medullary epithelial cells, even though the latter play an important role in the testing. The essence of negative selection is to weed out future T cells that would bind too strongly to self-antigens and cause autoimmune disease. In order to administer this test, the medullary cells of the thymus must be able to produce essentially all of the self-antigens found in the rest of the body. Through an exception to the normal rules of gene expression, the thymic medullary epithelial cells are able to do so, but they must pass on the antigens to the medullary dendritic cells for interaction with the thymocytes rather than doing so themselves. During the course of testing for negative selection, a thymocyte may interact with as many as 500 different dendritic cells. Thymocytes that weakly bind to the self-antigens presented by the dendritic cells are allowed to survive, and those that fail the test by binding too strongly are sentenced to death. The newly mature thymocytes next move back to the corticomedullary junction, where they originally entered to the thymus, but now they leave (Zone 7) as mature, but immunologically naïve helper or killer T cells ready for duty in the immunological periphery. Exactly what pathway they take in leaving the thymus is still not well understood.

Once inside the lymph node, lymph enters a **subscapular sinus** (see Fig. 8.15B), which contains many macrophages that act as cellular filters for incoming lymph. The macrophages phagocytize the foreign material and either destroy it or present antigenic fragments to other immune cells (for details, see below). Within the sinuses, fine fibers of reticular connective tissue form a very loose meshwork that both slows down the flow of lymph and serves as attachment sites for the resident macrophages. The endothelially-lined subscapular sinus then sends branches (trabecular sinuses) along fine connective tissue projections (trabeculae) that partially subdivide the outer cortex of the lymph node into segments. These sinuses then converge in the medulla, where they eventually join the efferent lymphatic vessel in the hilus.

The blood supply to a lymph node is important, because both B- and T-lymphocytes enter the node via the blood. This occurs in a region of the cortex where the endothelial lining cells of postcapillary venules take on a columnar shape and are called **high endothelial venules** (see Fig. 8.15A). In most lymphoid organs, such an endothelial configuration seems necessary for the exit of lymphocytes from the bloodstream (Box 8.4).

B-lymphocytes (resting and immunologically naïve) form dense aggregates known as **primary lymphoid follicles** within the cortex, whereas T cells migrate into a **paracortical region** closer to the medulla, where they also form dense aggregates alongside the B cells of the lymphoid nodules (see Fig. 8.16). There they await activation. Lymphoid follicles are not static structures. There is a constant inflow of new B-lymphocytes into follicles and T-lymphocytes into the paracortical regions and outflow away from these areas.

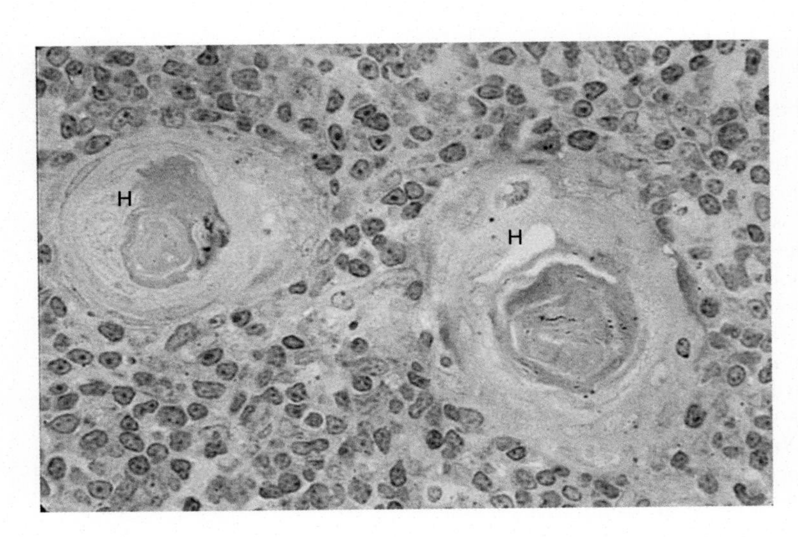

FIGURE 8.13 Photomicrograph of Hassall's corpuscles (H) in the thymus. *From Kierszenbaum and Tres (2012), with permission.*

After antigenic stimulation, lymphoid follicles, now called **secondary lymphoid follicles**, develop a lighter staining **germinal center** that consists predominantly of activated B cells with increased amounts of cytoplasm (Fig. 8.16). This area is composed of cells that are actively producing antibodies. Germinal center activity increases the overall size of the lymphoid nodule and contributes to the swelling of lymph nodes that occurs during an infection. Although the dominant cells are B-lymphocytes, lymphoid nodules also contain a substantial number of antigen-presenting cells called follicular dendritic cells and macrophages, which fulfill a phagocytic function.

Dendritic cells play an important functional role, especially within the cortex. Lymphoid nodules contain a meshwork of resident follicular dendritic cells, whereas the paracortical region tends to hold dendritic cells that have entered the node from its drainage area, although they can also be found within germinal centers. Within the medulla, tongues of cortical tissue (medullary cords) that contain many activated B cells or their plasma cell descendants alternate with medullary sinuses. Exiting B and T cells leave the lymph nodes via their efferent lymphatic vessels and ultimately enter the bloodstream through large lymphatic ducts.

Spleen

The **spleen** is the largest lymphoid organ in the body, and it plays a complementary role to lymph nodes. Whereas lymph nodes strain lymph for foreign pathogens, the spleen strains blood. Situated in the abdominal cavity in a niche bounded by the diaphragm, the stomach, and the pancreas, its simple ovoid external profile gives little hint of its complex internal structure. Internally, the spleen does not have a cortex or medulla, but rather contains two structurally and functionally different components—**white pulp**, which carries out traditional immune functions, and **red pulp**, which removes worn-out or abnormal erythrocytes from the blood (see p. 294). In the embryo and under certain extreme circumstances postnatally, the spleen also acts as a hematopoietic organ.

Key to understanding the functions of the spleen is its internal circulatory system. Its role in the white pulp and in immune functions is described here. The structure and function of red pulp are treated in Chapter 10.

Like lymph nodes and the kidneys, the spleen has a hilus through which a large splenic artery enters and a vein leaves. Much of the arterial blood becomes distributed throughout the spleen by branches that course through the dense connective tissue capsule and its internal projections (**trabecular arteries**). Some of the branches of these arteries (called **central arterioles**) that enter the splenic pulp are surrounded by a cuff (**periarteriolar lymphoid sheath**, or **PALS**) of lymphoid tissue (Fig. 8.17). Most of the cells of the PALS are helper T cells, along with macrophages. Although less directly associated with lymphatic nodules than in lymph nodes, the overall functional relationships remain much the same. Lymphoid nodules, representing the B cell component of the lymphoid system, are scattered along the PALS. When antigenically stimulated, they develop germinal centers. Surrounding the lymphoid nodules is a thin layer of T cells that tapers off into the red pulp without a clear line of demarcation. How lymphoid cells enter the spleen is not well understood, because high endothelial venules of the sort seen in lymph nodes or the thymus are not found in the spleen.

BOX 8.3 Early Stages in the Maturation of B Cells

Within the bone marrow, B cells (**pro-B cells**) in the earliest stages of development are heavily involved in rearrangements of their genes that form segments of antibody molecules (first the heavy chains and then the light chains). These genetic rearrangements produce **immature B cells** (see Fig. 8.14) with surface receptors that are capable of binding to many different antigens, even though a single cell can only react to a single antigen. Already a winnowing process begins, where cells that strongly bind to self-antigens encountered in the bone marrow are either internally instructed to commit suicide or they are sent back for more genetic rearranging to produce a different set of surface receptors. Even before leaving the marrow, about 70% of the immature B cells are either killed off or reprogrammed.

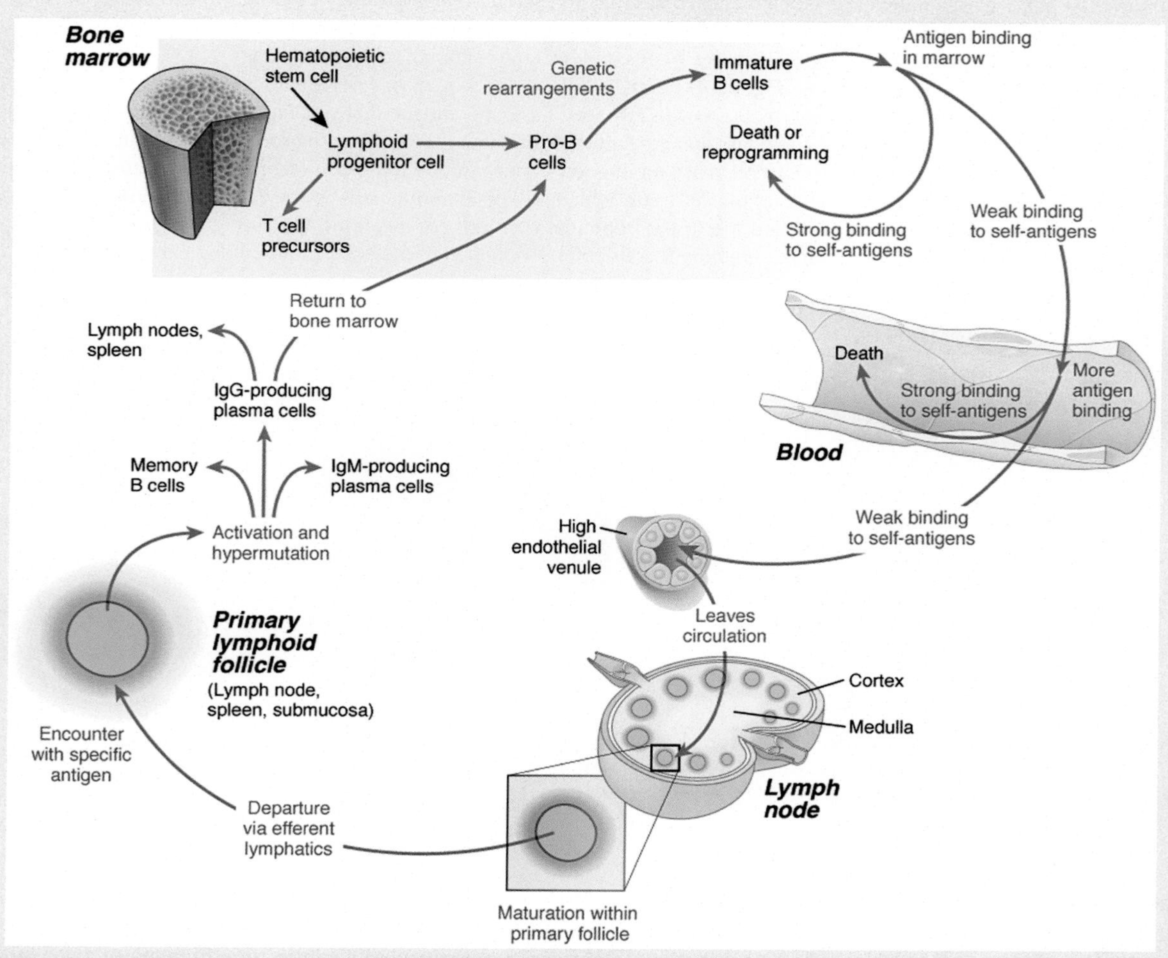

FIGURE 8.14 The life history of a B cell, starting with a hematopoietic stem cell in the bone marrow.

During the early maturation process, the immature B cells that are tolerant of the self-antigens present in the bone marrow gradually release their hold on the stromal cells of the marrow. Cells that are able to produce functional surface receptors (IgM, with a small amount of IgD) are allowed to leave the marrow and enter the general circulation, where they become exposed to other self-antigens of the body. Those cells that react strongly to them are also condemned to death, because at this stage genetic reprogramming is not an option. Once in the blood, immature B cells are swept into secondary lymphoid organs where the final stages of maturation occur.

BOX 8.4 Later Stages in the Maturation of B Cells

Responding to locally secreted chemokines, immature B-lymphocytes leave the circulation through the high endothelial venules in the lymph nodes (see Fig. 8.15A). They are then attracted to primary lymphoid follicles by other cytokines secreted by the follicular dendritic cells. While in the primary lymphoid nodules and associated with the follicular dendritic cells, they undergo further changes until they are ready to leave the follicle as mature, but immunologically naïve B cells.

Not all immature B cells are able to enter lymphoid follicles, and those that do not, die within a few days. Those mature B cells that leave the follicles travel around the body in blood or lymph searching for antigen that is complementary to the configuration of their surface antibodies. If these cells fail to encounter their complementary antibodies, they also die within about 100 days.

The normal place where naïve B cells encounter antigens is in primary lymphoid follicles, where the antigens are presented by macrophages or follicular dendritic cells. After the initial antigenic encounter, they proceed to the T cell area near the follicle, where they hook up with antigen-specific helper T cells and become immunologically activated.

Three fates await activated B cells (see Fig. 8.14). Some proliferate and quickly differentiate into plasma cells that produce secreted IgG, rather than membrane IgM, which is not highly specific to the antigen that was encountered. Most of the antigenically stimulated B cells enter a primary follicle and enlarge, while at the same time undergoing somatic hypermutation to produce more potent antibodies (see Box 8.1). Those cells with the highest affinity for the antigen undergo further maturation within the germinal centers. They finally differentiate into plasma cells, which will reside principally in the medullary cords of lymph nodes or the red pulp of the spleen, or make a return trip to the bone marrow. In these locations, they secrete high-affinity antibodies. Still other activated B cells turn into quiescent memory B cells and become part of the body's immunological memory bank.

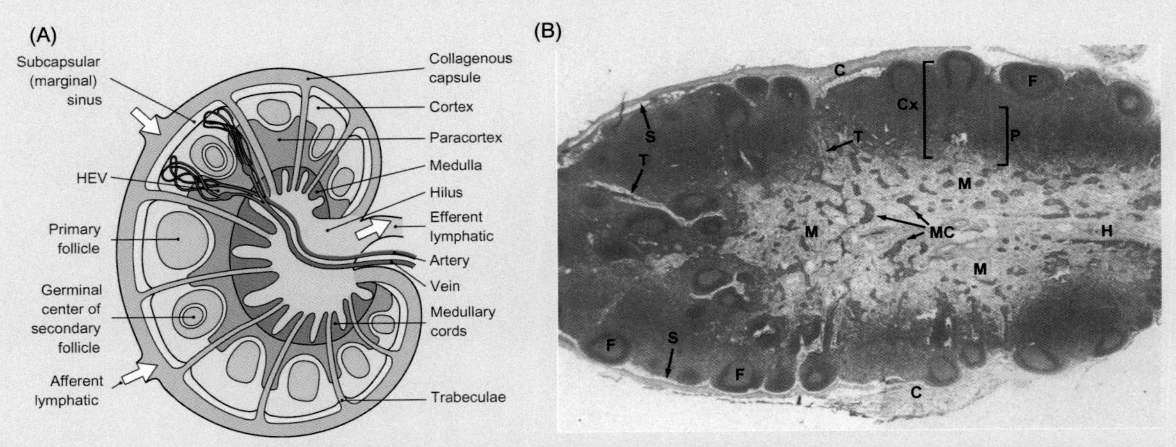

FIGURE 8.15 Structure of a lymph node. (A) Diagram of overall organization. HEV, high endothelial venule. (B) Photomicrograph. *C*, capsule; *Cx*, cortex; *F*, follicle; *H*, hilum; *M*, medulla; *Mc*, medullary cords; *P*, paracortex; *S*, subcapsular sinus; *T*, trabecula. *(A) From Male et al. (2013), with permission. (B) From Young et al. (2006), with permission.*

The principal function of the splenic lymphoid tissue is to screen the blood for blood-borne pathogens. The first line of screening is macrophages that are situated throughout the spleen. In many cases, the macrophages phagocytize bacteria directly, but they and dendritic cells also function as antigen-presenting cells to the lymphocytes of the white pulp. At this level, the splenic lymphoid tissue functions in a manner very similar to that of peripheral lymphoid tissue elsewhere in the body. In contrast to lymph nodes, lymphocytes leave the spleen through the blood, rather than through lymphatic vessels.

Skin and Mucosa-Associated Lymphoid Tissue

Although seemingly not impressive when one examines a small region, the lymphoid tissue associated with the skin and mucosal surfaces makes up a large component of the overall lymphoid system. Most of the lymphocytes of this system are scattered throughout the dermis of the skin or the layer of loose connective tissue (lamina propria) that immediately underlies the epithelium of a mucosa, such as the lining of the gut or respiratory system, but especially in internal

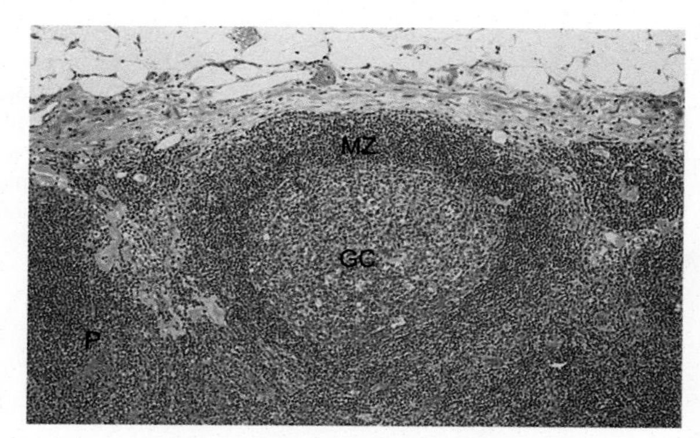

FIGURE 8.16 Secondary lymphoid follicle in a lymph node. *GC*, germinal center; *MZ* mantle zone; *P* paracortex. *From Young et al. (2006), with permission.*

(A)

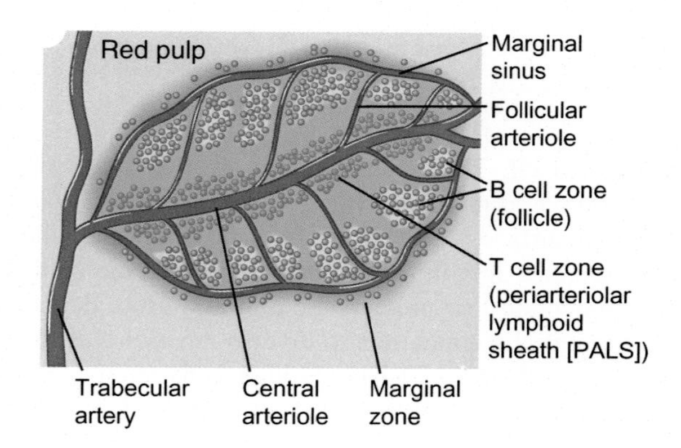

(B)

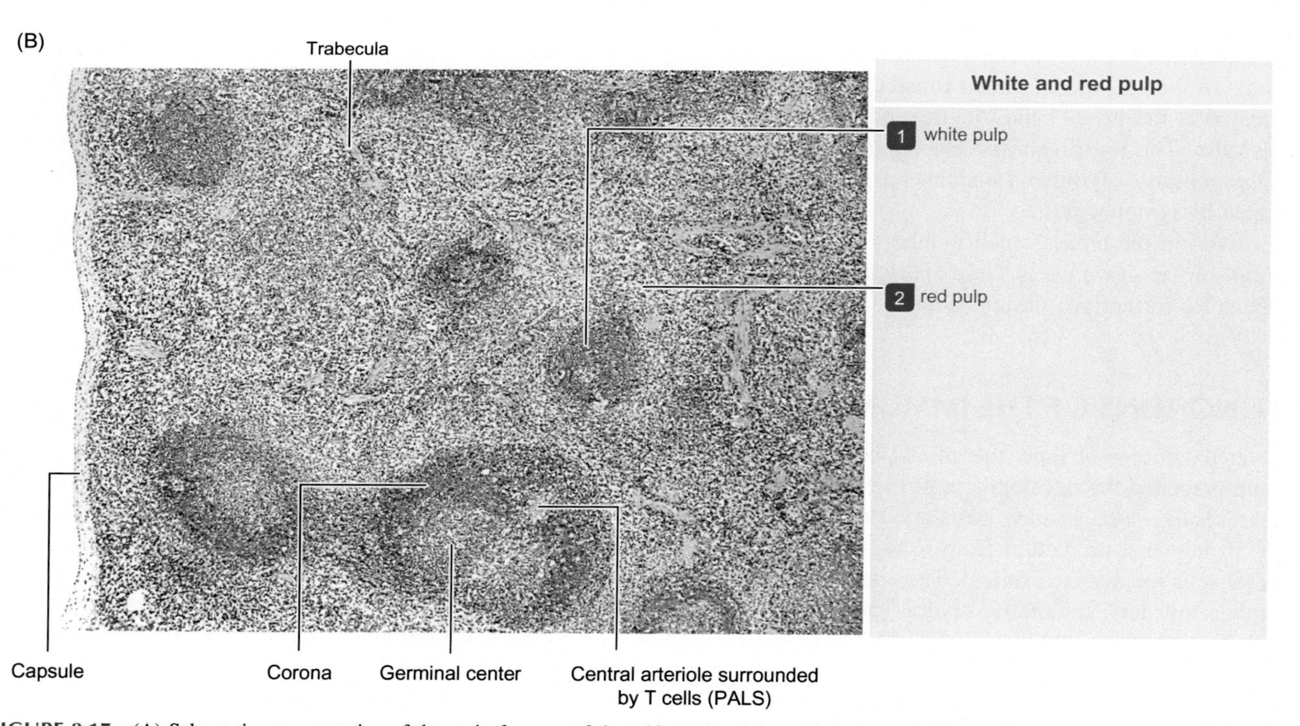

FIGURE 8.17 (A) Schematic representation of the main features of the white pulp of the spleen. (B) Histological structure of the spleen. *(A) From Abbas et al. (2016), with permission. (B) From Kierszenbaum and Tres (2012), with permission.*

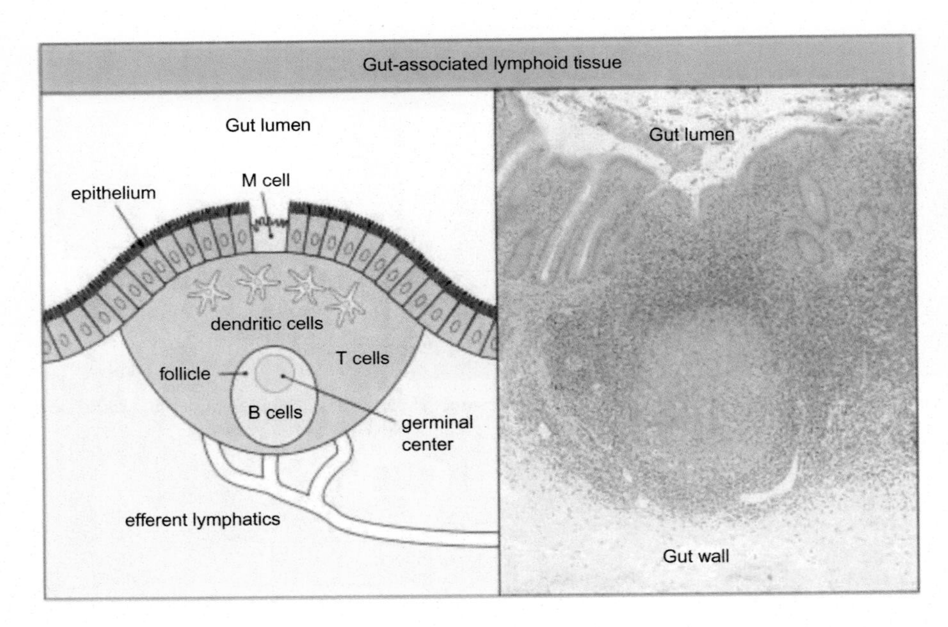

FIGURE 8.18 GALT (gut-associated lymphatic tissue). Left—schematic representation; right—photomicrograph. M cells (left) within the epithelium of the gut are specialized for the delivery of pathogens from the lumen of the gut to the lymphoid cells within the wall of the gut. *From Parham (2009), with permission.*

mucosal layers, lymphocytes are also present within the epithelium itself. These lymphocytes are principally T cells, but distributed among the scattered T cells are lymphoid nodules composed mainly of B cells (Fig. 8.18).

The main function of MALT is to protect the vulnerable internal surfaces of the body from invasion by pathogens. Such protection even begins outside the epithelium. Some of the secreted IgA in the gut passes through the epithelium into the lumen of the intestine, where it becomes bound to a carbohydrate that protects the antibody from degradation by proteolytic enzymes. If a pathogen or foreign antigen passes through the epithelium, a diverse repertoire of lymphocytes, dendritic cells, macrophages, and mast cells awaits to deal with the invaders through conventional immune responses. Lymphoid tissue in the gut is often called GALT (gut-associated lymphoid tissue) instead of the more generic term MALT.

Especially in the gut, aggregations of lymphoid nodules are very prominent. Surrounding the entrance to the pharynx is a ring of lymphoid tissue (**Waldeyer's ring**) that deals with pathogens as they enter the digestive tract. The components of the ring are all called **tonsils**. One pair, the palatine tonsils, is situated along the sides. A single lingual tonsil is located at the base of the tongue, and another (called an **adenoid** when swollen) is found on the dorsal surface of the pharynx. The tonsils consist mostly of lymphoid nodules, which are only partially encapsulated. Within the intestine, aggregations of lymphoid nodules are known as **Peyer's patches**. The epithelium above Peyer's patches is heavily infiltrated by lymphocytes.

Even in the breast, small numbers of lymphocytes and plasma cells underlie the epithelium of the mammary ducts. In all of the above cases, fluid and lymphoid cells drain into lymphatic channels that make their way to regional lymph nodes for further processing.

FUNCTIONS OF THE IMMUNE SYSTEM

Over the course of time, the animal body has evolved a variety of defenses against invading pathogens. Many of them long preceded the development of the adaptive immune system that is present in the human body. Even primitive invertebrates like sponges have developed some defense mechanisms. These defenses constitute the innate immune system.

In humans, protection from foreign invaders can be viewed as a multilayered system of defenses. The first line of defense is mainly mechanical. The epidermis covering the outer surface of the body is a highly effective barrier to pathogenic invaders. Internally, several systems—digestive, respiratory, and reproductive —are essentially open to the outside through their various orifices. The total surface area of these systems is enormous—in the range of 400 m^2. For the most part, the epithelial lining of these tracts are more delicate than the epidermis of the skin, but nevertheless, with some help, they are able to ward off most infections. Tight junctions between adjacent epithelial cells do not permit the ready passage of pathogens through the epithelium.

Epithelia employ a variety of strategies to increase their capacity to kill or repel invaders. Mucus, secreted by goblet cells or mucous glands, presents an effective medium for trapping microbes. In addition, some mucous epithelial cells secrete **lysozyme** or other powerful antibacterial agents, into the overlying mucus. One important family of defense molecules is the **defensins**. Defensins are produced by normal cells and consist of several types, each of which targets different pathogens. Defensins are secreted in an inactive form, which protects the nearby normal cells, but when enzymatically activated, they are able to penetrate bacterial membranes and kill the bacterial cells. The squamous epithelial cells of the vagina produce large amounts of glycogen which, when broken down into lactic acid by resident *Lactobacillus* bacteria, reduces the pH from neutral to ~4.3—sufficient to hold bacterial populations in check. In other cases, antibodies produced by cells in the submucosal tissues add an additional protectant to the surface layer.

Innate Immunity

The innate immune system comes into play when a pathogen breaches an epithelial barrier. It is designed to provide a rapid and local, but relatively nonspecific response to foreign invaders. For example, many innate immune receptors are somewhat generic, in that a single receptor may bind to a variety of foreign antigens.

One of the first responses to invading pathogens, e.g., bacteria, is activation of the complement system (see Fig. 8.9). C3 breaks down into two fragments, C3a and C3b. C3b fragments coat the surfaces of the bacteria through specific binding sites in a process that has been classically called **opsonization**. When opsonized, the C3b-coated bacteria are tagged for eventual destruction by phagocytes. In some cases, complement acts directly to destroy bacteria by forming membrane attack complexes (see Fig. 8.9), which kill the bacterial cells. In the meantime, C3a serves as an attractant that brings phagocytic cells into the immediate area of the invasion.

Resident tissue macrophages represent the first line of cellular defense against invading pathogens. Macrophages have surface receptors that bind to the C3b on the surfaces of opsonized bacteria. Macrophages are more efficient in binding to opsonized than nonopsonized bacteria. Once a bacterium is bound to a macrophage, folds of cytoplasm engulf the bacterium until it is completely enclosed in a membrane-bound vesicle, now called a **phagosome** (Fig. 8.19). Then lysosomes within the cytoplasm of the macrophage fuse with the phagosome and release their lytic enzymes, which fragment and dissolve the bacterium. Some small molecular fragments of the bacterium are moved to the surface of the macrophage for presentation to cells of the adaptive immune system (see below).

Other macrophage receptors send signals to the nucleus that stimulate the internal production of cytokines, which are secreted and organize other aspects of the immune response. A good example is the **Toll-like receptors**.[7] These receptors, of which there are at least 10 major types in humans, have different specificities for different types of microorganisms. Some are found on the plasma membrane, where they target external invaders; others, embedded in different internal membranes, target intracellular viral invaders. The messages conveyed by the Toll-like receptors allow the cells to tailor their responses, often the secretion of groups of cytokines, to different types of infection.

In some cases, local macrophages prove insufficient to ward off a pathogen invasion. Cytokines secreted by local macrophages, along with C3a and C5a complement fragments, are essentially signals for outside help. These signals have two broad functions. First, they stimulate white blood cells to leave the bloodstream in the area of the invasion. Second, through concentration gradients, they direct the white blood cells that have left the circulation to the exact site of infection.

The local emigration of white blood cells (**leukocytes**) from the blood into the area of an infection is a remarkable example of the coordination inherent in an immune response. Cytokines, especially IL-1 and TNF secreted by the local macrophages, diffuse to the nearest venules and cause the endothelial cells lining them to produce a type of surface receptor molecule called a **selectin** (Fig. 8.20). The selectins weakly bind to selectin ligands protruding from the plasma membranes of leukocytes, e.g., neutrophils, causing them to slow down and roll along the endothelial surface. Then stronger interactions mediated through integrins on the neutrophils and ICAM, a cell adhesion molecule on the endothelial cells, result in a firmer binding of the neutrophils to the endothelium. The neutrophils begin to squeeze their way between adjacent endothelial cells in a process called **diapedesis** and then enzymatically dissolve their way through the basal lamina that surrounds the endothelial layer of the venule. Finally, once in the tissues outside the venule, they migrate up concentration gradients set up by the cytokines until they reach the site of the infection.

In addition to stimulating leukocyte emigration from the blood, cytokines also act on local blood vessels, causing them to dilate and increase their blood flow. This results in local redness, heat, swelling, and some degree of pain—all

7. Toll-like receptors were first discovered in *Drosophila* in 1985 where they were found to be important in establishing the dorsoventral axis in the early embryo. Their name came from the German name "toll," which means great!, after the excitement of their discovery.

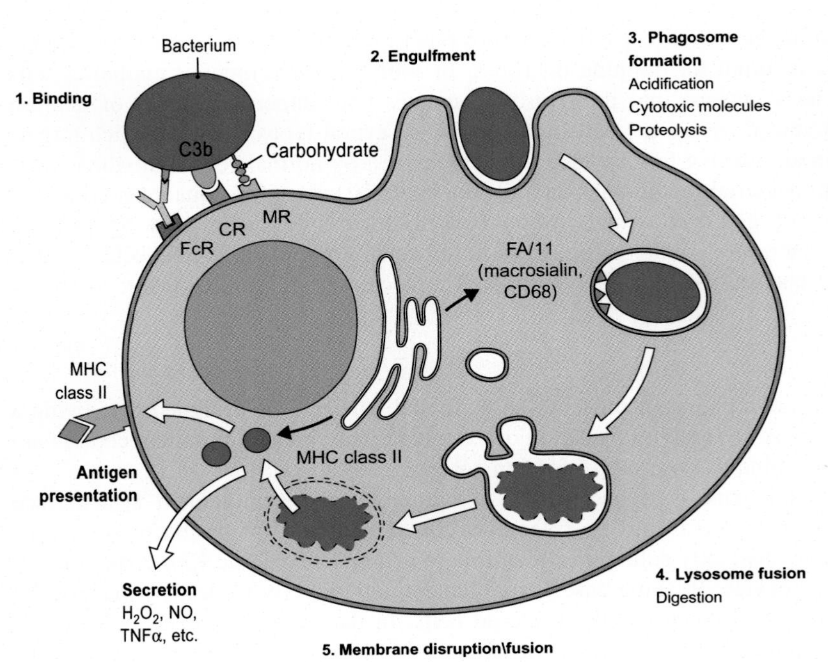

FIGURE 8.19 Stages of phagocytosis of an opsonized bacterium by a macrophage. 1. Binding of opsonized bacterium by surface receptors. 2. Engulfment of the bacterium. 3. Formation of an acidic phagosome. 4. Fusion of phagosome with lysosomes and digestion of the bacterium. 5. Disruption of phagosome membrane and release of breakdown products. *From Male et al. (2013), with permission.*

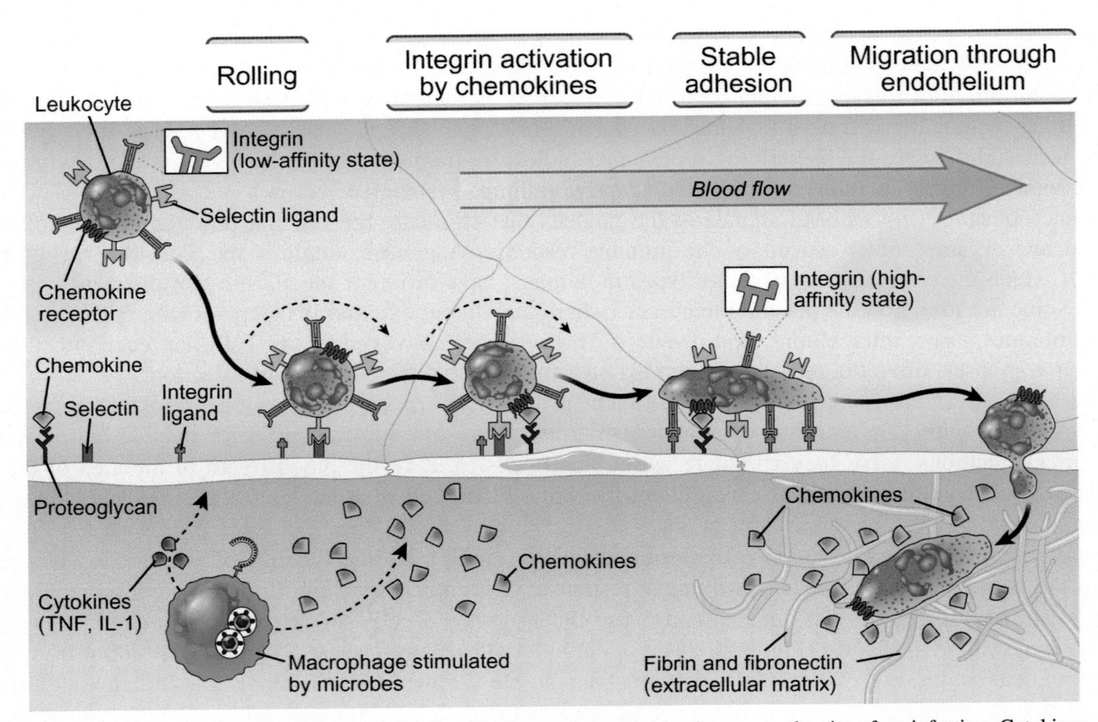

FIGURE 8.20 Sequence of events in the emigration of a blood leukocyte from the bloodstream to the site of an infection. Cytokines, produced by macrophages and other cells at the site of an infection, activate endothelial cells to express selectin and chemokines, which begin the process of tethering neutrophils from the blood onto the endothelium. This begins the process of emigration of the neutrophils from the blood through the endothelium and into the area of the infected tissue. *From Abbas et al. (2016), with permission.*

cardinal signs of inflammation. Swelling is the result of the extravasation of plasma from the dilated blood vessels. Some of the plasma proteins form local clots, which themselves act to contain the spread of bacteria. Others are protease inhibitors, which inhibit the actions of proteolytic enzymes secreted by certain types of bacteria, e.g. *Streptococcus*, that facilitate the spread of the bacteria.

Neutrophils are designed to kill things, and once they enter an area of infection, they immediately begin to phagocytize and degrade the pathogens. The tiny cytoplasmic granules of neutrophils contain many powerful destructive agents or their precursors, and when activated, they produce large amounts of **hydrogen peroxide** (H_2O_2) and **superoxide** radicals. If the infection is severe, these substances can cause significant damage to nearby cells of the body. Within a day or two, the neutrophils die and most are phagocytized by macrophages, which then salvage components of some of the bacterial molecules for presentation to cells of the adaptive immune system. If this system of removal is overwhelmed, dead neutrophils accumulate in the area of infection as pus. Several inflammatory cytokines, produced during the heat of this local battle, enter the bloodstream as **pyrogens** and stimulate a fever response through their actions on the hypothalamus. Fever is another antipathogenic mechanism because most bacteria and viruses proliferate best at lower-than-body temperatures. A fever causes them to divide more slowly.

Viruses, which inhabit cells, are dealt with differently by the innate immune system. One of the cytokines, IL-12, activates NK cells, which specialize in antiviral defense. In addition, virus-infected cells produce IFNs (see p. 219), which not only interfere with viral replication, but also make the infected cells more vulnerable to attacks by killer lymphocytes. Although most cells can produce IFNs, special IFN-producing lymphocytes in the blood can produce up to 1000 times more IFN than ordinary cells. NK cells have surface receptors that can distinguish between normal and virally infected or otherwise abnormal cells. After they have bound to a virally infected cell, they inject into it molecules that activate the intracellular apparatus that leads to **apoptosis** (intrinsic cell death) of the infected cell. Enzymes released during this process of cell death also inactivate or degrade the virus.

In the vast majority of cases, the innate immune system is capable of dealing with a local infection without further help. On some occasions, however, the number or virulence of the pathogens is such that it overwhelms the innate immune system. Then the adaptive immune system is called into action.

Adaptive Immunity

When the situation gets out of hand for local innate immune defenses, they call in elements of the adaptive immune system for support. In the case of foreign pathogens, the elements that are called on to respond to the problem depend upon the nature of the infection. Special cases in which the adaptive immune system is brought into play involve allergens, tumor cells and, in contemporary times, organ transplants. This section will use several examples that illustrate how the various components of the immune system work together in defending the integrity of the body. These examples are greatly simplified, especially in molecular detail, but they are designed to present an overview of what is involved in immune reactions to foreign invaders.

A Local Bacterial Infection

For this, we will use the example of a puncture wound of the skin that allows bacteria to penetrate the epidermal barrier and overwhelm the innate immune defenses outlined above. Even while the local innate defense system is combating the invaders, signals are being sent out to the adaptive immune system to be prepared to enter the fray, if needed.

Some of these signals take the form of soluble molecules; others require presentation by cells to other cells of the adaptive immune system. The essence of the signaling phase is capture of antigens and their presentation to lymphocytes.

As a prelude to this discussion, it is important to understand that within the normal body at all times are lymphocytes that have surface receptors specific for each of the tens of thousands of possible foreign antigens, but that each lymphocyte has receptors specific for only one antigen. This is the basis for the **clonal selection theory** of immunity, which was doubted when first proposed because of the huge numbers of cells that would have to be involved. Now, this theory is generally accepted.

In responding to a bacterial infection or some other foreign antigen, the big initial issue is presenting a foreign antigen to the relatively few lymphocytes that can recognize it. For a skin wound, some bacterial antigens may be freely suspended in tissue fluid and be transported to a regional lymph node through the lymphatic drainage of the area of infection. Other bacteria may be phagocytized by antigen-presenting cells—dendritic cells (Langerhans cells) in the epidermis or tissue macrophages in the underlying connective tissue. These cells are also carried to the regional lymph nodes via local lymphatic channels (Fig. 8.21).

Before they arrive in the lymph node the phagocytized bacteria are processed by the antigen-presenting cells. Processing involves first the encapsulation of the bacteria by plasma membrane vesicles (phagosomes) of the phagocytic cell (see Fig. 8.19). Through proton pumps, the interior of the phagosomes becomes acidified. This environment

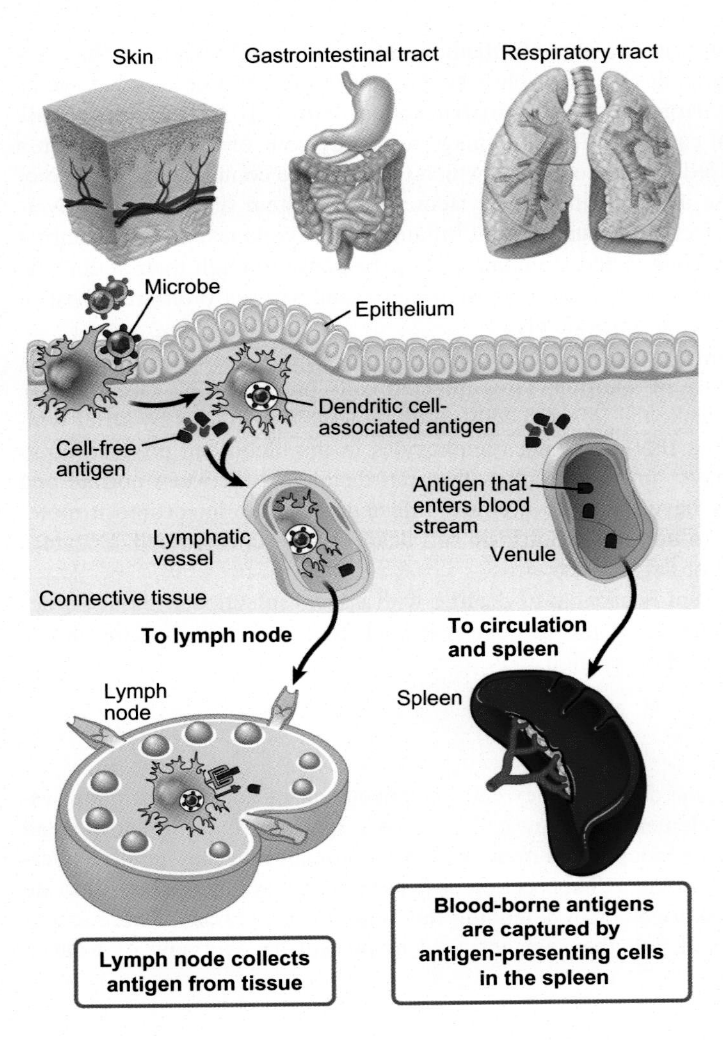

FIGURE 8.21 Early stages in the processing of a microbial infection that passes through an epithelium. Dendritic cells, which have captured the microbes, enter lymphatic vessels and are carried to nearby lymph nodes. Free antigen may enter the bloodstream and be taken to the spleen for processing. *From Abbas et al. (2016), with permission.*

activates the enzymes of the lysosomes, which have fused with the phagosomes. The bacteria are then fragmented, and peptide residues bind to class II MHC molecules with an affinity for each specific peptide. The class II MHC molecules with attached antigenic peptides are then taken to the surface of the cell, where end of the molecule containing the antigen is exposed.

When antigen-presenting cells enter a lymph node, they are carried to the paracortical T cell zone, where they encounter naïve T cells. Naïve T cells do not sit in one place waiting for antigens to come to them. They continuously move from one area to the next in search of an antigen that fits their particular antigen receptors. When by chance a naïve T cell meets with its antigen fit, it then begins a sequence of activation stages.

Activation begins with the linking of an antigen-presenting cell, for example a Langerhans cell from the epidermis, to a naïve T-lymphocyte (Fig. 8.22). This stimulates the T cell to produce cytokines, one of which (IL-2) acts on the cell that produces it (an autocrine effect) by binding to IL-2 receptors on that cell. This stimulates a burst of proliferation of that cell in a phase called **clonal expansion**, which produces large numbers of T cells with the same antigenic specificity as the founder cell of that clone. Many of these cells differentiate into helper T cells, which are critical for activation of other cells, for example B cells or macrophages, of the immune system. Other activated T cells become **memory cells** through processes that remain little understood. The memory cells are long-lived and are ready to respond quickly if the body encounters the same antigen in the future.

At the edges of lymphoid follicles within regional lymph nodes, an activated helper T cell may encounter a B cell that has bound the same peptide antigen to which the T cell is responsive. The joining of these two cells through mutual recognition of the same antigen then stimulates the clonal proliferation of that B cell and its further differentiation into plasma cells, which produce large amounts of antibodies against that antigen. These antibodies circulate in the blood

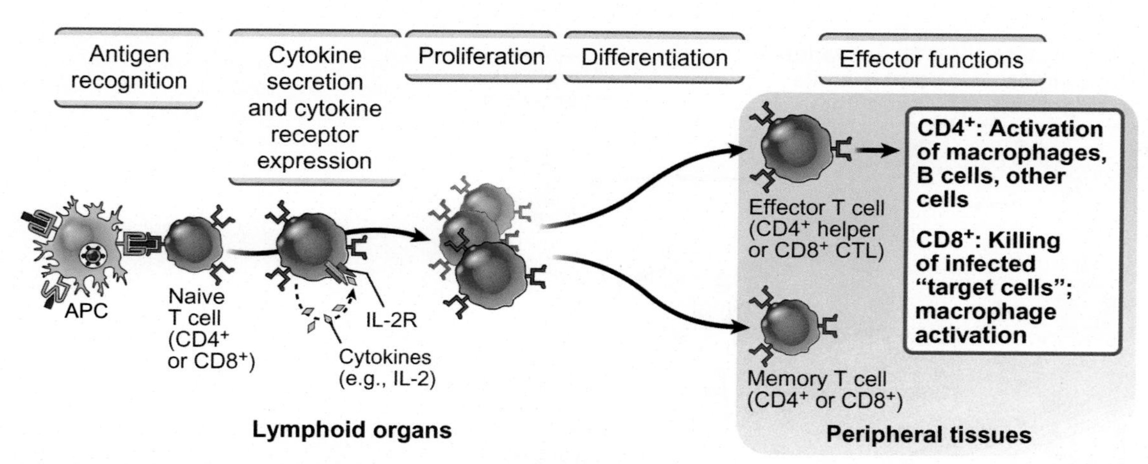

FIGURE 8.22 Steps in the activation of T-lymphocytes. *APC*, antigen-presenting cell; *CTL*, cytotoxic lymphocyte; *IL*, interleukin. *From Abbas et al. (2016), with permission.*

and lymph, and when they encounter the same antigen on invading bacteria, they coat the surface of the bacteria in a form of opsonization. This makes the bacteria more susceptible to phagocytosis and destruction by macrophages and neutrophils. Other B cells from the same clone become memory cells.

In the meantime, helper T cells with an affinity to the same antigen find their way to the site of the local infection and secrete IFN-γ, which activates the macrophages in the region to ingest the microbes more efficiently. After they break down the ingested bacteria, the macrophages present antigenic fragments of these microbes on their surfaces and become antigen-presenting cells themselves.

A Viral Infection

Combating a viral infection involves a different dynamic from that described above for a bacterial infection. The principal difference is that viruses invade normal cells of the body, and within these cells, they co-opt some of the normal cellular machinery to enable them to reproduce. As a result, they are essentially invisible to the surveillance mechanisms that patrol the body for bacteria and other extracellular invaders. The main task for infected cells is to present antigenic components of the viruses to the immune system for an appropriate response. Fighting viruses and other intracellular invaders is an important function of the cellular immune system.

Some of the cells (often dendritic cells) infected with a virus are able to break them down through their intracellular enzyme systems. These viral breakdown products are transported to the endoplasmic reticulum, where they are combined with class I MHC molecules (remember that in extracellular bacterial infections class II MHC is involved). Then the class I MHC with combined viral antigenic peptide is carried to the surface of the cell, where the antigen is presented to the outside environment. Cytotoxic T-lymphocytes (CTLs) recognize class I MHC molecules, and if one of these cells has surface receptors that fit with the presented viral antigen, they bind with the antigen-presenting cell (Fig. 8.23). They become activated by this contact and are prepared to kill the virally infected cells to which they are attached.

Cytotoxic T-lymphocytes, sometimes called killer T cells, kill infected cells by introducing a protein called **perforin** into the cell. As its name implies, perforin makes holes in membranes and allows the penetration of another cytotoxic T cell product, called **granzyme**, into the cytoplasm of the infected cell. Granzyme stimulates a sequence of intracellular enzymatic reactions that lead to death (apoptosis) of the infected cell.

Cancer Surveillance

Human cancers or tumors arise in many different ways and consist of a large variety of types. Common to most is their production of proteins that differ from those found in the normal body. Because of the ability of the immune system to distinguish between self and nonself, cancer cells are susceptible to attack by the immune system.

Several immunological defense mechanisms may be employed in the fight against tumor formation, but the dominant cell involved is the CD8 + **cytotoxic T-lymphocyte**. These cells screen the body for peptides that do not conform to the normal array of class I MHC molecules expressed on the surfaces of all nucleated cells in the body. Often,

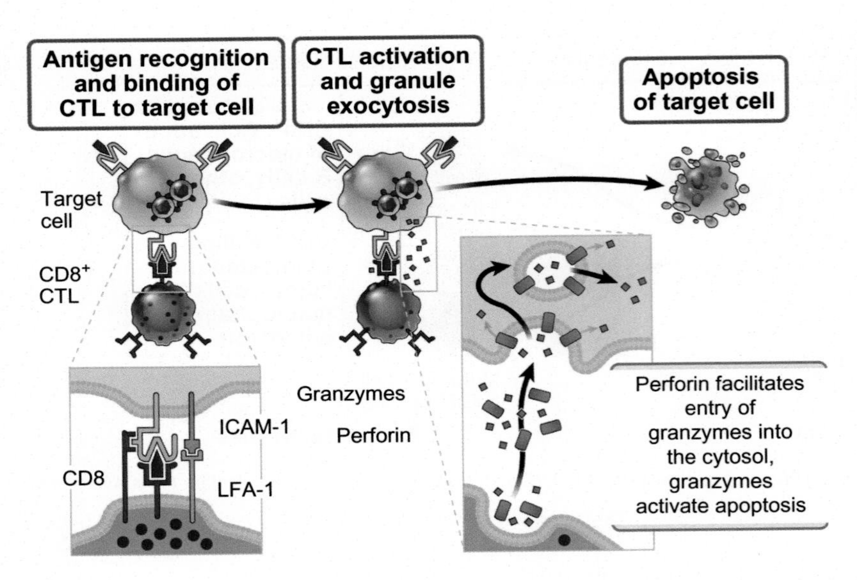

FIGURE 8.23 Killing of virally infected target cells by CD8$^+$ cytotoxic T-lymphocytes (CTLs). *ICAM-1*, intercellular adhesion molecule-1; *LFA-1* Leukocyte function-associated antigen-1. *From Abbas et al. (2016), with permission.*

however, tumor suppression begins with the ingestion of a tumor cell by a dendritic cell somewhere in the body (Fig. 8.24). The dendritic cell processes the foreign antigen of the tumor and displays the antigen on its surface, where it may come to the attention of a cytotoxic T-lymphocyte. Through the mediation of cytokines released from a helper T cell, the cytotoxic T-lymphocyte, which has attached to the tumor antigen on the dendritic cell, undergoes proliferation and differentiation. Its progeny are then released into the circulation, where they seek out other tumor cells of the same antigenicity. When such cells encounter tumor cells, they bind to them and kill them in the manner described above for their interaction with virally infected cells.

This is not the only way in which the body fights tumor cells. In some cases, NK cells of the innate immune system are able to bind to and destroy tumor cells. In addition, macrophages or dendritic cells can alone phagocytize and destroy tumor cells.

Not all tumor cells are immunologically destroyed. Because of their great variety, several ways of evading an immune response are possible. One simply involves evading detection. Early groups of tumor cells are small and often do not elicit an inflammatory response. As a result, very few cytotoxic lymphocytes pass their way, and their presence is undetected. Others lose or fail to produce tumor antigens or produce immunosuppressive proteins. Some tumors grow so rapidly that they simply overwhelm the immune response.

Graft Rejection

In a manner similar to that involved in the defense against tumor cells, the immunological repertoire designed to protect self from nonself is marshaled into action when a tissue or organ from a different individual other than an identical twin is transplanted into the body. The cells of a foreign tissue graft produce class I MHC molecules of a different configuration from those of the body of the recipient. This is quickly recognized by CTLs, which mount a massive response against the transplant (Fig. 8.25). Some of the most potent foreign antigens are found in the walls of blood vessels, and frequently a major reaction of the immune system of the recipient is directed against the blood vessels within the transplant, causing the transplant to be rejected within days or weeks.

The standard way of dealing with the rejection problem is the use of immunosuppressive drugs. Over the years, a number of immunosuppressive drugs have been developed. They act in different ways. Most are directed toward interfering with some aspect of T cell recruitment or function. Some block cytokine production or the proliferation of cytotoxic T cell precursors. Others, such as monoclonal antibodies, are targeted at specific T cells, and cause their destruction or inactivation. A side effect of immunotherapy is a weakening of the immune system of the recipient and a resulting greater susceptibility to viral and fungal diseases.

Protection from Parasitic Infections

For people in developed countries, the rise in allergic reactions and asthma represents an immunological paradox. Much of it centers around IgE and the cells with which this immunoglobulin interacts. Over many years of evolution, humans

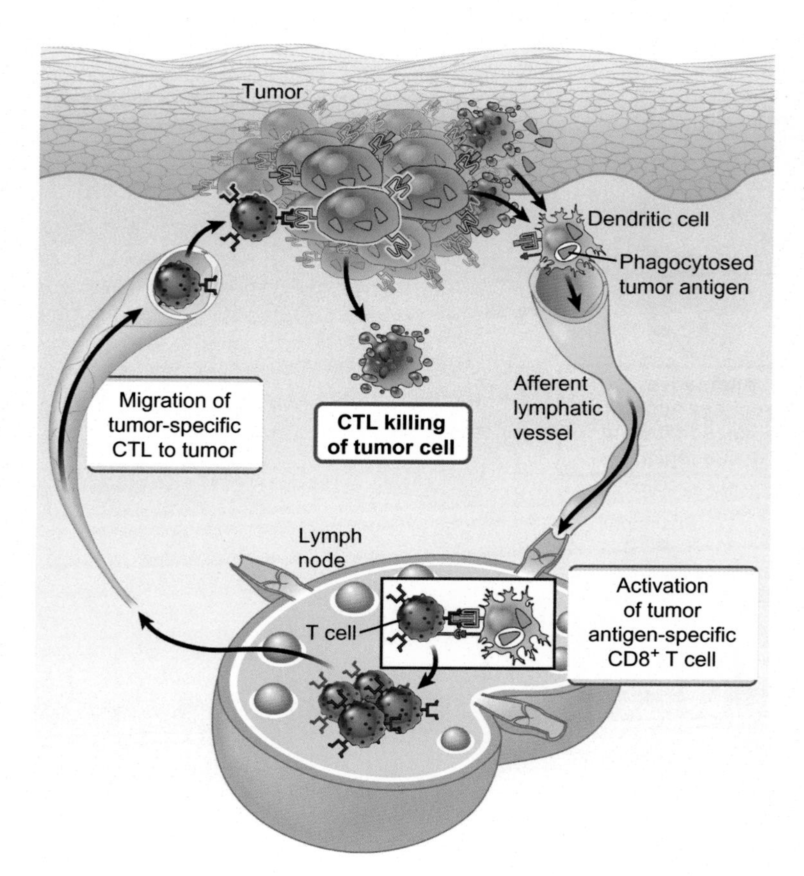

FIGURE 8.24 Induction of $CD8^+$ T cell responses against tumor cells. Dendritic cells pick up tumor antigens and take them to lymph nodes where tumor-specific CD8 + T cells are activated and released. From there they are transported do the site of the tumor, where they kill tumor cells. *From Abbas et al. (2016), with permission.*

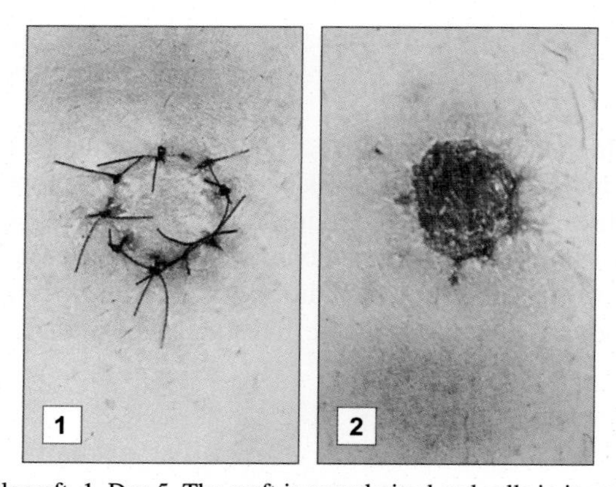

FIGURE 8.25 Rejection of a human skin allograft. 1. Day 5, The graft is vascularized and cells in it are dividing. 2. By 12 days, a rejection response has totally destroyed the graft. *From Male et al. (2013), with permission.*

developed a robust immunological defense mechanism involving both IgE and eosinophils, basophils, and mast cells to combat helminth (parasitic worms) infections. By improving sanitation, the threat of parasites in the developed world has greatly receded. Unfortunately, the lymphoid system does not understand that, and its considerable power to deal with parasitic worms now frequently results in misery for the many folks who suffer from allergic conditions.

The adaptive immune reaction to an intestinal helminth infestation begins with local dendritic cells picking up and displaying helminth antigens. In the case of **schistosome** parasites, antigens coming from the eggs are the most potent. Upon presentation of the egg antigens by dendritic cells, the principal responder to an intestinal helminth infection is a subset of helper T cells, called T_H2 cells. These cells respond to the parasitic antigens by secreting the cytokines IL-4,

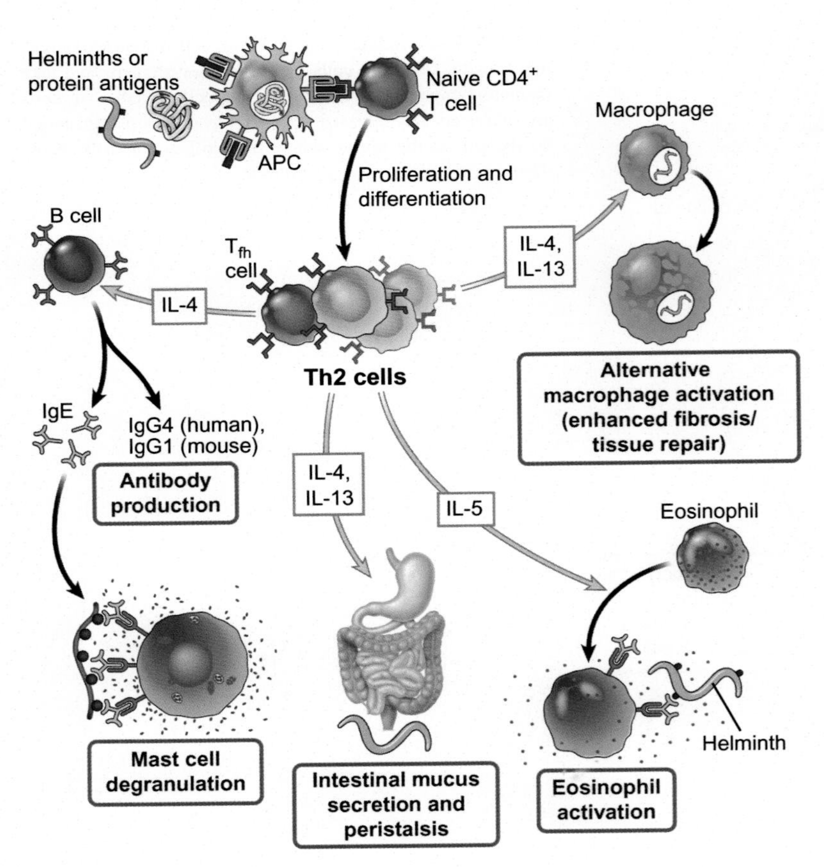

FIGURE 8.26 Functions of T_H2 cells in fighting a helminth infection. The T_H2 cells stimulate a variety of responses, ranging from degranulation of mast cells to stimulating mucus secretion and peristalsis by the intestine. *From Abbas et al. (2016), with permission.*

-5, and -13 (Fig. 8.26). IL-4 acts on B cells and causes them to switch from forming the more generic immunoglobulin IgM to producing and secreting IgE. IL-5 activates eosinophils and mast cells, which are direct participants in the fight against the parasites.

Most helminth parasites are too large to be phagocytized, and the IgE system is believed to have evolved as a mechanism for dealing with these larger invaders. IgE is the link that connects eosinophils to the outer surface of a parasitic worm. The large cytoplasmic granules of eosinophils contain a variety of proteases and toxic proteins that, in aggregate, can often kill larval intestinal parasites to which the eosinophils are attached. Mast cells, activated by IL-5 and coated by IgE also release molecules toxic to parasites, but in addition, the histamine released by expulsion of mast cell granules causes contractions of the smooth muscle cells of the intestinal walls as a physical mechanism for expelling the worms from the digestive tract.

Hypersensitivity

About 25% of the US population is afflicted with a variety of allergic conditions. Considerable evidence suggests that allergy results from the misdirection of the immune response would normally be directed against parasites. Epidemiological data suggest a correlation between parasitic infections and low allergy rates or high allergy rates in populations with minimal parasitic infestations. Of note, when groups of people have been treated for intestinal parasites, the rates of allergic symptoms have risen. Also, a number of the common human allergens are quite similar in structure to allergens found on parasites.

Although many aspects of human allergy continue to be debated, a common starting point is the recognition that allergic (called **atopic**) individuals produce far more IgE than do nonallergic individuals. This is because allergic individuals have a more active contingent of Th2 helper T cells than nonallergic people, who have a bias toward greater activity of Th1 helper T cells. As mentioned earlier, the cytokines secreted by Th2 cells activate eosinophils and stimulate B-lymphocytes to form IgE. In contrast, relatively greater Th1 activity results in the formation of greater numbers of IgG antibodies. What tips the balance between greater Th1 and Th2 activity is still under investigation.

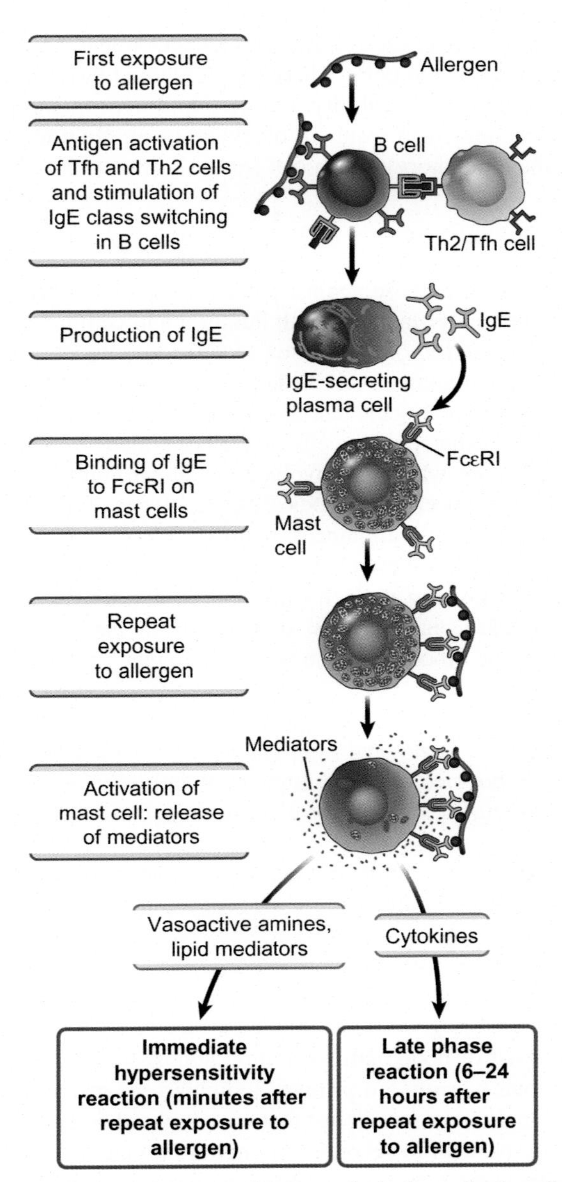

FIGURE 8.27 A hypersensitivity reaction. The first exposure to an allergen results in the production of a specific IgE. The IgE antibodies become bound to the surface of mast cells. Upon a repeat exposure to the allergen, the mast cells are quickly activated and release mediators of the hypersensitivity reaction. *From Abbas et al. (2016), with permission.*

In allergic individuals, overproduction of IgE has significant consequences. In the blood, IgE lasts only a day or so, but it also coats mast cells, where it remains on the surface for weeks. If an allergic person breathes in an allergen, for example, pollen, the allergen combines with the IgE antibodies on the surface of the mast cells residing in the mucosal lining of the nasal cavity and elsewhere (Fig. 8.27). This clusters the IgE receptors on the plasma membrane of the mast cells and results in the release of histamine and other substances from the mast cell granules. The immediate hypersensitivity reaction to histamine release in the nose is itching, swelling, a runny nose, and sneezing. Eosinophils and basophils take more time to react, since they are in the blood or the bone marrow. They respond to IL-5 secreted by the T_H2 cells by leaving the bone marrow or the blood over the course of days in what is called a delayed or chronic allergic reaction. The 2–3 day response to a tuberculin skin test (Mantoux test) is a good example of a delayed allergic reaction.

The worst case hypersensitivity scenario is anaphylactic shock, which can occur about after a second or later exposure to the same allergen, for example, a bee sting. Long-lived mast cells are already coated with IgE molecules that react to the allergen. If sufficient mast cells in the respiratory passages release their histamine, the smooth muscle surrounding these passageways can contract to the point where the individual has difficulty to breathing. In addition, mast cell secretions result in increased capillary permeability and the rapid accumulation of fluid in the tissues. This, combined with the smooth muscle contractions, can result in blockage of air flow to the lungs. On a long-term basis, the widespread removal of fluid from the circulatory system can seriously interfere with heart function.

SUMMARY

Innate immunity is the body's first line of defense against pathogens. If this defense is breached, the adaptive immune response comes into play. Adaptive immunity includes both humoral and cellular immune mechanisms.

Humoral immunity involves the formation of antibodies specific to antigens. In cellular immunity, lymphoid cells themselves act as defensive agents. An antigen is a molecule to which an antibody specifically combines.

The lymphoid system consists of cells and organs that are specifically designed to inactivate foreign antigens or pathogens. Lymphoid cells are derived from hematopoietic cells in the bone marrow. There are two types of lymphocytes—B cells, which produce antibodies, and T cells, which are involved in cellular immune responses. B-lymphocytes differentiate into plasma cells, which are the actual producer of antibodies. Several types of T-lymphocytes have been identified. Some are killer cells, and others serve as helper cells in immune responses.

Macrophages, derived from blood monocytes, engulf pathogens or foreign material by phagocytosis. They also secrete a large variety of cytokines (signaling molecules). Neutrophils, the most common type of white blood cell, migrate to sites of infection and release enzymes and oxidants from their cytoplasmic granules that kill bacteria. Eosinophils and basophils move from the blood into tissues and participate in hypersensitivity reactions and also defend against certain parasites.

Mast cells mediate allergic reactions. Dendritic cells present antigens to other cells of the lymphoid system.

Antibodies consist of identical pairs of heavy and light chains. There are several major classes of antibodies. IgM is attached to the surfaces of naïve lymphocytes. After immune stimulation, these cells produce IgG, the most prominent form of secreted antibody. IgA is associated with mucosal surfaces. IgE is secreted by B cells and binds to the surface of mast cells.

The complement system represents an evolutionary ancient component of the immune response. Complement proteins bind to surfaces of pathogens and make them more susceptible to destruction by other elements of the immune system.

Major histocompatibility proteins (MHC) present antigens to other members of the immune system. Class I MHCs are present on the surfaces of all normal cells. Infected cells present abnormal MHCs and are attacked by killer cells. Class II MHCs monitor the presence of foreign antigens in the extracellular environment.

Lymphoid tissues and organs consist of bone marrow, lymph nodes, spleen, thymus, and skin- or mucosa-associated lymphoid tissue. The thymus and bone marrow are primary lymphoid organs, and within them, immune cells undergo functional maturation. Secondary lymphoid organs house mature lymphoid cells that await antigenic activation.

Foreign antigens or pathogens are brought to secondary lymphoid organs by dendritic cells. This stimulates the final functional maturation of lymphocytes, which then either produce antigens or leave the secondary lymphoid organs and migrate to the site of infection.

In innate immune responses, complement fragments coat the surfaces of bacteria through a process called opsonization. Coated bacteria are then targeted for destruction by phagocytes, especially macrophages. In response to infections, macrophages produce cytokines that call other immune cells to the scene.

Inflammation is the result of vascular changes that cause lymphoid cells and fluid to accumulate at the site of an infection. Cardinal signs of inflammation are local redness, heart, swelling, and pain.

If a bacterial infection overwhelms the innate immune system, the adaptive immune system comes into play. The initial phase consists of capturing antigens and presenting them to other cells of the adaptive immune system. Individual lymphocytes have surface receptors specific for any form of antigen. Antigens are brought to local lymph nodes by dendritic cells. There is then a search for lymphocytes that have receptors specific for the particular antigen. Once that match has been made, the now-stimulated lymphocyte multiplies. Some daughter cells go on to produce antibodies; others remain as memory cells.

Viral infections affect individual cells and cause them to produce surface antigens that are different from normal. T-lymphocytes recognize these infected cells and destroy the infected cells. Similarly, cancer cells are recognized as foreign by the cellular immune system and are destroyed. A similar mechanism results in the destruction of foreign tissue grafts unless immunosuppressive agents are applied.

Basophils and mast cells evolved as a defense against certain helminth infections. Antigens from the parasites are presented by dendritic cells to T cells, which secrete cytokines that activate mast cells and basophils, which release histamine and other substances that result in the death or expulsion of the parasites.

In allergic individuals, overproduction of IgE coats mast cells, which then release large amounts of histamine when antigenically stimulated. Anaphylactic shock can occur after a second exposure to the antigens.

Chapter 9

The Endocrine System

The more we learn about the human body, the more we realize how interconnected and how integrated are most bodily functions. Obvious examples of whole-body interconnectedness are the nervous and circulatory systems because of their structural organization and distribution. Other interconnections are more subtle. The connective tissue linkages throughout the body are very structural, but their importance in the overall functioning of the body has only recently been generally recognized. Another set of linkages—the hormones produced by the endocrine glands—is principally chemical, but through the mediation of both the circulatory and nervous systems, endocrine secretions reach almost every cell of the body.

The endocrine system is often compared with the nervous system, both of which touch almost all components of the body, but aside from their wide distribution, their differences outweigh their similarities. Through its network of nerve fibers, the sensory component of the nervous system collects specific information from almost all areas of the body, but on the effector side, the direct action of the nervous system is limited to skeletal muscle fibers (through the motor end-plates), smooth and cardiac muscle, and glandular tissue. Endocrine glands, on the other hand, respond to signals carried to them by the blood, lymph, or nerve fibers, but their hormonal secretions are carried to cells throughout the body. Although almost all cells of the body are exposed to circulating hormones, only if they have appropriate receptors to specific hormones will they respond. Whereas signaling by the nervous system is measured in ms, hormonal effects can take from seconds to even days to be felt.

Classical endocrinology was focused on the grossly identifiable endocrine glands (Fig. 9.1), but researchers have increasingly recognized that many cells and tissues secrete powerful hormones that affect both nearby and distant cells. Hormonal effects are described as **endocrine** (working at a distance and carried mainly through the blood), **paracrine** (diffusion in tissue fluids to nearby cells and tissues), or **autocrine** (affecting the cells that actually produced the hormones) (Fig. 9.2). Characteristic of all endocrine glands (in contrast to exocrine glands) is the absence of a duct system for removal of the hormonal secretions from the gland. Instead, endocrine glands are endowed with an abundant microcirculation that surrounds the hormone-producing cells (Fig. 9.3). The secreted hormonal products are quickly and efficiently taken up by the local capillaries and from there are carried by the blood to all parts of the body.

Regardless of their mode of transport and their targets, hormones follow a similar generic pathway between their synthesis and ultimate degradation (Table 9.1). Box 9.1 describes in greater detail how hormones exert their effects at a cellular and molecular level. Hormones are of two fundamental biochemical types—either steroid or amino acid based. Those in the latter category range from single amino acids to proteins. In addition to the major hormones listed in Table 9.2, increasing numbers of local hormones, which often act in a paracrine manner, are being identified.

A significant number of the major hormones are products of a hierarchical system in which one organ sends a hormonal signal to another endocrine organ, which responds to the signal by producing its own hormone. The hierarchy begins in the cortical regions of the brain, where sometimes even conscious neural signals stimulate a hormonal response (Fig. 9.6). The overall pathway then proceeds to the hypothalamus, which produces a variety of peptide hormones that are carried to the pituitary gland (hypophysis). The pituitary, in turn, produces and secretes its own hormones, which stimulate the secretion of specific hormones by peripheral endocrine glands (e.g., adrenal cortex, gonads, thyroid).

In a hierarchical endocrine system, hormones from higher in the hierarchy stimulate or inhibit the synthesis or secretion of other hormones produced at the next lower level. Then through a complex series of feedback mechanisms, the hormones secreted by the peripheral endocrine glands at the lowest level of the hierarchy send secretory or inhibitory signals that control the synthesis and secretion of hormones upstream in the hierarchy. Especially among nonhierarchical endocrine glands, hormone secretion is strongly influenced by factors such as the concentrations of ions or metabolites (e.g., glucose) in the blood. Neuronal signals and mental activity modify the activity of many hormones. Even general environmental factors (e.g., light and temperature) can exert a strong influence on the activity of certain hormones.

The Human Body. DOI: https://doi.org/10.1016/B978-0-12-804254-0.00009-0

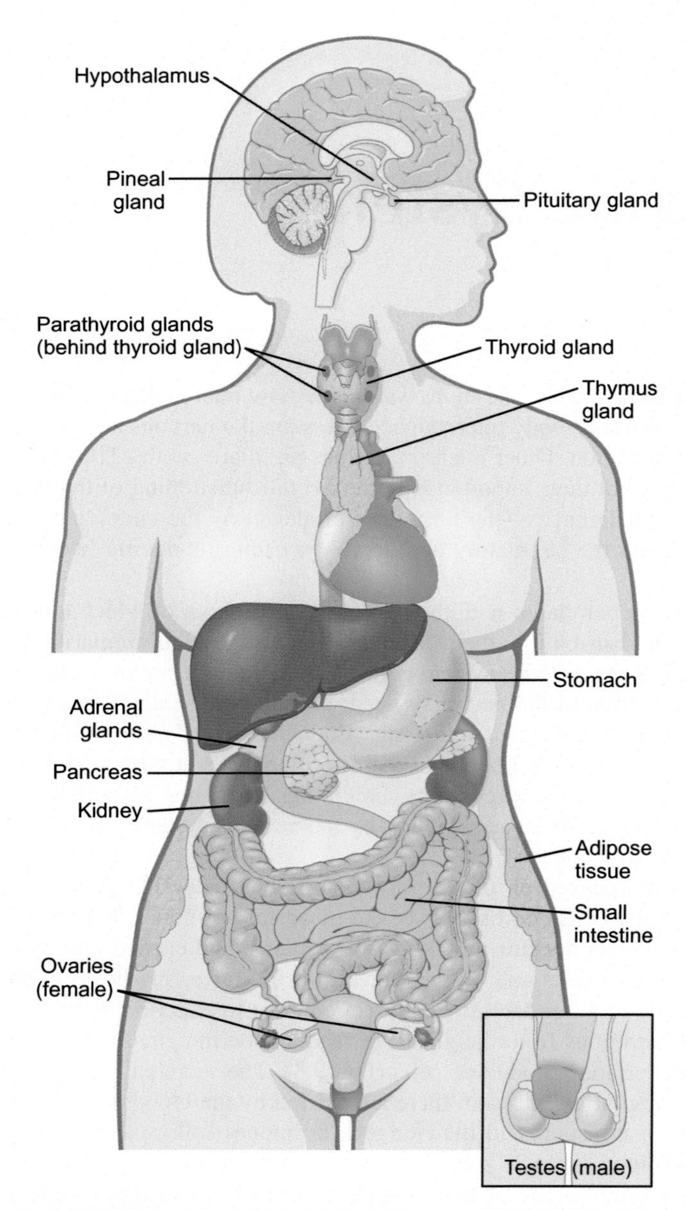

FIGURE 9.1 Locations of major endocrine glands throughout the body. *From Guyton and Hall (2006), with permission.*

EVOLUTION OF THE ENDOCRINE SYSTEM

Unlike many other organ systems of the body, the endocrine system did not evolve as a coordinated unit. Most endocrine secreting cells and glands arose from a need to integrate certain bodily functions or to maintain homeostasis. The biochemical roots of many of our hormones go back to the earliest single-celled organisms. For example, enzymes of the type used in the formation of steroids are even found in bacteria. Complex hormonal systems are found not only in vertebrates, but also in invertebrates and plants.

The essence of the evolution of endocrine systems is the repurposing of a variety of molecules and structures to meet the needs of bodies that are changing over the eons as they adapt to meet new environmental challenges. The more complex the body, the greater the need for internal coordinating functions, such as neural or endocrine controls.

Steroid and nonsteroid hormones followed separate evolutionary pathways. Steroids are made by most unicellular organisms—a testament to their early origins. Although many of the biochemical pathways leading to the formation of specific steroid molecules are quite similar, the functions of steroid hormones vary tremendously among different organisms. Some of the many steroids found in fungi and higher plants are involved in growth and flowering. In insects,

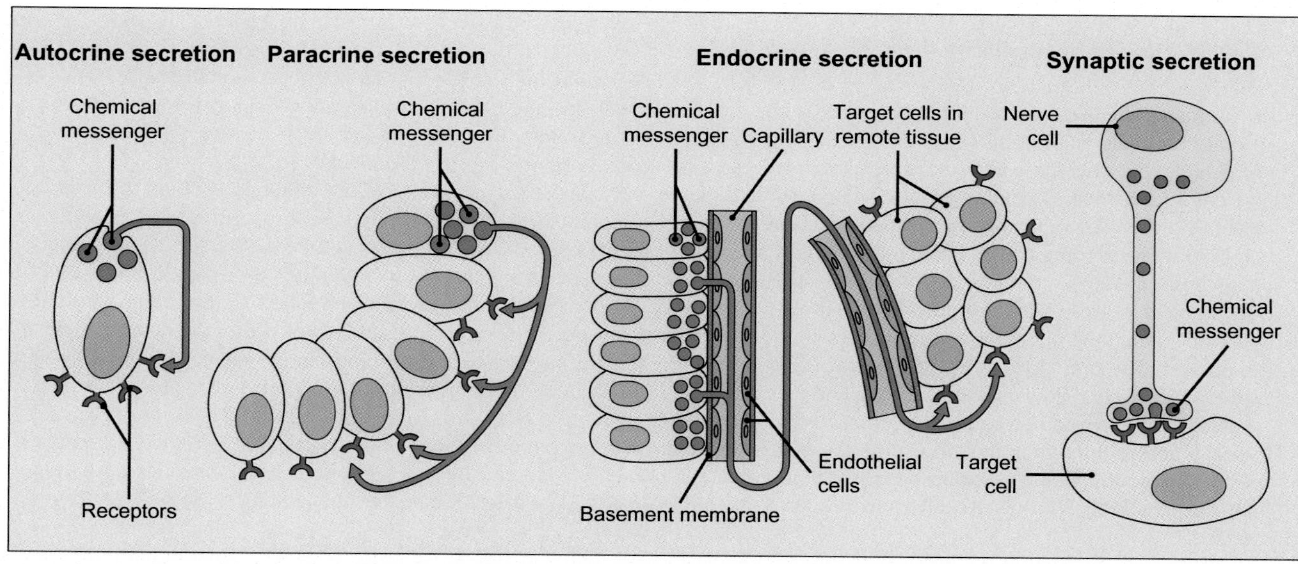

FIGURE 9.2 Types of chemical messaging among cells. *From Stevens and Lowe (2005), with permission.*

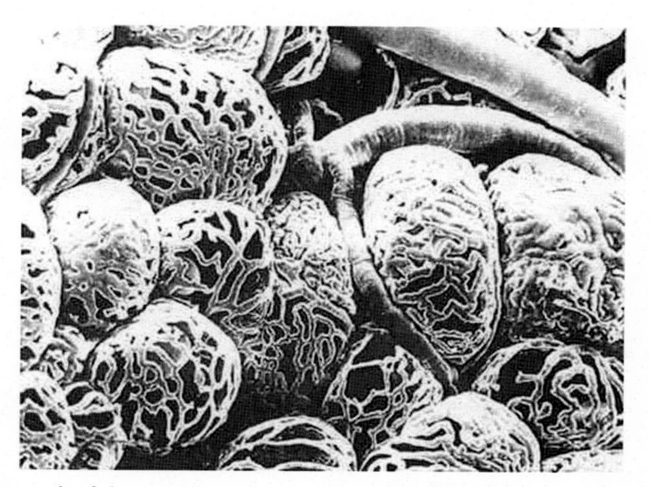

FIGURE 9.3 Scanning electron micrograph of dense capillary networks surrounding follicles in the thyroid gland. *From Erlandsen and Magney (1992), with permission.*

TABLE 9.1 The Life Cycle of a Generic Hormone
1. Biosynthesis
2. Storage
3. Secretion and activation
4. Transport to target cells
5. Binding to receptors—either on plasma membrane or in nucleus
6. Signal transduction and amplification of signal
7. Response by target cell
8. Degradation of hormone

BOX 9.1 How Hormones Exert Their Effects on Cells

A hormone can only exert an effect on a cell if it binds to a receptor molecule specific for that hormone. Even in the presence of the hormone, there can be no hormonal effect in the absence of such specific binding.[1] Individual cells may have receptors for several hormones, but not all cells have the same combinations of receptors. Blood-borne hormones are exposed to most cells of the body, but only those with the receptors specific for a particular hormone can react to them.

Cells interact with hormones in two fundamentally different ways. Those that are proteins or peptides must bind to receptors located on the plasma membranes of cells because they are hydrophilic and cannot readily pass through membranes without help. Hormone binding causes changes in the configuration of the receptor, which are translated through a series of molecular interactions, called signal transduction, into a variety of cellular responses, ranging from secretion of stored molecules to alterations in gene expression. These responses typically occur very rapidly. Steroid and thyroid hormones, on the other hand, are able to penetrate the plasma membrane. They then bind to hormone-specific receptors located either within the cytoplasm or in the nucleus. The hormone–receptor complex binds to specific sites on the DNA of the chromosomes, where it influences gene expression. This is a slower process, taking hours or days. Details of both modes of hormone action on cells are given later.

Steroid and thyroid hormones are transported through the circulation in two forms—free and bound. Free hormones are biologically active, but within the circulation the free forms deteriorate quite rapidly. Most of the circulating hormone is bound to specific hormone-binding globulins or to serum albumin. The bound, but inactive form of a hormone has a much longer half-life than the free form. Through equilibrium reactions, bound hormone is released when the concentration of the free form is reduced.

Both steroid and thyroid hormones pass through the plasma membrane of cells by diffusion (Fig. 9.4). Thyroid hormone goes directly to the nucleus, where it attaches to a nuclear hormone receptor. The receptor is bound to a segment of DNA known as a **hormone response element**, which is typically located upstream of a gene that is regulated by the hormone. Depending on the gene, the action of the receptor on the hormone response element can either activate or repress transcription (the formation of mRNA). Steroid hormones represent a variant on this theme in that they commonly bind to receptors located in the cytoplasm, rather than the nucleus. The bound receptors then move into the nucleus, where they also exert their effects on hormone response elements.

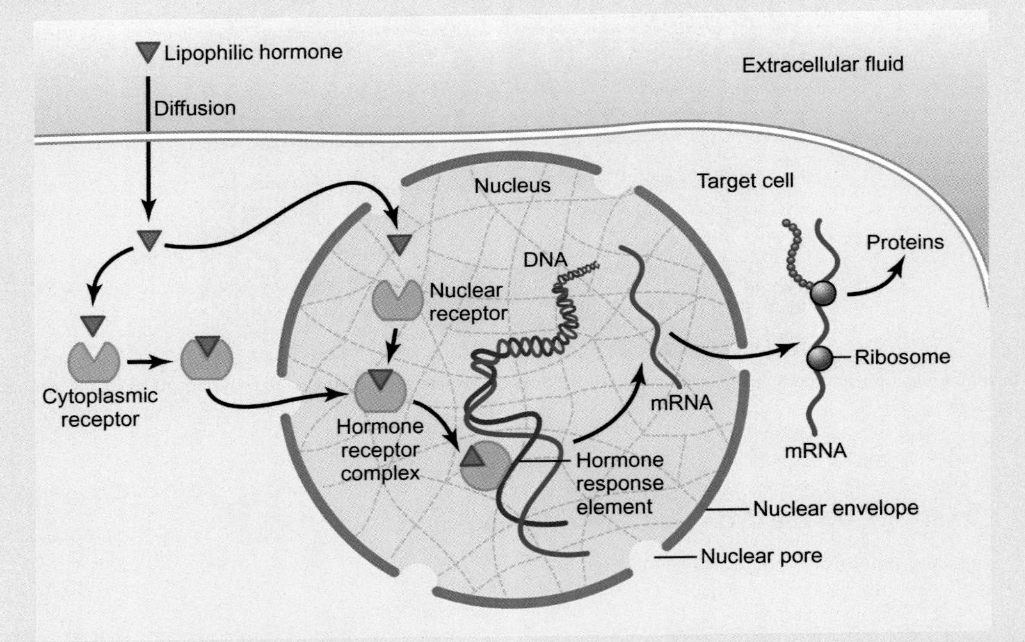

FIGURE 9.4 Steps in the uptake of a steroid hormone by a cell and the cell's response. After diffusion through the plasma membrane, the hormone could either bind to a cytoplasmic receptor and make its way into the nucleus or directly enter the nucleus and bind to a nuclear receptor. After it interacts with a specific region of DNA, the DNA then produces an appropriate mRNA, and synthesis of a new protein is initiated. *From Guyton and Hall (2006), with permission.*

(*Continued*)

BOX 9.1 (Continued)

Peptide and protein hormones do not readily cross plasma membranes. Instead, they bind to membrane-bound cell surface receptors (also proteins) that contain three domains: (1) an extracellular binding site, (2) one to seven transmembrane domains where the receptor molecule loops through the plasma membrane, and (3) a cytoplasmic domain, where the receptor is linked to various signaling proteins (Fig. 9.5). When a peptide hormone attaches to its receptor outside the cell, the configuration of the receptor commonly changes in all three of its domains. These changes activate a signal transduction pathway(s) in which the binding of the hormone is translated into some type of cellular response. Several such pathways exist, and more than one may be activated by a single hormonal signal. The net result may be a change in some cytoplasmic function, such as the stimulation of secretion, activation of enzymes, or even changes in the cytoskeleton. Other pathways lead to the nucleus, where they affect the function of transcription factors (see p. 8) that control the activity of specific genes. These latter nuclear responses take a longer time for implementation than most cytoplasmic responses.

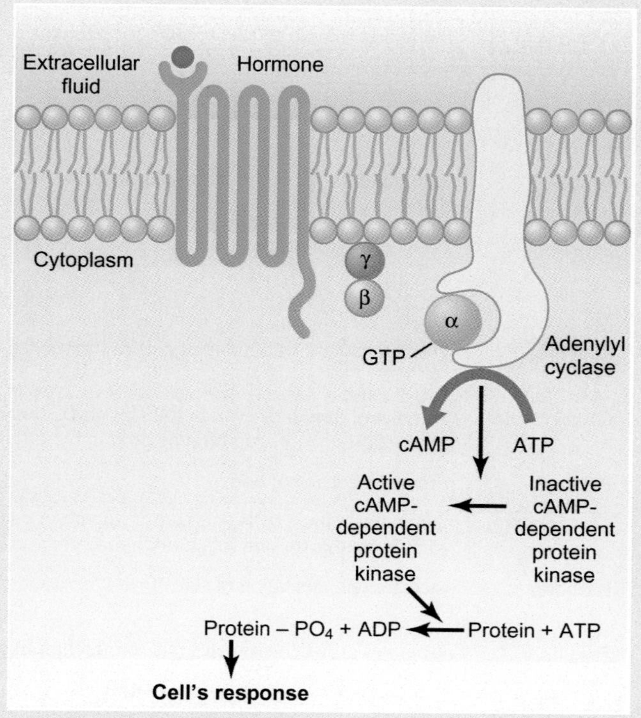

FIGURE 9.5 A typical interaction of a cell with a protein hormone. The hormone binds to a transmembrane receptor. This releases a molecular cascade that ultimately produces a cellular response. *ADP*, Adenosine diphosphate; *cAMP*, cyclic adenosine monophosphate; *GTP*, guanosine triphosphate. *From Guyton and Hall (2006), with permission.*

1. As an example, genetic males who are lacking in testosterone receptors (**testicular feminization syndrome**) develop the external configuration of a woman, including well developed breasts and fat deposit characteristics of females, even though they have high concentrations of testosterone in their blood.

the complex changes involved in metamorphosis are coordinated by a set of steroid hormones. Steroids in vertebrates are mostly involved in reproduction and homeostasis.

Nonsteroid hormones, usually amino acid based, have similarly been put to use in a large variety of functions. A classic example is the pituitary hormone prolactin, a protein hormone that comes in different sizes. Among the vertebrates, prolactin is known to have as many as 300 different functions, as diverse as osmoregulation in fishes, stimulating the drive to water in some terrestrial newts, regulating secretions from the crop in pigeons, and coordinating lactation in mammals. Similarly, thyroid hormone, which serves more generalized metabolic functions in humans, is required for stimulating metamorphosis in amphibians (whereas metamorphosis in insects is mediated by steroids).

How a hierarchical control system evolved for some hormones remains a matter of debate. One hypothesis involves "sequential capture." In this model, control of the secretion of hormones by peripheral endocrine glands, such as the

TABLE 9.2 Summary of Major Hormones

Name	Chemistry	Functions
Hypothalamus		
Corticotropin-releasing hormone (CRH)[a]	Peptide	Stimulates release of ACTH from anterior pituitary
Growth hormone-releasing hormone (GHRH)	Peptide	Stimulates release of GH from anterior pituitary
Gonadotropin-releasing hormone (GnRH)	Peptide	Stimulates release of FSH and LH by anterior pituitary
Somatostatin (SS)	Peptide	Inhibits secretion of GH and other hormones
Thyrotropin-releasing hormone (TRH)	Peptide	Stimulates release of TSH by anterior pituitary
Pituitary (Hypophysis)		
Posterior pituitary (neurohypophysis)		
Oxytocin (OT)	Peptide	Stimulates uterine contraction during labor; stimulates ejection of milk by mammary gland
Vasopressin (antidiuretic hormone (ADH))	Peptide	Causes vasoconstriction; increases water resorption by kidney
Anterior Pituitary (adenohypophysis)		
Adrenocorticotropic hormone (ACTH)	Protein	Promotes synthesis of adrenal cortical hormones
Follicle-stimulating hormone (FSH)	Glycoprotein	Stimulates growth of ovarian follicles and estrogen production; stimulates production of androgen-binding protein by testis
Growth hormone (GH)	Protein	Promotes growth of body
Luteinizing hormone (LH)	Glycoprotein	Stimulates follicle cells and corpus luteum in ovary to produce progesterone; stimulates testosterone secretion in males
Melanocyte-stimulating hormone (MSH)	Peptide	Increases pigmentation of skin
Prolactin	Protein	Promotes growth; stimulates milk production in mammary glands
Thyroid-stimulating hormone (TSH)	Glycoprotein	Stimulates secretion of thyroid hormones
Thyroid		
Triiodothyronine (T_3)	Amino acid	Regulates metabolic rate
Thyroxine (T_4)	Amino acid	Regulates metabolic rate
Calcitonin	Peptide	Causes decrease in blood Ca^{++} level
Parathyroid		
Parathyroid hormone (PTH)	Protein	Causes increase in blood Ca^{++} level
Endocrine Pancreas		
Glucagon	Protein	Increases blood sugar levels
Insulin	Protein	Decreases blood sugar level
Gonads		
Ovary		
Activin	Protein	Stimulates growth of ovarian follicles

(*Continued*)

TABLE 9.2 (Continued)

Name	Chemistry	Functions
Estrogen	Steroid	Stimulates cellular proliferation in reproductive tissues
Human chorionic gonadotropin (hCG)	Glycoprotein	Maintains activity of corpus luteum during pregnancy
Inhibin	Protein	Inhibits FSH secretion by pituitary
Progesterone	Steroid	Stimulates reproductive glandular secretions, maturation of female reproductive tissues
Testis		
Dihydrotestosterone	Steroid	Stimulates development of male reproductive tissues
Inhibin	Protein	Inhibits FSH secretion by pituitary
Testosterone	Steroid	Stimulates development of male reproductive tissues
Adrenals		
Cortex		
Aldosterone	Steroid	Control of ion balance by kidneys and other organs; blood pressure control
Androgens	Steroid	Similar functions to testosterone, but less potent
Cortisol	Steroid	Regulation of glucose; anti-inflammatory actions
Medulla		
Epinephrine	Amine	Rapid response to stressful situations
Norepinephrine	Amine	Rapid response to stressful situations
Heart		
Atrial natriuretic hormone (ANH)	Peptide	Vasodilation; diuretic effect on kidneys
Kidney		
Renin	Protein	Increases blood pressure and perfusion pressure in kidneys
Adipose Tissue		
Leptin	Protein	Promotes sense of satiety
Small Intestine		
Cholecystokinin (CCK)	Peptide	Mediates digestion in small intestine
Pineal Gland		
Melatonin	Amine	Entrainment of circadian rhythms

[a]*Commonly used abbreviations are given in parentheses.*

thyroid, adrenal, and gonads, was taken over by the anterior pituitary gland and the various stimulatory hormones that it produced. Then at a later time, the pituitary itself was captured by the hypothalamus, and the secretion of pituitary hormones was placed under the control of the hypothalamic releasing hormones. A second hypothesis suggests that preexisting ligand (hormone)-receptor systems can be pieced together over time to form new functioning endocrine networks. This hypothesis requires that organisms possess scattered, but not linked components that can form the basis of a new network.

Even at the organ level, complex endocrine glands show evolutionary mix-and-match patterns. Lampreys have scattered chromaffin tissue (precursors of the adrenal medulla), but the tissue source of adrenal cortical hormones is not clear. Fishes possess both adrenal cortical and medullary tissues, but they are scattered. Amphibians have multiple adrenal glands, but each component contains both cortical and medullary tissues. Not until reptiles, birds and mammals are

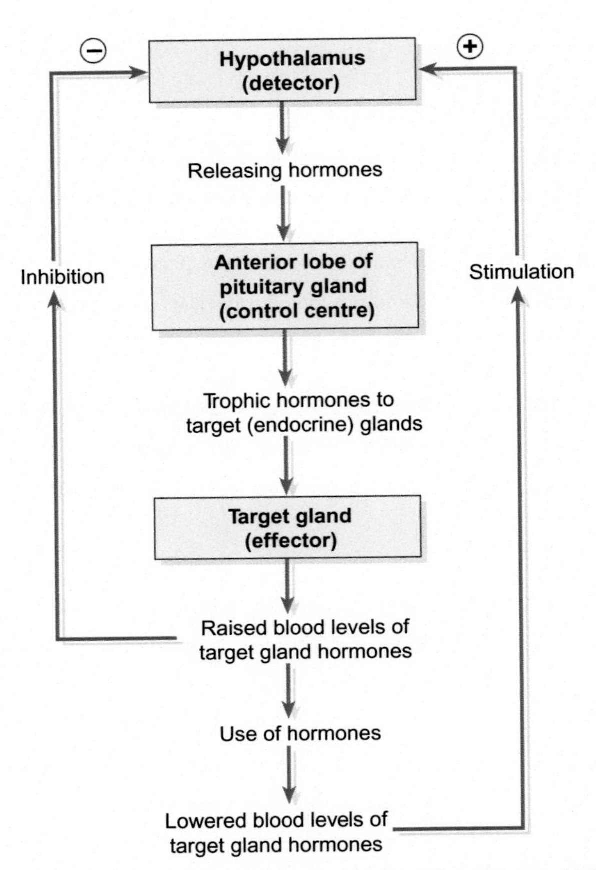

FIGURE 9.6 The endocrine gland hierarchy, including positive and negative feedback from lower to higher levels of the hierarchy. *From Waugh and Grant (2014), with permission.*

unified paired adrenal glands found. The pancreas follows a different pattern, where the separate endocrine and exocrine components found in fish give way to a unified pancreas in higher vertebrates in which the islets of Langerhans (endocrine) are scattered among the exocrine cells in the body of the gland

THE HIERARCHICAL HORMONE SYSTEM

Hypothalamus

The **hypothalamus** sits at the top of the hierarchical system of hormone producers. Yet it, too, relies on external signaling to modulate its functioning. Although not an endocrine gland, the cerebral cortex can exert powerful influences on the hypothalamus and its hormonal secretions. Two examples illustrate the effect of cortical input on the hypothalamus. Nursing mothers rely upon the hormone oxytocin, which stimulates the milk letdown reflex. Oxytocin is produced by neurons in the hypothalamus and is carried to the posterior pituitary gland, from which it is secreted into the blood. Although normally brought about by suckling, the milk letdown reflex can sometimes be elicited simply by seeing or hearing other babies. Conversely, anxiety can inhibit this reflex. Similarly, acute or chronic stress is mediated through the cerebral cortex and transmitted to a different part of the hypothalamus, which then produces a hormone that acts on the anterior pituitary gland, the next level down in the hierarchy and finally activates the release of hormones from the adrenal cortex. The hypothalamus also responds to blood levels of several hormones, which provide feedback upon which the hypothalamus produces more or less of its own intrinsic hormones.

Parts of the hypothalamus act as a neurosecretory organ, in which neurons themselves produce and secrete blood-borne hormones. Large neurons in the **paraventricular** and **supraoptic nuclei** of the hypothalamus synthesize **oxytocin** and **vasopressin** (see Fig. 6.15). After synthesis in the neuronal cell bodies within the hypothalamus, the hormones are transported down axons which lead into the posterior pituitary gland (neurohypophysis). Once in the axon terminals within the neurohypophysis, the hormones remain there until an action potential traveling down the axon stimulates their release into the blood. The production apparatus of oxytocin-producing neurons is strongly influenced by external events. After childbirth, these neurons hypertrophy and develop a more complex system of neuronal processes in order to accommodate the body's greater need for oxytocin during lactation.

Smaller neurons in other hypothalamic nuclei are specialized for the synthesis and secretion of releasing hormones that stimulate the anterior pituitary to secrete its own stimulatory hormones (see Fig. 6.15). In addition, other hypothalamic cells produce **dopamine**, which, instead of stimulating, represses the release of the pituitary hormone prolactin. In contrast to those that send axons directly to the posterior pituitary, the hypothalamic neurons that influence the anterior pituitary extend axonal processes only as far as the median eminence of the hypothalamus at the base of the neck of the pituitary gland. There they terminate, and their secretions enter a specialized venous system (**hypophythalamo-hypophyseal portal system**, see p. 152) that carries the releasing hormones to a specific target—the anterior pituitary—rather than broadcasting them throughout the entire systemic circulation. Such a confined system has the efficiency of allowing a smaller amount of releasing hormones to be produced. Like the hypothalamic hormones that are released from the neurohypophysis, control of release of those targeting the adenohypophysis is subject to many varieties of input into the hypothalamus. Some of the control is feedback in response to blood levels of circulating hormones. In other cases, notably corticotropin-releasing hormone, input from higher brain centers, especially under situations of stress, plays a major role.

Pituitary Gland (Hypophysis)

The **pituitary gland** is the next level down from the hypothalamus in the hierarchical system of endocrine control within the body. Even from its earliest embryonic origins, the **hypophysis** is a two-part gland, with one part (**neurohypophysis**) originating as a downgrowth from the brain and the other (**adenohypophysis**) as an outpocketing from the roof of the embryonic mouth. These different characteristics are retained throughout life. In many respects the neurohypophysis can be looked upon as an extension of the hypothalamus (see p. 152), where actual secretion of the neurohypophyseal hormones occurs. Its main structural specialization is a dense capillary network that surrounds the nerve terminals and allows the neurosecretory products (oxytocin and vasopressin) ready access to the blood.

The adenohypophysis is constructed more like a typical endocrine gland. About the size of a pea, it contains several types of cells, each of which responds to a different hypothalamic releasing factor and, in response, produces and secretes its own hormone (Fig. 9.7). Developmentally, three lineages of hormone-secreting cells arise from its embryonic precursors.

One lineage produces cells that secrete GH, prolactin and TSH (see Table 9.2). **Growth hormone** has a general anabolic effect on many cells in the body. Because of this property, GH analogs have been used—often illegally—as performance-enhancing drugs by athletes. An excess of GH during one's growth period results in gigantism, whereas a deficit results in **pituitary dwarfism** (Fig. 9.8A). After normal growth has ceased, an excess of GH (usually due to a pituitary tumor) causes **acromegaly**, a condition characterized by excessive growth of the hands, feet, nose, and jaw (Fig. 9.8B).

TSH has a more specific effect, namely stimulating the release of thyroid hormones from the thyroid gland. **Prolactin** is an ancient hormone that has many functions in various vertebrate animals. In humans, its principal target is the mammary glands, where it promotes milk production (see Fig. 14.22), but it has many other functions, as well. In fishes and amphibians it plays an important role in salt and water balance. Secretion of GH and TSH is stimulated by growth hormone-releasing hormone (GHRH) and thyrotropin-releasing hormone (TRH) that are carried to the adenohypophysis from their origin in the hypothalamus via the hypothalamohypophyseal portal system. Secretion of all three of the hormones in this pituitary lineage is inhibited by the hypothalamic hormone **somatostatin**. Prolactin is unique among the pituitary hormones in that its control by the hypothalamus is inhibitory, mainly through dopamine, and also by somatostatin.

A second pituitary lineage consists of cells that produce the gonadotropic hormones, **follicle-stimulating hormone** and **luteinizing hormone**. Production of both of these hormones is stimulated by gonadotropin-releasing hormone (**GnRH**) coming down from the hypothalamus. Especially in the female, the secretion of FSH and LH plays a critical role in the reproductive cycle (see Fig. 9.17). Stimulatory and inhibitory control occurs at a number of levels within the overall hierarchy of endocrine glands. In females, both FSH and LH target the ovaries, where they play complementary roles in the menstrual cycle (see Fig. 9.17). FSH exerts its effects mainly during the first half of the cycle, whereas a peak of LH secretion stimulates ovulation and drives changes in reproductive tissues during the postovulation phase of the cycle. In males, FSH and LH exert their effects on cells in the testes.

The third pituitary cell lineage includes cells that secrete **adrenocorticotropic hormone** and **melanocyte-stimulating hormone** (MSH). ACTH represents the intermediate hormone in what is often called the

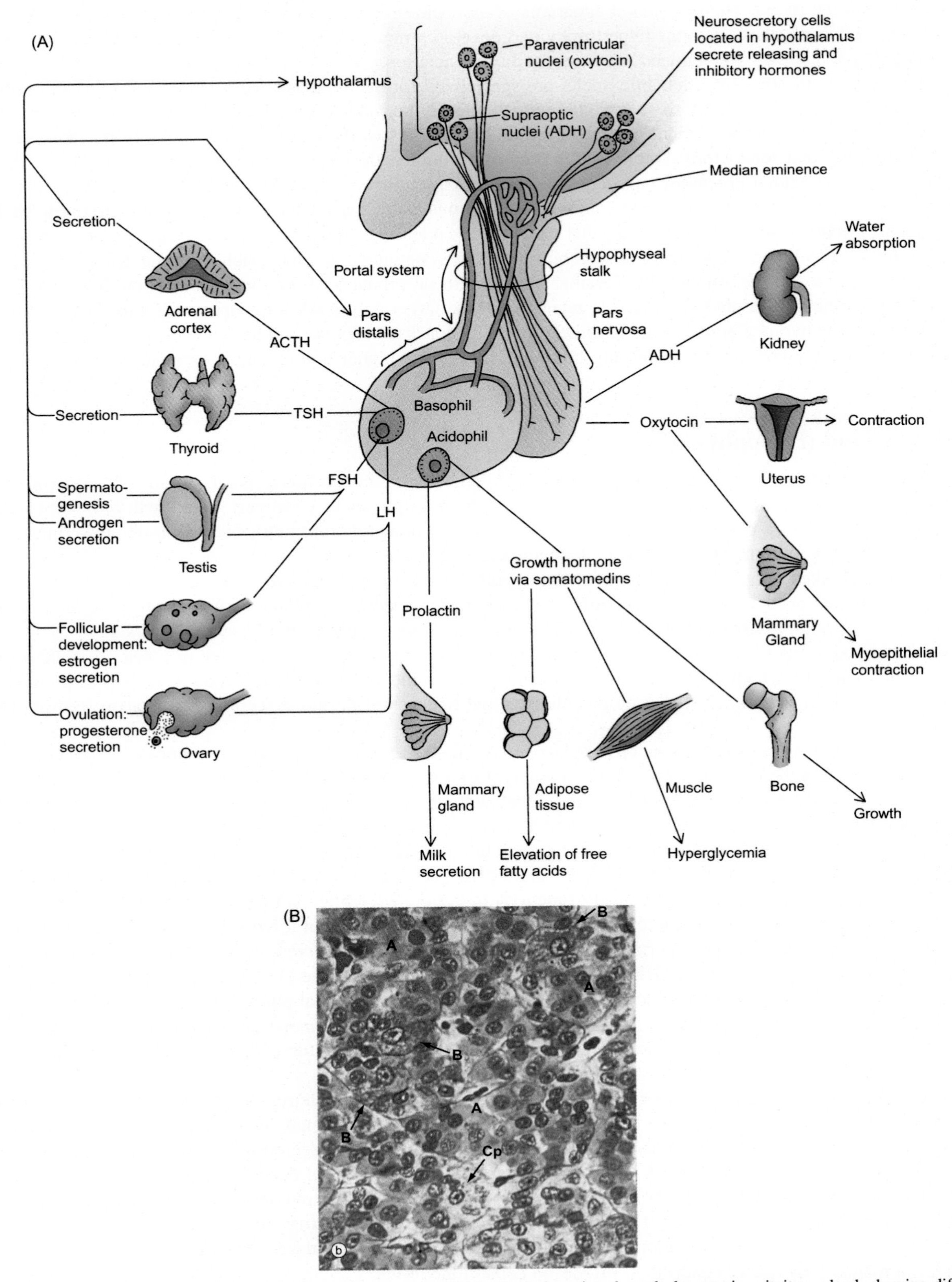

FIGURE 9.7 (A) Effects of the pituitary gland on other organs. (B) Histological section through the anterior pituitary gland, showing differential staining. *A*, Acidophils (red); *B*, basophils (purple); *Cp*, chromophobes. Immunostaining, rather than chemical dyes, is now used to localize specific hormones and their secretory cells. *(A) From Gartner and Hiatt (2007), with permission. (B) From Young et al. (2006), with permission.*

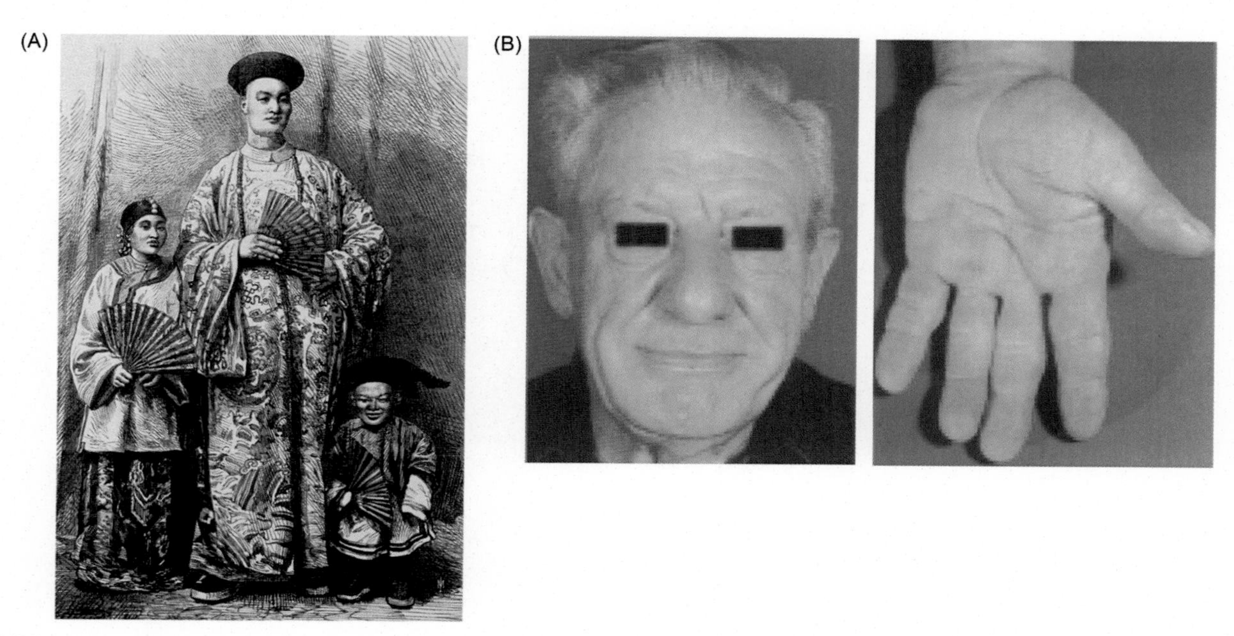

FIGURE 9.8 (A) Historical picture showing a pituitary giant (2.3 m tall) and a dwarf (0.9 m tall) alongside a person of normal height (left). (B) A case of acromegaly, showing coarsening of facial features through growth of the frontal and mandibular bones and enlargement of the bones and soft tissues of the hands. *(A) From Waugh and Grant (2014), with permission. (B) From Chew and Leslie (2006), with permission.*

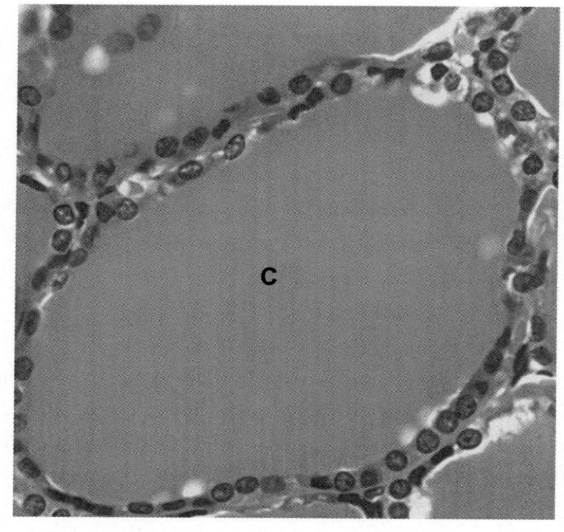

FIGURE 9.9 Histological section through a thyroid follicle. *C*, Colloid, representing stored thyroglobulin. The epithelial cells surrounding the follicle are cuboidal in form. *From Young et al. (2006), with permission.*

hypothalamus—pituitary—adrenal axis, which will be described in more detail on p. 255. MSH is a product of what in some animals is a distinct **intermediate lobe** of the pituitary gland. As its name implies, MSH stimulates the synthesis of the dark pigment melanin (see p. 75), and in amphibians and fishes it contributes toward the darkening of the skin when they live in environments with dark backgrounds. Although its functions are attenuated in humans, MSH levels rise during pregnancy and are responsible for the increased pigmentation in many pregnant women.

Thyroid Gland

Located in the neck and partially enwrapping the trachea, the richly vascularized thyroid consists of large numbers of small thyroid follicles embedded in a connective tissue stroma. Each follicle is lined by a simple cuboidal epithelium, and the cavity is filled with a material commonly called colloid (Fig. 9.9).

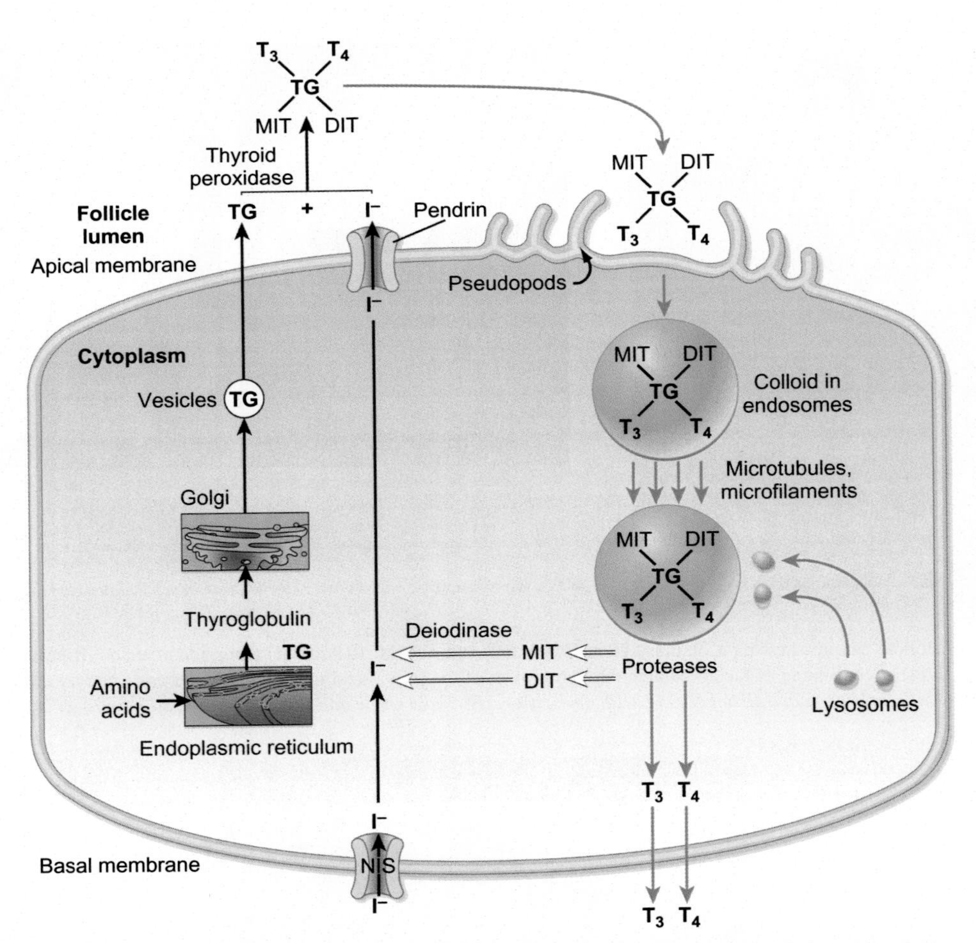

FIGURE 9.10 Synthesis (black arrows) and secretion (orange arrows) of thyroid hormones by a thyroid epithelial cell, starting with the synthesis of thyroglobulin from amino acids on the rough endoplasmic reticulum. *DIT*, Diodothyrosine; *I*, Iodide; *MIT*, Monoiodothyrosine; *T_3*, triiodothyronine; *T_4*, tetraiodothyronine; *TG*, thyroglobulin. *From White and Porterfield (2013), with permission.*

The synthesis and release of thyroid hormone follows an unusual pattern that includes both intracellular and extracellular events. **Iodine** is taken up from the circulation by the follicular epithelial cells and is passed through them into the **colloid** material in the lumen of the follicle (Fig. 9.10). At the same time, the **thyroglobulin** protein is synthesized in the rough endoplasmic reticulum, modified by the addition of sugar groups in the Golgi apparatus and then secreted into the lumen. Within the colloid, iodine is incorporated onto certain tyrosine (an amino acid) groups contained in the thyroglobulin molecule. The iodinated thyroglobulin remains in the colloid as a storage form until needed. When the need arises, iodinated thyroglobulin is taken up from the colloid in the form of membrane-bound endosomal vesicles, which soon fuse with lysosomes. The lysosomal enzymes degrade the thyroglobulin into individual amino acids. Some of these are iodinated tyrosine molecules, which have bound three (T_3) or four (T_4) iodines. **T_3 and T_4** (generically called **thyroxine**) are then passed into capillary networks that surround the thyroid follicles. An insufficiency of dietary iodine can lead to the formation of a **goiter** (Box 9.2).

Within the circulation, almost all of the thyroid hormone (>99%) is bound to plasma proteins. Only the small amount of free T_3 and T_4 in the circulation represents active thyroid hormone. The protein-bound thyroxine molecules represent a reserve that can be called into play when needed. T_3 is by far the most potent form of thyroxine, but 90% of the thyroxine released from the thyroid follicle is T_4. Mechanisms exist at the level of peripheral cells to convert T_4 into T_3. Thyroid hormone is carried to virtually every cell in the body, where it plays an important role in regulating metabolic activity.

Circulating T_3 also plays a role in modulating the chain of events that ultimately leads to hormone production. Acting on small hypothalamic neurons, T_3 inhibits the synthesis of TRH. T_3 also acts directly on thyrotropin cells in

BOX 9.2 Goiter

Goiter is an enlargement of the thyroid gland, which can be due to a number of causes. The most common worldwide (over 90%) is a simple goiter due to an iodine deficiency in the diet. In this case, the follicular cells sense the deficiency and to compensate, proliferate to try to increase the iodine-collecting efficiency of the thyroid gland. Some goiters seen in regions where iodine is deficient become very large (Fig. 9.11). Hyperthyroidism can also lead to goiter as a result of an oversecretion of TSH and a build-up of thyroid tissue. Other causes of goiter are less common and can involve tumors or autoimmune reactions.

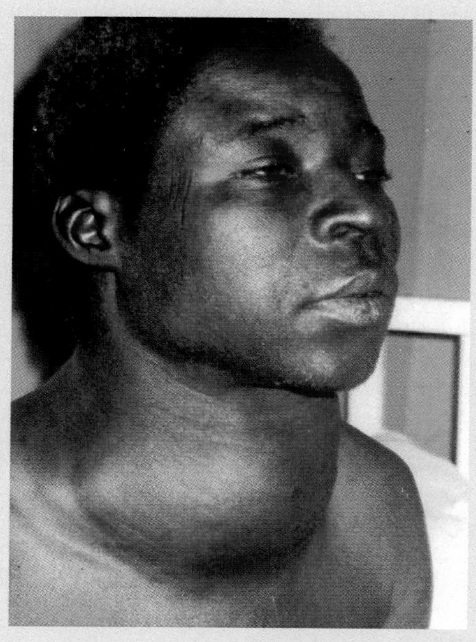

FIGURE 9.11 Iodine deficiency goiter in a young person from central Africa. *From Boron and Boulpaep (2012), with permission.*

the anterior pituitary. If in excess, it inhibits the production of TSH; a reduction in the concentration of free T_3 in the plasma results in a huge increase in TSH production by the pituitary.

Once inside a cell, much of the entering T_4 becomes enzymatically converted to T_3 so that their intracellular concentrations are about equal. T_3 and T_4 are then transported into the nucleus, where they bind to a thyroid hormone receptor which activates or represses specific genes that encode for a large number of intracellular functions.

Thyroid hormone affects almost all cells in some way, mainly by influencing internal metabolic processes involving proteins, carbohydrates, and lipids. All this increases the overall basal metabolic rate of the body. Too much or too little thyroid hormone results in profound general effects on the body. In humans, normal growth and development require a thyroid influence. If there is a deficit in thyroid hormone, physical growth of the body is considerably reduced, and mental development is severely retarded—a condition called **cretinism**. In amphibians, the process of metmorphosis from larval to adult forms is totally dependent on thyroid hormones. Temperature control is also strongly influenced by thyroid hormones. In excess, the body temperature and general metabolism rises, and a thyroid deficiency results in a lowered body temperature along with a reduced metabolism, in general. In newborn humans and in rodents, thyroid hormone specifically activates brown fat as a means of maintaining a normal body temperature.

Adrenal Glands

The adrenal gland consists of two separate glands that occupy the same general space. The **adrenal cortex** is a classical steroid-secreting endocrine gland that in the embryo arises from the same primordial tissue mass as the kidneys. Even in the adult, the adrenal glands are situated on the upper ends of the kidneys. The **adrenal medulla**, on the other hand, can be looked upon as an extension of the sympathetic nervous system with respect to both its embryological origins (neural crest) and its adult function.

Adrenal Cortex

Like the thyroid, the adrenal cortex is one step down in the hierarchy from the pituitary gland in terms of hormonal control. **Adrenocorticotropic hormone (ACTH)** from the pituitary is the prime stimulus for the synthesis and secretion of adrenal cortical hormones.

The adrenal cortex is divided into three structurally and functionally different zones (Fig. 9.12). Although each zone produces several steroid hormones, one is dominant for each. The dominant hormone secreted by the outer **zona glomerulosa** is **aldosterone**—a **mineralocorticoid**, which works on the tubules of the kidney to control the amount of Na^+ that is excreted or resorbed. Through this action, aldosterone is a major determinant of extracellular fluid volume in the body.

Cells of the middle **zona fasciculata** produce **glucocorticoids**, the major one of which is **cortisol**. Cortisol increases blood glucose by mobilizing amino acids from proteins and then, acting on the liver, causing it to convert the amino acids into glucose or glycogen. Both cortisol and aldosterone are derived from cholesterol through a series of enzymatic reactions within the cells of the adrenal cortex.

The inner **zona reticularis** secretes **androgenic steroids**. Like the other adrenal steroids, the hormones of the zona reticularis are synthesized from cholesterol. Although these hormones (dihydroepiandrosterone and androstenedione)

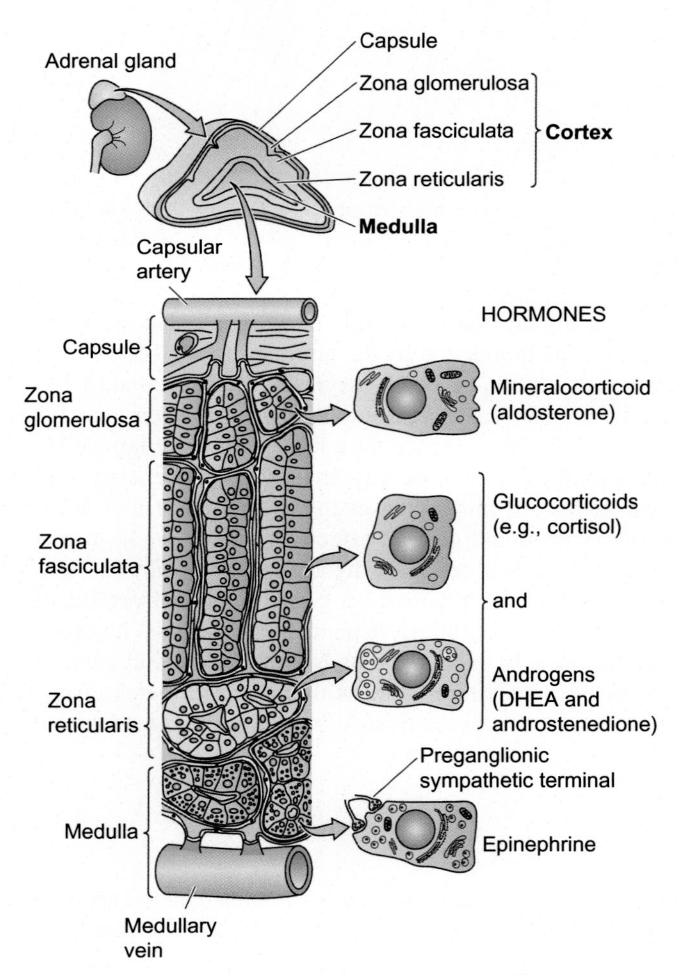

FIGURE 9.12 The cellular organization of the adrenal gland. *From Boron and Boulpaep (2012), with permission.*

have some androgenic activity in their own right, they also serve as metabolic intermediates in the synthesis of the more powerful testosterone and estrogen molecules. Tried and true mnemonic devices for remembering the zones of the adrenal cortex and their products are GFR (glomerular filtration rate) for glomerulosa, fascicularis, and reticularis and salt–sugar–sex for aldosterone, cortisol, and androgens.

The adrenal gland is one of the most richly vascularized structures in the body. Several arteries supply the capsule. They ultimately branch into an outer cortical capillary plexus, which then feeds into a series of thin-walled sinusoids that pass through the cortex and form a portal system. These, in turn, branch into a second set of capillaries within the medulla. Ultimately, the medullary blood collects into a central medullary vein and leaves the adrenal gland. Physiological evidence suggests a relationship between increased cortical blood flow and cortisol synthesis. In addition, hormonal secretion in the medulla is influenced by events in the cortex.

The **hypothalamic–pituitary–adrenal axis** is a classic example of hierarchical control within the endocrine system (Fig. 9.13). Cortisol is secreted by the adrenal cortex in response to acute or chronic stress. Stimuli for its secretion begin in the cerebral cortex, where stress is translated into several different manifestations (e.g., sleep

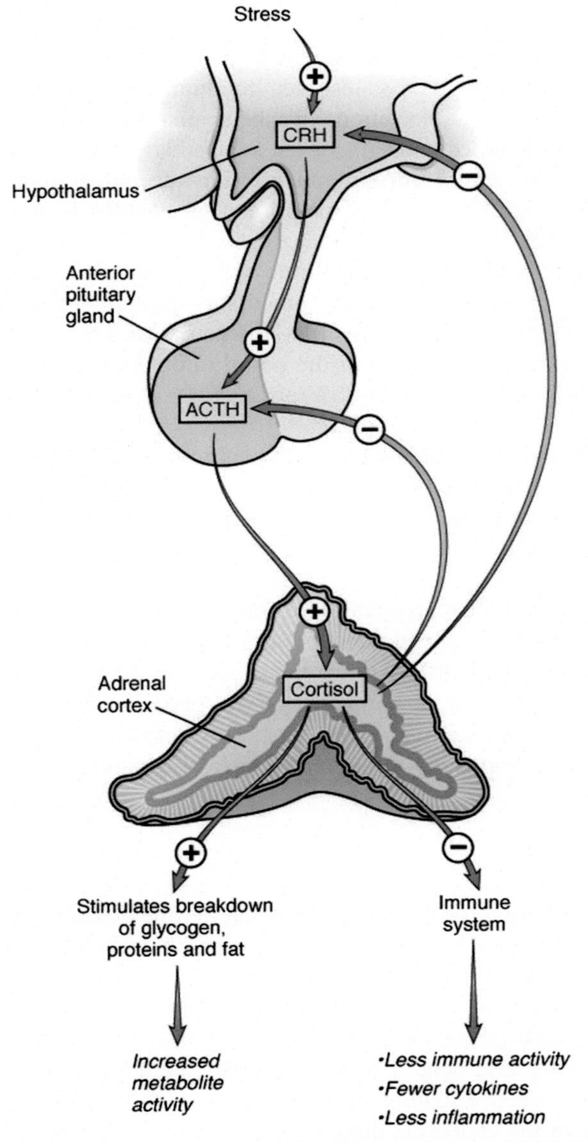

FIGURE 9.13 The hypothalamic–pituitary–adrenal axis and regulation of cortisol secretion. *From O'Neill and Murphy (2015), with permission.*

disturbances, emotional and physical tension, etc.) that stimulate the small neurons of the paraventricular nucleus of the hypothalamus to produce corticotropin-releasing hormone (CRH). This releasing hormone then acts on the anterior pituitary, resulting in the release of ACTH, which in turn stimulates the adrenal cortex to secrete cortisol. Cortisol is a powerful regulator of cellular metabolism—especially carbohydrate and protein breakdown. In addition, it is a potent suppressor of immune functions, with prominent anti-inflammatory activity. Negative feedback, resulting in dampening of the response, occurs in the form of increased blood cortisol levels, which act on both the hypothalamic and pituitary levels to reduce secretion of their hormones. In addition, increased blood levels of ACTH from the pituitary also feed back onto the hypothalamic neurons to repress their formation of CRH.

Aldosterone secretion in the zona glomerulosa is controlled in a more complex manner. Although its secretion is partially stimulated by ACTH and by an increased blood concentration of K^+, the renin–angiotensin–aldosterone axis represents the dominant means by which aldosterone secretion is regulated (see p. 369 and Fig. 13.10 for details). In essence, it represents a means of regulating body fluid levels and blood pressure through adjustments in ion exchange by the kidneys.

Adrenal Medulla

Cells of the adrenal medulla, which is not part of the hierarchical control system, fall into the general category of **chromaffin cells** because a pigment that they contain stains with chromium salts. Other masses of chromaffin cells are scattered around the abdominal cavity in close association with the sympathetic nervous system, and in aggregate they can be viewed as representing postganglionic components of the sympathetic nerves (see Fig. 6.4). All of these chromaffin cells secrete **epinephrine (adrenalin)** and/or **norepinephrine (noradrenalin)** upon being stimulated by acetylcholine released by preganglionic sympathetic neurons. This means that, in contrast to most endocrine-producing cells, these cells have a very rapid response time.

Cells of the adrenal medulla produce norepinephrine and epinephrine (collectively called **catecholamines**) through a series of enzymatic changes from their precursor—the amino acid tyrosine. These cells are not completely independent of the CRH–ACTH–cortisol axis because ACTH directly stimulates synthesis of certain epinephrine intermediates, and cortisol arriving from the portal circulation in the cortex also upregulates one of the enzymes involved in epinephrine synthesis. About 80% of the adrenal medullary cells secrete epinephrine and 20% secrete norepinephrine.

Before secretion, newly synthesized catecholamines are bound to granular proteins within the medullary cells. Being essentially components of the autonomic nervous system, adrenal medullary cells release their secretory contents by an electrical depolarization process upon nervous stimulation, namely actetylcholine release by the preganglionic sympathetic neurons that contact these cells. After release, the catecholamines are released from their binding granular proteins.

Once released, epinephrine, in particular, triggers the elements of the flight-or-fight response in the peripheral tissues. This response includes increasing the blood flow to muscles to allow more efficient function, relaxation of bronchial smooth muscle to provide for greater airflow to the lungs, and generation of energy by degradation of glycogen in muscle and lipids in fat. In an indirect effect, insulin levels are decreased, resulting in greater concentrations of glucose in the blood. The latter is important to maintain brain function. Control of secretion by the adrenal medulla resides almost entirely within the nervous system, with little or no feedback regulation by other endocrines.

Gonads

Despite their position on the bottom rung of the hormone hierarchy, the gonads are responsible for coordinating some of the most complex functional scenarios in the body. In females, their overlying purpose is preparing the body for pregnancy, then supporting the development of the embryo during pregnancy, and finally, nourishing the newborn infant. All of this requires a high degree of coordination of changes in many cells and tissues during the course of a single reproductive cycle even if pregnancy does not result. In addition to reproductive cycles, hormones generated from the gonads in both sexes exert a powerful control over the growth and development of the body during childhood and puberty. Finally, gonadal hormones exert a significant influence on emotions and mind-set. The ovaries and testes function downstream of the pituitary in the endocrine hierarchy, but like the other peripheral endocrine glands, the gonads also exert powerful feedback control on the upper levels of that hierarchy, namely the hypothalamus and the pituitary.

We are still in the early stages of understanding how hormonally influenced thought processes either directly or indirectly influence the upper levels of the endocrine hierarchy.

The gonads produce both androgens and estrogens, but different cells produce each hormone. The cells that produce either androgens or estrogens in the ovaries and testes are homologous in that they are descendants of the same type of precursor cell in the early embryo. As embryos develop into either males or females, the precursor cells differentiate into cells appropriate for the ovaries or testes. The similarities don't end with hormones produced. Cells in both sexes that produce testosterone respond to LH, and those that produce estrogens respond to FSH.

Ovaries

Hormone production in the **ovaries** takes place in the cells of the developing **follicles** (Box 9.3). Under the influence of LH from the anterior pituitary, it begins with the formation of androgens from the precursor molecule, cholesterol, in the cells of the **theca interna** (Fig. 9.15), which is part of the outer layer of the follicle (see Fig. 9.14). The newly synthesized androgens then diffuse through the basement membrane that separates the overlying thecal cells from the inner **granulosa cells** of the follicle. Within the granulosa cells, an **aromatase** enzyme, activated by FSH, catalyzes the conversion of androgens to estrogens, which are then released into the general circulation. Circulating estrogens are bound to a sex hormone-binding globulin and are distributed to cells throughout the body. As is the case with other circulating hormones, they influence only those cells that have specific receptors (in this case, cytoplasmic) for them.

Circulating estrogens influence overall growth and development of the female body, as well as the monthly reproductive cycles of postmenarchal females. Before puberty, estrogens promote bone growth, and after puberty, they cause fusion of the epiphyses (see p. 99), resulting in cessation of growth of the long bones. Estrogens are also largely responsible for the deposition of fat around the hips and in the breasts during and after puberty (Fig. 9.16). Contemporary research is also increasingly documenting the role played by estrogens in overall brain functions, such as learning, memory, and mood.

Estrogens play a dominant role in the first half of a woman's monthly reproductive cycle (see also p. 386). During the **menstrual phase** (days 1–5), when the uterine lining (endometrium) is shed, FSH and LH from the anterior pituitary stimulate the formation of estrogens from developing ovarian follicles (Fig. 9.17). The secreted estrogens support

BOX 9.3 The Life Cycle of an Ovarian Follicle

An **ovarian follicle** (an ovum with its coverings) has a long life history, with many twists and turns along the way. Postnatal ova are surrounded by a partial or full single layer of **follicular cells** (Fig. 9.14). Development of these early primary follicles is a lengthy process, measured in months, and is independent of pituitary hormones. Instead, it depends upon paracrine interactions between follicular cells and the ovum inside. The follicular cells increase in number and begin to form an additional layer of cells. At this point, they become minimally sensitive to pituitary hormones, but are independent of the hormonal fluctuations associated with the menstrual cycle. Over a period of about 2 months, the expanding primary follicle develops a small fluid-filled space within the follicular cells and is now known as a secondary follicle measuring about 2 mm in diameter. By this point, the follicular cells become acutely sensitive to FSH. During any menstrual cycle, about 15–20 follicles develop to this stage. Starting at the beginning of the proliferative stage of the cycle, these follicles, which are now 2–5 mm in diameter, become dependent upon FSH support. Through mechanisms still not completely understood, one of these follicles becomes dominant and grows rapidly until it reaches a diameter of 15–20 mm by the time of ovulation. The large secondary follicles also develop a covering of ovarian connective tissue, called the theca (both a **theca interna** and a **theca externa**), which is the tissue sensitive to LH and which begins the process of estrogen synthesis, although the synthetic process is handed off to the underlying granulosa cells for completion. A basement membrane (membrana granulosa) separates the granulosa cells from the theca interna, and estrogens formed by the granulosa cells must diffuse through the basement membrane to capillaries immediately overlying it.

Under the influence of the LH surge, two events of major import occur. One is ovulation, which releases the egg contained in the dominant follicle from the ovary. The other is the initiation of the first meiotic division, which is necessary for balancing the chromosomal content of the egg (and sperm) so that at fertilization the new embryo is endowed with the correct number of chromosomes (46 in the human).

(*Continued*)

BOX 9.3 (Continued)

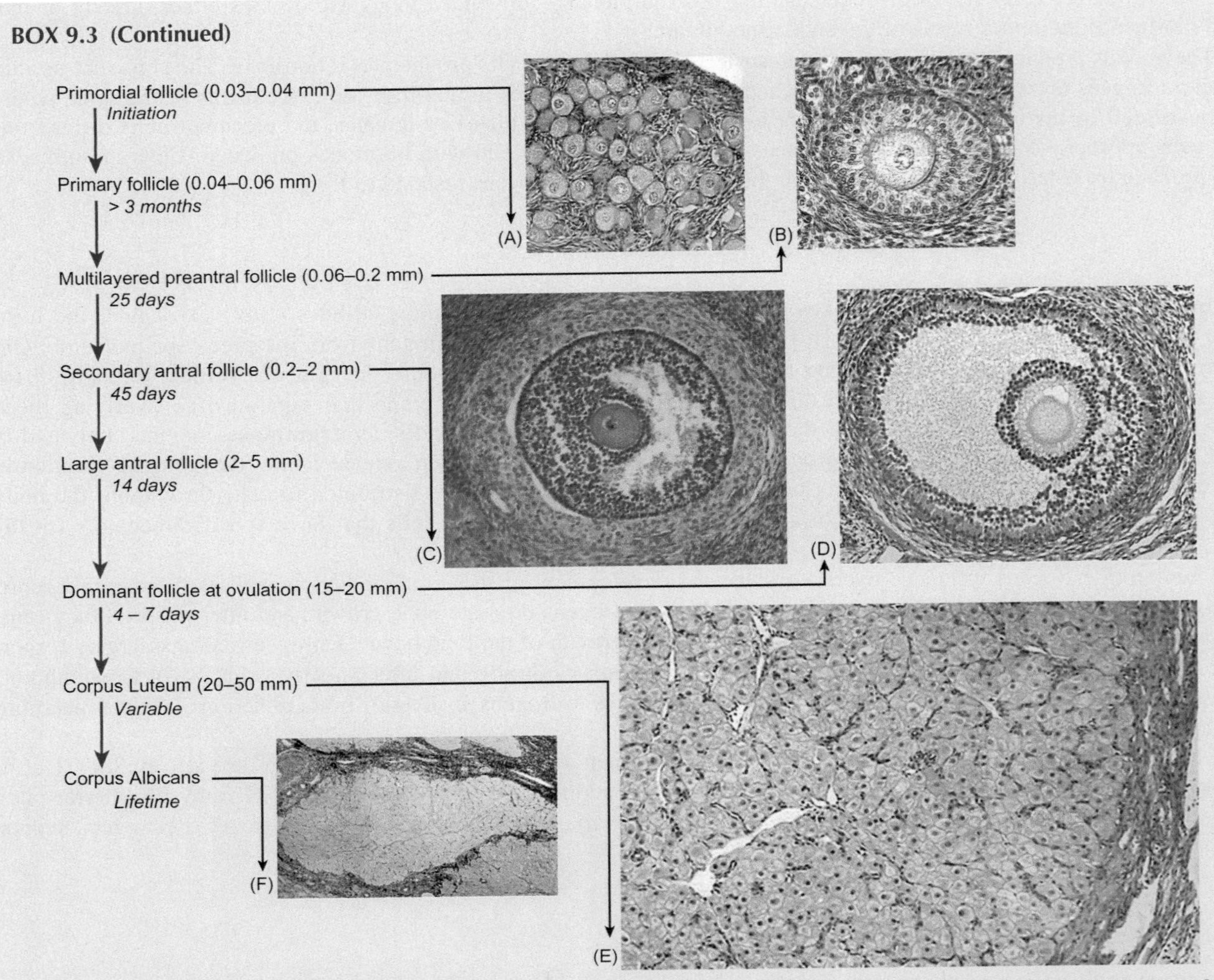

FIGURE 9.14 The life history of an ovarian follicle: (A) primordial follicles; (B) primary follicle; (C) secondary antral follicle; (D) dominant follicle; (E) corpus luteum; and (F) corpus albicans. *Photomicrographs from Erlandsen and Magney (1992); Kierszenbaum and Tres (2012), with permission.*

After ovulation, still under the influence of LH, the ruptured follicle undergoes a series of changes, called **luteinization**. The follicle enlarges greatly (up to 2–5 cm in diameter) and is now known as a **corpus luteum** (yellow body). The granulosa cells change their form and produce large amounts of estrogen. In addition, blood vessels now penetrate the basement membrane that surrounded the granulosa cells of the preovulatory follicle so that the granulosa cells become closely associated with a local vascular supply. If fertilization has not occurred, the inhibition of pituitary hormones by follicular hormones ultimately results in degeneration of the corpus luteum.

If fertilization has occurred, cells surrounding the early embryo produce a hormone called **chorionic gonadotropin**, which acts principally to stimulate LH receptors in the corpus luteum. Now called the **corpus luteum of pregnancy**, it continues to expand, and for the first 10 weeks of pregnancy, it produces the progesterone necessary for the maintenance of the pregnancy. Progesterone promotes a process called **decidualization** of the endometrium—a change needed to maintain the embryo within the endometrium.[2] After that time the placenta itself produces sufficient progesterone, and the corpus luteum becomes no longer indispensible. Over time it degenerates into a white mass of scar tissue, called the **corpus albicans**, which persists in the ovary for years.

2. During pregnancy many of the stromal cells of the uterine lining undergo substantial changes that enable them to support the implantation and later support of the embryo. Because the parts of the uterine lining containing these cells are shed at childbirth, they are called decidual tissues.

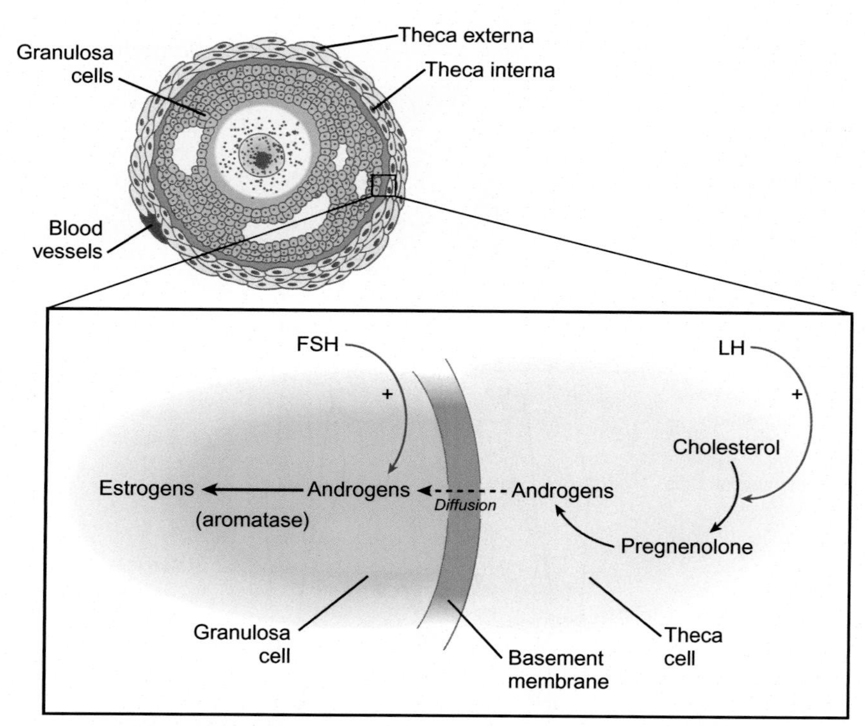

FIGURE 9.15 Estrogen synthesis in an ovarian follicle.

the regeneration and gradual build-up of the endometrium over the **proliferative phase** (days 5–14) of the menstrual cycle. As ovarian follicles mature during that time, one becomes dominant and produces so much estrogen that it becomes a positive feedback stimulus to the pituitary, causing it to secrete a massive amount of LH by about days 12 and 13 of the menstrual cycle. The spike of LH release (actually a series of sharp secretory spurts) is the stimulus for ovulation, which marks the transition from the proliferative to the **secretory phase** (days 15–28) of the menstrual cycle. Other changes in peripheral female reproductive tissues, which are designed to facilitate both insemination and gamete transport, are covered in Chapter 14.

The secretory phase of the menstrual cycle is focused on preparing the endometrium for implantation and nourishment of the early embryo. The endocrinological driving force behind it is the formation of the **corpus luteum** from the remains of the ovulated follicle and the production of **progesterone** from the former granulosa cells—now called **granulosa lutein cells** (see Fig. 9.15 and Box 9.3). While the secreted progesterone is stimulating further thickening of the endometrium, along with maturation of the endometrial glands and increasing the local vascular supply, the combination of progesterone, estrogen, and inhibin, secreted by the ovaries, acts upon the hypothalamus and anterior pituitary to greatly reduce the secretion of FSH and LH. If pregnancy has not occurred, the corpus luteum begins to deteriorate after a few days. The reduction of progesterone and estrogen secretion from the degenerating corpus luteum fails to support the endometrium so that by the end of the fourth week of the menstrual cycle, the endometrial microcirculation begins to shut down, resulting in local ischemia. Then patches of the ischemic endometrium begin to be shed as the menstrual period begins.

Testes

The **testes** have two main functions—producing viable spermatozoa and testosterone. Producing spermatozoa is discussed in Chapter 14; producing testosterone and maintaining a climate suitable for generating spermatozoa is discussed here.

A testis contains 600–1200 looped **seminiferous tubules**, immunologically protected environments within which spermatozoa develop (Fig. 9.18A). The endocrine component of the testes centers around two cell types—**Sertoli cells**, which line the insides of the seminiferous tubules, and **Leydig cells**, which lie outside them (Fig. 9.18B). Leydig cells are classic steroid-secreting cells, and under the influence of LH, they synthesize testosterone from cholesterol. Through negative feedback acting upon both the hypothalamus and the anterior pituitary, **testosterone** largely controls

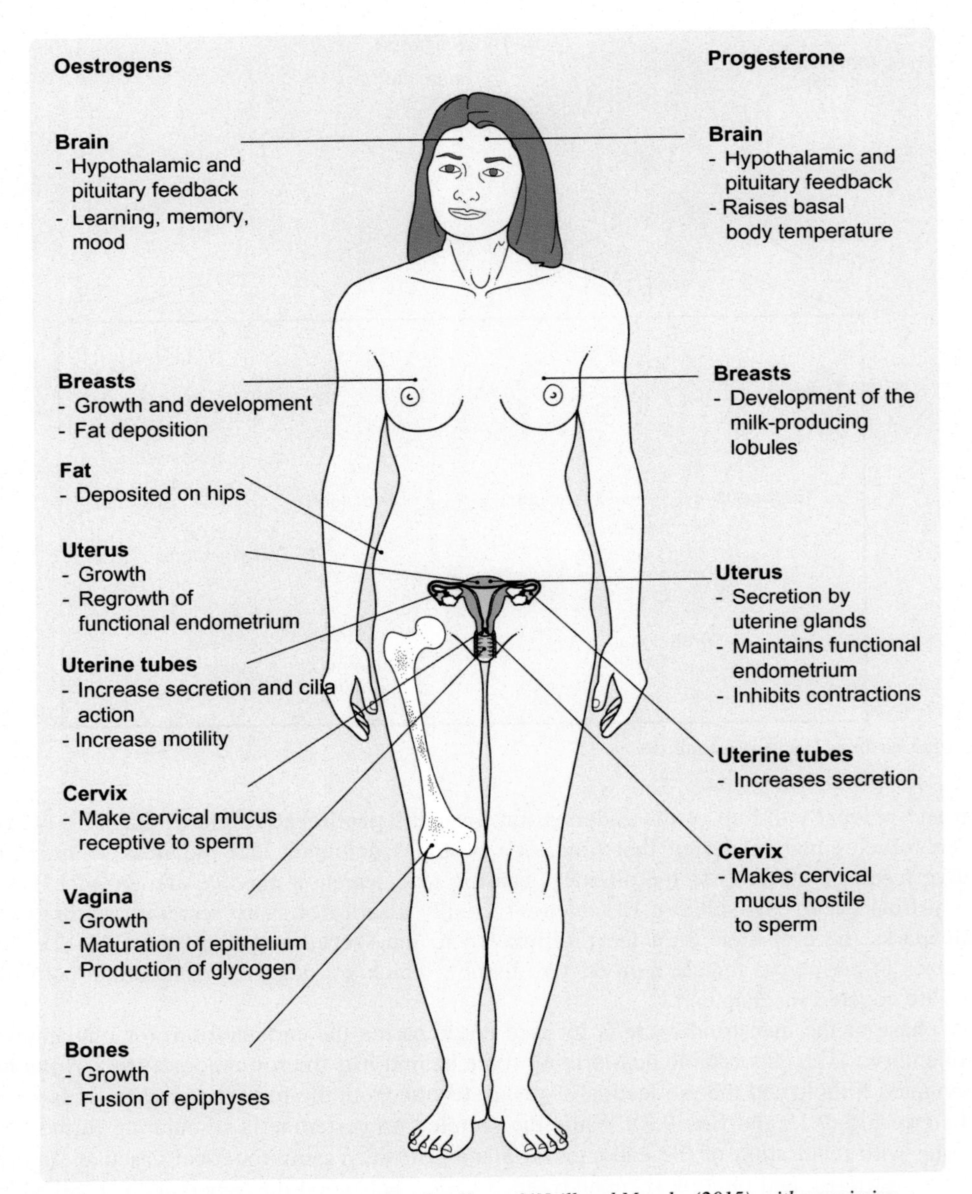

FIGURE 9.16 The actions of estrogens and progesterone on females. *From O'Neill and Murphy (2015), with permission.*

its own production by inhibiting the release of LH (Fig. 9.19). Sertoli cells are stimulated by FSH to generate **androgen-binding protein**, a molecule similar to that which binds to blood-borne testosterone. Some of the testosterone produced by the Leydig cells enters the blood for general distribution throughout the body, but testosterone also diffuses through the linings of the seminiferous tubules, where it attaches to the testosterone receptors on the Sertoli cells and accumulates in a concentration up to 100 times that in the blood. The high intratubular testosterone concentration is part of the environment promoting spermatogenesis that is provided by the Sertoli cells. Sertoli cells also secrete **inhibin**, which acts on the hypothalamus and anterior pituitary to inhibit FSH release, but with little effect on LH release.

Although potent in its own right, circulating testosterone is made even more powerful by being converted to **dihydrotestosterone** by an enzyme (**α-reductase**) found in tissues of the male reproductive tract, genital skin, hair follicles, and the liver. Dihydrotestosterone is the hormonal agent responsible for maturation of the external genitalia, the prostate and male hair patterns. Testosterone plays an important role in the development of the male skeleton, skeletal muscle, abdominal fat, the larynx, and in brain functioning (Fig. 9.20).

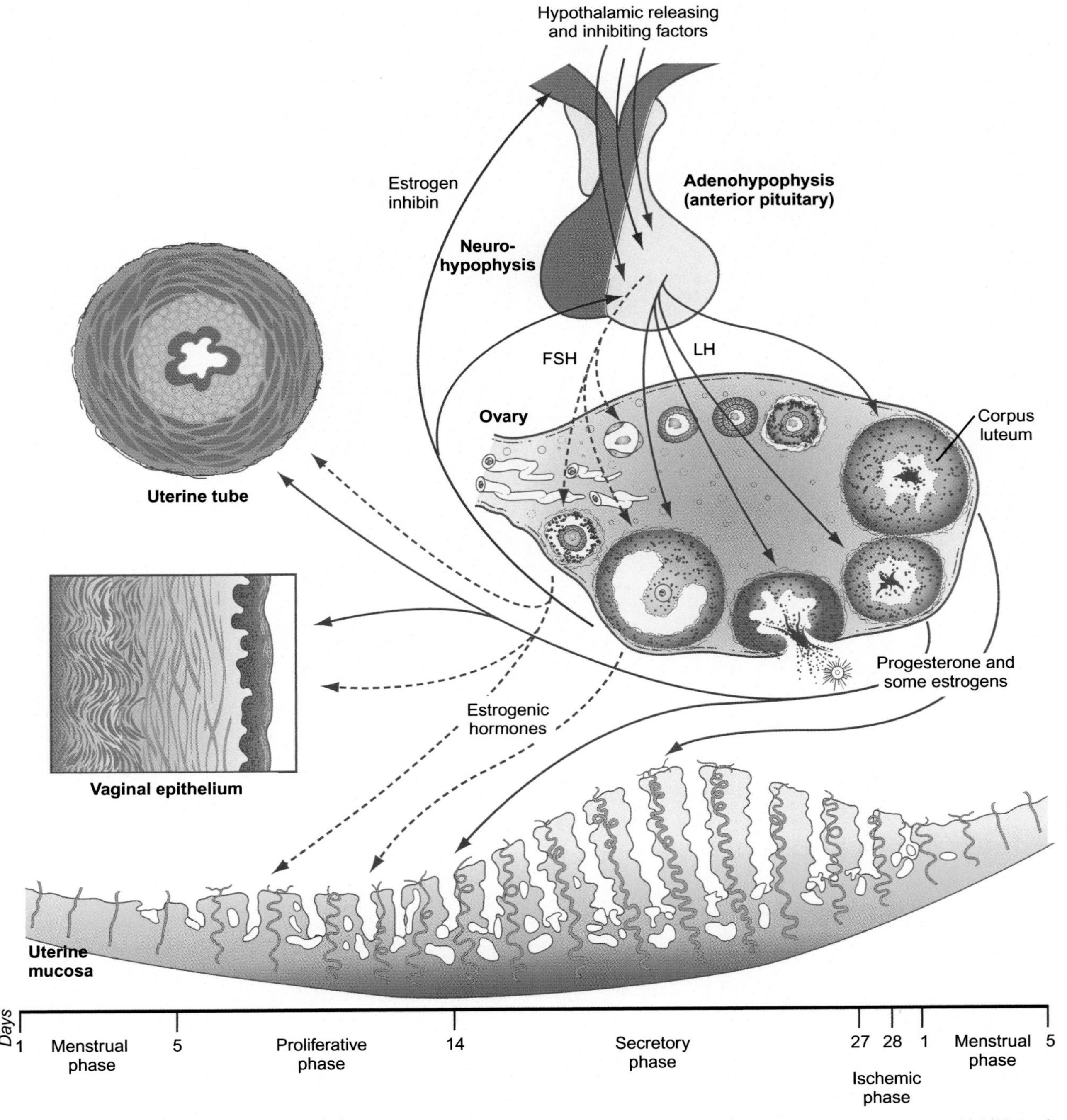

FIGURE 9.17 General scheme of hormonal control of reproduction in women. Stimulatory factors are represented by red arrows and inhibitory factors by purple arrows. Hormones involved principally in the proliferative phase of the menstrual cycle are represented by dashed arrows and those in the secretory phase by solid arrows. *From Carlson (2014), with permission.*

NONHIERARCHICAL ENDOCRINE GLANDS

Another category of endocrine glands consists of those that do not operate in a hierarchical manner, but are involved in maintaining the moment-to-moment homeostasis of the body. These glands take their cues from sensors of the body's internal environment (e.g., pH, osmolarity, and metabolite concentration) and respond by secreting hormones that act

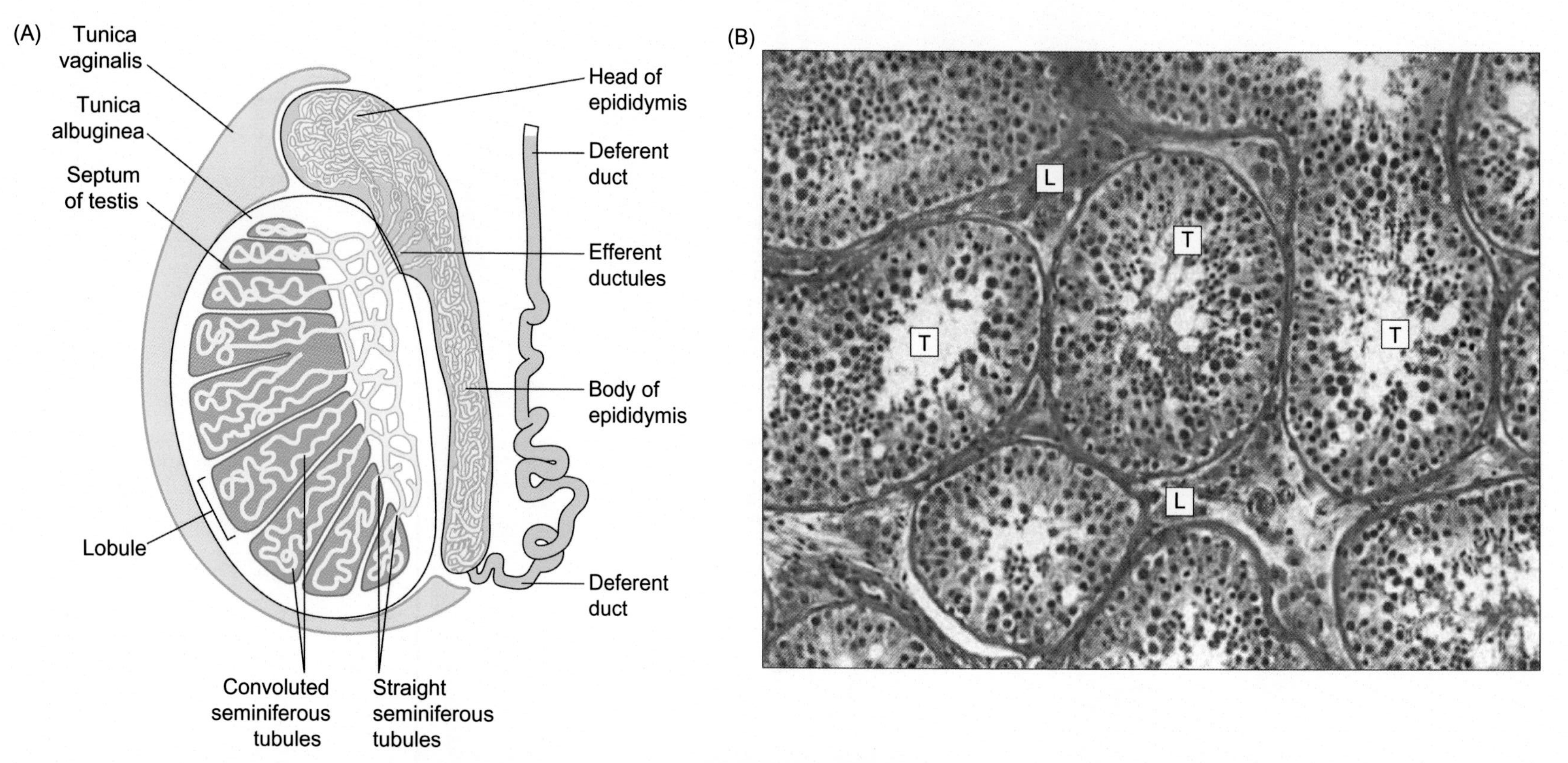

FIGURE 9.18 (A) Overall structure of the testis, epididymis, and ductus deferens. (B) Histological cross section through seminiferous tubules. *L*, Testosterone-secreting Leydig cells in the interstitium; *T*, seminiferous tubule. *(A) From Waugh and Grant (2014), with permission. (B) From Kierszenbaum and Tres (2012), with permission.*

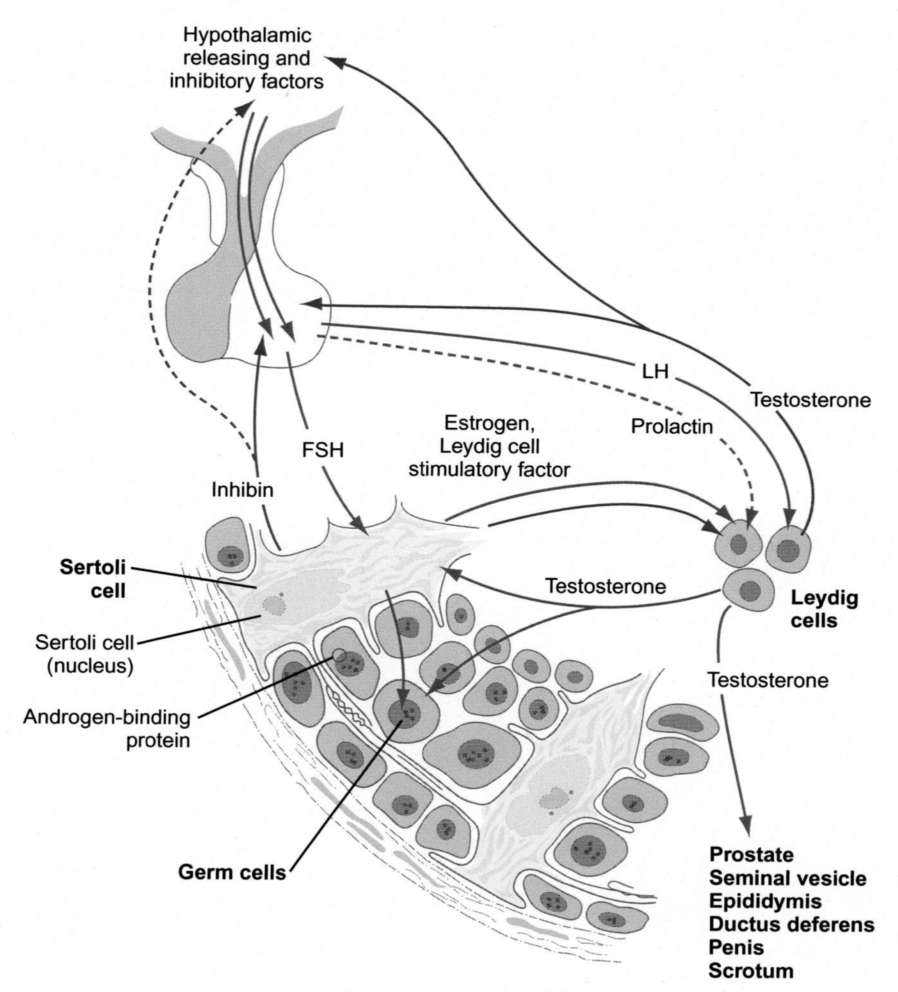

FIGURE 9.19 Hormonal control of reproduction in the male. Stimulatory influences are represented by red arrows and inhibitory influences by purple arrows. Suspected interactions are indicated by dashed arrows. *From Carlson (2014), with permission.*

within seconds or minutes, rather than hours or days. In some cases, especially involving the digestive system, the distinction between endocrine and paracrine action can be blurred. This section will cover those functions in which endocrine mechanisms predominate. Paracrine secretions of the digestive system are covered in Chapter 12.

Control of Calcium Balance

The control of **calcium balance** in the body involves three hormones (PTH, calcitonin, and vitamin D and its derivatives) acting on three major tissues—bone, the intestines, and the kidneys. Calcium balance is important for a variety of cellular functions, including the mineralization of bone, muscle contraction, exocytosis (release of intracellular components), intracellular signaling, and a variety of enzymatic reactions. Over 99% of the body's calcium is bound in bone matrix. About half of the calcium in the blood is bound to proteins; the remainder is unbound in ionized form. The latter is the form of extracellular calcium that is under hormonal influence, and Ca^{++} concentrations in the blood must be kept within a narrow range for normal homeostasis.

By far, the most important influence on calcium levels is **parathyroid hormone (PTH)**, which is secreted by the four small parathyroid glands that are embedded within the thyroid gland. PTH is made and secreted by epithelial **chief cells**—the dominant cell type in the gland. Another type of cell in the parathyroids, the oxyphil, is still little understood and has no well-defined function.

In response to low circulating Ca^{++} levels detected by a calcium-sensing receptor on the chief cells, these same cells secrete PTH (Fig. 9.21). PTH acts to increase Ca^{++} levels in the blood through increasing bone resorption and the

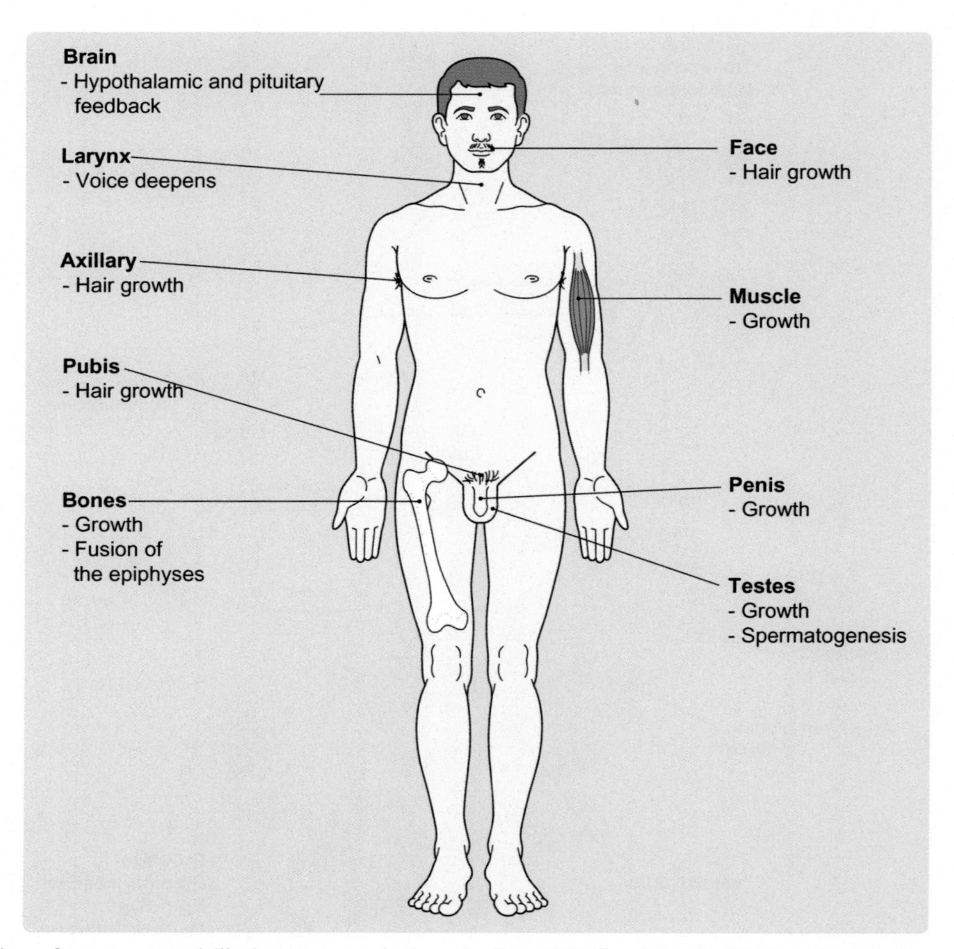

FIGURE 9.20 Actions of testosterone and dihydrotestosterone in the male. *From O'Neill and Murphy (2015), with permission.*

subsequent release of Ca^{++} and phosphate from the bone matrix. This is accomplished through the action of osteoclasts (see p. 93) on bone, but interestingly, PTH does not directly affect osteoclasts. Instead it binds to osteoblasts (see p. 91), which secrete factors that stimulate the conversion of osteoclastic precursors (monocytes) into osteoclasts.

In addition to stimulating the release of Ca^{++} from bone, PTH also acts directly on the kidneys to increase Ca^{++} resorption from the forming urine and to stimulate the formation of an enzyme that activates **vitamin D**. Activated vitamin D, in turn, acts to increase the absorption of Ca^{++} by the intestines. Vitamin D also plays a role in the formation and calcification of osteoid (see p. 91) in newly forming bone.

Calcitonin, the third member of this endocrine trio, plays a more enigmatic role in calcium homeostasis. Calcitonin, which is produced by parafollicular cells that migrate into the thyroid gland during embryogenesis, is known to lower calcium levels in the blood. However, complete removal of the thyroid gland including all of the parafollicular cells, produces no discernible effect on blood calcium levels. This suggests the presence of unidentified physiological backup mechanisms that can compensate for the loss of this hormone.

Although overall its effect appears to be minor in humans, calcitonin is an important Ca^{++}-regulating hormone in fishes, where exposure to Ca^{++} in the water is an important feature of their environment. At the cellular level, calcitonin targets osteoclasts through specific receptors on their surfaces, and it reduces their activity, thus reducing Ca^{++} loss from bone.

Glucose Regulation and the Pancreas

Glucose is almost universally used by cells as a metabolite and a source of energy, and its level in the blood must be kept within a narrow range, because either too much or too little can cause catastrophic acute and chronic effects.

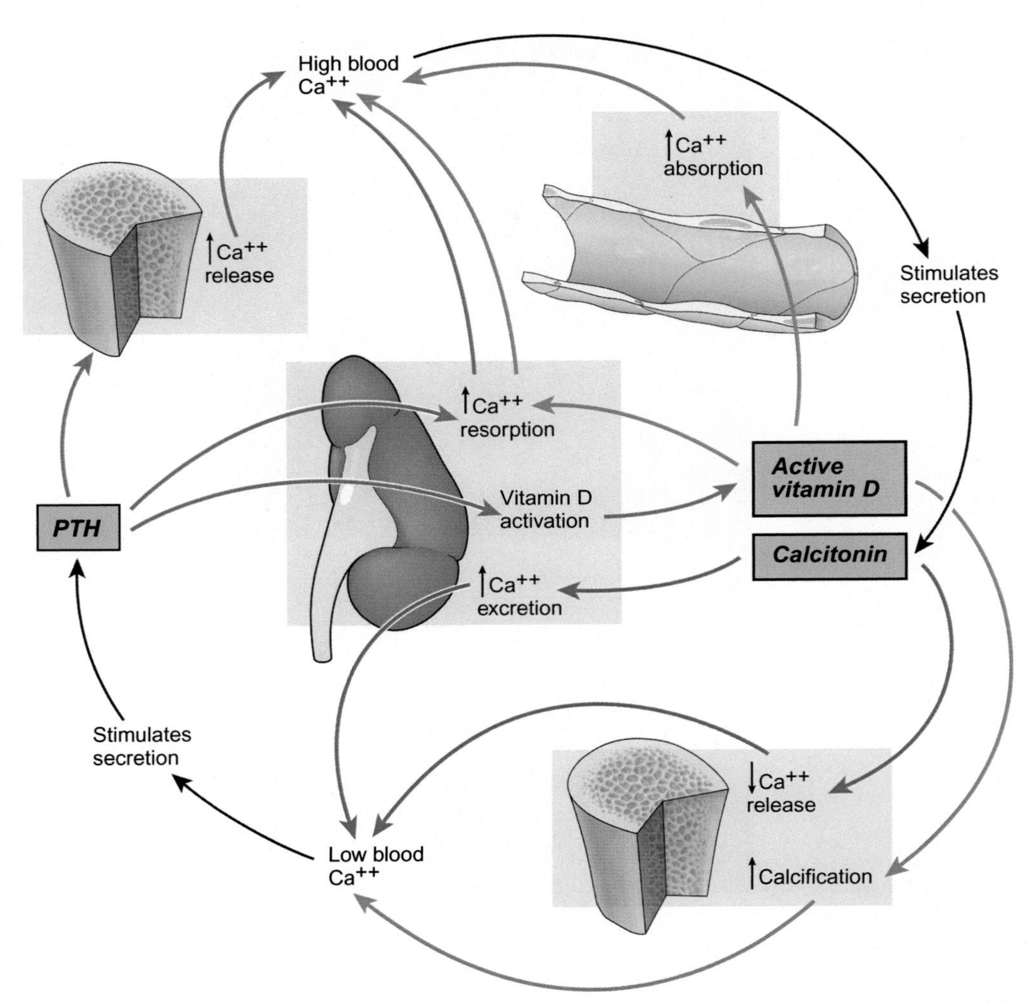

FIGURE 9.21 Interactions in the control of blood calcium. Orange arrows, PTH effects; green arrows, vitamin D effects; maroon arrows, calcitonin effects; black arrows, blood Ca^{++} effects.

Regulation of glucose levels cannot depend upon hierarchical control because the effects of imbalance can be felt within minutes. The body has dealt with this issue by evolving powerful hormonal mechanisms to either raise or lower glucose levels in the blood and its uptake into cells. The prime movers are two hormones produced by the **islets of Langerhans** within the pancreas—**insulin** which lowers and **glucagon** which raises blood glucose levels. Although these hormones are most influential, glucose levels and utilization are modified, often on a less acute basis, by a variety of secondary neural and hormonal influences.

Not surprisingly, because of its central metabolic importance, insulin or its equivalent has been around a long time in our evolutionary history. Insulin is a major hormone in all vertebrates, and insulin-like molecules have been found in life-forms as ancient as protozoa, bacteria, and fungi. In addition to its effects on sugar metabolism, insulin or its evolutionary equivalent is also often involved in growth processes.

In humans, the endocrine component of the pancreas consists of about three million microscopic islets of Langerhans scattered throughout the exocrine substance of the gland (Fig. 9.22). The exocrine component produces the digestive enzymes, which empty into a system of pancreatic ducts. The exocrine functions of the pancreas are discussed in greater detail in Chapter 12. Islets of Langerhans average about 0.1 mm in diameter. Despite occupying only between 4% and 5% of the pancreatic volume, they receive almost 15% of the total blood flow within the pancreas. High vascularity is typical of an endocrine structure, and within an individual islet, networks of capillaries are closely associated with the hormone-producing cells. Pancreatic islets contain five types of cells, each of which is associated with a different hormone (Table 9.3). The islets of Langerhans are innervated by the autonomic nervous system. Sympathetic

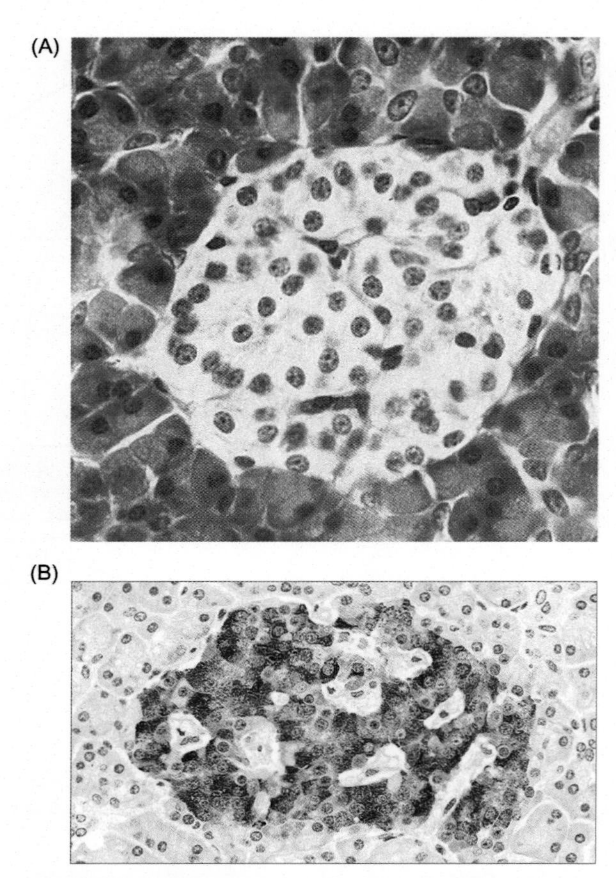

FIGURE 9.22 (A) Islet of Langerhans—standard histological preparation. The dark cells surrounding the islet are pancreatic acinar cells, which produce digestive enzymes. (B) Islet of Langerhans, with insulin-secreting calls stained brown. *(A) From Young et al. (2006). (B) From Stevens & Lowe (2005), with permission.*

TABLE 9.3 Pancreatic Islet Hormones

Cell Type	Hormone	Function
α-Cell (20%)	Glucagon	Increase blood glucose level
β-Cell (70%)	Insulin	Decrease blood glucose level
		Increase glucose uptake by cells
δ-Cell (8%)	Somatostatin	Inhibit insulin and glucagon release
		Inhibit GH
F(γ)-cell (2%)	Pancreatic polypeptide	Inhibit pancreatic exocrine functions
ε-Cell	Ghrelin (mainly in stomach)	Stimulates gastric functions, decreases insulin secretion

stimulation inhibits the release of insulin, whereas parasympathetic stimulation increases the secretion of both insulin and glucagon.

Insulin release is the result of a process that begins with the uptake of glucose into β-cells (Fig. 9.23). Its intracellular metabolic pathway results in the closure of ATP (adenosine triphosphate)-sensitive K^+ channels, leading to

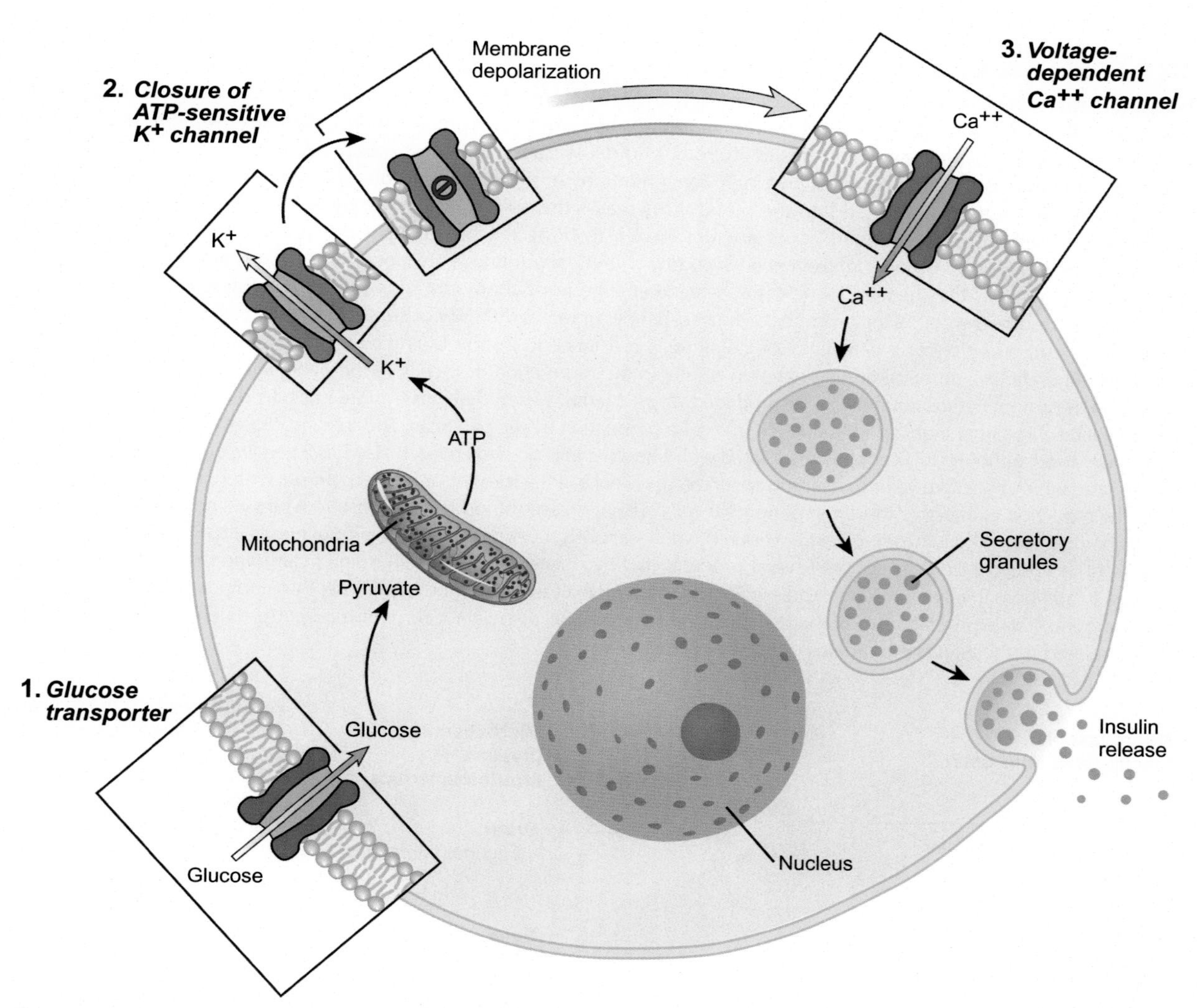

FIGURE 9.23 Steps in the control of insulin secretion after glucose enters an islet cell.

depolarization of the plasma membrane of the β-cell. This opens voltage-sensitive Ca^{++} channels and allows Ca^{++} to enter the cell. Increased intracellular Ca^{++} causes the release of already formed insulin molecules as well as a more sustained release of newly formed insulin. The secreted insulin finds its way into the blood and contacts the cells in the body, most of which have insulin receptors on their surfaces. Responsive cells increase their uptake of glucose and other metabolites and convert them to a stored form, e.g., glycogen. The overall effect is anabolic, and growth is a common result. Insulin has a short half-life (~5 minutes in the blood) and is broken down mainly in the liver and kidneys (or the placenta in pregnant women).

High levels of insulin, as would normally appear 30–60 minutes after a meal, inhibit the secretion of glucagon, but as insulin is broken down and blood levels of glucose fall, glucagon-secreting α-cells detect the low glucose concentration and release their stored glucagon. Glucagon helps to stabilize blood glucose levels by mechanisms such as breaking down glycogen to release glucose and the synthesis of glucose from amino acids (**gluconeogenesis**). Glucagon is not the only molecule that can stimulate an increase in blood glucose. Epinephrine, released in response to acute stress, inhibits insulin, and in response to longer-term stress cortisol reduces the sensitivity of cells to the effects of insulin. Inadequate blood insulin leads to the development of high blood glucose and **diabetes** (Box 9.4). Too little blood sugar produces symptoms of **hypoglycemia**.

BOX 9.4 Diabetes and Hypoglycemia

The word **diabetes** means an increased flow of urine. The disease normally assumed by the word is actually **diabetes mellitus**—an increased flow of sugary urine resulting from high blood sugar. Another type of diabetes, **diabetes insipidus**, is a symptom of a lack of antidiuretic hormone, which results in enormous fluid loss from the kidneys (see p. 367).

Diabetes mellitus is a condition involving high concentrations of glucose in the blood. Normally after a meal when blood glucose levels rise, insulin is secreted into the blood. It increases the uptake of glucose by cells, with a corresponding drop in blood glucose levels. In diabetes there is insufficient insulin to bring the blood sugar down to normal physiological values. When the concentration of blood glucose exceeds a certain level, the kidneys are not able to resorb all of the filtered glucose, and for osmotic reasons the sugar in the forming urine keeps the normal amount of water from being resorbed by the kidneys. Instead the nonresorbed water is lost from the kidneys—hence an increased flow of sugary urine.

Diabetes is caused by either a deficiency of insulin or a resistance to insulin by the responding cells. In **Type 1 diabetes** (often called **juvenile diabetes**), an autoimmune reaction results in the destruction of β-cells in the pancreatic islets and a corresponding reduction in the amount of insulin that can be produced. **Type 2 (adult-onset) diabetes** is more typically due to a resistance by cells to the effects of insulin, rather than a primary deficit in insulin production by the pancreas. The specific causes of Type 2 diabetes are still poorly understood, although it is certainly associated with obesity and lack of exercise. It also has a genetic component. In early stages of the disease, which may take years to become apparent, the β-cells in the islets of Langerhans produce more insulin to try to compensate for the reduction of effective insulin in peripheral cells. This increases blood levels of insulin while yet not producing the desired effect. According to one hypothesis, the extra work ultimately exhausts the β-cells and their production of insulin, leading to even higher levels of blood sugar. Fat cells secrete several products that modulate insulin secretion and action and may contribute to insulin resistance. In addition, the liver produces and sends into the blood excessive amounts of glucose, thus compounding the problem.

Regardless of the type or cause, diabetes can result in a host of associated chronic medical problems (Fig. 9.24). Many of these are related to the effects of diabetes on blood vessels or nerves.

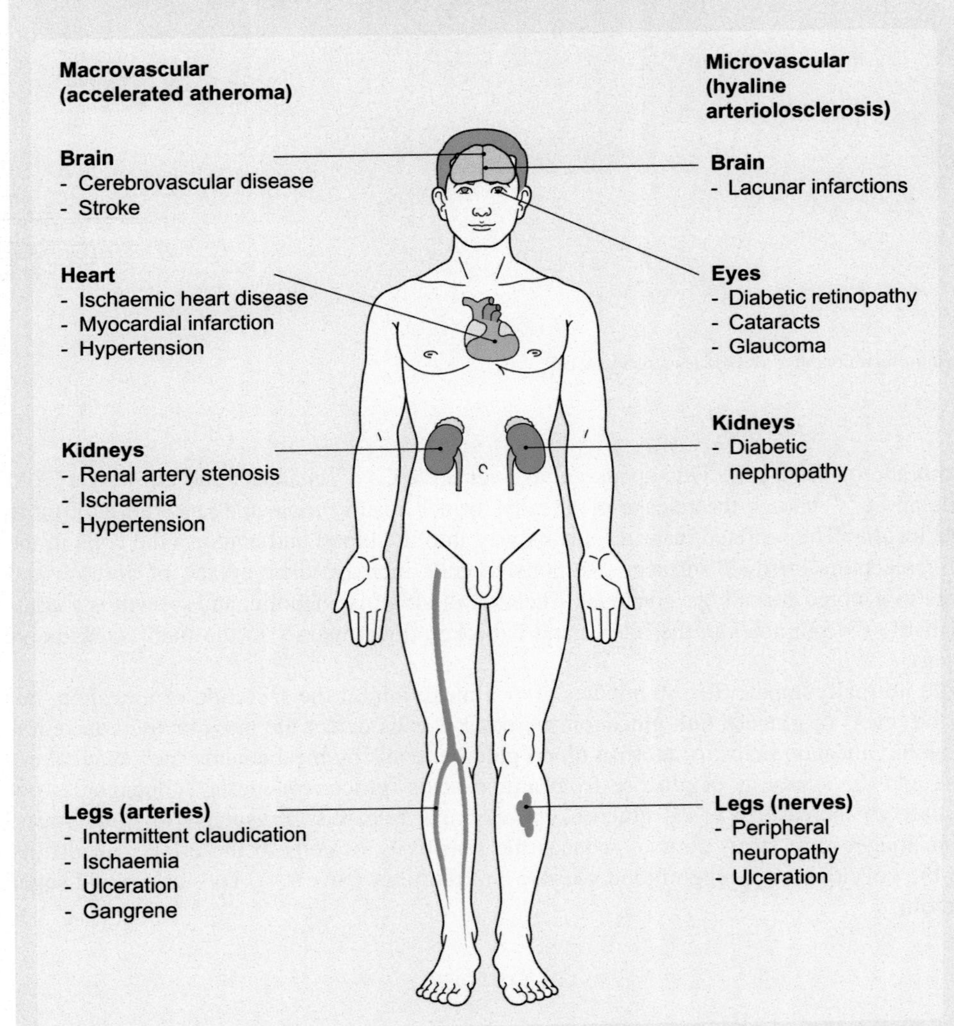

FIGURE 9.24 Complications of diabetes mellitus. *From O'Neill and Murphy (2015), with permission.*

(Continued)

BOX 9.4 (Continued)

The opposite condition from diabetes is **hypoglycemia**, characterized by excessively low blood sugar. On an acute level, hypoglycemia can be more dangerous than diabetes, because the brain requires a substantial amount of glucose to function properly. Hypoglycemia can have many causes, but the basis for most is a poor regulation of blood insulin or too much insulin or too little glucagon. An acute episode of hypoglycemia, which is often characterized by hunger, trembling, sweating, anxiety, rapid pulse, and even coma, can be counteracted simply by oral or intravenous intake of glucose.

SUMMARY

The endocrine system consists of structures that produce secreted hormones. Hormonal effects can be endocrine (working through the blood), paracrine (diffusion in tissue fluids to nearby cells and tissues), or autocrine (affecting the cells that produce the hormones). Most hormones are either steroids or peptides.

The hierarchical endocrine system begins with the hypothalamus, which sends releasing hormones via the hypothalamohypophyseal portal system to the anterior pituitary gland, which produces peptide hormones that affect a number of glands or tissues throughout the body. Growth hormone (GH) has a general anabolic effect and affects many tissues throughout the body. Thyroid-stimulating hormone (TSH) stimulates the release of thyroxin from the thyroid gland. Gonadotropic hormones (follicle-stimulating hormone (FSH) and luteinizing hormone (LH)) stimulate both ovaries and testes to produce characteristic androgenic and estrogenic steroid hormones. Adrenocorticotropic hormone (ACTH) stimulates the adrenal cortex to produce cortical steroids.

Follicles of the thyroid gland, located in the neck, produce thyroxine, which influences the metabolism of most cells of the body. Parafollicular cells in the thyroid produce calcitonin, but not under pituitary control. This hormone increases Ca^{++} retention in bone.

The adrenal cortex consists of three zones, each of which produces a different steroid hormone. The outer glomerular zone produces the mineralocorticoid, aldosterone, which influences Na^{+} exchange by the kidneys. Cortisol, a glucocorticoid, is produced by the middle zona fasciculata and increases blood glucose levels. The inner zona reticularis secretes androgenic steroids. The adrenal medulla functions independently of the pituitary and secretes epinephrine or norepinephrine upon stimulation by sympathetic nerve fibers that supply it.

The ovaries are stimulated in a cyclical manner by pituitary gonadotropins. FSH causes ovarian cells to secrete estrogens during the first half of the menstrual cycle, and LH stimulates other ovarian cells to secrete progesterone during the second half of the menstrual cycle.

In the testes, LH stimulates the Leydig cells to produce testosterone. FSH acts on Sertoli cells to generate androgen-binding protein.

The parathyroid glands, embedded within the thyroid gland, secrete parathyroid hormone (PTH), which results in the release of Ca^{++} from bone.

Islets of Langerhans, within the pancreas, secrete insulin, which decreases blood glucose levels, and glucagon, which increases them. Secretion of these is not under the direct control of the pituitary.

The posterior pituitary is a neuroendocrine gland, which releases oxytocin and vasopressin, both produced in the hypothalamus, into the blood. Oxytocin acts to cause milk letdown and also uterine contractions. Vasopressin affects water resorption by the kidney and also increases blood pressure.

Chapter 10

The Circulatory System

Very simply put, the function of the circulatory system is to circulate. With very few exceptions (e.g., cartilage, epithelia), the circulatory system penetrates all areas of the body. Only a few tens of micrometers, at most, separate cells of the body from terminal branches of the circulatory system.

In contrast to most arthropods, which have an open circulation, all vertebrates possess closed circulatory systems. A closed circulatory system consists of three principal components: a pump, a series of closed pipes leading from and back to the pump, and a medium—usually liquid—to be circulated. In vertebrates, the pump is the heart; the outflowing pipes are the arteries, which are connected to the inflowing veins by an extensive network of capillaries; the liquid medium is the blood. In addition, a separate lymphatic system drains tissue fluid from many parts of the body and carries it into the general circulation.

The main purpose of the circulatory system is to carry cells, molecules, and ions from one part of the body to another. Of most immediate importance is the transport of gases—oxygen taken up by the lungs—to all peripheral tissues and carbon dioxide, originating from these same tissues, back to the lungs for removal from the body. The circulation carries many small molecules—sugars, salts, amino acids, and lipids—to cells that require them for their metabolism. Conversely, the circulation picks up metabolic wastes from active cells and carries them to the appropriate excretory organs.

In addition to the more generic functions listed above, the circulatory system carries specialized cells and more complex molecules that are used only by specific cells and tissues within the body. Good examples are immune cells or antibodies, which are directed toward parts of the body invaded by pathogens, or hormones that only act on certain tissues. In these cases, it is important to note that the circulating blood is indiscriminate in carrying these cells or molecules to all parts of the body. The presence of local controls, such as cell surface receptors or humoral factors, for example, cytokines (see Chapters 8 and 9), determines whether or not the particular cells or molecules will be taken up where needed.

The pressures within the circulatory system are also used to transmit forces. One example on a micro scale is the filtration pressure in the kidneys that begins the process of urine formation. At the macro end, the pressure required to produce and maintain a penile erection is a requirement for reproduction in most higher vertebrates.

One of the fundamental aspects of physiological control is the maintenance of homeostasis—keeping many bodily processes operating within normal physiological limits. A classic example is body temperature. Most birds and mammals maintain their body temperatures within a range of just a few degrees. The circulatory system is exquisitely adapted to move heat from one part of the body to another so that overall body heat is either lost or retained, depending on the body's need at the time. Other components of the homeostatic environment, for example, O_2 levels, ionic balance, pH, or blood pressure, are tightly controlled through systems of sensors and effectors scattered throughout both the circulatory system and other organs.

At all levels in the circulatory system, structure is closely attuned to functional requirements. These structural adaptations range from the gross structure of the heart and the cardiac valves to the configuration of the cells lining some of the specialized capillary beds. Problems can arise when age or pathology interfere with the delicate balance between structure and function.

BLOOD

Blood is the medium through which the circulatory system operates. The adult human circulation contains approximately 5 L of blood, but the amount of circulating blood is in some form of equilibrium with reserves of noncirculating blood in the liver and spleen and with interstitial fluid in the tissues.

The Human Body. DOI: https://doi.org/10.1016/B978-0-12-804254-0.00010-7

Blood consists of two main parts—plasma and cells. Each of these is itself a complex mixture. All of the components within the blood—both cellular and molecular —have normal ranges of abundance, but the relative and absolute amounts of specific components in the blood also vary according to the needs of the body at the time. For example, numbers of certain types of white blood cells can rise precipitously during infective processes, and the number of red blood cells increases significantly in people who live at high altitudes.

Blood Cells

Blood cells, or their equivalents, have a long evolutionary history. Even the most primitive multicellular animals, such as sponges, contain cells, called amebocytes, which move about and perform phagocytic functions. These cells are not confined to vessels, but rather move about in spaces between tissue layers. Higher up the evolutionary scale, starting with worms, different types of blood cells appear. Some are still phagocytic, but others are involved in cell-mediated immune responses or in clotting reactions to injury.

Red blood cells, which carry oxygen, are not found in invertebrates. Some smaller species are able to supply oxygen to their tissues directly through small open channels to the outside. Larger species or more complex invertebrates often possess a poorly developed vascular system, but they distribute oxygen to their tissues by means of large oxygen-carrying molecules, which are directly dissolved in their tissue fluids. One such molecule is **hemocyanin**, which unlike vertebrate hemoglobin, contains copper rather than iron ions. The copper gives the fluid a bluish cast.

Red blood cells (erythrocytes) and their contents of hemoglobin have evolved with the vertebrates. Other than mammals, essentially all other vertebrate species possess nucleated red blood cells. Because they are not nucleated, mammalian erythrocytes have a limited life span and must be constantly renewed.

One feature common to both vertebrate and invertebrate blood cells is that they are typically produced from precursor stem cells (**hematopoietic stem cells**), rather than from division of mature blood cells. Vertebrate **hematopoiesis** is a complex process involving many stages and many cell types arising from a single common hematopoietic stem cell (see Fig. 8.2). Formation of each type of blood cell (Fig. 10.1, Table 10.1) follows its own unique pathway and is subject to different genetic and molecular controls.

In adult life, hematopoiesis (blood cell formation) takes place in the **bone marrow**—specifically the **red marrow**, which is located in the regions of spongy bone toward the ends of long bones or in flat bones, such as those of the

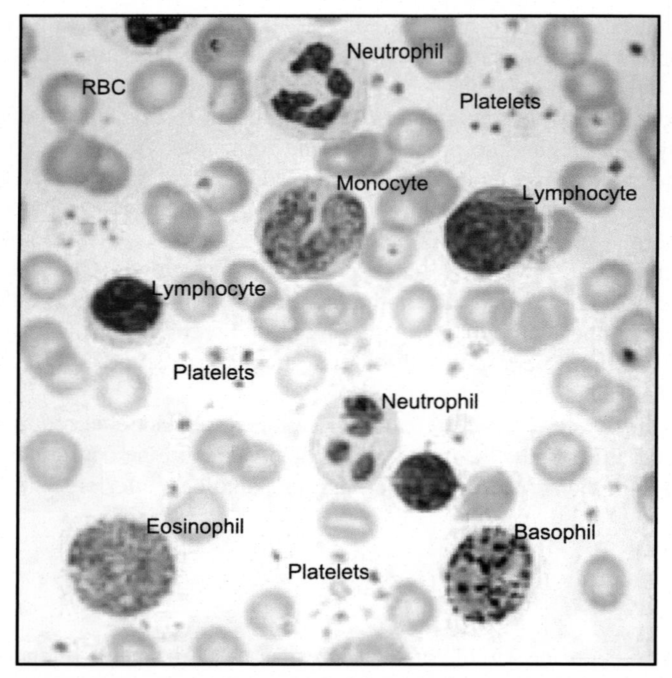

FIGURE 10.1 Cells found in the circulating blood, as seen in a blood smear. RBC— red blood cell. *From Pollard and Earnshaw (2004), with permission.*

TABLE 10.1 Properties of Mature Cells Found in Blood

Cell	Diameter (μm)	No. /μL	% of Leukocytes	Lifespan	Function
Erythrocyte	7.2	$4.5–5.5 \times 10^6$		120 days	Carry O_2 and CO_2
Platelet	2–3	$2.5–4 \times 10^5$		10–14 days	Agglutination and clotting
Lymphocyte	8–10	$1.5–2.5 \times 10^3$	10–30	Days–years	Immune responses
Monocyte	12–15	$2–8 \times 10^2$	3–8	3–4 days	Phagocytosis
Neutrophil	9–12	$3.5–7 \times 10^3$	50–70	1–2 days	Pathogen defense
Eosinophil	10–14	$1.5–3 \times 10^2$	1–3	1–2 days	Parasite defense
Basophil	8–10	$0.5–1 \times 10^2$	0.5	1–2 days	Pathogen defense

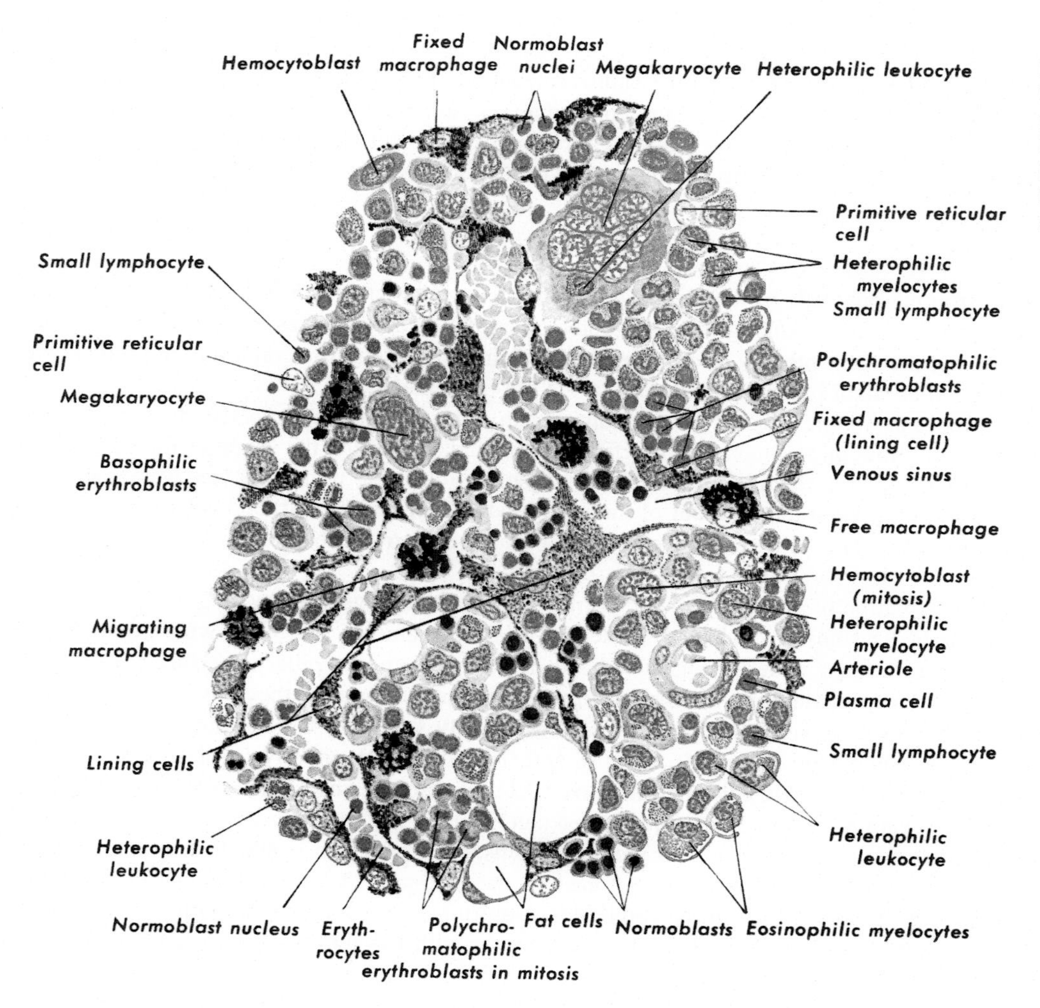

FIGURE 10.2 Drawing of a histological section through red bone marrow. Some of the phagocytic cells have taken up India ink particles (black) that had been injected in to the circulation. *From Bloom and Fawcett (1975), with permission.*

pelvis. When viewed under an ordinary microscope, the places in the marrow where hematopoiesis takes place look messy. It is difficult to define structural boundaries, and red and white blood cells in all stages of development, along with other connective tissue cells, are present (Fig. 10.2). Nevertheless, there is some degree of organization. Red marrow is technically called a **reticular connective tissue**, meaning that the three-dimensional space consists of large

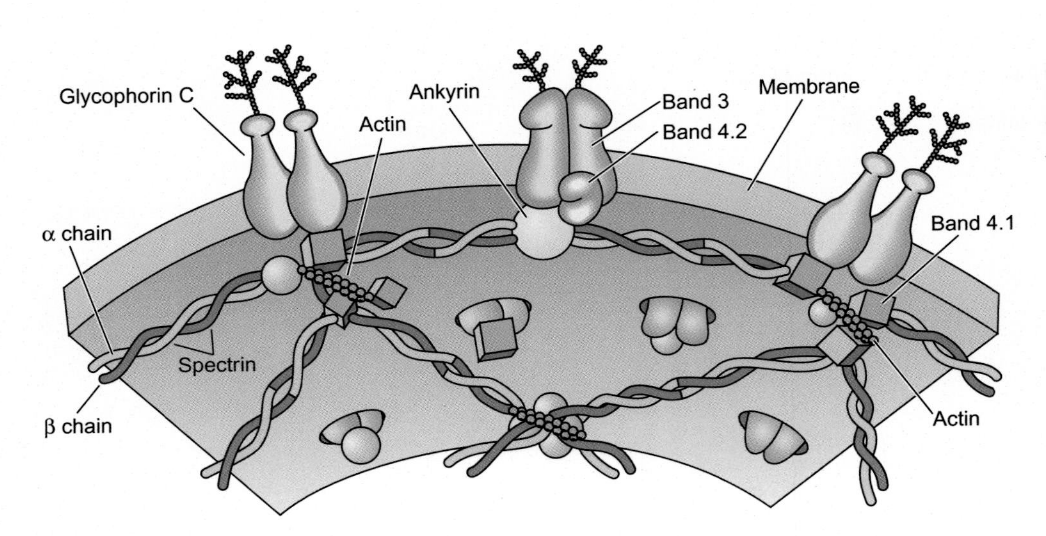

FIGURE 10.3 Proteins, especially spectrin, associated with the erythrocyte membrane. *From Gartner and Hiatt (2007), with permission.*

vascular sinusoids—large capillary-like vessels, with porous walls—surrounded by a loose meshwork of thin reticular fibers (see p. 41) and the reticular cells that produced them. The remaining space is mainly occupied by clusters of hematopoietic cells, which are in the process of forming maturing blood cells. The fluid milieu is awash with a variety of growth factors, often locally produced, that facilitate the maturation of the various types of blood cells.

Red Blood Cells (Erythrocytes)

Erythrocytes are by far the most numerous cells in the blood. Adult males have about 5–5.5 million erythrocytes per mm^3 of blood, and females about 4.5 million. Among the vertebrates, mammalian erythrocytes are unique in that they have no nucleus. Toward the end of their seventh day of development within the bone marrow, maturing red blood cells lose their nuclei and enter the blood as small biconcave disks about 7.2 μm in diameter.

Erythrocytes are often referred to as simple bags of hemoglobin, but in order to function properly, they need structural and molecular adaptations that allow these functions. The first of these adaptations is the shape of the cell, which by its design has a much greater surface area relative to volume than would be the case if erythrocytes were spherical. This greatly facilitates the diffusion of oxygen into and out of the red blood cells. Mature red blood cells are devoid of nuclei and most cytoplasmic organelles. This allows more space for the oxygen-carrying hemoglobin molecules to be packed into the cells.

Another important functional adaptation of erythrocytes is deformability. Roughly every 20 seconds a red blood cell must pass through a bed of capillaries or sinusoids, which are often 3–5 μm in diameter. For a cell with a diameter of 7.2 μm, this means that an erythrocyte must either change its shape or get stuck in a capillary. Erythrocytes have a remarkable capacity to undergo deformation and then to rebound to their original shape once they have passed through a capillary bed. Much of this is due to a network of protein molecules, mainly **spectrin**, which underlies the plasma membrane (Fig. 10.3). The spectrin network provides both strength and flexibility to the overlying plasma membrane. With aging or certain diseases, for example, **sickle cell anemia** (Box 10.1), erythrocytes become less flexible, and problems with erythrocyte flow through capillaries or cellular breakdown can occur.

Hemoglobin is a tetrameric molecule, consisting of four globin proteins, each of which is attached to a nonprotein heme group. One iron ion is jointly bound to the heme groups of each of the four subunits of the tetramer. Hemoglobin avidly binds oxygen in the lungs, but then releases it in the capillaries. Unfortunately, **carbon monoxide** (CO) binds to hemoglobin 250 times more avidly than does oxygen. It is the competition with oxygen binding in hemoglobin that results in death from CO poisoning.

Red blood cells contain many surface antigens, which collectively determine one's **blood type**. The two most common types are the **ABO system** and the **Rh system**. Erythrocytes in some individuals have type A antigens (agglutinogens) on their surfaces(type A blood), whereas others have B (type B blood), A and B (type AB blood) or no AB antigens (type O). People also have corresponding antibodies (agglutinins) against the other antigens (Table 10.2). For example, people with type A blood have antibodies against the B antigen, and conversely, people with type B blood have antibodies against the A antigen. This is of importance in blood transfusions. It is critical not to transfuse blood of

BOX 10.1 Sickle Cell Anemia

Sickle cell anemia is a genetic recessive condition caused by the mutation of a single nucleotide of the gene coding for the β-globin component of the hemoglobin molecule. This mutation, in which the normal GAG is substituted by GTG at a specific location in the DNA chain, results in the insertion of the amino acid valine instead of glutamic acid into the β-globin protein of hemoglobin. This simple change causes the hemoglobin to form abnormal protein fibers in the red blood cells when the oxygen concentration in the blood is low. This change is reversed after the oxygen levels return to normal, but after several episodes, the accumulated effects result in a stiffening of the membrane of the red blood cell and its assuming a sickle shape. Such cells lose the ability to undergo the deformation required to allow them to pass easily through capillaries. As a result, masses of sickle cells plug the capillaries in certain parts of the body. The cells and tissues supplied by these capillaries are then starved of oxygen, resulting in severe ischemic pain or even death of cells (necrosis) in the oxygen-starved areas. Many such episodes can result in the premature death of the afflicted individual. Anemia occurs when the rate of degeneration of affected erythrocytes exceeds their rate of production.

Because sickle cell disease is genetically recessive, each parent must have at least one copy of the defective gene in order to transmit the full-fledged disease to their offspring. If each parent has one copy of the defective gene, their progeny would have a 25% chance of having two copies of the defective gene (homozygous), a 50% chance of having only one copy of the gene (heterozygous) and a 25% chance of having two normal copies of the gene.

The sickle cell mutation is most commonly seen in sub-Saharan Africa and in locations around the Mediterranean Sea—areas where malaria is common. There is considerable evidence that heterozygous individuals who have only one copy of the sickle cell gene (**sickle cell trait**) might be afforded some degree of protection from the malaria parasite, which lives in red blood cells. In this case, some sickling of red cells occurs—enough to interfere with lodging of the parasite, but not enough to cause a serious health problem. Such protection seems to provide a sufficient evolutionary advantage for the group of heterozygous individuals to compensate for the disadvantage of full-fledged sickle cell disease in homozygous individuals.

TABLE 10.2 ABO Blood Types

Blood Type	Frequency	Antigens[a]	Agglutinins[b]	Blood Types Agglutinated
A	41%	A	Anti-B	B, AB
B	9%	B	Anti-A	A, AB
AB	3%	A, B	None	None
O	47%	None	Anti-A, Anti-B	A, B, AB

[a] *Also called agglutinogens.*
[b] *Or antibodies.*

the wrong type into an individual, because the reactions against the red cell antigens of the recipient would cause a transfusion reaction and the clumping (agglutination) of the blood cells. Individuals with type AB blood have no AB antibodies in their plasma and are therefore universal recipients. People with type O blood have both A and B antibodies in their blood and therefore can be transfused with only type O blood. On the other hand, type O blood contains no A or B antigens and is therefore a universal donor for recipients with any ABO blood type.

The other major blood type is Rh. Blood of about 85% of Americans contains one of three major **Rh antigens** (C, D, and E), and such individuals are considered to be **Rh-positive**. The remaining 15% do not possess these antigens and are **Rh-negative**. Problems can arise if an Rh^- mother is pregnant with an Rh^+ fetus. If some fetal blood leaks into the maternal circulation, as often happens, the mother initially produces IgM antibodies against the Rh antigens of the fetus. IgM antibodies are membrane-bound (see p. 216) and would normally not pass back to that fetus through the placenta. However, if the mother had a second pregnancy with another Rh^+ fetus, smaller IgG antibodies (the successors to IgM), which had formed earlier, could be able to leak through the placenta and react against the blood cells of the fetus, causing the condition of **erythroblastosis fetalis**. This condition results in the breakdown of the fetal blood cells and can even be fatal.

White Blood Cells (Leukocytes)

In contrast to the enormous numbers of erythrocytes, **white blood cells** quantitatively constitute only a minor cellular constituent of the blood (6,000–10,000 cells/mm^3 of blood). Whereas erythrocytes remain within the bloodstream, leukocytes function outside the bloodstream, and the circulatory system merely represents the means by which they are carried to the extravascular locations where they fulfill their duties.

Leukocytes arise from common hematopoietic stem cell precursors in the bone marrow (see Fig. 8.2). Very early in the process, however, the branch leading to the formation of lymphocytes splits off from the branch that forms all the other types of red and white blood cells. Their maturation is described in Chapter 8.

Other leukocytes (neutrophils, eosinophils, and basophils, see Fig. 10.1) are characterized by containing prominent cytoplasmic granules in their cytoplasm. These granules contain a variety of substances used in the defense against pathogens and are extruded from the circulation at sites of infection. The other type of leukocyte is the monocyte. Monocytes, which do not contain large granules, leave the bloodstream and make their way into the local connective tissues, where they act as macrophages.

Platelets

The other major cellular component of the blood is the **platelets**. Platelets are small (2–3 μm) fragments that are pinched off of projections extending from large cells called **megakaryocytes**, located within the bone marrow. Megakaryocytes have very large nuclei, which are the product of many rounds of chromosomal replication (up to 16 times) without corresponding cell division —a process called **endomitosis**. Each of their cytoplasmic processes buds off up to several thousand nonnucleated platelets, which enter the general circulation and survive for about 2 weeks. Platelets play a prominent role in stemming blood loss, first by forming aggregations (**platelet plugs**) in breaches of the vascular wall, and then by playing a prominent role in the blood-clotting process. One mm^3 of blood contains 250,000–400,000 platelets.

Blood Plasma

Plasma represents the liquid component of blood and all that is dissolved in it. About 90% of plasma is water; the rest is dissolved proteins, ions, nutrients, and gases. The most prominent plasma protein is **albumin**, which like many other proteins in the blood is produced in the liver. A major function of albumin is to maintain the osmotic pressure of the blood—of particular importance in capillary exchange (see p. 288) —but it also acts as a carrier protein for many substances (e.g., cations, some hormones, fatty acids, and even some pharmaceutical agents). Other important plasma proteins belong to the globulin family. Some **globulins** transport ions and lipids, much as do the albumins, but the γ-globulins are the main circulating antibody molecules (see p. 216). Complement proteins (see p. 217) are also important components of the overall immune defense of the body. The various protein precursors of the cascade of blood-clotting proteins constantly circulate in the blood as they await a possible call to stem bleeding after a wound (Box 10.2). Finally, several lipoproteins are heavily involved in the distribution of cholesterol and other lipids to sites throughout the body.

BOX 10.2 Hemostasis

When a wound breaches the walls of a blood vessel, several mechanisms quickly come into play to stem the flow of blood (**hemostasis**). The first is almost immediate constriction of the injured vessel by contractions of the smooth muscle cells found in its walls. The second is formation of a plug, consisting principally of platelets, that adheres to internal vascular surfaces where the endothelium has been disrupted. For small vascular ruptures, plugging the leak by platelets usually suffices. Larger breaks in a blood vessel call into play the blood-clotting mechanism.

Blood clotting is a complex chain of events, with over 50 molecular components involved. It operates on much the same principle as the complement system (see p. 217), with one factor acting as a catalyst to stimulate the activation of the next clotting factor down the line. The ultimate objective of blood clotting is to form a dense meshwork of fibers of a protein called **fibrin**. The fibrin meshwork envelops red blood cells, platelets, and other components of blood to form a tough plug that prevents blood leakage while the injured blood vessel repairs itself.

(Continued)

BOX 10.2 (Continued)

Stimuli for the formation of a blood clot appear from two different sources and lead down two separate pathways—an **intrinsic** and an **extrinsic pathway** (Fig. 10.4). The intrinsic pathway involves a large number of clotting factors produced by the liver and always present in the blood. The release of **thromboplastin** by injured tissues sets off the extrinsic pathway. Through the sequential activation of many other blood-clotting factors, these two pathways ultimately stimulate the conversion of **prothrombin** to **thrombin**, which in turn stimulates the conversion of the precursor molecule **fibrinogen** to fibrin (see Fig. 10.4). **Vitamin K** plays a vital role in several steps of the clotting process.

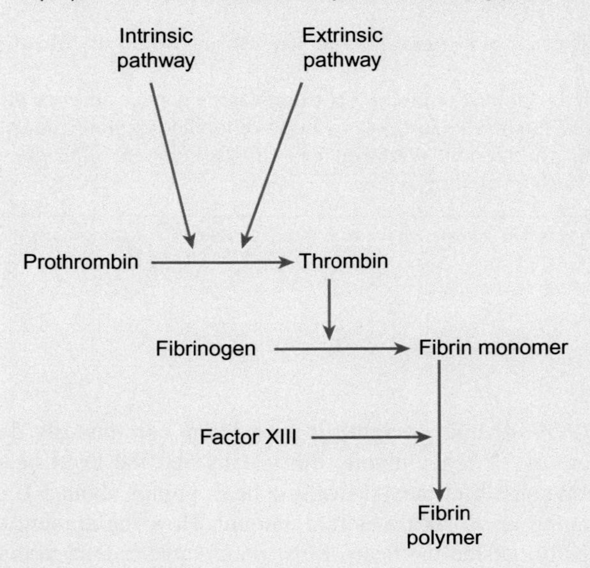

FIGURE 10.4 Abbreviated summary of major steps in blood clotting.

The fibrin meshwork initially adheres to the areas of damage and then entraps red blood cells and platelets to form the clot (Fig. 10.5). Shortly after the initial clot forms, it begins to contract and within an hour, it has squeezed out most of the serum[1] that it contains. The platelets contribute a variety of factors that help to stabilize the clot. As the clot contracts, it pulls the edges of the vascular wound together, thereby facilitating ultimate healing of the vascular walls.

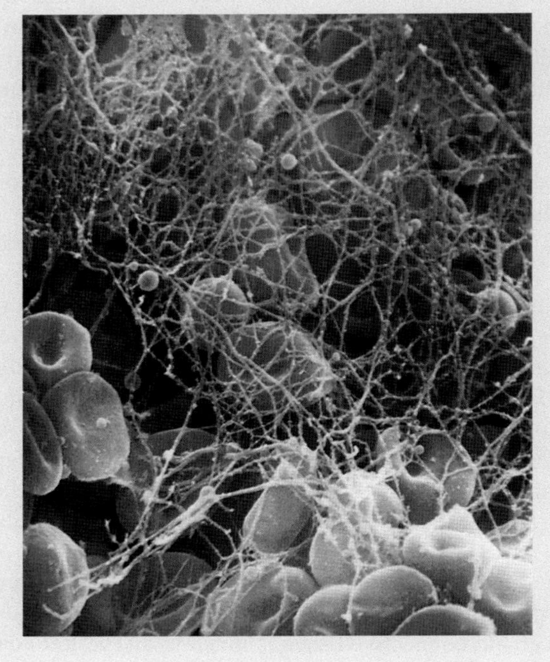

FIGURE 10.5 Scanning electron micrograph of a blood clot, showing red blood cells immobilized within a network of fibrin threads. *From Levy and Pappano (2007), with permission.*

(Continued)

BOX 10.2 (Continued)

A negative aspect of our highly efficient clotting mechanism is its potential for the formation of clots at inappropriate times and locations. Normally, clots do not form in mechanically uninjured blood vessels, but certain types of pathology can lead to the formation of a **thrombus** (blood clot) in the absence of gross injury to a vessel. A thrombus by itself can disrupt the fluid dynamics of flowing blood, but if a thrombus forms and breaks free, it is carried as an **embolus** within the blood until it reaches a branch of the vessel too small to accommodate it and subsequently blocks the flow of blood to the tissues served by that vessel. If this occurs in the brain, it can cause a **stroke**. If in the lungs, a **pulmonary embolus** can cause death in over 20% of the cases.

To maintain balance, the body produces a variety of factors that can dissolve clots. Some are resident within the blood. Others, such as **heparin** (see p. 214) are produced within the tissues.

The absence of any one of the many clotting factors interferes with the clotting of blood. For example, the genetic absence of **factor VIII** results in **hemophilia**, a condition formerly rampant in royal families in Europe as a result of inbreeding. Individuals prone to stroke are administered anticoagulants to prevent the formation of thrombi. **Warfarin**, for example, competes with vitamin K in the liver for binding sites and interferes with the formation of several clotting factors.

1. Serum refers to the liquid component of blood that no longer contains any clotting factors. Plasma, on the other hand, contains all the components of whole blood (including clotting factors) except for cells.

HEART

The **heart** is a most remarkable organ. Weighing about 0.3 kg (0.5% of body weight), it must pump continuously day and night in order for us to live. Assuming an average heart rate of 72 beats/minute, this means that the heart beats almost 104,000 times per day or almost three billion times in an average lifetime. At rest, the heart pumps about 5 L of blood per minute, but it can adapt to strenuous exercise by pumping up to six times that amount. How the amount of blood pumped is regulated is subject to a variety of controls, mostly outside the heart. Most remarkable is that cardiac muscle cells are rarely renewed. Therefore, the same cells must continue to maintain their function over a lifetime, without being replaced by new cells. This is in marked contrast to the epithelial cells covering the skin or lining the gut, which have lifetimes measured in days.

Although only one organ, the heart is really two pumps operating in series. Poorly oxygenated venous blood enters the **right atrium** and is then pumped into the **right ventricle** (Fig. 10.6A). The right ventricle then pumps all of the blood through the **pulmonary artery** (the only artery that carries poorly oxygenated blood) to the lungs. Within the lungs, the blood becomes oxygenated (see p. 317) and is returned to the left atrium of the heart through the **pulmonary vein**, the only vein in the body to carry highly oxygenated blood. Receiving blood from the **left atrium**, the **left ventricle** then pumps the oxygenated blood into the **aorta**, which distributes it to smaller vessels throughout the body. Veins collecting blood from capillary beds throughout the body then bring the blood, laden with carbon dioxide and metabolites, back to the right atrium of the heart, thus completing the circuit.

In order to function effectively to pump and distribute blood as described above, the heart must at a minimum possess two major properties—the ability to pump and to ensure that the pumped blood always moves in a forward direction. The former is accomplished through the coordinated actions of the cardiomyocytes (the muscular cells of the heart wall) and the cardiac valves.

Cardiomyocytes are muscle cells. Their contractile apparatus is organized and functions much as does the contractile apparatus of skeletal muscle fibers (see p. 118). Significant differences are (1) that the individual cardiomyocytes are not supplied with motor nerve endings, as are skeletal muscle cells, and (2) cardiomyocytes are functionally connected to one another at their ends through **gap junctions** (see Fig. 2.25). Because of their need to pump continuously, cardiac muscle cells are richly supplied with mitochondria —a parallel with slow skeletal muscle fibers. As befitting cells that have high metabolic requirements, cardiomyocytes are surrounded by robust networks of capillaries that carry oxygen to them and remove carbon dioxide and metabolic wastes. Despite the fact that there is little or no turnover of actual cardiomyocytes, their internal components are continuously renewed to keep them in good working condition.

The relative thicknesses of the walls of the four chambers of the heart reflect the functional demands placed upon them. Little force is required to move blood from the two atria to the ventricles. Therefore, the atrial walls are much thinner than those of the ventricles, and the ratio of cardiac muscle cells to connective tissue is much less. The left ventricle must pump blood at a significant pressure (mean pressure ~95 mm Hg) to reach all of the peripheral tissues and

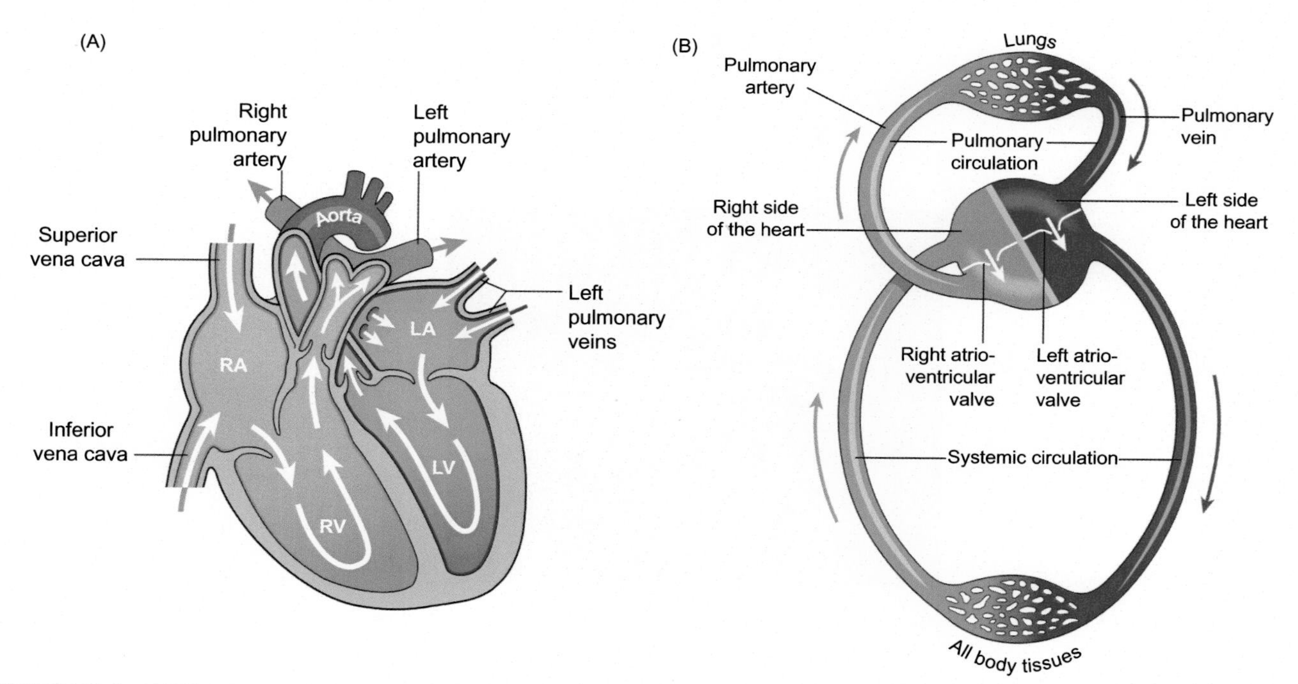

FIGURE 10.6 (A) Direction of blood flow through the heart. (B) General scheme of blood circulation throughout the body. *From Waugh and Grant (2014), with permission.*

therefore has a very robust muscular wall. On the other hand, much less pressure is required to perfuse the lungs with blood (mean pressure ~15 mm Hg), and the thickness of the right ventricular wall is correspondingly less. All parts of the heart from inside out consist of three major layers—an inner **endocardium**, the lining cells of which are very similar in structure and function to the endothelial cells that line blood vessels; a middle **myocardium**, the muscular layer; and a thin outer **epicardium**. The epicardium is covered with a simple epithelium that acts as the interface with the lining of the **pericardium**, the sack surrounding the heart. The space between the epicardium and the pericardium contains a thin layer of fluid that acts as a lubricant. Under normal circumstances, one never notices the displacements between the outer surface of the heart and its overlying pericardium. Only if there is inflammation or the leakage of blood or tissue fluid into the pericardial cavity is one painfully aware of this space.

For the heart to function effectively as a pump, it must be able both to reduce its internal volume in order to expel blood and to ensure that the expelled blood is all moving in the same direction. The former is a function of the overall shape of the heart and the arrangement of the muscle fibers within it. The latter depends upon the presence of a good system of one-way valves.

Most cardiomyocytes are arranged in an end-to-end fashion and are both structurally and functionally connected to one another by intercalated disks (for strength) and gap junctions (for electrical connectivity). The muscular component of the ventricular walls originates at a series of rings of dense connective tissue, sometimes called the **cardiac skeleton**, that surrounds the heart valves. Then the interconnected heart muscle cells proceed as bands that spiral toward the apex of the heart. Coordinated contractions of these muscle fibers reduce the size of the ventricular cavities. Interestingly, a surprisingly small amount of shortening of the cardiomyocytes (6%−10%) is required to reduce the volume of the left ventricle to the point where it squeezes out the typical amount of 70 ml of blood per beat at rest.

The directionality of blood flow throughout the circulatory system is maintained by a system of valves in both the heart and the veins. Within the heart, four valves ensure that blood is efficiently transferred out of each of the four chambers (Fig. 10.7). At the bases of the aorta and pulmonary arteries are aortic and pulmonary valves (the **semilunar valves**), each of which contains three overlapping leaflets (Fig. 10.8). The valves are designed to be open when blood is being pumped out from the ventricles, and they snap closed when the ventricles relax and the blood pressure in these major arteries is greater than that in the ventricles. A three-leaflet **tricuspid valve** between the right atrium and right ventricle and a two-leaflet **mitral valve** between the left atrium and left ventricle are constructed to prevent blood from back-flowing into the atria when the ventricles contract. Critical to the long-term maintenance of their ability to hold

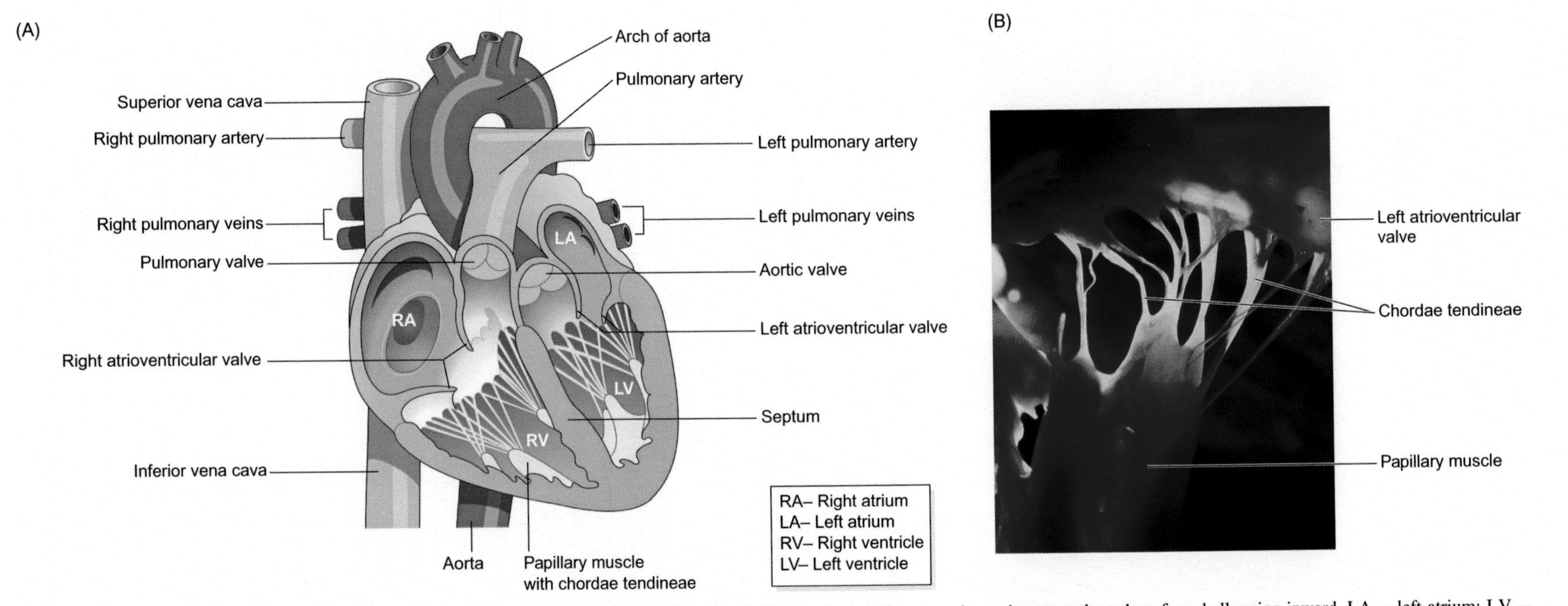

FIGURE 10.7 Heart valves. Note the chordae tendinae, which connect the edges of the valve leaflets to the papillary muscles and prevent the valves from ballooning inward. LA – left atrium; LV – left ventricle; RA - right atrium; RV – right ventricle. *From Waugh and Grant (2014), with permission.*

(A)

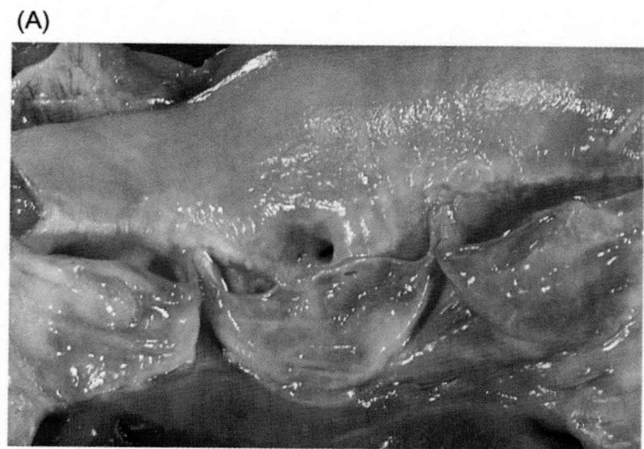

(B)

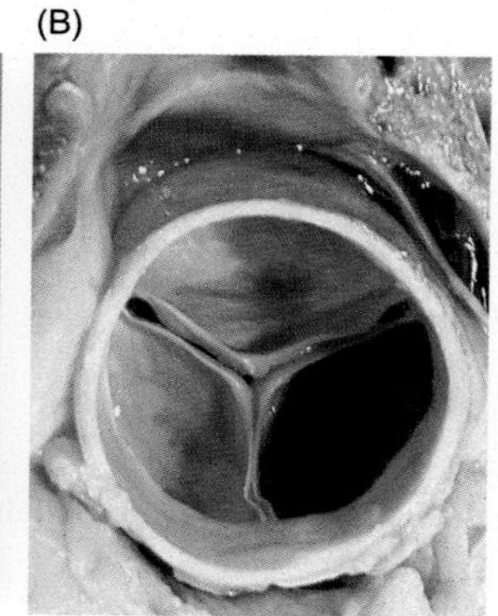

FIGURE 10.8 The aortic valve. (A) The opened aortic valve. *A*, aorta; *C*, the three cusps of the valve; *CA*, origin of one of the coronary arteries; *FR*, fibrocollagenous ring around the valve; *LV*, left ventricle. (B) The closed aortic valve. *From Stevens and Lowe (2005), with permission.*

back blood under pressure are groups of thin cords of dense connective tissue (**chordae tendineae**) that connect the ventricular sides of the valve leaflets to **papillary muscles**—projections of cardiac muscle into the ventricular cavities (see Fig. 10.7B). Without the chordae tendineae, the leaflets of the tricuspid and mitral valves would be blown backward into the atria when the ventricles contract.

The four-chambered heart with valves is a product of later vertebrate evolution. Fishes do not need such a complicated arrangement because of the relationship between the heart and gills. In a fish, venous blood enters the single-chamber atrium of the heart and passes through to a single ventricle. An atrioventricular valve prevents backflow into the atrium when the ventricle contracts. Blood ejected from the ventricle enters a ventral aorta, which sends branches and ultimately a capillary network to the gills, where the blood is oxygenated. The oxygenated blood is then collected from the gills into a large dorsal aorta, which distributes the blood throughout the body. Venous blood then returns to the heart through a system of veins. With the development of lungs in amphibia, the overall pattern of the circulation became more complex, and incomplete partitions of the heart into left and right chambers appeared. Despite being anatomically incomplete, the partitions nevertheless effectively separate pulmonary and systemic blood. The embryological development of the human heart and major blood vessels, especially in the area of the pharynx, reflects the evolutionary origins from the simple pattern of blood vessels that supply the gills in fishes.

Cardiac Rhythms

Control of the heartbeat occurs at many levels. Slow rhythmic beating is an intrinsic property of cardiomyocytes. If heart muscle cells are grown in tissue culture or are transplanted into other regions of the body, they spontaneously contract and relax at regular intervals, often as slowly as 20 beats per minute. One interesting property of heart muscle cells is that they are very responsive to signals telling them to beat faster. Some of these signals are intrinsic to the heart. Others come from the outside. The general rule is that the cells that generate the fastest rate of contraction determine the frequency of the overall heartbeat.

The generation and conduction of intrinsic signals occur through two groups of **pacemaker cells** and a specialized system of modified cardiac muscle cells that carry the signals to all regions of the heart (Fig. 10.9). The master pacemaker is the **sinoatrial (SA) node**, which is located in the upper part of the right atrium near the region where the superior vena cava enters the atrium. Action potentials arising from the highly modified cardiac muscle cells that constitute the node spread to adjoining cells through gap junctions and determine the rate of their contraction. Normally, the rate generated by the SA node is greater than 60/min, in contrast to intrinsic beats in other regions of the heart, which range from 30–40/min. Specialized **conduction pathways**—also of highly modified cardiomyocytes—lead from the SA node. One pathway leads to another pacemaker, the **atrioventricular (AV) node**, also located in the right ventricle, but close to the atrioventricular junction. Another pathway leads to the wall of the left atrium. Within both the right and left atria, cardiomyocytes not part of any established pathways rapidly pass action potentials from one cell to the next so that each atrium contracts as a whole. The intercellular contractile signals are prevented from entering the ventricular heart muscle by the nonconducting fibrous rings of the cardiac skeleton that divide the atria from the ventricles and also serve as anchors for the cardiac valves.

The contractile signal travels throughout the atria and to the AV node at a rate of about 1 m/second. Within the AV node, the signal slows down. Otherwise, the normal speed of these contractile signals would not allow the heart to fulfill its mechanical functions in a coordinated manner. The cells at the lower end of the AV node merge smoothly into

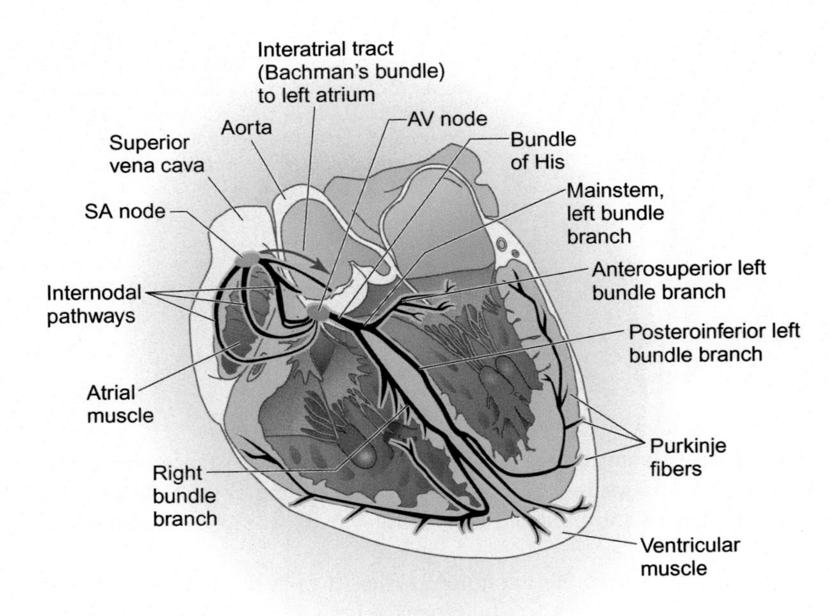

FIGURE 10.9 The conduction system in the heart (black lines). *From Boron and Boulpaep (2012), with permission.*

another bundle of modified cardiac muscle fibers (**bundle of His**), which passes through the fibrous tissue of the cardiac skeleton and into the region of the **interventricular septum**. Normally, this is the only means by which the contractile signal can pass from the atria to the ventricles. The bundle of His soon splits into two branches that proceed down the muscular part of the interventricular septum toward the apex of the heart. When it approaches the apex, the components of the right bundle branch fan out in many directions throughout the right ventricular wall (see Fig. 10.9). The left bundle branch follows a similar pattern. Within the ventricular walls, the cells of the conduction system become enlarged and are filled with glycogen. Called **Purkinje fibers**, they contain few myofibrils and conduct the contractile impulse more rapidly than do ordinary cardiac muscle cells. Purkinje fibers have an intrinsic beat of 15–40/minute, but under normal circumstances, this beat is overridden by the more rapid signals coming down the conduction system from the SA and AV nodes. As with the atria, the wide distribution of conducting tissue and the rapid spread of the contractile stimulus throughout the ventricular cardiomyocytes produces coordinated contractions of each of the ventricles.

The overall function of the heart can be well illustrated by following a drop of venous blood from the time it enters the right atrium until it finally leaves the left ventricle as arterial blood. This sequence of events is often called the **cardiac cycle**, which begins with an action potential arising in the SA node and spreading throughout the muscular wall of the right atrium (events in the left side of the heart will be covered later). The electrical depolarization of the muscle of the right atrium results in a uniform contraction (**atrial systole**) and a small increase in right atrial pressure (up to 5 mm Hg). Meanwhile, the right ventricle is relaxing (**ventricular diastole**) after its last contraction. The combination of atrial contraction and ventricular relaxation results in the flow of blood from the right atrium through the open atrioventricular (tricuspid) valve and into the right ventricle. While this is occurring, the contractile stimulus has progressed from the SA node to the AV node and into the ventricles via the bundle of His and the bundle branches. When the right ventricular volume is at its maximum, the ventricular musculature is stimulated to contract (**ventricular systole**) as the result of the spread of the action potential from the bundle branches and terminal conduction pathways into all of the ventricular cardiomyocytes, thereby increasing the intraventricular pressure to about 35 mm Hg. This pressure forces the tricuspid valve to close, while at the same time opening the semilunar valve at the base of the pulmonary artery to open. The restraining effects of the chordae tendineae and the papillary muscles to which they are attached prevent the leaflets of that valve from ballooning into the right atrial cavity as the valves prevent backflow into the right atrium.

Once in the pulmonary artery, the poorly oxygenated venous blood flows to the lungs, where it becomes reoxygenated as it flows around the pulmonary alveoli (see p. 318). The relatively low resistance within the pulmonary

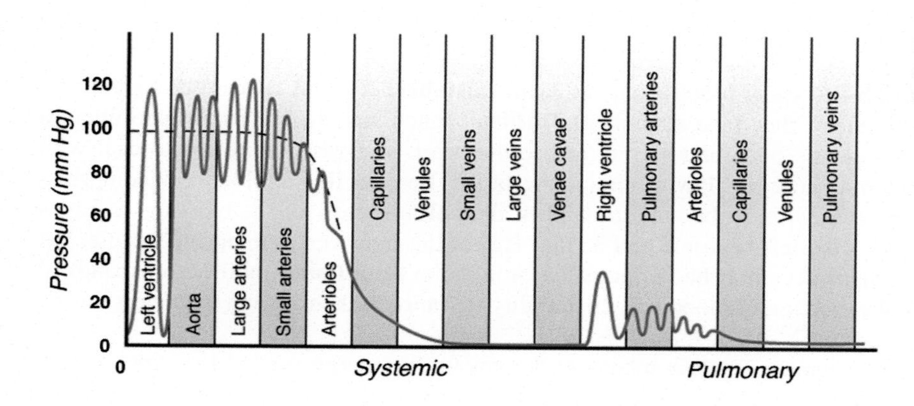

FIGURE 10.10 Typical blood pressures in various parts of the circulatory system. *Modified from Guyton and Hall (2006), with permission.*

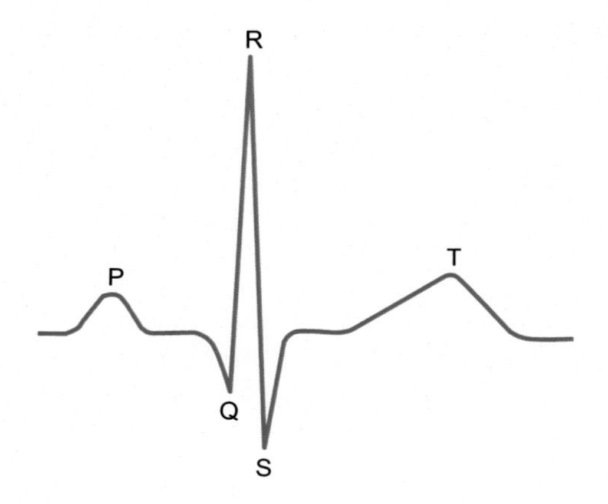

FIGURE 10.11 Components of an electrocardiogram tracing. *From Levy and Pappano (2007), with permission.*

vasculature reduces the requirement for a higher blood pressure in the pulmonary arterial system (Fig. 10.10). The well-oxygenated blood returns to the left atrium through four pulmonary veins.[2] The left atrium contracts simultaneously with the right, but the pressure of the left atrial blood is a few mm Hg higher than that in the right atrium. This pressure differential can be significant in cases of congenital persistence of open channels in the interatrial septum. In such circumstances, some blood from the left atrium flows into the right atrium, leading to a gradually worsening circulatory imbalance within the heart.

As the left atrium contracts in synchrony with the right, the oxygenated blood flows through the **mitral valve** into the left ventricle while both the left and right ventricles are relaxed in diastole. The left ventricle contracts during the systole that it shares with the right ventricle, and the intraventricular pressure rises as high as 120 mm Hg. This elevated pressure forces the leaflets of the mitral valve to close while blood is pumped from the left ventricle into the aorta through the **aortic semilunar valve** (see Fig. 10.8). Approximately 70 ml of blood leaves the left ventricle and enters the aorta with each heartbeat.

The general principle in the cardiac cycle is that electrical stimulation precedes the mechanical response. The pattern of electrical stimulation is well reflected in a typical electrocardiogram (ECG or EKG) tracing (Fig. 10.11). The P wave represents the electrical depolarization that sweeps over the atria from its origin in the SA node, and it immediately precedes the atrial systolic contraction. The QRS component of the tracing represents the electrical excitation that stimulates the systolic contraction of the of the ventricles along with an increase in intraventricular pressure, and the interval between the P wave and the QRS complex corresponds to the time involved in carrying the electrical signal from the SA to the AV node. The return of the ventricles to the resting state (repolarization) is indicated by the T wave, which is coincident with relaxation of the ventricular muscles and the lowering of intraventricular pressure. The cycle then repeats itself.

---

2. The reason for four pulmonary vein openings into the left atrium is that during embryonic development the wall of the expanding left atrium engulfs the single terminal trunk of the pulmonary vein. The process of engulfment continues until the left atrial wall has swallowed up the first and second branching points of the pulmonary vein, leaving four secondary branches separately entering the left atrium.

THE VASCULATURE

Blood vessels have a seemingly simple job—to carry blood from the heart, distribute it to all tissues throughout the body and finally, return it to the heart. In reality, they face a number of difficult functional challenges. It takes about a minute for blood leaving the heart to traverse the body and be returned to the heart. The vasculature has to deal with both propelling blood forward and maintaining an even flow of circulating blood. Being a liquid, blood is incompressible, and the vascular system (mainly the arteries) must be able to dampen the pulsatile flow of the blood as it leaves the ventricles at pressures of 120 mm Hg for the left ventricle and 35 mm Hg for the right ventricle during systole, but then drops to 0 mm Hg during diastole. This is accomplished principally through the **compliance** (ability to expand) of the large arteries (see below) during systolic ejection of blood and their ability to return to their resting diameters during diastole.

Another design issue that the vascular system has to face relates to the amount of energy needed to propel a fluid through pipes. If a pipe with a given cross-sectional area branches into two equal smaller pipes, which yet have the same total cross-sectional area as the single larger pipe, it takes eight times as much energy to move the liquid through the two smaller pipes. The body deals with this dilemma by expanding the number of branched pipes (smaller arterial branches) to the point where their total cross-sectional area is sufficient to overcome their individual resistances to fluid flow. By the time, the arterial system has completed its branching from the single aorta to its final capillary bed, the total cross-sectional area of the lumens of the capillary bed is about 700 times greater than that of the aorta. Without getting into specific details, suffice it to say that the overall pattern of branching and the respective diameters of the various arterial branches well accommodates the physical requirements of fluid flow dynamics. Constriction of the arteries in one region of the body must be accompanied by relaxation of arteries somewhere else in order to accommodate the full flow of blood through the vasculature.

Another important consideration is the characteristics of blood flow. In most parts of the vascular system, blood flow is smooth (laminar) instead of turbulent. The dynamics of fluid flow show that actual flow is greatest in the center of the lumen of a pipe or blood vessel and that it slows greatly toward the vascular wall in a paraboloid fashion (Fig. 10.12). In the case of blood vessels, this pattern of flow reduces stress on the endothelial cells that line the vessels and makes it easier for white blood cells to exit the circulation.

To meet the mechanical and metabolic demands of the circulation, the walls of the blood vessels must be structured in a manner that meets these demands, which vary considerably from one region to another. This is accomplished by regional modifications. The basic structure of a blood vessel is straightforward. The wall consists of three layers, each of which is constructed to meet the local functional demands. The inner layer, called the **tunica intima**, consists of the endothelial lining and a backing of connective tissue. The middle layer (**tunica media**) contains smooth muscle cells intermingled with varying amounts and types of connective tissue fibers. The outer **tunica adventitia** consists of connective tissue. In larger vessels, this layer also contains small blood vessels (**vasa vasorum**), which are needed to nourish the cells of the thick vascular walls.

In order to maintain homeostasis, the vascular system must be able to make fine adjustments to accommodate local or systemic needs. Much of this is accomplished through small and often ill-defined sensors that are distributed in various locations throughout the cardiovascular system.

Arteries

The main function of the arterial system is to conduct blood from the heart to the terminal capillary beds. An important mechanical challenge, especially for those large arteries closest to the heart, is to buffer the pressure changes caused by the heartbeat. Blood leaves the heart at high pressure during ventricular systole, but during diastole, no blood leaves the heart. The large arteries must be able to accommodate the 70 ml of blood that is pumped out of the heart with each beat. Mechanically, this is only possible if the large arteries are extensible when that bolus of blood is pumped into

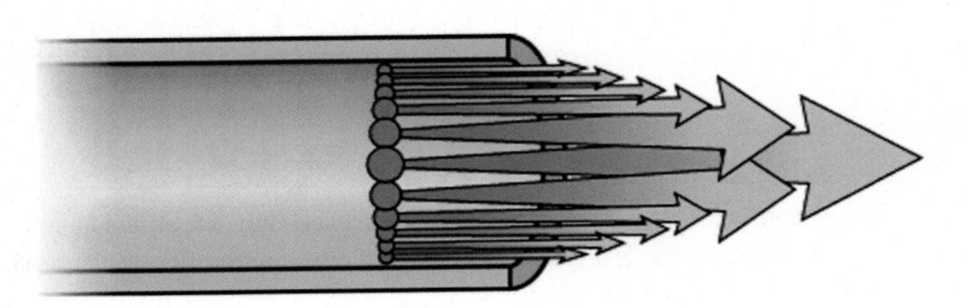

FIGURE 10.12 Longitudinal section through a blood vessel, showing the parabolic distribution of speeds within the vessel, with the fastest speed (*longest arrows*) in the center and the slowest speeds (*short arrows*) near the edge of the lumen.

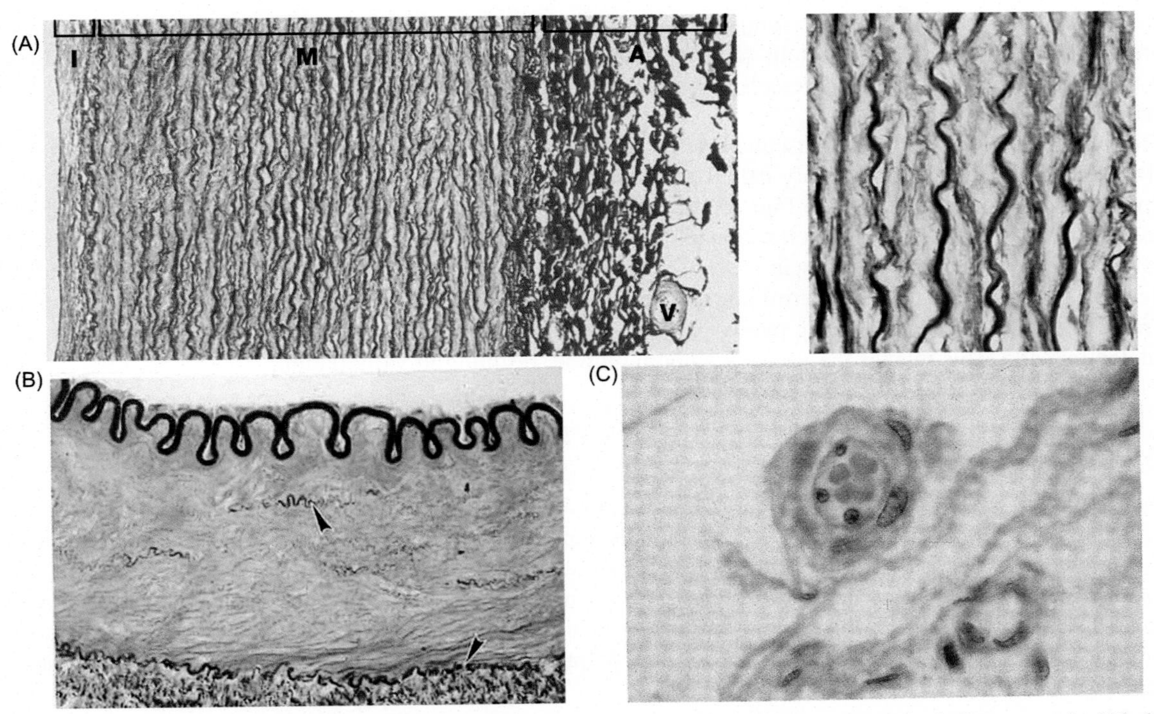

FIGURE 10.13 Histological sections through arterial walls of different dimensions. (A) Elastic artery (aorta). Elastic fibers are stained black, smooth muscle pink, and collagen fibers are blue. (B) Muscular artery, stained to demonstrate elastic fibers (black). The lumen of the vessel is on top. Note the prominent internal elastic lamina (convoluted black line) and the thick layer of smooth muscle (yellowish brown) beneath it. (C) Small artery (arteriole) with four pink erythrocytes in the lumen. The three dark circles are cross-sectioned endothelial cell nuclei. A single layer of smooth muscle cells surrounds the endothelium. *From (A) Young et al. (2006), with permission, (B) and (C) Erlandsen and Magney (1992), with permission.*

them. The aorta and other large arteries (often called **elastic arteries**) accomplish this largely through the structure of the tunica media. In these vessels, the tunica media consists of concentric layers of elastic fibers, smooth muscle cells, and collagen fibers (Fig. 10.13A). As blood is pushed out of the ventricles during systole, the elastic arteries expand to accommodate the extra volume of blood. The elastic fibers are for the most part oriented in a straight-line fashion in the unstretched artery, and as the artery expands, they become stretched under tension. Then, as the heart enters diastole, the elastic fibers in the walls of the expanded arteries spring back to their resting length, much like a rubber band—a property called **resiliency**. This action propels the blood forward and evens out the blood pressure, so that instead of going from 120 to 0 mm Hg with each heartbeat and then back again, a typical arterial pressure might be 120/80 mm Hg.[3] The collagen fibers in the tunica media and tunica adventitia provide the mechanical strength that prevents the excessive stretching or bursting of the arteries. In the unstretched artery, the collagen fibers, in contrast to elastic fibers, are kinked. As the artery expands, the inelastic collagen fibers stretch out, and after a certain point, the wall of the stretched artery becomes stiffer (up to 50 times) due to the resistance of the collagen. Elastic arteries contain relatively few smooth muscle fibers in their tunica media. Smooth muscle becomes more prominent and functionally more important in smaller arteries farther down the arterial tree. The wall of the aorta and other elastic arteries is so thick that diffusion alone cannot supply the cells of the wall with required nutrients and oxygen. These larger vessels are supplied with small vessels (**vasa vasorum**) that pass through their walls to serve the needs of the cells of the tunica media and adventitia.

As the large elastic arteries branch into smaller vessels, the character of the arterial walls changes, especially in the tunica media. These medium-sized arteries are often called muscular arteries (typically the smallest named arteries in the body) because smooth muscle cells become the most prominent component of the tunica media. Instead of being scattered in layers among other components of the media, the elastic fibers consolidate into a robust internal elastic lamina on the inner border of the muscular tunica media and a thinner layer along its outer border (Fig. 10.13B). The

3. As one ages, the compliance (extensibility) of arterial walls decreases due to changes in both the elastic and collagen fibers. Because of this, the walls of the aorta and larger elastic arteries cannot expand to accommodate the systolic pulse of blood as efficiently as they did at a younger age. The result is an increase in blood pressure.

smooth muscle layer is innervated by sympathetic nerve fibers. Although thanks to their internal and external elastic membranes, muscular arteries still retain some elasticity, the increasing prominence of smooth muscle allows active constriction and relaxation of these vessels, whereas the reactions of elastic arteries to the pulses of blood are more passive.

With further branching, smaller muscular arteries give way to arterioles, the smallest members of the arterial system (Fig. 10.13C). Arterioles are commonly called **resistance vessels**, and it is in this segment of the arterial system that blood pressure drops most steeply (see Fig. 10.10). Relative to the size of their lumen, these vessels have a thick smooth muscle wall, and by contraction or relaxation of the smooth muscle, they exert the most powerful effect on overall blood flow. Because their smooth muscle is well innervated by sympathetic nerve fibers, local contraction of these vessels plays a very important role in determining the peripheral distribution of arterial blood. Relaxation of most arterioles is not nerve-mediated. Instead of being purely passive, arteriolar relaxation can be mediated by nitric oxide released by the endothelial cells. This relaxes the vascular smooth muscle, resulting in vasodilation.

The Microcirculation

Terminal arterioles have diameters of roughly 10–100 μm and are barely visible to the naked eye. They are the vascular gatekeepers to the network of capillaries that actually supply cells and tissues with oxygen and nutrients and remove waste materials (Fig. 10.14). Arterioles most commonly branch into capillaries, which then collect into small venules, but in some tissues, such as the skin, terminal arterioles connect directly to small venules (**arteriovenous anastomoses**), allowing blood to bypass the capillary network. In the skin, this can be useful in regulating body temperature. When blood reaches the capillaries, the pressure is less than 35 mm Hg, and by the time it leaves the pressure has fallen to ~15 mm Hg.

Capillaries are very thin vessels, with diameters between 3 and 8 μm. They consist of a thin inner layer of endothelial cells, an underlying basement membrane and a few strands of surrounding collagen fibers (Fig. 10.15). Occasionally a generic type of cell, called a **pericyte**, lies alongside capillaries. The function of pericytes has long been obscure. Although thought to be contractile and in some manner influencing capillary diameter, pericytes are now also suspected of being adult stem cells, whose capabilities for differentiation into other cell types is presently under investigation.

Even with their very thin walls, capillaries can withstand relatively high intravascular pressures. This can be explained by **Laplace's law**, which relates tension in the vessel wall to the transmural pressure and the radius of the vessel, while taking wall thickness in to account. Seemingly counterintuitively, at the same intraluminal pressure, the wall tension of a larger tube is greater than that of a smaller one. With respect to human blood vessels, the wall tension of the aorta is about 12,000 times greater than that for capillary walls, even though the difference in blood pressure is only about fivefold. A practical example is that of a partially blown-up balloon. Inside the balloon, the air pressure is

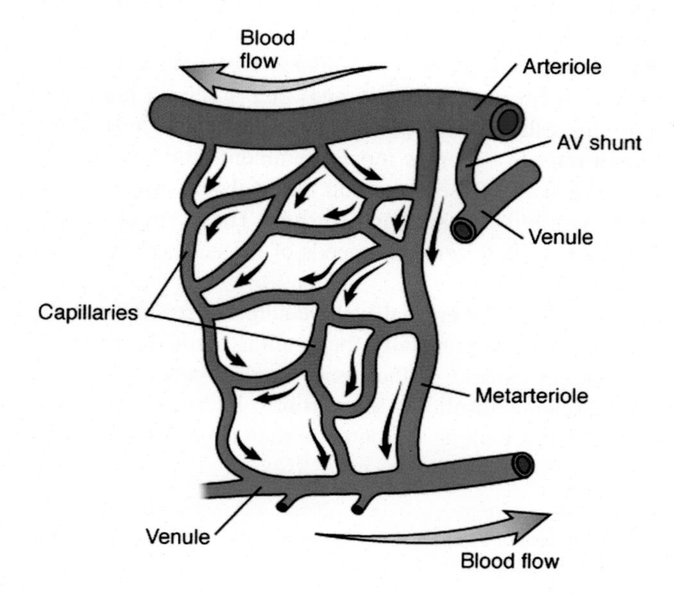

FIGURE 10.14 Schematic diagram showing the main elements of the microcirculation. An arteriovenous anastomosis is represented by the Metarteriole label. Arrows indicate the direction of blood flow. *Modified from Levy and Pappano (2007), with permission.*

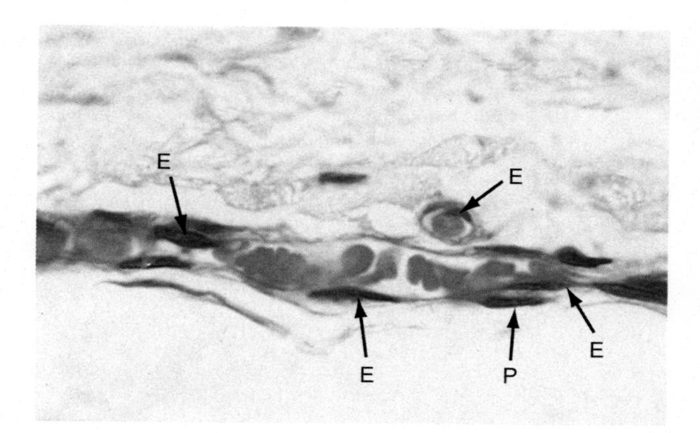

FIGURE 10.15 Longitudinal and cross-sectional views through capillaries. The cross-sectioned capillary is above the longitudinally sectioned one. The red blood cells (pink) within the capillaries are 7 μm in diameter. *E*, endothelial cell nuclei; *P*, pericyte. *From Young et al. (2006), with permission.*

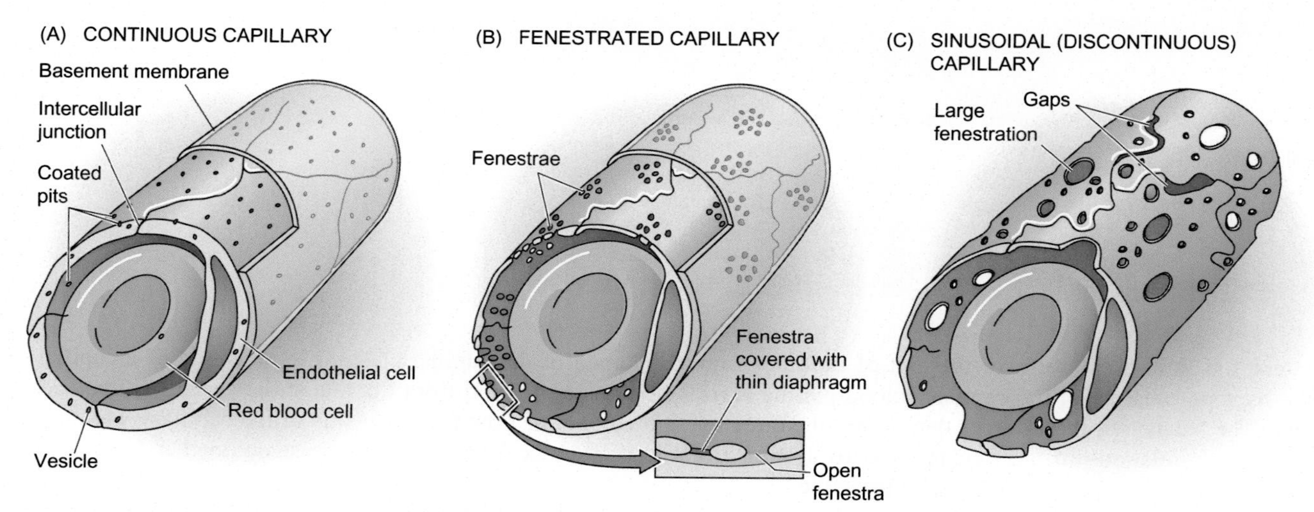

FIGURE 10.16 The three major types of capillaries. (A) Continuous capillary; (B) Fenestrated capillary; (C) Sinusoidal (Discontinuous) capillary. *From Boron and Boulpaep (2012), with permission.*

the same all throughout the interior, but the tension in the wall of the blown-up part is easily felt, whereas the unblown-up tip may be actually flaccid.

Three morphological types of capillaries have been described (Fig. 10.16). The most common type is the **continuous capillary**, which is found in muscle, skin, and most other tissues (Fig. 10.17A). The endothelial cells are continuous and are tightly bound to one another by intercellular junctions. Gases and small molecules can pass through these cells by diffusion, whereas larger molecules are taken up and transported out by **pinocytotic vesicles** (Fig. 10.17B). The surrounding basement membrane acts only minimally as a filter in this type of capillary.

Fenestrated capillaries are found in tissues with high levels of metabolism and molecular exchange, such as the kidneys (see Fig. 13.5C), small intestine and some endocrine glands. The endothelial cells of these capillaries contain pores (fenestrae, from the Latin, *fenestra*—window). Some of these pores are completely open; others are covered with thin molecular diaphragms. In some cases, the basement membrane acts more as a selective filter than does its equivalent in continuous capillaries.

The third variety of capillary is the **sinusoidal** or **discontinuous type**. Found in the liver, bone marrow, and spleen (see later), these sinusoidal structures allow the ready passage of essentially all molecules, as well as cells.

Most exchange between blood and surrounding tissues in the body takes place through the walls of continuous capillaries. Exchange of fluid by filtration between capillary blood and interstitial tissues involves four main variables—the pressure of the capillary blood, the interstitial fluid pressure, the osmotic pressure of the capillary blood and the osmotic pressure of the surrounding interstitial tissue. Overall, fluid leaves the blood in the arterial ends of capillaries and returns at the venous ends. The blood pressure, the osmotic pressure of the interstitial fluid, and a small negative

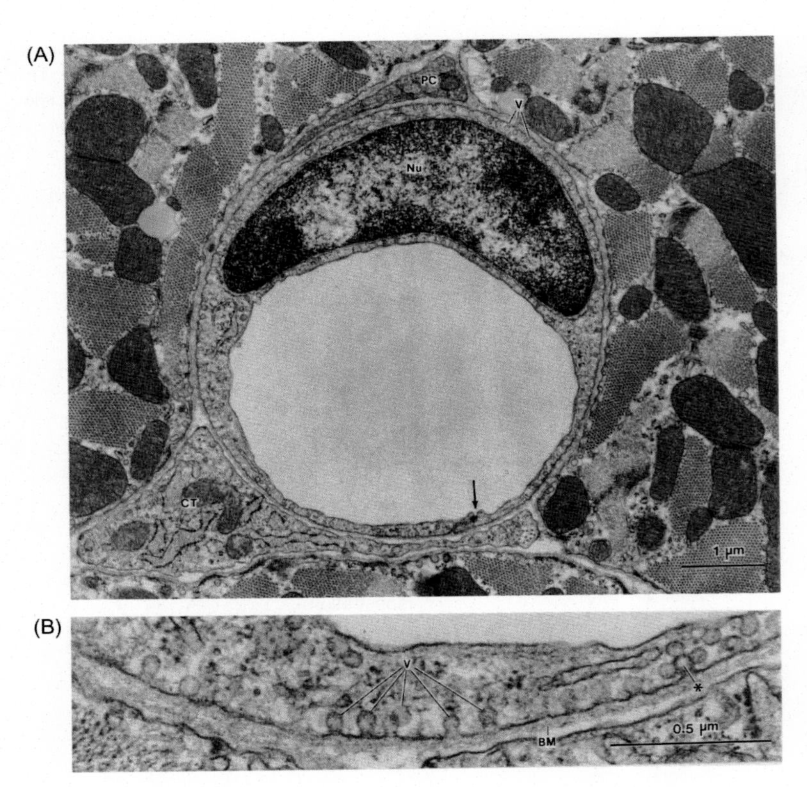

FIGURE 10.17 (A) Electron micrograph of a continuous capillary. *Nu*, endothelial cell nucleus; *V*, vesicles; *arrow*, intercellular junction. (B) Higher power micrograph through the thin cytoplasm of an endothelial cell. the lumen is on top. *BM*, basement membrane; *V*, pinocytotic vesicles; *asterisk*, GET. *From Levy and Pappano (2007), with permission.*

interstitial fluid pressure favor the exit of fluid out from the lumen of the capillary. At the venous end of a capillary, the osmotic pressure of the capillary blood plasma favors the transfer of fluid from the interstitial space into the capillary. The balance of these forces determines the direction and amount of flow from one side of the capillary wall to the other. This relationship is often referred to as **Starling's law** after the English physiologist outlined it in the early 1900s. Table 10.3 outlines the general principle.

The passage of small solutes, gases, and metabolic wastes through capillary walls is largely accomplished by diffusion. Larger lipid-insoluble molecules cannot readily penetrate the membrane of the endothelial cell. These must rely upon uptake by pinocytotic vesicles (see Fig. 10.17B) to bring them inside the endothelial cell.

After blood has passed through a capillary network, the capillary meshwork consolidates into small (10 μm diameter) postcapillary venules, the first element of the venous collecting system. These terminal venules are constructed much like capillaries, with an endothelium underlain by a basement membrane and surrounded by some wisps of collagen fibers and a scattering of pericytes. The walls of venules do not contain smooth muscle cells. The pressure of blood entering the venules is ~15 mm Hg.

Veins

The main function of the venous system is to collect blood from the capillary networks and to return that blood to the heart. Less commonly recognized, a major function of the venous system is to serve as a reservoir for the blood. About 75% of the systemic blood is found on the venous side of the circulation, especially the smaller veins.

Blood pressure in the venous system drops to close to zero past the level of the venules. Given the lack of pressure, one of the challenges of the veins in dependent parts of the body is countering the gravitational effects of hydrostatic pressure in order to move blood back to the heart. The presence of one-way valves in larger veins and mechanical pressures from muscular movements are major factors that maintain the even flow of venous blood. While standing, the pressure in foot veins can be as high as 80 mm Hg (Fig. 10.18). This is due to not only gravitational pressure, but also to the influx of liquid entering from the capillary circulation and from tissue fluids. When the intravenous pressure in the lower body exceeds that farther up the venous system, the blood is pushed up the veins until it passes through one of the one-way intravenous valves. At this point, it is prevented from falling back. The muscular movements of walking or running considerably reduce venous pressure in the legs because the muscular contractions assist in pushing the

TABLE 10.3 Example of Starling's Law of Capillary Exchange[a]

	Arterial End (mm Hg)	Venous End (mm Hg)
Forces Moving Fluid From Capillary		
Capillary pressure	30	10
Interstitial fluid pressure[b]	3	3
Interstitial fluid osmotic pressure	8	8
Total force moving fluid outward	41	21
Forces Moving Fluid into Capillary		
Plasma osmotic pressure	28	28
Total force moving fluid inward	28	28
Sum of Forces		
Outward	41	21
Inward	28	28
Net force	13 mm outward	7 mm inward

[a]*Adapted from Guyton and Hall Medical Physiology 11/e (2006).*
[b]*Interstitial fluid pressure is actually a negative force leading fluid out from capillaries because interstitial fluid draining into lymphatics creates what amounts to a slight fluid vacuum.*

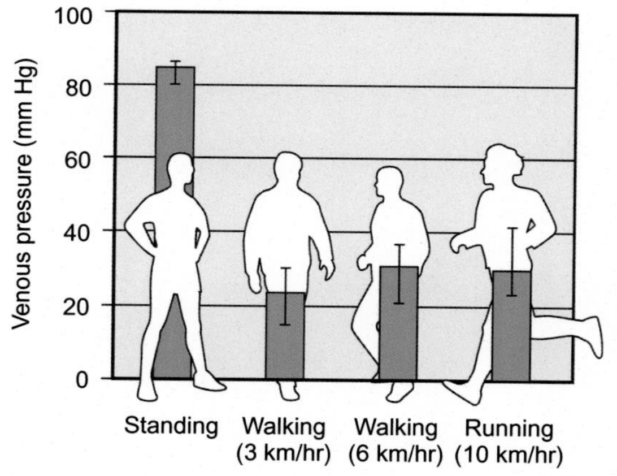

FIGURE 10.18 Effect of various activities on mean pressures in foot veins. *From Levy and Pappano (2007), with permission.*

blood along. If the venous valves become incompetent, **varicose veins**, due to the chronic pooling of blood, can result (Fig. 10.19).

Because of the lower intravenous blood pressure, the walls of veins are not as robust as the walls of arteries of the same internal diameter. As postcapillary venules merge into venules (Fig. 10.20) and larger venous channels, increasing amounts of connective tissue, containing both elastic and collagen fibers, appears as a layer surrounding the endothelial basement membrane. A thin, but well-defined layer of smooth muscle cells, innervated by sympathetic nerve fibers, appears next as the small veins approach 1 mm in diameter. This is the level in the venous system where major adjustments in blood volume occur. At the level of medium and large veins, valves appear and the walls consist of a more robust intermingling of smooth muscle and connective tissue. Like large arteries, the largest veins contain vasa vasorum in their walls. Because of our habitual erect posture, veins draining the head are minimally, if at all, equipped with valves. Gravity normally suffices to drain blood from the head.

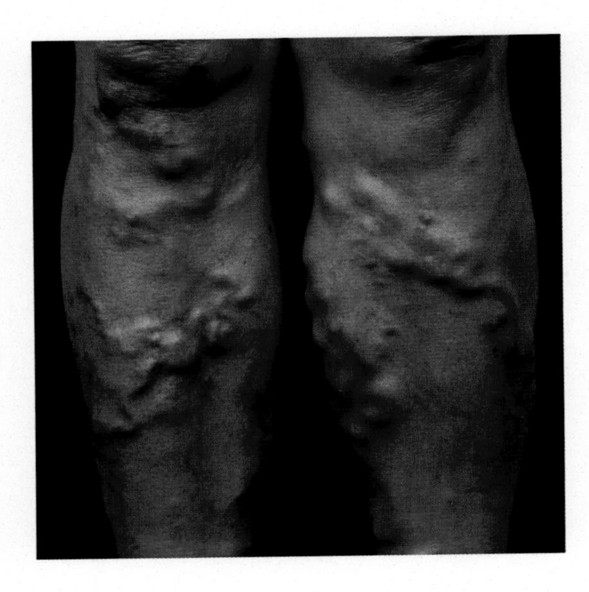

FIGURE 10.19 Varicose veins in the legs. *From Waugh and Grant (2014), with permission.*

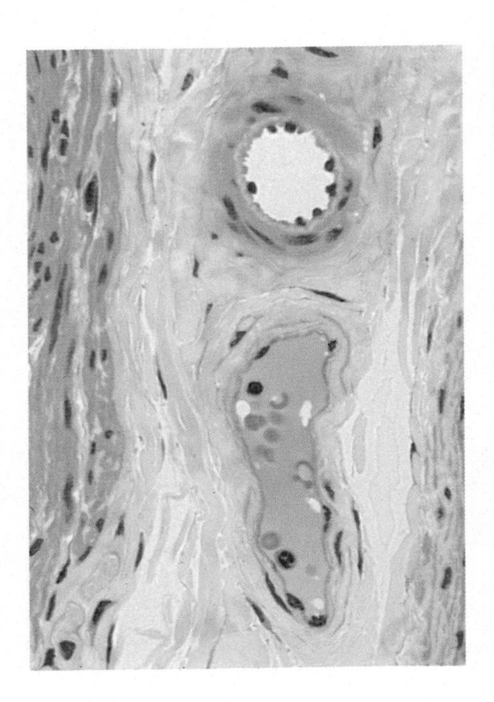

FIGURE 10.20 Photograph of an arteriole (above) and a venule (below). The arteriole has a thicker wall, mainly because of the presence of smooth muscle in the tunica media. *A*, arteriole; *L*, lumen; *N*, nucleus of endothelial cell; *RBC*, red blood cell; *TM*, tunica media; *Ve*, venule. *From Gartner and Hiatt (2007), with permission.*

Special Circulations

Different parts of the body have different functional requirements, and the vascular system often shows adaptations designed to meet these requirements. These arrangements range from the gross topographical relationships of large vessels to the fine structure of individual endothelial cells. Several of these special circulations are discussed here.

Limbs

In limbs, the major veins often closely parallel the deeply seated main arteries as well as nerves, but this juxtaposition is rarely illustrated in anatomy textbooks (Fig. 10.21) because the arterial and venous systems are often described separately. Limbs also contain sets of superficial veins, and the deep and superficial veins are interconnected. This arrangement is the anatomical basis for heat retention or loss in the limbs. When it is cold, the superficial veins collapse, and

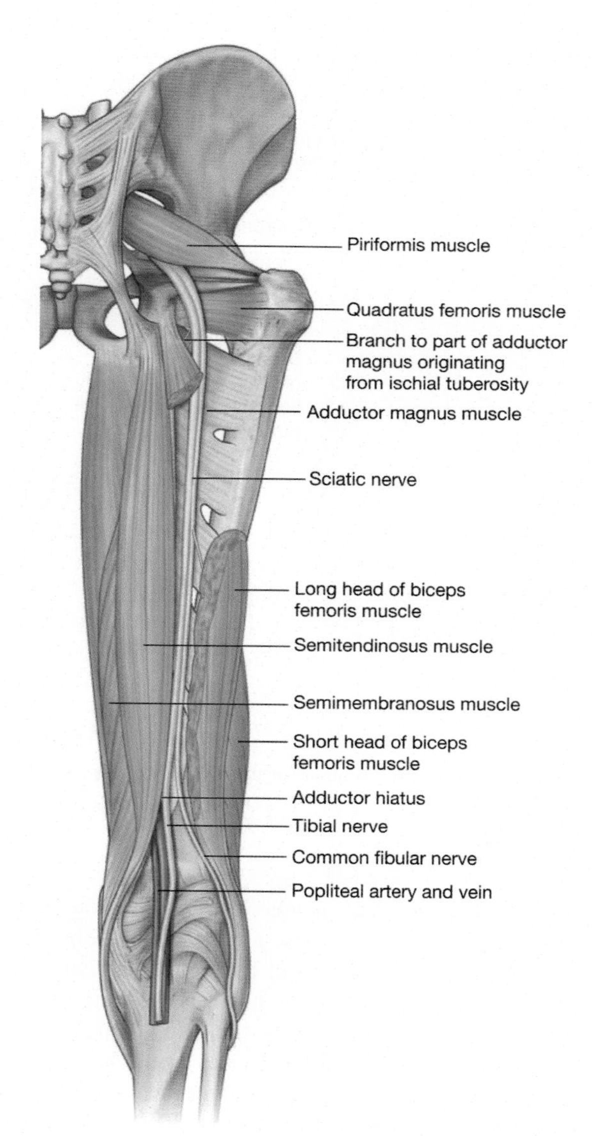

FIGURE 10.21 Dissection of the thigh, showing the close parallel relationship between the popliteal artery (red) and vein (blue), as well as nerves (yellow). *From Drake et al. (2005), with permission.*

the venous return from the limbs flows through the deep veins. Conversely, when it is warm, or during exercise, the need to remove excess body heat allows the superficial veins to become engorged through the deep-superficial vein anastomoses.

This system of close approximation of deep arteries and veins is the basis for a **countercurrent system**, which is critical for survival for many animals living in cold environments. Imagine a duck swimming or a heron standing in ice-cold water. Its feet would radiate so much body heat into the water that the bird would soon become hypothermic. This complication is avoided by the arrangement of the arteries and veins supplying the legs and feet. Right alongside the central artery (or a group of parallel arteries) bringing blood into the limbs is a set of veins running parallel and very close to the central artery as they drain the limbs. When warm arterial blood enters the artery of the leg, much of its heat is transferred directly to the vein(s) returning blood into the body. Thus by the time, the arterial blood gets to the feet, its temperature is greatly reduced from 40°C to slightly over freezing in the winter, and very little body heat is lost. In addition to heat control, countercurrent mechanisms play important roles in functions, such as salt concentration (e.g., the kidney [see p. 368] or salt-excreting glands of many ocean animals) or oxygen exchange in the gills of aquatic animals.

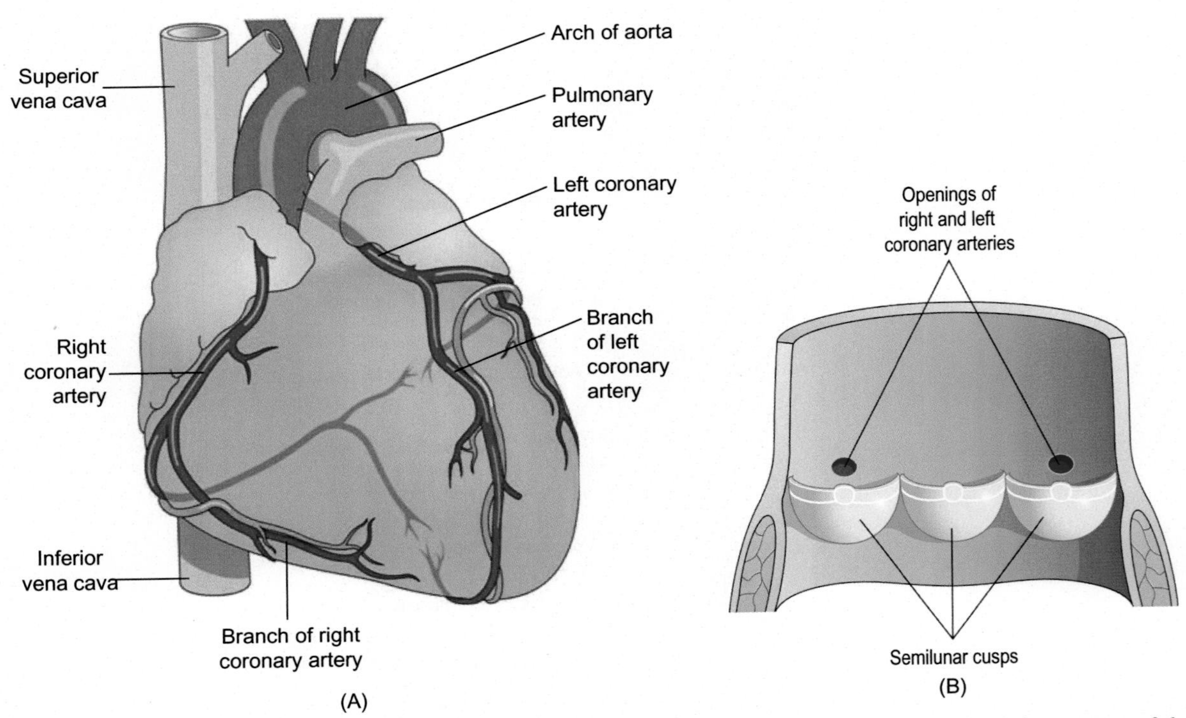

FIGURE 10.22 (A) Distribution of the coronary vessels. (B) Locations of the openings of the coronary arteries in relation to cusps of the aortic valve. *From Waugh and Grant (2014), with permission.*

Coronary Circulation

The heart is too large to be supplied with oxygen and metabolites by diffusion from the cardiac cavities. Instead, it is supplied by two coronary arteries (right and left), which arise as the first branches from the ascending aorta located just behind the leaflets of the aortic valve (Fig. 10.22). Because of its constant activity, the heart must be well supplied with blood. Although comprising only 0.5% of the body mass, the heart uses 5% of the total oxygen consumed by the body. Fortunately, the nature of the flow pattern of blood leaving the left ventricle does not compress the valve leaflets directly onto the aortic wall, or the openings of the coronary arteries would be obstructed.

The flow pattern in the coronary arteries is highly unusual in that it pretty much ceases in the intramuscular branches during ventricular systole and then resumes during diastole. The reason for this is that during systole, the vessels become compressed as the heart muscle contracts. In fact, sometimes the flow in these branch arteries can even become briefly reversed. Because the main coronary arteries are located on the epicardial surface of the heart, they are not compressed during systole.

In many parts of the body, connections exist between neighboring arterial trees so that if one artery becomes blocked, blood from its neighbor can flow into the branches of the blocked artery (this is called **collateral circulation**). Such connections are minimal between the right and left coronary arteries. Because of this, blockage of a coronary artery or one of its main branches results in a deficit of blood flow to the area (**ischemia**). If the blockage is complete, the unsupplied heart muscle dies (**myocardial infarct**). If the patient survives, the muscle of the infarcted region is replaced by scar tissue, and that part of the wall of the heart is weakened. In the normal heart, most of the arterial branches end in capillary networks, which lead into a system of coronary veins that ultimately empty into the right atrium. Another type of small veins, called **Thebesian veins**, empty directly into the heart chambers, especially the right atrium. Those that empty into the left side of the heart dilute the well-oxygenated blood with their small contribution of venous blood.

Skin Vasculature

The skin has a much greater blood flow than is dictated by its metabolic requirements. The reason for this is due to its large surface area and the role of the dermal vasculature in regulation of body temperature. If the body needs to get rid of

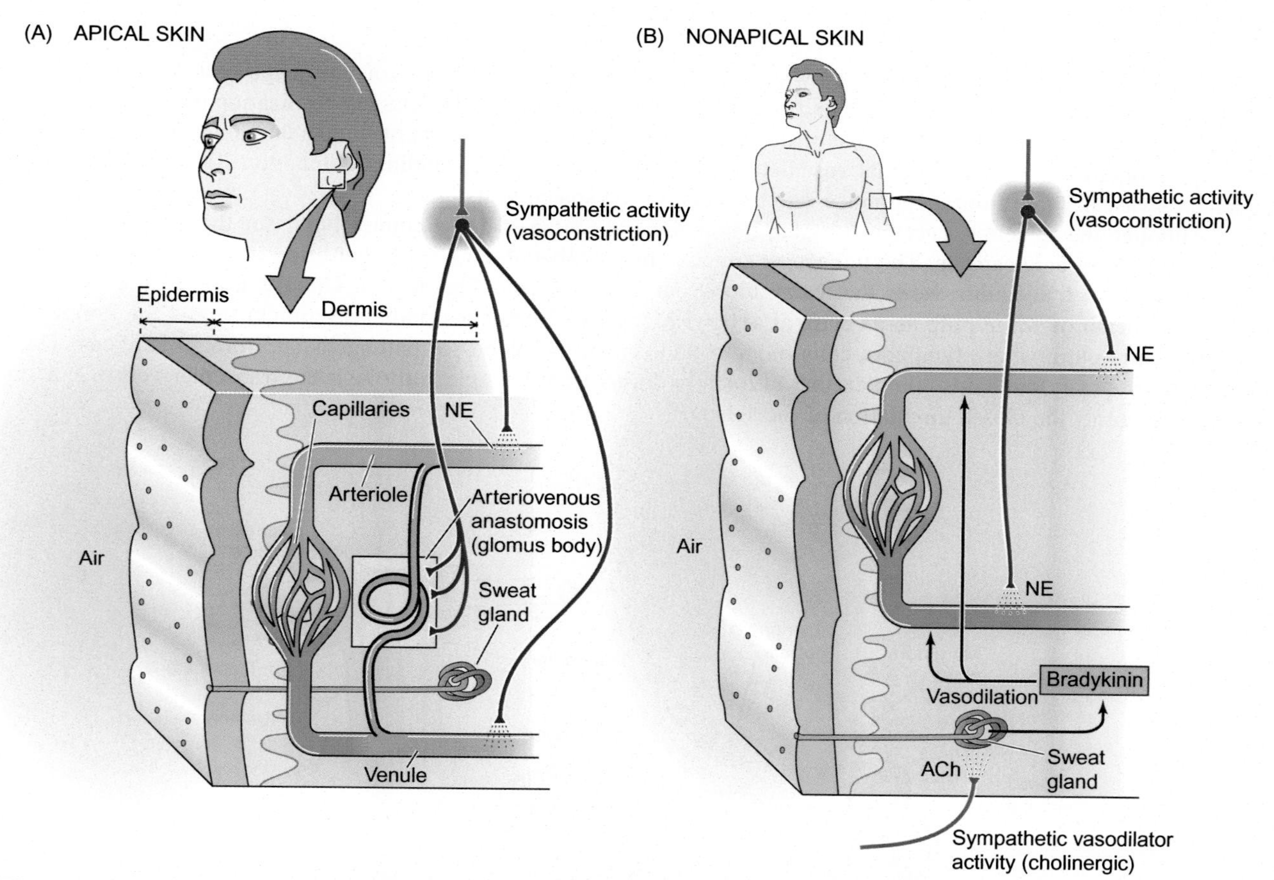

FIGURE 10.23 Two different patterns of blood flow in the skin. (A) Apical skin, with glomus bodies in addition to capillary networks. (B) Nonapical skin lacks glomus bodies. *From Boron and Boulpaep (2012), with permission.*

excess heat, blood flow in the skin increases; on the other hand, if the body needs to conserve heat, blood flow in the skin decreases. Depending upon needs, blood flow in the skin can range from almost nothing to 30% of the total cardiac output. Blood flow to the skin can be influenced by sympathetic innervation, local metabolites, or local warming or cooling.

Two types of skin vasculature, each with a different structure and different control mechanisms, exist. **Apical skin**, as the name implies, covers terminal parts of the body—the hands and feet, the nose, ears, and lips. The remainder of the skin is considered to be nonapical. Both types of skin contain an extensive subdermal venous plexus through which much of the superficial blood flows.

The vasculature of **nonapical skin** is characterized by standard capillary beds that parallel the surface of the skin (Fig. 10.23). Local cooling can cause blanching of the skin, and heating can cause reddening through the constriction and dilation of the skin vasculature. Stimulation of the sympathetic innervation of the arterioles and venules causes vasoconstriction due to the release of norepinephrine from the nerve terminals. Paradoxically, other sympathetic nerve fibers cause vasodilation through the release of acetylcholine. Exactly how the latter occurs is not certain. During embryonic development, the sympathetic nerve fibers innervating sweat glands undergo a transmitter switch from norepinephrine to acetylcholine. One hypothesis is that the secreted acetylcholine directly affects sweat glands rather than the microcirculation and that vasoactive products of stimulated sweat glands act on the small vessels of the skin, causing them to dilate.

Apical skin contains capillary networks of the same sort as nonapical skin, but in parallel with the capillary beds are **arteriovenous anastomoses** (see Fig. 10.23), which directly connect arterioles and venules without intervening capillaries. Called a **glomus body**, the arteriovenous anastomosis has a thick wall of contractile myoepithelial cells enmeshed in connective tissue. In contrast to the overlying capillaries, which are subject to local metabolic control, glomus bodies respond to norepinephrine secreted by their sympathetic innervation. Sympathetic stimulation causes them to constrict, sometimes completely. Dilation after sympathetic stimulation is withdrawn is passive.

Spleen

The **spleen** is a dual-purpose organ. Its white pulp is an important component of the lymphoid system (see Chapter 8). Its red pulp is involved in the storage and processing of red blood cells. The tissue organization of the spleen, especially that of the red pulp, has remained one of the most poorly understood aspects of human anatomy, mainly because of the complex organization of its internal vasculature. One approach to understanding its functional anatomy is to follow the spleen's vascular tree from its entrance into the body of the spleen to its exit.

Although the spleen is a very soft organ, it contains trabeculae of dense connective tissue that penetrate into its interior from the splenic capsule. The trabeculae contain small **trabecular arteries** that bring blood into the parenchyma of the spleen and **trabecular veins** that carry blood out of the spleen (Fig. 10.24). Upon exiting a trabecula, the artery enters a sheath of white pulp as its **central artery**. Many small branches (radial arterioles) supply the lymphoid tissue of the white pulp, where lymphoid cells monitor and screen the blood for pathogens and foreign antigens. This blood collects in a marginal sinus complex that ultimately drains from the spleen through the trabecular veins. This arrangement represents the closed circulation of the spleen.

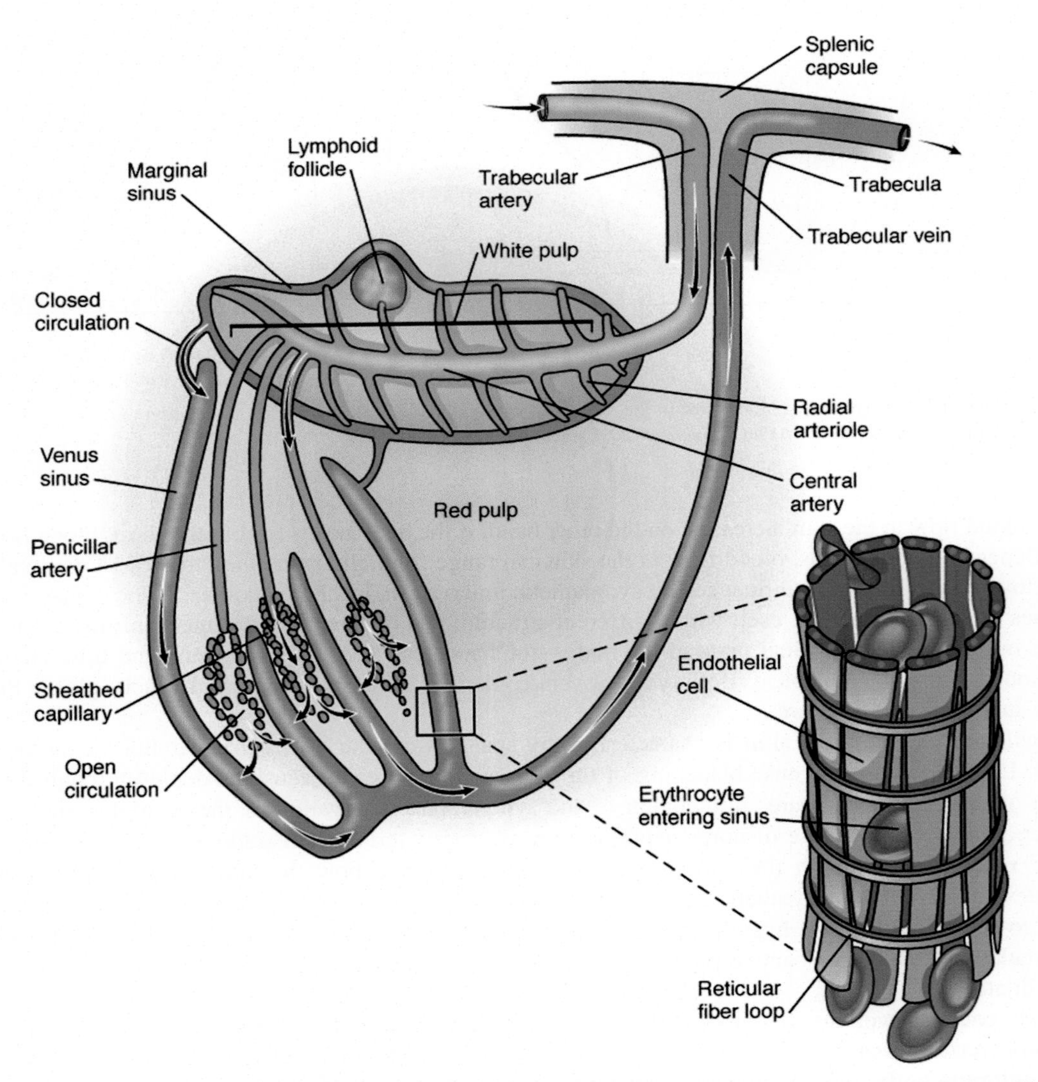

FIGURE 10.24 Vascular architecture of the spleen. Blood entering through a trabecular artery passes through the white pulp and leaves through a series of penicillar arteries, which then open up, releasing blood into an open circulation in the red pulp. The blood is then collected into venous sinuses and leaves the spleen via trabecular veins. The closed circulation of the spleen consists of blood leaving the white pulp via the marginal sinus and from there connecting directly with venous sinuses. The inset shows the structure of the wall of a venous sinus.

Other branches of the central artery leave the white pulp as straight **penicillar arteries**, which are located in the red pulp (see Fig. 10.24). Blood from the penicillar arteries flows into sheathed capillaries—unique microvessels that are lined by macrophages. Blood from the sheathed capillaries drains into splenic cords, which is a very loose version of connective tissue containing fibroblast-like cells called **reticular cells**, a loose meshwork of fine reticular fibers, and macrophages. Blood leaves the splenic cords by entering a network of **venous sinuses**. The terminal branches of the venous sinuses are lined by longitudinally oriented endothelial cells that are arranged as loosely fitting barrel staves. Surrounding these endothelial cells are thin bands of reticular fibers (see Fig. 10.24 inset). The endothelial cells lining the venous sinuses are not tightly fitting, and red blood cells pass through the chinks in that lining to reenter the closed circulation of the spleen. The venous sinuses consolidate into larger venules and ultimately leave the parenchyma of the spleen via trabecular veins.

While the blood is in the open circulation within the splenic cords, it is screened for bacteria and foreign particles by the macrophages, which then phagocytize them. Healthy erythrocytes are able to pass through the interstices between the endothelial cells of the venous sinuses, but old or unhealthy erythrocytes are unable to undergo the necessary deformation to enter. These cells are then broken down in the splenic cords, and the cellular residue is taken up by the macrophages. The hemoglobin from these cells is first broken down into heme and globin components. Iron (Fe^{++}) from the degraded heme component is combined with the iron transporter protein **ferritin** and is either stored in the macrophages or sent via the bloodstream into the bone marrow for its reutilization in erythropoiesis.

Hepatic Portal Vein

In many respects, the hepatic portal vein resembles many systemic veins. Its branches arise from capillary networks in the intestines (see Fig. 12.29) and the stomach. In addition, it receives venous blood from the spleen (Fig. 10.25). The blood entering the hepatic portal system from the intestines is rich in digestive products and hormones secreted by enteroendocrine cells of the digestive tract. From a functional standpoint, it is more efficient to bring this enriched blood directly to the major metabolic center (the liver) than to empty it into the general circulation, where it would go the heart and then be distributed generally to all parts of the body. Instead, the hepatic portal vein brings all of this blood directly to the liver, where it branches into smaller caliber veins which finally empty into the **hepatic sinusoids** (see Fig. 12.30). The open nature of the sinusoidal walls allows free access of all the contents of the portal blood to the hepatic parenchymal cells (hepatocytes) for their metabolic use. From the sinusoids, the blood, now depleted of

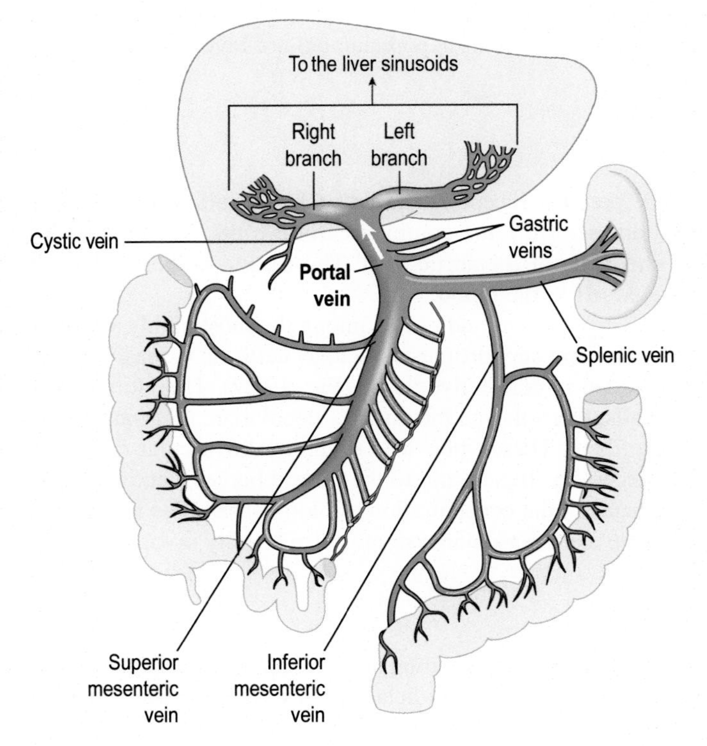

FIGURE 10.25 The hepatic portal venous system. *From Waugh and Grant (2014), with permission.*

metabolites, but enriched in newly synthesized molecules, travels into branches of hepatic veins and ultimately empties into the inferior vena cava and into the heart.

ELEMENTS CONTROLLING CIRCULATORY DYNAMICS

Because of its wide distribution to almost all parts of the body, the circulatory system must be highly regulated in order to maintain bodily homeostasis. One could write a large book on this topic, but some of the more important elements of circulatory control are covered below. Circulatory controls can be systemic or local, short- or long-term. They can affect the heart or the blood vessels or both; they can arise from within or without the circulatory system. This section examines how several important properties of the circulation are controlled.

Blood Pressure

Blood pressure is sensed by specific baroreceptor structures embedded within the large blood vessels. The main pressure receptors are nerve endings contained in the **carotid sinuses**, slight enlargements at the bases of the internal carotid arteries (Fig. 10.26). Other less sensitive pressure receptors are located in the wall of the arch of the aorta. An increase in blood pressure increases the rate of firing of the nerve fibers in the baroreceptor. The nerve endings actually respond to mechanical stretch, which occurs when increased blood pressure dilates the carotid sinus and stretches the nerve endings. These neural signals are carried up the fibers of two cranial nerves (IX [glossopharyngeal] for the carotid sinus and X [vagus] for aortic arch receptors) to a **medullary cardiovascular center**. The medullary cardiovascular center then integrates the afferent information and coordinates a response that consists of the reduction of vascular resistance (by inhibiting sympathetic constriction of smooth muscle) and a reduction in heart rate via stimulation of parasympathetic nerve fibers.

Blood pressure itself is influenced by many factors, but the main determinants of arterial blood pressure are cardiac output in relationship to peripheral arterial resistance. **Cardiac output** is measured by the heart rate times the stroke volume (blood ejected per heartbeat). Cardiac output is heavily influenced by exercise, with increases up to sixfold in elite athletes. **Peripheral resistance** depends heavily upon arterial compliance (stretchability) to accommodate the pulses of blood leaving the left ventricle during systole. These factors in aggregate, plus the arterial blood volume, all impact final arterial blood pressure.

Except for some short-term phenomena, such as blood loss or blood transfusions, the control of blood volume occurs over more extended periods of time. The kidney is the ultimate determinant of blood volume through its control of water loss, but many other systems of the body (e.g., liver, adrenal cortex, and hypothalamus) are involved in the coordination of water loss or retention by the kidney (see Chapter 13).

Blood Gases

Peripheral receptors for dissolved blood gases exist close to the pressure receptors. In the neck, a small **carotid body** is located at the bifurcation of the internal and external carotid arteries from the common carotid trunk (see Fig. 10.26). Groups of smaller **chemoreceptors** are scattered along the underside of the aortic arch. The carotid bodies are highly sensitive to reduced oxygen, increased carbon dioxide or reduced pH in the blood.

In keeping with their function of sampling blood, the small (2 mg) carotid bodies are among the most highly vascularized tissues in the body. Internally, they consist of sensor (glomus) and support (sustentacular) cells, an extensive capillary network and terminals of autonomic nerve fibers, all embedded in connective tissue (Fig. 10.27). The glomus cells are the chemosensors, and when the O_2 pressure around them falls, their plasma membranes depolarize, releasing neurotransmitters, which then activate the afferent sympathetic nerve fibers. These fibers lead to the cardiovascular medullary center through a pathway that follows the nerves from the baroreceptors. In addition to the carotid bodies, central sensors in the brain also play a role in orchestrating the response to changes in the concentration of blood gases. The response to hypoxia is principally an increase in the depth of breathing, but the cardiovascular system is also heavily involved.

Heat Management

Heat management involves many aspects of bodily physiology, including sweating and metabolic mechanisms (see Fig. 3.6), but the circulatory system plays a major role in the distribution of body heat. Much of this control resides in the peripheral vasculature. To save body heat, the vasculature near body surfaces constricts, keeping more of the blood

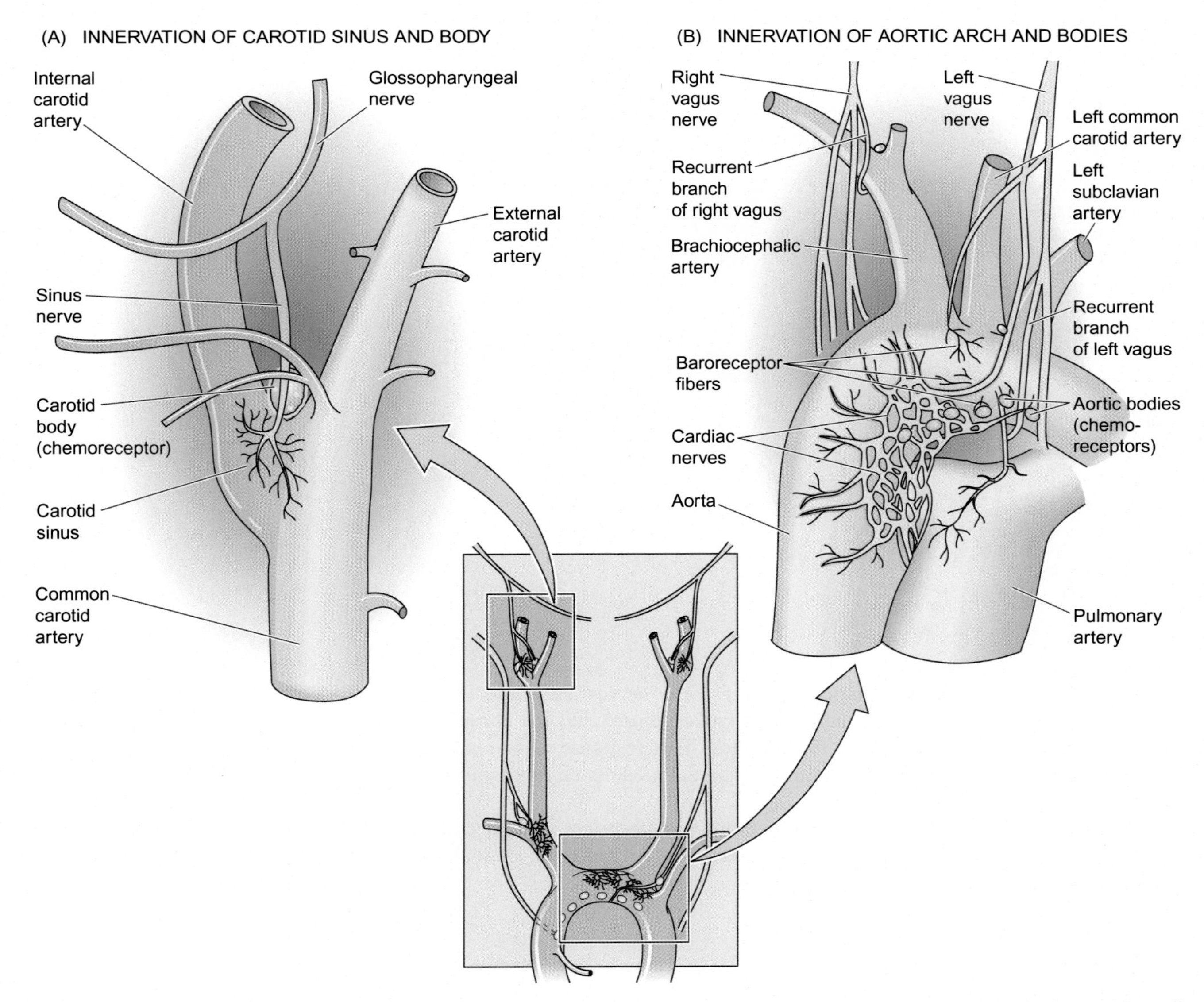

FIGURE 10.26 Baroreceptors and chemoreceptors in and near the walls of large arteries. Baroreceptors are located in the carotid sinus and the wall of the aortic arch, whereas chemoreceptors are located near the bifurcation of the internal and external carotid arteries (carotid body) and alongside the aortic arch. Afferent nerve fibers from the carotid sinus/carotid body reach the brain via the glossopharyngeal (IX) nerve, and those from the aortic arch region use the vagus (X) nerve as their conduit. *From Boron and Boulpaep (2012), with permission.*

in the core of the body. To lose excess heat, the reverse occurs. Heart rate tends to increase with excessive body heat, and it decreases when core body temperature decreases. These responses are based on information gained by thermal sensors in the skin and brain. The sensory information is transferred to the brain, which then responds through the sympathetic nervous system (see Fig. 10.23).

In addition to systemic control mechanisms, the microcirculation to small areas of the skin is responsive locally to applied heat or cold. This is accomplished by constriction or dilation of the microcirculatory vessels, especially the arteriovenous anastomoses, in the affected area and probably involves local metabolic as well as neural control.

Emotional Factors and Cardiovascular Function

The dynamics of the circulatory system can also be strongly influenced by cognitive centers within the central nervous system. The most dramatic example of this is the **fight-or-flight mechanism**, which is a response to perceived danger or anger. Pathways from the cerebral cortex lead to the hypothalamus and from there to the medullary cardiovascular

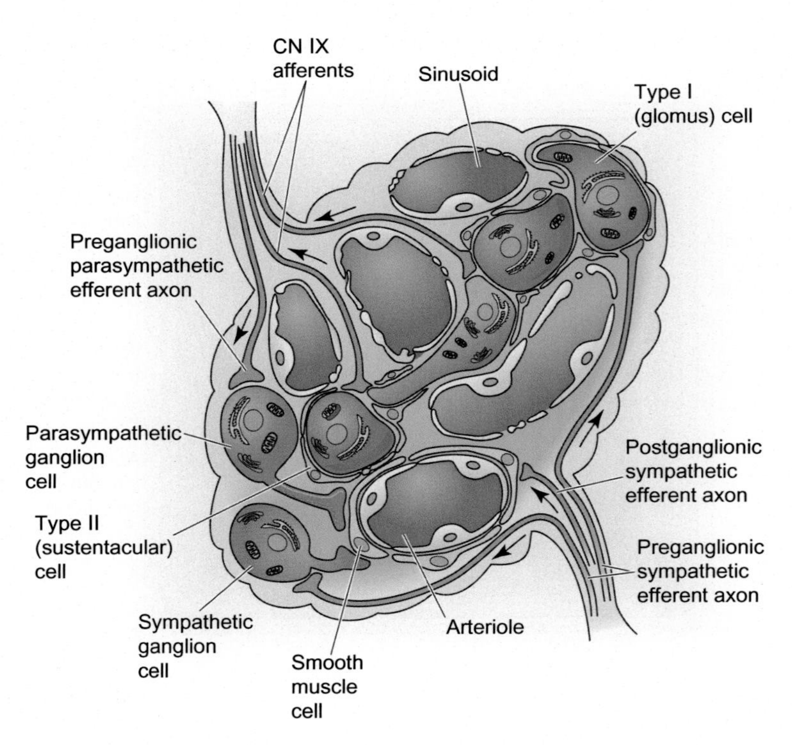

FIGURE 10.27 Microscopic structure of the carotid body. *From Boron and Boulpaep (2012), with permission.*

center, which controls the stimulation of sympathetic nerves leading to the heart and blood vessels. This results in an acceleration of heartbeat and an increase in stroke volume and the constriction of blood vessels to the viscera, along with an increase of blood flow to the muscles. This response also involves many other body systems. Epinephrine released from the sympathetically stimulated adrenal medulla circulates all over the body through the blood and accentuates the response.

At a chronic level, emotional stress is translated from the cerebral cortex through the hypothalamus and anterior pituitary to the adrenal cortex. Chronically increased secretion of adrenal cortical steroids contributes to a syndrome that affects many tissues of the body—increased blood pressure being among the most pronounced symptoms.

THE LYMPHATIC SYSTEM

The **lymphatic system** plays two important functional roles in the body—transporting fluid to the heart and defending the body against pathogens through its immune functions. The latter function is discussed in Chapter 8. This section discusses the circulatory functions of the lymphatic system.

The bulk of the lymphatic system consists of a network of channels, which in parallel with blood vessels, drain essentially all parts of the body except the brain, cartilage, bone, and epithelia (Fig. 10.28). At critical collecting points, the lymphatic channels lead into and out of clusters of lymph nodes. Due to the vagaries of embryological development, the pattern of lymphatic return to the heart is highly asymmetrical. Lymphatics from the entire lower half of the body and the left side of the upper half drain into the venous circulation (the left subclavian vein) through the thoracic duct (Fig. 10.29). Those draining the right arm and right half of the thorax, along with the right side of the head and neck, enter the right subclavian vein via the right lymphatic duct.

The lymphatic circulation begins as small blind endings (**lymphatic capillaries**) located in interstitial spaces (Fig. 10.30). The blind endings are essentially pouches of endothelial cells that are loosely connected with the surrounding tissue elements by anchoring filaments. The endothelial cells are not tightly joined, so tissue fluid that is not taken up by venous capillaries enters the blindly ending lymphatic capillaries through spaces between endothelial cells. This loose arrangement is important, because not only fluid enters the lymphatic channels, but also any tissue debris, pathogens and proteins that are too large to make their way back into the blood vessels. When the pressure inside the lymphatic capillaries exceeds that of the surrounding tissue fluid, the spaces between endothelial cells close in the fashion of a one-way valve. As much as 2–3 L of fluid is picked up by the lymphatic channels each day.

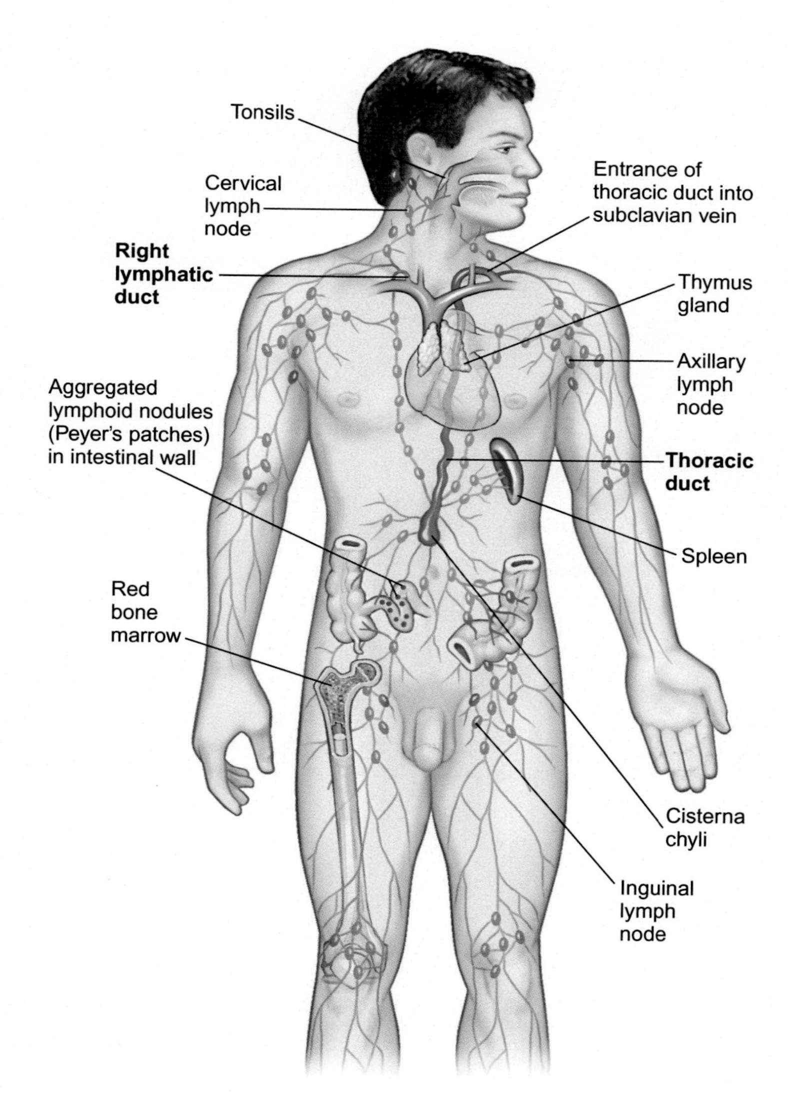

FIGURE 10.28 Overview of the lymphatic system. *From Thibodeau and Patton (2007), with permission.*

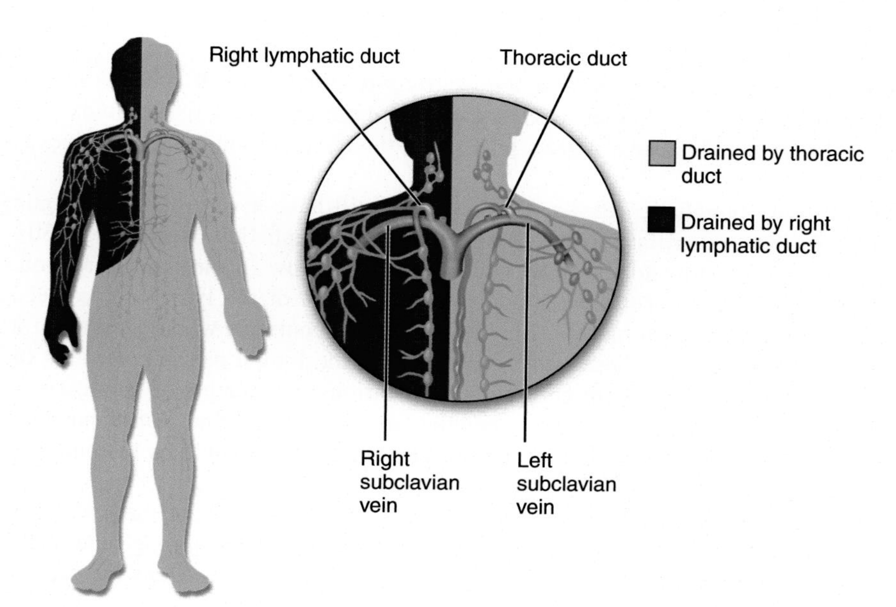

FIGURE 10.29 Lymphatic drainage regions in the body. *From Thibodeau and Patton (2007), with permission.*

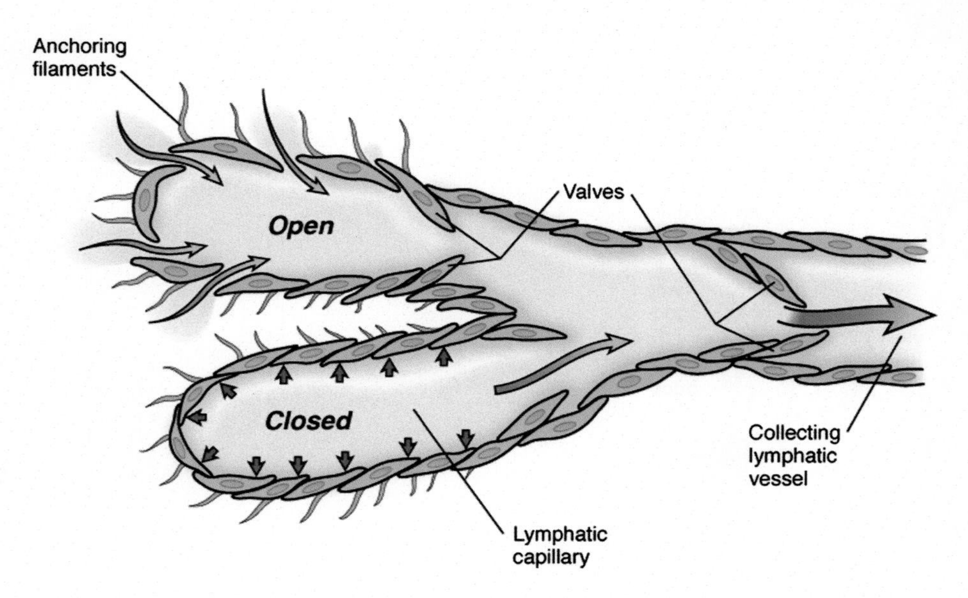

FIGURE 10.30 The structure of a terminal lymphatic duct. Blue arrows show the direction of lymph flow. Tissue fluid enters the lymphatic capillary through spaces between the endothelial cells. Internal lymphatic pressure closes these spaces. The valves prevent the backflow of lymph.

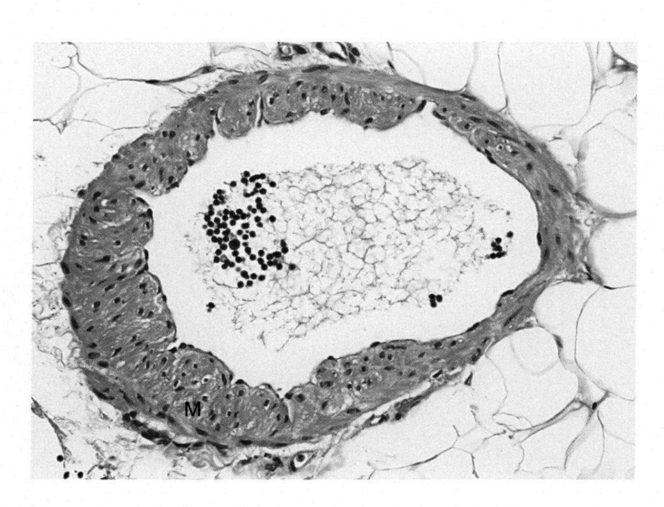

FIGURE 10.31 Histological section through a medium-sized lymphatic vessel. M – smooth muscle layer. *From Young et al. (2006), with permission.*

Initially, the lymphatic fluid has as much the same composition as the nearby interstitial fluid. It contains no red blood cells or platelets, but it does pick up lymphocytes and other leukocytes that are present in the tissue fluids. The protein content of lymph is much less than that of blood plasma. After a fatty meal, the lymph draining the intestines is very rich in fat.

The lymphatic capillaries drain into collecting lymphatic vessels. These vessels still have a very thin and fragile wall. Internally, they contain one-way valves that are critical for the maintenance of lymph flow (see Fig. 10.30). Like venous valves, they are open to lymph flowing toward the heart and prevent backflow in the other direction. The movement of lymph is caused by both contractions of the walls and by compression of the lymphatic vessels. Larger lymphatic vessels are lined by a thin layer of smooth muscle (Fig. 10.31), which contracts when the wall of a lymphatic vessel is stretched. Called the **lymphatic pump**, these contractions propel the lymph past the next set of valves. In addition to internal contractions, lymphatic vessels are compressed by various movements of or pressures on the body and by skeletal muscle contractions. The pooling of fluid (**edema**) around the feet that can occur with prolonged standing or sitting reflects the importance of contractions of leg muscles or leg movements on maintaining a proper flow of lymph.

The lymphatic vessels converge upon clusters of lymph nodes at a variety of key points throughout the body (see Fig. 10.28). They enter the lymph nodes through the capsular region, and the lymph then percolates through the node. Lymph leaves the node through a single efferent vessel in the hilar region. Whatever pathogens or antigens are carried in the lymph are then exposed to immune cells within the lymph nodes for an appropriate response (see Chapter 8).

SUMMARY

The circulatory system consists of the heart, blood and lymphatic vessels, and blood.

Blood is a liquid made up of both cells and plasma. Red blood cells (erythrocytes) are anuclear and contain hemoglobin, which carries oxygen to all parts of the body. White blood cells are components of the lymphoid system. Platelets are also anucleated cellular fragments that are an integral part of the blood-clotting mechanism. There are several antigenic blood types, the most important are types A, B, AB, and O, along with a set of Rh antigens. Blood plasma contains abundant albumin, globulins, and clotting proteins.

The heart is a muscular organ that beats continuously. It contains two atria and two ventricles. Venous blood enters the right atrium and proceeds to the right ventricle, from which it is pumped to the lungs for reoxygenation. Oxygenated blood then enters the left atrium and then the left ventricle, which pumps it into the aorta and the coronary arteries, which supply the heart itself. Cardiac valves prevent the backflow of blood. Contraction of the heart is called systole and relaxation called diastole.

Although cardiomyocytes have an intrinsic beat, cardiac rhythm is controlled by a pacemaker (the sinoatrial node). From there, the beat is transferred to the atrioventricular node and then down the bundle of His and the Purkinje fibers to meet individual cardiomyocytes. The beat is also transferred from one cardiomyocyte to another through connecting gap junctions.

Blood vessels consist of arteries and veins. Arteries have thick muscular walls that dampen the pressure of the cardiac output. Their walls contain collagen and elastic fibers for strength and elasticity, and smooth muscle fibers, which control their diameters.

The microcirculation extends from small arterioles to capillaries and small venules. The walls of capillaries are adapted for the exchange of gases and various dissolved substances within the blood. Much of the exchange is controlled by blood and osmotic pressure differences between the blood in the capillaries and the surrounding tissue fluids. Under some circumstances, immune cells (leukocytes) leave the bloodstream to fight local infections.

Veins collect blood from the periphery and transport it to the heart. Relatively lacking in smooth muscle within their walls, veins depend upon muscular movements and valves for the transport of blood.

Several organs have special adaptations in their local circulation. These include the kidney, spleen, skin, and liver.

Special sensory structures influence circulatory function. The carotid bodies and chemoreceptors measure the oxygen pressure, and if it is too low, they stimulate the sympathetic nervous system to increase breathing and circulatory dynamics.

The lymphatic system is a network of vessels that collect tissue fluids and move them toward the heart. Lymph channels begin as blind sacs in the periphery, and like veins, become successively larger as they collect more inflowing branches. They contain valves to prevent backflow like veins. Two large lymphatic channels, draining asymmetrically right and left sides of the body, empty into the subclavian veins.

Chapter 11

The Respiratory System

Virtually all animals need to take in oxygen and remove carbon dioxide in order to survive. Some are small enough that gas exchange through their outer surfaces is sufficient, but almost any animal big enough to be seen without a microscope requires some form of specialized respiratory system. Most aquatic animals have evolved external sets of feathery gills that extract oxygen from the water and allow carbon dioxide to escape from the body. Vertebrate land animals, on the other hand, developed very complex internal lungs that are subdivided into enormous numbers of tiny air sacs lined with a thin moist epithelium that allows the efficient passage of gases to and from the networks of capillaries that underlie the epithelium. The epithelium of the lungs is very delicate, and in order to protect it, mammals have evolved a long channel, consisting of the nostrils, the pharynx, the trachea, and bronchial passages, that conditions the air before it reaches the lungs. Without this, the cell lining of the lungs would become seriously damaged by breathing in air that is too cold, or hot, or dry.

All of the structures from the nose to the lungs make up the respiratory system, but as it evolved, certain other body functions became incorporated into the fabric of the respiratory system. The two most prominent are smell (**olfaction**) and sound production (**phonation**). From the phylogenetic standpoint, the nose first arose as the site of the sense of smell (as well as magnetoreception in some animals). As air-breathing evolved and animals came onto the land, the olfactory organ remained in the nose, because out of water the inner environment of the nose provides the moist conditions most conducive for the sense of smell. Most forms of sound production by vertebrates require both vibrating parts and the movement of air. For this reason, the respiratory system was the logical site for the development of the larynx, the principal organ of sound production.

At every level, from the outermost part of the nose to the smallest air sacs in the lungs, the structure of the respiratory system is designed to serve the function required of that component. In this chapter, we will follow a breath of inspired air from the nose to the lungs (Fig. 11.1) and see how the structures along the way condition the air so that when it finally reaches the lungs gas exchange between the air and the body can be efficiently accomplished.

BREATHING

In a broad sense, respiration begins with taking a breath and filling the lungs with air. Normal quiet breathing is an automatic process that begins with a signal generated from three pairs of **respiratory centers** located within the brainstem (pons and medulla). A **pneumotaxic center** in the pons acts as a switch between inspiratory and expiratory signals, which are generated by **inspiratory** (dorsal respiratory group) and **expiratory** (ventral respiratory group) **signaling centers** located in the medulla. Nerve fibers emanating from these signaling centers connect with other nerve fibers that directly supply the respiratory muscles.

Inspiration is accomplished by the contractions of a group of inspiratory muscles, chief among them being the **diaphragm** (Fig. 11.2). When the diaphragm—the dome-shaped muscle separating the thoracic from the abdominal cavities—contracts, the dome flattens toward the abdomen, thereby increasing the volume of the thoracic cavity. At the same time, the upper external intercostal muscles elevate the ribs, further increasing the size of the chest cavity. The lungs themselves follow the contours of the expanding chest wall because of a negative pressure between the outer layer (**visceral pleura**) of the lungs and the inner layer (**parietal pleura**) lining the chest wall. As the lungs expand, a second negative pressure builds up within them, and outside air rushes into the respiratory system through the nose or mouth. The tissues of the lungs are inherently elastic, and once the inspiratory muscles relax, the lungs begin to contract in a manner reminiscent of a stretched rubber band. As a result, air passively moves from the lungs to the outside.

Forced breathing can result from exercise, inadequate oxygen in the air or a variety of pathological conditions. In this case, voluntary contractions of muscles respond to cerebral signals and add an extra dimension to the breathing process

The Human Body. DOI: https://doi.org/10.1016/B978-0-12-804254-0.00011-9

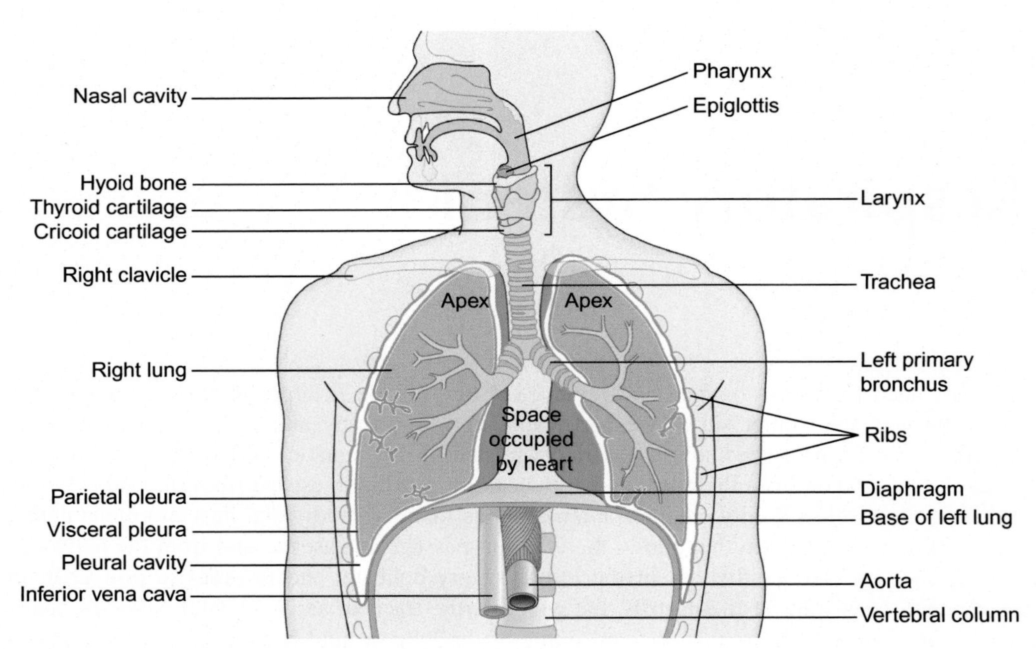

FIGURE 11.1 The respiratory system. *From Waugh and Grant (2014), with permission.*

by recruiting other groups of muscles. Prominent in forced inspiration is the action of some neck muscles—the scalenes and sternocleidomastoids (see Fig. 11.2), which further elevate the upper ribs and the sternum. Adding to these actions are contractions of muscles that elevate the pectoral girdle and extend the back; thus further increasing the volume of the chest cavity. Forced expiration is accomplished by adding contractions of abdominal and some of the internal intercostal muscles to the normal elastic recoil of the lungs. These supplemental aids to breathing are needed by individuals with many forms of **chronic obstructive pulmonary disease** (COPD), such as emphysema, asthma, and chronic bronchitis.

The most forcible exhalation of air occurs during a sneeze. Following stimuli as diverse as sudden exposure to an allergen, cold or even bright light in some individuals, all of the muscles involved in forcible exhalation react in rapid synchrony. A cough or sneeze begins with momentary closure of the glottis, which blocks the airway. Contractions of the intercostal muscles then rapidly build up the intrathoracic pressure. Then the glottis quickly opens, allowing the air from the lungs, now under high pressure, to be expelled from the lungs. Recent data suggest that the velocity of air expelled during a sneeze is much less (~5 m/second) than commonly cited earlier estimates of 100 m/second. Surprisingly, the air velocity of a sneeze is not much faster than that of a cough. Up to 40,000 aerosol droplets are spread per sneeze. The difference between a cough and a huff is that in the latter, the glottis never closes.

THE NOSE

In the aquatic vertebrates that breathe by means of gills, the nose is an organ designed to serve the function of smell. As land animals evolved, however, gills were abandoned in favor of lungs, and a means of getting air to the lungs had to be developed. Up to a point, simply taking in air through the mouth can work, but humans and other mammals were faced with a new problem—how to suckle milk and breathe at the same time. This problem was solved by the development of the **secondary palate**—the thin shelf of bone and soft tissue at the roof of the mouth that forms in the embryo and separates the nasal cavity from the oral cavity. With breathing separated from eating, the nose was free to develop into an organ that has as one of its main functions the conditioning of inspired air. This, plus the primitive function of olfaction, determines both the anatomical configuration and physiological functions of the human nose.

The Nose during Breathing

In order to appreciate what happens to a breath of air that enters the nose, it is useful to recall the main functions of a good air conditioner, which (1) brings air to the proper temperature (32–35°C), (2) humidifies air (to ~80%), and (3)

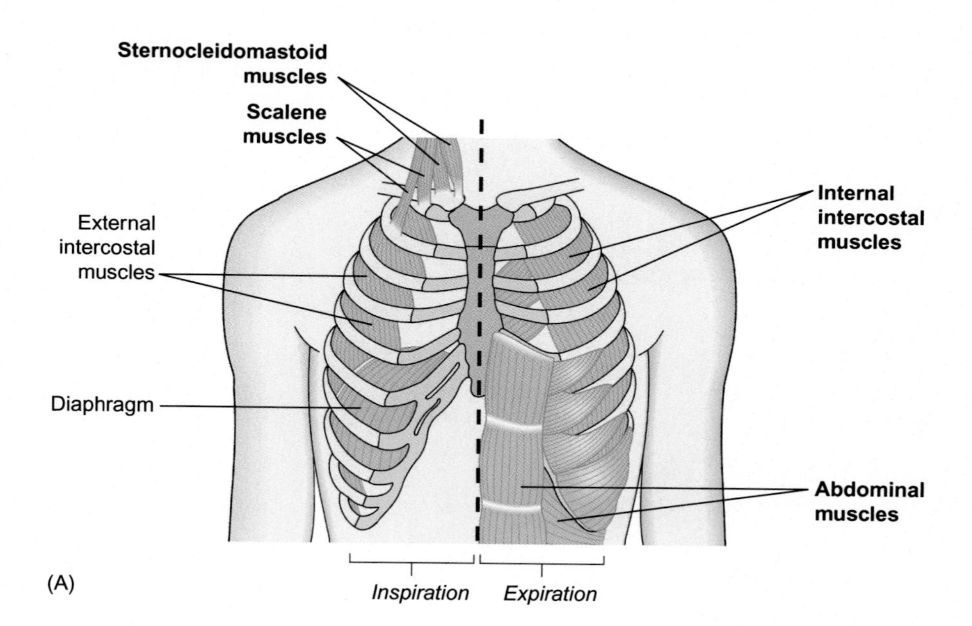

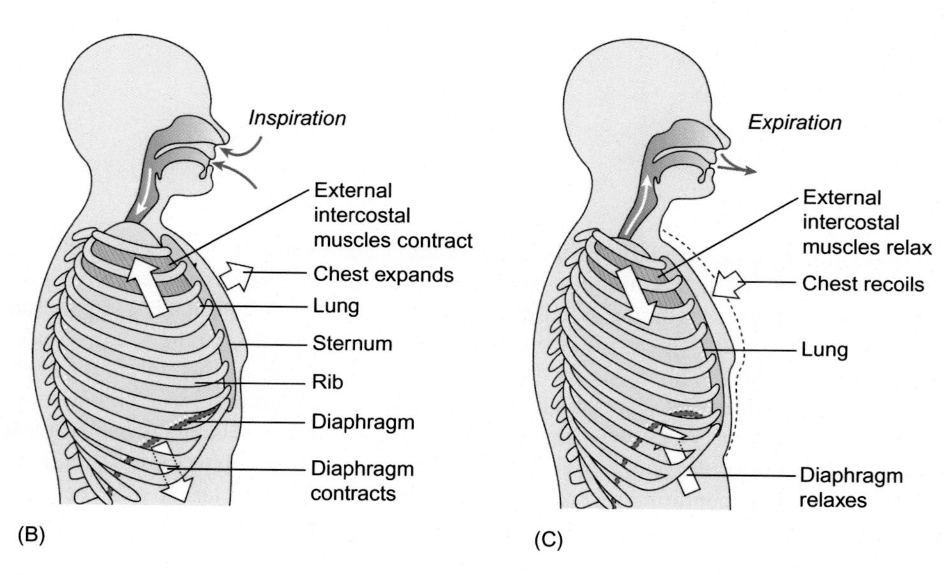

FIGURE 11.2 The chest during breathing. (A) Muscles involved in inspiration and expiration. (B) The chest during inspiration. (C) The chest during expiration. *From Waugh and Grant (2014), with permission.*

removes particulate matter and allergens from the air. The nose accomplishes all of these functions through some finely tuned structures and processes.

When we inspire, our expanding chest cavity creates a partial vacuum, which draws a rush of outside air into the nose. The air could be frigid and dry in a northern winter or very dusty, with even a few flying insects thrown in, during the summer. By the time, the air reaches the lungs, its temperature approaches that of the body, and it is nearly saturated with humidity. Most particulate matter in the air has been removed, as well.

The process of biological air conditioning begins as soon as the air enters the nose (**external nares**, Fig. 11.3). The first structures that the air encounters are the coarse nasal hairs, which are rooted in the outer **vestibule** of the nasal chambers (**internal nares**). These hairs act as a filter to prevent insects or large airborne objects from entering the nose. A second element involves the pathway of the inspired air. When air is first drawn into the vestibule, it follows a nearly vertical path. Then it has to bend in a more horizontal manner in order to reach the main nasal chambers. The bending causes air turbulence, which increases the contact between the air and the moist nasal mucosa. This is a first step in the process of temperature and humidity control.

Once past the vestibule, the air enters the main paired nasal chambers. It is here where the main air conditioning occurs. An examination of the structures lining the nasal chambers shows how efficiently the nasal chambers are

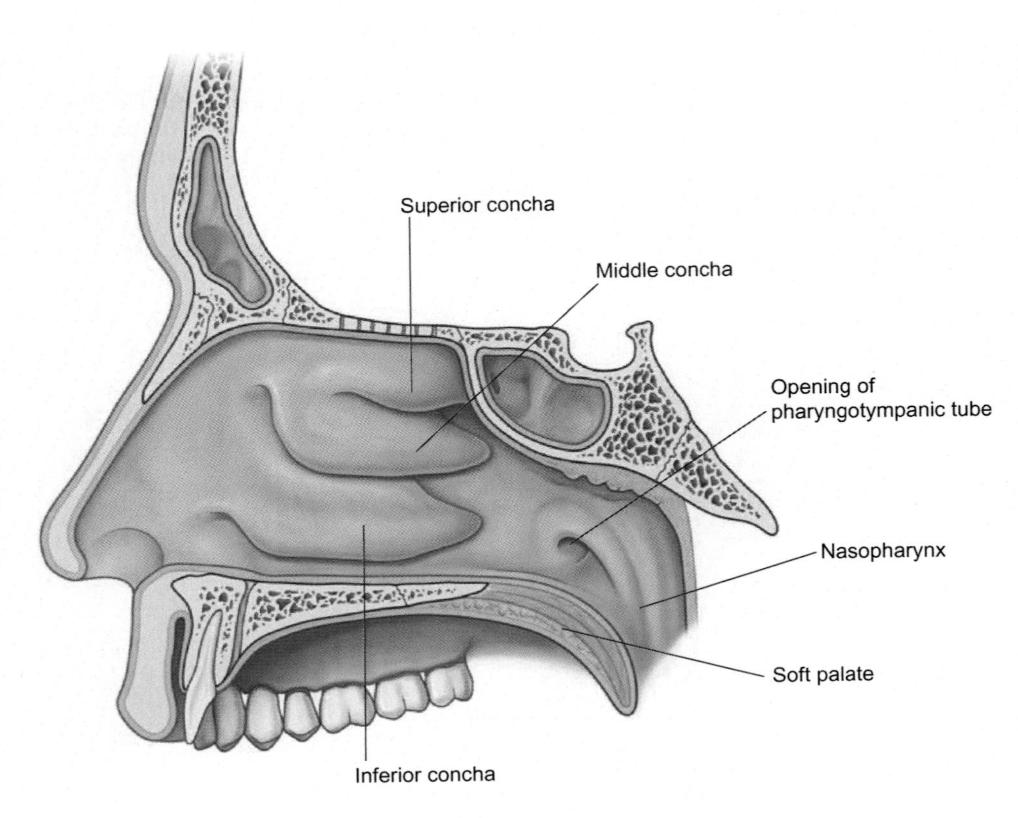

FIGURE 11.3 The lateral wall of the nasal cavity and pharynx. *From Drake et al. (2005), with permission.*

designed for conditioning air. A section cutting perpendicularly to the internal nares shows that there are right and left chambers with a thin **nasal septum** separating them in the midline. The lateral walls of the nasal chambers are highly irregular and are dominated by three drooping shelves of tissue, called the superior, middle, and inferior **nasal conchae** (Fig. 11.4A), which act as baffles. The nasal conchae greatly narrow the width of the nasal chambers and greatly increase the surface area of the walls of the chambers. This alone facilitates the warming of the air that enters the nose, but the configuration of the nasal conchae also causes the inspired air to form eddy currents in the nose. The swirling allows the air to remain longer in the nose, and that, too, allows for more effective warming. Nasal conchae in most other mammals are much more complex than those of the human (Fig. 11.4B). Although the reason for these differences remains controversial, both the need for greater air conditioning while pursuing prey and a much larger olfactory surface have been postulated. The other feature that greatly contributes to warming the air is the presence of a complex of dilated veins beneath the nasal lining, especially around the nasal conchae (see Fig. 11.5). Warmth from the flowing blood is dissipated into the air, and changes in the amount of blood flow regulate the degree of warming the air. Temperature sensors in branches of the trigeminal nerve (cranial nerve V) are involved in the control of blood flow.

The epithelium that lines the nasal cavities plays a critical role in the overall function of the nose. Except for the vestibule and the region near the pharynx, the nasal chambers are lined by a highly specialized **ciliated pseudostratified columnar epithelium** (Fig. 11.5). The outer surfaces of most of these epithelial cells are covered by large numbers of microscopic cilia. Scattered among these cells are other cells that form and secrete mucus. Because of their appearance under the microscope, they are called **goblet cells**. The goblet cells, along with small glands located beneath the epithelium, produce a thin layer of nasal mucus that covers the entire surface of the nasal chambers. In addition to mucus, some of the cells in the mucous glands are specialized to produce a watery (serous) secretion that enters the nasal chambers and plays an important role in humidifying the air.

Located within the cheeks are the large **maxillary sinuses**, which empty into the nose (Fig. 11.6). These sinuses are covered by a similar epithelium, but with fewer mucus-producing cells, and they contribute to both the warming and humidification of the air. Several other sinuses (frontal, ethmoid, and sphenoid) also open to the nasal cavity. There is no agreement on the functions of the sinuses. From a mechanical perspective, they reduce weight in the bones of the face and also probably provide architectural strength, and they serve as a resonance chamber for vocalization. Because they are blind chambers, airflow to and from them is very slow, but they may provide some thermal and humidity

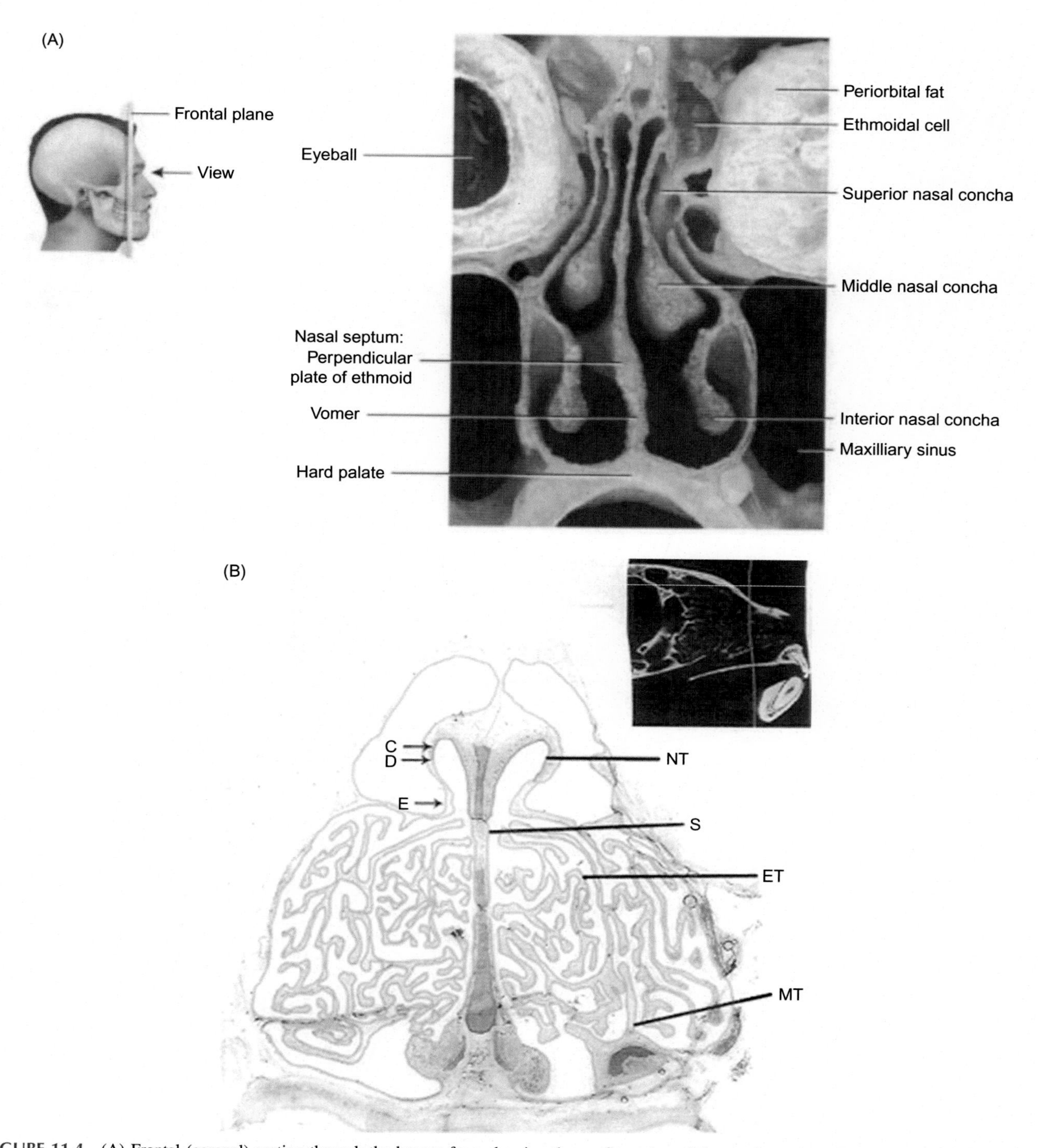

FIGURE 11.4 (A) Frontal (coronal) section through the human face, showing the configuration of the nasal conchae. (B) Coronal section through the nose of a domestic cat. Note the great complexity of the turbinate bones compared with that of the human. *C*, olfactory epithelium; *D* and *E* respiratory epithelium; *ET*, ethmoturbinals; *MT*, maxilloturbinals; *NT*, nasoturbinals; *S*, – nasal septum. *From Tortora and Derrickson (2009), with permission.*

buffering of inspired air. It has commonly been assumed that their interior is nearly sterile, but if they do become infected and the local immune cells are overwhelmed, resolution of the infection is often a lengthy process.

The principal role of nasal mucus is to trap tiny particles in the air to prevent them from getting to the lungs. The trapping function is facilitated by the swirling of air currents in the nose, as well as by electrostatic charges on the mucus,

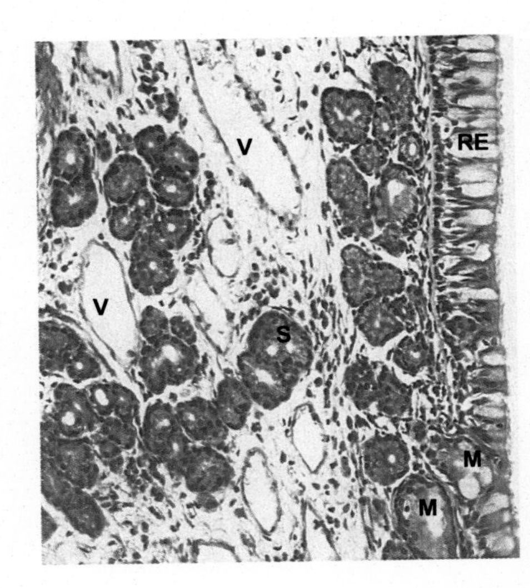

FIGURE 11.5 The nasal mucosa. *M*, mucous glands; *RE*, respiratory epithelium; *S*, serous glands; *V* blood vessels. *From Young et al. (2006), with permission.*

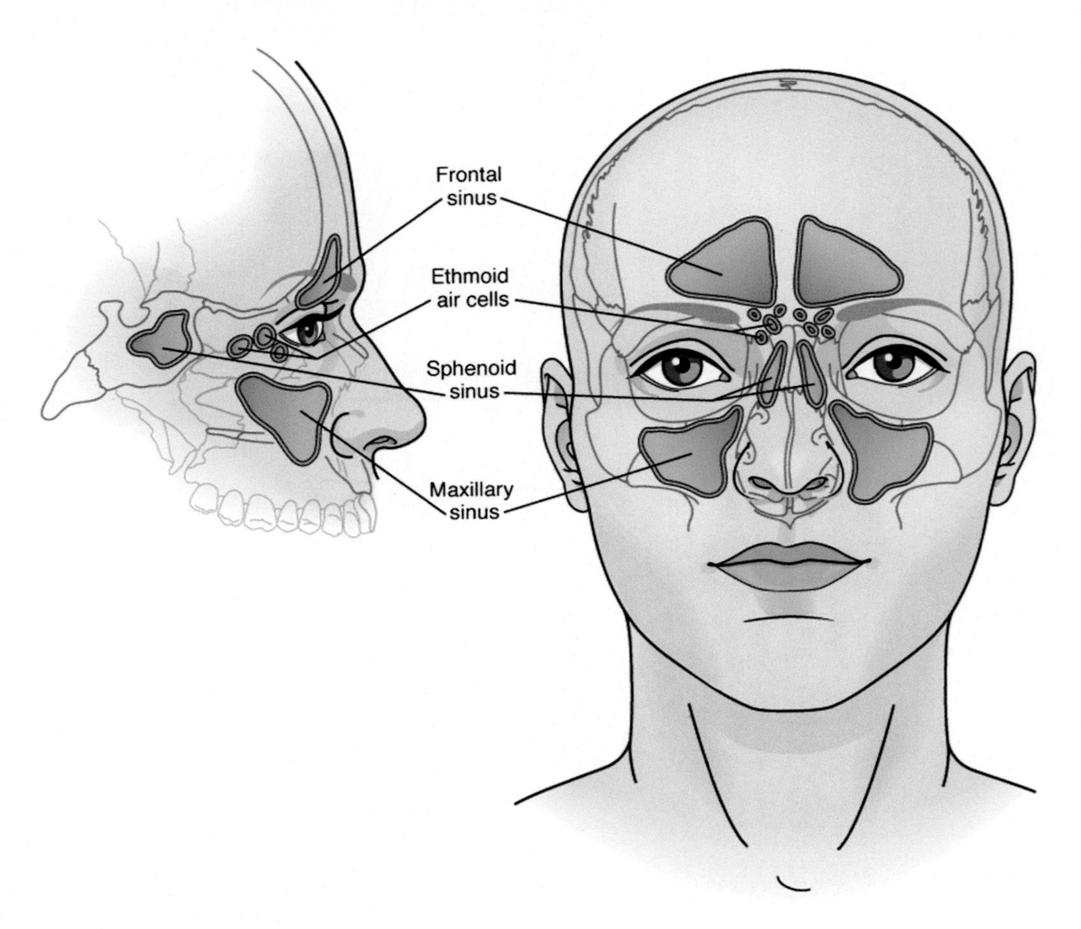

FIGURE 11.6 The paranasal sinuses.

which attract bacteria, pollen, or dust to the mucus. Certain serous cells within the nasal glands also form an enzyme, called **lysozyme**, which joins the nasal mucus and kills trapped bacteria by breaking down polysaccharides in their cell walls. In the tissues, immediately beneath the nasal epithelium are specialized immune cells that further protect this vulnerable inner body surface from foreign invaders (see Chapter 8). A side effect of this protective mechanism is the sneeze. When some immune cells come into contact with allergens trapped in the nasal mucus, they secrete histamine, which stimulates nerve endings in the nose. Their signals to the brain result in the explosive expulsion of air from the nose and mouth and the release of many thousand potentially infectious droplets into the air.

The continuous beating of the fields of cilia lining the surface of the nasal epithelium pulls the sheet of nasal mucus toward the nasopharynx and into the oropharynx at the back of the mouth. Cilia have an active beat, followed by a slower recovery beat. In the nose, the frequency is 4–10 beats per second. A wave of ciliary beats runs over the nasal epithelial surface toward the pharynx. One can visualize the power of the cilia in the laboratory by placing a small piece of cork onto the surface of the roof of the mouth of a frog and watching the cork being carried to the back of the mouth and into the esophagus. In humans, as much as a pint of nasal mucus is swallowed per day, usually without our being aware of it. This process serves as a natural trash disposal system for the upper respiratory tract.

The ciliated epithelium that lines the nasal chambers is very sensitive and can be badly damaged by harsh conditions, such as extremely cold, dry air, mechanical trauma, and smoke. Under these conditions, the ciliated columnar cells die and are replaced by a different type of cell—flat pavement-like cells that form patches of stratified squamous epithelium. This process of transformation of one type of cell to another in a tissue is called **metaplasia**. Under some circumstances, such transformations lead to the development of tumors, but in this case, the metaplasia is benign. The formation of patches of stratified squamous epithelium within the nose is very disruptive to the physiology of the nose. This type of epithelium is not ciliated, and therefore mucus that becomes pushed onto these cells is not transported further, and it tends to dry onto the epithelium. If such a patch of crusted mucus is blown or picked off, it may pull off some of the epithelial cells and even cause minor bleeding at the site. Because it is directly exposed to the outside air, the epithelium lining the outer vestibule of the nose is normally of the stratified squamous variety. The location of the border between the stratified squamous epithelium of the vestibule and the ciliated epithelium of the inner nasal chambers is not constant and depends upon the conditions to which the nose is exposed.

The nose has developed a unique mechanism for the protection of the nasal epithelium. One can justifiably wonder why there are two side-by-side nasal chambers instead of just a single larger one. One of the reasons is that constant exposure to flowing air results in metaplasia of the nasal epithelium. The nose deals with this problem through the actions of the **swell bodies**—the complex of blood vessels beneath the nasal epithelium. The swell bodies literally do swell upon an alternating cycle of every 20–30 minutes. When the swell bodies on one side of the nose are expanded, those on the other side are typically contracted, allowing less resistance to the inspired air. This is the side of the nostril through which most of the air intake occurs. Then roughly a half an hour later, these swell bodies expand and those on the other side of the nose contract. The absence of strong airflow on the side of the nose with the expanded swell body gives the nasal epithelium time to recover and prevents the occurrence of squamous metaplasia. An unfortunate side effect of the rich vascular supply of the nasal lining is the tendency of the nose to bleed (**epistaxis**) when subjected to trauma, infection, excessive drying, or allergies.

Another mechanism that protects the nasal epithelium from damage is increased watery secretions from the serous glands. This is particularly noticeable when one is outside in very cold, dry weather. An unrelated source of a runny nose occurs when one is crying. In this case, tears from the eye enter tiny nasolacrimal ducts that run from each eye and enter the lateral sides of the nasal cavity just beneath the inferior nasal concha (see Fig. 11.4A).

Olfaction

Olfaction is one of the most basic of all sensory functions, and in vertebrates, this sense is concentrated in the nose. Fish have purely external nasal canals, through which water passes. **Odorants** (molecules that elicit an olfactory response) present in the water bind to receptors and excite neural pathways that process and store the olfactory stimulus. The olfactory sense of some fishes is exquisitely sensitive. Salmon swimming in the ocean can identify the stream in which they were born and return to the exact site in the stream where they were hatched. Land-dwelling vertebrates have retained the nose as the seat of smell, but the nasal canals evolved into dual-use organs for both respiration and olfaction and became connected with the oral cavity. Land animals normally detect odorants that are present in the air, which explains to a great degree why the olfactory organ still resides in the nose. In order to an odor to be detected, the odorant molecules must still become incorporated into a liquid before they can be detected.

The human nose can detect thousands of odorant molecules, all of which are less than 300 Mw. Most perceived odors represent combinations of many odorant molecules. For example, the smell of a rose represents a mixture of 275 different odorants. These molecules enter the nose with the inspired air. Some of this air reaches the site of the olfactory organ, which is located in the roof of the nose above the superior nasal conchae (Fig. 11.7). With normal quiet respiration, the main air currents barely reach that level. This is why when one wants to detect a faint odor one sniffs several times. Sniffing directs an airstream to the upper reaches of the nasal cavities, where the odorant

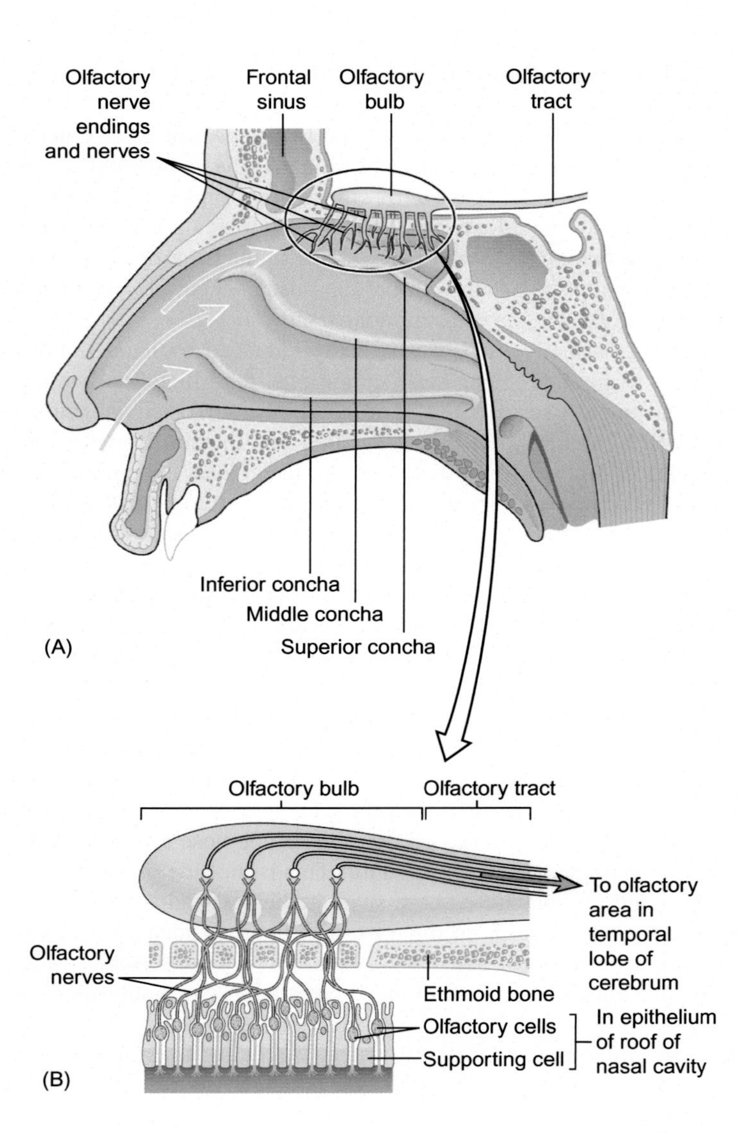

FIGURE 11.7 The olfactory system. (A) Gross anatomy. (B) Cellular structure of the olfactory apparatus. *From Waugh and Grant (2014), with permission.*

molecules in the air can directly contact the olfactory organ. In order to be detected, the odorant molecules must enter a watery mucous film (produced by the local Bowman glands) that covers the olfactory epithelium (Fig. 11.8). The mucous film also contains lysozyme and immunoglobulin A, which inactivate pathogens and keep them from entering the brain via the olfactory nerves.

Some classes of odorants are lipid-soluble, whereas others are more soluble in water. Lipid-soluble odorants, such as those present in cucumbers and bell peppers, can be perceived in extremely low concentrations (in the range of hundreds of nanomoles), whereas water-soluble odorants, for example, ethanol cannot be detected until their concentration reaches as much as 2 mM. Within the mucus are odorant-binding proteins, which serve to concentrate the odorant molecules and carry them to olfactory receptors. The olfactory mucus turns over every 10 minutes. This prevents the accumulation of odorants and allows better discrimination of individual odors in time and space.

The bound odorants come into contact with highly specific receptor molecules, which are located on specialized **olfactory receptor neurons** (see Fig. 11.7). These bipolar neurons terminate in a knob-like structure from which up to 20 long (30–200 μm) cilia extend over the surface of the olfactory organ. Each neuron expresses only one of the roughly 1000 types of receptor molecules, but there are as many as 25,000 olfactory neurons in the olfactory epithelium. The many thousands of discrete odors can be recognized through a combinatorial process. An individual odorant can be recognized by multiple receptors, and a single receptor can recognize multiple odorants. By a process involving the participation of multiple receptors, the nose can detect and recognize specific odors.

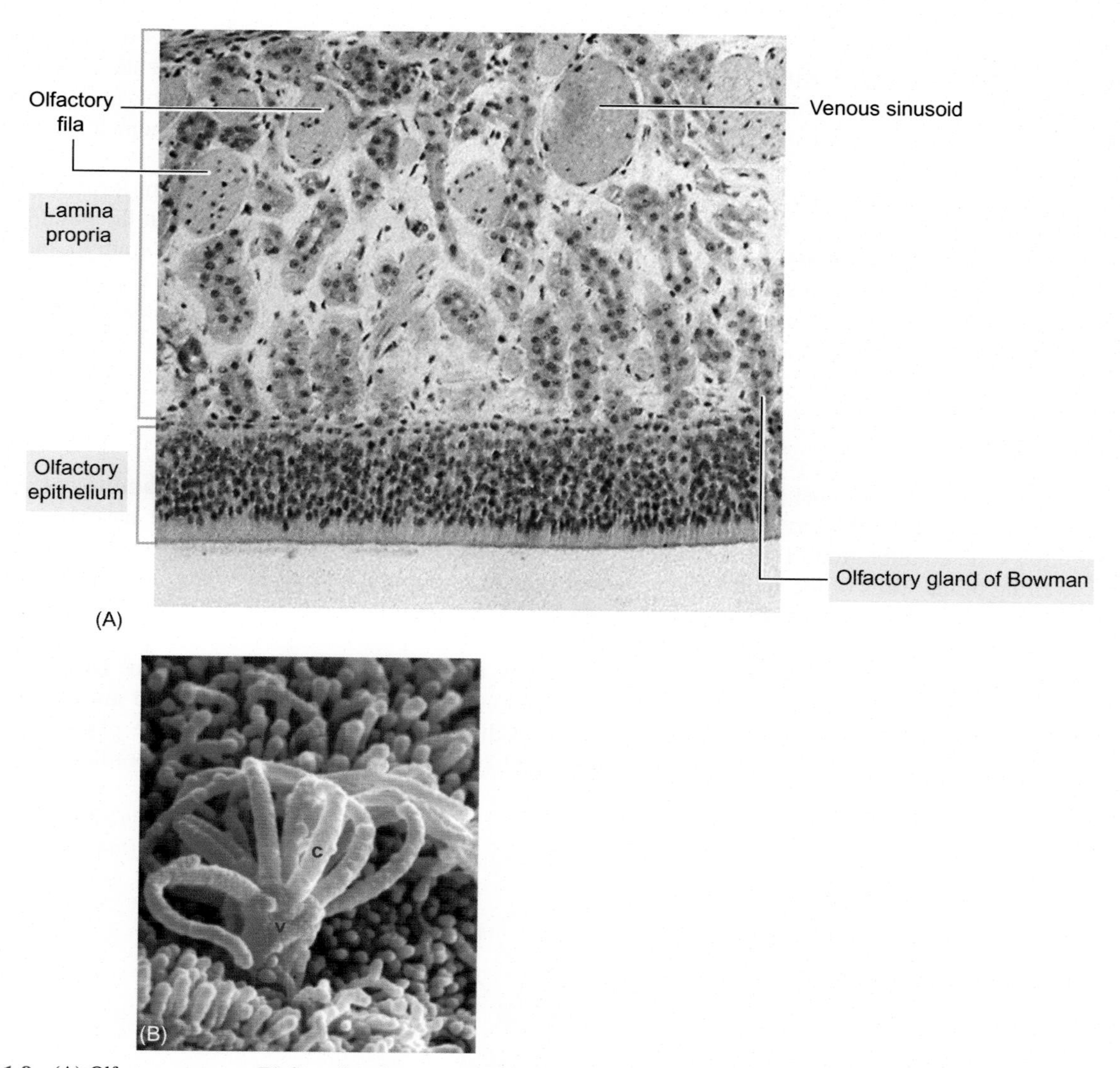

FIGURE 11.8 (A) Olfactory mucosa. (B) Scanning electron micrograph, showing elongated chemosensory cilia (c) protruding from a single olfactory epithelial cell. *From (A) Kierszenbaum and Tres (2012). (B) Nolte (2009), with permission.*

Olfactory acuity depends to a large part on the size of the olfactory organ and the number of olfactory neurons contained within it. The human olfactory organ, which is not particularly sensitive as organs of smell go, covers a surface area of approximately 5–10 cm^2, whereas that of some dogs with long noses covers an area as great as 170 cm^2 and has a density of olfactory neurons several times greater than that seen in humans. The nose of a bloodhound is 10–100 million times more sensitive than that of a human, and the nose of a grizzly bear is several times more sensitive than that.

The binding of as few as four odorant ligands to the receptor molecules in an olfactory neuron triggers a set of pathways leading to the depolarization of the neuron. The neurons themselves send axonal processes through the cribriform plate of the ethmoid bone (see Fig. 11.7) to the olfactory bulb of the brain, where individual neuronal signals are integrated with others and are then distributed to various parts of the brain. Some of the pathways lead directly to the amygdala area, a primitive part of the brain and one of the seats of basic long-term and emotionally charged memory.

Olfactory neurons are among the few in the body that have a normal turnover cycle. The normal lifespan of a single olfactory neuron is approximately 40 days, after which it dies and is replaced by a new one generated from olfactory stem cells. Despite the turnover, over the course of a lifetime the number of olfactory neurons declines, and along with it one's olfactory acuity. By age 70, the ability to distinguish different odorants has decreased by almost 50% from that of a 20-year old.

The olfactory region also contains another sensory system, which involves the trigeminal nerve—the large cranial nerve that serves most sensory functions in the facial area. When exposed to chemical stimulants, such as menthol, chlorine, ammonia, and capsaicin, sensory components of the trigeminal nerve produce hot or cool, tingling or irritating sensations. The trigeminal nerve also reacts to other odorants, but is much less sensitive to them than are olfactory neurons.

More controversial is the question of whether or not humans have retained a functional system for detecting **pheromones** (nonodorous substances that can affect sexual or social behavior in many animals). Within the nasal septum is a small structure called the **vomeronasal (Jacobson's) organ**, which is recognizable in the human embryo, but then atrophies during development. In rodents and many other mammals, the vomeronasal organ plays a major role in social communication. Present evidence suggests that traces of the vomeronasal organ can be found in many adult humans, but positive evidence for functional pheromone receptors in them is still lacking.

THE PHARYNX

Once past the nose, inspired air must pass through the pharynx before entering the trachea. The **auditory (Eustachian) tubes** empty into the upper pharynx (**nasopharynx**) (Fig. 11.9). The continuity between the nasopharynx and the middle ear allows the equilibration of air pressure between the atmosphere and the middle ear. A pharyngeal infection can cause swelling of the walls of the auditory tubes and make the equilibration of air pressure more difficult, especially for air travelers. In addition, an infection can travel up the tubes, resulting in middle ear pathology. The nasopharynx is lined by a ciliated epithelium similar to that of the nose.

The **oropharynx** is a double duty structure that must allow the passage of both food and air. During inspiration, the loose walls of the pharynx tend to collapse, but coordinated contractions of certain pharyngeal muscles, such as the pharyngeal constrictor and genioglossus muscles, stiffen the walls of the pharynx and allow free passage of air. Like those of the diaphragm and the intercostal muscles, contractions of the pharyngeal muscles are coordinated through the

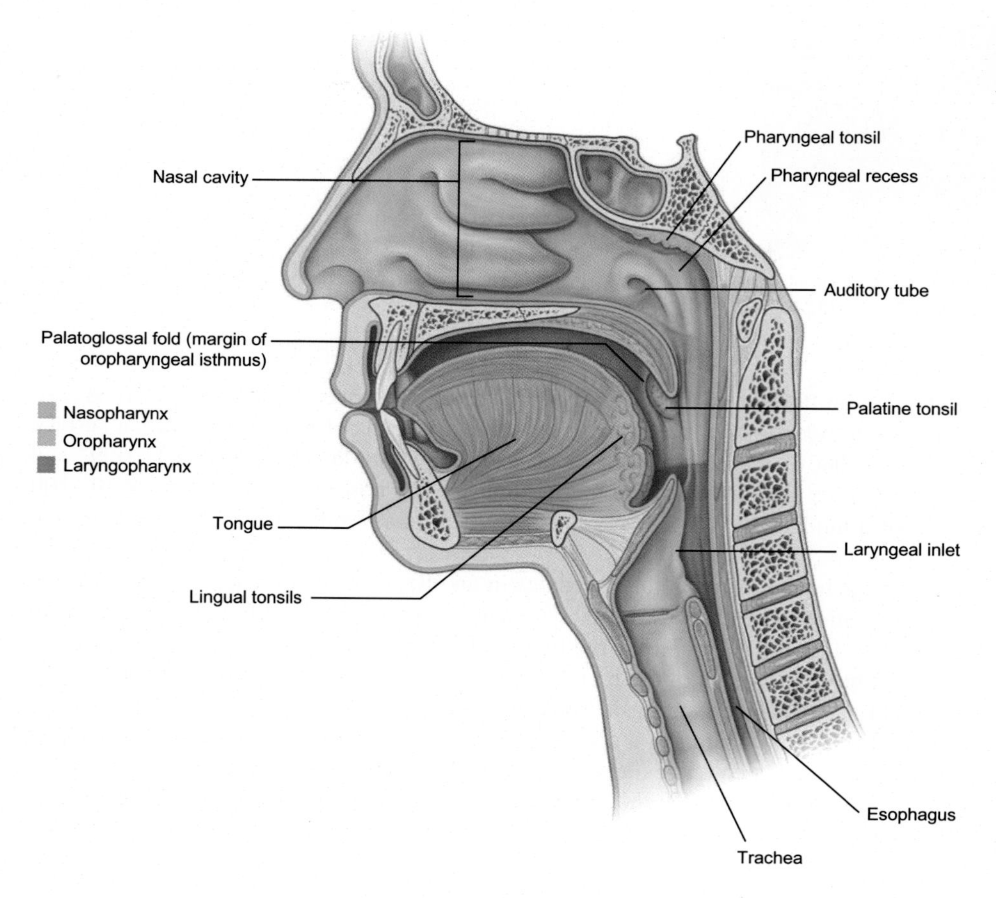

FIGURE 11.9 Anatomy of the pharynx. *From Drake et al. (2005), with permission.*

respiratory centers in the medulla. The oropharynx is lined by a stratified squamous epithelium to protect it from the mechanical abrasion produced by contact with ingested food and warm liquids.

The lowest region of the pharynx is called the laryngopharynx. It runs from the base of the tongue to the esophagus, and a prominent feature is the entrance to the larynx and lower respiratory tract.

THE LARYNX

The **larynx** (see Fig. 11.9) is a structure that has made use of moving air for a completely different purpose from respiration. Although an integral part of the respiratory tract, it has evolved a complex structure uniquely adapted for producing sound. In addition, it functions to prevent ingested food from entering the trachea.

The food-blocking function is served by the **epiglottis**, a flap-like structure with a cartilaginous core. During swallowing, the larynx becomes elevated through the actions of extrinsic laryngeal muscles, and the downward movement of the tongue causes the epiglottis to fold over the narrow glottis—the narrow upper opening of the larynx. This allows swallowed food to slide over it and into the esophagus. Within the upper part of the larynx, a pair of relatively inelastic **vestibular folds** (sometimes called **false vocal cords**) also plays a major role as a barrier to the entry of food or large foreign objects into the respiratory tract. In the absence of the epiglottis, the vestibular folds alone serve as an effective barrier to the passage of food and liquids into the trachea.

The pair of **vocal folds** (cords), located just below the vestibular folds (Fig. 11.10), represents the main engine for sound production in humans. Within each vocal fold is a band of highly elastic connective tissue that is suspended between the thyroid and arytenoid cartilages. The vocal folds tighten or relax, depending upon the actions of intrinsic

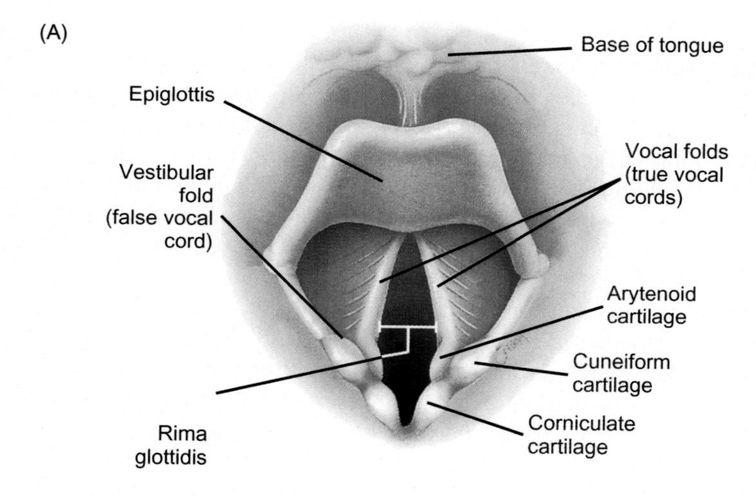

(B)

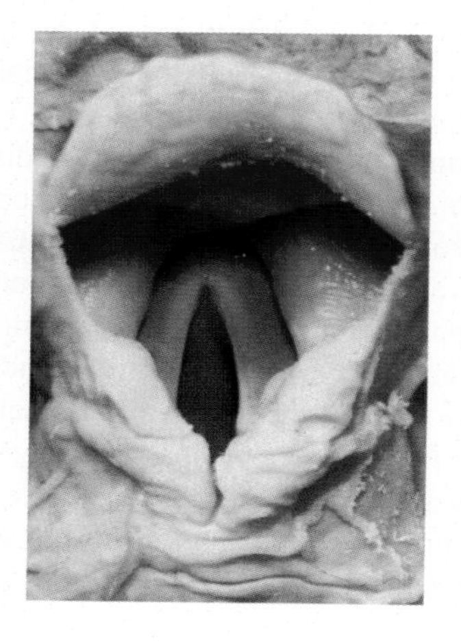

FIGURE 11.10 (A) The vocal cords, as viewed from above. (B) Photograph of vocal cords as viewed from above. *From (A) Thibodeau and Patton (2007). (B) Stevens and Lowe (2005), with permission.*

laryngeal muscles that position the arytenoid cartilages. As befits a mechanically active mucosa, the vocal folds are lined by a stratified squamous epithelium.

Sound production by the vocal cords is called **phonation**. Pitch is determined by the length and tension of the vocal cords, and volume is a function of the amount of air passing over them. When the vocal cords are stretched and brought close together (adducted) by the inward pivoting of the arytenoid cartilages through contractions of the lateral cricoarytenoid muscles, a high pitched sound is produced. A lower-pitched sound results from the relaxation and moving apart (abduction) of the vocal cords due to the swinging out of the arytenoid cartilages through contractions of the posterior cricoarytenoid muscles. At puberty, androgenic hormones stimulate the larynx of males to enlarge relative to that of females, resulting in the lower-pitched male voice.

The nature of the pharynx itself is of considerable importance in phonation. The human larynx is positioned lower down in the pharynx than is the larynx of other primates or even proto-humans. This increases the length of the air passage through the pharynx, with the result that humans have more control of the airflow coming out of the larynx. As a result, it is possible to form vowel sounds more precisely than can any other mammal.

Phonation is only the first step in human sound-making. Modulation of the sound produced by the vocal cords is accomplished by an intricately controlled set of actions by the muscles surrounding the oral cavity. The tongue and cheek muscles, along with the lips, are the principal players in what is sometimes called the articulation of sound (see Chapter 12). The resonant tones in singing and speaking are also greatly influenced by the shape and size of the paranasal sinuses.

THE LOWER RESPIRATORY TRACT

The Respiratory Tree (Conducting Zone)

The lower **respiratory (bronchial) tree** consists of the **trachea** and the 23 generations of successive tubes that branch in a classic fractal pattern. Inspired air must traverse all these branches in order to reach the millions of alveoli in the lungs, where actual gas exchange between air and the body occurs. At each level of branching, the structure of the respiratory tract changes in order to maintain the most efficient functional characteristics (Fig. 11.11). These functions are very similar to those of the nasal passages. In addition to allowing the passage of air, conditioning of that air (heating and humidification) continues in the lower respiratory tract. The extensive surface of the lower respiratory tract facilitates the removal of both particulate matter and pathogens, and the production and transport of mucus is an important function in the removal process.

A main function of the respiratory tree is to provide an open conduit for the passage of air. In the trachea, measuring about 12 cm in length and 2.5–3 cm in diameter, this is accomplished through a series of 16–20 C-shaped cartilages distributed along its entire length. Like flexible metal conduit, these cartilages maintain the structural stability of the tracheal wall while allowing bending. The open ends of the C's, which run parallel to the esophagus, are filled in by a layer of smooth muscle (**trachealis muscle**) (see Fig. 11.11). This band of soft tissue allows some distensibility, enabling a large object passing down the esophagus to intrude into tracheal space as it goes along.

As the trachea enters the chest (mediastinum), it branches into two mainstem **bronchi**—one going to each lung. At this level, without the need for flexibility to accommodate the needs of the esophagus, the cartilaginous rings in the walls of the trachea undergo a transition into large irregularly shaped plates that surround the mainstem bronchi. The trachealis muscular band disappears. Over the next several levels of branching, as the diameters of the bronchial segments gradually decrease, the closely spaced plates of cartilage become more broadly dispersed, before disappearing entirely as the bronchi branch into up to twelve branching sets of **bronchioles**. With the gradual disappearance of the cartilaginous support structure as the bronchi become narrower comes the appearance of a layer consisting of helical bands of smooth muscle, which completely encircle the bronchioles (see Fig. 11.11). The smooth muscle plays a powerful role in regulating the diameter of the bronchiolar segments. In an **asthma** attack, the smooth muscle strongly constricts, making breathing very difficult. The administration of epinephrine counteracts the parasympathetic signals that cause the bronchiolar smooth muscle to constrict the airways in an asthma attack. At the terminal ends of the respiratory tree, smooth muscle is represented only by isolated groups of cells in the walls of the respiratory bronchioles.

The other, mainly air conditioning functions of the respiratory tree are served principally through structures located in the inner mucosal layer. These structures produce a watery mucus that provides an interface between themselves and the moving air. The main function of the airway mucus is to trap and sometimes inactivate particulate matter and pathogens that enter the lower respiratory tract. In addition to its inherent sticky nature, the mucus contains protective

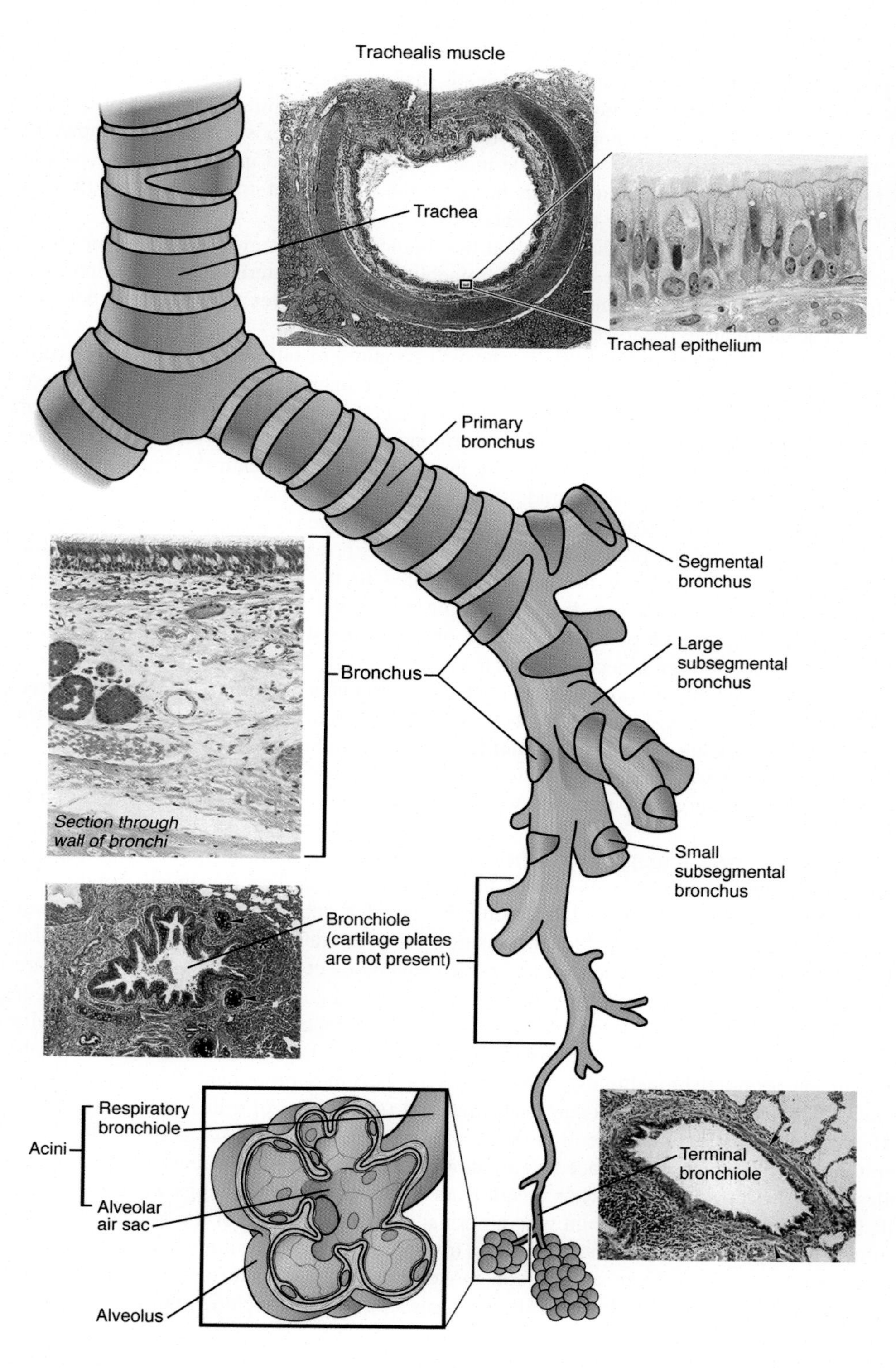

FIGURE 11.11 The respiratory tree, with photomicrographs illustrating the microscopic structure of key sections. *Photomicrographs from Erlandsen and Magney (1992), Kierszenbaum and Tres (2012), Stevens and Lowe (2005), and Young et al. (2006), with permission.*

molecules like those of the nasal mucus (lysozyme, immunoglobulins) which inactivate pathogens and certain toxic molecules.

The cells of the respiratory mucosa are designed to both produce and move the mucus. Respiratory mucus is bilayered, with the surface layer being more mucoid (sticky) and the lower layer, which contacts the cells of the respiratory epithelium, being more watery. Much of the actual mucus is produced by goblet cells, which are liberally interspersed (6500 cells/mm^2) among the cells of the respiratory epithelium. The watery part of the respiratory mucus, as well as additional mucus, is produced largely by small mucoserous glands that penetrate deeply into the walls of the trachea and bronchi. Most of the surface of the lower respiratory tree is covered by an epithelium composed of ciliated cells. Much like in the nose, these ciliated cells beat in a coordinated fashion, with the power stroke directed upward, in the direction of the pharynx. The waves of ciliary beating move the mucus film toward and into the pharynx, where it enters the digestive tract to be broken down in the stomach. In humans, the velocity of tracheal mucus flow is 4–5 mm/min, and 10–100 ml of liquid is transported up the trachea each day.

The character of the respiratory epithelium changes as the bronchial tree branches into progressively smaller channels. A dominant overlying principle is the need to clear the lungs and bronchioles of any material that would interfere with air exchange. This means that ciliated cells must extend farther down the respiratory tree than mucus-producing cells, because otherwise mucus would accumulate in the lower reaches of the bronchi.

The epithelium of the trachea is robust (ciliated pseudostratified), consisting mainly of tall columnar epithelial cells studded with numerous cilia on their outer surface. Interspersed among these are numerous goblet cells that produce specific mucus molecules. Embedded at the base of the tracheal epithelium are stem cells that replenish shed epithelial cells. Also present are scattered cells that produce a variety of peptide hormones. The ducts of the serous glands pass through the surface epithelium to empty their watery secretions into the mucus film. Just beneath the tracheal epithelium is a thin layer of connective tissue (lamina propria) that, in addition to providing support, contains a scattering of immune cells, which protect the interior of the body from foreign molecules or pathogens. Such cells, which among other things produce immunoglobulin A and cytokine protective agents, line any exposed body surface as part of the body's overall immunological defense system. A defense mechanism unrelated to the respiratory epithelium is an aggregation of sensory nerve endings located in the **carina**, the site of splitting of the trachea into the two primary bronchi. When these nerve endings sense the presence of a large foreign object at the site of splitting of the trachea, they stimulate a cough reflex, which expels the foreign object from the trachea.

The splitting of the trachea into right and left primary bronchi and the subsequent splitting of these into three secondary bronchi in the right lung and two in the left, followed by the branching of tertiary bronchi from the secondary bronchi, subdivides the lungs into discrete **bronchopulmonary segments**—10 in the right and 8 in the left lung. These segments are separated from one another in a way that tends to localize infections or tumors within a segment. If necessary, a bronchopulmonary segment can be surgically removed without causing major damage to neighboring parts of the lung.

As the bronchial segments become narrower through their repeated branching, the surface epithelium becomes progressively simpler in its construction, and mucus flow is slower than in the trachea. The frequency, height, and complexity of the ciliated cells decrease, trending from a high columnar toward a cuboidal shape in the **terminal bronchioles**. Most of these cells are ciliated. As the bronchioles transition into respiratory bronchioles and alveolar ducts, the epithelial cells lose their cilia and become flattened single-layered squamous epithelial cells. Goblet cells are present in diminishing numbers through the system of small bronchi, but only scattered goblet cells remain in the larger bronchioles. The arrangement of ciliated epithelial cells extending farther down the bronchial tree than the goblet cells enables the mucus coating to be swept upwards toward the pharynx without its accumulating within the lungs. At levels where the goblet cells disappear in the terminal bronchioles, a new cell type becomes prominent. These are **Clara cells**—dome-shaped cells without cilia. Clara cells secrete a surfactant material (see below) of a different composition from that secreted by the type II cells lining the terminal alveoli in the respiratory tree.

Smoking destroys ciliated epithelial cells, which are replaced by squamous nonciliated cells. Because of this, mucus produced by the goblet cells piles up in the lungs. The smoker's cough is an attempt to clear the lungs when the distribution of ciliated epithelial cells has been disrupted to the point where normal ciliary propulsion of the mucus is unable to do the job. The bronchial cilia not only move mucus, but they also transport dust cells (pulmonary macrophages located within the alveoli) up the respiratory tract. Even among nonsmokers, mucus clearance in the bronchial tree becomes slower with increasing age. This leaves elderly individuals more susceptible to bronchitis and lower respiratory infections.

Within the walls of the branches of the respiratory tree is a network of both arterial and venous channels, which comprise the **bronchial circulation**. The bronchial circulation supplies the tissues of the respiratory tract with oxygen and nutrients and removes metabolic waste products. The blood in the bronchial circulation is not involved in the respiratory gas exchange that occurs in the alveoli. Also present within the walls of the respiratory tree is a complex system of lymphatic channels, which arise as blind endings near the respiratory areas of the lungs. Excess tissue fluid and immune cells (mainly lymphocytes) collect in these lymphatic capillaries, which then drain into larger lymphatic vessels that ultimately leave the lungs. Lymph nodes scattered along the pathways of these lymphatic vessels process the lymphocytes passing through in the lymph as part of the overall strategy of immune defense for the body.

Gas Exchange in the Lungs (Respiratory Zone)

Gas exchange, the ultimate function of the respiratory system, takes place in the end portions of the respiratory tree, where the terminal bronchioles give way to **respiratory bronchioles**. The significant difference between these two is that terminal bronchioles are purely tubes, whereas respiratory bronchioles begin to lose their tube-like character and are also studded with occasional air sacs (**alveoli**), where gas exchange actually occurs (Fig. 11.12). The respiratory bronchioles feed into a delicate network of **alveolar ducts**, which terminate in **alveolar sacs** composed of aggregations of individual alveoli. According to some estimates, the lungs contain approximately 600 million individual alveoli with a total surface area, through which gas exchange occurs, of 140–150 m^2 (somewhat smaller than that of a tennis court). This respiratory surface accommodates the 350 ml of inspired air that reaches the respiratory portions of the lungs during a normal breath. (Of a typical inspired volume of 500 ml, 150 ml of air remains in the bronchial tree and is not available for gas exchange.)

The tissue interface between the air in the alveoli and the capillary blood in the lungs is extremely thin ($\sim$0.1–0.5 μm) and is well adapted for the passage of gases. It consists of a very thin squamous epithelium (**type I alveolar cells**) lining the alveolus, an equally thin layer of capillary endothelial cells and a common basement membrane interposed between the two (Fig. 11.13). This interface represents the majority of the wall of an alveolus, but other elements are also involved. Another prominent component of the alveolar wall is the **type II alveolar cell**. These cells protrude into the alveolar cavity and are important in lung function by producing **pulmonary surfactant**, a thin layer that lines the inner surface of the alveolus. A principal function of surfactant is to reduce surface tension, thereby making it easier for an alveolus to inflate during inspiration. Through its surface tension-reducing properties, the surfactant coating also prevents the alveoli from collapsing upon expiration. Type II alveolar cells also serve a stem cell-like function.

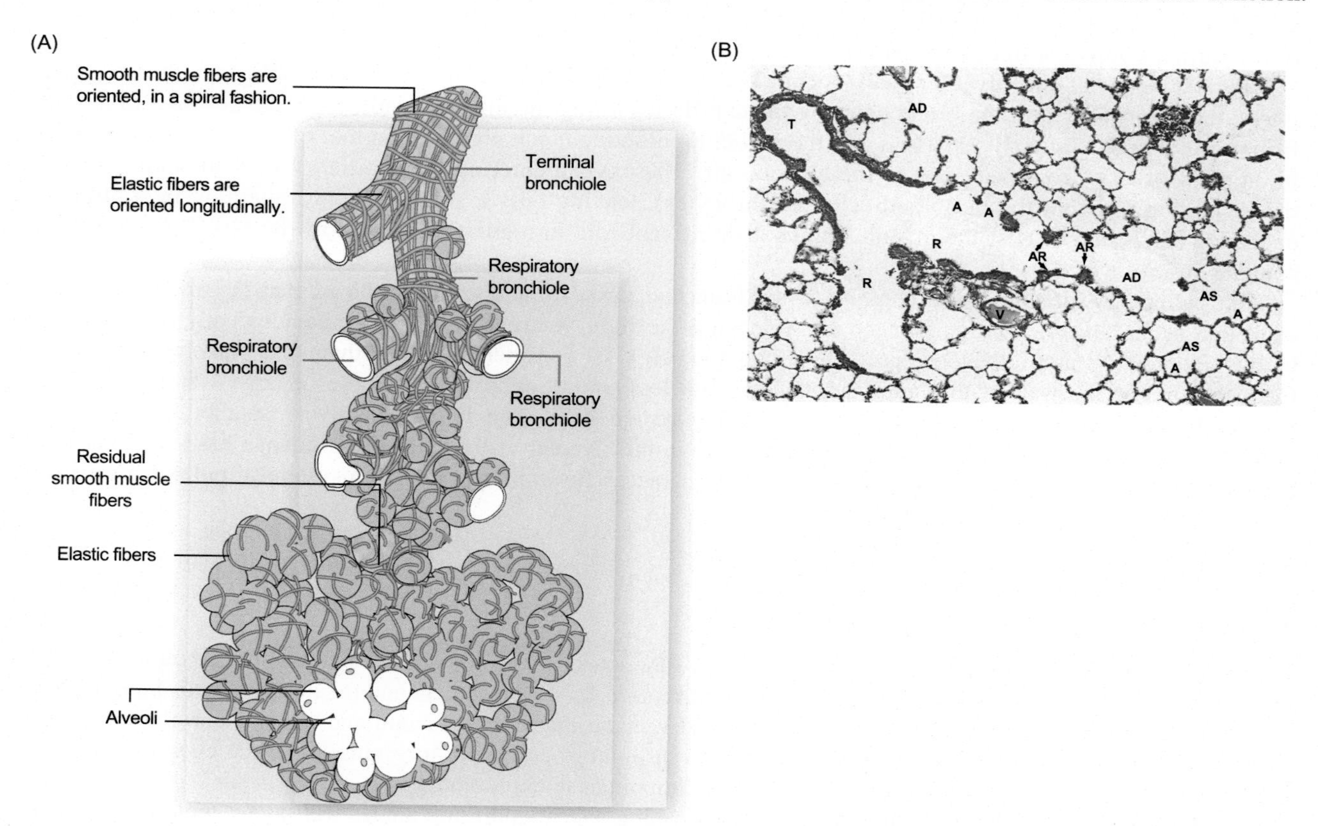

FIGURE 11.12 The terminal respiratory tract. (A) Anatomy, (B) microscopic structure. *From (A) Kierszenbaum and Tres (2012). (B) Young et al. (2006), with permission.*

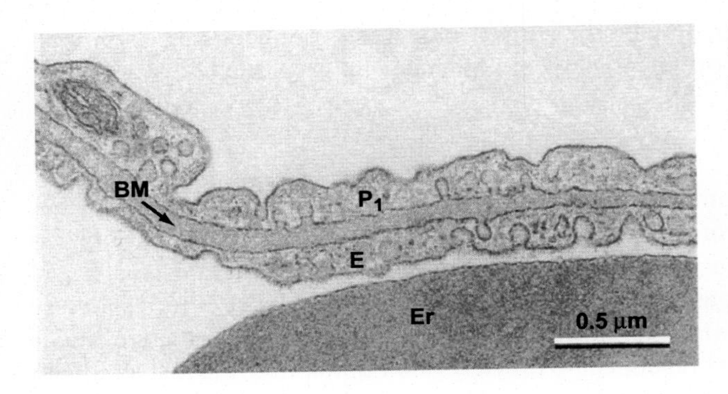

FIGURE 11.13 The barrier between air and blood—electron micrograph. The interior of an alveolus is on top, and the lumen of a capillary is at the bottom. *BM*, common basement membrane; *E*, – endothelial cell; *Er*, erythrocyte; P_1 type I pneumocyte. *From Young et al. (2006), with permission.*

Following damage to the alveolar lining, type II cells can divide and produce type I lining cells. The wall between two adjacent alveoli may be punctuated by a small pore that serves to equalize air pressure. It can also serve as an avenue for the spread of pathogens in the lung.

An additional type of cell found within the alveolus is the blood-derived (from monocytes) dust cell, which acts to take up dust and other particulate matter that might have penetrated as deeply as the respiratory part of the lungs. Dust cells (pulmonary macrophages) also take up asbestos particles that are present in inspired air. These cells, along with their phagocytized contents, make their way up the respiratory tree. When they contact the mucous layer of the bronchioles they are moved up toward the trachea until they are expelled into the pharynx and are then subsequently swallowed.

The thin connective tissue of the alveolar wall, mainly at the corners between adjacent alveoli, contains two other types of cells. One is the **mast cell**, which responds to foreign antigens by releasing histamine (see p. 237). The other is the fibroblast, which in the alveolar part of the lung produces a matrix rich in elastic fibers. The recoil of the elastic fibers plays a significant role in the passive exhalation process.

Gas exchange at the alveolar wall occurs because the partial pressure of oxygen ($\mathbf{P_{O_2}}$) in the inspired air is greater than that within the blood entering the capillaries lining the alveolar wall. Thus the net flow of oxygen is from air to blood. Similarly, the ($\mathbf{P_{CO_2}}$) in the incoming blood is much greater than that of the inspired air, resulting in the flow of CO_2 from the blood to the alveolar air. The oxygen entering the capillary blood soon combines with the hemoglobin within the red blood cells (erythrocytes), leaving only a small amount remaining in solution. The exchange of O_2 and CO_2 at the alveolus plays a significant role in regulating the pH of the blood, mainly through changes in levels of CO_2.

The pulmonary circulation, in contrast to the bronchial circulation, carries the blood that is enriched by the gas exchange in the alveoli. The pulmonary artery leaving the right ventricle of the heart paradoxically contains venous blood. Pulmonary arterial blood pressure (25/8 mm Hg) is considerably less than that of systemic arteries. Correspondingly, the wall of that artery is thinner, with less smooth muscle, than those of large systemic arteries. In parallel to the bronchial tree, the pulmonary arterial system breaks up into successively smaller branches, which ultimately become distributed as capillary networks within the alveolar walls. Once gas exchange has taken place in the alveolar capillaries, the venous blood, now arterial in character, flows into a collection system of pulmonary veins that ultimately empty into the left atrium of the heart.

SUMMARY

The main function of respiratory system is gas exchange—the intake of oxygen and the removal of carbon dioxide. To do so requires the nose for air conditioning, the trachea, and bronchi for air movement and the lungs for actual gas exchange.

Breathing movements are required to create a partial vacuum that brings air into and sends it out of the respiratory system. Respiratory centers in the brain send signals to inspiratory and expiratory muscles in the chest, as well as the diaphragm. Elastic properties of the lungs and chest are important in expiration.

The nose functions as an air conditioner by both warming and humidifying air, as well as by removing particulate matter. This is accomplished by the overall structural configuration of the nasal chambers, a rich submucosal vascular supply, and secretions of the nasal epithelium. The nasal epithelium is very sensitive and is protected by the alternation of air intake between the two sides of the nose through expansion of the swell bodies.

A secondary function of the nose is olfaction. Olfactory receptors in the nose can detect thousands of different odorants. After reception of an odorant, olfactory neurons send signals to the olfactory bulb of the brain for further processing.

Inspired air passes through the pharynx, the larynx, the trachea, and bronchi. Phonation is another secondary function of the respiratory system. Air passing over the vocal folds produces sounds, which are later modified by structures within the oral cavity.

The respiratory tree is lined with a ciliated epithelium, which moves mucus that is secreted by goblet cells up to the trachea toward the esophagus. This action removes particular matter from the lungs. Respiratory mucus also has antibacterial properties. Smooth muscle in the walls helps to control the diameter of elements of the lower respiratory tree.

Gas exchange occurs in the alveoli of the lungs. The epithelium of the alveoli is very thin and is underlain by a rich capillary network. Actual gas exchange occurs when the partial pressure of oxygen is greater than that in the capillary blood that enters the alveolar wall. Pulmonary surfactant functions to maintain the patency of the alveoli. Dust cells within the alveoli function as macrophages and engulf foreign particulate material or bacteria.

In addition to the pulmonary circulation that serves gas exchange, the lungs also have a bronchial circulation that supplies oxygen and nutrients to the tissues of the lungs.

Chapter 12

The Digestive System

The digestive tract is one of the most remarkable organ systems in the body. It is essentially a long tube through which food and liquid enter at one end and, after a lot of processing, emerge from the other end as feces (Fig. 12.1). Through this entire journey, the ingested food is subjected to a large number of mechanical and chemical influences that help to extract the maximum amount of nutrients from the food and absorb these nutrients into the body. As is also the case with the respiratory system, the entire interior of the alimentary canal is continuous with the external environment and is technically not part of the body.

Because the lumen of the gastrointestinal tract is continuous with the external environment, it is necessary to protect the body from pathogens or toxic substances that enter it. These strategies range from killing pathogens by secreted enzymes or antibodies to subjecting them to a pH in the stomach that is low enough to kill most pathogens. In addition, the immune system maintains a robust outpost of phagocytic and antigen-presenting cells in the connective tissue layer just beneath the epithelium of the gut. Nevertheless, the lower regions of the digestive tract are home to countless bacteria (probably three to four times as many bacterial cells as there are cells within the human body) which are essential to the normal functions of the bowel. Collectively called the **microbiome** and weighing over 1 kg, these bacteria exert profound effects on a surprisingly large number of bodily functions, even including cognitive ones. The overall composition of the estimated 500–1000 different species of bacteria living in our gut determines whether we have normal gut function or live with a variety of chronic gastrointestinal diseases.

The adaptation of structure to function in the digestive system covers a wide range of dimensions—from the gross anatomical level to the molecular—and at all of these levels, the adaptations change from one region of the gut to another. Two examples should suffice. The length of the intestine is closely correlated to the type of diet. Herbivorous animals have extremely long intestines and very complex stomachs in order to digest the cellulose products that make up most of their diets. Carnivores, on the other hand, have much shorter intestines. An interesting example is seen in frogs. Tadpoles are herbivorous, and their greatly coiled intestines are readily visible through their thin abdominal skin. Yet after their metamorphosis into carnivorous frogs, their intestines are greatly reduced in relative length. At the other end of the dimensional spectrum, appropriately distributed ion exchange channels in the small intestine ensure that a proper balance of ions and water is maintained in the body.

This chapter will follow the course of a meal consisting of proteins, carbohydrates, and fat (e.g., a hamburger) from the time, it first touches the lips until its remains are extruded from the anus.

ORAL CAVITY

The oral cavity is the place where food (our hamburger) first enters the digestive system. It is first brought into and retained in the mouth through actions of the lips. It is then chewed by the contractions of the muscles of mastication acting through the teeth. While this is taking place, enzymes brought in with the salivary secretions begin the process of digestion. Once mastication is completed, the tongue propels the bolus of food into the pharynx.

Lips

The **lips** are the guardians of the digestive system. Nothing can enter the mouth unless the lips are opened, and if the lips are closed, they prevent the contents of the mouth from being expelled. For structures as simple as they may seem, human lips have evolved a remarkably diverse set of functions. In addition to their role in food intake, lips are of vital importance in sound production, specifically in making labial sounds and vowel rounding during ordinary speech and in singing (Fig. 12.2). They are indispensable in whistling and playing wind instruments. Lips serve as a general tactile

The Human Body. DOI: https://doi.org/10.1016/B978-0-12-804254-0.00012-0

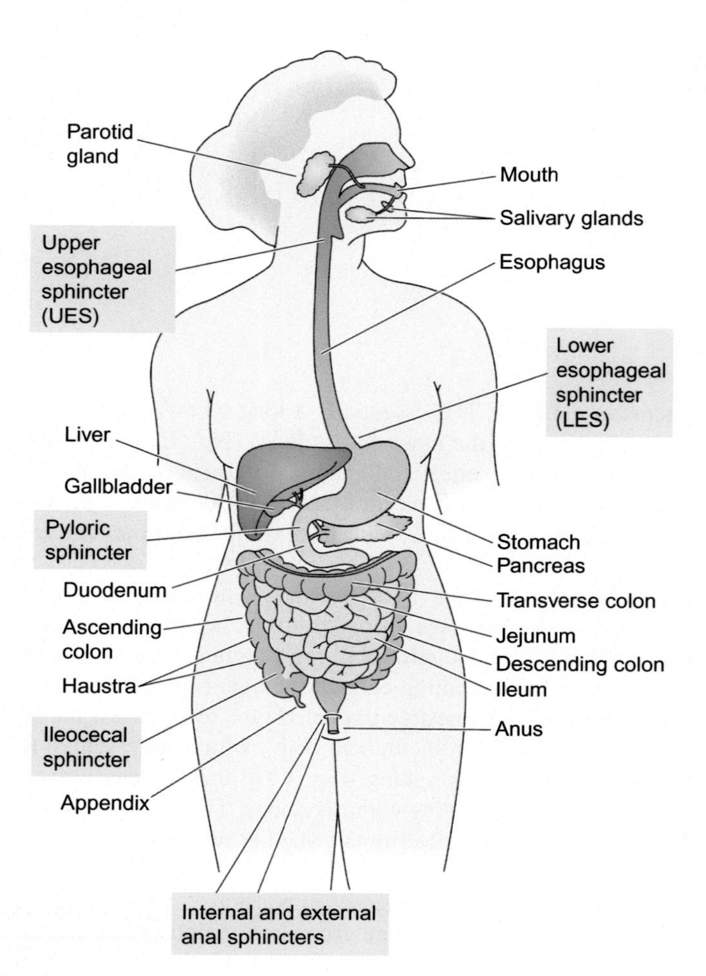

FIGURE 12.1 The digestive trace and associated glands. *From Boron and Boulpaep (2012), with permission.*

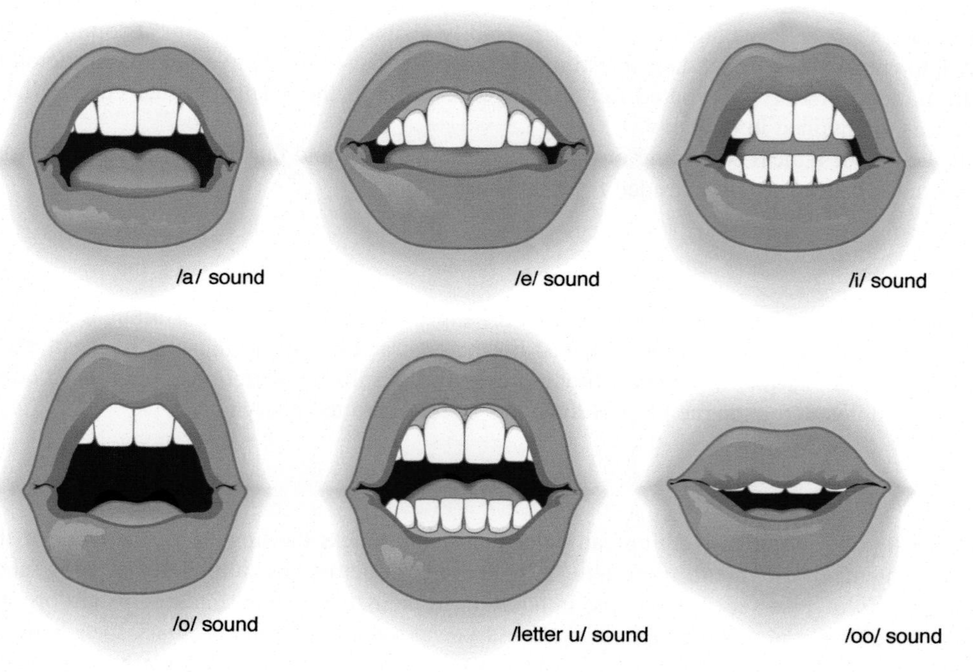

FIGURE 12.2 Lip positions while making vowel sounds. *Adapted from Berkovitz et al. (2009), with permission.*

organ, their erogenous function being one specialized aspect of this modality. Estrogens promote fuller lip development, which many psychologists believe serves as an attractant—hence the popularity of lip augmentation procedures. Finally, in humans (and many other primates) the lips play a central role in facial expression and the display of emotions.

For a number of these functions, the musculature within and attached to the lips is critical. Overall, both upper and lower lips are flaps of soft tissue, with thin skin on the outside surface, an underlying layer of muscle, and an inner mucous membrane (Fig. 12.3). Connecting the inner and outer layers is the vermillion border, consisting of very thin hairless skin overlying dense networks of blood vessels and nerve endings.

Many of the functions of the lips are mechanical, and most of these are the result of interactions of the intrinsic lip musculature (**orbicularis oris muscle**) and the variety of muscles of facial expression that attach to it (Fig. 12.4). The orbicularis oris muscle consists of two bands of muscle fibers that merge at the corners of the mouth. This muscle acts as a sphincter, causing the lips to pucker. The sphincter function is assisted by contractions of a deep cheek muscle, the **buccinator** (from the Latin for trumpeter). Other functions of the lips are elevation or depression of the corners of the mouth, each of which function is assisted by at least seven and five muscles, respectively (see Fig. 12.4). It is the presence and disposition of all of these facial muscles that allow the almost infinite means of expressing emotion by means of the mouth and lips, not to mention other parts of the face, especially the eyes, which are controlled by other sets of facial muscles. One of the most important mechanical functions of the lips is transferring food into the oral cavity. Between the inside edge of the muscle and the oral epithelium is a thin layer of connective tissue containing a scattering of small salivary glands.

Mastication

Once a bite of food has passed from the lips into the oral cavity, the process of mastication begins. Chewing involves an amazing number of moving parts. The most obvious motion is the forceful elevation of the lower jaw through contractions of the **muscles of mastication** (Fig. 12.5), all innervated by the trigeminal nerve (cranial nerve V). The effectors of this motion are the teeth, which sink into the food. The teeth play two important roles in chewing. The initial motion of the teeth is oriented vertically and is used in crushing or cutting. This is followed by a second lateral motion that is used for shearing purposes, especially with tough food, such as meat. Lateral motion is accomplished through the action of other facial muscles and is facilitated by the structure of the temporomandibular joint, which allows some lateral movement of the jaw.

Chewing obviously involves breaking a bite of food down into smaller pieces by the teeth, but the teeth are not the only players in this process. Getting food between the teeth in the first place is a function of the tongue, which not only places food there, but also later involved in the chewing process also selects the pieces to be chewed. Muscles of the cheek, especially the buccinator muscle, are also involved in chewing. They automatically contract during biting and keep food particles from escaping into the buccal (cheek) side of the teeth during the chewing process. The lips keep food in the mouth during chewing. In humans and other mammals, chewing is normally done by the teeth on only one side of the mouth.

Chewing is a rhythmic activity that is coordinated by an oral rhythm center (central pattern generator) in the brainstem. It involves three phases—(1) a closing stroke (initial contact), (2) a power stroke in which the food is mechanically broken down, and (3) an opening stroke, in which the jaw relaxes. Chewing is a very powerful action. A typical bite force ranges from 5–15 kg, and maximum bite force is ~50 kg. To prevent excessive damage to tooth enamel, a reflex action relaxes the biting action just as the teeth are about to come into contact.

As chewing proceeds, the tongue becomes more heavily involved by forming a bolus of food. It compresses the bolus by pressing it against the hard palate. The hard palate is also a passive player in the chewing process by separating the oral from the nasal cavity and allowing breathing while chewing. While all these mechanical events are taking place, the salivary glands are emptying their secretions into the oral cavity. These secretions play both a moistening and lubricating role. In addition, their contribution of salivary amylase begins the initial stages of digestion of carbohydrates. As an example of this activity, try chewing a piece of bread over 30 times before swallowing it. The bread begins to taste sweet, because the salivary amylase converts starches in the bread to more simple sugars, for example, maltose.

(A)

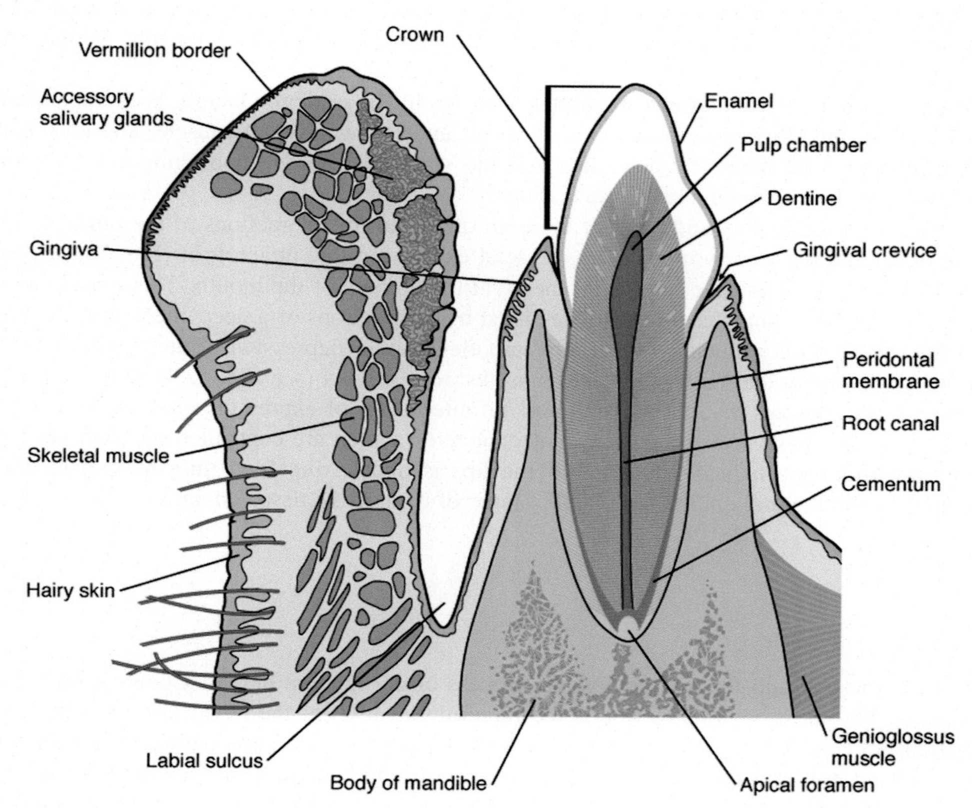

(B)

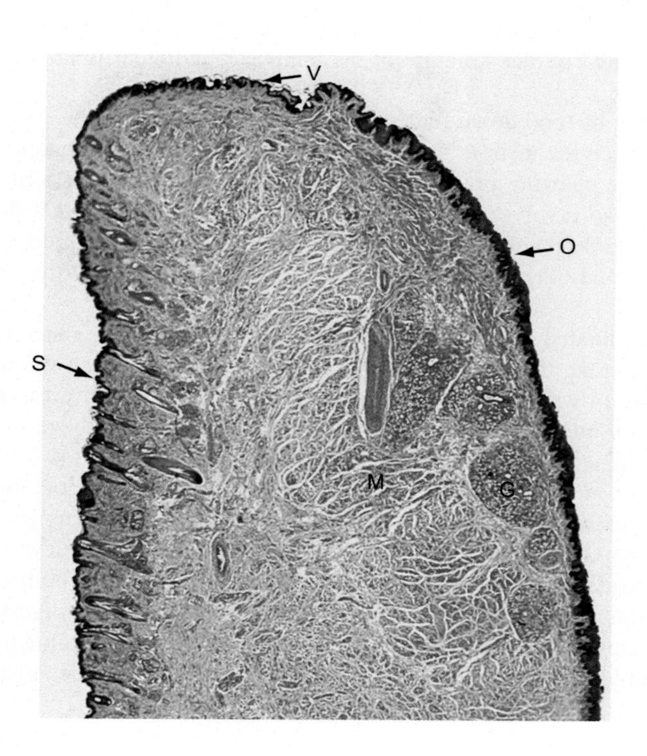

FIGURE 12.3 (A) Section through lip and jaw, showing major components. (B) Histology of the lip. *G*, accessory salivary glands; *M*, orbicularis oris muscle; *O*, oral mucosa; *S*, skin; *V*, vermillion border of lip. *From Young et al. (2006), with permission.*

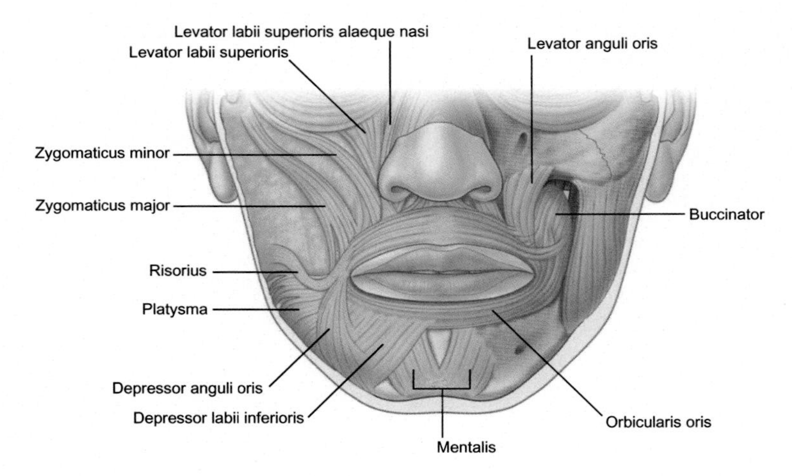

FIGURE 12.4 Muscles of facial expression. *From Drake et al. (2005), with permission.*

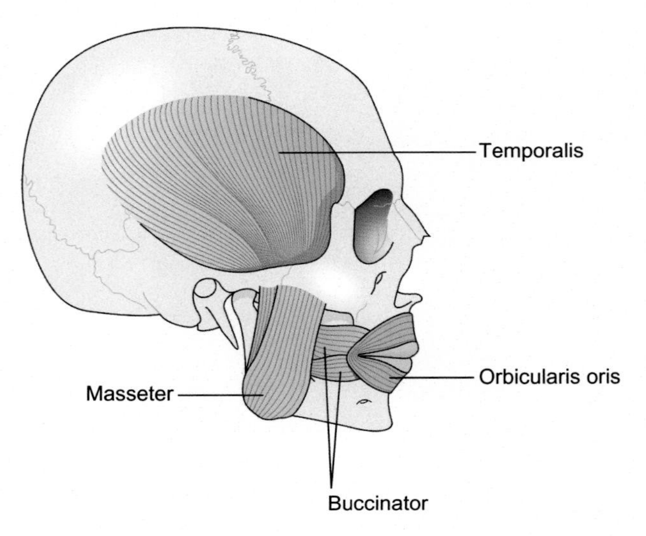

FIGURE 12.5 Muscles used in chewing. *From Waugh and Grant (2014), with permission.*

Teeth

Teeth are the body's instruments for breaking up ingested food into manageable dimensions, although for very soft food the tongue can also fulfill this role. Our teeth can trace their evolutionary origins to the spiny scales (placoid scales) of sharks which, in contrast to other types of fish scales, possess the same components as human teeth, that is enamel, dentine, and a pulp cavity.

Humans and other mammals have four main categories of teeth—incisors, canines, premolars, and molars (Fig. 12.6). Incisors are sharp-edged and are used for cutting. No incisors can match cutting power against the incisors of a beaver. Canine teeth are used for tearing. The overgrown canine teeth of saber-toothed tigers represent the utmost development of this type of tooth. Molars are used for grinding, and herbivores have strongly developed molars. With our omnivorous diet, humans do not emphasize one type of tooth over the others.

Teeth consist of **crowns** and **roots**. Their roots are deeply embedded within the upper and lower jaw bones and the gums, and their crowns are exposed (see Fig. 12.3A). In keeping with its mechanical role, the outer part of a tooth is a layer of enamel, the hardest, and most durable component of the human body. This is necessary to prevent excessive wear, since teeth must last a lifetime. The enamel layer, which is up to 2.5 mm thick in places, is arranged in many microprisms with only a small amount ($\sim$2%) of organic matrix interspersed among them. Even within the enamel, variations in its inner structure and composition give it the qualities necessary to withstand a lifetime of chewing.

Beneath the enamel is a layer of softer **dentine** (but still the second hardest tissue in the body after enamel), and inside the dentine is a soft core of **dental pulp**, containing loose connective tissue, blood vessels, and nerve fibers (see Fig. 12.3A). The enamel covering of the dentine ceases at about the gumline. Below that, covering the dentine in the roots of the tooth, is another hard material, **cementum**. Both the enamel and dentine are noncellular materials

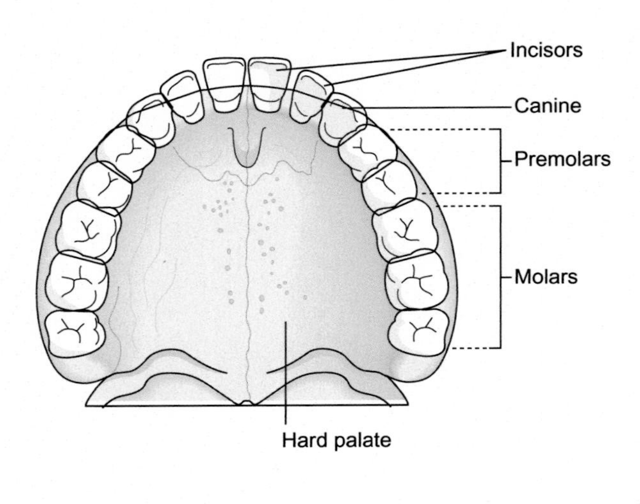

FIGURE 12.6 Types of teeth. *From Waugh and Grant (2014), with permission.*

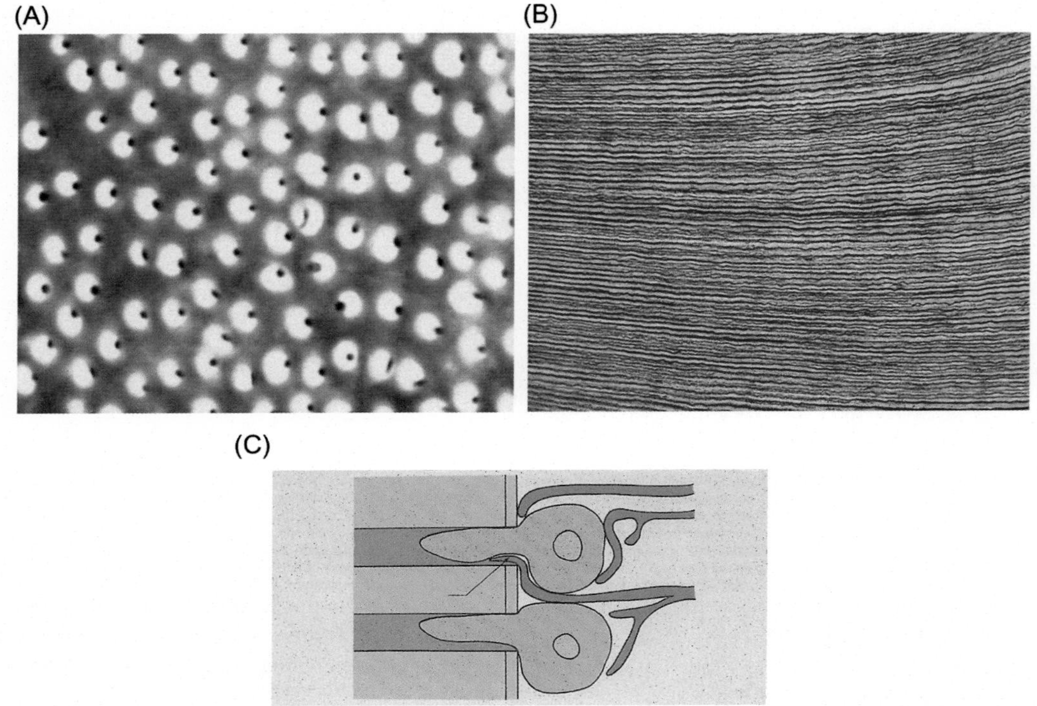

FIGURE 12.7 Dentinal tubules. (A) Cross-sectioned. The dark-stained odontoblast processes occupy only a small part of the tubules (white circles). (B) Longitudinal section through dentine, showing the parallel dentinal tubules. (C) Diagram, showing odontoblast processes (orange) and sensory nerve fibers (green) within dentinal tubules. *From Berkovitz et al. (2009), with permission.*

that represent extracellular secretions of two layers of cells found in developing teeth. In contrast to enamel, cells (**odontoblasts**) on the inner surface of the dentine produce new dentine throughout life. In a manner similar to bone formation by osteoblasts (see p. 91), odontoblasts secrete the dentinal matrix around them, and it then calcifies. As the predentine matrix becomes thicker, the odontoblasts retreat, leaving long cellular processes behind them. In mature dentine, the regions originally occupied by these processes form tiny (2.5 μm in diameter) dentinal tubules (Fig. 12.7A, B) that are partially filled by fluid and partially by the retreating cellular processes (Fig. 12.7C). In addition to odontoblastic processes, tips of sensory nerve fibers are also found in the dentinal tubules. The pain that one feels from a dental drill appears when the drill bit hits the dentine and is likely caused by the movement of fluid within the dentinal tubules acting on the tips of the sensory nerve fiber endings.

Dental pulp is a loose connective tissue, through which runs a network of vascular plexuses and a rich array of sensory nerve terminals. It also contains a surprisingly large number of dendritic antigen-presenting cells (see Fig. 3.9).

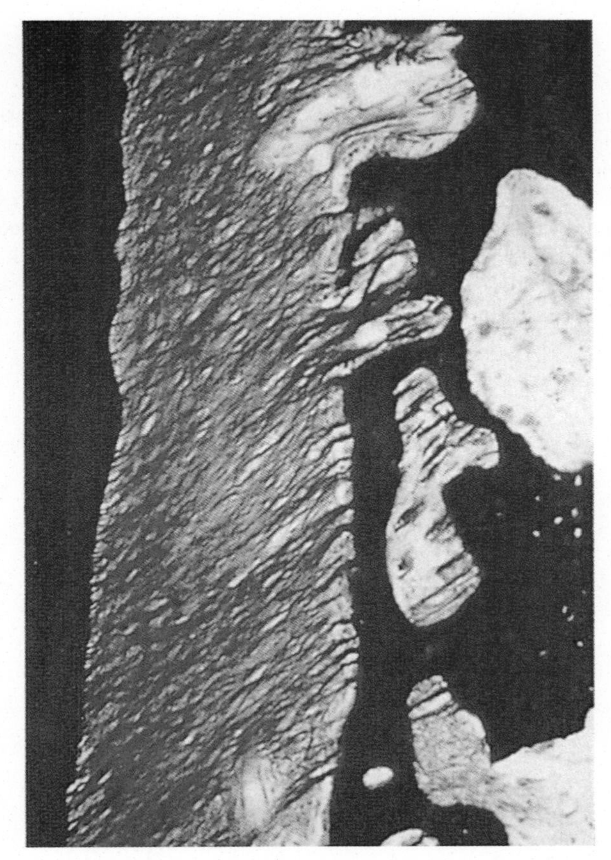

FIGURE 12.8 Oblique fibers of the periodontal ligament, connecting the dentine of a tooth (left) with alveolar bone (right). *From Berkovitz et al. (2009), with permission.*

A small canal at the apex of the roots of the teeth represents the connection between the pulp and the surrounding connective tissue. The interface between the pulp and the dentine is the single layer of odontoblasts.

In order to remain stable in the face of the powerful biting forces, teeth must be anchored firmly to a substrate. This is accomplished in several ways. Teeth are firmly embedded within sockets in the alveolar ridges of the maxillary and mandibular bones. Between each pair of teeth is another ridge of bone. The teeth are attached to the alveolar bone through the many strong collagen fibers of the **periodontal ligament**, which connects the cementum surrounding the roots of the teeth to the bony alveolar ridges (Fig. 12.8, see Fig. 12.3A). The periodontal ligament is continuous with the dental pulp. The periodontal ligament is richly supplied with blood vessels and sensory nerve endings that transmit information about mechanical forces acting on the teeth. The soft tissue that covers the alveolar ridges and ends at the borders of the teeth is called the gingiva. This area and the spaces between the gingiva and the teeth are home to up to 250 species of bacteria, which play a prominent role in the formation of both **dental caries** (cavities) and plaque.

The periodontal membrane and associated structures ensure stability of the teeth when biting. Nevertheless, the teeth are not immutably stable. When pressure or tension are applied to individual teeth, their relative position with respect to their neighbors can slowly be changed through remodeling of both the periodontal membranes and alveolar bone. This is how teeth adapt to the growth forces of the maturing jaws. Remodeling serves as the basis for orthodontic treatment, where artificial forces, such as braces, are applied to the teeth.

Tongue

The **tongue** is a multipurpose structure and is certainly the most complex muscular organ in the body. Not only does it play an important role in chewing and swallowing, but it is essential for the articulation of consonants during speech. It is also responsible for our sense of taste through the taste buds that are located on its surface (Box 12.1). In addition, it (especially the tip) is very sensitive to touch (see the homunculus in Fig. 6.24).

BOX 12.1 Taste

Taste, like olfaction, is one of the fundamental chemical senses, whose origins go back to the earliest forms of life. Single-celled animals and even bacteria respond to chemicals in their aquatic environments by moving toward or away from a chemical stimulant. In humans, taste is a function of an array of taste buds located in various areas on the surface of the tongue. Taste per se is perceived by the activation of specific receptors on the taste buds by tastants (Table 12.1), but the more subtle appreciation of taste occurs through processing of individual taste signals within the central nervous system. Even more subtle than pure taste is flavor, which represents a combination of taste plus odor. This is accomplished by the process of **retronasal olfaction**, which means bringing odorants from the food into the olfactory area of the nose through the back door, namely the upper pharynx (Fig. 12.9). Again, central processing melds the stimuli from both the olfactory area and taste buds to produce the subtle appreciation of flavor.

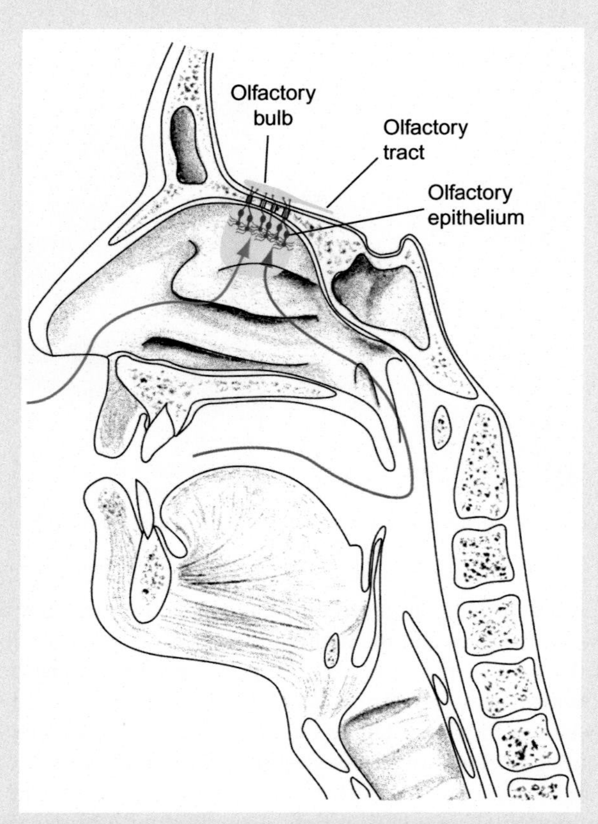

FIGURE 12.9 Orthonasal (*green arrow*) and retronasal (*red arrow*) pathways by which air containing odorants reache the olfactory epithelium. *From Nolte (2009), with permission.*

The tip of the tongue is most sensitive to sweet, salt, and **umami**[1] tastes. When this area is stimulated, other feeding mechanisms, such as salivation, mouth movements, and swallowing, are activated. In contrast, sour tastes cause puckering of the cheeks. Strong bitter tastes, which are sensed mainly at the base of the tongue, sometimes stimulate the gag reflex. This has evolutionary survival value, because many toxic substances are bitter. Ejecting them quickly in this case can be a matter of life or death. Like olfactory memory, gustatory memory, especially of unhealthy tastes, is very persistent over time and may also have considerable survival value. Such memory has been well demonstrated in experiments on rats.

Taste buds are found on three kinds of papillae located throughout the surface of the tongue (Fig. 12.10). The most numerous papillae on the tongue, **filiform papillae**, do not contain taste buds. **Fungiform papillae**, named because they are somewhat mushroom-shaped, are scattered throughout the anterior two-third of the tongue and contain about 25% of all taste buds. Another variety of papilla is the ridge-like **foliate papillae**. Located along the posterolateral edges of the tongue, they also contain about 25% of the total number of taste buds. The remaining 50% of taste buds are found along the walls of nine **circumvallate papillae**, which form a broad chevron at the base of the tongue. About 250 **taste buds** line the walls of each circumvallate papilla. The total number of taste buds on the tongue varies from person to person, but the common range is 2000–5000.

(*Continued*)

BOX 12.1 (Continued)

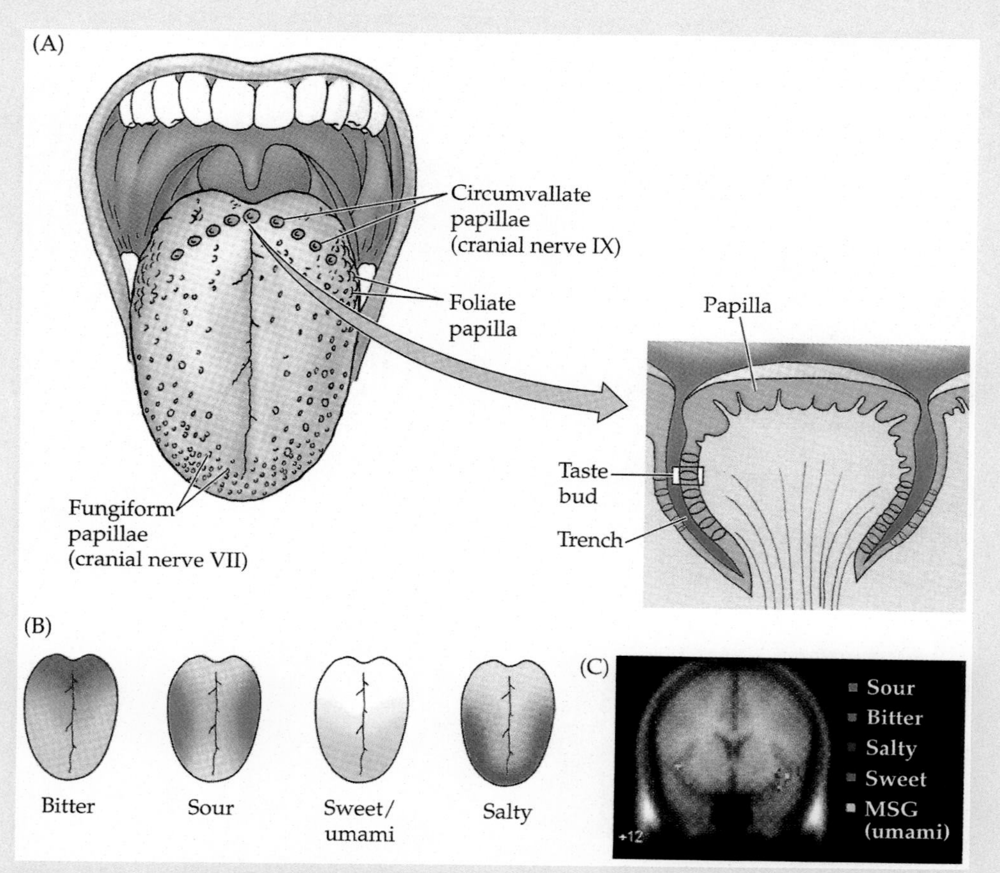

FIGURE 12.10 Taste buds and their innervation. (A) distribution of taste buds over the surface of the tongue. (B) Responses of different regions of the tongue to four fundamental tastants. (C) Composite fMRI image, showing location of parts of the brain responding to five different tastes. *From Purves et al. (2008), with permission.*

Taste buds are not only found on the tongue, but in humans small numbers are also scattered on the surfaces of the pharynx, the larynx, and upper esophagus in human. Some animals that rely heavily on taste have a much broader distribution of taste buds. Catfish, which locate food almost entirely by taste, have numerous taste buds (over 100,000) scattered all over their skin, which is how they can effectively feed at night in heavily stained or muddy water. They are literally swimming tongues. Crayfish and lobsters have taste buds on their antennae, whereas flies have them on their feet. At the other end of the spectrum, many birds and snakes have minimal senses of taste. This might be a distinct advantage to carrion feeders!

An individual taste bud looks something like an onion embedded in the epithelium of the tongue (Fig. 12.11). It opens up to the surface through a small taste pore. The taste bud is made up of around 100 banana-shaped cells with an apical surface studded with microvilli, which contain the taste receptor molecules. The basal ends of these cells synapse with terminal processes of afferent **gustatory neurons**. At the bottom of a taste bud is a collection of basal cells, which probably act as stem cells. Like olfactory cells, taste cells have a short lifespan—less than 2 weeks, so they must be constantly renewed from progeny derived from the basal cells.

(*Continued*)

BOX 12.1 (Continued)

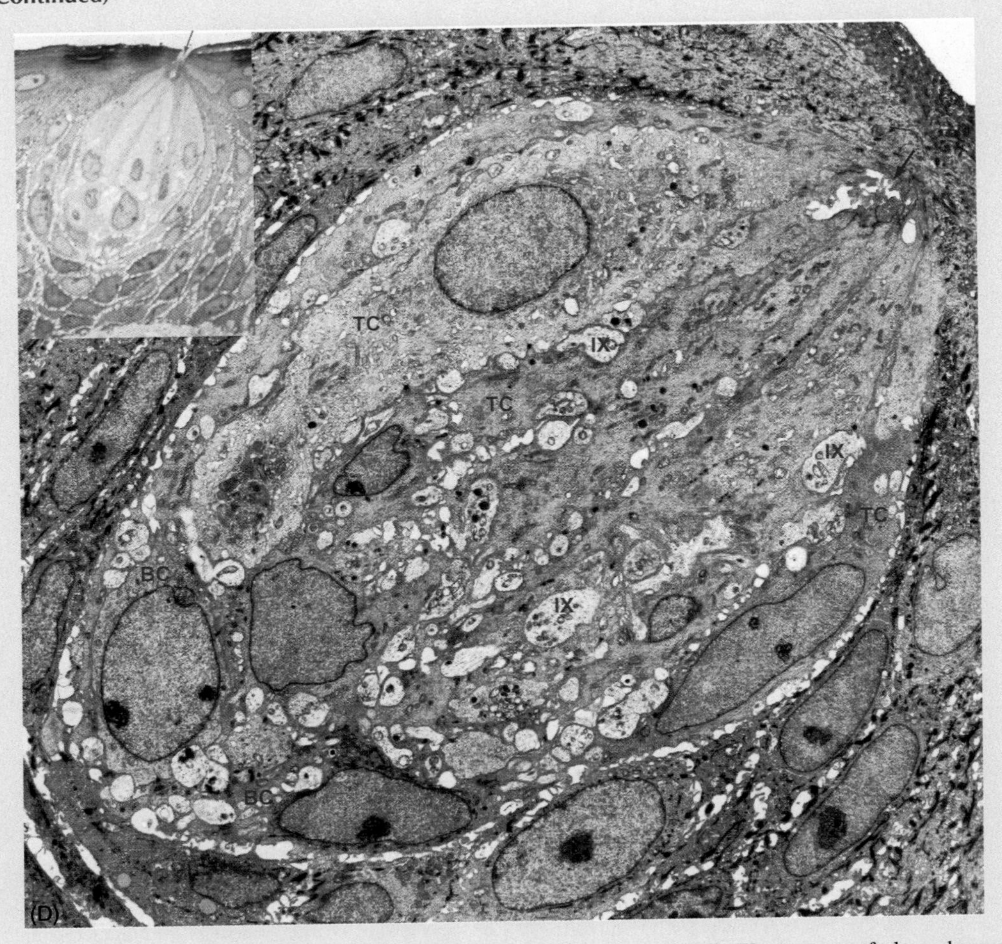

FIGURE 12.11 Light and electron micrographs of a taste bud. *BC*, basal cell; *TC*, taste cell; *IX*, processes of glossopharyngeal (IX) nerve fiber. *Arrow*, taste pore. *From Nolte (2009), with permission.*

When stimulated by a tastant, the excited taste cells release a transmitter that fires the gustatory afferent nerve fibers. Signals carried along individual nerve fibers pass up the dendrites and go to a gustatory nucleus in the medulla. From there, they pass through the thalamus and on to the primary gustatory center in the cerebral cortex, where various taste and olfactory signals are finally integrated.

1. In addition to the classic four basic tastes (salt, sweet, sour, and bitter), umami is now recognized as a fifth basic taste. Coming from Japanese words, meaning pleasant savory taste, umami is a taste based on specific receptors for glutamate. It is represented in the tastes of monosodium glutamate (MSG), meat and fish broths, and some fermented products.

TABLE 12.1 Fundamental Types of Taste

Taste Type	Principal Tastant
Salt	Na^+
Sweet	Sugars, artificial sweeteners
Sour	Low pH, H^+
Bitter	K^+, quinine, many toxins
Umami	Sodium glutamate

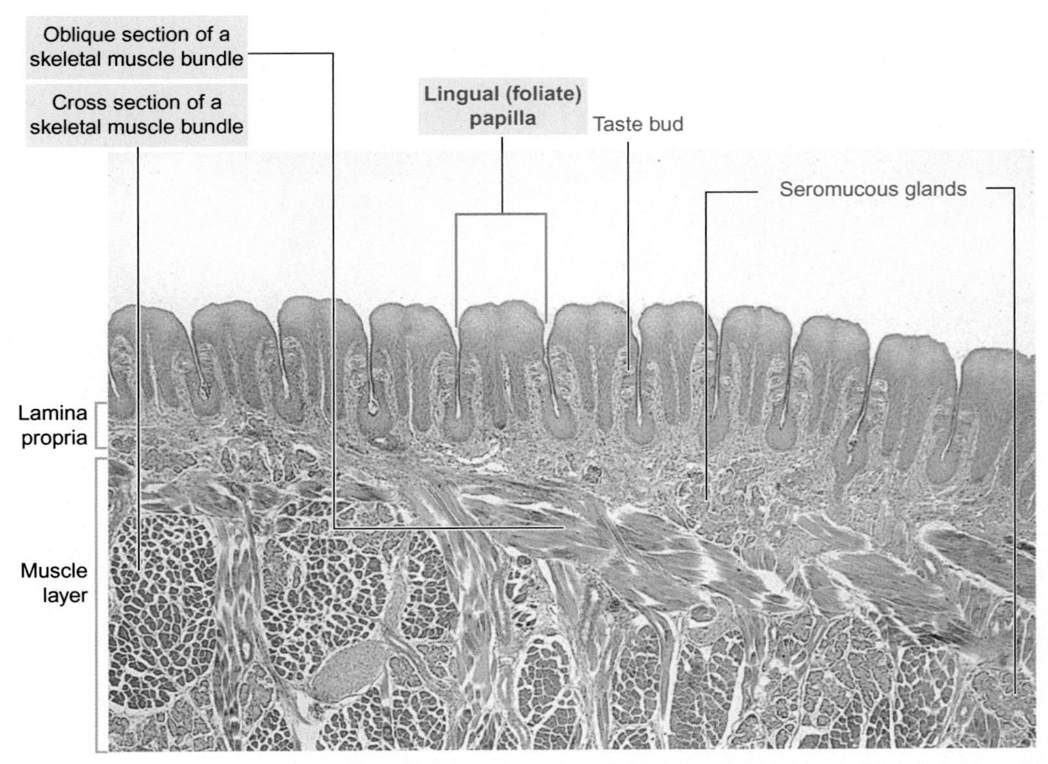

FIGURE 12.12 Histological section through the tongue, showing the three-dimensional orientation of the underlying skeletal muscle bundles. *From Kierszenbaum and Tres (2012), with permission.*

The incredible range of movements of the tongue is made possible by the arrangement of the muscle fibers contained within it. Starting from their first appearance in the embryo, most of the muscle fibers in the tongue are organized in small bundles that are perpendicular to the others in all three dimensions (Fig. 12.12). This is a characteristic shared by only a few other structures in the animal kingdom, such as the tentacles of an octopus or squid or the trunk of an elephant. This allows the elaborate and fine-tuned motions that are involved in forming a bolus of chewed food or in swallowing. Examination of Fig. 12.13 shows how many shapes the tongue has to assume in order to articulate consonant sounds.

Salivary Glands

The mouth contains three main **salivary glands**, plus a large number of very small mucous glands scattered throughout the oral mucosa. In aggregate, the salivary glands produce about 1 L of **saliva** per day, without which almost all oral functions would cease. First and foremost, saliva is a moistening and lubricating agent. Forming a bolus of chewed food would be almost impossible in the absence of saliva, and its lubricating function smoothes the passage of any solid object through the mouth and reduces damage to the oral mucosa by hard food. Saliva also begins the process of digestion through its content of **amylase**, which breaks starch down into maltose, and also **lingual lipase**, which begins the breakdown of lipid molecules. Saliva is important in cleaning the mouth of food particles after eating. It also brings into solution the **tastants** that stimulate the taste buds.

Defense against pathogens is an important function of saliva, and normal saliva contains a number of protective molecules. One is immunoglobulin A (see p. 216). Other more generic defenders against pathogens are **lysozyme**, which breaks down the cell walls of bacteria, and **lactoferrin**, which chelates iron, an ion necessary for the growth of many pathogens. Less saliva is produced at night. This allows more rapid growth of bacteria at night; these bacteria can account for bad breath in the morning. Inorganic ions (especially Ca^{++}) present in saliva are of some importance in maintaining the hard components of the teeth, but they also contribute to the formation of unwanted **tartar** (**calculus**) at the base of the teeth.

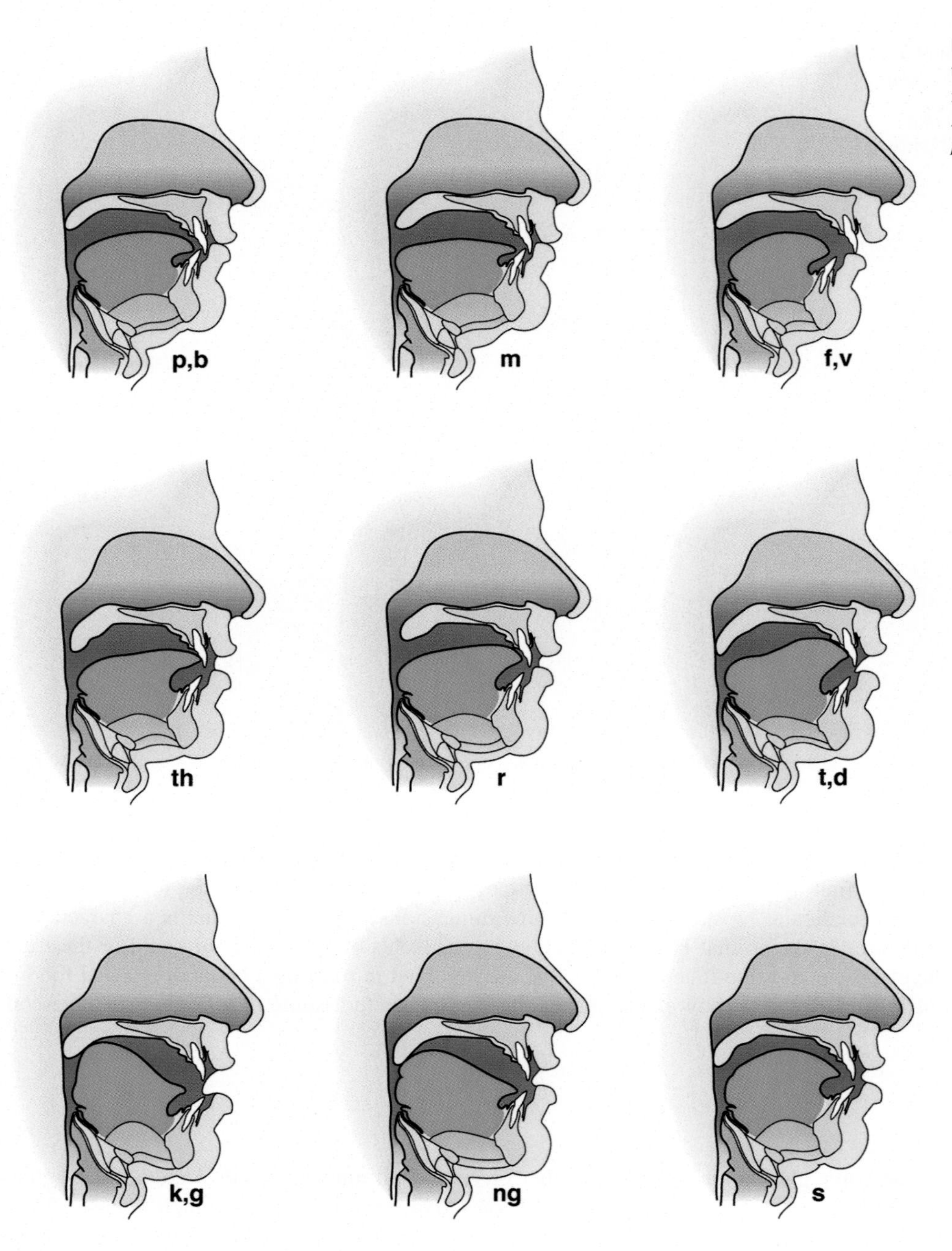

FIGURE 12.13 Configurations of the tongue and other oral structures in making consonant sounds. *From Berkovitz et al. (2009), with permission.*

Another function of saliva is protection of the oral mucosa during vomiting or when noxious substances are introduced into the mouth. A reflex action results in the copious secretion of saliva just before vomiting. Otherwise, the gastric acid would have a corrosive effect on both the teeth and the oral mucosa.

Of the salivary glands, the **parotid glands** are the largest (Fig. 12.14). The parotid is called a serous gland because it produces a watery secretion that accounts for about 25% of the daily volume of saliva. The parotid glands empty their secretions into the oral cavity through ducts opening into the middle of the cheek lining at about the level of the molar teeth. Located along the lower jaw, the **submandibular glands** produce a mix of watery and mucous secretions, but in greater amounts than the parotid gland (~70% of all salivary secretion). These glands empty into the mouth through duct openings on either side of the thin midline frenulum that is found under the tongue. The **sublingual glands**

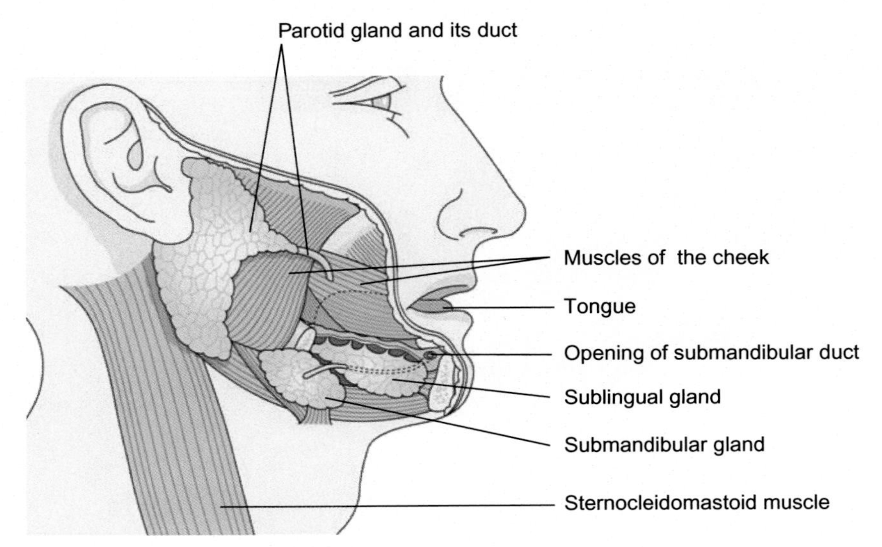

FIGURE 12.14 Locations of the three major salivary glands. *From Waugh and Grant (2014), with permission.*

produce small amounts of a mainly mucous secretion, which leaves the glands through a number of small ducts located beneath the tongue.

The salivary glands are organized much like bunches of grapes, with **secretory acini** the equivalent of the grapes themselves and duct systems organized like the stems. Each salivary gland has its own structure and mode of secretion (Fig. 12.15). Acini of the parotid gland are purely serous, and the epithelial cells lining the acini are constructed to secrete ions and water, along with some protein products. The sublingual gland is purely mucous, and its epithelial cells are all mucus-secreting. The submandibular gland is a mixed gland that can secrete both a serous and mucous form of saliva. The inner part of a submandibular acinus is constructed like a mucous gland, but capping it is a **serous demilune**, which produces a watery secretion. This secretion is able to pass through the mucous part of the acinus by means of tiny canaliculi, which are continuous with the excretory ducts (see Fig. 12.15). Surrounding the acini is a layer of fingerlike **myoepithelial cells** (Fig. 12.16). As their name implies, these cells can contract and assist the acini in expelling their secretions. Also associated with the acini are plasma cells which produce immunoglobulin A, as well as sympathetic and parasympathetic nerve endings that play important roles in salivary secretion.

The duct system (see Fig. 12.15) consists of ever larger ducts, much like tributary creeks emptying into a stream. Smaller ducts are lined by a low epithelium; as the ducts get larger, the height of the epithelium correspondingly increases. A functionally important part of the duct system, especially in the parotid gland, is a segment called the **striated duct**. The epithelial cells lining the striated ducts play an important role in determining the final composition of the saliva. Salivary glands are richly vascularized. In fact, the blood flow in a salivary gland is 20 times that of the equivalent mass of a muscle.

Salivary glands are unique among exocrine glands in that the flow of saliva responds almost exclusively to neural rather than to hormonal stimulation. Both parasympathetic and sympathetic nerves stimulate secretion, but most important is parasympathetic. Parasympathetic stimulation promotes a watery secretion, whereas sympathetic stimulation results in a more viscous saliva. Because of its dependence on innervation, salivary secretion is greatly influenced by higher brain functions. The famous conditioning experiments on dogs by Pavlov over a century ago well established the brain-salivary gland connection.

Secretion in the parotid gland begins when the epithelial cells of the acini pump Cl^- into the cells from the blood and tissue fluids. Na^+ and K^+, along with H_2O, follow. As the myoepithelial cells propel the secretion into the duct system and into the region of the striated ducts, further active changes modify the amount and composition of the saliva. Within the striated ducts, Na^+ and Cl^- are actively absorbed and K^+ and HCO_3^- are secreted. Water also leaves the duct, but with relatively more ions leaving, the resulting saliva is hypotonic. The concentration of electrolytes in saliva is strongly dependent upon its rate of secretion. When the secretory rate is low, saliva is quite hypotonic, but as the flow increases there is not sufficient time for the normal exchange of ions, and the composition of the saliva more closely resembles that of plasma and the initial fluid in the acini.

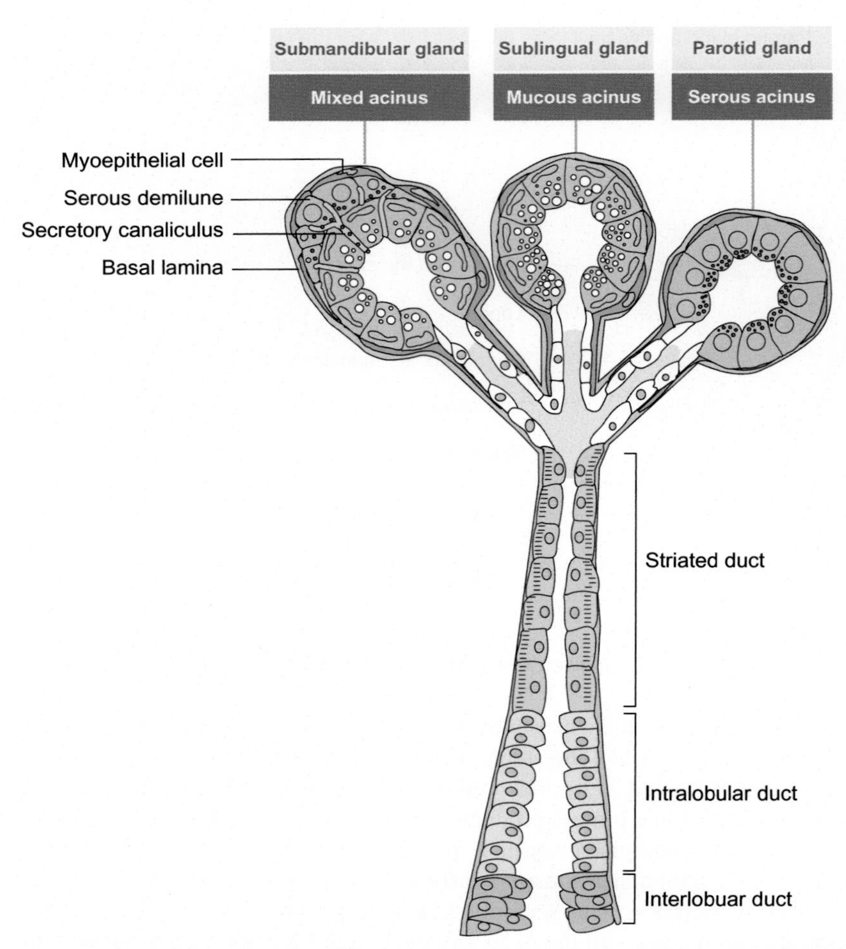

FIGURE 12.15 The three different types of acinar structure characteristic of the three types of salivary glands. *From Kierszenbaum and Tres (2012), with permission.*

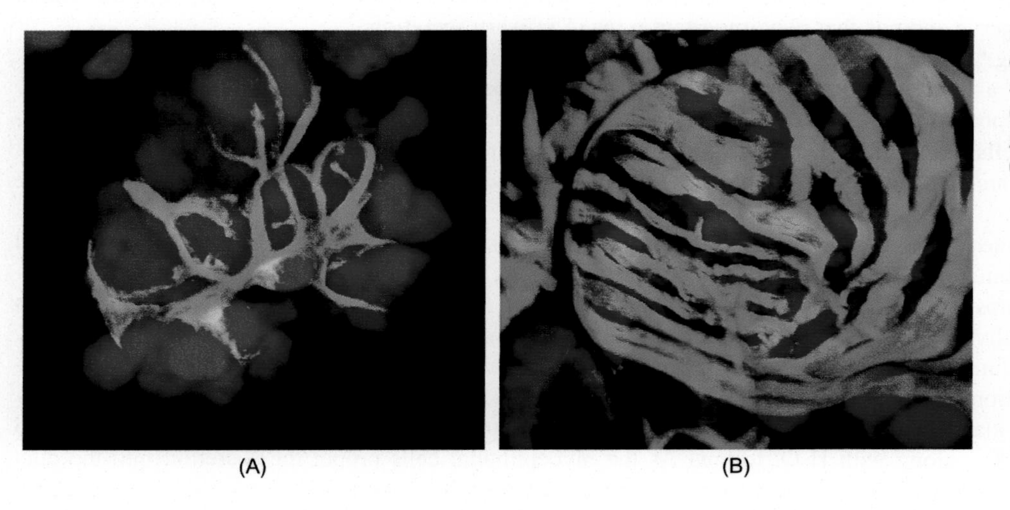

FIGURE 12.16 Myoepithelial cells (green) surrounding secretory acini (red) of the submandibular (A) and sublingual (B) salivary glands. *From Berkovitz et al. (2009), with permission.*

PHARYNX AND ESOPHAGUS

The pharynx and esophagus are all about swallowing. Because swallowing is so rapid, digestion in these areas is minimal. Swallowing actually begins in the brain, where conscious signals passing through the **swallowing center** in the reticular formation proceed to the musculature involved in swallowing. Swallowing begins when contractions of the tongue propel a bolus of chewed food into the upper (oro-) pharynx. The tongue forms a longitudinal groove and pushes the food against the roof of the mouth (hard palate). This motion initiates an automatic reflex that involves contractions of the superior constrictor muscles of the pharynx. (Fig. 12.17). These contractions close off the nasopharynx so that the ingested food cannot enter the nasopharynx. The swallowing reflex also inhibits breathing and causes elevation of the larynx and closing off of the glottis by the epiglottis, preventing any food or liquid from entering the trachea. Within a second, the bolus of food, lubricated by saliva, enters the pharynx and is next propelled into the esophagus by a peristaltic contraction of the upper pharyngeal muscles.

Although once initiated swallowing occurs in a reflex manner, the musculature of the pharynx and upper esophagus is striated and under voluntary control. Starting in mid-esophagus, smooth muscle becomes increasingly prominent in the esophageal wall until by the lower third, skeletal muscle fibers disappear and only smooth muscle is present. Like the entire gut, the musculature of the esophagus is arranged with an inner layer of circularly oriented muscle and an outer layer of longitudinal muscle (Fig. 12.18).

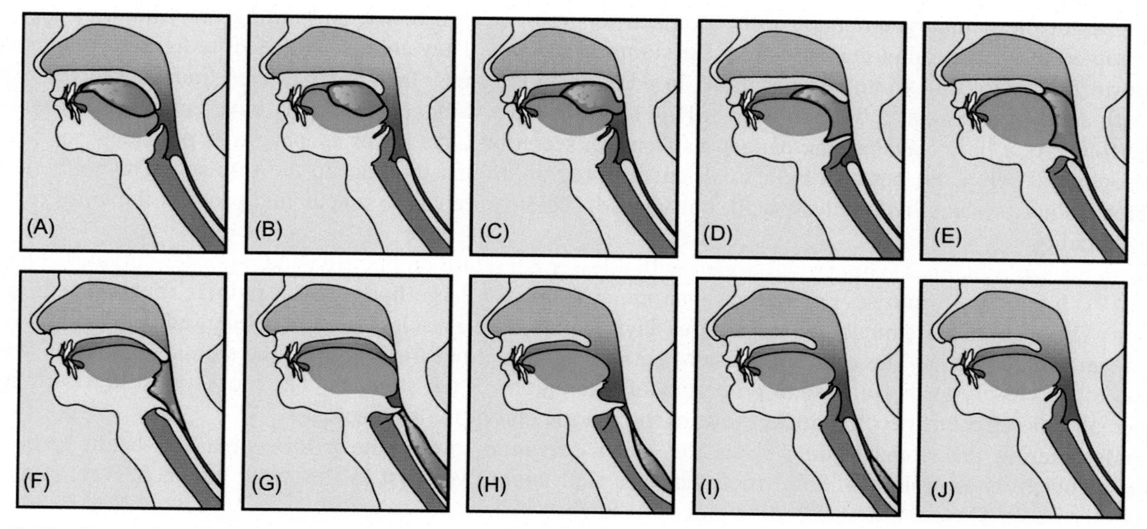

FIGURE 12.17 Stages of swallowing. (A)–(C), oral stage; (D)–(F), pharyngeal stage; G–J, esophageal stage. *From Berkovitz et al. (2009), with permission.*

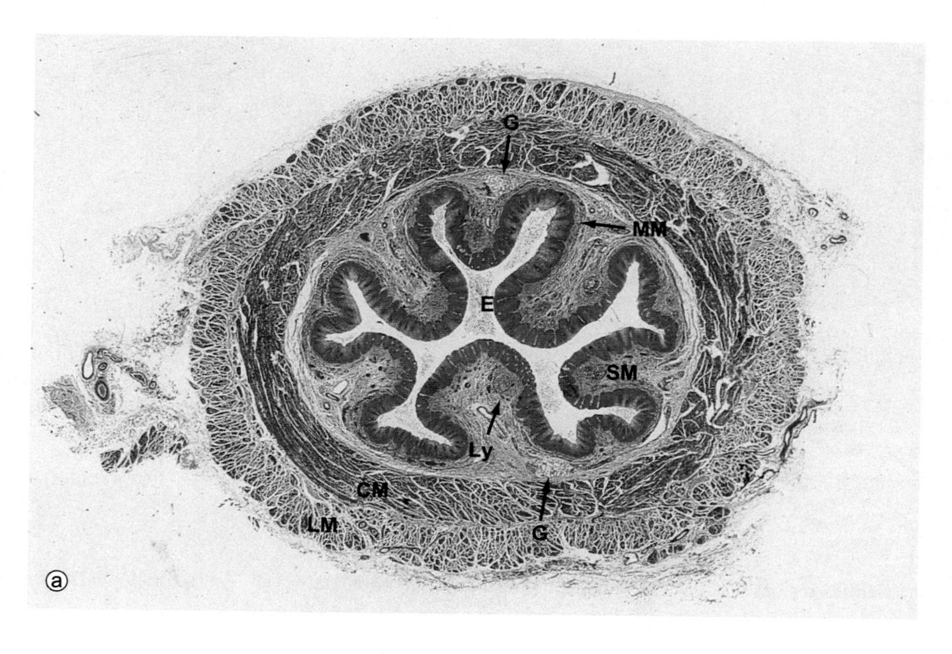

FIGURE 12.18 Cross-section through the esophagus. *CM*, circular smooth muscle; *E*, stratified squamous epithelium; *G*, seromucous glands; *LM*, longitudinal smooth muscle; *Ly*, lymphoid aggregate; *MM*, muscularis mucosae; *SM*, submucosa; *TP*, tunica propria. *From Young et al. (2006), with permission.*

Passage of food from the pharynx into the esophagus occurs by contractions of the lower pharyngeal muscles. Then the upper esophageal muscles, which together act as a sphincter, relax, allowing the bolus of food to enter the esophagus. At this point, a series of peristaltic contractions, beginning with the upper esophageal musculature, moves down the esophagus at a range of 2–6 cm/s and carries the bolus of food down with it, until the food reaches the lower end of the esophagus (see Fig. 12.17). During the esophageal phase of swallowing, the soft palate, tongue, and glottis return to their normal positions, and the airway is reestablished. Entry into the stomach is permitted by relaxation of the **lower esophageal sphincter**, which consists of the smooth muscle surrounding the lower 1–2 cm of the esophagus. At the same time, the musculature of the upper stomach relaxes as part of the overall swallowing reflex, lowering the pressure within the stomach and allowing the food to enter.

STOMACH

Once the piece of chewed and swallowed hamburger enters the **stomach**, it encounters a completely different environment. Pieces of the hamburger remain in the stomach for up to several hours, and during this time they are mechanically disrupted by the churning movements of the stomach muscles. They are also exposed to digestive enzymes, which break down macromolecules into simpler units. In addition to enzymes, the food particles find themselves exposed to fluids with a pH of as low as 1.0. Such a low pH is lethal to most pathogens that may have entered the stomach along with the food. As a buffer against the dangerous stomach secretions, the lower esophagus is protected not only by the lower esophageal sphincter, but also by a small **cardiac region** around the inlet to the stomach. The epithelial cells in this region do not produce hydrochloric acid, but instead, secrete an alkaline mucus that protects the entry region from excessive acidity.

The stomach is a highly asymmetrical organ, with a lesser curvature on one side and a greater curvature on the other (Fig. 12.19). It is divided into several regions—an upper **fundus**, a large **body**, and a **pyloric (antral) region** near its exit point. The wall of the stomach contains three layers of smooth muscles—an outer longitudinal, a middle circular, and an inner oblique layer. These layers are not continuous throughout the length of the stomach, but their extent and thickness relate closely to the functional properties of the regions of the stomach. The most robust layers (longitudinal and especially the circular) become thicker toward the pyloric end of the stomach.

Liquids entering the stomach slide down the lesser curvature toward the pyloric area and begin to be emptied from the stomach within minutes. Solid food particles find themselves first in the upper stomach, where contractions are relatively weak. There the first digestive processes begin.

Because of the lesser amounts of muscle in the fundic part of the stomach, ingested food tends to remain there for some time—often over an hour. This allows the salivary amylase, which accompanies the bolus of swallowed food, to

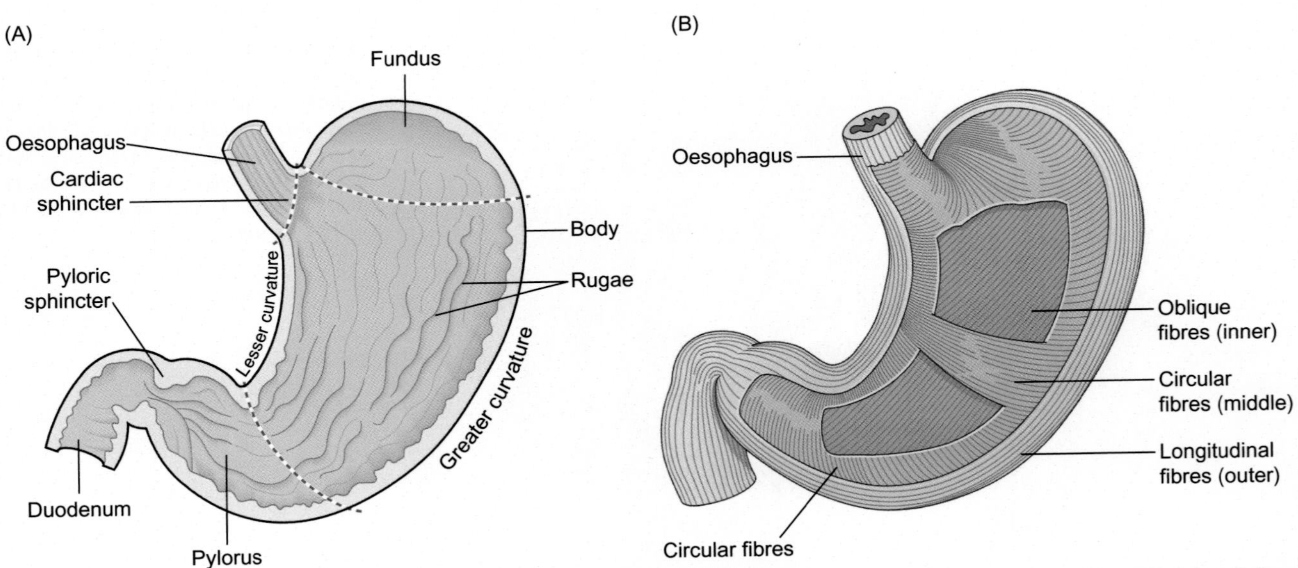

FIGURE 12.19 Structure of the stomach. (A) General regions and internal lining. B. Muscular layers of stomach wall. *From Waugh and Grant (2014), with permission.*

begin breaking down the large carbohydrate molecules in the food before the amylase is inactivated by the low pH of the gastric juice. In addition, the presence of food in the stomach is detected by sensory nerve endings, setting up a reflex action that, via efferent fibers of the vagus nerve, stimulates both the muscular movements and secretions of the stomach. The rhythmic contractions of the stomach musculature (about three per minute) both mix and mechanically break down the food particles. They also move the food in the general direction of the pyloric end of the stomach. During this movement through the body of the stomach, major digestive processes are also taking place. As the peristaltic contractions approach the pylorus, they increase in speed and intensity. Some of the stomach contents pass through the pylorus into the intestine. Other contents are bypassed by the contractions and are pushed backward toward the fundic part of the stomach by a process called **retropulsion**.

Digestion occurs mainly through secretions from cells of the gastric mucosa, which can be divided into oxyntic and pyloric areas. The **oxyntic mucosa**, comprising 80% of the stomach lining, is studded with several million **gastric pits**. These are openings draining tiny **gastric (oxyntic) glands** (Fig. 12.20). Each pit/gland combination represents in a microcosm the functional workings of the stomach. Most of the gastric mucosa, including the gastric pits, is lined by mucus-producing **surface epithelial cells**. A dominant cell type within the glands is the **parietal cell**, which produces hydrochloric acid (HCl). The other major cell type is the **chief cell**, the source of the peptide-degrading enzyme pepsinogen. At the base of a gastric gland is a scattering of **G cells** (a subset of a group of cells called **enteroendocrine cells**) that produce local gastric hormones. The junction between the gastric pit and gland is occupied by stem cells, called **mucous neck cells**. Progeny of these cells migrate into the glands and differentiate into parietal, chief or G cells. Other progeny of these stem cells migrate toward the surface of the stomach lining and become **surface epithelial cells**.

Parietal cells are remarkably adapted to form and secrete **hydrochloric acid**. Scattered among the walls of the gastric glands, these cells are characterized by two dominant morphological features (Fig. 12.21). The first is a large number of mitochondria, which are necessary to produce the ATP required to move H^+ against a millionfold concentration gradient. The second is a large meshwork of cytoplasmic tubulovesicles which, when active, empty into an intracellular canaliculus that leads to the outside of the cell. The membranes of the tubulovesicles contain the enzymes (carbonic anhydrase and H^+, K^+, and ATPase) that are critical for the production of HCl. Within 10 minutes of some form of stimulation, the parietal cells begin to produce HCl, largely through the passage of H^+ from inside the cell to the exterior in exchange for the movement of K^+ from the gastric surface through the parietal cell and into the blood. At the

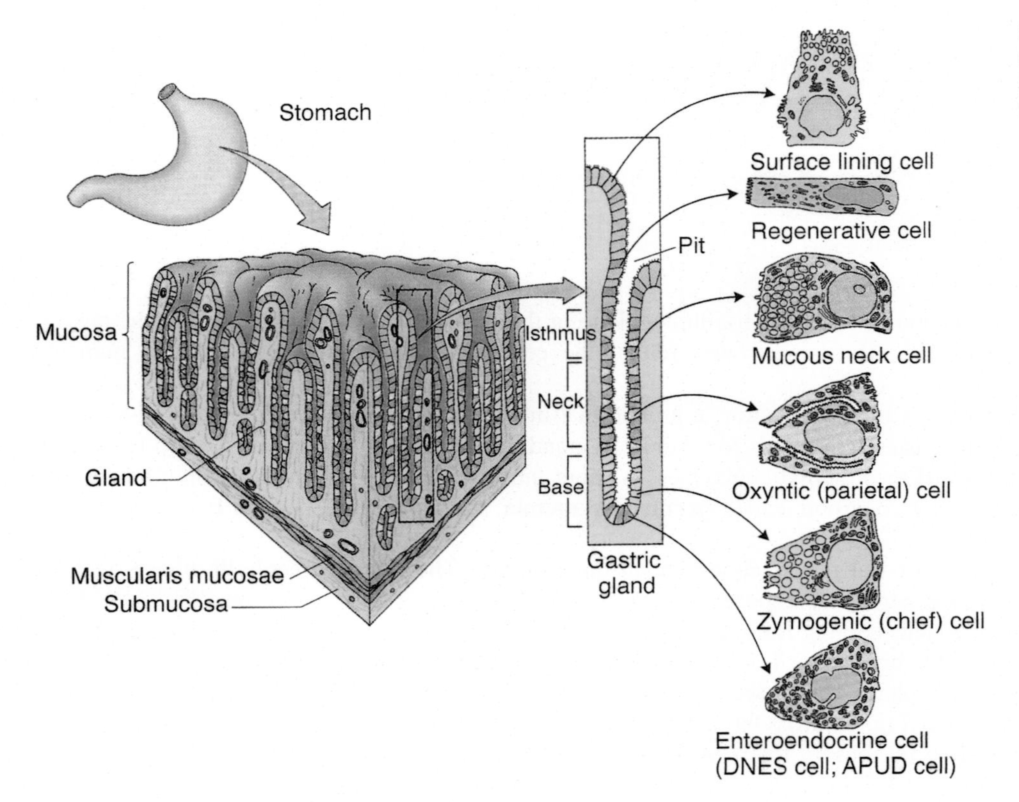

FIGURE 12.20 Histological structure of the stomach lining. *From Gartner and Hiatt (2010), with permission.*

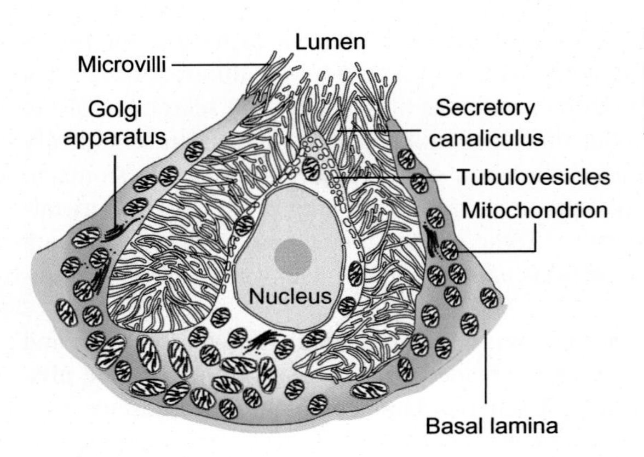

FIGURE 12.21 Drawing of a parietal cell. Note the abundant mitochondria, which supply energy for HCl formation and the numerous microvilli, which increase the surface area for secretion of HCl. *From Kierszenbaum and Tres (2012), with permission.*

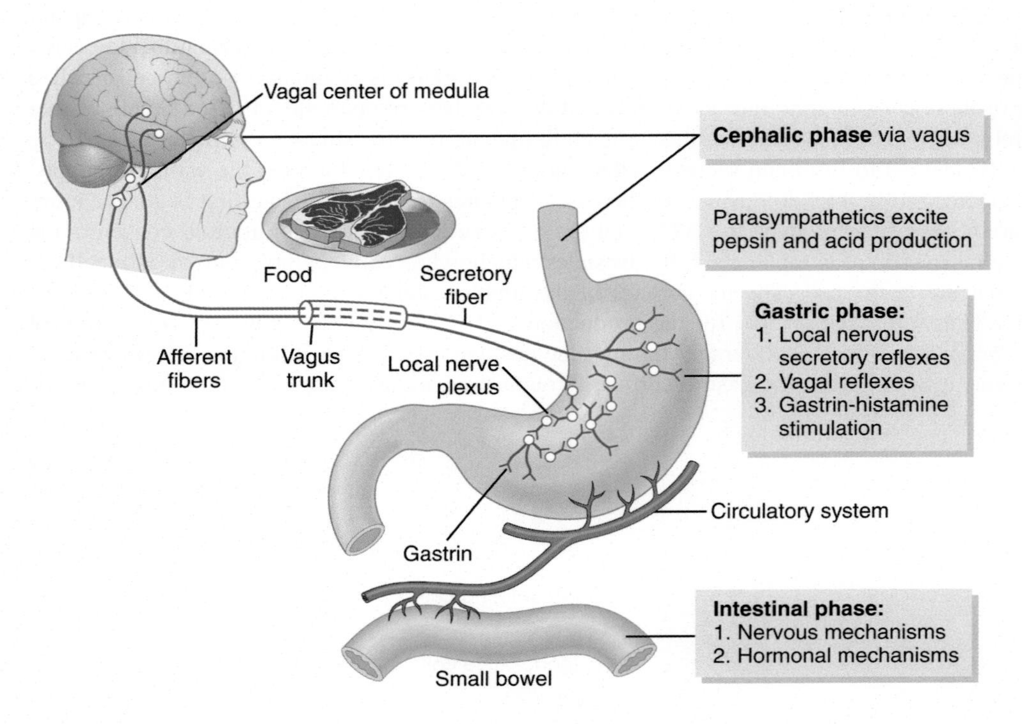

FIGURE 12.22 Phases in gastric secretion (cephalic, gastric, and intestinal) and their controls. *From Guyton and Hall (2006), with permission.*

same time, Cl^- ions follow the H^+ ions into the gastric lumen. During this time, the tubulovesicles link to the intracellular canaliculus, thus increasing the effective surface area of the cell, which is already studded with numerous microvilli.

Another product of parietal cells is **intrinsic factor**, a protein that in humans is required for the absorption of vitamin B_{12} by the small intestine. Intrinsic factor binds to vitamin B_{12}, and this combination survives the trip through the stomach unscathed by the gastric pH or proteolytic enzymes. Without intrinsic factor, vitamin B_{12} cannot be absorbed by the intestine and a life-threatening condition, called **pernicious anemia**, results because vitamin B_{12} is required for normal red blood cell formation.

Secretion of gastric acid above a basal rate is controlled in three phases. The first is a **cephalic phase**, which is mediated by signals from the vagus nerve, a component of the parasympathetic nervous system (see Fig. 6.4). Either psychological factors or afferent signals coming from areas of the mouth or pharynx stimulated by chewing or swallowing stimulate the vagus nucleus in the brainstem. Terminal branches of stimulated vagal neurons directly stimulate parietal cells to secrete HCl. The second phase of acid secretion is called the **gastric phase**. Two principal factors stimulate the release of HCl during this phase (Fig. 12.22). One is mechanical distension of the stomach from the ingested food. The sense of distension is picked up by sensory fibers of the vagus nerve, and their afferent signals set up a vagovagal reflex that acts on both the parietal cells and cells of the pyloric area and duodenum that produce the local hormone

gastrin. Gastrin, the second stimulant, is carried to the body of the stomach and stimulates the secretion of HCl by the parietal cells. Gastrin secretion is inhibited when the pH of the stomach falls below 3.0, but when food first enters the stomach, it neutralizes the gastric pH, which can rise as high as 6.0. This provides an appropriate condition for gastrin release, which is controlled through local positive and negative feedback loops. A minor third phase of gastric secretion (**intestinal phase**) is influenced by the presence of protein digestion products in the duodenum. This stimulates the release of gastrin from endocrine-producing cells in the upper duodenum. In addition, amino acids entering the blood may have a stimulatory effect on the parietal cells.

Another important modulator of HCl secretion is **histamine**, which is produced by specialized cells just beneath the gastric lining. When stimulated by gastrin, these cells produce histamine, which then acts directly on the parietal cells. This secondary action potentiates the direct action of gastrin on the parietal cells and increases their secretion of HCl.

The secretion of HCl is inhibited in several ways that provide partially redundant backup systems. With a substance as potentially harmful as HCl, it is important not to rely upon only one means of control. When the pH of the gastric fluid falls below 3.0, the hormone **somatostatin** in addition to being a pituitary hormone (see p. 249), which is released from cells in the pyloric region and acts in a paracrine manner, inhibits the production of HCl by the parietal cells. The presence of fatty acids in stomach contents releases **gastric inhibitory peptide**, another locally produced hormone, which depresses the secretion of HCl by parietal cells.

Concomitant with the secretion of acid, other cells of the stomach, the chief cells, secrete **pepsinogen**, the inactive form of the protein-digesting enzyme **pepsin**. The stimulus for the conversion of pepsinogen to pepsin is a low pH environment, and at a pH of 2.0 or less, the conversion is extremely rapid. Secretion of pepsinogen is closely tied to the formation of gastric acid, and the stimulatory influences of the vagus nerve act on the chief cells, as well as on parietal cells. In addition, the increasingly acid gastric environment after the intake of food also results in greater levels of secretion of pepsinogen. Other local hormonal influences also play a minor role in pepsinogen secretion. The closely coordinated timing of acid and pepsinogen secretion provide the conditions for the initial breakdown of dietary proteins into smaller peptide fragments while the food particles are being jostled about within the stomach.

Another product of the chief cells is **gastric lipase**, which is involved in the breakdown of fatty substances. Unlike salivary lipase, gastric lipase functions best at an acid pH (3.0–6.0). Most fat digestion, however, occurs in the small intestine, and neither salivary nor gastric lipase is critical to the digestive process.

To compensate for its ability to produce extremely strong acid secretions and proteolytic enzymes, the stomach must also have the means to protect itself from self-digestion. In the absence of such protection, **ulcers** can form (Box 12.2). Protection of the gastric mucosal lining is accomplished by the production of a layer of insoluble

BOX 12.2 Gastric Ulcers

When I was in medical school years ago, one of the most common surgical procedures was designed to ameliorate the symptoms of gastric ulcers. Called vagotomy and pyloroplasty, the procedure was based upon the best understanding of gastric physiology at the time. As the name implies, a gastric ulcer is an erosion of the stomach lining. As soon as the erosion begins, patches of the stomach lining are left without the protective layer of mucus and the epithelial cells that produce it. Over time, the acidic and proteolytic environment of the gastric juices further erodes the wall of the stomach. If the erosive processes eat through the wall of a blood vessel or through the wall of the stomach itself, the resulting gastric bleed can be a medical emergency.

During the era of surgical treatment of ulcers, it was commonly believed that ulcers were caused by stress. **Vagotomy** (severing the vagus nerve) was performed to eliminate the connection between the source of stress (the brain) and the source of gastric acid (the parietal cells). We now know that this treatment was directed at the cephalic phase of gastric secretion. The **pyloroplasty** component of the operation was designed to counter a serious side effect of vagotomy, namely paralysis of the pyloric sphincter, which prevented the release of stomach contents into the small intestine. (The pyloric sphincter is the last outpost of central control of gastrointestinal movements until such control is restarted at the anus.) Pyloroplasty disrupts the integrity of the smooth muscle layers in the pylorus and allows the passage of food from the stomach. Eventually, control of gastric emptying is taken over by the enteric nervous system and reasonable pyloric sphincter function is restored.

The whole paradigm of gastric ulcer treatment changed in the early 1980s, when Dr. Barry Marshall in Australia realized that an acid-loving bacterium, *Helicobacter pylori*, was almost uniformly associated with gastric ulcers. Initially, his hypothesis that ulcers were essentially consequences of a bacterial disease was met with great skepticism, but he convinced the doubters by drinking a solution containing *H. pylori* and coming down with a gastric disturbance himself. For this groundbreaking discovery that improved the lives of millions, Dr. Marshall was awarded the Nobel Prize in Medicine in 2005. Now the standard treatment for gastric ulcers is antibiotics.

mucus by the surface epithelial cells that line the inner surface of the stomach. This mucus contains alkaline products also secreted by the surface epithelial cells and is able to neutralize the HCl in the immediate vicinity of the gastric mucosa. In addition, it prevents pepsin from reaching the mucosal cells. Mucous neck cells secrete a different, soluble form of mucus that helps to lubricate the food particles that are being digested and moved around in the stomach.

After the food in the stomach has been mechanically and chemically broken down by the actions of the stomach muscles and gastric juices, peristaltic-type contractions of the stomach muscles move the food toward the pyloric end of the stomach. Only particles less than 1 mm^3 are normally allowed to pass through the pyloric sphincter.

Proper operation of the **pyloric sphincter** depends upon coordination between the central and enteric nervous systems. Sensors in the stomach lining stimulate afferent nerve fibers in the vagus nerve, and after central processing, signals traveling down efferent nerve fibers in the vagus nerve stimulate the brief opening of the pyloric sphincter. This allows a small amount of the highly acidic gastric contents to pass into the duodenum. Once the pyloric sphincter has opened, peristaltic contractions controlled by the enteric nervous system move the gastric contents along. Duodenal receptors feed back on the stomach musculature to inhibit excessive gastric emptying.

With gastric emptying, the immediate problem for the duodenum is to neutralize the gastric acid in order to prevent the formation of **duodenal ulcers**. As is the case with the stomach, the body has evolved redundant mechanisms to ensure that the duodenal lining is protected from gastric acidity. The first line of defense is a dense aggregation of **Brunner's glands** that occupy much of the duodenal submucosa for about 5–10 cm below the pyloric opening. Brunner's glands produce an alkaline secretion (pH 8.5–9.0) that helps to neutralize stomach acids. The other main player in neutralizing gastric acids entering the duodenum is a watery alkaline fluid secreted by epithelial cells lining the ducts leading out of the pancreas. Together these two glandular fluids effectively neutralize the pH of the gastric fluids that enter the duodenum. Yet by feedback from poorly understood local sensors, they do not oversecrete. Too alkaline an environment would also be highly damaging to the wall of the duodenum. Once the buffering of the gastric fluids is completed, the duodenal contents find themselves in a neutral or slightly alkaline environment that is ideal for the next set of digestive enzymes that the food will encounter.

SMALL INTESTINE

Structure

By the time that fragments of the partially digested hamburger reach the **small intestine**, they enter an environment that is specialized for digestion, absorption, and propulsion. Key to absorption is a very large surface area. This is accomplished by both the overall length of the small intestine (somewhat over 5 m long) and structural adaptations that increase its cross-sectional area. In addition to the simple diameter of the small intestine (~2.5 cm), modifications at several levels greatly increase its effective absorptive area. At a grossly visible level are mucosal **folds (plicae)**. Protruding from the plicae are densely packed almost microscopic **intestinal villi** from 0.5 to 1.0 mm long (Fig. 12.23A) that have a huge (30 times) multiplier effect on the effective surface area of the small intestinal lining. This is even further increased (600-fold) by the presence of tiny projections (**microvilli**) from the surfaces of the intestinal epithelial lining cells (Fig. 12.23B). The apical surface of each epithelial cell contains 2000–3000 microvilli, which are barely distinguishable at the light microscopic level.

Like almost all regions of the digestive tract, the wall of the small intestine contains many layers, each designed to fulfill one or more specific functions (Fig. 12.24). The innermost region of the intestinal wall is called the **mucosa**, and it consists of three layers. Adjacent to the lumen is the **intestinal epithelium**, a simple columnar epithelium. The vast majority of these epithelial cells are designed for absorption, but scattered among them are mucus-producing goblet cells. Immediately beneath the epithelium is the **lamina propria**, a thin layer of loose connective tissue that serves as the campground for cells of the peripheral immune system that have been deployed to guard against any ingested pathogens that might have survived the hostile environment of the stomach. These cells constitute the gut-associated lymphatic tissue (GALT, see p. 228). The lamina propria also contains dense networks of small blood vessels and lymphatics, whose job it is to transport absorbed nutrients and any hormones produced by deep epithelial endocrine-producing cells (Fig. 12.25). In addition, branches from the autonomic nerve plexuses in the intestinal wall traverse the lamina propria to reach target blood vessels or epithelial cells. Beneath the lamina propria is a very thin layer of smooth muscle, the **muscularis mucosae**. Nearly continuous contractions of the intertwined muscle cells of the muscularis

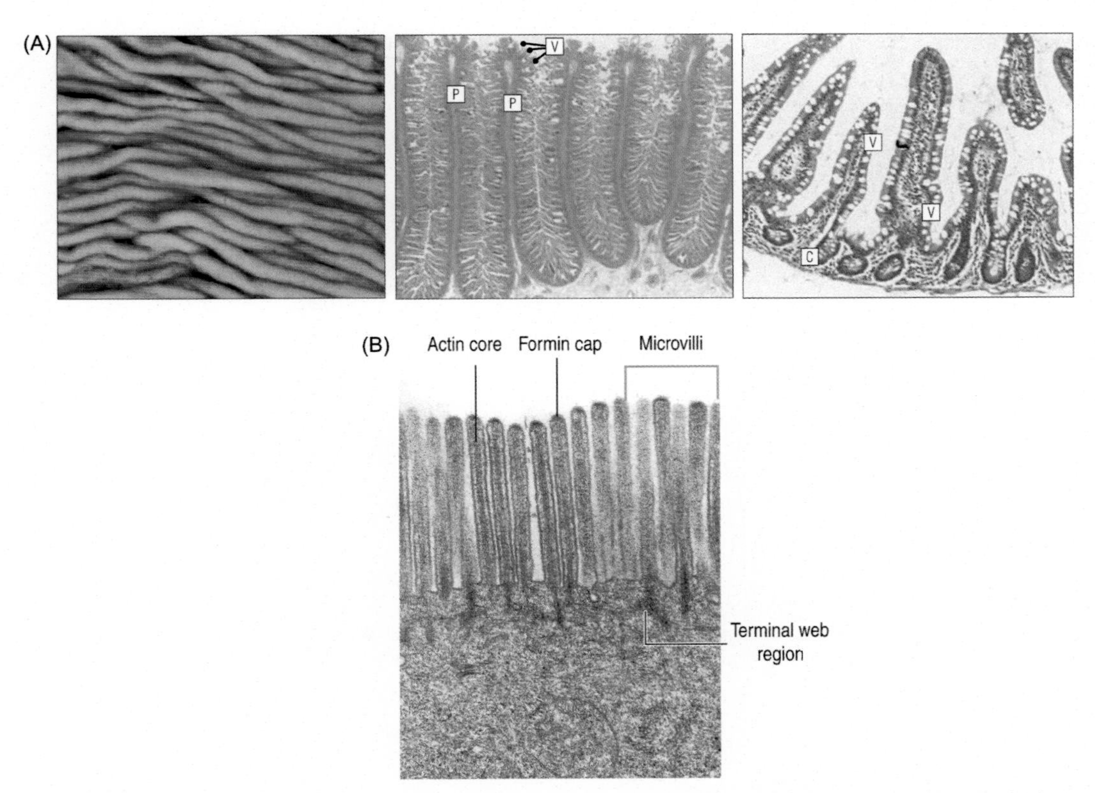

FIGURE 12.23 Structural modifications for increasing the absorptive area of the small intestine. (A) Light micrograph of plicae and villi. (B) Electron micrograph of microvilli on the surfaces of the epithelial cells. *From (A) Kerr (2010), with permission. (B) Kierszenbaum and Tres (2012), with permission.*

mucosae help to expel the contents of mucosal glands and are assumed to enhance contact between the mucosa and intestinal contents.

Surrounding the three layers of the mucosa is a layer of tough connective tissue called the **submucosa**. This is the material (catgut) that was formerly used for surgical sutures and tennis racquet strings before the advent of more uniform artificial materials. Of note, the submucosa contains a set of autonomic nerve plexuses (**Meissner's** or **submucosal plexuses**), which are important components of the enteric nervous system (see below).

The submucosa is surrounded by a robust muscular layer consisting of two layers of smooth muscle. The inner layer contains circularly oriented smooth muscle fibers, whereas those of the outer layer run longitudinally. This arrangement provides the mechanical basis for the peristaltic movements that propel the contents of the small intestine toward the large intestine. Interspersed among the individual smooth muscle cells is an interconnected network of **interstitial cells of Cajal**. These cells communicate with smooth muscle cells by gap junctions and coordinate the production and propagation of slow waves (3–12 cycles/min) of smooth muscle contraction. These slow waves themselves do not account for peristaltic movements, but they provide the physiological substrate for the more powerful phasic contractions that actually move the intestinal contents along. Between the two layers of smooth muscle is another set of prominent nerve plexuses called **Auerbach's plexuses**.

Meissner's and Auerbach's plexuses are the most visible components of the enteric nervous system of the gut. The enteric nervous system is so extensive that it is often called the second brain. The number of nerve cells in the small intestine is estimated to be over 100 million, and the number of neurons in the gut exceeds the number of neurons in the spinal cord. Although the enteric nervous system receives input from parasympathetic fibers running down the vagus nerve and nerves from the sacrum, many of the neurons are not directly connected to either the brain or spinal cord. As a result of this and their numerous interconnected neuronal processes, the enteric nervous system operates with a high degree of autonomy and can direct peristaltic function in the absence of any central connections. Independent of any connection to the central nervous system, the enteric nervous system can receive signals from sensor cells in the gut, process them, and then direct an appropriate response by effector cells.

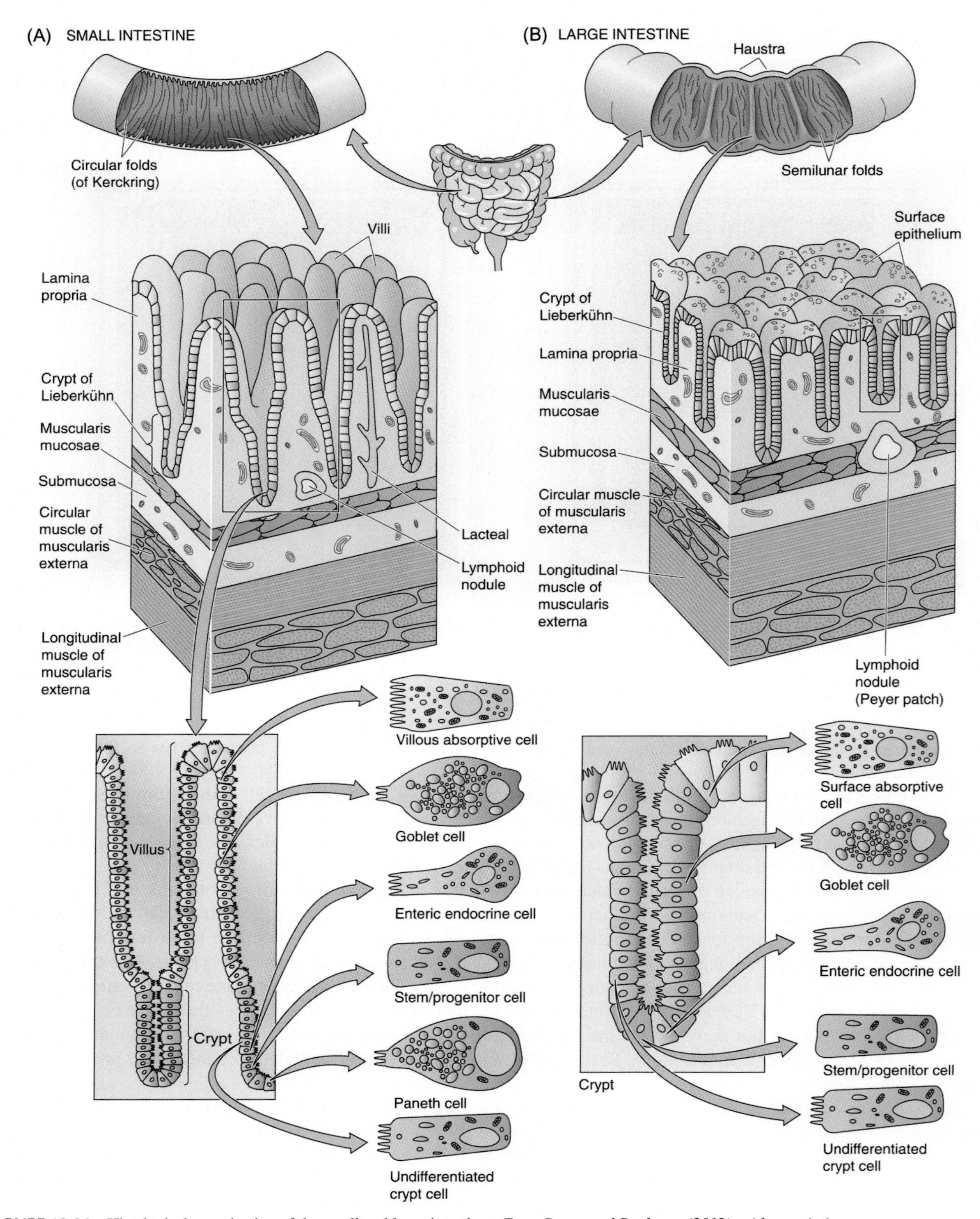

FIGURE 12.24 Histological organization of the small and large intestines. *From Boron and Boulpaep (2012), with permission.*

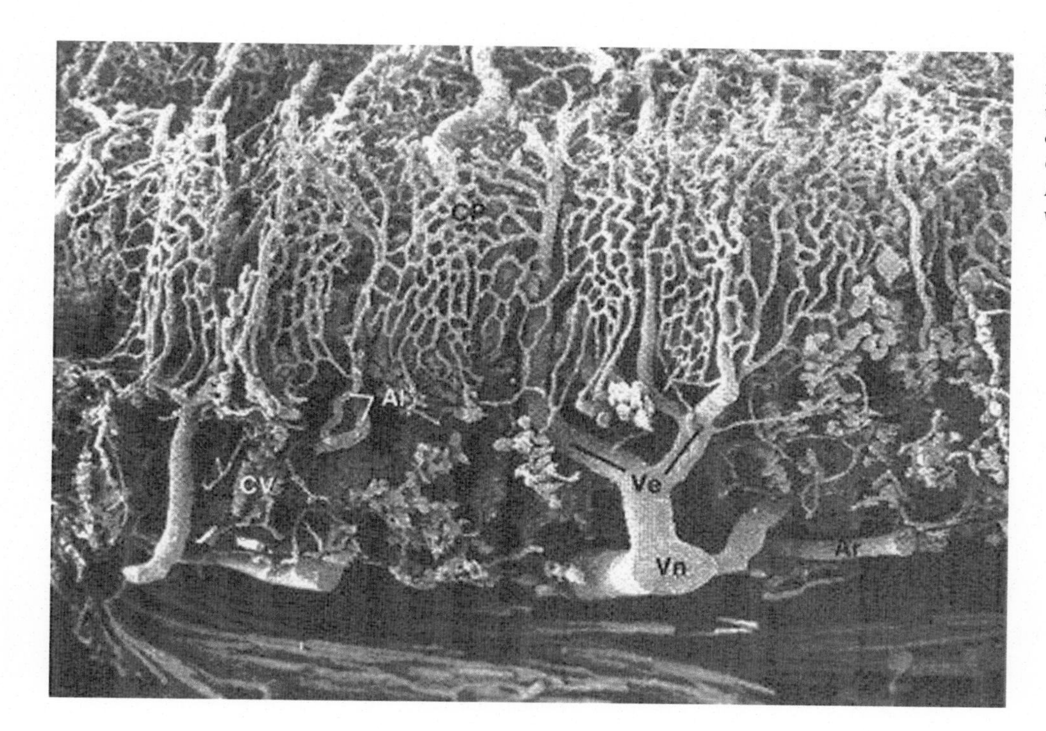

FIGURE 12.25 Scanning electron micrograph of the vasculature of the wall of the jejunum. *Al*, arteriole; *Ar* artery; *CP*, capillary plexus; *CV*, capillary; *Ve*, venule; *Vn*, vein. *From Kessel and Kardon (1979), with permission.*

Surrounding the muscularis layer of the intestine is a thin layer of connective tissue, called the **adventitia**. The connective tissue is covered on the outside by a single-celled epithelium (technically called a **mesothelium**) that is continuous with the peritoneal lining of the abdominal cavity. The intestine, as a whole, is suspended from the body wall by the mesentery—a thin sheet of connective tissue covered by the peritoneal lining. The blood vessels and nerves that supply the intestine travel through the mesentery.

On the inner side of the small intestine, the projecting villi alternate with crypts that penetrate into the lamina propria layer (see Fig. 12.24). The epithelial surfaces of the villi and crypts contain a number of cell types, each with its unique function. Covering the villi are tall columnar epithelial cells, called **enterocytes**. With their apical surfaces studded with microvilli, these cells are specialized for absorption. Scattered among the enterocytes on the villi is a scattering of **goblet cells** that produce mucus to facilitate the movement of intestinal contents as they make their way down the digestive tract. Goblet cells steadily increase in number as the small intestine approaches the colon.

Two types of cells are dominant in the crypts. One is **Paneth cells**, which produce large amounts of various proteins involved in the defense from and coexistence with intestinal bacteria. Their functions remain poorly understood. Scattered enteroendocrine cells at the base of the crypts produce hormones, such as secretin and cholecystokinin, which exert paracrine effects on various parts of the digestive system.

Also located within the crypts are **stem cells** that give rise to replacements for all of the other types of cells within the intestinal epithelium. A typical enterocyte has a lifespan of only 4 days (Fig. 12.26). A villus loses about 400 enterocytes per day. These are replaced by other cells derived from the stem cells. The replacement cells move up the villi according to a regular schedule. Less is known about the loss and replacement cycles of the other intestinal epithelial cells.

Both the overall structure and function of the small intestine change along its length, and it has been subdivided into three named segments—the **duodenum**, the **jejunum**, and the **ileum**. The short duodenal segment[2] is the region where stomach acids are neutralized and most digestive processes are begun. This is the region where products of the pancreas and gallbladder enter the intestine. The bulk of digestion and absorption of food occurs in the jejunum, which occupies about 40% of the length of the small intestine. Correspondingly, the jejunum has a greater diameter, more prominent folds and longer villi than other parts of the small intestine. The terminal segment of the small intestine, the ileum, accounts for about 60% of its total length. The ileum plays a mopping-up role by completing any digestive and

2. The word duodenum is derived from the Latin word duodactylum, meaning 12 finger breadths, which approximates its length.

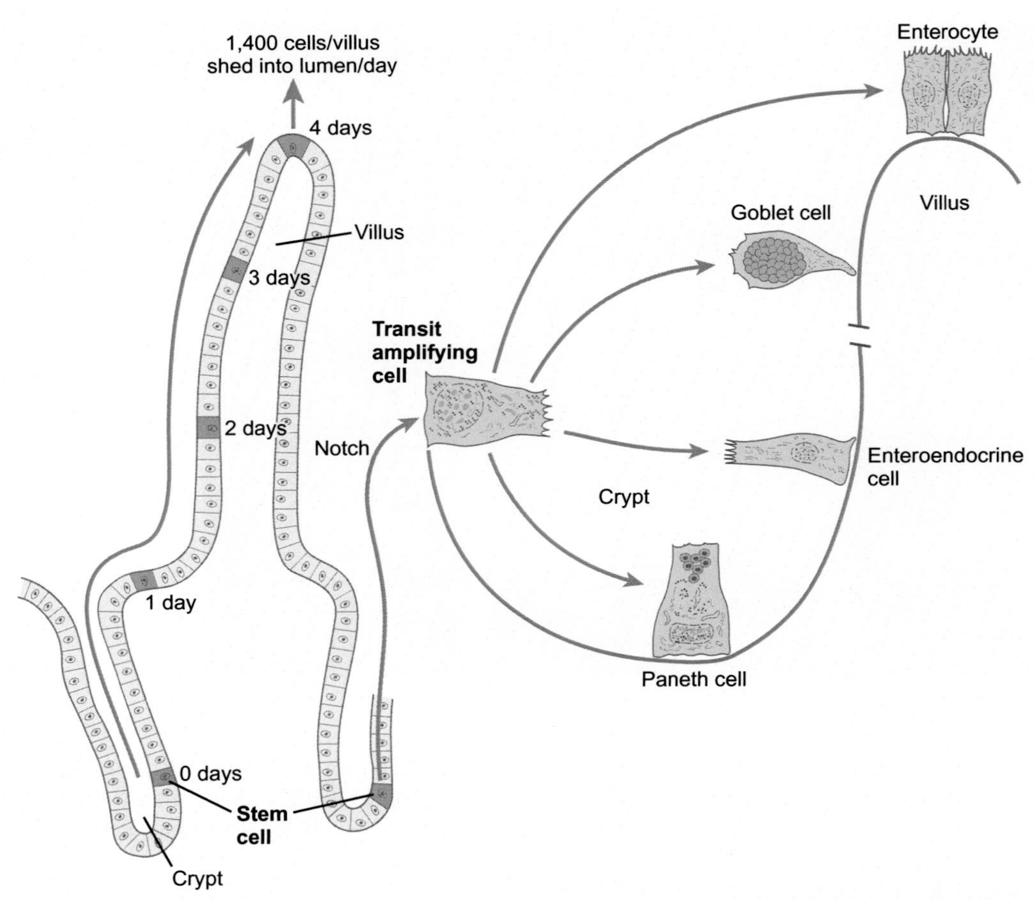

FIGURE 12.26 Differentiation of intestinal epithelial cells from stem cells located in the crypts. The time scale shows the typical courses of migration of daughter cells starting with their generation from the stem cell population to their being shed form the villus into the intestinal lumen. *From Carlson (2014), with permission.*

absorptive process that did not take place in the jejunum. A major unique function of the ileum is absorption of vitamin B_{12} and bile salts. The degree of folding of the mucosa and size of villi decrease, and the number of mucus-secreting goblet cells increases along the length of the ileum. The ileum is an area where prominent lymphatic nodules, Peyer's patches, predominate. These represent an important component of the immune defense of bodily surfaces. The remains of the ingested food become more concentrated and, on the substrate of the secreted mucus, are moved along toward the colon. An **ileocecal valve** at the terminal part of the ileum plays an important role in the movement of food along the intestines.

Digestion and Absorption

The bulk of digestion and absorption of food takes place in the small intestine. Critical to the overall digestive process are secretions entering the upper small intestine through the pancreatic and bile ducts (Box 12.3). Digestion, which was begun in the stomach, is largely completed in the small intestine with the help of a large number of enzymes released from both the pancreas and the intestinal epithelium (Fig. 12.28). In addition, bile salts entering the intestine from the gallbladder are important in the breakdown of complex lipid molecules.

Although most small molecules are readily absorbed, a gut-vascular barrier, composed of pericytes and glial cells associated with the enteric nervous system, prevents the passage of very large molecules and microorganisms from the lumen of the gut into the blood. Certain bacteria, such as *Salmonella typhimurium*, which cause intestinal disease, are

BOX 12.3 The Pancreas and Digestive Enzymes

The pancreas is a mixed gland with both endocrine and exocrine functions. In Chapter 9, the endocrine portion (islets of Langerhans) and its production of insulin are discussed. The bulk of the pancreas, however, is an exocrine gland, producing a wide array of digestive enzymes and fluids that drain from the gland into the intestine via a duct system, rather than into the blood. Phylogenetically, the pancreas has not always been so organized. In many fishes, the islets are separated from the exocrine pancreas. Early diabetes researchers often experimented on an ugly marine fish, the toadfish, because its islets were located in a discrete lump of tissue that was not connected to the cells producing pancreatic digestive enzymes.

Overall, the exocrine pancreas is organized in a manner very similar to that of the salivary glands. In fact, in the embryo, the same epithelial cells can form either pancreatic or salivary tissue depending upon the signals that they are given. The pancreatic acini are lined by protein-producing cells (Fig. 12.27). In fact, their rate of protein synthesis is the highest of any cells in the body. At their base is a dense accumulation of rough endoplasmic reticulum, where the peptide backbones of the enzymes are formed. These peptides then move into an extensive Golgi apparatus, where they are further modified by the addition of carbohydrate side groups. The dormant enzymes (**proenzymes**) are then stored as prominent granules in the apical cytoplasm while awaiting release.

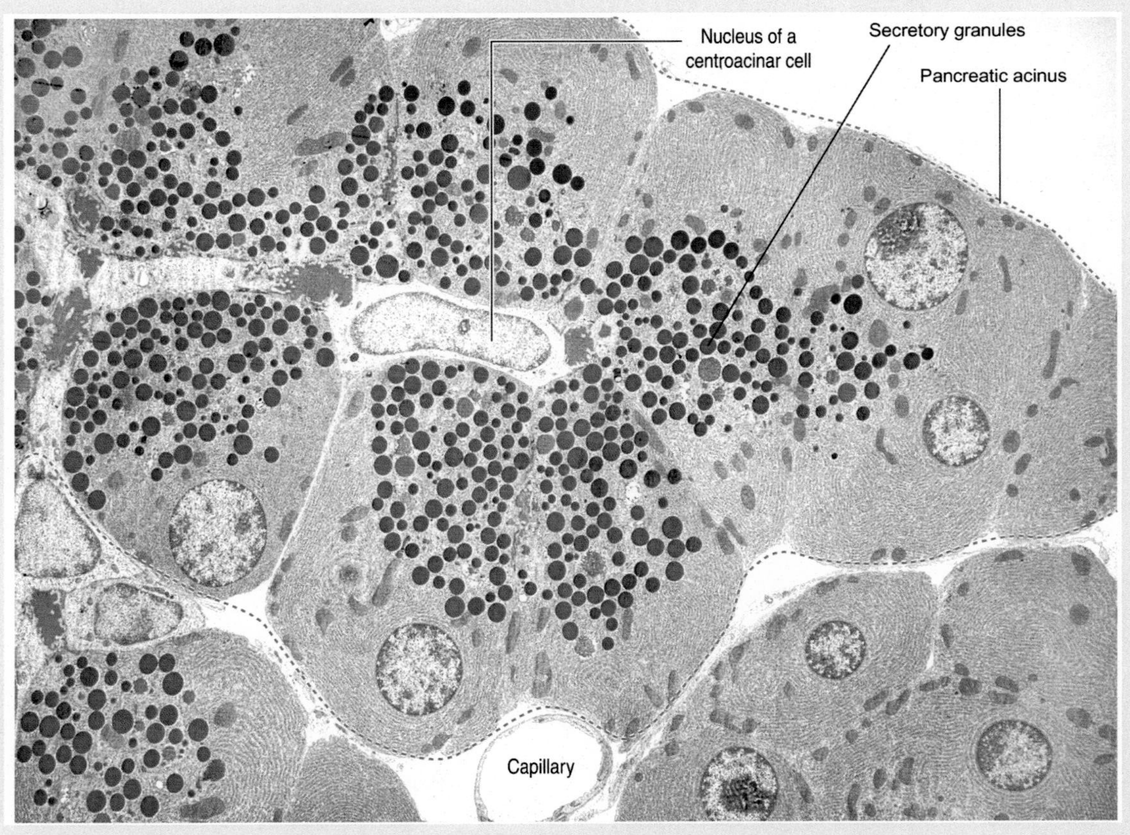

FIGURE 12.27 Low power electron micrograph of a pancreatic acinus (outlined by dashed line). *From Kierszenbaum and Tres (2012), with permission.*

Like salivary glands, the pancreas is well supplied by both sympathetic and parasympathetic nerve fibers. In contrast to the salivary glands, however, neural stimulation is not the only mechanism that regulates pancreatic secretion (see below). Overall, parasympathetic nerve firing stimulates and sympathetic inhibits the secretion of pancreatic enzymes.

Protein-containing secretions from the pancreatic acini flow into a duct system, which finally empties into the upper duodenum. The upper levels of the duct system contribute large amounts of water (up to 1 L/day) and HCO_3^- to the mix. Although pancreatic ducts do not have the morphological equivalent of the striated ducts of salivary glands (see p. 333), the cuboidal epithelial cells lining the ducts are equipped with the appropriate enzymes and membrane channels for exchanging Na^+, H^+,

(Continued)

BOX 12.3 (Continued)

and Cl^-. Toward the end of the pancreatic duct system, the epithelial cells lining the ducts become columnar and consist of increasing numbers of mucus-secreting goblet cells, probably to ensure that the lining of the duct is protected from the acidity of the newly arrived gastric contents.

Like the stomach, secretion in the pancreas is responsive to three phases of stimulation. The first is the **cephalic phase**, which depends almost entirely upon stimulation by parasympathetic nerve endings on the acinar cells. Stimuli for this phase are smell, taste, chewing, and swallowing. For all of these, the initial stimuli are processed in the brain, with the resulting efferent signal traveling down the vagus nerve (cranial nerve X) to the pancreas. Once food has entered the stomach, distension of the stomach initiates a vagovagal reflex (**gastric phase**) that stimulates further secretion of pancreatic exocrine products. The **intestinal phase** is most important and accounts for about 75% of all pancreatic secretion. Secretion in this phase is set off by the presence of food and gastric acid in the upper duodenum.

Gastric acid and fatty acids emptying into the duodenum from the stomach stimulate the secretion of the local hormone **secretin** from S cells located in the base of the intestinal glands. The lower the duodenal pH, the greater amount of secretin is released. As a protective agent, secretin fulfills two important functions. First, it is carried to the stomach by the blood, where it inhibits the formation of HCl by the parietal cells. Second, it is also carried to the acini of the pancreas and stimulates the secretion of digestive enzymes by the acinar cells and HCO_3^- by the ductal cells.

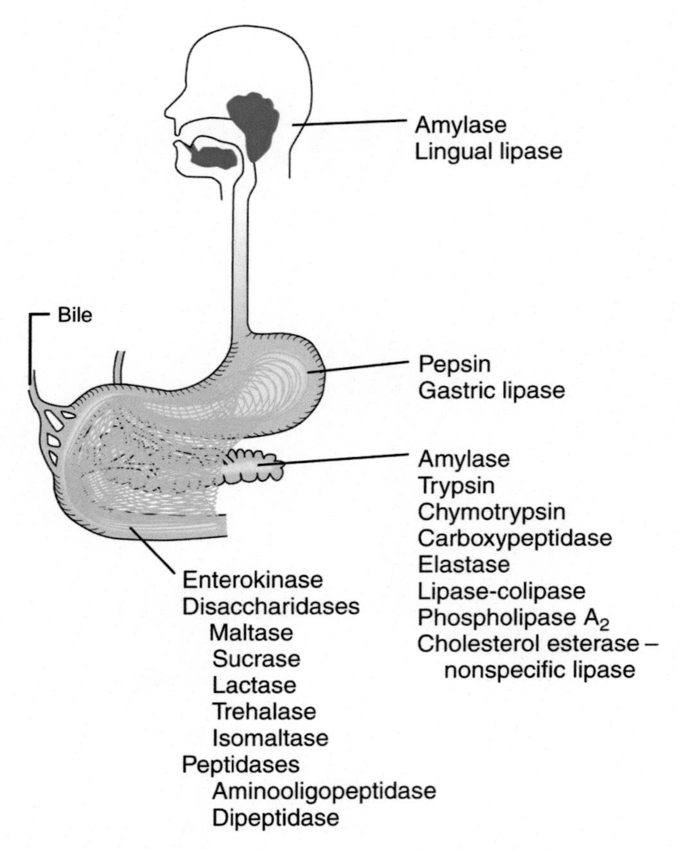

FIGURE 12.28 Sources of major digestive enzymes and bile. *From Johnson (2014), with permission.*

able to break down the gut-vascular barrier. Another bacterial source of food-borne illness (*Clostridium*) causes problems by disrupting the tight junctions between intestinal epithelial cells. The pharmaceutical industry is currently looking at using the disruptive agent (**claudins**) as a means of improving the delivery of drugs designed to be absorbed in the small intestine.

Two general types of digestion are recognized. One is **luminal digestion**, in which food material freely distributed within the lumen of the intestine is further broken down by digestive enzymes or bile. The other is **membrane digestion**. Here enzymes that are an integral part of the plasma membrane of the enterocytes break down disaccharides or dipeptides into simple sugars or amino acids that are then directly absorbed into the enterocytes.

Carbohydrate Digestion and Absorption

Although digestion of starch begins in the mouth and the upper stomach under the influence of salivary amylase, by far the greatest amount of carbohydrate digestion occurs in the small intestine. Here the breakdown of starch into smaller di- and trisaccharides continues in the lumen of the duodenum and jejunum through the action of **pancreatic amylase**. These smaller sugars are then further broken down into simple sugars (glucose, galactose, and fructose) through brush border enzymes on the microvilli of the enterocytes. The small intestine has an enormous capacity to both break down and absorb sugars, so its full capacity is almost never used. Most simple sugars are absorbed in the proximal parts of the jejunum. **Cellulose** cannot be digested by the human digestive tract, so it passes through as undigested fiber. Ruminant mammals, such as cattle, are able to utilize cellulose because of symbiotic bacteria that live within their complex stomachs. These bacteria produce enzymes that break down the cellulose walls of grasses and provide nutrition both for the cow and for themselves. Like humans, cattle do not have their own enzymes that have the capacity to digest cellulose.

Protein Digestion and Absorption

Proteins enter the small intestine partially broken down by stomach acid and pepsin. These protein products are then bombarded by a formidable array of proteolytic enzymes released from the pancreatic duct. At least five major enzymes attack proteins from both well within the molecule and at the ends. These enzymes enter the duodenum as inactive precursor molecules. Were this not the case, they would begin digesting the pancreas as they descend down the pancreatic duct system. In fact, this is the basis of **pancreatitis**, a chronic inflammation of the pancreas. The first pancreatic enzyme to be activated is **trypsinogen**, which is acted upon by enterokinase, an enzyme released from the microvilli of the intestinal wall through the action of bile salts. **Trypsin**, the activated form of trypsinogen, then activates the precursors of the other proteolytic enzymes. The enzymes are rapidly inactivated, but before this occurs, they have broken down proteins into simple peptides and their constituent amino acids. These molecules are absorbed by enterocytes in much the same manner as simple sugars. The proximal small intestine has a greater capacity to absorb di- and tripeptides, whereas farther distally amino acids are preferentially absorbed.

Lipid Digestion and Absorption

The processing of lipids in the digestive tract is complex. First of all, lipids are not chemically as homogeneous as are proteins or carbohydrates. Some lipids, such as fatty acids and triglycerides, are basically linear in construction, whereas others, such as cholesterol and the fat-soluble vitamins A, D, E, and K, have a much more complex structure. Because lipid molecules are not soluble in water, different strategies must be used to break them up and get them absorbed from the digestive tract.

Lipid digestion begins in the stomach, but it consists mostly of mechanical breaking up of lipid droplets from the churning movements of the stomach. When gastric contents are released into the duodenum, lipid droplets are emulsified through the actions of bile salts released from the gallbladder. Emulsification reduces surface tension at the lipid/water interface and allows smaller lipid droplets to form and persist. At this stage, they are then susceptible to further chemical breakdown through the action of **pancreatic lipases**, which, for example, reduce triglycerides to simpler glycerol and fatty acid molecules. Although the stomach also produces a gastric lipase, its effect in adults is insignificant in relation to that of pancreatic lipase. Other pancreatic enzymes reduce cholesterol and fat-soluble vitamins to simpler structures.

With the help of bile salts, the lipid breakdown products are aggregated into small water-soluble structures, called **micelles**. The inside of a micelle is hydrophobic and is the region that concentrates the lipids. The outer surface of a micelle is hydrophilic and keeps the micelle in solution. Micelles are taken up by enterocytes. Simpler lipid metabolites, such as glycerol and small fatty acids are water soluble and can enter the enterocytes independently. Within the cytoplasm of the enterocytes, more complex lipids, for example, triglycerides, are reconstituted. Then various reconstituted intracellular lipids become coated with lipoprotein molecules and form mixed aggregates called **chylomicrons**.

Transport of Absorbed Digestive Products

The central core of a villus contains the means by which absorbed nutrients begin their journey from the intestinal epithelium to the liver, where they are either further metabolized or distributed to other parts of the body. The core of a villus is essentially lamina propria tissue that contains a capillary network and a central blind-ending lymphatic channel—a **lacteal** (Fig. 12.29). Sugars, amino acids, and simple lipid molecules pass from the enterocytes into the capillaries within the villi. From there, the capillaries empty into a prominent system of veins—the hepatic portal system—that drains the intestines, bringing all of the absorbed nutrients into the liver. As befits a portal system, within the liver the hepatic portal vein then breaks down into a network of smaller veins and ultimately into an extensive system of venous sinusoids that bring the nutrient-rich blood into direct contact with the epithelial cells of the liver (hepatocytes). The hepatocytes are in essence biochemical factories that perform an amazing variety of biochemical functions (Box 12.4).

The lipid micelles that enter the lacteals become part of the lymphatic fluid. Instead of draining directly into the liver, the intestinal lymph vessels converge upon a central lymphatic collecting system, called the **cisterna chyli**, which sends lymph from the lower part of the body through the thoracic lymphatic duct. The thoracic duct empties directly into the blood vascular system through the left subclavian vein, the main vein draining the left arm.

Transport of Water and Electrolytes

One of the most important functions of the small intestine is the exchange of water and electrolytes. Throughout the digestive tract, the water that passes into the lumen and is then removed is immense (Fig. 12.32). Each day from 7 to 10 L of water enters the small intestine via the stomach, bile, and pancreatic juice and through the wall of the intestine itself. Of that amount, all but about 1.0–1.5 L is absorbed into the small intestinal wall. Most of the remainder is absorbed by the colon. The movement of water in either direction across the intestinal epithelium is closely tied to the transport of ions. Proceeding from the duodenum to the end of the ileum, there is a net loss of Na^+ and Cl^-, and

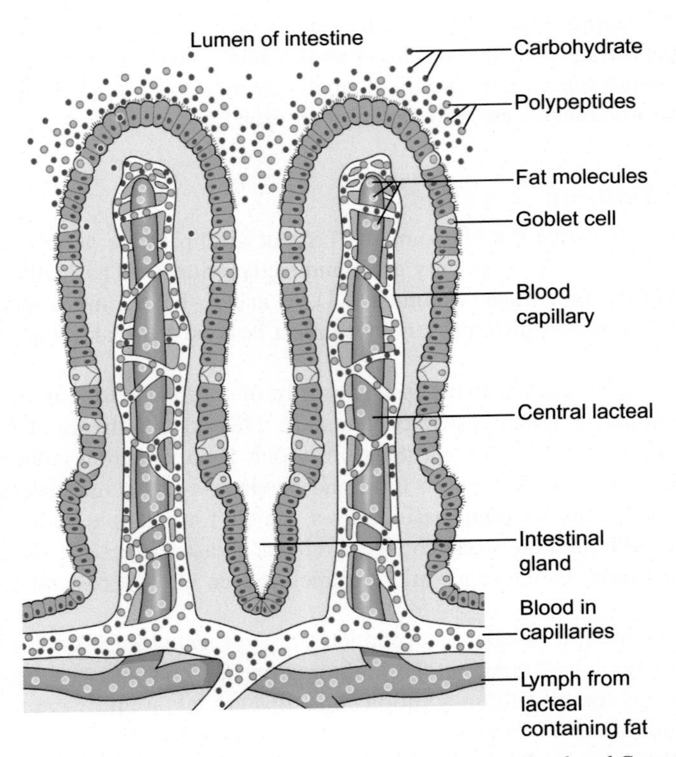

FIGURE 12.29 Internal structure of an intestinal villus and the absorption of nutrients. *From Waugh and Grant (2014), with permission.*

BOX 12.4 The Liver

The liver, weighing from 1–2 kg, is the largest glandular structure in the body. It is a multifunctional organ, basically a biochemical factory, and is absolutely essential to life. Located in the right upper quadrant of the abdomen, it consists of two major lobes that are closely associated with the inferior vena cava.

Venous blood rich in nutrients enters the liver from the hepatic portal venous system. The hepatic portal vein drains the digestive tract and efficiently transports metabolic building blocks (sugars and amino acids) directly to the liver, where they are reconstituted into more complex molecules.

Internally, the liver is made up of many small **hepatic lobules**—the functional units of the liver—that are barely visible to the naked eye. Hepatic lobules are roughly hexagonal in cross-section and are characterized by a central vein in the middle and a triad of vessels (small branches of the hepatic artery and portal vein, and a bile duct) at each outer corner (Fig. 12.30). Radiating outward from the center are irregularly shaped sheets, a single cell thick, of **hepatocytes**, the epithelial cells of the liver. On one side of the sheet is a sinusoid, lined by a loose layer of macrophages called **Kupffer cells**, through which blood from the hepatic

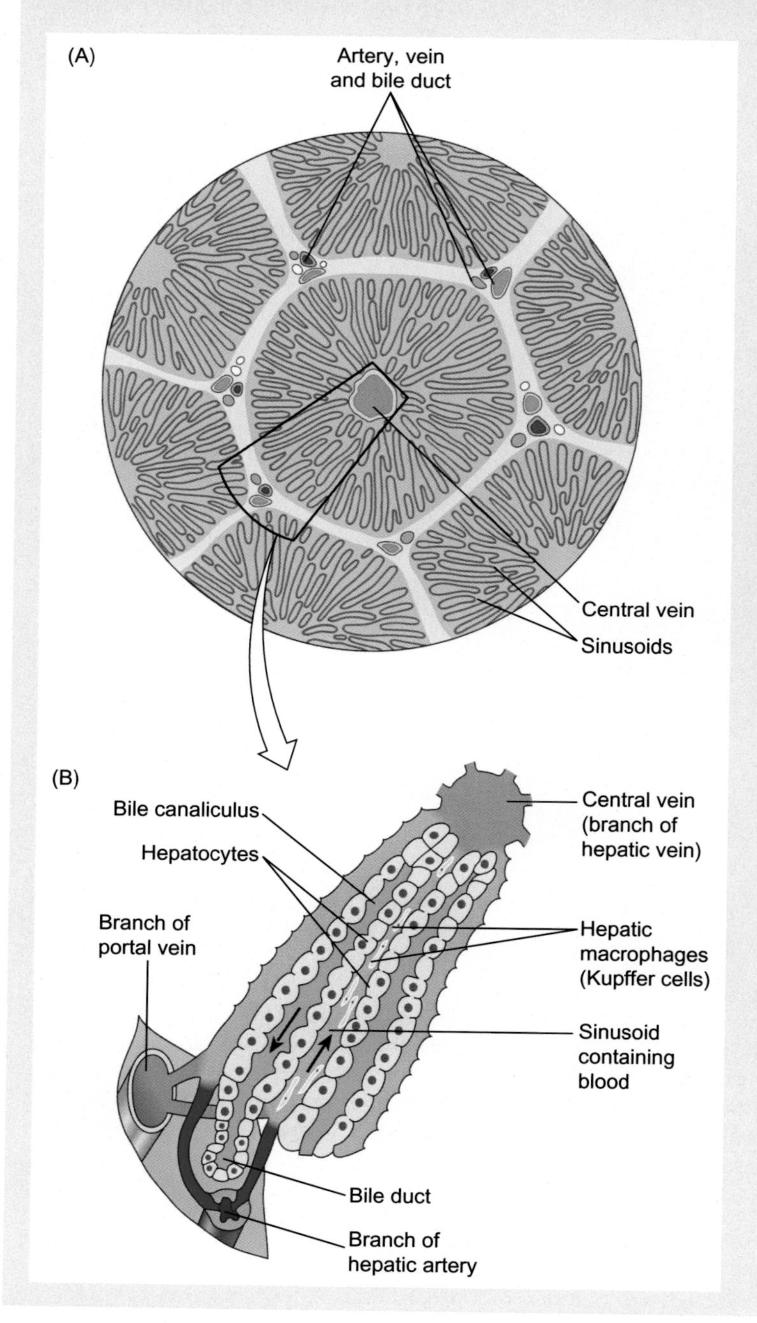

FIGURE 12.30 Architecture of a liver lobule. In (B) the black arrows indicate the direction of blood flow (in violet) and bile (in green). *From Waugh and Grant (2014), with permission.*

(Continued)

BOX 12.4 (Continued)

artery and portal vein flows from a triad toward the central vein. This blood brings in both oxygen and the raw materials for the metabolic functions of the liver. These materials are taken up by the hepatocytes as the blood slowly flows past them. The same blood also picks up the molecules produced by the hepatocytes and transports them from the central veins into larger hepatic veins and finally into the general circulation via the inferior vena cava. On the other face of the sheet are tiny **bile canaliculi**, into which waste products and newly formed bile flow in the opposite direction from the blood. The biliary fluid collects into successively larger channels and ultimately empties into the **gallbladder**.

In one sense, hepatocytes can be viewed as very generalized cells because they are capable of so many different functions (Table 12.2). Their cytoplasm contains a wide variety of organelles—mainly those involved in the synthesis and breakdown of larger molecules (Fig. 12.31). Despite their generic nature, hepatocytes are capable of responding to metabolic demands, and various cytoplasmic components become prominent, depending upon the need. A good example of this is the smooth endoplasmic reticulum, which expands considerably when it is necessary to detoxify drugs or to deal with excess exposure to alcohol. The rough endoplasmic reticulum is also highly developed, because hepatocytes produce large amounts of albumin, coagulation proteins, and complement, all of which are secreted into the blood.

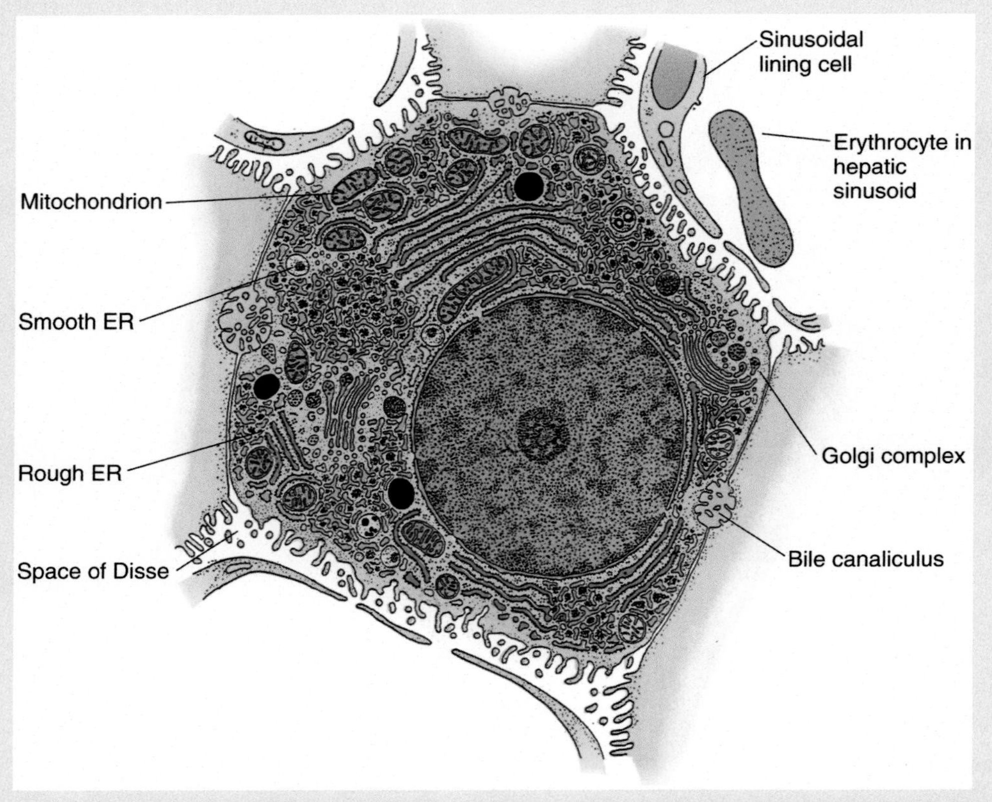

FIGURE 12.31 The cellular structure of a hepatocyte—a highly generalized cell with many functions. *From Gartner and Hiatt (2007), with permission.*

The Kupffer cells lining the sinusoids play a role, along with the spleen and bone marrow, in the breakdown of worn red blood cells. Heme is separated from the globin protein of hemoglobin, and the iron core of heme is separated from the remainder of the heme molecule. The iron gets recycled, and the remainder of the heme makes its way into the bile via the bile canaliculi.

Within a hepatic lobule, there are zones of different metabolic activity. The nature of the activity is largely based on the oxygen concentration of the zones. Closest to the area of the triads, where the hepatic arteries bring oxygenated blood into the liver, hepatocytes focus on the synthesis of plasma proteins and glycogen. Close to the central vein, where the oxygen concentration is low, detoxification processes are more prominent.

TABLE 12.2 Functions of Hepatocytes

Functions of Hepatocytes
Metabolism and storage of carbohydrates (as glycogen)
Protein metabolism—breakdown and reuse of amino acids, synthesis of many plasma proteins
Lipid metabolism—Breakdown of fatty acids, formation of triglycerides
Heat production as a by-product of metabolic activity
Detoxification—Breakdown of drugs and toxins, metabolism of ethanol
Inactivation of hormones—for example, steroids, insulin, glucagon, and thyroid hormone
Formation and secretion of bile components
Storage—for example, glycogen, iron, copper, fat-soluble vitamins A, D, E, K, and vitamin B_{12}

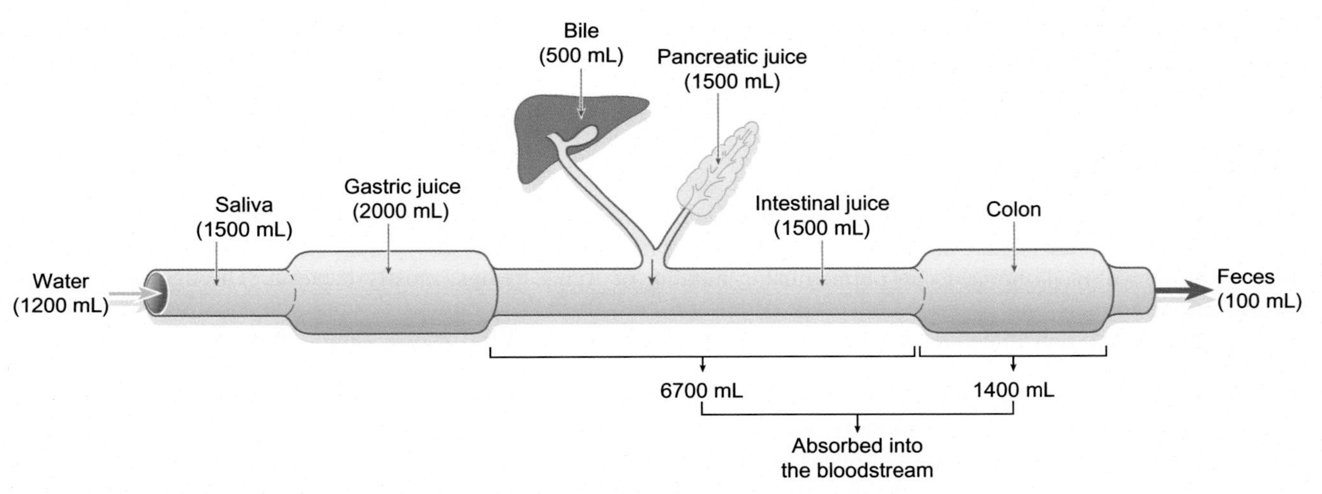

FIGURE 12.32 Fluid exchange in the digestive tract. *From Waugh and Grant (2014), with permission.*

the fluid in the distal ileum is somewhat hypotonic. Final processing of intestinal contents, called **chyme**, takes place in the colon.

LARGE INTESTINE

When the ileocecal valve opens, the small intestine empties its mostly liquid contents into the large intestine (**colon**) at its origin in the lower right-hand quadrant of the abdomen (see Fig. 12.1). The site of emptying, however, is not at the very end of the colon. The colon ends in a blind pouch, called the **cecum** (from the Latin word for blind), which extends a few centimeters beyond the site where the ileum empties into the colon. In many mammals, the cecum is a site containing many anaerobic bacteria, which assist in the overall digestive process.

Protruding from the cecum and emptying into it is the thin wormlike **(vermiform) appendix**. The human appendix is constructed like a lymphoid organ, with many lymphoid nodules, but its normal function in the human has remained obscure. Long-considered a vestigial or rudimentary structure, more recent research has suggested that it plays a significant role in the immune defense of the lower digestive tract and that it might also be involved in the normal relationship between nonpathogenic (commensal) bacteria and the body. According to one hypothesis, the appendix serves as a reservoir for normal gut bacteria. After bouts of diarrhea or cholera, bacteria present in the appendix that are not washed out during the episodes of diarrhea can then emerge from the appendix and repopulate the gut with beneficial bacteria.

The colon itself follows a 1.5-m course as it frames the small intestines (see Fig. 12.1). Its ascending segment rises along the right abdominal wall before making a sharp turn and becoming the transverse colon. Then at the upper left-hand corner of the abdomen, it makes another sharp turn while becoming the descending colon and passing downward along the left abdominal wall. The descending colon then becomes S-shaped as the short sigmoid segment before terminating as the rectum and the anal canal.

Digestion and Absorption in the Large Intestine

For all practical purposes, digestion is completed by the time chyme enters the large intestine. Absorption of organic molecules is usually limited to minor amounts of small lipid fragments, including vitamins, that either pass through the small intestine or may be produced by intestinal bacteria. The colonic mucosa is lined with cells that are adapted for absorption of water and ions or for the production of mucus (goblet cells, see Fig. 12.24).

The major absorptive activities of the large intestine involve inorganic ions and water. Na^+ is actively absorbed from colonic fluid and is largely replaced by K^+. similarly, Cl^- is replaced by HCO_3^-. If metabolically necessary, the adrenal hormone, aldosterone, can assist the colonic epithelium in further increasing the absorption of Na^+ until virtually none is lost from the body. The replacement of Na^+ by K^+ in the large intestine leaves the body susceptible to K^+ deficiency after prolonged bouts of severe diarrhea. Eating foods, like bananas, which are rich in K^+, is a good source of replenishment in cases of diarrhea.

Of the roughly 1500 ml of water that enters the colon from the small intestine, 1400 ml are absorbed into the colonic wall (see Fig. 12.32). As in other areas of the intestine, water leaves the colon by following an osmotic gradient, and the chyme entering the colon is hypo-osmotic.

The Colonic Microbiome

If this book had been written 10 years ago, this section would not have been included. We now know that the phrase "no man is an island" is much truer than the person who coined that phrase had ever imagined. Our gut, especially our colon, is home to an unimaginable number of living beings, mostly bacteria. Throughout the course of evolution, animals and microbes have coexisted and coevolved in a mutually supportive relationship. We need some of the things that intestinal bacteria can provide, and they need the warm stable environment that our intestines can provide.

Although estimates vary greatly, a good guess is that our gut, mainly the colon, is home to 3–4 times as many bacteria as there are cells in the human body. Molecular biology studies indicate that the aggregate number of bacterial genes in the gut exceeds the number of genes in our own cells by a factor of 100. To date, between 500 and 1000 species of bacteria have been found in the digestive tracts of humans, although not that many are found in a single person. In addition, increasing numbers of intestinal viruses that relate to both intestinal bacteria and their human host are being found. With all of this quantitative information, the main question is why do we, as humans, house so many creatures within our bodies[3]. Because of their own unique metabolisms, bacteria can perform biochemical operations that we cannot—possibly because over the eons, we have ceded these responsibilities to the bacteria. In turn, the human digestive tract is a very favorable environment for the growth of these same bacteria, which have undoubtedly evolved to be dependent upon this environment.

Even though they are not essential for our survival, intestinal bacteria play several important roles in our normal physiology. Certain of these bacteria are able to digest cellulose compounds (called **xyloglucans**), which are found in lettuce and onions, into nutrients—mainly for the bacteria themselves. Other bacteria break down dietary fiber into short-chain fatty acids that provide nutrition to cells of the colonic mucosa. A very important area of interaction is that between intestinal bacteria and the immune system. The immune system must learn to recognize the normal intestinal bacterial flora as being part of "self," but yet be able to mount an immune response against pathogens. Intestinal bacteria influence the differentiation of regulatory T cells, which modulate immune responses. Some species synthesize vitamin K and folic acid, functions that our bodies have ceded to the microbiome many years ago. In addition, there are strong indications of relationships between intestinal bacteria and growth control, obesity, and even mental functions, such as depression. As a by-product of their metabolic activity, intestinal bacteria produce varying amounts of gases, most commonly hydrogen, carbon dioxide, methane, and sometimes hydrogen sulfide.

---

3. In the mouth, on the skin, in the respiratory tract, and in the female reproductive tract, especially in the vagina. As in the gut, these relationships are often beneficial to both the bacteria and to the human hosts. With improvements in molecular technology, it is now possible by analyzing their RNA (specifically 16 S ribosomal RNA) to identify forms of bacteria that have never been cultured. Biomedical scientists have been amazed at the diversity of species that constitute the human microbiome.

In cases of a number of chronic gastrointestinal diseases, such as irritable bowel syndrome, or after diarrhea, or the overuse of antibiotics, populations of colonic bacteria become seriously depleted or skewed toward unfavorable species. The use of probiotic supplements, now readily available, to readjust one's intestinal bacterial populations is becoming increasingly commonplace In some cases of severe imbalances of the microbiome, researchers, and physicians have used fecal transplants from normal individuals to introduce normally healthy bacterial strains into the depleted colon.

Motility of the Colon

Like the small intestine, the large intestine has two layers of smooth muscle—an inner circular and an outer longitudinal layer. Unlike the small intestine, the outer longitudinal layer is not continuous, but rather exists as three strips of longitudinal muscle called the **teniae coli**. The mechanics of these muscles cause the colon to take on a puckered appearance, with bulges called **haustra** (see Fig. 12.1) found along its length. At the level of the rectum, the longitudinal smooth muscle reforms a continuous layer, and the haustra disappear.

Motility in the colon differs from that in the small intestine. Instead of relatively frequent (every few seconds) peristaltic movements, colonic contractions are segmental in nature and may last up to a minute. Several times a day **mass movements** take over and supersede the local segmental contractions. These mass movements propel the intestinal contents significant distances along the length of the large intestine each time.

At the levels of the descending and sigmoid colon, the intestinal contents have changed from a liquid to a semi-solid state. Local segmental contractions become more frequent, but they do not propel the intestinal contents along. Instead, they retard the movement of the material into the rectum. Retention in this area further reduces the fluid content of the now-forming fecal material. Propulsion of intestinal contents throughout the colon is facilitated by the abundant mucus secreted by goblet cells, the dominant cell type in the colonic epithelium.

Formation of Feces and Defecation

By the time the intestinal contents reach the **rectum**, they have been transformed into feces. Most of the water has been absorbed, but about 100 ml of water per day is lost with the feces. In addition to water, feces contain a variety of solid components, especially bacteria, undigested fiber, shed intestinal epithelial cells, and some fatty acids. The typical brown color of feces is due to bile pigments (breakdown products of hemoglobin), especially **stercobilin**, that intermingle with the intestinal contents.

Normally, the rectum is empty, but when a colonic mass movement pushes fecal material into it, the distended rectum stimulates a **rectosphincteric reflex** in the anal region. The **anal canal** is a complex structure that marks a sharp transition from colonic epithelium to a stratified squamous epithelium characteristic of skin. Underlying the lining of

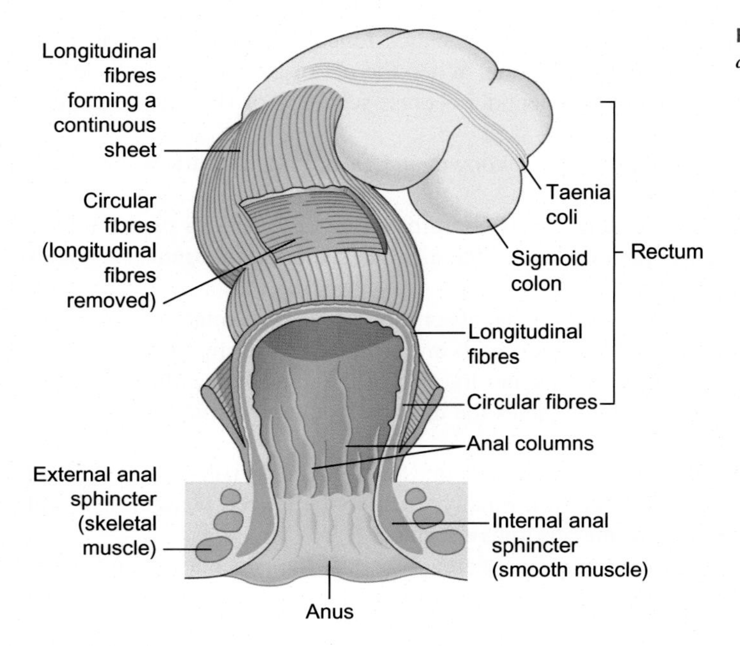

FIGURE 12.33 Musculature of the rectum and anus. *From Waugh and Grant (2014), with permission.*

the anal canal are two muscular sphincters. The internal sphincter is composed of smooth muscle and is under involuntary control. The external sphincter, under voluntary control, consists of skeletal muscle fibers.

Relaxation of the internal sphincter in response to rectal filling can be transient. The external sphincter is normally in a state of tonic contraction, which accounts for bowel control. Under these circumstances, the relaxation of the internal sphincter is replaced by contraction because receptors in the rectal wall become accommodated to the state of filling. In normal adults, defecation is a voluntary action, which involves relaxation of the external anal sphincter and increasing the intraabdominal pressure. This is accompanied by involuntary contractions of the rectal musculature and relaxation of the internal sphincter.

Lining the upper anal canal are 6–10 mucosal folds, called **anal columns** (Fig. 12.33). Each column contains a branch of the rectal artery and its corresponding vein. In addition, two or three hemorrhoidal cushions are present in the anal canal. These structures are thin-walled sinusoids that do not contain smooth muscle in their walls. Normally, they play a small role in maintaining fecal continence. Not uncommonly, however, chronic increased intraabdominal pressure causes the weak walls of the hemorrhoidal sinuses to bulge, forming **hemorrhoids**.

SUMMARY

The digestive system is a continuous tube beginning with the mouth and ending at the anal opening. Associated with the digestive tube are important glands—salivary glands, the pancreas, and liver.

Food is brought into the digestive system and initially broken down by the lips, teeth and the oral cavity. Breakdown of food in the oral cavity is largely mechanical, through the teeth and chewing, but some digestion of carbohydrates begins through enzymes released by the salivary glands. There are three salivary glands—the parotid, the submaxillary, and sublingual glands. Each has a slightly different function, and the saliva that they secrete has different properties. The tongue is covered with taste receptors and is also the means by which food is moved into the pharynx and esophagus. The oral cavity and lips are also very important in vocalization.

The pharynx and esophagus function to facilitate swallowing. Swallowing is accomplished by skeletal muscles attaching to or lining the walls of these structures. during swallowing, elevation of the palate seals off the nasal cavity, and the epiglottis seals off the larynx so that food or liquid will not enter the respiratory tract. The swallowing reflex inhibits breathing during swallowing. Entry of food into the stomach is permitted by relaxation of the lower esophageal sphincter.

With a three-layered muscular wall, the stomach mechanically churns the swallowed food and begins the digestive process. The lining of the stomach contains parietal cells that produce hydrochloric acid, which reduces the pH of the interior of the stomach to as low as 1.0. Other epithelial lining cells produce specialized mucus that protects the lining of the stomach from the acid.

The production of HCl is controlled by three phases—the cephalic, the gastric, and the intestinal phase. In the first phase, secretion is stimulated by activity of the vagus nerve. The other two phases function through stimulation of secretion of the local hormone gastrin, which also stimulates HCl production. When the pH of the stomach falls below 3.0, other local hormones, such as somatostatin, and gastric inhibitory peptide, depress secretion of HCl by the parietal cells.

The locally produced enzyme, pepsin, begins the process of protein breakdown, and gastric lipase initiates the breakdown of fats.

The pyloric sphincter controls emptying of the stomach. Its function is coordinated through actions of the central and enteric nervous systems. As acid stomach contents enter the duodenum, the alkaline secretion of Brunner's glands in the duodenum protects the duodenal lining from ulceration.

The small intestine is the segment of the gut where the most digestion and absorption occur. The small intestine contains three segments—the duodenum, the jejunum, and the ileum. Stomach acids are neutralized and digestive processes begun in the short duodenal segment. Secretory ducts from the pancreas and liver enter the duodenum. Most digestion and absorption of digested food occurs in the jejunum. The ileum completes the digestive and absorptive processes and functions uniquely to absorb vitamin B_{12} and bile salts.

Several specializations greatly increase the absorptive surface of the small intestine. Grossly visible folds (plicae) constitute the first level. On them are numerous fingerlike intestinal villi. Finally, each epithelial cell possesses several thousand microvilli, which account for the final magnification of the absorptive surface.

Digestion of food particles depends heavily upon bile salts and enzymes secreted by both the pancreas and intestinal glands. Digestion can be intraluminal or by membrane digestion by enzymes intrinsic to membranes of the intestinal epithelial cells.

Initial digestion of carbohydrates is accomplished through the action of pancreatic amylase. Breakdown of di- and trisaccharides into monosaccharides is accomplished by brush border enzymes of the intestinal epithelium. Most simple sugars are absorbed in the jejunum.

Proteins partially digested in the stomach are broken down into amino acids in the small intestine by at least five enzymes produced by the pancreas. One of the most powerful is trypsin, which is derived from trypsinogen after activation by intestinal enterokinase.

Lipid digestion is complex because of the chemical diversity of lipids. Not being soluble in water, they are first emulsified through the action of bile salts released from the gallbladder. Pancreatic lipases break down triglycerides into glycerol and fatty acids. Aggregations of lipid breakdown products, called micelles, are taken up by enterocytes in the small intestine.

Absorbed sugars, amino acids, and simple lipid molecules enter the capillary circulation within the intestinal villi. From there, they are carried to the liver via the hepatic portal venous system. More complex lipids are taken up by lacteals—blind endings of the lymphatic system in the villi—and then make their way through the lymph channels.

Another important function of the small intestine is the absorption of water and electrolytes. The small intestine absorbs up to 7 L of water per day.

Movement of digested food along the small intestine is accomplished by rhythmic contractions of the circular and longitudinal smooth muscle layers in the intestinal wall. These contractions are largely controlled by the enteric nervous system.

Upon opening of the ileocecal valve at the distal end of the small intestine, ileal contents empty into the colon. Most digestion has been completed by the time intestinal contents (chyme) enter the colon. The colonic epithelium is specialized for the absorption of water and ions and for the production of mucus. About 95% of the water entering the colon is resorbed there. The colon resorbs any remaining vitamins or lipid molecules that enter it.

The colon contains a large and functionally important microbiome, with hundreds of species of bacteria within a single individual. Some bacteria can break down certain cellulose compounds or dietary fiber into smaller molecules that can be absorbed through the wall of the colon. Intestinal bacteria also produce varying amounts of gases, such as carbon dioxide, methane, and hydrogen sulfide.

The colon contains two layers of smooth muscle in its wall, but instead of regular peristaltic movements, several times per day mass movements take place and supersede local segmental contractions.

By the time intestinal contents reach the end of the colon, most of the water has been removed. As they reach the rectum, they have been transformed into feces. The solid content of feces includes bacteria, undigested fiber, and shed intestinal cells. The usual brown color of feces is due to the presence of stercobilin, a breakdown product of hemoglobin.

The anal canal is bounded by two sphincters. The internal sphincter is of smooth muscle and is under involuntary control. The external sphincter is made of skeletal muscle and operates under voluntary control.

Chapter 13

The Urinary System

At first glance, the purpose of the urinary system would seem to be quite straightforward—the production of **urine**. A major function of the urinary system is the removal of nitrogenous waste (mainly in the form of urea) and other metabolic by-products from the blood. However, like most body systems, other functions are intimately tied to the structures serving the formation and excretion of urine. The process of making urine is part of the body's drive toward maintaining **homeostasis**—the maintenance of a stable internal environment. Homeostasis involves preserving an appropriate volume of blood and tissue fluids. It also encompasses maintaining a healthy balance of inorganic ions (e.g., Na^+, K^+) as well as keeping the pH of the blood within a healthy range. In addition, groups of cells within the kidney are heavily involved in the control of blood pressure. Even the male urethra plays a shared role as a common conduit for the outflow of urine and semen.

In brief, the story of urine formation begins with the flow of arterial blood into the kidney through the renal arteries (Fig. 13.1). Then, within the kidneys, the blood undergoes a complex round of processing, as many components pass from the circulating blood into the tubules of the kidney. Most of these components are then returned to the circulation before the blood leaves the kidney. Meanwhile, the filtrate of blood that has passed into the renal tubules is selectively processed until it ultimately leaves the kidney as fully formed urine. The urine passes through the ureters into a storage reservoir, the urinary bladder, where it remains until it is voluntarily voided during urination.

One of the principal functions of kidneys is the removal of nitrogenous metabolic waste products resulting from the breakdown of large molecules such as proteins and nucleic acids or their precursor molecules. About half of the final nitrogenous waste is in the form of urea. The rest of the waste nitrogen is in the form of uric acid and creatinine phosphate. These molecules are very toxic if allowed to accumulate in the body, resulting in the condition of **uremia**.

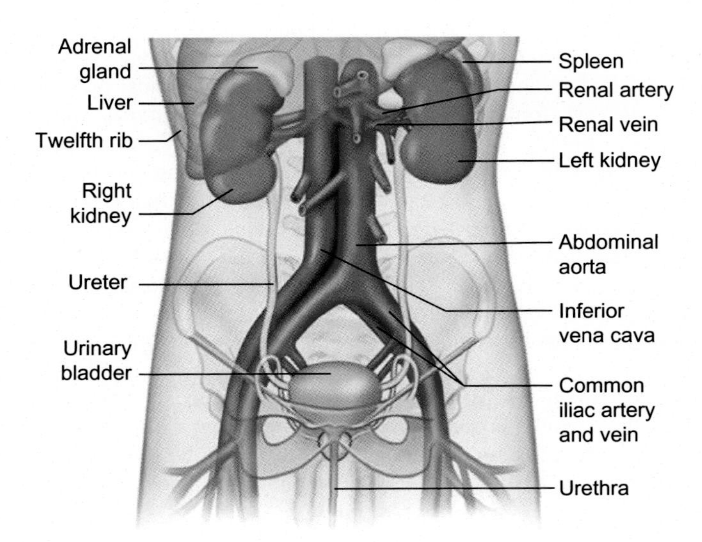

FIGURE 13.1 Overview of the urinary system. *From Thibodeau and Patton (2007), with permission.*

The Human Body. DOI: https://doi.org/10.1016/B978-0-12-804254-0.00013-2

EVOLUTION OF THE URINARY SYSTEM

The kidneys have had a long and complex evolutionary history. The earliest precursors of the vertebrates didn't have actual kidneys. These segmented animals likely had simple tubes in each segment that drained fluids from the body cavity directly to the outside. As the body became more complex, a mechanism (the glomerulus) developed for removing water and solutes from the blood for eventual processing into urine. Kidneys originally arose in aquatic fish-like animals. These kidneys, called **pro-** and **mesonephric kidneys**, were highly elongated strips that stretched out along the entire length of the body cavity. Because in freshwater there is no need for the body to preserve water, the early kidneys evolved with a means for both filtering the blood and then returning to the body most of the needed solutes, while at the same time eliminating metabolic wastes, although much of the urea left the body by secretion through the gills. In contrast to the kidneys of land-dwelling vertebrates, there was little need to recover filtered water.

With their emergence onto land, vertebrate animals were faced with a new problem—conserving water to avoid dehydration. This issue was met by the evolution of a new component of the renal tubule, the loop of Henle, and an expanded inner region of the kidney, the medulla, which act in concert to provide a remarkably efficient mechanism for preserving both water and solutes. These structures evolved along with a new type of kidney, the **metanephros**, the form of kidney seen in humans (Box 13.1). Without gills, which served as accessory excretory organs, land animals were also forced to adapt their kidneys to secrete metabolic wastes, such as urea or uric acid.

BOX 13.1 A Primer on Kidney Structure

The fundamental functional unit of the kidney is called the **nephron**, and each human kidney contains about a million nephrons. The location of both entire nephrons and components of individual nephrons within the kidney is critical to their function. This box is designed to provide a structural framework for understanding specific details of urine function presented throughout the chapter.

Although there is considerable variation in size, a typical kidney is a bean-shaped structure approximately 12 cm long, 6 cm wide, and 4 cm thick. At the center of the concave face is the **hilum**—a region through which a renal artery enters the kidney and from which a renal vein drains it (Fig. 13.2A). Much of the hilum is occupied by the **renal pelvis**, which serves as the collecting chamber for the urine formed by the individual nephrons which empty their contributions into the pelvis. The renal pelvis then merges smoothly into the **ureter**, which carries the urine into the **urinary bladder**.

The substance of the kidney (the parenchyma) is divided into an outer cortex and an inner medulla. As seen later in the chapter, a concentration gradient of solutes in the medulla is critical to the formation of a concentrated urine. The human kidney is subdivided into 8–18 lobes, each of which includes a region of cortex and pyramidal region of medullary tissue. The tip of a pyramid, called the papilla, empties newly formed urine into a system of minor and major renal calyces, which are continuous with the renal pelvis.

Individual nephrons have the same general structure, but their specific function depends upon where their various components are located. All nephrons consist of the same components (Fig. 13.2B). The front end of a nephron is called the renal corpuscle, a structure consisting of a glomerulus (a tangled mass of small capillaries) embedded within a cup-like structure called **Bowman's capsule**. Bowman's capsule drains into a proximal convoluted tubule that leads into a hairpin-shaped loop of Henle. Both the descending and ascending limbs of the loop of Henle contain thick and thin segments—so called because of the nature of the epithelium that lines the loops. The loop of Henle continues into a distal convoluted tubule, which empties into a system of collecting ducts. All of these segments have specific functional properties, all of which are critical to the formation of normal urine.

Of considerable importance is where the components of nephrons are located. Nephrons themselves sort out into two main topographic types. About 25% of them (called **juxtamedullary nephrons**) begin close to the corticomedullary border and send their loops of Henle deep into the medulla. These are the nephrons that feed the strong osmotic gradient in the medulla. The remainder of the nephrons (**cortical nephrons**) arise farther away from the medullary border, and they do not send their loops of Henle as deeply into the medulla as do the juxtamedullary nephrons.

(Continued)

BOX 13.1 (Continued)

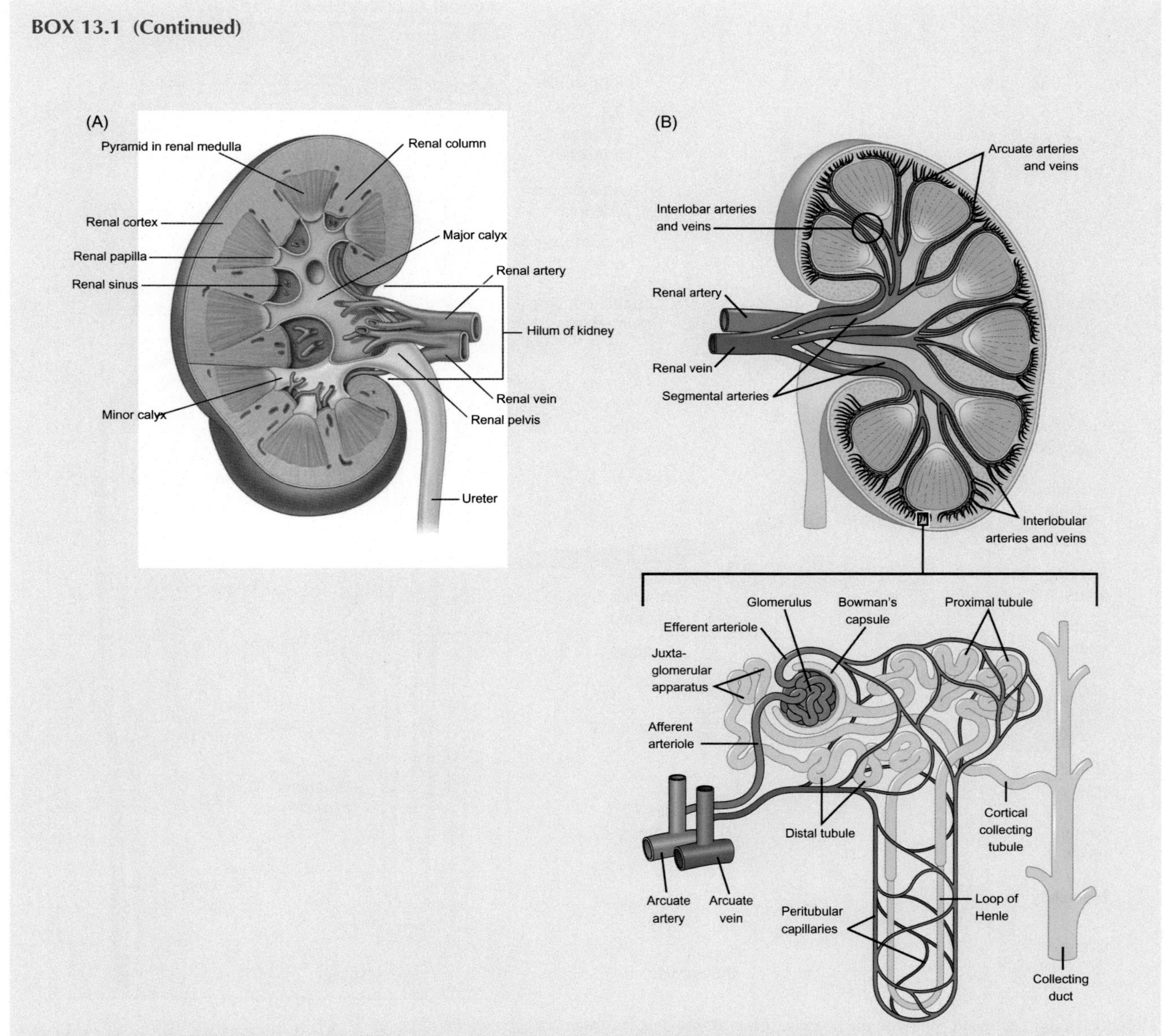

FIGURE 13.2 (A) Structure of the kidney. (B) Topography of the nephron—the functional unit of the kidney. Two major types of nephrons—cortical and corticomedullary—have different functional properties, mainly relating to the depth of penetration of the loop of Henle into the medulla. *(A) From Drake et al. (2005), with permission. (B) From Guyton and Hall (2016), with permission.*

THE FORMATION OF URINE

Filtration

Through the **renal arteries**, a disproportionate share of the circulatory volume flows to the kidneys. Although their combined mass is less than 0.5% of the total body mass, the kidneys receive approximately 20% of the total cardiac output and account for approximately 8% of the body's oxygen consumption. As a renal artery enters the kidney, it

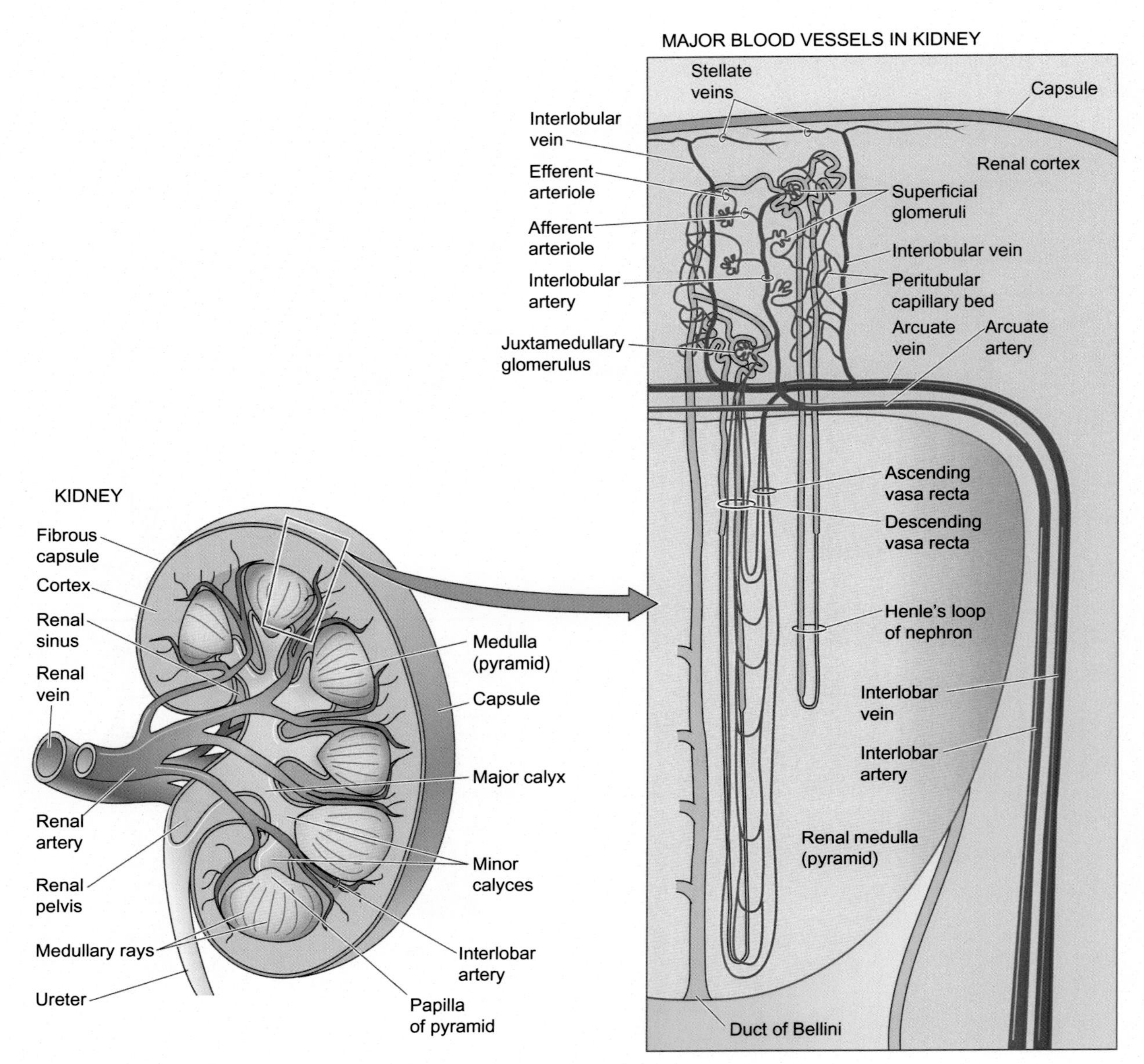

FIGURE 13.3 The blood supply to the nephron. *From Boron and Boulpaep (2012), with permission.*

splits into progressively smaller branches until tiny **afferent arterioles** reach each of the roughly 0.5–1.0 million glomeruli within a kidney (Fig. 13.3). At the glomerulus, the afferent arteriole breaks up into about 30 highly contorted glomerular capillary loops (Fig. 13.4).

The blood reaching the **glomerulus** arrives under arterial pressure (~90 mm Hg) and remains under relatively high pressure as it passes through the glomerulus because of the large combined cross-sectional area of the glomerular capillaries, which are arranged in parallel. The glomerulus is embedded in a double-walled cup-like **Bowman's capsule** (with a diameter of 300 μm)—the terminal end of an individual renal tubule system, or **nephron**. The collective term **renal corpuscle** is applied to Bowman's capsule along with its embedded glomerulus. At the glomeruli, about 20% of the blood plasma flowing through the kidneys is filtered and passes from the blood into the hollow space within Bowman's capsule. What passes from blood to nephron depends upon both physical characteristics of the circulating blood (hydrostatic and colloid osmotic pressure) and the structural filtration barrier between the circulating blood and the lumen of the nephron.

The **filtration barrier** between the blood and the lumen of Bowman's capsule consists of three elements—the endothelium of the glomerular capillary, the inner (visceral) layer of Bowman's capsule, and a thick basal lamina between

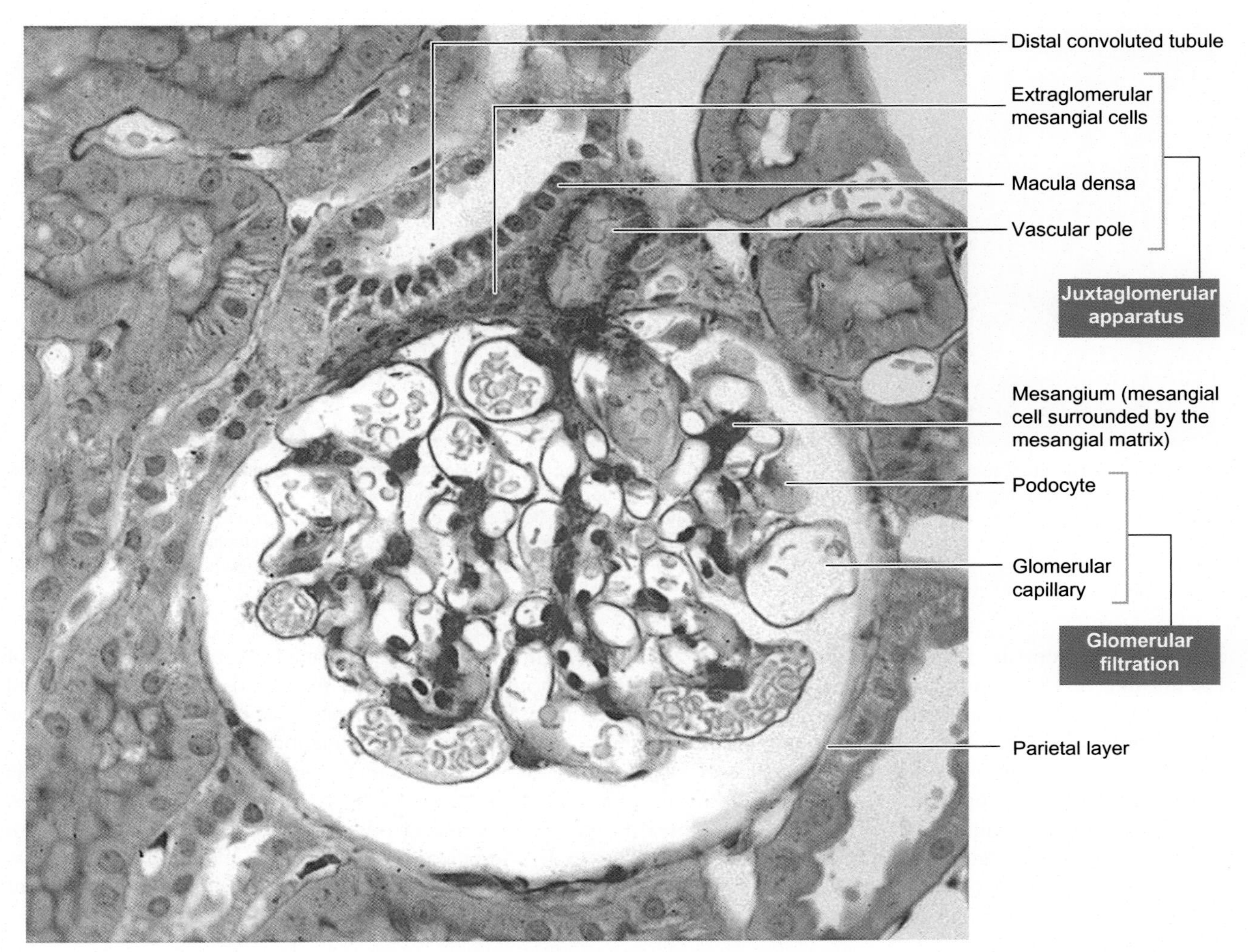

FIGURE 13.4 Histological organization of the nephron and juxtaglomerular apparatus (JGA). *From Kierszenbaum and Tres (2012), with permission.*

them (Fig. 13.5). Each of these components has characteristics that play a role in the filtration properties of this barrier. In contrast to most tissues, the capillary endothelium is not continuous, but is studded with many 70 nm fenestrations, allowing plasma proteins direct access to the basal lamina. On the other hand, the surfaces of the endothelial cells are lined with negatively charged glycoproteins, which do not permit negatively charged plasma proteins, e.g., albumins, to pass through the endothelial cells themselves. The fenestrations, however, allow free passage of essentially all other components of blood except for cells. The epithelial cells of the visceral layer of Bowman's capsule are very complex in structure, with long processes extending from the cell body. The basal lamina, representing a fusion between the basal laminae secreted by both the endothelial cells and the podocytes, filters out molecules of a weight greater than 1000 Da. Because it is also negatively charged, it also repels large negatively charged proteins.

The final, and most structurally complex, component of the glomerular filtration barrier is the inner layer of Bowman's capsule. This layer consists of specialized octopus-shaped epithelial cells, called **podocytes**, which drape their arms over the glomerular capillaries (see Fig. 13.5B). Extending from the arms are regularly spaced podocyte foot processes, which interdigitate with those from neighboring podocytes. The spaces between podocyte foot processes (filtration slits) are occupied by thin membranes consisting largely of linked protein molecules, called **nephrins**. These represent the most selective filter, which permits passage of water, ions, urea, glucose, amino acids, and small proteins and peptides.

Driving much of the filtration process is the hydrostatic pressure (60 mm Hg) of the glomerular capillaries. This is opposed by the hydrostatic pressure of the fluid within the lumen of Bowman's capsule (~ 20 mm Hg) and the colloid osmotic pressure within the capillary blood. The aggregate result of these is a net pressure of approximately 12 mm Hg

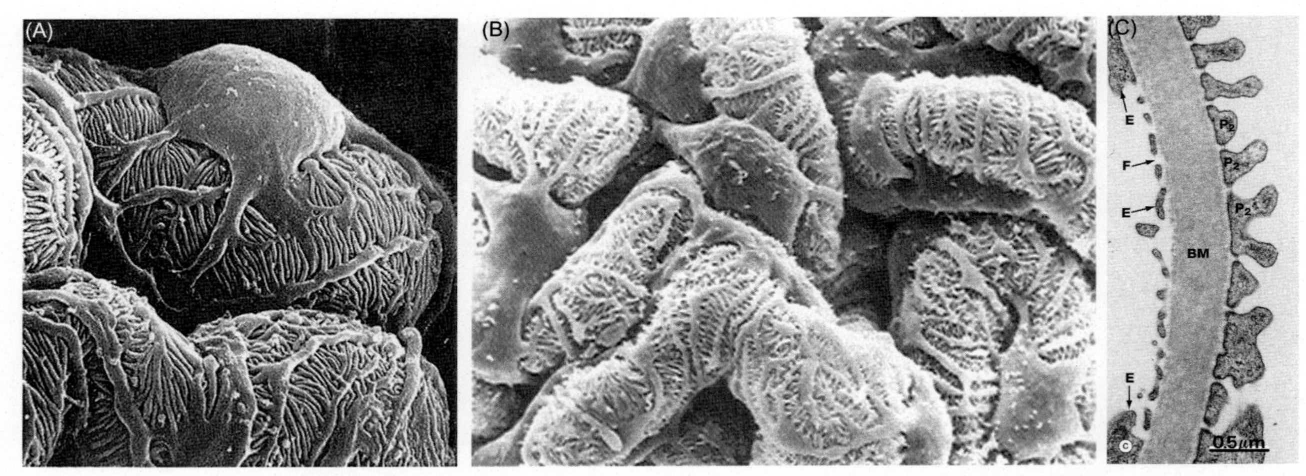

FIGURE 13.5 The glomerular filtration barrier. (A) Scanning electron micrograph of podocytes surrounding glomerular capillaries. (B) Higher power scanning electron micrograph of interdigitating foot processes of podocytes. (C) Transmission electron micrograph showing the discontinuous endothelium of the glomerular capillary, the thick basement membrane intervening between the endothelium and the podocyte layer and the foot processes of the overlying podocytes. *BM*, basement membrane; *E*, endothelial cells; *F*, fenestrations in endothelial cell cytoplasm; P_2, foot processes of podocytes. *(A) From Erlandsen and Magney (1992), with permission. (B) From Guyton and Hall (2016), with permission. (C) From Young et al. (2006), with permission.*

driving liquid from the renal circulation into the nephron. Overall, this leads to a total glomerular filtration rate of 125 mL/min or 180 L/day. Except for most proteins, the filtrate is remarkably similar in composition to that of blood plasma. The task confronting the tubular portion of the nephron is to return the vast majority of these constituents to the blood, while selectively retaining or secreting urea and other metabolic wastes for elimination in the urine.

One other element within the renal corpuscle plays a significant role in regulating glomerular filtration. That is the **mesangium**, a group of cells that abut onto the afferent and efferent arterioles and the glomerular capillaries. Mesangial cells are a component of a structural complex called the **juxtaglomerular apparatus** (JGA) (see Fig. 13.6), which has important regulatory functions (Fig. 13.6; see Fig. 13.4). Mesangial cells can reduce glomerular blood flow by smooth muscle-like constriction and putting pressure on the microvessels. They also secrete **endothelin**, a potent protein that induces constriction of the smooth muscle walls of the arterioles.

Secretion and Reabsorption

The filtrate entering the lumen of Bowman's capsule must undergo a tremendous amount of processing before it emerges from the kidney as urine. About 180 L of filtrate is generated every 24 hours, and of that amount 178.5 L is reabsorbed and returned to the circulation. In addition, most of the ions and small molecules are also removed from the forming urine, while other waste products are added. All of this is accomplished by the remainder of the nephron. The nephron is subdivided structurally and functionally into a proximal tubule, a loop of Henle, and a distal tubule before connecting with the system of collecting ducts, which convey the mostly formed urine toward the exterior of the kidney (Fig. 13.7).

The Proximal Tubule

The **proximal tubule** (see Fig. 13.7) is an intensely active region of the nephron where two-thirds of the glomerular filtrate, including water, important ions, and small molecules, is reabsorbed. Assuming a glomerular filtration rate of 130 mL/min, this means that the proximal tubules reabsorb about 85 mL/min of water. In addition almost 100% of the filtered glucose, amino acids, and small peptides are reabsorbed in this region. Reabsorption of solutes requires large amount of energy, which is why this region accounts for most of the oxygen consumption in the kidney.

The cuboidal epithelial cells of the proximal tubules are well designed for their functional role (Fig. 13.8). Their apical surface, facing the lumen of the tubule, is covered with densely packed microvilli to form what is often called a **brush border**. The microvilli greatly increase the surface area available for absorption. The other side of the cell, the basolateral membrane, is also highly folded, again increasing its surface area available for exchange. In keeping with their high metabolic activity and need for oxygen, these cells contain large numbers of elongated mitochondria, which,

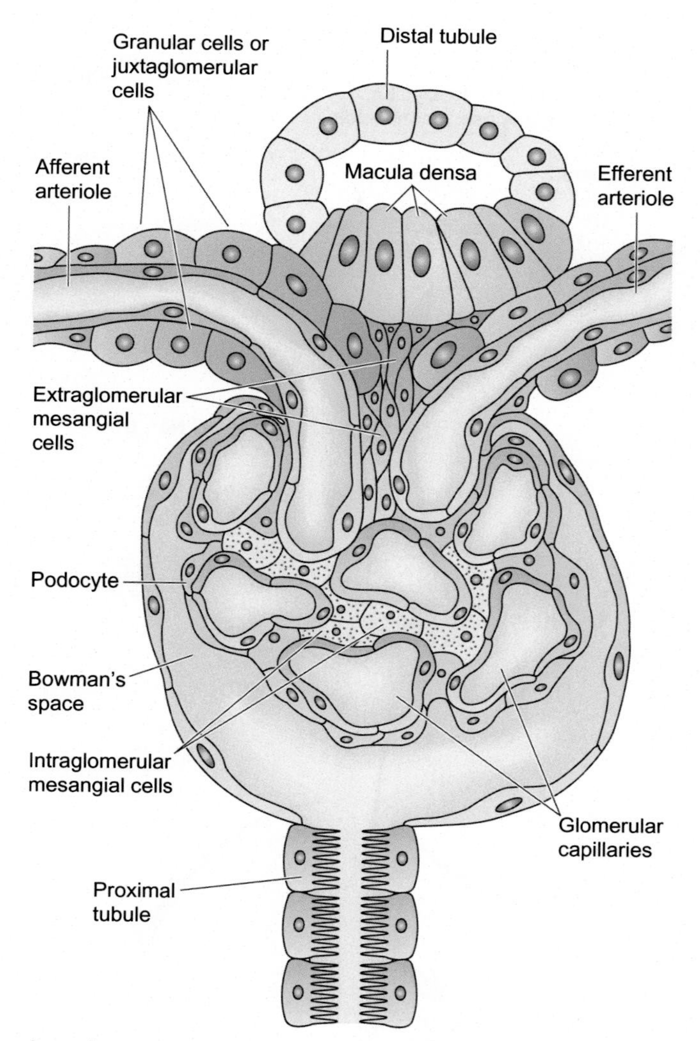

FIGURE 13.6 Structure of the juxtaglomerular apparatus (JGA). *From Boron and Boulpaep (2012), with permission.*

through their production of ATP (adenosine triphosphate), provide the energy required for ion transport. The mitochondria are concentrated in the basolateral region of the cell, where the active ion pumps are located. Near the apical surface of the cell are accumulations of vesicles and lysosomes. These structures facilitate the ingress of solutes into the cell and the breakdown of larger molecules, such as peptides, into smaller amino acid units. Between epithelial cells are tight junctions that maintain the continuity of the epithelium. All tight junctions are not the same. In the proximal tubules (and the descending limb of the loop of Henle just following), the tight junctions are physiologically leaky, whereas throughout the rest of the nephron the tight junctions are much more resistant to the passage of water and other solutes.

Structural adaptations facilitating reabsorption do not end at the cellular level. In contrast to most tissues, the renal tubular epithelium is underlain by a minimal amount of connective tissue. This allows a network of capillaries to enwrap the proximal tubule much more closely than would otherwise be the case, and it reduces the distance between the fluids within the lumen of the proximal tubule and the lumen of the capillaries. Such an arrangement is reminiscent of that in the lung, where oxygen diffusion from alveolus to blood is facilitated by juxtaposition of capillary wall to alveolar epithelium.

The initial stages of urine formation within the proximal tubules utilize the exchange of many different substances, including various ions, sugars and amino acids, and water, between the glomerular filtrate within the lumen of the proximal tubule and the interstitial fluid on the other side of the tubular epithelial cells. For each of these, special

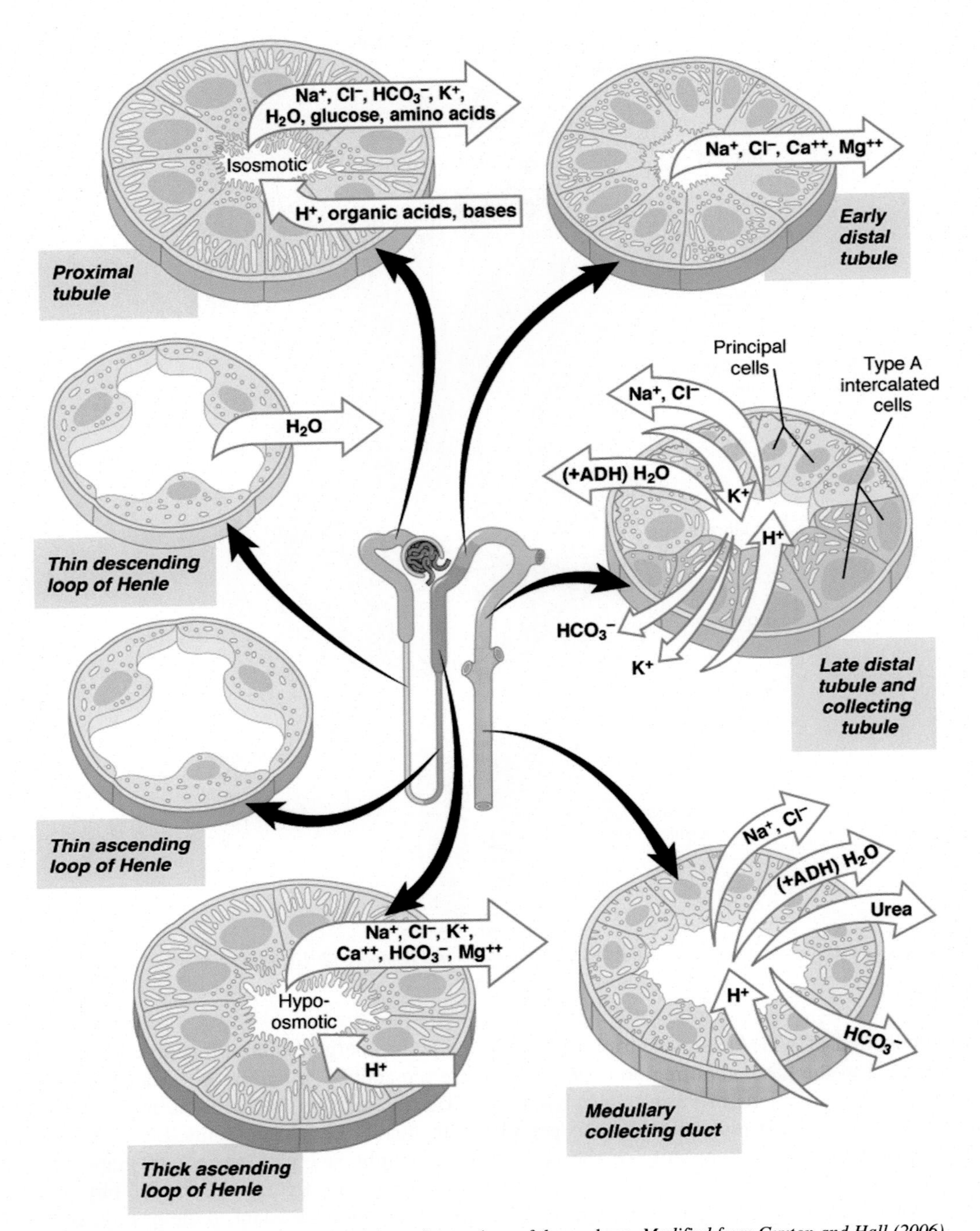

FIGURE 13.7 Structural and functional characteristics of the major sections of the nephron. *Modified from Guyton and Hall (2006).*

molecular mechanisms are employed, often using enzymes and channel proteins that are common to other parts of the body, as well. Box 13.2 summarizes some of the most important processes involved in the initial processing of the glomerular filtrate.

The proximal tubule, especially the distal straight portion, is responsible for the elimination of urea, excess metabolites, and toxic substances. Urea is filtered in the glomerulus and half of that is reabsorbed by the epithelial cells of the proximal tubule. It is also secreted by the epithelium of the proximal tubule and, especially, farther down the nephron (loop of Henle). The net result of a sequence of secretory and reabsorptive processes is that about 40% of the amount of filtered urea is ultimately eliminated in the urine.

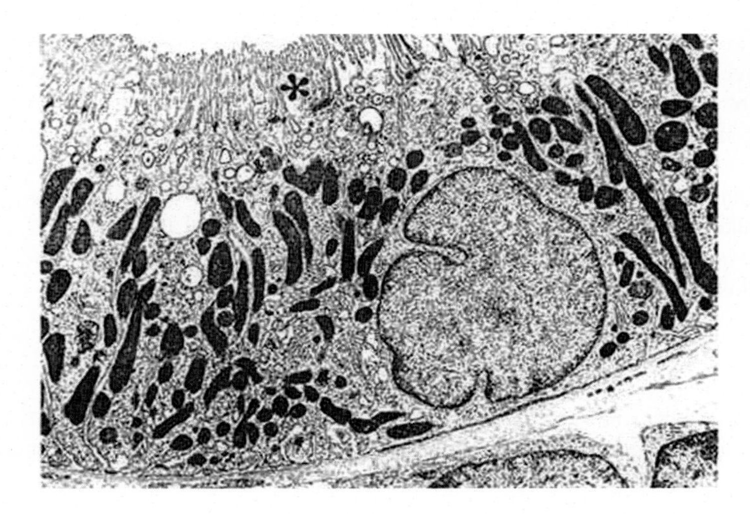

FIGURE 13.8 Electron micrograph of a cell from the proximal convoluted tubule, showing prominent mitochondria and microvilli. *From Erlandsen and Magney (1992), with permission.*

BOX 13.2 Mechanisms of Secretion and Absorption in the Proximal Tubule

The driving force for most exchange processes in the proximal tubule is the active transport of Na^+ from the ultrafiltrate in the lumen of the tubule into the blood. This process accounts for most of the oxygen consumption in the kidney. The key is the presence of Na^+, K^+-ATPase, located in the basolateral membranes. Using energy derived from ATP produced in the basolateral mitochondria, this ATPase pumps three Na^+ ions out of the cell in exchange for the ingress of two K^+ ions into the cell. This sets up an electrochemical gradient through which Na^+ passively enters the apical surface of the epithelial cell from the lumen of the tubule. Active transport of ions represents such a fundamental property of life that the molecular homologies of the transport proteins are preserved across taxonomic lines from bacteria to humans. Along with the active Na^+ uptake, Cl^- is also reabsorbed, mainly passively, through the junctional complexes. The reabsorptive processes within the luminal fluid also result in an increased K^+ concentration, which becomes reduced by the passive movement of K^+ ions into the tubal epithelium.

The large amount of water reabsorption in the proximal tubule takes place across both the epithelial cells themselves and their intercellular junctions. Water passes through the epithelial cell membrane through special water channels lined by aquaporin-1 proteins. Although most water passes through the cells, some also makes its way through the leaky intercellular junctions. Water removal from the ultrafiltrate within the proximal tubule is powered by a small osmotic gradient that sucks the water molecules into the interstitial fluid. From there, it flows to the network of capillaries that surrounds the tubule. Water entry into the capillaries is based largely on a small difference in hydrostatic pressure between the interstitial fluid and that within the capillaries. The pressure of the renal interstitial fluid is slightly elevated over that of most interstitial fluid, and the hydrostatic pressure within the capillaries is lower than that of most other capillary beds.

More complex is the relationship between H^+ and HCO_3^- in the proximal tubule (see Fig. 13.7). While Na^+ is removed from the luminal fluid, H^+ is actively transported into it. As a buffering mechanism, HCO_3^- combines with the H^+ to form carbonic acid (H_2CO_3), which is then broken down into H_2O and CO_2 by the enzyme **carbonic anhydrase** bound to the brush border of the epithelial cells. The removal of HCO_3^- from the tubular fluid slightly acidifies the forming urine. Within the epithelial cells, a cotransporter removes HCO_3^- along with Na^+ from the cell into the interstitial fluid. These movements of H^+ and HCO_3^- between kidney tissue and the blood play a significant role in maintaining a proper acid–base balance within the body.

Many low molecular weight solutes, such as glucose, amino acids, and organic acids are brought into the epithelial cells along with Na^+ through a cotransport mechanism involving a binding protein. Each type of compound seems to have its own set of transport molecules. Normally there is sufficient capacity to completely clear the urine of these compounds and return them to the body, but if through some metabolic disturbance the concentration of one of these in the blood is too high, the amount in the glomerular ultrafiltrate exceeds the capacity of the cotransporter to remove it all, and that compound then appears in the urine.

A good example is blood glucose in diabetics. When the concentration of glucose in the blood rises more than approximately 2¼ times above normal, the cotransport mechanism becomes saturated, and the extra glucose remains in the urine. This is the basis for many tests for diabetes.[1]

1. An amusing anecdote regarding sugar in the urine involved A.J. (Ajax) Carlson, a well-known professor of physiology at the University of Chicago in the early 1900s. He emphasized to his class the importance of accurate observations. In a lecture on diabetes, he was said to have told the class about the ability to diagnose the condition (**diabetes mellitus** = sweet diabetes) by tasting its sweet presence in the urine. He brought a beaker of urine from a diabetic to class and put on his desk. Then he told the class to do exactly what he was doing. He put one finger into the urine and then putting a finger to his mouth, he told the class that if one tasted the urine on the fingertip, one could tell by its sweetness if the patient had diabetes. He then told the class to come up and try it. One by one, the members of the class dipped a finger into the urine and then made disgusted faces as they put the fingertip into their mouths, but they did notice the sweet taste. When this class had finished the exercise, Prof. Carlson then said to them, "For much of this class I have tried to impress on you the power of acute observation. You may have noticed that I put my middle finger into the urine, but my index finger into my mouth!"

Other substances enter the forming urine mainly through the proximal tubule. Metabolic by-products and many toxic substances that have passed through the liver are actively absorbed across the basolateral regions of the epithelial cells of the proximal tubule through linkage with a Na^+ cotransport mechanism or similar means. Thus, their entry into the cell is limited by the ratio of the molecule/transporter sites. These substances leave the cells and enter the lumen of the nephron along decreasing concentration gradients.

Most common drugs are metabolized in the liver and exit the body via the kidneys as metabolites. These also pass through the proximal tubules. Special care must be taken on drug dosages with patients who have renal insufficiency, because of the reduced secretory capacity of the kidneys. It is often necessary to lower the intake of the drug so that the ability of the kidneys to excrete it is not exceeded and an effective overdose occurs.

The Loop of Henle

From the proximal tubule, the fluid on its way to becoming urine next passes into the **loop of Henle**. In doing so, it also enters the medulla of the kidney, a region with a dramatically different appearance and properties from the outer cortex (see Fig. 13.2). The cortex contains all of the glomeruli and the convoluted tubules, both proximal and distal, and upon gross examination it is rather homogeneous in appearance. The medulla, on the other hand, is subdivided into around a dozen medullary pyramids, whose apices point toward a central area, called the renal pelvis, where fully formed urine is collected before leaving the kidney. The vast majority of structures within a medullary pyramid are straight (the limbs of the loop of Henle, the system of collecting ducts [see later] and the blood vessels). This linearity is quite evident even at very low magnification. In addition to these structural elements, another characteristic of major functional importance is the character of the interstitial fluid, which becomes much more concentrated from the corticomedullary junction inward toward the tip of the pyramid. These features all play a role in the ongoing formation of urine within the kidney.

The fluid leaving the proximal tubule is greatly reduced in volume, but little changed in concentration or ionic composition (except for Cl^-) because of the relatively nonselective reabsorption taking place in that area. On the other hand, it is almost completely cleared of sugars, amino acids, but it contains more waste products than the glomerular filtrate. Within the loop of Henle, significant changes in the composition take place. In contrast to the proximal tubule, many of the changes within the loop of Henle are quite selective for a single component of the tubular fluid.

At the beginning of the descending loop of Henle, the epithelium is low cuboidal in character and has some properties similar to those of the proximal tubule. The epithelium then undergoes an abrupt change in morphology to that of a simple squamous epithelium with few surface specializations and few mitochondria. Because of numerous aquaporin-1 channels in both the apical and basal surfaces in these cells, this epithelium is freely permeable to water. Loops of Henle are of different lengths, with the hairpin ends of some situated relatively close to the corticomedullary junction and others ending far down into the medulla—close to the tip of the pyramid. In the human kidney, the osmolality of the medullary interstitial fluid is near normal (280 mOsmol/L) at the corticomedullary junction, but it increases to 1200 mOsmol/L at the tip of the pyramid. As the tubular fluid passes through the descending limb of the loop of Henle, water passively diffuses out until at the tip of the hairpin loop the volume of fluid is only approximately 10% of that of the glomerular filtrate in those loops that extend deep into the medullary pyramid. Only small amounts of NaCl and urea diffuse into the descending limb. The net result is that the osmolality of the tubular fluid also reaches 1200 mOsmol/L.

As the fluid rounds the bend of the loop and begins to pass through the ascending limb of the loop of Henle, it encounters an epithelium that looks very much like that of the thin descending limb but has quite different physiological properties. A principal difference is that this epithelium no longer contains aquaporin-1 water channels, and the tight junctions between cells are no longer leaky. The epithelium is essentially impermeable to water.

Due to the egress of water from the descending limb of the loop of Henle, the concentration of NaCl within the tubular fluid increases greatly until it is almost twice that of the interstitial fluid. Through passive diffusion, much of the NaCl leaves the ascending limb and enters the interstitial fluid. The interstitial fluid of the medulla contains a high concentration of urea, but the epithelial cells permit little urea to enter the tubule at this point. Closer toward the cortex, the cells of the ascending loop of Henle thicken to a low cuboidal shape. Within an apical membrane containing a high concentration of microvilli are NaKCl2 cotransporter molecules, and near the basolateral border are numerous mitochondria, although not so numerous or as complex in structure as those of the proximal tubules. The cotransporter actively removes both Na^+ and Cl^- from the tubular fluid, with the result that by the time the fluid reaches the upper end of the loop of Henle it is actually hypoosmotic to the interstitial fluid.

Distal Convoluted Tubule

By the end of the thick ascending limb of the loop of Henle, the nephron has again reentered the cortex. Its fluid consists of only 10% of the volume filtered at the glomerulus and less than 10% of the filtered solutes; it is thus hypotonic to normal body fluids. The fluid then enters the **distal convoluted tubule**. The epithelial lining cells of the distal tubule are specialized for ion exchange. Although they have few apical microvilli, they have a highly complex basolateral membrane with many folds and the highest density of mitochondria and basal Na^+, K^+-ATPase of any segment of the nephron. The epithelium is still impermeable to water, and it continues to remove NaCl from the tubular fluid, thus making the fluid even more hypotonic. Within the cortex, the distal convoluted tubule then empties into the first limb of a system of collecting tubules and ducts.

Collecting Tubules and Ducts

Like tributaries to a stream, the individual collecting tubules in the cortex merge into a larger straight **collecting duct** that drops through the medullary pyramid and leaves through its apex as the **duct of Bellini**. Most of the collecting duct system is lined by a cuboidal epithelium containing two types of cells—**principal cells** and **intercalated cells** (see Fig. 13.7). In addition to a basolateral membrane adapted for Na^+,K^+ exchange, the apical surface of a principal cell contains a **primary cilium**, which acts as a mechanosensor of fluid flow. Principal cells are also heavily involved in K^+ secretion into the lumen. Intercalated cells have apical microvilli and contain many mitochondria. They engage in K^+ and HCO_3^- secretion out of and H^+ secretion into the collecting duct.

An important characteristic of the epithelium of the collecting duct is that its secretory activity is heavily influenced by the adrenal cortical hormone **aldosterone**. Aldosterone, which acts through the nucleus of the cells, increases K^+ secretion and Na^+ reabsorption by the principal cells and H^+ secretion into the lumen by the intercalated cells. Overall, aldosterone serves to protect the composition of body fluids, and under conditions of dehydration the adrenal gland secretes more aldosterone, which causes the cells of the collecting ducts to secrete more K^+ into the forming urine. Because of their responsiveness to external agents, such as aldosterone, the collecting duct is also the target of diuretic medications.

The collecting duct system is also the region where urine is acidified. This is accomplished by the secretion of H^+ by the intercalated cells into the collecting duct.

One important process in the final production of urine is a last phase of removal of water in the deep medullary part of the collecting duct system. This is accomplished by the principal cells, operating under the influence of **antidiuretic hormone (ADH)**, produced by the posterior pituitary gland (Box 13.3). In addition to maintaining bodily homeostasis through the selective resorption and secretion of many critical solutes, the kidney serves the vital role of regulating the vast majority of water that is retained by or lost from the body. Understanding the ins and outs of water flow in the kidney requires a knowledge of the osmotic gradient found in the medullary pyramid and the basis for it.

BOX 13.3 ADH and Diabetes Insipidus

The basolateral surfaces of the principal cells within the collecting ducts contain aquaporin-based water channels and are always permeable to water. Permeability of the apical surfaces, however, depends upon the presence or absence of ADH. In the absence of ADH, the apical surfaces of the principal cells of the collecting ducts are impermeable to water. When ADH is present, aquaporin-2-based water channels that have been stored in vesicles close to the apical membrane are inserted into the membrane. This allows water to flow from the urine into the cells. The water then flows through the basolateral water channels and into the medulla of the kidney because of the high osmotic pressure of the interstitial fluids of the medulla. Upon removal of ADH, the water channels are taken up by the sub-membrane vesicles and are stored in them until needed. In addition to its effect on water uptake, ADH also increases the removal of urea from the tubular fluid deep in the medulla. The removed urea becomes concentrated in the deep renal medulla.

If ADH is absent or is secreted at less than 80% of normal levels, water removal from the collecting ducts fails, and the body excretes massive amounts of urine—typically more than 4 L/day. This condition is called **diabetes insipidus** and is accompanied by great thirst and dehydration in the absence of increased fluid intake. Less common causes of diabetes insipidus are genetic defects affecting the water channels in the collecting ducts. In these cases the patients may have normal levels of circulating ADH, but in the absence of receptors or with defective water channel genes, the hormone is without effect on the kidneys.

The Renal Medulla and Its Countercurrent Multiplier System

A key element in urine formation is the gradient of osmolality present in the renal medullary pyramids (Fig. 13.9). The formation and maintenance of this gradient is the result of a countercurrent mechanism that involves both the loop of Henle and its accompanying vasculature.

As described earlier, the loop of Henle is a tight hairpin loop that extends from the corticomedullary junction to the inner medulla. Accompanying the loop is a similar hairpin-looped vascular configuration, called the **vasa recta**. Both the loop of Henle and the vasa recta work in concert to maintain the osmotic gradient.

The prime mover in the **countercurrent multiplier system** is the loop of Henle, where NaCl is actively removed from the ascending loop and enters the medullary interstitium. This, along with a high concentration of urea, coming from both passive reabsorption from the proximal tubule and removal in the deep medullary collecting ducts, results in a fourfold concentration of solutes (NaCl and urea) in the deep medulla over that of the cortex. The increasing concentration gradient within the medullary interstitium provides the basis for the large amount of water removal from the descending limb of the loop of Henle. With water entering the system, one would expect that it would dilute the concentration gradient, causing it ultimately to disappear. This is where the hairpin loop of the vasa recta comes in. As blood flows down the loop, water leaves and NaCl and urea enter the blood, but as the blood passes the tip of the loop, the situation becomes reversed. Water now reenters the blood within the ascending loop and NaCl and urea leave. A greater inflow of water from the medullary interstitium into the ascending limb of the vasa recta results in a net increase of about 15%–25% of the volume of water per minute leaving the medulla in comparison to the volume of blood that entered. By leaving much of the solutes behind, this removal of water allows the osmotic gradient of the medullary interstitium to be maintained.

Countercurrent mechanisms are important in other aspects of physiological control, as well. Another example in humans involves the maintenance of a temperature gradient in the testes, where a lower than normal body temperature is required for proper sperm production. A closely spaced arrangement of parallel straight segments of small arteries and veins with blood flowing in opposite directions (the pampiniform plexus) removes heat from the blood and keeps the testes several degrees cooler than the rest of the body. A similar mechanism operates in the feet and legs of aquatic birds to prevent them from losing valuable body heat while they are swimming in freezing water.

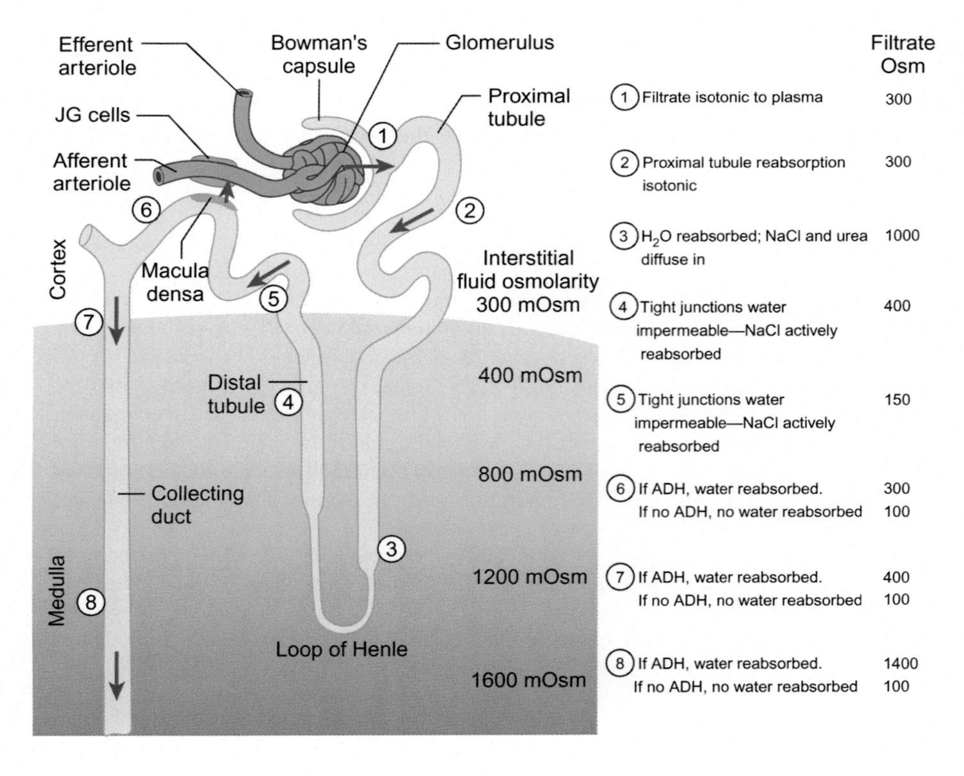

FIGURE 13.9 Secretion/absorption functions in various regions of the nephron in relationship to the gradient of osmolarity within the renal medulla. *From Carroll (2007), with permission.*

OTHER FUNCTIONS OF THE KIDNEYS

The JGA and the Control of Blood Pressure

A small area situated near the glomerulus plays a vital role in the modulation of blood pressure throughout the body. Called the **juxtaglomerular apparatus** (**JGA**, see Fig. 13.6), it consists of a small region of the distal convoluted tubule (macula densa), which is situated very close to the afferent arteriole of its nephron. The characteristic cells of the macula densa lie adjacent to a segment of the afferent arteriole whose modified smooth muscle cells are called **granular cells**. Completing this cellular aggregate is a group of **mesangial cells** which are connected to each other and the other components of the JGA by gap junctions (which facilitate intercellular communication). Sympathetic nerve terminals in the area complete the structure of the JGA.

The granular cells of the afferent arteriole make and secrete into the blood **renin**, an enzyme that, as will be seen later, is critical in the control of blood pressure. Several stimuli can trigger the release of renin. One is the mechanical sensing of low blood pressure by the wall of the afferent arteriole itself, which acts as a baroreceptor. A second is a signal from the macula densa of the distal convoluted tubule. Cells of the macula densa sense the concentration of NaCl in the fluid passing through the distal convoluted tubule, and if the concentration is low, they signal to the granular cells to release renin. A third stimulus comes from the sympathetic nerve terminals, which fire if reduced blood pressure causes the baroreceptor reflex to stimulate medullary control centers in the brain stem to fire the sympathetic neurons supplying the JGA.

Regardless of the nature of the stimulus, once renin is released into the general circulation it stimulates the activation of the **renin–angiotensin–aldosterone system** (Fig. 13.10). Renin acts as an enzyme that breaks down **angiotensinogen** (a large protein produced by the liver) into **angiotensin I**, a decapeptide that is still biologically inactive. Within the vascular epithelium of the lung and some other locations, another enzyme, angiotensin-converting enzyme, cleaves off two more amino acids, leaving octopeptide **angiotensin II**, a highly biologically active molecule. Angiotensin II acts indirectly through the hypothalamus of the brain and directly upon the adrenal cortex to stimulate the production of aldosterone. In addition, angiotensin II directly causes vasoconstriction, thereby increasing blood pressure. It also acts on the kidney by stimulating the cells of the proximal tubule to enhance their reabsorption of NaCl, and it indirectly affects the kidney by stimulating the release of ADH from the posterior pituitary.

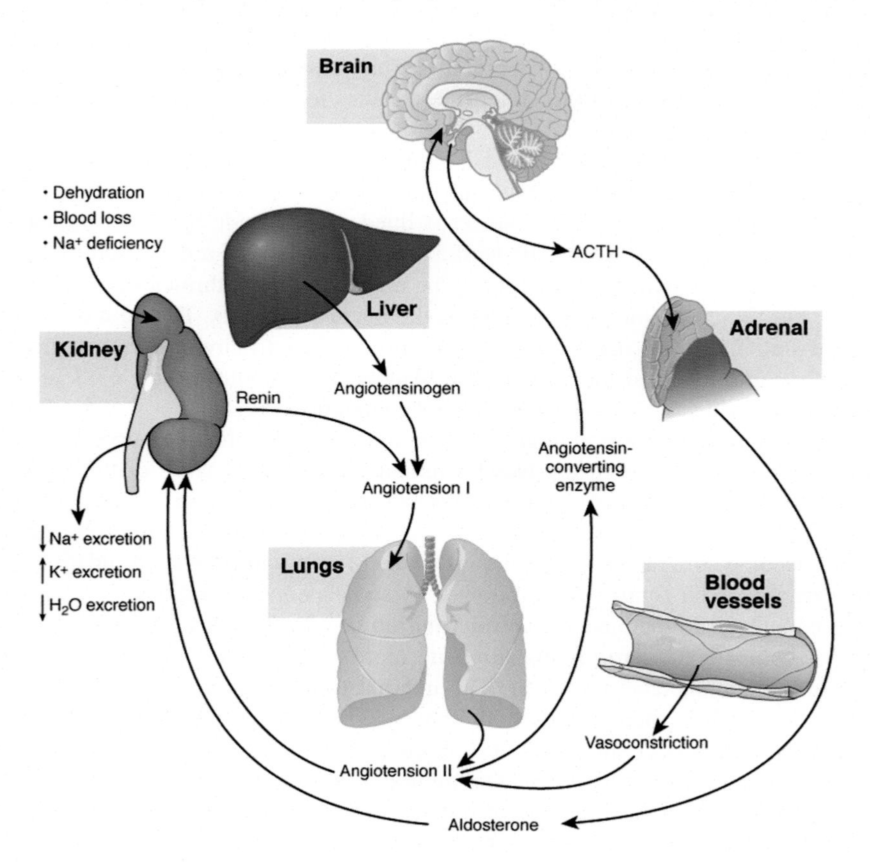

FIGURE 13.10 Interactions among the renin–angiotensin–aldosterone system in the maintenance of blood pressure in the face of blood loss, dehydration, or Na^+ deficiency.

THE STORAGE AND TRANSPORT OF URINE

By the time the urine has reached the end of the collecting duct, its final volume and composition of solutes is set and does not change as it is transported through the remainder of the urinary system. From the collecting ducts, the urine passes through a series of progressively larger channels (**minor** and **major calyces**) until it reached a common collecting area, the renal pelvis, which funnels the urine in to the ureter (see Fig. 13.2A).

Ureter

The **ureter** is about 25 cm long and less than 5 mm in diameter, and it transports urine from the kidney into the urinary bladder. Like all the urine transport channels, as well as the urinary bladder, the lumen of the ureter is lined with a special type of epithelium, called **transitional epithelium** (see p. 32), that is highly adapted for stretch. From the calyces in the kidney to the bladder, layers of smooth muscle underlie the mucosa. This smooth muscle acts as a functional syncytium, which produces peristaltic waves of contraction. Arising from electrical pacemakers in the renal pelvis, these waves of contraction propel urine down the ureters in series of squirts with a frequency of 2–6/min. In addition, mechanical stretch anywhere along the transport pathway can elicit smooth muscle contractions. The smooth muscles of the ureteral walls are well supplied with autonomic nerve endings, which modulate the smooth muscle contractions. Parasympathetic stimuli enhance contractions, whereas sympathetic signals inhibit them. Pressure receptors in the walls of the ureters convey a strong sense of pain to the central nervous system, which is why the passage of a **kidney stone** causes such excruciating pain. Such an event triggers a **ureterorenal reflex**, in which increased pressure within a ureter stimulates a sympathetic discharge that not only reduces the smooth muscle contractions, but also reduces the rate of glomerular filtration, thus reducing urine buildup in an obstructed ureter.

Urinary Bladder

The ureters enter the **urinary bladder** through its back wall, in an area called the **trigone** (Fig. 13.11). Several features of the trigone work in concert to prevent the backflow of urine from the bladder into the ureters. The ureters pass through the wall of the trigone at a shallow angle, allowing the bundles of smooth muscle in the wall of the bladder (sometimes called the **detrusor muscle**) to press on the terminal part of the ureter. This pressure flattens the lumen of the ureter and, along with a flap of bladder mucosa at the orifice of the ureter, these forces together effectively act as a sphincter sealing off the ureters from backflow of urine.

As urine enters the urinary bladder, the bladder begins to fill. The bladder is remarkably well adapted for stretching and contraction. The bladder rests in a cavity partially bounded by the symphysis pubis on three sides and partially resting on a layer of skeletal muscle, the pelvic diaphragm. The wall of the bladder is lined on the inside by a transitional epithelium that can range from 6 cells thick in an empty bladder to 2 cells with a squamous shape in a distended bladder. Minimal transport of either water or solutes occurs through this epithelium. The mucosal layer is thrown into prominent folds, called rugae, in the empty bladder, but these become stretched out as the bladder expands. The bulk of the wall of the bladder consists of several layers of smooth muscle, the cells of which are innervated individually by terminals of autonomic neurons. Yet in the detrusor muscle wall at the back of the bladder, the smooth muscle cells are interconnected by low resistance electrical pathways, which allow more coordinated contraction.

As the empty bladder begins to fill, the wall stretches correspondingly to accommodate the volume of urine contained within it. At low bladder volumes, the fluid pressure within the bladder remains very low because of the high degree of compliance (stretchability) of the bladder, and sympathetic stimuli relax the smooth muscle fibers within the wall. When the volume of urine rises to over 200 mL and the internal pressure of the bladder correspondingly rises, stretch receptors within the wall of the bladder become stimulated. Under the influence of parasympathetic stimulation, the muscular wall of the bladder begins to undergo periodic contractions, followed by spontaneous relaxations.

By the time 400–500 mL of urine have accumulated in the bladder the **micturition (voiding) reflex** becomes activated. Signals from the spinal cord through parasympathetic nerve fibers stimulate more persistent contractions of the detrusor muscle. Then neural control of two urinary sphincters takes over. The internal urinary sphincter is based on slips of smooth muscle that extend from the bladder wall to the upper urethra. Two layers of longitudinal smooth muscle surround an internal circular layer of smooth muscle and elastic fibers. The circular muscle functions as a constrictor. When the longitudinal smooth muscle contracts, it shortens and widens the urethra, allowing urine to pass.

As the bladder is filling, signals from sympathetic nerve fibers keep the internal sphincter constricted, but stretch signals that go from the bladder to the pons in the brain stem stimulate inhibitory signals to the sympathetic nerve

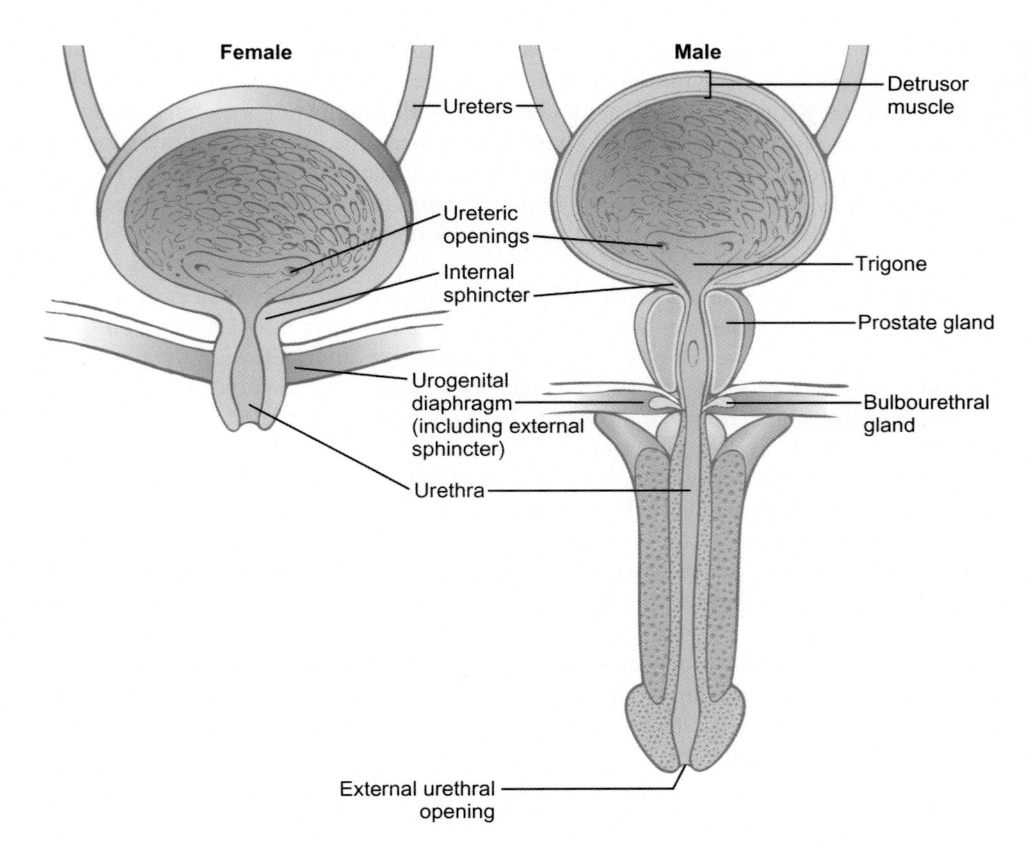

FIGURE 13.11 Structure of the bladder and urethra in males and females. *From Guyton and Hall (2016), with permission.*

fibers, resulting in relaxation of the muscles of the internal sphincter. If a person wishes to urinate, then the external urinary sphincter comes into play. This sphincter consists of skeletal muscle fibers that are under conscious cerebral control and are kept constricted by signals coming from the pons. Upon the desire to urinate, these signals cease, and the external sphincter relaxes. At almost the same time, parasympathetic stimulation causes the detrusor muscle of the bladder wall to contract and the muscle of the internal sphincter to relax. At this point, urine leaves the bladder and enters the urethra.

Urethra

The **urethra** in the male is about 20 cm long (Fig. 13.11). It arises from the trigone of the bladder and consists of three segments: the prostatic, membranous, and penile urethra. Within the short (3–4 cm) prostatic segment, the male urethra receives the ejaculatory ducts as well as many small ducts leading form the prostate gland. The even shorter membranous urethra passes through the external urinary sphincter as it continues within the shaft of the penis as the penile urethra. The ducts of the bulbourethral glands (see p. 380) empty into the penile urethra close to its origin. The epithelium of the urethra changes from transitional within the prostate to stratified columnar throughout the rest of its length. In males, the last drops of urine are forced through the urethra by contractions of the bulbocavernosus muscle at the base of the penis. The female urethra is much shorter ($\sim$4 cm), and like that of the male, its epithelium changes from transitional to pseudostratified columnar, before changing again to a stratified squamous form close to the external opening (meatus). In the female urethra, small mucous glands empty into the urethra. This may be a mechanism that protects against urinary tract infections, which are nevertheless more common in females than in males.

SUMMARY

The urinary system consists of the kidneys, the ureters, the bladder, and the urethra. The process of urine production is part of an overall maintenance of bodily homeostasis.

Vertebrate kidneys have evolved through three stages. Pro- and mesonephric kidneys are found in fish-like aquatic animals and are present for a time in human embryos. The metanephric kidney is the adult form of the mammalian kidney. A major function of the metanephric kidney is to preserve water and solutes while eliminating metabolic waste.

Within the kidney, urine formation begins with filtration of blood within the glomerulus. Blood enters the glomerulus through afferent arteriolar branches of the renal artery and reaches the glomerulus under arterial pressure. The glomerulus is embedded in a double-walled Bowman's capsule, which is the terminal end of the nephron. Because of fenestrations in the glomerular capillary endothelium, essentially all components of blood except for cells leave the capillaries.

The epithelial cells (podocytes) of Bowman's capsule are structurally very complex, with many projections extending from the cell body. The basal lamina filters out molecules with a molecular weight greater than 1000 Da. Filtration slits between processes extending from the foot processes of podocytes act as a selective filter, allowing the passage of water, ions, urea, glucose, amino acids, and small proteins and peptides into the lumen of the nephron. Over 24 hours, 180 L of fluid is filtered through the glomerulus and Bowman's capsule. Mesangial cells associated with the afferent and efferent arterioles can reduce the flow of blood to the glomerulus.

The nephron consists of Bowman's capsule, the proximal convoluted tubule, the loop of Henle, and the distal convoluted tubule. When the filtrate enters the proximal convoluted tubule from Bowman's capsule, two-thirds of the filtrate, including water, ions, and small molecules, is resorbed. Almost 100% of the filtered glucose, amino acids, and small peptides are absorbed in this region. Resorption is an active process that is facilitated by high microvillous projections from the apical surfaces of the epithelial cells and by metabolically active absorptive processes. Urea is both secreted and absorbed in the proximal tubule.

The loop of Henle, which extends deeply within the hyperosmotic medulla of the kidney, is a region where significant water loss back into the tissues occurs. The epithelium of the ascending loop of Henle is impermeable to water. Through passive diffusion, much of the Na^+ and Cl^- ions leave the ascending loop of Henle. A countercurrent multiplier system involving a vascular network surrounding the loop of Henle plays a very important role in ion exchange.

The epithelium of the distal convoluted tubule remains impermeable to water, and its contents become hypotonic with the continued removal of ions. Selective ion exchange continues in the collecting tubules and collecting ducts. The nature of this exchange is heavily influenced by aldosterone, secreted by the adrenal cortex. Urine is acidified in the collecting duct system. Under the influence of antidiuretic hormone (ADH) (vasopressin) from the posterior pituitary, the last phase of water removal from the urine occurs in the deep part of the collecting duct system.

In addition to making urine, the kidneys also play an important role in controlling blood pressure through the juxtaglomerular apparatus (JGA), located within the epithelium of the distal convoluted tubule. Next to this region is a group of mesangial cells associated with the afferent arterioles. Granular cells of the afferent arterioles secrete renin into the blood. Renin enters the liver via the blood and converts angiotensinogen into angiotensin, which both stimulates the production of aldosterone by the adrenal cortex and directly causes vasoconstriction and a resulting increase in blood pressure.

Newly formed urine is transported from the kidney to the bladder through the ureters by waves of peristaltic contraction by smooth muscle elements in the ureteral walls. The urine then enters the bladder, where it remains until it is voided.

The transitional epithelium of the bladder is designed to accommodate stretching, and it returns to a nonstretched configuration after the bladder has emptied. A full bladder activates the micturition reflex. This reflex relaxes the muscle of the urinary sphincters, and urine leaves the bladder through the urethra. The female urethra is much shorter than that of males, which passes through the prostate gland at the base of the bladder.

Chapter 14

The Reproductive System

The reproductive system exists for the sole purpose of perpetuating the species. For purely genetic reasons, hundreds of million years ago, animals abandoned fission as a strategy for reproduction in favor of sexual reproduction. This change in strategy necessitated the evolution of separate sexes. The main function of females is to produce eggs; the main function of males is to produce sperm, which fertilize the eggs. Beyond these basic functions, animals have evolved an amazing variety of strategies for not only producing gametes, but bringing them together.

One of the simplest modes of reproduction is for females to produce huge numbers of eggs and then spread them around a favorable aquatic environment. The male then distributes its sperm in the same general area. Quantitatively, this is a very inefficient mode of reproduction, and in many species, a 1:1000 chance of an egg's producing a viable offspring would be considered good. Nevertheless, this reproductive strategy makes sense for a lot of aquatic species. With the appearance of land vertebrates, reproductive strategies changed dramatically. Internal fertilization eliminated the need for the female to produce large numbers of eggs, although males still need to emit huge numbers of sperm cells in order to ensure fertilization. Reptiles and birds lay eggs, so the female's direct role in reproduction ends when the fertilized eggs are deposited into a nest. With the exception of the platypus and a few other species, reproduction for female mammals does not end with fertilization. Instead, the reproductive process continues with intrauterine embryonic development and does not end until birth. Even after birth, the reproductive system in females continues a vital function in nursing the infant. For all stages in the reproductive process, the structure of the male and female reproductive systems is exquisitely adapted for the production of gametes and, in the female, for harboring and nourishing the developing embryo.

THE MALE REPRODUCTIVE SYSTEM

The male reproductive system has two principal functions—producing large numbers of spermatozoa and depositing them in a location where they will have access to the egg(s). Sperm production occurs in the testes, which are connected to a series of ducts that carry mature spermatozoa to the outside during the act of copulation. The penis evolved as a means for introducing the sperm into the female reproductive tract. In embryos, many of the same molecules that are involved in the formation of limbs are also utilized in formation of the penis.

Sperm Production (Spermatogenesis)

Each day, tens of millions of spermatozoa are produced in the testes. The testes themselves are located in the **scrotum**, a sac-like structure that in most mammalian species suspends the testes outside the body cavity (Fig. 14.1). Some exceptions to the rule are elephants and marine mammals. The scrotum is divided into two halves, each containing a testis, separated by a central connective tissue septum. Why so many mammals have a scrotum at all is a significant question. According to one hypothesis, the scrotum is an adaptation to protect the testes from being jostled by intra-abdominal movements and prematurely releasing sperm.

We do know that within the scrotum the temperature surrounding the testes is approximately 2°C cooler (35°C) than that of the abdominal cavity and that this lower temperature is more favorable for sperm production than the normal body temperature of 37°C. What we do not know is why a lower ambient temperature is necessary for ideal sperm production. Did this requirement lead to the evolution of the scrotum, or is it an evolutionary consequence of the testes' residing in the cooler scrotum?

Regardless of the evolutionary uncertainty, three structural and functional adaptations promote temperature control within the scrotum. The first is the **cremaster muscle**, a series of slips of the internal oblique muscle, which surround

The Human Body. DOI: https://doi.org/10.1016/B978-0-12-804254-0.00014-4

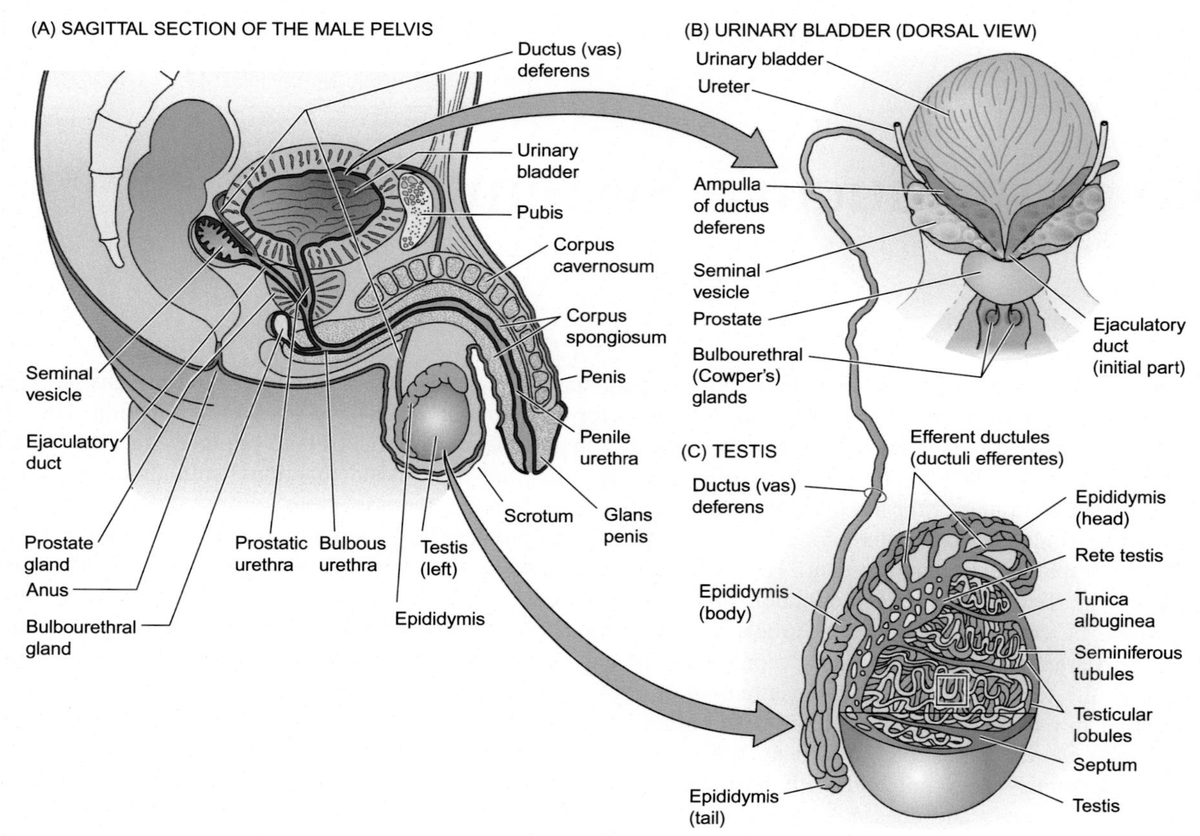

FIGURE 14.1 Anatomy of the male genitalia and accessory sex organs. *From Boron and Boulpaep (2012), with permission.*

the spermatic cord. This muscle contracts in cold temperatures, bringing the testes closer to the abdomen, and relaxes when the outside temperature is warm, causing the scrotum to sag and the testes to hang farther from the abdomen. The cremaster muscle also brings the testes closer to the abdomen during sexual arousal. A second muscle, the dartos, is a smooth muscle that forms a thin layer beneath the skin of the scrotum. This muscle, which reacts to temperature both directly and through its sympathetic innervation, is slower to contract than the cremaster. The result of its contraction is shriveling of the scrotal skin, thus reducing the surface area of the scrotum in a temperature-saving response to cold. The third temperature control mechanism within the scrotum is a vascular network called the **pampiniform plexus** (Fig. 14.2). This plexus forms a countercurrent system (see p. 368) in which a network of veins is situated closely parallel to the testicular artery. Through this configuration warm arterial blood coming from the abdomen transfers some of its heat to the cooler blood of the veins returning from the scrotal structures. As a result of the transfer, the arterial blood bathing the testes is cooler. Such a countercurrent heart exchange plays a major role in maintaining the lower environmental temperature needed for optimum sperm development.

Situated within the scrotum, the testes are the site of sperm production. Each testis is covered by a thin layer of dense connective tissue (**tunica albuginea** [see Fig. 14.1C]). The nonexpansibility of this layer is the main reason why mumps in mature males can be such a painful condition, because swelling within the testis (**orchitis**) itself cannot be accommodated by a corresponding expansion of its connective tissue covering. Internally, each testis is subdivided into 250–300 lobules, consisting of a connective tissue lining that contains one to four tightly packed seminiferous tubules (Fig. 14.3). These tubules are the sites of sperm production. The U-shaped **seminiferous tubule** is about 80 cm long, with each end emptying into the rete testis, a meshwork of small channels leading out from the testis. Scattered among the seminiferous tubules are small clumps of **Leydig (interstitial) cells**, which produce testosterone.

The highly complex internal structure of a seminiferous tubule is a reflection of the many important cellular events that take place within it. Its main function is to produce spermatozoa from ordinary-looking stem cells. Unseen within this remarkable transformation is the process of **meiosis**, the genetic resetting of the chromosome number so that a mature spermatozoon contains only half the number of chromosomes of the other cells that make up the body (Fig. 14.4). The same set of meiotic events occurs within the developing egg. As a result, when the sperm and egg

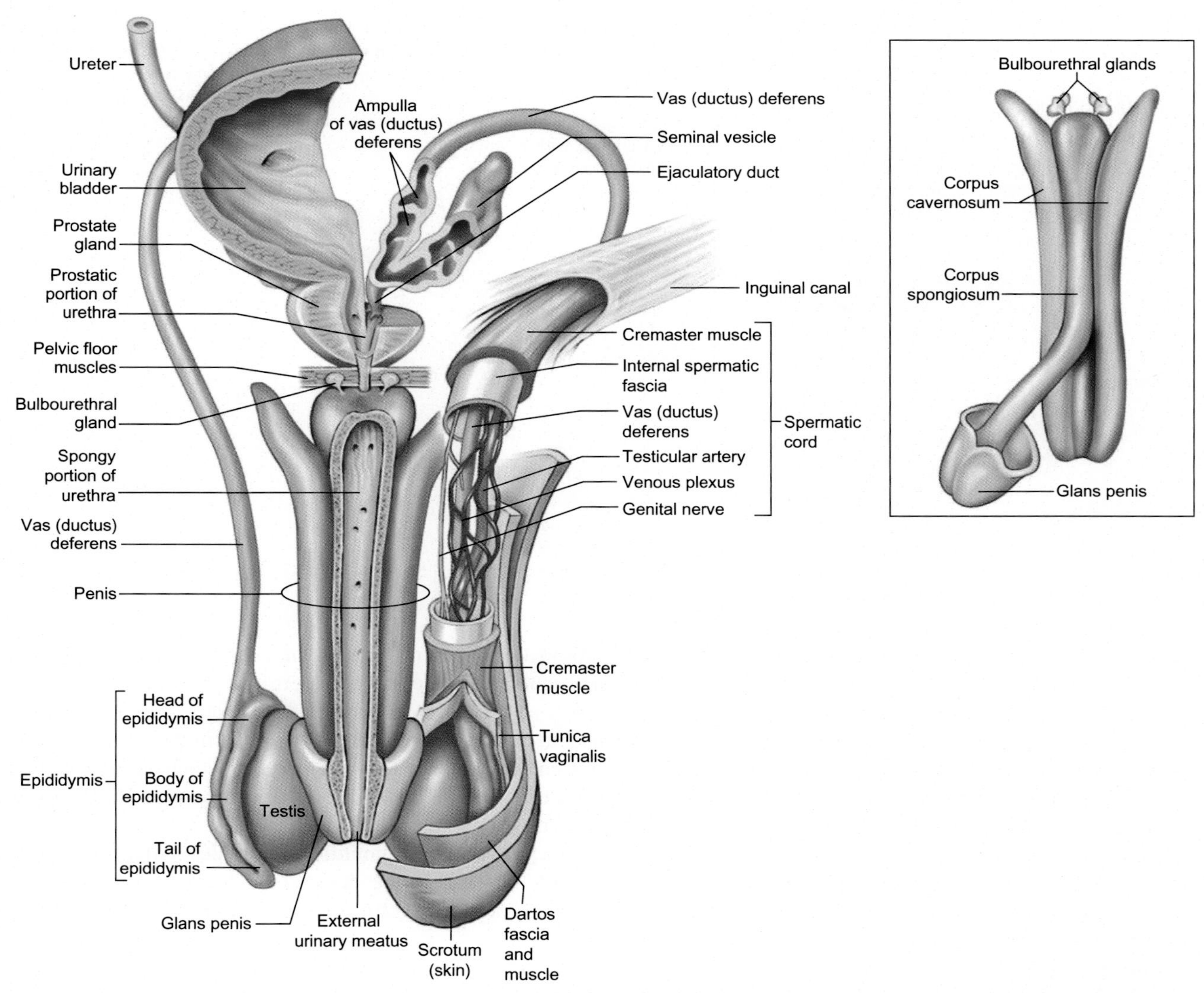

FIGURE 14.2 The pampiniform plexus (venous plexus) surrounding the testicular artery and forming a countercurrent heat-removal system within the spermatic cord. *From Thibodeau and Patton (2007), with permission.*

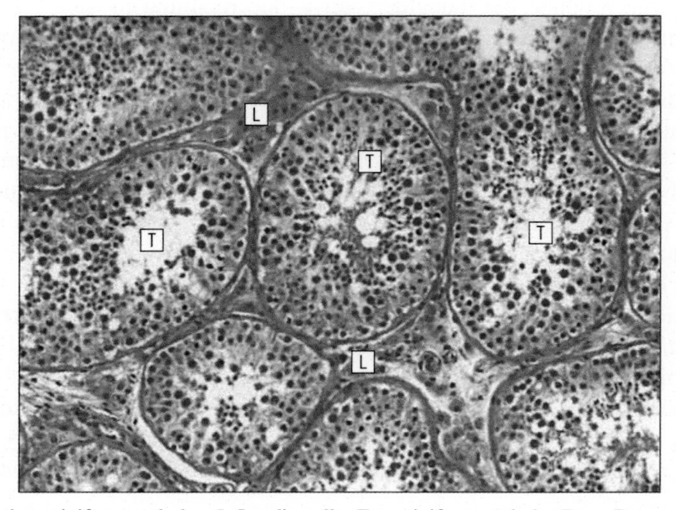

FIGURE 14.3 Cross-section through seminiferous tubules. *L*, Leydig cells; *T*, seminiferous tubule. *From Stevens and Lowe (2005), with permission.*

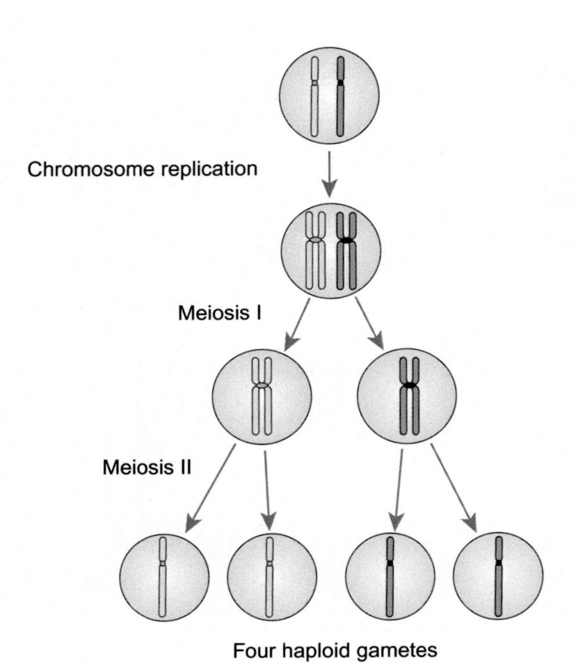

FIGURE 14.4 The essential steps in meiosis. The process begins with one round of chromosome replication. During the first meiotic division, one replicated chromosome goes to one daughter cell, and the other one goes to the second daughter cell. The second meiotic division involves the splitting of each chromosome resulting from the first meiotic division, and one half would go to each of the daughter cells. At the end of this process, each cell contains only half the number of chromosomes of the premeiotic cell. Compare with a mitotic division, as seen in Fig. 1.22A. *From Nussbaum et al. (2016), with permission.*

come together during fertilization the resulting **zygote** (a fertilized egg) again contains the normal number of chromosomes. In the absence of meiosis, the number of chromosomes would double every time fertilization occurred.

A cross-section through a seminiferous tubule (see Fig. 14.3) shows a thin covering of connective tissue, consisting of collagen fibers, fibroblasts, and myoid cells, which are capable of contraction and are involved in the initial stages of sperm transport (see later). Inside the covering are two main categories of cells—**germ cells**, which ultimately form spermatozoa, and **Sertoli cells**, specialized somatic cells that are vital to the process of sperm formation. In the center of the seminiferous tubule is a lumen, which contains liquid and newly formed spermatozoa.

Before puberty, Sertoli cells constitute the predominant cell type within the seminiferous tubules. After puberty, the ratio is reversed, and about 90% of the cells lining the seminiferous tubule are devoted to sperm production. Until puberty, male germ cells are relatively quiescent. Only after puberty does the germ cells begin to multiply in earnest—first by mitosis and then by the specialized process of meiosis.

Spermatogenesis (sperm formation) is a complex, but very orderly process taking about 64 days in the human (Fig. 14.5). Spermatogenesis begins with local stem cells (sometimes called **prespermatogonia**), located at the outer edge of the seminiferous tubule. These cells are diploid (containing the normal number of chromosomes [23 pairs] seen in somatic cells) and have few distinguishing morphological characteristics. Through the process of ordinary cell division (mitosis), prespermatogonia produce two daughter cells—one retaining its stem cell characteristics, whereas the other is committed to forming spermatozoa. The latter cell, now called a **spermatogonium**, divides mitotically several times to form eight progeny—all connected by cytoplasmic bridges. The cytoplasmic bridges allow all these cells to develop and divide in a synchronous fashion throughout their entire transformation into spermatozoa.

The next stage in spermatogenesis is a critical one. Each spermatogonium enters the process of meiosis, and while doing so, it becomes situated inside a **blood-testis barrier** formed by interlocking Sertoli cells. Within this barrier developing sperm cells can become immunologically foreign without eliciting an autoimmune rejection response from the body. While undergoing the first meiotic division, a process taking several weeks, the former spermatogonium produces two **primary spermatocytes**. Under a microscope, they still look very much like ordinary somatic cells, but their number of chromosomes has been reduced by half.

The second meiotic division, which takes only 8 hours, produces two **secondary spermatocytes** from each primary spermatocyte. During these meiotic divisions, the spermatocytes are pushed farther toward the central lumen of the seminiferous tubule. By the completion of the second meiotic division, the two daughter cells from each secondary spermatocyte, now called **spermatids**, now contain only 23 single nonduplicated chromosomes. This represents the genetic complement allotted to each spermatozoon. A spermatid still looks like a small ordinary somatic cell, but over a few weeks, each spermatid undergoes a profound morphological transformation (**spermiogenesis**) into a tadpole-shaped

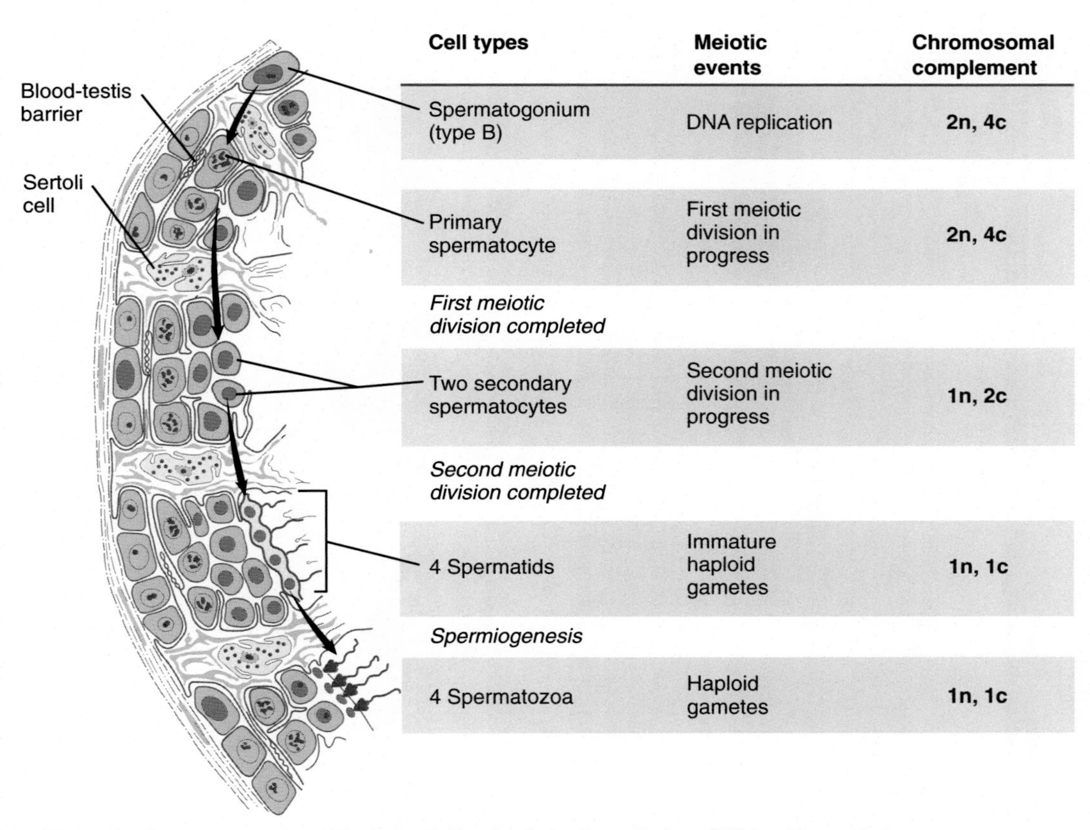

FIGURE 14.5 The cycle of spermatogenesis within the seminiferous tubule. *From Carlson (2014), with permission.*

spermatozoon, which is released into the lumen of the seminiferous tubule. During spermiogenesis, much of the cytoplasm of the spermatids is shed and subsequently phagocytized by the Sertoli cells.

Spermatozoa are remarkable cells. They have been stripped of anything that would interfere with their function of fertilizing an egg. A spermatozoon consists of a **head**, a **midpiece,** and a **tail** (Fig. 14.6). The head contains a nucleus with tightly packed chromosomal material and a cap-like structure, called the **acrosome**, which contains enzymes that are needed for penetrating the coverings of the egg during the fertilization process. The midpiece is a region dominated by a tight coil of mitochondria, which generate the energy needed for the swimming movements of the sperm. Interestingly, a mature spermatozoon does not possess an internal energy supply. For this, it must rely upon external energy sources present in the seminal fluid. The tail consists of a long flagellum constructed much like a cilium and which energetically beats to propel the sperm. Although morphologically mature, a newly formed spermatozoon is incapable of fertilizing an egg because it is not functionally mature. Functional maturation occurs as the sperm are transported from the testis into the epididymis (see later).

During the entire process of spermatogenesis, developing germ cells are intimately bound to Sertoli cells. Individual Sertoli cells have a complex morphology. Each Sertoli cell is firmly connected to the outer wall of the seminiferous tubule, but yet extends almost to the central lumen of the tubule. Close to the outer wall of the tubule, adjacent Sertoli cells tightly interconnect to form the **blood-testis barrier**. Sandwiched between the Sertoli cells and the outer wall of the tubule are the spermatogonia and the stem cells, all located outside the blood-testis barrier. As the spermatogonia become transformed into primary spermatocytes, the blood-testis barrier (cytoplasm of the Sertoli cells) shifts from the luminal side to the outside of these cells, at this point, segregating them from the immune environment of the body. At all remaining stages of sperm formation, the developing sperm cells are partially embedded in the Sertoli cells, which both support and nourish them.

Sertoli cells play an important role in male reproductive endocrinology. Stimulated by **follicle-stimulating hormone** (FSH), produced by the pituitary, they secrete **androgen-binding protein** and also influence testosterone production by Leydig cells in the testes. They also produce **inhibin**, a hormone that inhibits FSH production by the pituitary in a negative feedback fashion.

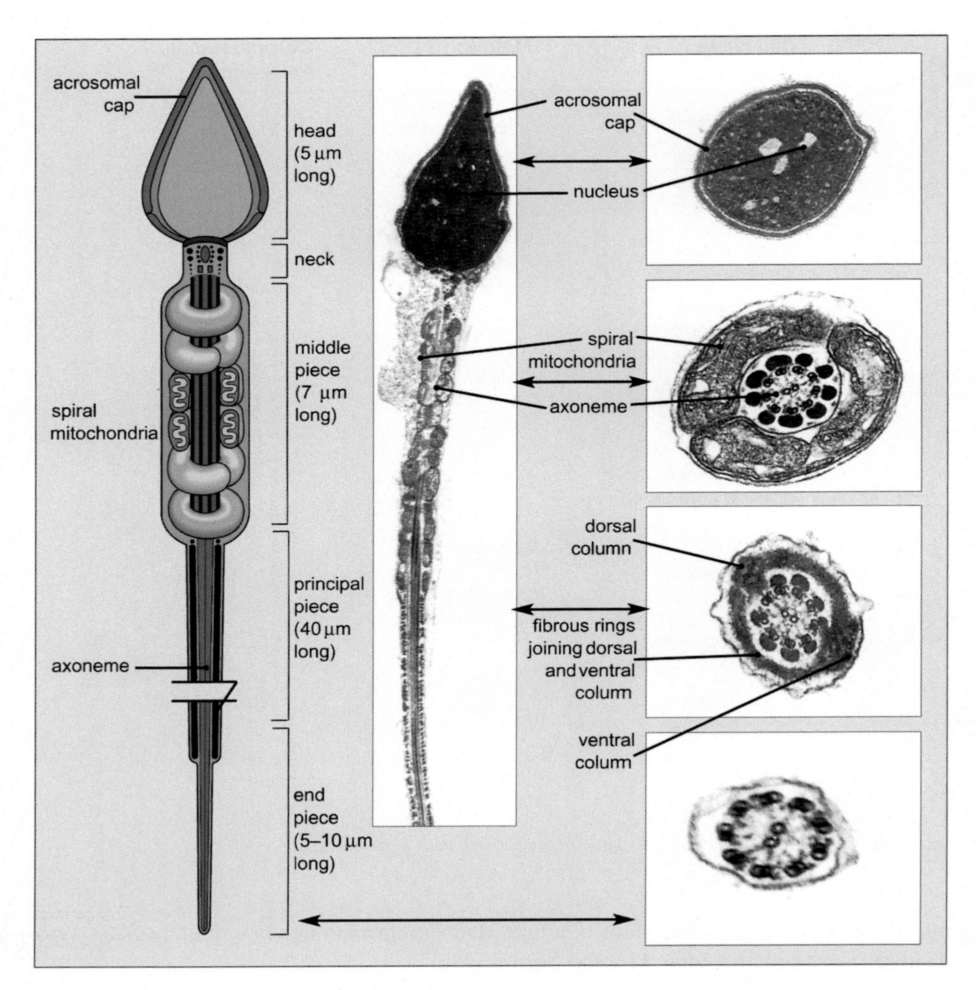

FIGURE 14.6 Diagram of a human spermatozoon. *From Stevens and Lowe (2005), with permission.*

Sperm Transport

Forming spermatozoa is the first stage in the male reproductive process. Next, the sperm must be transported from their site of origin to a place where they can encounter an egg. Three phases of sperm transport can be identified—two occurring in the body of the male and the third in the body of the female (Fig. 14.7). The first is a **slow phase**, during which morphologically mature spermatozoa are carried passively from the seminiferous tubules within the testis to the epididymis, a coiled tube lying upon the testis, where the spermatozoa undergo functional maturation. The structural substrate for this phase is sometimes called the **sperm maturation pathway**. A second **rapid active phase** is that of **ejaculation**, which rapidly propels the sperm from the epididymis through the ductus deferens and out of the erect penis. During the course of ejaculation, the sperm are joined by the secretions of several accessory reproductive glands, which provide nutrients for sperm motility and fulfill other functions within the female reproductive tract. In mammals, the third phase of sperm transport takes place within the female reproductive tract, where the sperm must travel from the site of insemination to the region containing the ovulated egg.

Slow Passive Sperm Transport

At the completion of spermiogenesis, spermatozoa are released, as the cellular interconnections among spermatids are broken and the residual bodies, consisting of cytoplasmic remnants of the spermatids, become disconnected from the spermatozoa and are phagocytized by the Sertoli cells. The newly minted spermatozoa float in the fluid within the lumen of the seminiferous tubule and are slowly propelled from it by the pressure of fluids secreted by the seminiferous tubules and by rhythmic contractions of the **myoid cells** that are part of the outer lining of the seminiferous tubule. Still within the testis, the immobile spermatozoa, along with testicular fluid, pass from the seminiferous tubules into a meshwork of channels called the **rete testis** (see Figs. 14.1 and 14.8). The cuboidal epithelial cells lining the rete testis look

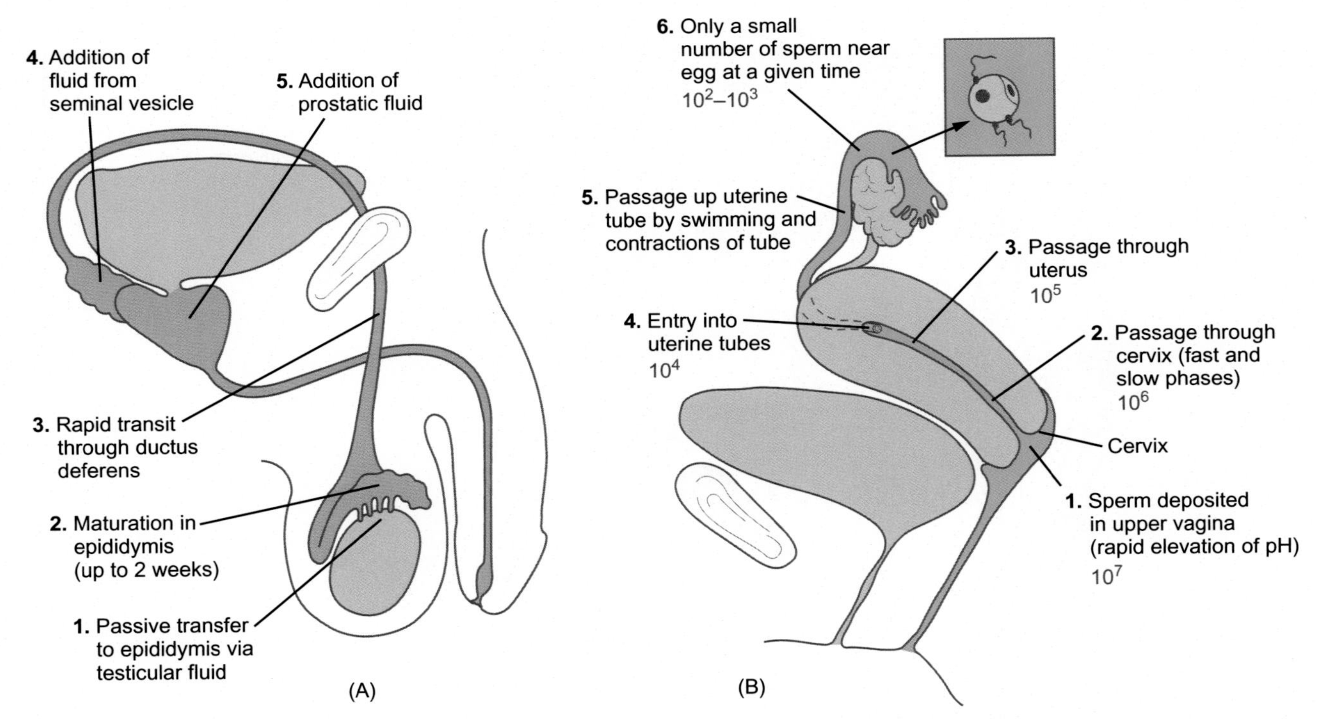

FIGURE 14.7 Pathways of sperm transport in (A) the male and (B) the female reproductive tracts. In B, numbers of spermatozoa typically found in the various parts of the female reproductive tract are indicated in red. *From Carlson (2014), with permission.*

rather unremarkable, but they are tightly interlinked on their apical border to maintain the blood-testis barrier that immunologically protects the sperm cells. From there, they leave the testis itself and enter one of about a dozen or more tiny **efferent ductules**, all of which empty into the epididymis. Assisting the passive transport of the spermatozoa through the efferent ductules is the coordinated action of cilia, which protrude from many of the epithelial lining cells. Other nonciliated cells within that epithelium are specialized for absorbing some of the testicular fluid.

Upon gross examination, the **epididymis** is a roughly cylindrical structure about 7–8 cm in length. In reality, it consists of a highly coiled duct that, if stretched out, would measure 6–7 m in length. It is divided into a **head**, a **body,** and a **tail**. During their ~20-day residence in the epididymis, the spermatozoa undergo functional maturation as they slowly pass from the head toward the tail. The epithelium lining the epididymal duct is a highly characteristic high columnar variety (Fig. 14.8). Protruding from the apical surface of these cells are long **stereocilia**, which by increasing the surface area of the epithelial cells facilitate the removal of excess testicular fluid from the duct. At the same time, glycoproteins, produced by the epididymis become attached to the surface of the heads of the spermatozoa. These surface molecules play a role in the fertilization process of humans and many other mammals. During their period of residence within the epididymis, spermatozoa gradually acquire the capacity for forward motility so that by the time they have progressed to the tail of the epididymis, they are fully capable of fertilizing an egg. Of the estimated 200 million sperm stored in the epididymis, half are in the head and body. These portions of the epididymal duct are lined with a thin layer of sparsely innervated smooth muscle. Through slow peristaltic waves, contractions of the smooth muscle fibers move the sperm and fluid down the epididymal duct toward the tail. The caudal (tail) portion of the epididymis is also lined with smooth muscle, but it is richly innervated by sympathetic nerve fibers, allowing a much faster and coordinated contraction of the smooth muscle in this region.

Fast Active Sperm Transport (Ejaculation)

The fast active phase of sperm transport has little to do with the swimming capabilities of spermatozoa since they are still swept passively along the male reproductive tract by external forces. Ejaculation consists of two phases—emission and expulsion. **Emission** carries a package of sperm from the tail of the epididymis through the **ductus deferens (spermatic duct)** and into the urethra. **Expulsion** propels semen (sperm plus glandular secretions) through the urethra and out from the penis.

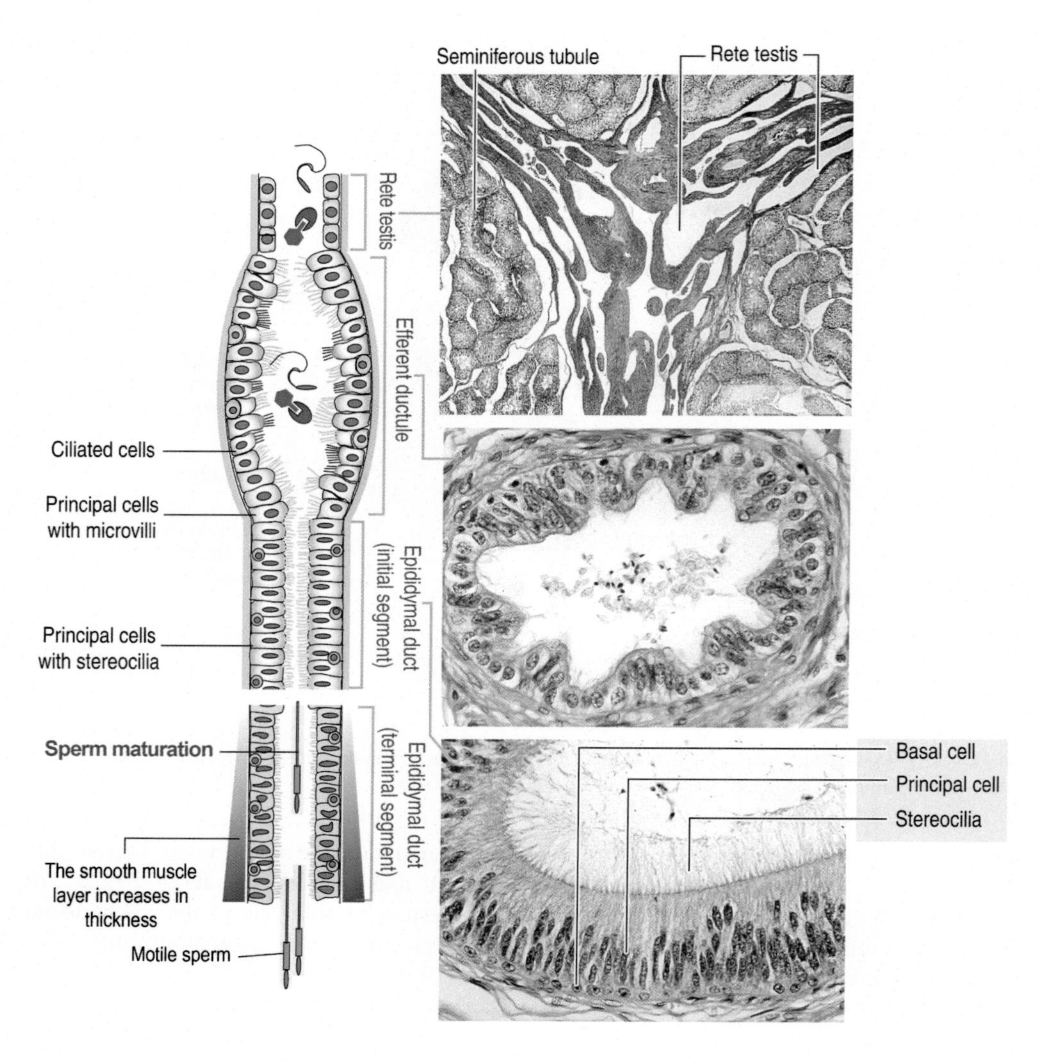

FIGURE 14.8 The male reproductive tract from the rete testis to the epididymis. Left, diagram; right, photomicrographs. *From Kierszenbaum and Tres (2012), with permission.*

Emission is initiated through stimulation of the smooth muscle cells lining the tail of the epididymis by sympathetic nerve fibers. This represents a shift in neural input from the early arousal stage of sexual excitement, which is based on signals from parasympathetic nerve fibers and results in erection of the penis. Contraction of the epididymal smooth muscle moves the sperm into the base of the ductus deferens, which is lined by thick layers of smooth muscle innervated by sympathetic nerve fibers (Fig. 14.9). Both the gross structure of the ductus deferens (straight and not highly coiled like the epididymal duct) and its thick muscular wall (inner and outer layers of longitudinally and a middle layer of circularly oriented muscle fibers) are adaptations for rapidly moving its contents (sperm and fluid) a relatively long distance—~40–50 cm.

By the time, the sperm reach the distal end of the ductus deferens, which ends in an expanded ampullary region, another element is introduced into the sperm transport process—namely, providing both appropriate nutrition and a favorable environment for sperm motility. This is accomplished by the secretions of two major accessory reproductive glands—the **seminal vesicles** and the **prostate gland** (see Fig. 14.1). Another pair of small **bulbourethral glands** plays a minor environmental role. These glands function in response to signals from the autonomic nervous system. Under the influence of parasympathetic stimulation during sexual arousal, their epithelia empty their secretions into the lumens of the glands. Then at the time of the sympathetically controlled ejaculatory process, they expel those secretions into the male reproductive ducts. **Semen** is the collective term for the bolus of sperm, testicular fluid. and secretions of the seminal vesicles, prostate and bulbourethral glands that is expelled from the penis.

The distal end of the ampulla of the ductus is joined by the duct of the seminal vesicle to form the **ejaculatory duct**. As the sperm and remaining testicular fluid pass from the ductus deferens, the seminal vesicle contributes a secretion rich in fructose, prostaglandins, and seminal coagulating proteins. The ejaculatory ducts join the urethra within the

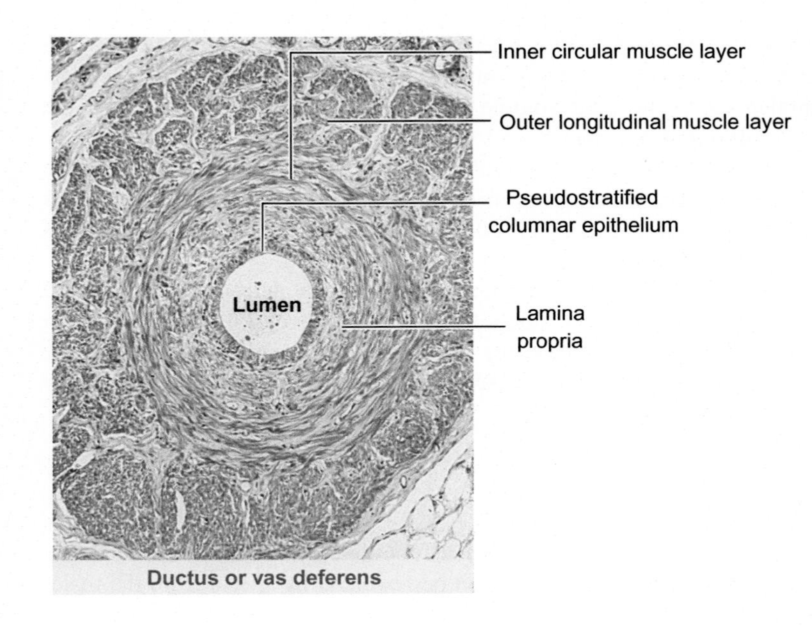

FIGURE 14.9 Cross-section through the vas (ductus) deferens. Note the thick layer of inner circular and outer longitudinal smooth muscle around the lumen. *From Kierszenbaum and Tres (2012), with permission.*

BOX 14.1 Erectile Dysfunction

Erectile dysfunction is common among elderly males, but most cases can now be treated by medications that have a specific physiological effect. Normal erection is initiated by nitric oxide (NO) secreted by parasympathetic nerve terminals. This activates the enzyme guanylate cyclase, which, in turn, converts guanosine triphosphate (GTP) into cyclic guanosine monophosphate (cGMP). cGMP causes relaxation of the smooth muscle layers surrounding the arteries supplying the corpora cavernosa and corpus spongiosum and allows them to become engorged with blood. A normal erection subsides when an enzyme, phosphodiesterase, breaks down the cGMP, thus reducing the amount of inflowing blood. Erectile dysfunction is often a function of too much phosphodiesterase activity, which results in a premature reduction of blood flow into the penis. Erectile dysfunction medications inhibit phosphodiestererase activity and allow more blood to enter and be retained within the erectile tissues of the penis.

body of the prostate. The prostate contributes a multifunctional alkaline mixture of proteins, polysaccharides, and ions. The bulbourethral glands secrete a viscous mucopolysaccharide-rich lubricating fluid.

Ejaculation (expulsion) is a very rapid phase of sperm transport that takes semen from the area of the prostatic urethra and ejects it from the **penis**. The penis has already achieved an erect condition during the arousal phase through parasympathetic-mediated neural input resulting in the engorgement of the paired **corpora cavernosae** and the single median **corpus spongiosum**, which surrounds the urethra. Nitric oxide (NO), secreted by the parasympathetic nerve endings, stimulates relaxation of the arteries supplying the erectile tissue, causing the erectile tissue to become engorged with blood and stiffen (Box 14.1).

Actual ejaculation is a response that involves a sympathetic nerve-driven response by the smooth muscle of the reproductive duct system and a motor nerve-driven response by the muscles of the pelvic floor. The sympathetic response involves a highly coordinated series of contractions of the smooth muscle coats of the ampullae of the ductus deferens, the seminal vesicles, and the prostate. It begins with the emptying of secretory products from the seminal vesicles and prostate and is followed by the relaxation of the external urinary sphincter and their propulsion, along with sperm, down the urethra. At the same time, the internal sphincter at the neck of the bladder closes, preventing urine from mixing with the semen and sperm from being retrogradely transported into the bladder. A major driver of the ejaculatory response is the reflex contraction of striated muscles of the pelvic floor, as well as the **ischiocavernosus** and **bulbocavernosus muscles**, in response to coordinated motor nerve signals emanating from the spinal cord. The result is a series of several spasmodic contractions of the **bulbospongiosus muscle** surrounding the base of penile urethra that propel the semen through the penile urethra in a series of spurts.

Ejaculated semen is a complex mixture of cells and secretory products. Spermatozoa themselves constitute 2%–5% of the total volume of semen. Of the remainder, 65%–75% consists of secretions from the seminal vesicles, 25%–30% from the prostate and <1% from the bulbourethral glands. The total volume of semen is typically 2–5 ml, and the usual concentration of sperm ranges from 50 to 150 million/ml of semen.

During ejaculation, the way is prepared by secretions from the bulbourethral glands, which both lubricate the penile urethra and reduce the acid pH, caused by residual urine. The first portion of semen emitted from the penis consists principally of prostate secretions and a majority of the spermatozoa. This is followed by a component rich in secretions from the seminal vesicles.

Soon after ejaculation, the semen becomes very viscous and coagulates through the actions of a seminal vesicle-derived protein, **semogelin**. This process is presumed to play a beneficial role in the early phases of sperm introduction into the female reproductive tract. Shortly thereafter, proteases in the prostatic secretions decoagulate the semen, rendering it more liquid again. The prostaglandins within the semen activate smooth muscle and may facilitate early transport of sperm within the female reproductive tract.

THE FEMALE REPRODUCTIVE SYSTEM

The female reproductive system (Fig. 14.10) in mammals is designed to be multifunctional. First, it must produce a sufficient supply of eggs. Next, it must transport the eggs to a site where they can be fertilized by sperm from the male. If fertilization does occur, it must provide a site where the resulting embryo is both protected and nourished until it is ready to be born. Lastly, it must expel the fetus through the birth canal at the appropriate developmental stage. Even the skeletal structure of the female pelvis differs from that of males to allow easier passage of the term fetus

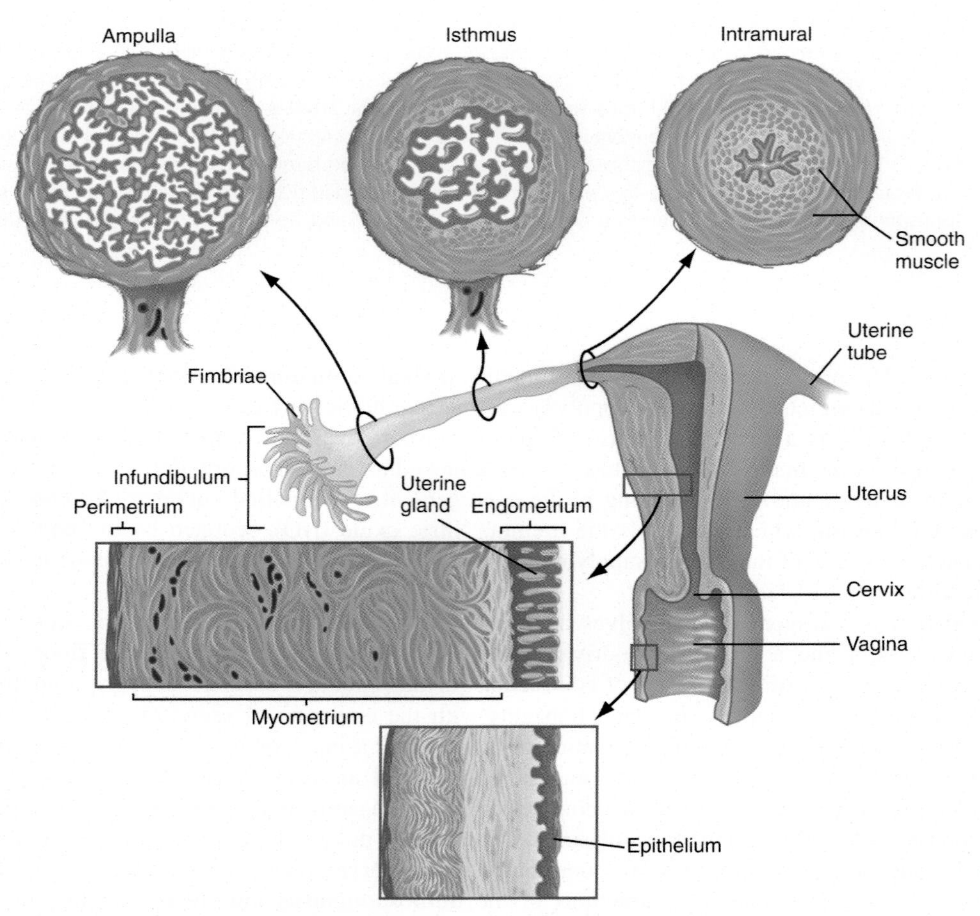

FIGURE 14.10 Internal anatomy of the female reproductive tract. *From Carlson (2014), with permission.*

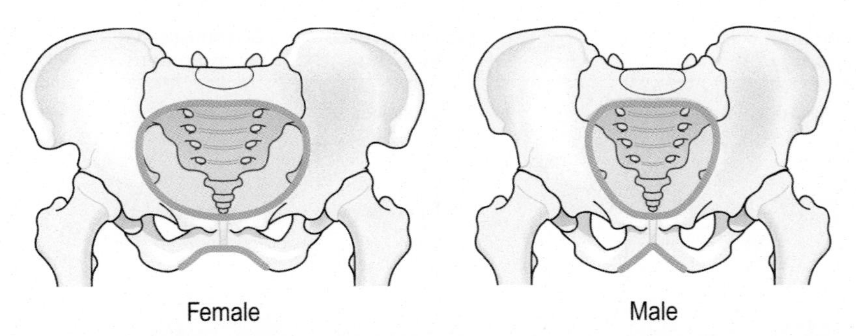

FIGURE 14.11 Differences in the shape of male and female pelvic skeletons. *From Waugh and Grant (2014), with permission.*

(Fig. 14.11). The female reproductive system is subject to complex cyclic endocrine controls, which affect all stages of the reproductive process from egg production to childbirth. Reproductive cycles in the female are predicated upon the possibility that pregnancy will occur. In the case of humans, the vast majority of reproductive cycles (menstrual cycles) do not result in pregnancy.

Egg Production (Oogenesis)

Egg production in females differs greatly from sperm production, even as early as embryonic life (Fig. 14.12). In males, serious sperm production does not begin until puberty, whereas by the time of birth, a female has as many eggs in her ovaries as she will ever have. From the second to the fifth month of pregnancy, the number of **oogonia** (egg-forming cells) in the ovaries increases by mitosis from a few thousand to about seven million. After that, the number of egg-forming cells progressively decreases through a natural degenerative process called **atresia** until by the time of birth, the ovaries contain about two million eggs. By puberty, that number is further reduced to about 40,000. Of this number, only about 400 are actually ovulated during the reproductive lifetime of the woman.

In contrast to sperm-forming cells, **primary oocytes** (genetically equivalent to primary spermatocytes) enter meiosis late in embryonic life, but then the process of meiosis is arrested. The block occurs early in the first meiotic division and remains until puberty or later. Not until an egg is ready to be ovulated in the first meiotic division completed. Since only ~400 eggs are ovulated, this means that none of the remaining oocytes ever complete the first meiotic division. In eggs that are not ovulated until late in a woman's life, the first meiotic block may persist for almost 50 years.

Oocytes require hormonal stimulation to mature. Indispensable for such maturation to take place is the development of cellular coverings around each egg, a process that begins after birth. Over time, a single layer of epithelial cells, homologous to the Sertoli cells lining the inside of the seminiferous tubules, forms a complete layer around the oocyte, which is still in a state of **meiotic arrest** (see Fig. 9.14). The combination of egg and its cellular coverings is called a **follicle**. Meiotic arrest is maintained by a high concentration of **cyclic adenosine monophosphate (cAMP)**, which is produced by both the follicular cells and the oocyte itself.

The follicular cells, called **granulosa cells**, secrete a thin noncellular basement membrane called the **membrana granulosa**, which surrounds the layer of granulosa and serves as a barrier to the penetration of capillaries into the follicle itself. The pituitary hormone, FSH, acts on the follicle cells, stimulating them to form estrogens, which act both on the oocytes and the woman's body, in general (see Fig. 9.15). For reasons still not completely understood, only about 50 follicles per menstrual cycle respond to hormonal stimulation by undertaking further development (Fig. 14.13). These follicles produce more estrogen-producing granulosa cells, which form a follicular covering several cell layers thick. On top of that, another layer of cells forms a discrete layer outside the membrana granulosa. These cells are **thecal cells** and are the female equivalent of Leydig cells in the testis. Like Leydig cells, they produce androgens after being stimulated by another pituitary hormone, luteinizing hormone (LH). While these changes are occurring, the egg protrudes into a cavity that begins to form within the follicle. Under multiple hormonal influences, one follicle enlarges greatly and produces large amounts of hormones, whereas the other follicles that had begun to enlarge all undergo atresia. The mature follicle develops a large fluid-filled antrum, into which the ovum, surrounded by a layer of cells called the cumulus oophorous, protrudes (Fig. 14.14).

Responding to several spurts of LH production by the pituitary, the oocyte in the chosen follicle rapidly completes the first meiotic division and enters the second. In contrast to sperm development, however, the first meiotic division of

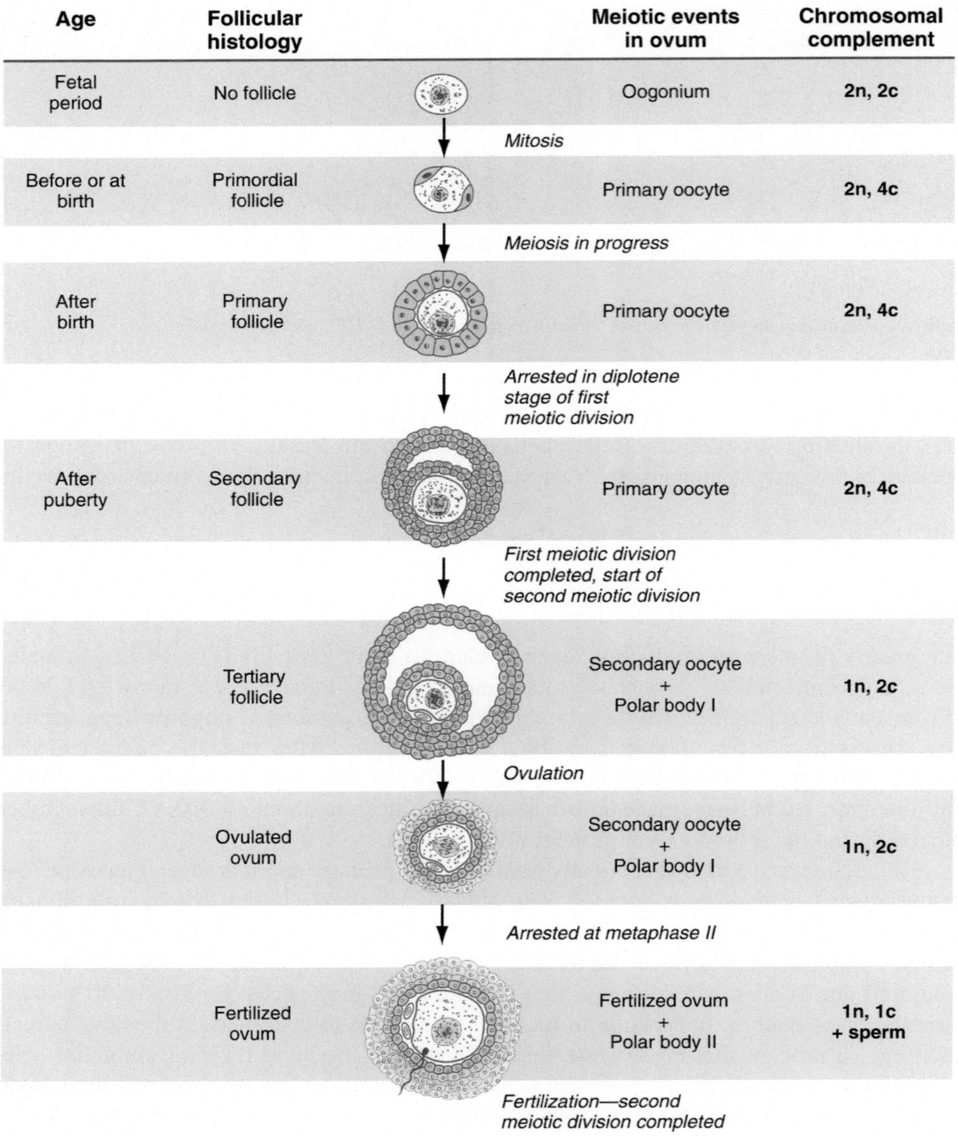

Age	Follicular histology	Meiotic events in ovum	Chromosomal complement
Fetal period	No follicle	Oogonium	2n, 2c
		Mitosis	
Before or at birth	Primordial follicle	Primary oocyte	2n, 4c
		Meiosis in progress	
After birth	Primary follicle	Primary oocyte	2n, 4c
		Arrested in diplotene stage of first meiotic division	
After puberty	Secondary follicle	Primary oocyte	2n, 4c
		First meiotic division completed, start of second meiotic division	
	Tertiary follicle	Secondary oocyte + Polar body I	1n, 2c
		Ovulation	
	Ovulated ovum	Secondary oocyte + Polar body I	1n, 2c
		Arrested at metaphase II	
	Fertilized ovum	Fertilized ovum + Polar body II	1n, 1c + sperm
		Fertilization—second meiotic division completed	

FIGURE 14.12 Summary of the major events in human oogenesis and follicular development. *From Carlson (2014), with permission.*

a primary oocyte produces unequal progeny—one **secondary oocyte** and a tiny **polar body**, which has no known function (see Fig. 14.12). This meiotic activity occurs within the **zona pellucida**, a thick basement membrane that has been forming between the oocyte and the surrounding granulosa cells. The secondary oocyte then enters the second meiotic division, but again this division is arrested and will not be completed unless the egg is fertilized.

Ovulation itself is stimulated by the surge of LH that is secreted by the pituitary gland midway through the menstrual cycle. The follicle begins to protrude from the surface of the ovary, and the follicular cells begin to produce molecules stimulating the ingrowth of capillaries into the follicular wall. Largely through the activities of enzymes that weaken the follicular wall, the follicle ruptures and releases the egg along with some of its cellular coverings from the ovary.

Female Reproductive Cycles

Maturation of the egg does not occur in isolation. In response to hormonal stimulation, the entire female reproductive tract undergoes cyclic changes designed to prepare it for the potential fertilization of an egg and a subsequent pregnancy. The **menstrual cycle**, normally 28 days in length, is most prominently represented by changes in the uterine

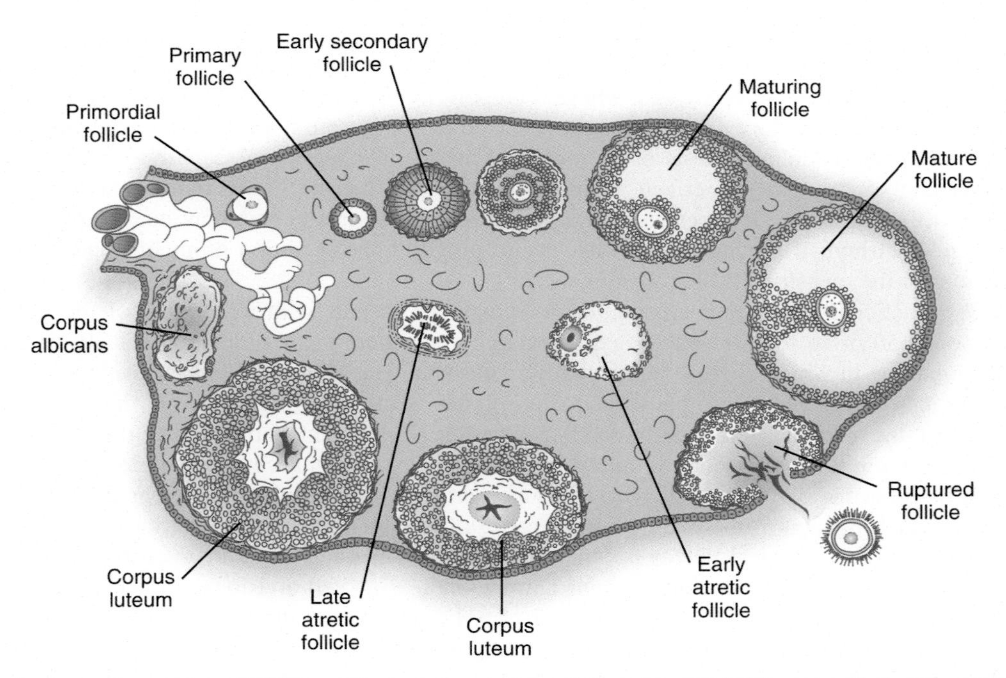

FIGURE 14.13 Sequence of maturation of follicles within the ovary, starting with the primordial follicle, going clockwise, and ending with the formation of a corpus albicans. *From Carlson (2014), with permission.*

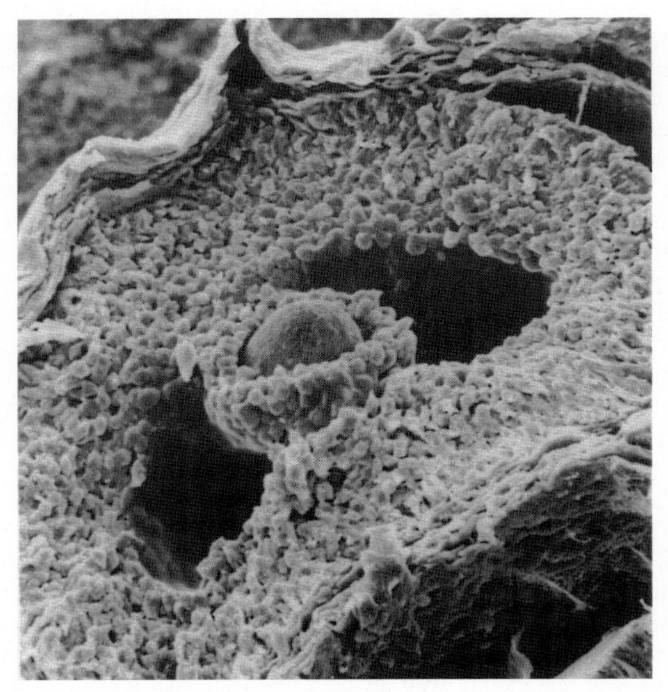

FIGURE 14.14 Scanning electron micrograph of a mature follicle in the ovary of a rat. The smooth spherical egg in the center is surrounded by a thin layer of cells, the cumulus oophorus, which protrudes into the antrum. *From Carlson (2014), with permission.*

lining, but all tissues of the female reproductive tract, the breasts, and even the brain show evidence of the monthly cycles.

The menstrual cycle is divided into two major phases, with ovulation constituting a discrete midpoint in the cycle (see Fig. 9.17). The first two weeks (often called the **proliferative** or **follicular phase** of the cycle) are involved with recovery from the previous cycle and preparing the tissues of the reproductive tract for ovulation, possible fertilization, and transport of the resulting embryo from the site of fertilization to a properly prepared uterine lining. The proliferative phase is dominated by the actions of **estrogens**, whose secretions by the ovaries are stimulated by the pituitary hormone FSH. The second phase (the secretory or luteal phase) is driven by the steroid hormone **progesterone**, which is secreted by the **corpus luteum**, the cellular remains of the ovulated follicle. Progesterone secretion is stimulated by LH from the pituitary, and one of its strongest effects is preparation of the uterine lining (**endometrium**) for receiving an embryo, should fertilization occur. If pregnancy does take place, the corpus luteum continues to pour out progesterone to maintain the endometrium. If pregnancy does not occur, progesterone secretion drops off, and the uterine lining is shed in preparation for the menstrual cycle to begin anew.

From a purely functional standpoint, preparation for pregnancy begins with maturation of the egg and ovulation. The above-mentioned 50 or so follicles that begin to enlarge under hormonal stimulation actually begin to enlarge in the cycle prior to the one in which one of them actually ovulates an egg. While follicular development is taking place, many changes are also taking place in other reproductive tissues. Following menstruation, during which much of the uterine lining is shed, the endometrium heals, much like a healing skin wound, and begins to thicken.

The cervix secretes a relatively viscous **G mucus** that effectively blocks access to the uterine cavity from the outside. This mucus block of the cervical canal plays an important role in preventing pathogens from entering the female reproductive tract, which is open to the body cavity because of the open ends of the uterine tubes. Not until the days just before ovulation does the cervical mucus become more watery (**E mucus**) to allow penetration by sperm. Early in the proliferative phase, the vaginal epithelium is thin. Through increasing cell divisions, it thickens as the time of ovulation approaches. The uterine tubes also reflect a period of reduced function during the early proliferative phase. The smooth muscle layer is relatively inactive, and the inner lining is dominated by generic columnar epithelial cells. As the time of ovulation approaches, increasing numbers of the epithelial cells respond to estrogen stimulation and begin to develop cilia. Smooth muscle activity also increases, with peristaltic contractions of increased strength.

Around the time of ovulation, the tissues of the female reproductive tract become finely tuned for the transport of both the egg and sperm (see below). Then the secretion of progesterone by the corpus luteum quickly brings the reproductive tissues to a state where a fertilized egg can be efficiently transported into the uterus, where it makes contact with a prepared endometrial lining and undergoes implantation. In anticipation of the possible implantation of an embryo, the endometrium becomes much thicker due to the rapid development of endometrial glands and blood vessels. The cervical mucus becomes much thicker, effectively sealing off the uterine cavity from the outside. Hormones secreted from an early embryo are necessary to maintain the integrity of the thickened uterine lining. When an embryo is present, a hormone called **chorionic gonadotrophin**, which is secreted by embryonic tissues, maintains the corpus luteum and its secretion of progesterone.

In the absence of an embryo, the corpus luteum in the ovary regresses and ceases to produce progesterone. Late in the nonpregnant menstrual cycle the reduction in progesterone results in the infiltration of the endometrium by white blood cells, loss of local fluids and the spasmodic constriction of the small arteries within the uterine lining. This causes tissue damage (ischemia) and results in small pieces of endometrium to be shed, along with some attendant bleeding. During the several days of the menstrual period, the uterine lining is shed, along with roughly 30 ml of blood. By the end of the menstrual period, the inner surface of the uterus resembles a raw wound, which begins to be recovered with epithelial cells arising from remnants of the uterine glands. With this, a new reproductive cycle begins.

Egg Transport

At the moment of ovulation, the egg is expelled from the ovary and floats freely within the woman's pelvic cavity. Two things must happen in order to get the egg into a proper location for fertilization. First, the ovulated egg must find its way into a uterine tube. Next, it must be transported pathway down the tube to a region where it is likely to encounter a sperm (see Fig. 14.17).

The female reproductive tract is remarkably well adapted to ensure that these two functions occur. First, however, we need to step back a bit to examine the ovulated egg itself. What is expelled at the time of ovulation is not simply a naked egg. Instead, the egg is surrounded by a thick basement membrane (the zona pellucida) and a layer of cells from the cumulus oophorus, called the **corona radiata**. This, plus a few other adhering cells, is the remnant of the granulosa cells that nourished the egg as it was maturing in the ovary. Once ovulation has occurred, these same cells play a different role in the reproductive process—namely providing bulk.

At the time of ovulation, the egg complex, which has no intrinsic motility, must be captured by the **uterine tube**. This is facilitated by several structural adaptations at both the gross and microscopic level. The open end of the uterine tube is funnel-shaped, and its rim consists of a row of finger-like **fimbriae**, which stick out from the tube much like petals of a flower. The fimbriae, however, are not passive structures. They contain strips of smooth muscle cells, which, as they contract and relax, cause slow, almost writhing movements of the fimbriae. The epithelium lining the inner surfaces of the fimbriae is also adapted for participation in the process of egg capture. Many of the epithelial cells are ciliated, and the synchronous beating of the collected cilia creates subtle inward currents in the liquid film that covers the epithelium.

Further adaptations are set in place to make egg capture by the tube more likely. The phase of the menstrual cycle leading to ovulation is characterized by the influence of estrogenic hormones. Estrogens act at several levels to make egg capture more likely. As the time of ovulation approaches, the fimbriated end of the uterine tube moves closer to the ovary, which reduces the distance between the site of ovulation and the entrance to the tube. In some mammals, such as the cat, the fimbriae partially surround the ovary. In addition to moving closer to the ovary, individual fimbriae increase their respective movements. This, plus the fluid currents generated by the cilia, creates the equivalent of a slow-moving biological whirlpool, which under normal conditions sucks the ovulated egg complex into the open end of the uterine tube.

The process of egg capture is remarkably efficient, and it operates in ways that are still not completely understood. One of the best demonstrations of this efficiency is seen in situations in which a woman has had one ovary and the contralateral uterine tube surgically removed. In some cases, these women have had normal pregnancies. The only way that this could happen would be for an egg ovulated by the left ovary, for instance, to be picked up by the right uterine tube.

After the egg complex has entered the funnel end of the uterine tube, it becomes transported down the tube. This is where the corona radiata that surrounds the egg comes into play. The corona cells appear to provide the bulk and possibly some surface characteristics that allow egg transport to occur. In experiments done on rabbits, beads the size of the egg alone without its surrounding cells were placed into the uterine tube, and instead of being transported, they merely spun around in place without going anywhere.

In a different dimension, the inner lining of the open (ampullary) end of the uterine tube is incredibly convoluted and is seemingly filled with epithelial folds to the point where the lumen of the tube is almost obscured. This increased surface area also increases the amount of liquid film that is propelled by the cilia, and it increases the efficiency of egg transport down the tube. On the negative side, if the uterine tubes have been inflamed through pelvic diseases, such as tuberculosis or gonorrhea, the resulting scarring of the tube can create pockets in which the egg can get stuck. If such an egg is fertilized, it won't get to the uterus, and the embryo implants into the uterine tube. This is an **ectopic (tubal) pregnancy**. As the embryo grows, it becomes constrained by the confines of the tube and causes intense pain, usually around a couple of months following fertilization. This condition requires surgery to correct.

Once within the uterine tube, the egg is transported about one-third of the way toward the uterus, where it awaits the arrival of some spermatozoa. Here again, the functions of the uterine tube are enhanced by the hormonal environment that precedes the time of ovulation. The uterine tubes are lined by a simple columnar epithelium that is partially ciliated. During much of the menstrual cycle, considerably fewer than 50% of these epithelial cells are ciliated, but under the influence of the estrogenic environment that leads up to ovulation, many more of the epithelial cells become ciliated (Fig. 14.15). The collective action of the cilia creates a pronounced fluid current leading down the tube in the direction of the uterus.

Beneath the epithelial lining of the uterine tube are layers of circular and longitudinal smooth muscle. When they contract, they create peristaltic-type movements that facilitate the transport of both the egg and the spermatozoa. These movements are also accentuated by estrogens at the time of ovulation. Both ciliary currents and smooth muscle action

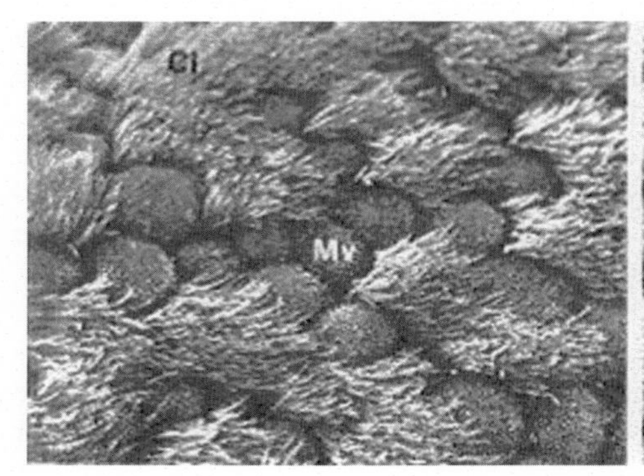

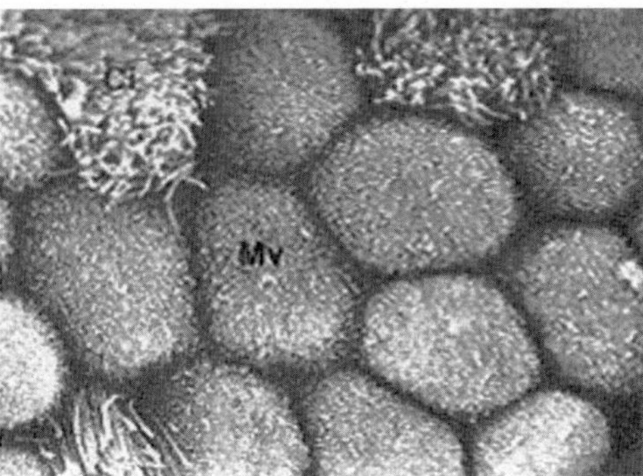

FIGURE 14.15 Scanning electron micrographs of the epithelial lining of uterine tubes of hormonally stimulated (left) and nonstimulated (right) rabbits. *From Kessel and Kardon (1979), with permission.*

work together to transport the egg. There is some redundancy in these functions, because pregnancies can occur in individuals in the absence of either ciliary action or smooth muscle movements. For important functions, it is common for the body to have evolved back-up mechanisms that can carry out the same broad function, although not necessarily as efficiently as the two working together.

SPERM TRANSPORT IN THE FEMALE REPRODUCTIVE TRACT

Sperm transport within the female reproductive tract is a cooperative effort between the functional properties of the sperm and seminal fluid on the one hand and cyclic adaptations of the female reproductive tract that facilitate the transport of sperm toward the ovulated egg. Much of the story of sperm transport in the female reproductive system involves the penetration by the sperm of various barriers along their way toward the egg (see Fig. 14.7B).

During **coitus** in the human, semen is deposited in the upper **vagina** close to the cervix. The normal environment of the vagina is inhospitable to the survival of sperm, principally because of its low pH (<5.0). The low pH of the vagina is a protective mechanism for the woman against many sexually transmitted pathogens, because no tissue barrier exists between the vagina (outside) and the peritoneal cavity (inside). The acidic pH of the vagina is bacteriocidal and is the reflection of an unusual functional adaptation of the vaginal epithelium. Alone among the stratified squamous epithelia in the body, the cells of the vaginal lining contain large amounts of **glycogen**. Anaerobic lactobacilli within the vagina break down the glycogen from shed vaginal epithelial cells, with the production of **lactic acid** as a byproduct. The lactic acid is responsible for the lowered vaginal pH.

Direct measurements have shown that within 8 seconds from the introduction of semen the pH of the upper vagina is raised from 4.3 to 7.2, creating an environment favorable for sperm motility. Another rapid event is the coagulation of human semen through the actions of semogelin by a minute after coitus. The coagulative function is incompletely understood, but it may play a role in keeping sperm near the cervical os. Thirty to 60 minutes after it coagulates, **prostate-specific antigen** (PSA), a proteolytic enzyme, degrades the coagulated semen. Within the semen and altered vaginal fluids, the sperm have begun to swim actively. A critical element in sperm motility is the availability of **fructose**, a nutrient provided by the seminal vesicles, within the semen. Because of their paucity of cytoplasm, spermatozoa require an external energy source. Unusually for most cells, spermatozoa have a specific requirement for fructose rather than glucose, the more commonly utilized carbohydrate energy source.

The next barrier facing sperm is the **cervix**. The cervical entrance (os) is not only very small, but it is blocked by cervical mucus. During most times in the menstrual cycle, cervical mucus is highly sticky (G mucus) and represents an almost impenetrable barrier to sperm penetration. Around the time of ovulation, however, the estrogenic environment of the female reproductive system brings about a change in cervical mucus, rendering it more watery and more amenable to penetration by sperm (E mucus).

Considerable uncertainty surrounds the question of passage of sperm through the cervix. The swimming speed of human sperm in fluid is approximately 5 mm/min, so in theory, sperm could swim through the cervical canal in a matter of minutes or hours. In reality, some sperm have been found in the upper reaches of the uterine tubes within minutes of coitus. These pioneers are likely to have been swept up the female reproductive tract during muscular contractions occurring at the time of or shortly after coitus. Research on rabbits has indicated that most of these sperm have been damaged and would not be able to fertilize an egg. The functional status of early-arriving human sperm is not known. On the other end of the spectrum, viable sperm have been taken from the cervix as long as 5 days after coitus. Between these two extremes, over the course of hours or even days, most of the spermatozoa make their way through the cervical mucus and up the cervical canal and into the uterus, where even less is known about the course of sperm transport in the human. Whether or not sperm are stored in the cervix is still not entirely certain. Sperm transport into and through the uterus is assumed to be assisted by contractions of its thick smooth muscle walls. There may or may not be subtle influences that favor the transport of sperm toward the opening of the uterine tube that contains the ovulated egg.

Of the huge numbers of sperm that enter the female reproductive tract, almost all fail to reach the uterine tubes. The unsuccessful sperm are removed by the infiltration of white blood cells into the cavities of the vagina, cervix, and uterus. These cells, along with certain immunoglobulins, inactivate and degrade foreign invaders, in this case, the excess sperm. Fortunately, the uterine tubes are not subject to this sort of cellular infiltration.

The openings of the uterine tubes into the uterus (**uterotubal junction**) represent another barrier to sperm transport. With two uterine tubes and usually only one ovulated egg, any spermatozoon that enters the empty uterine tube is automatically doomed to reproductive failure. Roughly 10,000 or fewer sperm cells of the millions in the ejaculate enter the correct tube. These sperm cells collect in the lower part of the uterine tube and attach to the epithelium of the tube for about 24 hours.

Two critical events occur during this period of attachment. The first is called **capacitation**, a reaction necessary for a spermatozoon to be able to fertilize an egg. The first phase of the capacitation reaction is the removal of cholesterol from the surface of the sperm. Cholesterol was introduced onto the sperm head to prevent premature capacitation. The next phase of capacitation is the removal of many of the glycoproteins that were deposited on the sperm head within the epididymis. After their removal, the spermatozoon is now capable of fertilizing an egg. It is likely that covering the sperm cells with glycoproteins and then cholesterol is done to prevent the sperm from prematurely attempting to fertilize other somatic cells that they encounter on their way to meeting the egg. Capacitation removes the molecular shield.

A second phenomenon occurring while the sperm are attached to the distal tubal lining is **hyperactivation** of the sperm. Hyperactivation is manifest by the increased vigor in their swimming movements and allows the sperm to break free from their binding with the tubal epithelial cells. Hyperactivated sperm are more efficient in making their way up the uterine tube and penetrating the coverings of the egg.

Once capacitated sperm break away from the tubal epithelium, they make their way up the uterine tube through a combination of their own swimming movements, peristaltic contractions of the smooth musculature of the tubal wall and the movement of tubal fluids directed by ciliary activity. In the upper third of the uterine tube, a few hundred sperm approach the ovulated egg. Only one of them out of the millions that left the male reproductive tract will attain is ultimate goal of fertilizing that egg.

FERTILIZATION

Fertilization represents the culmination of the journeys taken by both the sperm and the egg (Fig. 14.16). The usual site of fertilization is the upper third of the uterine tube, where the egg, by purely passive means, and the sperm, by a

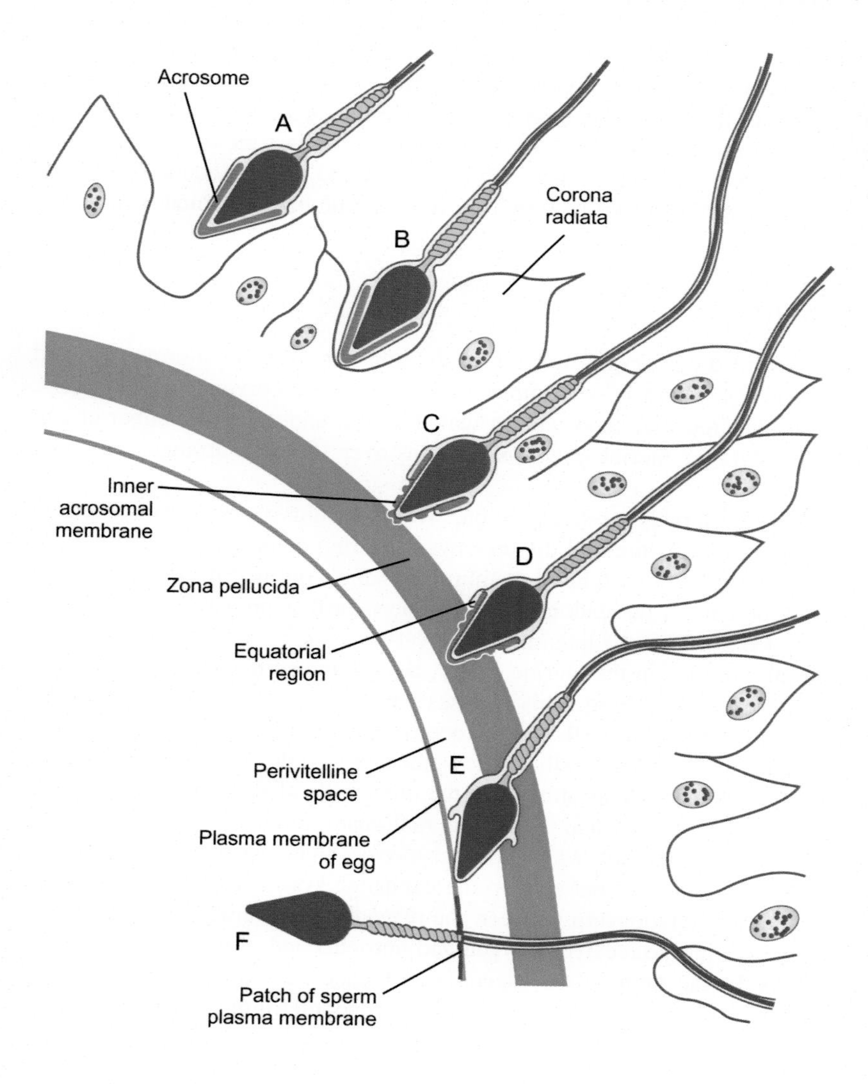

FIGURE 14.16 Sequence of events in the fertilization of an egg, starting with penetration of the corona radiata, passage through the zona pellucida and making contact with and penetration of the plasma membrane of the ovum. *From Carlson (2014), with permission.*

combination of passive transport and active swimming, find themselves in the same tubal compartment. At most, only a few hundred sperm have reached this part of the tube. What draws sperm and an egg together is still not completely known. Pure chance is the most commonly held explanation, but there remains the possibility that some form of chemical attraction of sperm to the egg may also play a role.

The story of sperm transport in fertilization remains the penetration of barriers between the sperm cell and the egg, but now in a microscopic dimension. In fertilization, the barriers consist of egg coverings, namely the cells of the **corona radiata** and within that, the membranous **zona pellucida** that invests the egg.

The swimming movements of the sperm are most important in penetration of egg coverings. Sperm make their way through the corona radiata by means of both swimming and the effects of **hyaluronidase**, an enzyme produced by sperm that breaks molecular connections between cells of the corona radiata. Once through the corona radiata, the sperm encounter yet another barrier—this time the zona pellucida. Two important events are necessary for a spermatozoon to penetrate this layer. The first is molecular binding to the zona. This is accomplished by the interaction of specific binding proteins on the head of the sperm and a glycoprotein component (ZP-3) of the zona pellucida. Such binding is species-specific, and it represents a defense mechanism against the fertilization of an egg by the sperm of another species. The second event necessary for penetration of the zona is a consequence of the **acrosomal reaction** in the sperm cell. Covering the nucleus of the sperm cell like a cap is a vesicle, called the **acrosome**. The acrosome contains many enzymes. When a capacitated sperm encounters the corona radiata, the outer membrane of the sperm cell begins to break down, allowing the release of the many enzymes contained within the acrosome. Some of these enzymes digest a hole through the zona pellucida, allowing the spermatozoon to wiggle its way through the zona and into the narrow space between the egg and the zona.

The last step in the sperm transport process is fusion between the sperm and the outer membrane of the egg. Within seconds after this event, electrical changes in the egg membrane do not allow other sperm to attach to the egg. A minute later, a reaction just below the cell surface results in the release of enzymes that degrade the sperm receptors in the zona pellucida. This eliminates the option of the penetration of the egg by other sperm, so that only one sperm actually fertilizes the egg Within 24 hours, the chromosomes of the sperm combine with those of the egg, resulting in the formation of a zygote—a genetically new individual. During those 24 hours, the egg completes its last meiotic division and extrudes another polar body with a redundant set of chromosomes. Since both the sperm and the egg contain only half of the normal complement of chromosomes in an ordinary cell, fertilization restores that number to normal, and development can proceed.

EMBRYO TRANSPORT AND PREGNANCY

Transport does not cease with fertilization. The fertilized egg (zygote) is still in the upper third of the uterine tube and needs to be carried to the uterine cavity (Fig. 14.17), where it will implant and continue embryonic development. Movement of the embryo down the uterine tube takes about five days, during which time it undergoes a number of cell divisions (the **period of cleavage**). By the time it enters the uterine cavity, it contains over 100 cells arranged around a central cavity (Fig. 14.18).

The trip down the uterine tube is accomplished in a largely passive fashion, with the embryo's being propelled by means of peristaltic contractions of the tubal smooth musculature and the downward beating of the cilia that cover the epithelial surface. During its entire period of descent through the tube, the embryo remains confined to its surrounding zona pellucida. When the embryo reaches the uterine cavity, the endometrium has been further prepared to receive an embryo by the stimulation of progesterone coming from the corpus luteum of the ovary.

At about the fifth day after fertilization, the embryo, now in the uterine cavity, begins to enzymatically dissolve the surrounding zona pellucida in a process commonly called **hatching**. This exposes embryonic cells to the environment of the uterus. Both the outer cells of the embryo and the epithelial cells of the endometrium express cell adhesion molecules on their surfaces. At some point, usually during the sixth day, cells of the outer layer of the embryo make contact with the endometrial surface and then become firmly attached. This stimulates the outer cells of the embryo to begin a relentless invasion of the endometrial tissues—equal to or greater than that of any malignant tumor.

As the embryonic cells continue their invasion, they erode the walls of local uterine blood vessels. These vessels spill blood into spaces surrounding the embryo—a phenomenon that sets up the exchange system between the mother and the embryo. The exchange system is based upon the relationship between the placenta of the embryo and the uterine tissues that surround it. Through it, the developing embryo receives oxygen and nutrients and in return, sends metabolic waste products into the maternal circulation. In the human, the **placenta** is bathed in a pool of free-floating maternal blood that flows from the open ends of small spiral arteries located within the endometrium (Fig. 14.19).

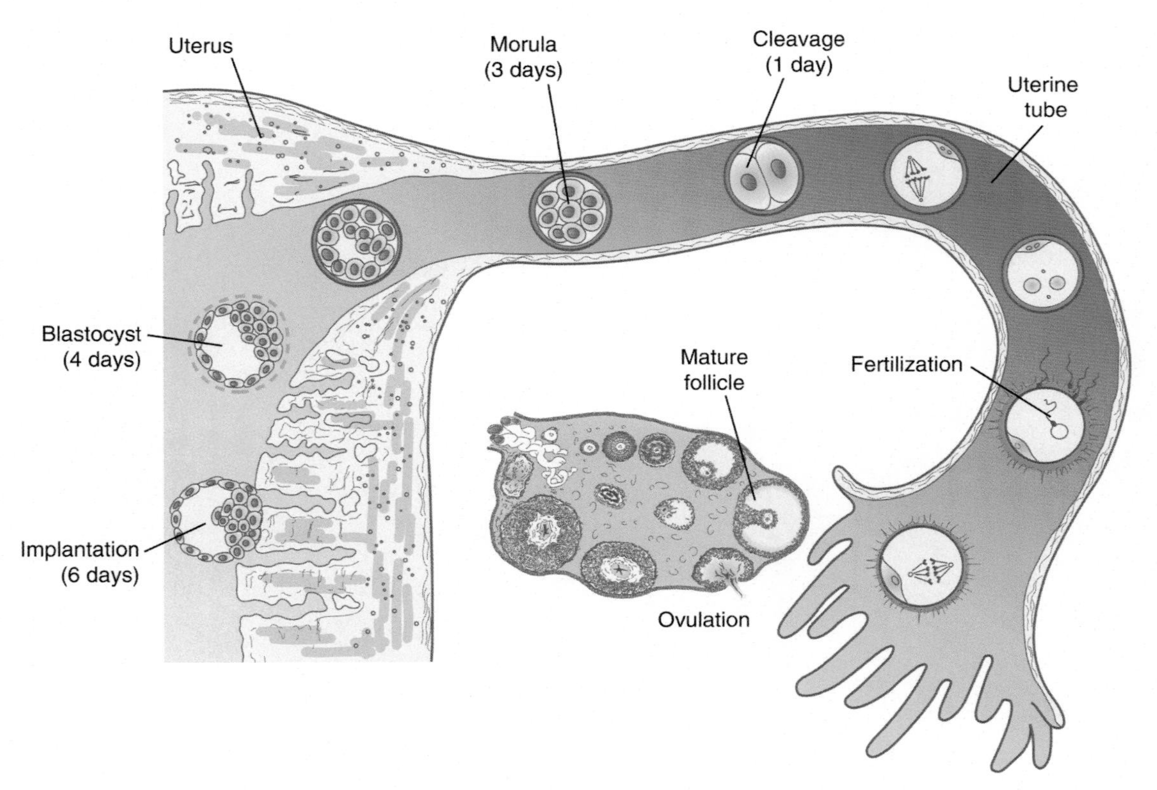

FIGURE 14.17 Transport of the ovulated egg and fertilized embryo through the uterine tube and uterus until the time of implantation. *From Carlson (2014), with permission.*

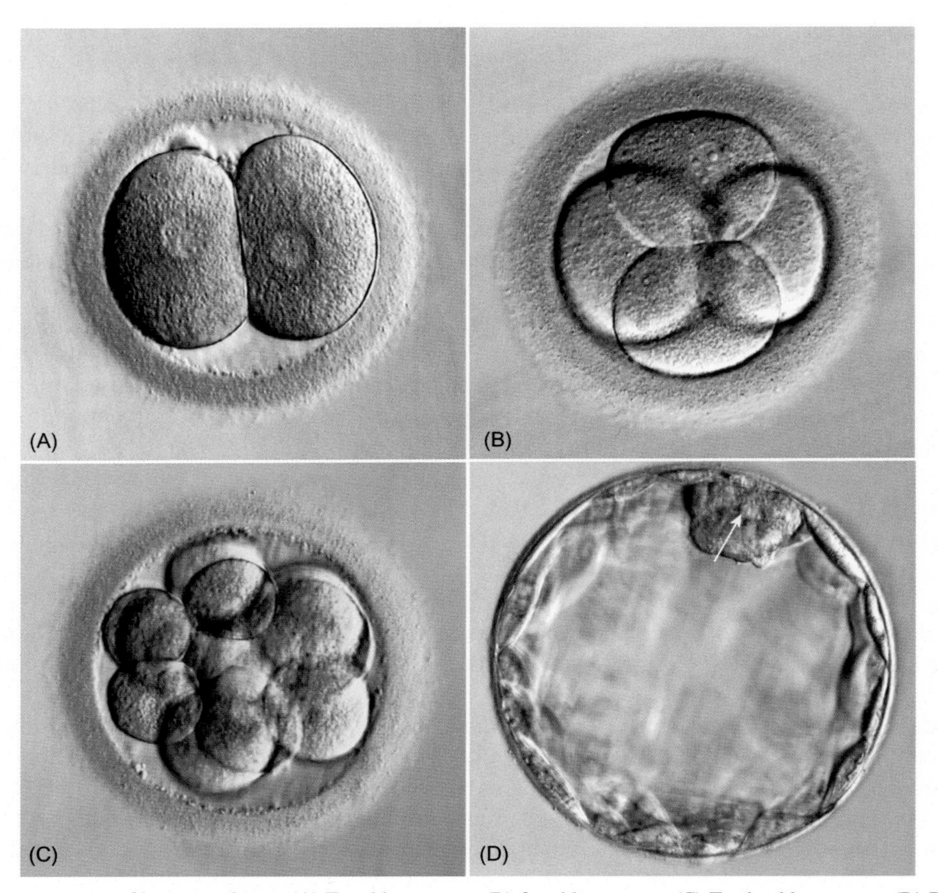

FIGURE 14.18 Cleavage stages of human embryos. (A) Two blastomeres; (B) four blastomeres; (C) Twelve blastomeres; (D) Blastocyst, showing a well defined inner cell mass (arrow), the cells from which the actual body of the embryo arises. *From Carlson (2014), with permission.*

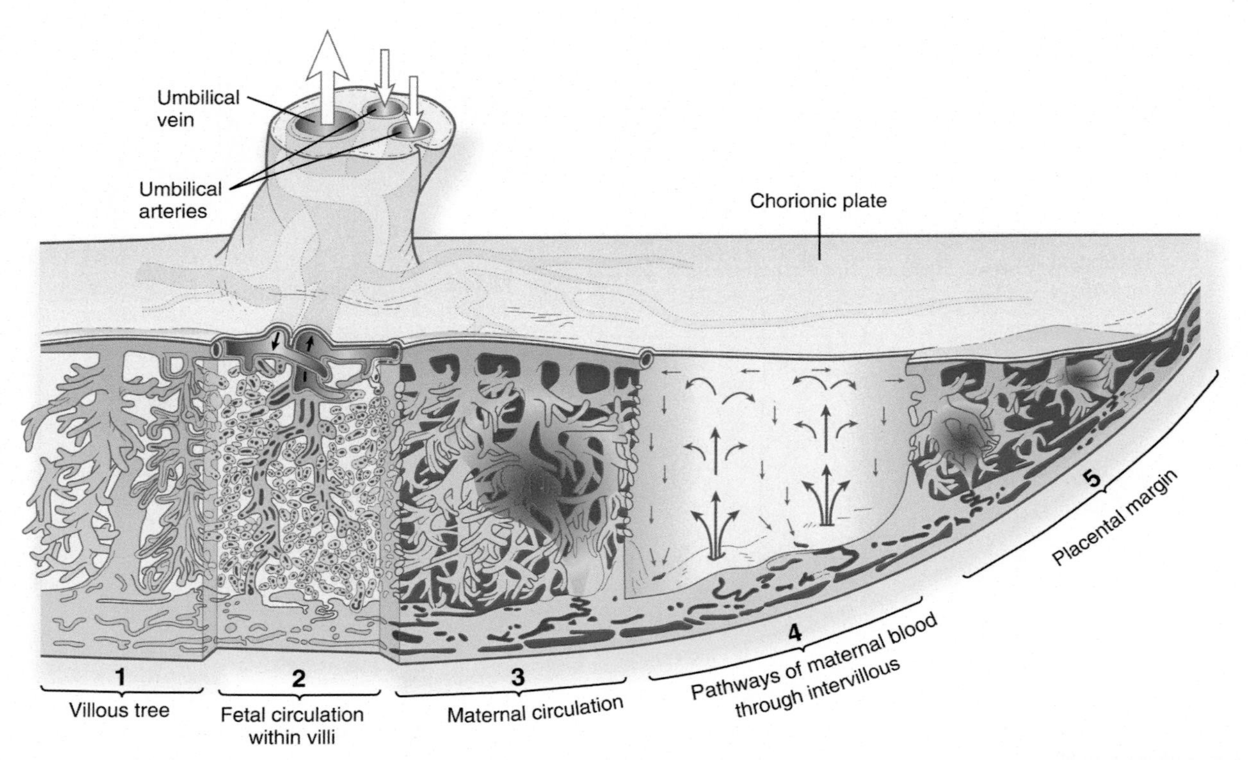

FIGURE 14.19 The structure and circulation of the mature human placenta. Blood (red) enters the intervillous spaces from the open ends of uterine spiral arteries. After bathing the placental villi, the blood (blue) is drained via the endometrial veins. Embryonic blood (blue) enters the placenta through the umbilical arteries, remains within microvessels in the placental villi, where it becomes oxygenated and receives nutrients before returning to the embryo via the umbilical vein (red). *From Carlson (2014), with permission.*

Throughout pregnancy, the uterus expands to provide a container for the embryo developing within. Keeping up with growth of the embryo, smooth muscle cells within the thick uterine wall divide and enlarge throughout pregnancy. Strong ligaments (the broad ligaments and the round ligaments) support the expanding uterus and anchor it to the internal body wall.

Much of the structure of the uterus, especially the muscular walls, is designed for the ultimate purpose of expelling the fetus when it has developed to the point where it can survive outside the mother's body. Interestingly, we know much more about what initiates **parturition** (childbirth) in sheep than we do in humans.

The signals leading to parturition are largely hormonal, and hormones, especially cortisol, from the adrenal glands of both the fetus and mother, are heavily involved. A final major signal for the initiation of labor is the hormone **oxytocin**, secreted by the posterior lobe of the pituitary. Oxytocin stimulates the strong periodic contractions of the uterine smooth musculature, which is the signal that labor has begun. While these changes are beginning, important changes are also taking place in the cervix. During most of pregnancy, a major role of the cervix is keeping the developing fetus within the uterus. This is accomplished largely by the presence of strong bands of collagen that encircle the cervical canal. As the body is preparing for the initiation of labor, the cervical canal undergoes a complete reconfiguration from a firm rigid canal to a soft, distensible structure that blends in with the overall functional architecture of the uterus and allows passage of the fetus. Although there remains much that we don't yet understand about the basis for these cervical changes, they are accomplished largely through the actions of **collagenase**, an enzyme that quickly dissolves the constraining bands of collagen that line the cervical canal. One of the stimuli for these changes is the production of a **prostaglandin** by the fetus. Most of these changes take place during the **first stage of labor**, which commonly lasts around 12 hours.

The **second stage of labor** is that of expulsion of the fetus. In addition to powerful directed contractions of the uterine smooth muscle, much of the abdominal musculature of the mother is also involved in pushing the fetus from the uterine cavity through the **birth canal** (cervix and vagina). Only recently has it been recognized that during a normal vaginal birth the baby is exposed to the mother's vaginal bacterial flora. These maternal bacteria represent the seed culture for the skin microbiome of the newborn infant. Research studies have shown a great similarity between the

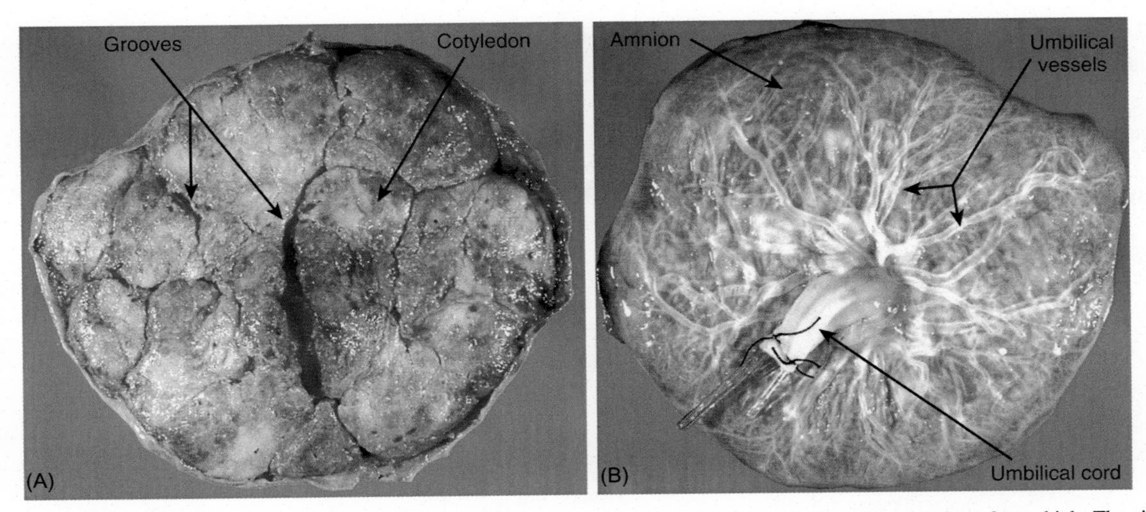

FIGURE 14.20 Photographs of a human placenta after delivery. A typical placenta is 20–25 cm in diameter and about 2 cm thick. The side facing the uterine wall (A) is bathed in uterine blood and is dull in color. It consists of closely spaced clumps of placental villi, called cotyledons. The cotyledons are separated by thin grooves. The other side (B) is covered with a shiny membrane (the amnion), which forms a fluid-filled bubble around the fetus throughout embryonic development. The cut umbilical cord, which connects the fetus to the mother, protrudes from the placenta in B. Rootlike fetal blood vessels are prominent on the fetal surface of the placenta. *From Moore et al. (2016), with permission.*

composition of the skin biome of a newborn and the vaginal biome of the mother. The **third stage of labor**, expulsion of the placenta (Fig. 14.20), must also be followed by oxytocin-induced strong contractions of the uterine smooth muscle, which greatly reduce the amount of maternal blood loss that results from tearing the placenta away from the naked uterine wall.

BREAST DEVELOPMENT AND LACTATION

Breast development and **lactation** are closely coordinated with those factors that control other aspects of female reproduction. Initial development and maturation of the breasts and even their further development during pregnancy are largely under the control of estrogens and progesterone, but after birth, other hormones take control of milk production and letdown.

The mature **mammary gland** consists of groups of epithelially-lined alveoli that make up secretory lobules (Fig. 14.21). Each alveolus empties into a ductule, which leads into progressively larger ducts. Ductules from about 15–20 secretory lobules empty into a single **lactiferous duct**. Several lactiferous ducts open into the nipple. Surrounding each alveolus is a layer of myoepithelial cells, which play an important role in lactation. The secretory lobules are separated from one another by fat and sheets of fibrous connective tissue. Sensory neurons, especially in the area of the nipple, represent an important functional link in lactation.

In prepubertal girls, estrogens play a prominent role in the initial development of the duct system of the breast. Then as puberty approaches, estrogens, acting on a base of growth hormone and insulin-like growth factor activity, stimulate further development of both the mammary ducts and the fatty tissue that surrounds them (Fig. 14.22). During pregnancy, the high levels of progesterone and the secretion of prolactin by the anterior pituitary starting in the second month stimulate further development of the duct system, as well as development of the secretory alveoli. Even in non-pregnant women, the effects of progesterone during the last half of the normal menstrual cycle can often be felt in the breasts. During pregnancy, actual milk production, which is stimulated by prolactin, is inhibited by the estrogenic and progesterone-rich hormonal environment.

After birth, the inhibitory effects of estrogens and progesterone are removed, and prolactin becomes the dominant hormone involved in the promotion of milk production. For the first few days after birth, the mammary glands produce **colostrum**, a secretion closely resembling milk except that it lacks the high-fat concentration of normal milk, which begins to be produced in increasing amounts late in the first post-partum week. **Milk** is a mixture of a large variety of constituents that must be sufficient to sustain the life and growth of a baby for an extended period. It is an emulsion of fats superimposed on a solution containing many other constituents (Table 14.1). Human milk contains about 20% more fat than does cow's milk. Within the aqueous solution, the major sugar is **lactose**, and the major protein is **casein**,

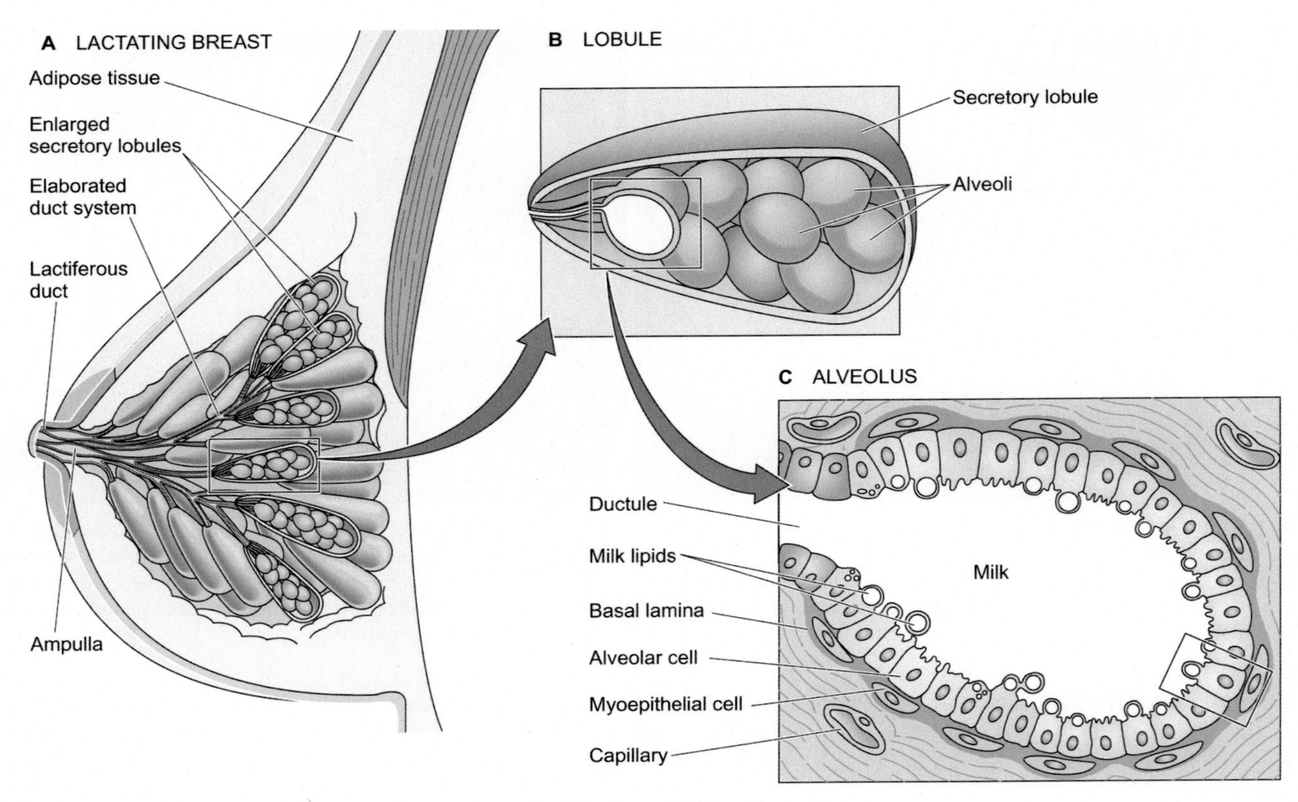

FIGURE 14.21 Structure of the human breast. *From Boron and Boulpaep (2012), with permission.*

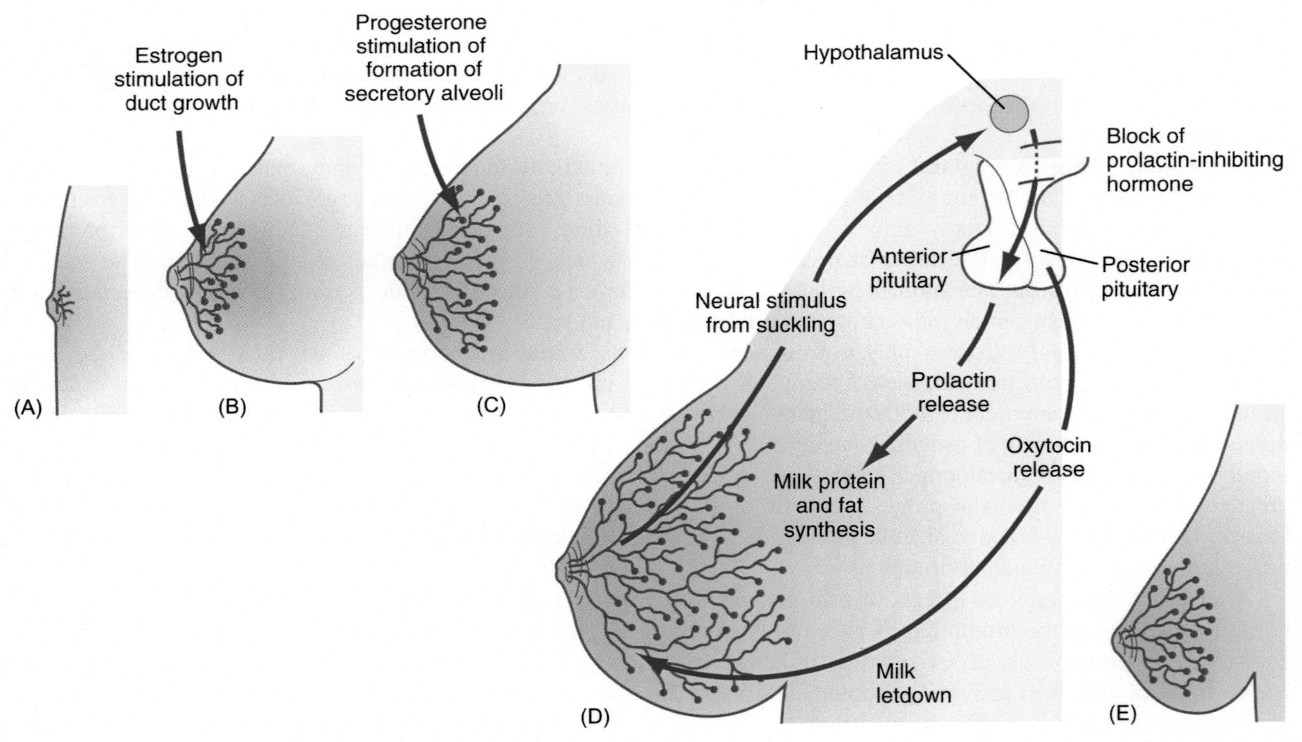

FIGURE 14.22 Hormonal control of development of the breast and mammary ducts. (A) Newborn; (B) young adult; (C) adult; (D) lactating adult; and (E) adult after the cessation of lactation. *From Carlson (2014), with permission.*

TABLE 14.1 Composition of Milk

Component (per 100 ml)	Human colostrum	Human milk	Cow's milk
Total protein (g)	2.7	0.9	3.3
Casein (% of total protein)	44	44	82
Total fat (g)	2.9	4.5	3.7
Lactose (g)	5.7	7.1	4.8
Caloric content (kcal)	54	70	69
Calcium (mg)	31	33	125
Iron (μg)	10	50	50
Phosphorus (mg)	14	15	96

From Boron and Boulpaep (2012), with permission.

another milk-specific component. Among the proteins present in milk, maternal antibodies are a very important constituent because the immune system of the newborn is not yet fully functional. It is now known that breast milk also contains bacteria. These bacteria provide seed colonies for parts of the intestinal microbiome of the infant who is drinking the mother's milk.

Prolactin is the principal hormone involved in milk production. After birth prolactin levels begin to fall, but each time the infant nurses, the production of prolactin by the pituitary spikes to a high level for about an hour after nursing. This ensures the production of sufficient milk for the next nursing episode. Milk is produced within the alveoli through a mechanism involving the pinching off of the apical portions of the alveolar cells (**apocrine secretion**). Apocrine secretion is the main mechanism by which fatty material enters the milk. The contribution of proteins, small molecules, ions and water to the mix is accomplished by the usual means of secretion and transport through a cellular layer.

The actual let-down of milk begins with suckling of the nipple by the infant. This stimulates afferent sensory nerves to send signals to the hypothalamus, causing the release of the posterior pituitary hormone **oxytocin** from the supraoptic and paraventricular nuclei (see p. 153). Oxytocin causes contraction of the myoepithelial cells surrounding the alveoli. These contractions squeeze the milk from the individual alveoli into the duct system. For the first 30–60 seconds after the initiation of suckling no milk comes out. This period represents the remarkably rapid time during which the milk let-down reflex is completed. Suckling on one breast stimulates milk let-down in both breasts. Emotional signals can also stimulate milk let-down. Some nursing mothers experience milk let-down when they hear other babies crying. Similarly, milk ejection can be inhibited by the mother's emotions.

Nursing has a broader effect on the female reproductive system. For many months during nursing, the production of gonadotropin hormones by the hypothalamus is inhibited. This suppresses ovulation and serves as a natural form of temporary birth control. Over time, even if the mother continues nursing, normal reproductive rhythms reemerge and ovulation begins again.

SUMMARY

The main function of the male reproductive system is to produce sperm and to get them into the vicinity of the egg. Tens of millions of spermatozoa are produced daily within the seminiferous tubules of the testes. The testes are located within the scrotum. The exposed location of the scrotum and a countercurrent vascular system leading to the testes reduce the temperature at which sperm cells develop.

Outside the seminiferous tubules, Leydig cells respond to luteinizing hormone (LH) from the pituitary to produce testosterone.

Spermatogenesis begins with spermatogonia located near the outer margin of the seminiferous tubule. Sertoli cells are closely associated with spermatogenesis. As spermatogonia become transformed into primary spermatocytes, they enter the first meiotic division and pass through the blood-testis barrier created by interdigitation of the Sertoli cells. Secondary spermatocytes have entered the short second meiotic division. After this has been completed, they become spermatids, which then undergo a transformation into tadpole-shaped spermatozoa. Although fully formed, spermatozoa within the seminiferous tubules are functionally inactive.

Newly formed spermatozoa are carried out of the seminiferous tubules by a slow passive transport through the rete testis and the efferent ductules. They then enter the highly coiled epididymis, where they undergo functional maturation.

A fast phase of sperm transport occurs during ejaculation, which consists of two phases— emission and expulsion. Emission carries a package of sperm from the tail of the epididymis through the ductus deferens and into the urethra. Secretions of two glands, the seminal vesicles and prostate, are added to the sperm to form the bulk of the ejaculate. Expulsion carries the ejaculate from the prostatic urethra and through the erect penis. This process is mediated by the parasympathetic nervous system, followed by sympathetic nervous signals that activate the smooth muscle of the reproductive tract. This is accompanied by reflex constriction of certain skeletal muscles of the pelvic floor.

Ejaculated semen is about 2–5 ml in volume and consists of 76%–75% of secretions from the seminal vesicles, 25%–30% from the prostate, <1% from the bulbourethral glands, and 2%–5% spermatozoa themselves.

The female reproductive system must first produce eggs. Then it must provide an environment where eggs and sperm can meet. Finally, it must be able to house a developing embryo until it is capable of survival when expelled at the end of pregnancy.

Egg production occurs in the ovaries and is under tight cyclical hormonal control. In the ovary, meiosis of oocytes begins in the embryo and is not completed until the time of fertilization. Normally, only one mature egg is produced per menstrual cycle. Maturation of oocytes occurs within ovarian follicles which, in response to follicle-stimulating hormone, produce estrogens.

Midway through the menstrual cycle a single oocyte is ovulated and makes its way into the female reproductive tract. The female reproductive tract consists of the uterine tubes, the uterus, the vagina, and the external genitalia.

The entire female reproductive system undergoes hormonally induced cyclical changes (menstrual cycle) over a 28-day period. The menstrual cycle is roughly divided into two phases, with ovulation occurring in the middle. The first two weeks (proliferative phase) operate under the influence of pituitary follicle-stimulating hormone, which causes follicular cells in the ovaries to produce estrogens. Shortly before ovulation on the 14th day, thecal cells within the ovaries respond to stimulation by pituitary LH by producing progesterone, which prepares the uterine lining for pregnancy. The second half of the menstrual cycle (15–28 days) is dominated by the influence of progesterone. In the absence of fertilization, the cycle ends and a new one begins.

Both the ovulated egg and ejaculated sperm must be transported to the site of fertilization in the upper end of the uterine tube. The ovulated egg first floats freely within the peritoneal cavity, but is soon captured by the fimbriated open end of the uterine tube. Under the influence of estrogens, the ends of the uterine tube move closer to the ovary and smooth muscle within the walls of the tube increases its contractions, which facilitates both entry of the egg into the tube and transport down the uterine tube. In addition, the tubal epithelium becomes more highly ciliated during the follicular phase. Ciliary currents assist in moving the egg down the tube.

Sperm transport within the female begins with deposition of semen in the upper end of the vagina during intercourse. Alkaline properties of the semen increase the normally low pH of the upper vagina in order to promote survival of the sperm. To enter the uterus, sperm must first traverse the cervix, the entrance of which is normally blocked by a thick mucus. Under the influence of estrogens just before ovulation, the character of the mucus changes and it becomes more watery, allowing the passage of sperm. Through poorly understood mechanisms, spermatozoa cross the uterine cavity and enter the lower ends of the uterine tubes. Peristaltic contractions of the uterine tubes ultimately result in the meeting of an egg and spermatozoa.

Fertilization first involves passage of spermatozoa through egg coverings—the cellular corona radiata and the thick zona pellucida. Penetration of these coverings is accomplished by both swimming movements of the sperm and enzymes secreted by the head of the sperm.

Once an egg is fertilized, transport down the uterine tube continues until by about five days, the embryo enters the uterine cavity and begins to undergo implantation within the uterine mucosa (endometrium). Implantation is made possible because of progesterone-induced changes of the endometrial epithelium. After implantation, the embryo develops in close association with the mother for the next 9 months. At that point, through mechanisms not yet well understood in humans, the process of labor begins, and the fetus is expelled from the mother.

Breast development is also closely tied to the overall hormonal environment of the woman. Development of mammary ducts is initiated through the actions of estrogens, but during pregnancy further development of the duct system is stimulated by high levels of progesterone and prolactin, produced by the anterior pituitary. After birth, prolactin stimulates milk production, and the actual letdown of milk upon a suckling stimulus is mediated by the posterior pituitary hormone oxytocin.

Glossary

Acinus A small grape-shaped group of secretory cells that empty their product into a small duct.

Acute inflammation The first few days of an inflammatory process, characterized by the presence of large numbers of neutrophils.

Adrenergic Referring to nerve fibers that use epinephrine or norepinephrine as a transmitter.

Afferent Leading to, e.g., an artery supplying an organ is afferent.

Ala (of nose) The soft, fleshy side of the nose.

Amino acid A simple organic acid in which an NH_2 replaces one of the hydrogen atoms.

Anastomosis A direct connection between an arteriole and a venule without an intervening capillary bed.

Antibody A globulin protein produced by immune cells that combines with antigenic compounds.

Antigen A substance (protein or carbohydrate) capable of eliciting an immune response, either antibodies or reactive cells

Apoptosis Cell death brought about by an internal cellular program.

Axon A long cellular process that carries signals away from the cell body of a neuron.

Basophilic A property of cells that attracts basic histological dyes, such as hematoxylin.

Bouton A tiny rounded ending of a terminal nerve process.

Carbohydrate Organic molecules composed of simple sugars. They can be simple sugars themselves or polymers containing large numbers of them.

Channel An protein-lined opening in the plasma membrane that allows small molecules or ions to pass through the cell membrane.

Cholinergic Referring to nerve fibers that use acetylcholine as a transmitter.

Chromatid One half of a paired chromosome after DNA replication. The members of the pair are joined at the centrosome before they separate during mitosis.

Chromosome A long chain of DNA and associated proteins located in the nucleus that contains genetic information.

Chronic inflammation An inflammatory process lasting more than a few days and characterized by the presence of macrophages.

Cloning The process of making identical copies of an individual.

Colloid osmotic pressure A type of osmotic pressure, generated by proteins, which tends to pull water into a cell or tissue compartment from surrounding compartments.

Compliance The ability of a tissue to stretch or become deformed under pressure.

Cytokine Any of a variety of substances secreted by immune cells that influence the behavior of other cells.

Dendrite

Depolarization A rapid change of the electrical charge in the inside of a cell from negative to positive as the result of a sudden inrush of sodium ions.

Differentiation The process of structural and functional specialization of a cell over time.

Disaccharide A molecule consisting of two simple sugars chemically linked together.

Disulfide bond A direct chemical linkage of two sulfide groups within a single peptide or protein or between different proteins.

DNA (deoxyribonucleic acid) A long self-replicating chain of linked nucleotides that contains genetic information.

Efferent Leading away from some structure.

Endocrine A substance secreted into the blood, rather than into a duct system.

Endocytosis A process by which a cell takes in a large molecule by engulfing it and surrounding it with a membranous covering.

Enzyme A molecule, usually a protein, that catalyzes a biochemical reaction, changing one substance into another.

Eosinophilic (acidophilic) A histological term for cellular components that bind to acidic dyes (eosin being the most common example).

Exocrine A gland that empties its secretory product into a system of ducts, rather than directly into the bloodstream.

Exocytosis The process of moving membrane-bound groups of large molecules (vesicles) out of a cell by the fusion of the membrane of the vesicle with the plasma membrane and its opening to the exterior of the cell, thus releasing its contents.

Fascia Sheets of fibroelastic connective tissue that separate parts of the body from one another, for example between individual muscles.

Gene A sequence of DNA that contains the information needed for the construction of an individual protein.

Glabrous Smooth; a term applied to regions of hairless skin.

Growth factor A naturally occurring molecule that produces an effect, such as proliferation or differentiation, in other cells.

Hilus The concave part of an organ that serves as the entry and exit point for blood and lymphatic vessels.

Histology The study of tissues with the aid of a microscope.

Homeostasis The process of active regulation of bodily components so that their concentration remains within very tight limits.

Hydrolysis A chemical reaction in which ionic components of H_2O are bound to cleavage sites in a molecule.

Hydrostatic pressure The physical pressure of a fluid within a vessel or some container.

Hyperosmotic A condition (e.g., renal medulla) in which the dissolved components in one compartment are more highly concentrated than those in another compartment.

Hypertonic A term describing a concentration of solutes that is greater than normal for the body.

Hypoosmotic A condition in which the dissolved components in one compartment are less concentrated than those in another compartment.

Hypotonic A term describing a concentration of solutes that is less than normal for the body.

Induction Embryonic signal-calling, in which one group of embryonic cells emits a signal that influences the development of another group of cells.

Interstitial In between structures, or spaces within a structure.

Ischemia A reduced blood supply to a tissue or organ, leading to pathological changes.

Lamina propria A thin layer of connective tissue that underlies an epithelium.

Ligand A molecule that binds to a receptor on the surface or the interior of a cell.

Lipid A carbon-based molecule that is insoluble in water, but is soluble in many nonpolar liquids.

Lumen The interior of a tube or vessel.

Metaplasia A change in the properties of a cell or tissue from one type to another.

Molarity (molality) The number of moles of a substance per liter of solution. For practical purposes, there is little difference between molarity and molality.

Mucous membrane (mucosa) The moist lining (epithelium plus underlying connective tissue) of a passageway or vesicle in the body.

Myoepithelial cell A cell found in glands between the base of the epithelium and the basal lamina. They contain myosin and can contract, assisting the expulsion of glandular material.

Necrosis The pathological death of cells or tissues.

Neural crest A cellular component of early embryos that gives rise to a wide variety of adult tissue derivatives, ranging from pigment cells, to sensory nerve cells, to bone.

Noradrenergic	Neurosecretory elements that produce norepinephrine as a transmitter.
Nucleotide	Organic molecules that are fundamental building blocks of nucleic acids. They contain a nitrogenous base, a five-carbon sugar, and a phosphate group.
Opsonize	The process coating a foreign object or molecule with molecules, e.g., antibodies, that facilitate phagocytosis of the foreign object.
Osmosis	The movement of water or a solvent across a semipermeable membrane from a region of lower to higher concentration of solute.
Oxidant	An agent that oxidizes another.
Oxidize	The process of adding oxygen or removing hydrogen to a molecule.
Paracrine	A mode of action of secreted substances that influences neighboring cells through local channels, rather than through the blood.
Parenchyma	The epithelial component of a gland or organ.
Pathogen	Any organism that causes disease.
Peptide	A compound molecule composed of several amino acids linked by peptide bonds.
Perineum	The region between the anus and the vulva in females and the scrotum in males.
Phagocytosis	The process by which a cell ingests a foreign object by engulfing it.
Piloerection	The erection of a hair through the contraction of the smooth muscle at its base.
Posttranslational processing	Modification of the structure of a polypeptide after the basic polypeptide chain has been assembled.
Primary cilium	A single cilium projecting from a cell that may play a number of sensory and/or mechanical roles in the cell.
Prokaryote	A single-celled organism, such as a bacterium, that does not contain a nuclear membrane or other specialized organelles.
Protein	A macromolecule consisting of many amino acids joined together by peptide bonds.
Receptor	A molecule with a special site that combines with a specific ligand(s). After binding, the receptor initiates a cellular response to the ligand.
Redundancy	The presence of two or more genes that perform essentially the same function.
RNA (ribonucleic acid)	A macromolecule containing the sugar ribose (in contrast to the deoxyribose of DNA) that is involved in protein synthesis.
Self-antigen	Antigens normally found on cells of one's own body.
Serous	A watery secretion of exocrine glands.
Solute	A material dissolved in a liquid solvent.
Strain	The amount of deformation after stress has been applied.
Stress	The mechanical force applied to an object, e.g., a bone.
Stroma	The connective tissue component of a tissue or structure.
Symbiosis	Two organisms living together to their mutual advantage.
Synapse	The space between the terminal of a neuron and its end organ.
Trabecula	A small spicule of bone.
Umami	One of the five basic tastes, meaning a pleasant savory taste.
Vesicle	A space within the body filled with fluid or air and surrounded by an epithelial lining.

Index

Note: Page numbers followed by "*f*" and "*t*" refer to figures and tables, respectively.

B

C

D

E

F

G

H

M

P

Q

R

U

V

W

X

Y

Z

Printed in the United States
By Bookmasters